PROFESSOR JÁNOS JÁKÓ
1st Department of Medicine
Postgraduate Medical School
Szabolcs u. 35
1135 Budapest, Hungary

TRANSLATION OF: GAMMOPATHIÁK


ISBN-13: 978-3-642-84931-2 e-ISBN-13: 978-3-642-84929-9
DOI: 10.1007/978-3-642-84929-9



Springer-Verlag Berlin Heidelberg 1993
Softcover reprint of the hardcover 1st edition 1993



Typesetting: Data conversion by TRÜKK Bt.

27/3130-5 4 3 2 1 0 - Printed on acid-free paper

János Jákó

# Gammopathy

With 325 Figures,
104 in Color, and 21 Tables

Springer-Verlag
Berlin Heidelberg New York
London Paris Tokyo
Honk Kong Barcelona
Budapest

# Contents

# Introduction

*"For the place gives, but also take from the spirit – none can deny, And the verse bears the imprint of the place where it was born."*
*(Janus Pannonius)*

It is generally expected that any author of a book of value in everyday clinical work in a highly specific area will have based his work not on a summary of the literature and the "urge to communicate" arising from a certain amount of experience acquired with routine, but that he will bring an individual approach distilling the results of the setbacks and achievements of many years. Obviously such an approach can only be shaped by the correct interpretation of the positive and negative responses coming from the environment, assisted by the selective use of the literature. However, it must be stressed that "selective" reference to the literature at once means not "only" selection, but within this also the assertion of a clearly defined trend which serves to support the given and explained interpretation. This leaves the reader free to exercise his critical faculties if he should happen to see, or think he sees, the facts differently on the basis of contrary data in the literature, which he too must prove if he wishes to refute the authors illustrative data. What is involved here is practice, not the confrontation of theories. These latter sentences explain why, in departure from would be theoretically desirable - although this is not always the practice everywhere - the book is consistent in not giving arguments referring to the opposite of its claims. In a certain sense this can be regarded as a (retrospective) summary and evaluation of work carried out for a specific goal. In other words, the author should only have an opinion worth passing on to others on matters in which he has a great deal of experience in many directions, acquired at bedside and/or in laboratory work. However, on analysis it may turn out that this opinion does not always coincide in all respects and at all times with the views of the different theoretical experts. The question immediately arises of why this is so, and with it also the answer: because, for technical reasons, there is always a time lag in practice. This lag is the result of the gap in time between the flow of information and its application in practice, naturally together with material considerations of a decisive nature. To put it even more briefly: a technical book inevitably reflects the stage of development, the level of economic and scientific development of the country, in which it was written. Parallel with this statement, the fact of the time lag gives rise to the greatest dilemma: should we write or rather translate? Although it is not possible within the scope of an introduction to provide a reassuring answer to this difficulty, honesty requires that an attempt be made, even if only in passing, to dispel the doubts.

One of directions of development in our age giving rise to a great of debate is the fragmentation of medicine into countless tiny, highly specialized fields caused by the growing demand to strive for increasingly precise diagnosis and for more effective forms of treatment tailored to the individual. However, this obligation represents a task full of pitfalls for the doctor working in a small specialized area and wishing to satisfy this demand. What is so big and difficult about this task? One must remain, in a certain sense a general physician well trained and capable of synthesis, that is, having a horizontal approach, and at the same time also becoming an expert moving vertically and thoroughly aquainted with a special task, a specialist who is also at home in the related disciplines. It is in possession of this type of knowledge that we can become capable to help the patients, even despite the handicaps outlined above. In other words, our treatment results should not differ greatly from those of the leading big powers. "Ut desint vires tamen est laudanda voluntas", and here forces are to be understood principally as material forces. Of course, when I began this work thirty years ago I did not begin to take my steps in a planned, linear way, after thinking through what has been said above, but was forced by necessity to do many things and to follow what at that time still appeared to be sidetracks. It was only after evaluating the work of this period that I became committed to writing down this train of thought as I have done and that I felt myself authorized to write this book. The sidetracks later proved to be important routes and they meant the involvement of indispensable colleagues whose self-sacrificing help appears all the more important the more distant the point from which it is viewed.

I cannot fail to mention my contemporaries either, since it was their direct influence, inspiration, criticism and not least of all their active help that created the environment in which those who dealt with immunology in Hungary worked and developed as self-made men: Richard Backhaus, Lóránt Kesztyűs, János Gergely, István Földes, György Medgyesi, Katalin Merétey and the most versatile expert in the field of immunochemistry, György Kocsis. Among those working in other countries, I had contacts with Professors Hitzig, Barandun, Skvaril, Moghadam, Kohn, Damacco and Bonomo. The list is not complete, even if we consider that it contains only the names of those members of the older generation with whom I came into contact.

The history of immunology as a discipline can be divided up into periods, but since there is not much point in discussing these stages I would rather draw attention to the change in character of the present which is actually of importance not only for immunology research but for basic research in medicine in general. Those means and - what is perhaps far more important - principles for the observation of the effect of biologically active substances, which consisted in the use of suitable methods based on rapid changes can no longer be regarded as the path to follow since the focus of attention will inevitably be directed towards the importance of perceiving the slow, and not always striking, changes, even though at present we do not possess the methods required for this. This means that we must urgently find the method and technique needed to monitor these processes and, within them, the pathological changes. In this way the two together, observation and the method elaborated, will lead to important new findings.

There are disciplines which can be cultivated, if not fully, but nevertheless almost in themselves. Immunology is not one of these but rather, a typical example of the eclectic disciplines. It deals with protein chemistry, citology, membrane biology, eukaryotic genetics, haematology and oncology in order to clarify the processes within immunology, or at least to throw more light on them. As a consequence, the synthesis of different terminologies, which are inevitably present is leading to the creation of a specific language. It is very important though that this separate language should not become esoteric but should be comprehensible for all doctors, since immunology and the way of thinking at the bedside are indispensable in everyday care of patients. Precisely for this reason, my aim has been to give a readily understandable description of this field, even if it is well known that the processes mentioned are not yet precisely studied or are perhaps often almost entirely unknown.

# History of the study of gammopathies

The first indirect proof of the importance of immunoglobulins and their role in preserving the homeostasis of the organism came from Jenner who vaccinated children againts the dreaded smallpox in a tiny house in the county of Berkeley in 1978 (Figure 1).

The direct proof of the presence of some kind of protein, later discovered to be part of the immunoglobulins, was provided by Henry Bence Jones in 1846. The story is in itself an excellent example of how the possibility provided by clinical practice was utilised to produce a result for which no explanation could be provided at the time but for many decades the phenomenon reported aroused the interest of researchers and of experts dealing with it (13).

Edward Jenner (1749–1823) Antoine Maxime Monsaldi Med. History 22, 335. 1978.

Fig. 1

In January 1846 a 45-year-old man in Cavendish Square, at the corner of Harley Street in London, was terminally ill.* The patient was treated by a Doctor MacIntyre, who is not recorded in the literature but who was actually the one who observed the urine described as "urine of a very high specific gravity". He reported his observation in a not internationally recognized journal in 1850 (18) some four years after Jones lecture on the same subject (11) and three and two years after the publication of his results, of which, in retrospect, the Lancet was the most important forum of publication (12, 13).

Henry Bence Jones migrated from the region of Transylvania in the early 1800 s and judging from his name he must have been from a mixed family of Romanians and Szeklers. He was primarily a physician-chemist, the contemporary and friend, as well as the doctor of Huxley, Darwin, Spencer and Faraday.**

H. Bence Jones received a sample for examination of the urine of the MacIntyre's patient - who observed that it contained a substance which precipitated when heated to 50-60 °C and again dissolved at boiling point, then precipitated again when cooled back to 60 °C - Watson, who was also one of the doctors treating the patient also sent him a urine sample. The letter Watson attached to the urine sample is of special interest since it contained the following (Figure 2).**

The name of the patient whose urine with such mysterious properties had been studied remained in obscurity for over a century until, in 1967, John Clamp found the death certificate attesting that Thomas McBean died of "atrophy from albuminuria" (27), (Figure 3).

* It is interesting that this part of the city is still the prestige district of English doctors.

** I take this opportunity to express my gratitude to Professor J.Kohn of London for placing the data and manuscript at my disposal.

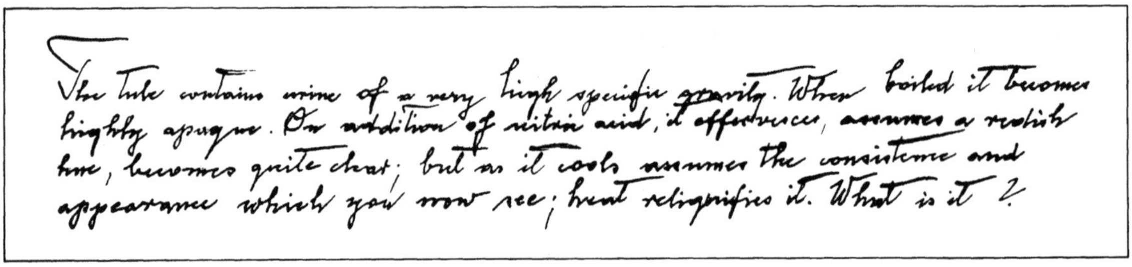
The tube contains urine of a very high specific gravity. When boiled it becomes highly opaque. On addition of nitric acid, it effervesces, assumes a reddish hue, becomes quite clear; but as it cools assumes the consistence and appearance which you now see; heat reliquifies it. What is it?

Fig. 2

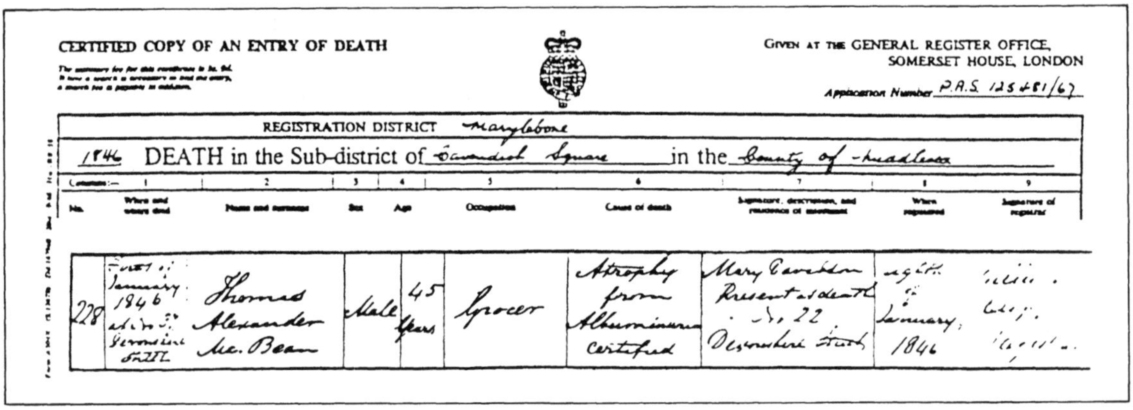
CERTIFIED COPY OF AN ENTRY OF DEATH

GIVEN AT THE GENERAL REGISTER OFFICE, SOMERSET HOUSE, LONDON

Application Number P.A.S. 125481/67

REGISTRATION DISTRICT Marylebone

1846 DEATH in the Sub-district of Cavendish Square in the County of Middlesex

| No. | When and where died | Name and surname | Sex | Age | Occupation | Cause of death | Signature, description, and residence of informant | When registered | Signature of registrar |
|---|---|---|---|---|---|---|---|---|---|
| 228 | First of January 1846 at No. 22 Devonshire Street | Thomas Alexander Mc. Bean | Male | 45 Years | Grocer | Atrophy from Albuminuria Certified | Mary Cavendish Present at death No. 22 Devonshire Street | Eighth of January 1846 | [illegible] |

Fig. 3

This diagnosis is of exceptional note, for it was less than a decade since the term albumin, and protein in general, had won general acceptance.

Five persons took part in the autopsy. The three doctors already mentioned were among those present. Together they observed the deformities caused by what we now call pathological fractures, the softness of the bones, and the "islands" in the bone marrow differing in colour from the normal bone marrow. The histological examinations were carried out in the laboratory of Dalrymple who found that there were droplets of fat and a mass of nucleated cells in the islands (6). He did not form an opinion on the classification of the cells, but his contemporary, Solly, considered on the basis of the pathological report that a malignant process was probably involved (31). Taking into account the histological characteristics observed, Rustizky created the expression "myeloma" in 1873 (29).

The merit of describing normal plasma cells and hypothesizing certain functional relations goes to Tamás Marschalkó, who was a university professor in Kolozsvár (now Cluj, Romania), where he was also buried (19, 20), (Figures 4, 5).

The Marschalkó family came from Germany and settled in Hungary in the 12th or 13th century, making their home in the Szepesség region (now in Slovakia) (5) although it must also be stressed that the family - as a name - had two coats-of-arms,only one of which - judging by its features - could have originated in Northern Hungary, while the other is clearly of Transylvanian origin.*** The original name was "Marschalkó" and it

*** Marschalkó family. The coat of arms is a two-tailed lion rampant bearing a vert branch on a vert field in an azure ground on the escutcheon. On the crown of the helm above the escutcheon, between two spread eagle wings stands a crane on one leg, holding a ball in its other raised foot. The escutcheon is surrounded by mantling. Nagy Iván: Magyarország családai czimerekkel és nemzedék-rendi táblákkal (The families of Hungary, with coats of arms and genealogical tables). 4, 337. 1860

Marschalkó family. Coat of arms: on an azure escutcheon a lion emerging from a crown bears a sword with a Turkish head; crescent moon and star in the chief; crest: armoured arm bearing a sword with a Turkish head; charges: azure-gold, red-silver. Cf.: NI. VII/337; T. III/140, Siebm. 404, Andr. 52-3, rcsk. Kempelen Béla: Magyar Nemes Családok. 7, 90. 1913

Tamás Marschalkó (1862–1915)

**Fig. 4**

Otto Kahler (1849–1893)

**Fig. 6**

*Harminczkilenczedik évfolyam.* 4. szám. *Budapest, 1895. január 27.*

ORVOSI HETILAP.

A HAZAI ÉS KÜLFÖLDI GYÓGYÁSZAT ÉS KÓRBUVÁRLAT KÖZLÖNYE.

Alapította Markusovszky Lajos dr. 1857-ben.

KIADÓ TULAJDONOS ÉS FELELŐS SZERKESZTŐ: HŐGYES ENDRE EGYETEMI TANÁR.

EREDETI KÖZLEMÉNYEK.

Az úgynevezett „Plasmasejtekről", adatok a lobos beszürődés sejtjeinek eredetéhez.[1]

(Egy szinnyomatú tábla melléklettel.)

*Marschalkó Tamás* dr.-tól.

**Fig. 5**

could have acquired the "ó" either in the Szepesség region or in Transylvania.

However, it was Otto Kahler who first thought of a link between the plasma cells, the pathological plasma cells and Bence Jones protein in the urine and in the close to a century since the publication of his opinion (1889) the literature has used the diagnosis of Kahlers disease as a synonym for myeloma multiplex (14).

Kahler worked in Prague and later became professor of internal medicine in Vienna (4), (Figure 6).

The period up to the beginning of the twentieth century could perhaps be called the primitive age in the history of monoclonal gammopathies when nothing was known as yet even about the protein molecule itself. This was followed by the period, in which the immunoglobulins became known, which could be called the middle ages. However, this period itself is also divided into two sections:

1. the study of the solubility of the globulin molecules, followed by the investigation of the link between the appearance of the proteins and the disease (1, 2, 7, 8, 9, 15, 21, 22, 23, 30, 32, 33, 35, 36, 37).

2. the period between 1940 and 1965 when these globulins were called gammaglobulins and scientists came closer to the true explanation of the processes which represent the relationship of cause and effect in the symptoms of the disease (16, 17, 25). It is during this time, moreover, that we encounter the term paraprotein coined by Apitz (3) and gammopathy created by Waldenström (34).

The present age has identified the five classes and the subclasses of human immunoglobulins and their complete primary structure - including also the three-dimensional structure of the light chains and Fab fragments - and has studied their morphological structure, (24, 28); furthermore, the specific functions, as well as the biosynthesis and genetic control (26) were analysed. Of these molecules while Porter and Putnam carried out research on the structural and physical properties of the immunoglobulins, Kunkel reported more and more results from the serological, genetic and clinical side which became decisive in shaping our present conception.

Naturally, the scope of the problem covered by the monograph cannot be considered as closed with this, not only because the genetic background of the gammopathies originating from heavy and light chain disease, as well as gammopathies due to the deletion or duplication of the immunoglobulins needs further investigation but also in view of the countless indefinable unknown factors which it is imperative to clarify especially in order to reveal the cause of the failures observed in therapy.

## References


1. Adams, W.
Mollities ossium.
Trans. Path. Soc. of London 23, 186. 1872.
2. Anders, J. M., Boston, L. N.
Bence Jones albumosuria, with report of three cases.
Lancet. i. 93. 1903.
3. Apitz, K.
Die Paraproteinosen.
Virchows Arch. f. pathol. Anat. 306, 631. 1940.
4. Berkessy, S.
Otto Kahler és a myeloma multiplex.
Orv. Hetil. 129, 1933. 1988.
5. Berkessy, S.
A plasmasejtek származásának felfedezése.
Dr. Marschalkó Tamás élete és munkássága.
B.-A.-Z. megyei tanács kórház nyomdaja.
Miskolc. 1989.
6. Dalrymple, J.
On the microscopical character of mollities ossium.
Dublin Qu. J. Med. Sci. 2, 85, 1846.
7. Geschickter, C. F., Gopeland, M. M.
Multiple myeloma.
Arch. Surg. 16, 807. 1928.
8. Howell, G. M., Alexanian, R.
Multiple myeloma in young persons.
Ann. Inter. Med. 84, 441. 1876.
9. Jackobson, V. C.
A case of multiple myeloma with chronic nephritis showing Bence Jones protein in urine and blood serum.
J. Urology. 1, 167. 1917.
10. Jenner, E.
An inquiry into the causes and effects of the variolae vaccinae, a disease discovered in some of the western counties of England, particulary Gloucestershire, and known by the name of the cow pox.
Sampson Low, Soho, London. 1798.
11. Jones, H. B.
Papers on chemical pathology: prefaced by the Gulstonian Lectures, read at the Royal College of Physicians, 1846.
Lancet. ii, 88. 1847.
12. Jones, H. B.
On a new substance occuring in the urine of a patient with mollities ossium.
Proc. R. Soc. 5, 673. 1847.
13. Jones, H. B.
On a new substance occuring in the urin of a patients with mollities ossium.
Philos. Trans. R. Soc. 55. 1848.
14. Kahler, O.
Zur Symptomatologie des multiplen Myeloms. Beobachtung von Albumosurie. I–II.
Prager med. Wschr. 14, 33. 1889.
15. Krauss, E.
Studien zur Bence-Jones'schen Albuminuria.
Deutsch. Arch. F. Klin. Med. 137, 257. 1921.
16. Kunkel, H. G., Tiselius, A.
Elektrophoresis of proteins on filter paper.
J. Gen. Physiol. 35, 89. 1951.
17. Kunkel, H. G., Slater, R. J., Good, R. A.
Relation between certain myeloma proteins and normal gammaglobulin.
Proc. Soc. Exp. Biol. Med. 76, 190. 1951.
18. MacIntyre, W.
Case of mollities and fragilitas ossium, accompanied with urine strongly charged with animal matter.
Med. chir. Trans. (London). 33, 211. 1850.
19. Marschalkó, Th.
Ueber die sogenannte Plasmazelle, ein Beitrag zur Kentniss der Herkunft der entzündlichen Infiltrationszelle.
Arch. Derm. Syph. 30, 3. 1895.

20. Marschalkó, T.
Az úgynevezett „Plasmasejtekről", adatok a lobos beszűrődés sejtjeinek eredetéhez.
Orv. Hetil. 39, 36. 49. 63. 1895.
21. Michels, N. A.
The plasma cell: a critical review of it's morphogenesis, function and development capacity under normal and under abnormal conditions.
Arch. Pathol. 11, 775. 1931.
22. Parkes-Weber, F.
A case of multiple myeloma with Bence-Jones protein in the urine.
Lancet. i. 731. 1903.
23. Pelzweig, W. A., Delrue, G., Geschickter, C.
Hyperproteinaemia associated with multiple myelomas: report of an unusual case.
JAMA. 90, 755. 1928.
24. Porter, R. R.
The structure of the heavy chain of immunoglobulin and it's relevance to the nature of the antibody-combining site.
Biochem. J. 105, 417. 1967.
25. Putnam, F. W., Hardy, S.
Proteins in multiple myeloma III. Origin of Bence Jones protein.
J. Biol. Chem. 212, 361. 1955.
26. Putnam, F. W.
Immunoglobulins I. II. III. structure.
In: The plasma proteins.
Ed.: by Putnam, F. W.
Academic Press, New York. 3, 1. 155. 223. 1977.
27. Putnam, F. W.
The post, present, and future of the immunoglobulins.
In: Clinical immunochemistry. The molecular basis of immunochemistry.
Section II.
Molecular factors in the immune response.
Ed.: by Natelson, Pesce, Dietz.
American Association for Clinical Chemistry. W. 45. 1978.
28. Putnam, F. W.
From the first to the last of the immunoglobulins.
Perspectives and prospects.
Clin. Physiol. Biochem. 1, 63. 1983.
29. Rustizky, J.
Multiples myelom.
Dtsch. Z. Chir. 3, 162. 1873.
30. Shar, A. F.
A case of plasma cell myeloma.
J. Path. Bacteriology. 26, 125. 1923.
31. Solly, S.
Remarks on the pathology of mollities ossium: with cases.
Med. Chir. Trans. 27, 435. 1844.
32. Tiselius, A.
Electrophoresis of serum globulin. II.
Electrophoresis analysis of normal and immune sera.
Biochem. J. 31, 1464. 1937.
33. Tiselius, A., Kabat, E. A.
An electrophoretic study of immune sera and purified antibody preparations.
J. Exp. Med. 69, 119. 1939.
34. Waldenström, J.
Studies on conditions associated with disturbed gammaglobulin formation (gammopathies).
Harvey Lect. 56, 211. 1961.
35. Weber, F. P.
A case of multiple myeloma (myelomatosis) with Bence Jones proteid in the urine (myelopathic albumosuria of Bradshaw Kahler's disease) and summary of Bence Jones albumosuria: with a report on the chemical pathology by R. Hutchison and J. J. R. Macleod.
Med. Chir. Trans. 86, 395. 1903.
36. Wintrobe, M. M., Buell, M. W.
Hyperproteinaemia associated with multiple myeloma: with report of a case in which an extraordinary hyperproteinaemia was associated with thrombosis of the retinal veins and symptoms suggesting Raynaud's disease.
Bull. Johns Hopkins Hosp. 52, 156. 1933.
37. Wright, J. H.
A case of multiple myeloma.
Johns Hopkins Hosp. Rep. 9, 359. 1900.

# Abbrevations

| | |
|---|---|
| α-ANF | alpha-atrial natriuretic factor |
| α-TdT | alpha-terminalis-deoxynucleotidyl-transferase |
| AcP-T | Acid phosphatase tartarat |
| ADA | Amyloid degrading activity |
| ADCC | Antibody dependent cell-mediated cytotoxicity |
| ADF | Amyloid degrading factor |
| AEF (1) | Allogenic effect factors |
| AEF (2) | Amyloid enhancing factor |
| AFC | Antibody forming cell |
| ALL | Acute lymphoid leukemia |
| ALS/PD | Amyotrophic lateralsclerosis/ Parkinson's dementia |
| AML | Acute myeloid leukemia |
| ANA | Anti-nuclear antibody |
| APC | Antigen presenting cell |
| ARDS | Acut respiratory distress syndrome |
| ASF | Acid soluble fraction |
| ATL | Adult T-cell lymphoma |
| AUO | Amyloid of unknow origin |
| beta-APP | beta-amyloid praecursor protein = A 4. |
| BCDF | B-cell differentiation factor |
| BCGF | B-cell growth factor |
| BMPC-LI | Bone marrow plasma cell labeling index |
| BSF-2 | B-cell stimulatory factor |
| BUdR | Bromdeoxyuridin |
| BUN | Blood urea nitrogen |
| CAH | Chronic active hepatitis |
| CALLA | Common acute lymphoblastic leukaemia antigen |
| CAPD | Continous ambulant peritoneal dialysis |
| CBE | Clonal B-cell excess |
| CDR | Complementarity determining regions |
| $C_L^P$ | Light chain identified protein |
| $C^+$ | Light chain identified protein positivity |
| CLL | Chronic lymphocytic leukemia |
| CNS | Central nerveous system |
| CTL | Cytotoxic Tlymphocyte |
| CTX | Cyclophosphamide |
| DCF | 2'-deoxycoformycin |
| EBNA | Epstein-Barr nuclear antigen |
| EBV | Epstein-Barr virus |
| ECF-A | Eosinophilic chemotactic factor of anaphylaxis |
| EGF | Epidermal growth factor |
| FAC | Familial amyloid cerebral localisation |
| FAEF | Fibril derivated amyloid enhancing factor |
| FAP | Familial amyloidotic polyneuropathy |
| FFP | Fresh frozen plasma |
| FGF | Fibroblast growth factor |
| FMF | Familial mediterranean fever |
| FR | „Framework"-region |
| GALT | Gut-associated lymphoid tissue |
| GMB | Glomerular basal membrane |
| G-CSF | Granulocyte-colony stimulating factor |

| | |
|---|---|
| GVHD | Graft versus horst disease |
| HAN-PC$_1$ | Human antigen normal plasma-cell-1 |
| HC-protein | Human complex-binding protein |
| HCHWA | Hereditary cerebral haemorrhage with amyloidosis |
| HES | Hydroxy-ethyl-Starch |
| HLA | Human leukocyte antigen |
| HLBI | Human lymphoblastoid interferon |
| HNE | Human neutorphil elastase |
| HPLS | High pressure liquid chromatography |
| HSF | Hepatocyte stimulating factor |
| IAA | Isolated atrial amyloidosis |
| IAPP | Islet amyloid polypeptid |
| ICS | Immunulogical component system |
| IDC | Interdigitating reticulum cell |
| IEL | Intraepithelial lymphocytes |
| IFN | Interferon |
| IgBF | Immunoglobulin binding factor |
| IL | Interleukin |
| IL-R | Interleukin receptor |
| J-Ch | Joining chain |
| LAK | Lymphokine activated killer |
| LCD | Light chain disease |
| LCIS | Light chain izotype suppression |
| LDH | Lactate dehydrogenase |
| LGL | Large granular lymphocyte |
| LI | Labeling index |
| LP | Lamina propria mucosae |
| LPP | Lipoprotein |
| LPS | Lipopolysaccharid (pl.: endotoxin) |
| LT | Lymphotoxin |
| M | Melphalan |
| MABs | Monoclonal anitbody B-cell surface |
| MDS | Myelodysplasia syndrome |
| MDR (1) | Maligmant drug resistance |
| MDR (2) | Multidrug resistance |
| MG | Monoclonal gammopathy |
| MGUS | Monoclonal gammopathy of undetertermined significance |
| MHC | Major histocompatibility complex |
| MLN | Mesenteric lymph nodes |
| MM | Myeloma multiplex |
| MP | Melphalan + Prednisolon |
| MoAb | Monoclonal-antibody |
| NC | Natural cytotoxic cell |
| NCF | Neutrophil chemotactic factor |
| NK | Natural killer cell |
| OAF | Osteoclast activating factor |
| P | Prednisolon |
| PAF | Platelet activating factor |
| PAG (PAGE) | Polyacrylamid-gel electrophoresis |
| PBA | Polyclonal B-cell activator |
| PBC | Primary biliary cirrhosis |
| PBL | Peripheral blood lymphocyte |
| Pc/Ig | Polyclonal immunoglobuline |
| PCA | Passive cutan anaphylaxis |

| | |
|---|---|
| PCL | Plasma cell leukemia |
| PDGF | Platelet derivate growth factor |
| PFC | Plaque-forming cell |
| PG | Prostaglandin |
| POEMS syndroma | Polyneuropathy, organomegaly, endocrinopathy, monoclonal gammopathy, skin changes |
| PPD | Purified protein derivate |
| PPPP | Post proctoscopic periorbital purpura |
| PSP | Progressiv supranuclear palsy |
| PTH | Parathormon |
| RC | Railed cell |
| SC | Secretory component |
| SDS | Sodium dodecyl sulphat |
| SMM | Smoldering multiple myeloma |
| SRS-A | Slow-reacting substance of anaphylaxis |
| SSA | Systemic senilic amyloidosis |
| TAC (tac) | T-cell antigen activation with concanavalin-A. = IL-2R |
| TAT (tat) | Transcriptional activator factor |
| Tdt (TDT) | Terminal deoxynucleotidyl transferase |
| TGF | Transforming growth factor |
| THF | T helper cell factor |
| TNF | Tumor necrosis factor (antigen specific suppressor factor) |
| TPA | 12-o-tetradecanoyl-phorbol-13-acetate (phorbol ester) |
| TRAP | Tartarat resistent acid-phosphatase |
| TRF | T-cell replacing factor |
| TRITC | Tetraethyl-rhomadin-iso-thio-cyanat |
| VIP | Vasoactive intestinal polypeptide |

# Diseases

# Index

# I. The structure of the immune system and the relationship between its components

(with special emphasis on antibody production)

Afferent, central and efferent phases comprise the three functional steps of the immune response.

The afferent response involves two types of immune reactions, non-specific response, mainly mediated by granulocytes and a specific one, involving monocytes and lymphocytes. These cells arise from the pluripotent cells of the bone marrow, fetal liver and possibly the spleen. But in the fetus the most important source of pluripotent cells is the yolk sac. Though the count and morphology of the progenitor cells is still unknown, extensive data exist on the function and morphology of their mature successors. Consequently, an overview of the available information, which is similar to the fragments of a mosaic, will provide a basic outline necessary for the comprehension of these relationships.

## I. Granulocytes

The three types of granulocytes do not show any specificity for antigens or memory.

- Neutrophils. Their role in antigen processing is described later. It is of prime importance, however, that their lysosomal enzymes can penetrate into the tissues eliciting local reactions and inflammation.
- Basophils. These cells are found in very small numbers in the circulation. They enter the circulatory system only for a short period necessary for their migration from the bone marrow to the periphery. Basophilia is due to characteristic blue staining of their cytoplasmic granules, containing histamine, serotonin, and heparin. Therefore, these cells are sometimes identified as plasmatic mast-cells. Functionally they are indistinguishable from the mast cells.
- Eosinophils. The elevated eosinophil count occurs as a reaction of immunity against helminthic infection. By binding to the pathogenic agents eosinophils induce cytolysis. Eosinophil granules contain the following active substances:

1. Major basic protein basic peptide with molecular weight of 10,000 Daltons released during binding of eosinophils to the pathogenic agent for its elimination.
2. Membrane-toxic cationic proteins.
3. Active substances of hypersensitivity promoting the control function of the specific immune cells.
4. Extract of eosinophils inhibits the inflammatory reaction in vitro (no data are available on its effect in vivo ).
5. SRS-A (Slow Reactive Substance of Anaphylaxis) decreases the cytotoxic effect of eosinophils.

## II. Monocytes

A wide variety of cells between bone marrow monoblasts and tissue macrophages. These bone marrow-derived cells show continuity in their development:

- Monoblasts are rarely found in the circulation, hence they can mainly be observed in monoblast leukemia.
- Promonocytes: due to their capability of adhesion to glass surfaces, promonocytes can be studied, albeit with some difficulties. These cells with high proliferation rate are of 10-20 microns in diameter, and exhibit certain phagocytosis. Promonocytes practically contain no lysosomal enzymes.
- Monocytes: the diameter of the circulating cells is smaller, compared to those in the other tissues; the DNA synthesis of these cells is low. The lysosomes are well developed and can be easily detected. The nuclear to cytoplasmic ratio, compared to other cells with similar functions, is lower. The nucleus has a typical "horseshoe" or "bean" shape. Phagocytosis is

markedly higher, compared to the immature forms.
- Monocytes: monocytes migrating into tissues increase in diameter and differentiate into macrophages and histiocytes. The term "macrophage" usually suggests free cells, for example, peritoneal macrophages, while "histiocyte" refers to cells fixed in different tissues, for example Kupffer cells. The latter are the final differentiated cell forms with no capability for further proliferation; their main role is phagocytosis and presentation of antigens to antigen-sensitive T-cells with subsequent release of monokines. According to available evidence these two functions can be either interrelated, or independent.

A system of so-called dendritic cells also belongs to this group.
- Four main groups of dendritic cells exist. Common characteristics include numerous extensions of protoplasma and an irregular nucleus. They show minimum phagocytic or endocytic activity, furthermore, carry no markers typical of T or B lymphocytes. They, however, do have MHC II receptors responsible for their antigen presenting activity.

  1. Lymphoid dendritic cells. Their relationship to lymphocytes, monocytes or reticulocytes is debated. Bone marrow gives rise to these cells, but they are also observed in the lymphatic nodes and spleen. These dendritic cells comprise a part of the spleen adherent cells. They have numerous mitochondria, but their plasma lacks free ribosomes. Dendritic cells carry Fc and MHC II antigen markers and can bind the complement components. They are insensitive to gamma-, but sensitive to ultra-violet irradiation. These cells play an active part in graft rejection.

  2. Follicular dendritic cells. They occur in lymphoid follicles. They have low cytoplasm content and few subcellular organelles. Their main function is "capturing" or "retinealization" of antigens, which might be a mechanism of developing cellular memory. This type of dendritic cells rather exhibits relationship with B lymphocytes. These cells display MHC II, Fc and $C_3$ binding activity. They differ from the lymphoid dendritic cells by having more and longer cytoplasmic processes which contain granules of unknown function.

  3. Interdigital dendritic cells are located in the T-dependent areas of the lymphoid tissue. They might show relationship with the lymphoid dendritic cells. Their main task is maintenance of communication with the T lymphocytes. These cells have a developed system of cytoplasmic granules, termed Birbeck granules. They do not possess Fc receptors, but according to recent data carry MHC II antigens. Bone marrow mononuclear progenitors give rise to these cells.

  4. Langerhans cells are the specific elements of skin epidermis. Histologically they can be stained with gold. Langerhans cells are in functional contact with the T lymphocytes. The main task of these cells carrying MHC II antigen and Fc receptors is antigen presentation. They also contain Birbeck granules, thus resembling to the interdigital cells.

## III. Lymphocytes

Lymphocytes recognize foreign molecules and develop cellular memory. They are also responsible of antibody production and cytotoxicity.
- Lymphocytes are the most common cellular elements of the lymphoid organs. Their diameter varies between 7 and 10 micrometers. Isolated lymphocytes or small groups of cells occur in the epithelium (for example in the mucosal lamina propria). These cells possess a high nuclear to cytoplasmic ratio and their cytoplasm is free of organelles. Antigenic challenge induces blast transformation. Three distinct categories of lymphocytes can be distinguished concerning their functions and cell surface markers:

  *A)* T lymphocytes are produced in the thymus. T lymphocytes express typical Thy-1 antigen markers and do not carry immunoglobuline receptors on their surface. It is noteworthy that some cells of the central nervous system also possess the same surface antigen marker, however its functional implication remains unknown. The following T cell subpopulations are defined: helper cells ($CD_4$), suppressor cells ($CD_8$), $T_{DTH}$ and specific killer cells.

*B)* B cells are capable to transform into antibody producing plasma cells following an antigenic challenge. B cells are identified by the endogenously produced immunoglobuline markers typical of this B cell.

*C)* Null cells. These cells do not carry either cell surface T antigens, or immunoglobuline B markers. Functionally they represent the so-called NK cells killing tumour cells.

- Mast cells. Play an important part in allergic reactions. They are commonly found in connective tissue, thus remaining beyond the immune system and circulation. During parasite infections, however, a marked accumulation of these cells is observed in the lymphatic nodes, especially in the cortical regions. Though the histologic origin of the mast cells is debated, it is assumed that they develop from pluripotent stem cells. Phylogenetically these primitive cells are identified in vertebrates having no developed system of blood cells. Mature cells can not proliferate. Their plasma contains numerous basophil granules releasing histamine, heparin, zinc and hydrolytic enzymes on degranulation initiated by allergens. These released mediators cause typical tissue effects and allergic symptoms.
- Platelets. In addition to their role in blood clotting, the vasoactive amines and enzymes of platelets are also involved in allergic reactions.

The above mentioned cells and tissues are organized into the organs of the immune system.

- Bone marrow is a system of hemopoietic and adipose tissue embedded into the vascular and reticular matrix. Both flat and long bones contain bone marrow. Bone marrow comprises about 2-5% of the body weight, its volume is about 2-3 litres. It produces precursors of all the blood cells from the stem cells. B cells in different stages of maturation are detected in the bone marrow, however, it contains no T cells. Bone marrow has no afferent lymphatics, the new blood cells leave the bone marrow entering the circulation via a system of efferent lymphatics. Bone marrow is a major site of lymphopoiesis and, therefore, is considered as a primary lymphoid organ.
- The thymus. It develops from the embryonic third pharyngeal sac. Thymus is a Greek word, originally meaning "thyme" because its structure resembles the leaves of this plant. Thymus has a typical bilobed structure. This first lymphoid organ develops by the 9th week of gestation. Referred to the whole body mass thymus shows a constantly decreasing tendency. Its absolute size reaches its maximum at puberty and after that period thymus exhibits continuous involution and fatty degeneration. However its potential function still remains utmostly important. The lobules of the thymus are separated from each other by connective tissue. The lobules consist of cortical and medullar areas. Cortical lymphocytes are very prone to steroids, while the medullar are not. Lymphocytes pass from the thymus into circulation mainly via the medullar veins. During their brief stay in the circular system more than 90% of lymphocytes are destroyed. The overall effect of lymphocyte traffic shows that the total weight of all the lymphocytes produced every two-three days by the thymus equals to the weight of this organ. The mature subpopulations and/or newly produced T lymphocytes still migrate to the thymus during its involution. Thymus drains these cells through the efferent lymphatics. Therefore, in the later stages thymus can not be considered as a primary immune organ. Epithelial cells of the thymus secrete thymosine, a a growth factor promoting T lymphocyte maturation both in the thymus and peripheral lymphatic organs. Thymectomization in neonatal periods greatly depletes the lymphocytes of the paracortical and medullar regions of the lymphatic nodes and those localized in the marginal zones of the germinal regions in the lymphatic nodes and the spleen (white pulp). These are the so-called thymus-dependent zones. Thymectomization in adults reduces

the resistance against infections, but is neither pronounced, nor lethal.

- The bursa of Fabricius. The bursa is a lymphoid epithelial organ in avian and some reptilian species. It develops from the intestinal epithelium, its structure shows great similarity to that of the thymus. The only difference is that the bursal cortical and medullar follicles are separated from each other by epithelial cells and not by the connective tissue. Following an early removal of the bursa (hatch period) the germinal centres fail to develop, the lymphatic nodes and the spleen remain underdeveloped and the maturation of the antibody producing plasma cells does not occur. Removal of the bursa before hatching exhibits no such sequellae. The above results indicate the functional importance of the bursa and its relationship with immune system.
- The spleen. It is surrounded by a collagenous capsule. A matrix of connective tissue trabecules and cells belonging to RES penetrate into the organ dividing it into parts containing active cells. The so-called white pulp contains small and large lymphocytes, monocytes, macrophages, reticular cells and lymphatic stem cells. The lymphocytes are organized in follicles possessing germinal centres which contain mature plasma cells. Despite its important immune function, the spleen can be removed. However splenectomy performed in children and adults differently affects their immunoreactivity. Spleen is included in the continuous active traffic of lymphocytes via the circulatory system. Lymphocytes leave through the efferent channels into the splenic veins. The terminal branches of the splenic artery end in the red pulp, at the region of venous sinuses.
- Lymph nodes. Lymph nodes are less than 1 cm in diameter. They form a network comprising a part of lymphatic system. Lymph nodes are surrounded with a collagenous capsule. Lymphocytes are embedded in a system of reticular cells. Lymph nodes filter and capture antigens (small particles and molecules). Not only lymphatics, but also blood vessels (artery and vein) enter and leave lymph nodes at the hilus, maintaining lymphocyte traffic between these two systems of "tubes". Afferent lymphatic penetrates the capsule of a lymph node at its convex surface. This cortical part contains aggregates of lymphocytes. The medulla forms sinuses which are drained into the efferent lymphatics at the hilus.
- GALT (Gut-associated lymphoid tissue) is a dispersed non-encapsulated lymphoid tissue specialized to perform uniform functions. This system comprises the most primitive elements of the immune system and occurs in similar forms in all vertebrates. Diffuse cells and small lymphocyte aggregates in the lamina propria, Peyer's patches with complex specialized structure, tonsils and the appendix, as well as the system of associated lymph nodes comprise the GALT. GALT activity is important in the maintenance of the homeostasis and will be described later (for details see Immunology of Gastrointestinal Tract).
- Reticuloendothelial system (RES) is a system of phagocytic cells and tissues. Due to their degrading metabolic activity, they effectively metabolize and eliminate foreign particles. RES cells (including fixed macrophages and dendritic cells) possessing surface antigens (epitopes), exhibit high antigen presenting activity (APC). The RES cells are the basic elements of the bone marrow, lymph node and spleen matrix. They are supported by collagenous trabecules and blood vessels. Such cells embedded in spongious structure are called histiocytes. Germinative centers and the red pulp of the spleen also contain RES cells. The suggestion that the erythroid and lymphoid stem cells are derived from these cells needs further evidence. The task of the RES cells of the liver and spleen is to remove the "old" erythrocytes. Its cellular mechanism is well documented, however the molecular mechanism needs further investigation. Evidence exist that these cells synthesize and secrete bile.
- Lymphatic circulation. The cells of the immune system at certain periods move through one lymphoid organ to another via lymphatics and blood stream, transferring information between the organs, performing actual tasks and control function.

This migration resembles that of the passengers taking various means of transport. Therefore, only a minor part of lymphocytes circulates in the blood vessels. The tissue lymphatic system starts with microscopic sinuses, whose walls contain cells without any apparent structure. Lymphatic vessels recirculate blood-born proteins, carbohydrates, and water, as well as different substances produced by the cells from the interstitium into the blood stream. The normal 24-hour lymph flow (including lymphocytes) is 1-2 L. Lymphatics will also transport tumor cells, which is important in spreading metastases. A small ratio of lymphocytes is derived from the interstitium, the rest migrate from the primary lymphoid organs. The ratio of T lymphocytes is higher because B cells comprise only 15-20%. Under normal conditions lymph contains no red blood cells, monocytes or granulocytes. The paracortex and cortex of a lymph node is a gate between the lymphatic and blood systems. Specialized post-capillary venules lined with typical high endothelium (High Endothelium Venules) or its accumulates are the main exit route for the lymphocytes. High endothelium is not a simple passive barrier between lymph and blood, but is an important regulatory element controlling lymphocyte migration. Thus, for example, the non-differentiated pre-T lymphocytes derived from bone marrow carry surface receptors determining their migration via the vascular endothelium of the thymus: endothelial identification in the first stage retains the cells preventing them from passing, later, however, endothelium promotes their migration into the thymus. Similar mechanisms might regulate the T lymphocyte migration from the bone marrow or thymus into the spleen or lymph nodes, or the traffic of the B cells into the Peyer's patches. Endothelial cells produce a vast amount of factors regulating this process, as well as the growth and proliferation of lymphoid cells. It is noteworthy that the high endothelial cells disappear from the postcapillar veins of the lymph nodes in rats free of T lymphocytes.

In summary, we can conclude that the lymphocyte traffic is not a passive mechanical process, but is an example of a complex active regulation.

After a brief overview of the immune cells and organs, the analysis of the afferent phase of immune reaction with a special emphasize of its humoral response is presented. The cellular immune response is only mentioned to show the relationship between the two mechanisms. Therefore, it seems logical to start with the mechanisms of immunization, thus enlightening the processes leading to formation of immunoglobulins and the whole humoral immune response.

## Immunization

Neonates have no immune protection of their own. They develop no specific antibodies to a variety of antigens if they were not infected during intrauterine life. Thus, adequate protection, defined as physiological passive immunization, is acquired via passive immunization: in some mammals it occurs through the placenta, in others - with colostrum, via the gut. (An other form of passive immunization is a passive artificial vaccination with a parenteral transfer of planned or "spontaneously" produced antibodies).

Similarly, two forms of active immunization exist:

- immunization employing natural factors with subsequent synthesis of antibodies;
- exposure to artificial antigens followed by a planned, maximum rise of antibody titers at desired periods. Antigen itself makes a long way before it stimulates antibody production.

### *Afferent phase*

Antigens entering the organism are first phagocyted either by the non-specific cells exhibiting no APC activity, or by the APC cells, thus activating specific immune response. As a rule opsonization occurs prior to phagocytosis. Genetics of molecular and cellular parameters of immune reactions including opsonization needs further investigation: opsonization suggests that an antibody could not react with its specific antigen, which has been a trigger of its formation, but also with other, less specific or similar molecules. It also implies the existence of non-specifically binding "janitor" immunoglobulins. Following opsonization, antigens enter the phagocytes cells and get entrapped in small vesicles - "phagosomes". Later they fuse with lysosomes forming "suicide bags" full of lysosomal enzymes. Such fusion is an energy-requiring process. Due to proteolytic effect of the enzymes, free antigen determinants are presented on the cell surface in a combination with MHC II determinants. This form is referred to as an APC cell. Apart from passive antigen presentation, APC cells also release IL-1, a growth factor locally activating the $T_4$ cells. $T_4$ lymphocytes carry $CD_4$, $CD_3$ and TCR determinants on their surface. These receptors are responsible for recognition of the MHC II products and antigen determinants on the macrophage surface. This help signal stimulates the B cells to produce antibodies. However, it is not certain whether antigen forms a bridge between $T_h$ and B lymphocytes, because the probability of such cell cooperation is extremely low. If it occurs, the velocity of this reaction must change in time. Experimental evidence indicate that $T_h$ - B cooperation is only a first specific step. Another non-specific factor - a "signal" is lso required. According to the first possible mechanism the related $T_h$ and B cells cannot be separated from each other, otherwise the non-specific factor would activate other adjacent cells.

The second possibility is based on the above experimental facts; it suggests that the $T_h$ cells bind antigen and deliver a signal promoting contact with the B cells and their activation.

The third model implies "remote" cooperation between the T and B cells. Experiments performed in culture systems, in which $T_h$ cells were separated from the B cells by a membrane permeable to soluble factors, but impermeable to cells indicate that the two factors, TRF and AEF (MW exceeding 10,000 D) elicit the immune response. TRF is not antigen-specific and is responsible for differentiation of B lymphocytes into plasma cells. AEF, a protein with a molecular weight of 30,000-40,000 D, promotes both primary and secondary immune responses to a considerable amount of antigens.

Therefore, there is number of ways in which the $T_h$ and B cells can cooperate:

- actual interaction between the $T_h$ and B cells;
- $T_h$ and B cells are close to each other both binding to the APC;
- $T_h$ and B cells are close to each other; the non-specific TRF and AEF factors derived from the T cells stimulate the B lymphocytes, promoting their differentiation.

Similarly to other biological processes, immune reactions also exhibit regulatory mechanisms. These systems elicit their regulatory function in the central phase of the immune response.

*Central phase*

- MHC II restriction is the first, but not the only mechanism regulating the $T_h$ - B cell cooperation; that means that the cooperating cells only recognize each other provided that both cells share same MHC II determinants. In this sense the cooperation between the T cells and macrophages (APC) is also a regulatory step - lack of similarity of the antigen determinants inhibits further delivery of information. In macrophage (APC)-free systems pure T-cell populations (without memory-cells) are not activated by the presence of antigen. Cell-free supernatant obtained after a several day long incubation of antigens with macrophages substituted the macrophage - T cell cooperation, activating pure T lymphocytes. It is noteworthy that this factor exhibited its effect only if the Ia products of the H-2 region (MHC II determinants) were identical in the macrophages and the T lymphocytes.
- Not only APC, but other cells, for example keratinocytes and synovial cells can produce IL-1. IL-1 is a growth factor also capable of inducing fever and inflammation. This factor removes rare metals from the plasma, suggesting that it is not a homogenous factor, but a series of peptides. Genetic studies proved existence of at least two different IL-1 molecules. The primary task of IL-1 is to activate the $T_h$ lymphocytes. B cells do not require IL-1; the presence of antigen is sufficient for their differentiation into antibody-producing cells. However evidence exists suggesting that IL-1 can affect the B lymphocytes.

- The second step of the MHC restriction is the binding of the activated $T_h$ cells to the surface immunoglobulins of the B lymphocytes. Such binding occurs only if the B cells possess MHC II determinants and are activated by IL-2, as well as BCGF and BCDF released from the T cells. The latter two factors are essential for differentiation of the B cells into antibody-producing plasma cells. The reqirement of similar MHC antigens on the interacting cells indicates that it is the second site of MHC restriction. BCGF is a B-cell growth factor causing proliferation and expansion in the presence of IL-2. Differentiation into plasma cells is induced by BCDF. This factor triggers, for example secretion of IgM antibodies by the cells carrying IgM receptors on their surface (Fig. I/1).

Other regulatory molecules, such as gamma IFN, are capable to enhance or slow down the whole process. Tolerance also comprises the central phase of the immune response. Tolerance means non-reactivity towards particular specific antigens. It differs from immunosuppression discussed later. The latter reflects immune disfunction of all the phases of the immune response with subsequent no-reaction. Antigen-specific tolerance can affect both T and B cells. Tolerance can be considered as one of the forms of immune memory: exposure of the immune system to antigens leads to specific unresponsiveness instead of a usual immune response. A primitive form of tolerance is unresponsiveness to self-antigens. Current data suggest that some T cells of the thymus are potentially self-reactive, but are very susceptible to local hydrocortisone treatment. Though there is no direct experimental evidence, self-reactivity can develop at any later stage of ontogenesis. Independently of the nature of the elimination (or suppression) of self-reactivity during the period of fetal development, the self-reactivity will not disappear, because the maturing T cells might later develop reactivity to the self antigens. It is noteworthy that the self-reactivity or tolerance of the B cells depends on T cell functions. There has been doubt about individual self-reactivity of the B cells. Graft experiments performed on dizygotic animals and inbred strains imply that the mainten-

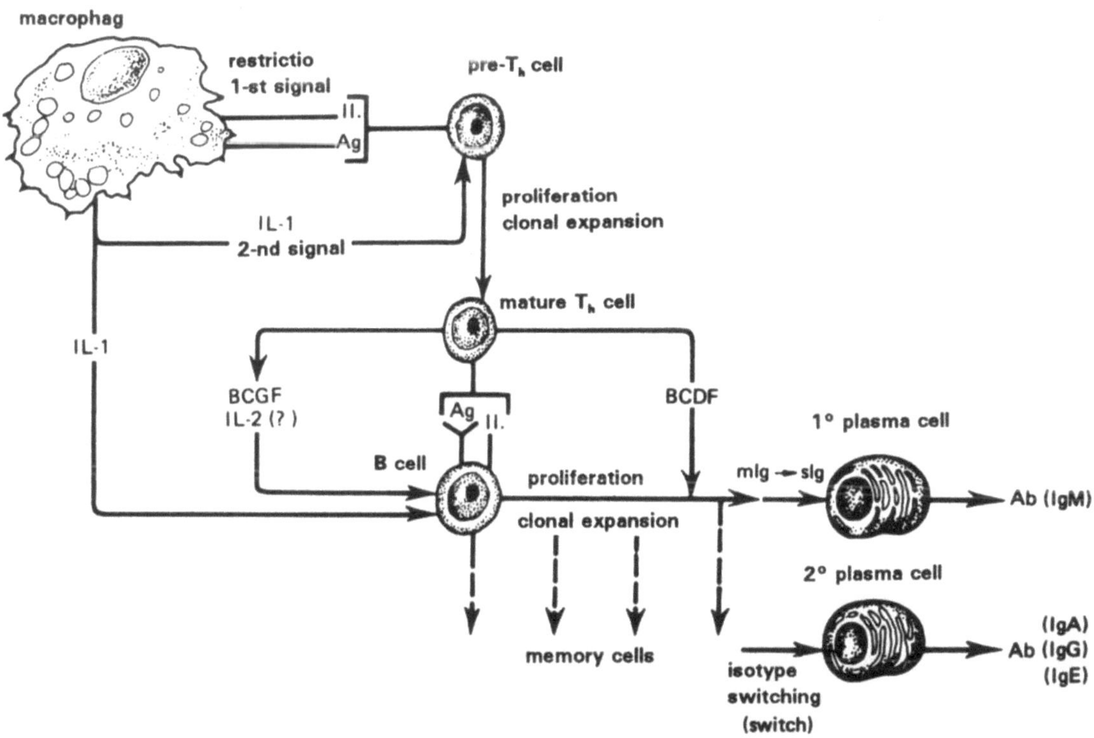


Fig. I/1

ance of the antigen is the basic prerequisite for tolerance. The ageing T cells can not transfer tolerance information to the young cells; this new cell generation must acquire this information individually which is impossible in the absence of antigen.

Immune exhaustion after immunization with pneumococcus polysaccharide is another form of tolerance. Such non-reactivity remaining for the whole life can not be elicited with proteins probably due to the above mechanisms.

Induction of tolerance depends on the dose of antigen, both low-dose and high-dose tolerance exist. Different features of T and B lymphocytes greatly contribute to the development of non-reactivity. Tolerance in T cells, compared to B lymphocytes, can be induced with antigens at much lower (by an order of magnitude) concentrations. The presence or absence of the μ and δ surface markers greatly affects the behaviour of B lymphocytes.

Tolerance is sometimes defined as clonal depletion, meaning that non-reactivity to certain antigens is due to elimination of the antigen-sensitive T or B cells. Later experiments, however, have proved that this is not the case and rather clonal suppression would cause tolerance. There is also evidence that functional non-reactivity of APC leads to a direct interaction between antigen and the T cell. Under these conditions T lymphocytes exhibit suppressor activity. Another antigen-specific factor ($T_sF$, Feldmann factor, M.W. 50,000 Daltons) might also contribute to the development of tolerance. Production of this factor is encoded by the If genes.

Cooperation between $T_8$ suppressors and APC is mediated via the MHC I and the expression of antigen determinants on the membrane. Suppressor lymphocyte carries $CD_8$, $CD_3$ and TCR surface antigens. IL-1 promotes growth of these cells, but later their function is also controlled by IL-2 and IFN.

*Efferent phase*

Following primary antigenic challenge and cell stimulation antibodies are produced. In the analysis of the antibody response it is necessary to consider first the destiny of the activated cells, the intracellular changes and cell migration processes.

B lymphocytes carry surface immunoglobulins behaving as antigen receptors as well. Bone marrow is a primary immune organ producing lymphocytes; lymph nodes, the spleen, lungs, gut and urogenital tract are considered as the secondary organs. Most of the lymphocytes are not fixed, but perform perpetual migration. In the course of this migration they mature, transfer information, and differentiate. Antibody formation is a final step of the differentiation process

In other words, antigen challenge activates the B cells, they begin to proliferate turning into immunoblasts. In several days immunoblasts would leave the locus of their activation and further differentiate either into plasma cells or lymphoplasmacytoid cells, they can also become small lymphocytes (memory cells) or, if differentiation occurs in the germinal centres, turn into centroblasts. Under normal conditions plasma cells function for several hours or days and decay (Fig. I/2).

The metamorphosis of B lymphocytes into plasma cells occurs in regions other than the site of the antigen challenge. Bone marrow plasma cells are derived from B lymphocytes that were activated in the secondary lymphoid organs following recycling between various organs. This finding indicates that the lymphoid cells are derived from the hemopoietic stem cells of the bone marrow.

B cell activation consequently selects and reorganizes the immunoglobulin genes to

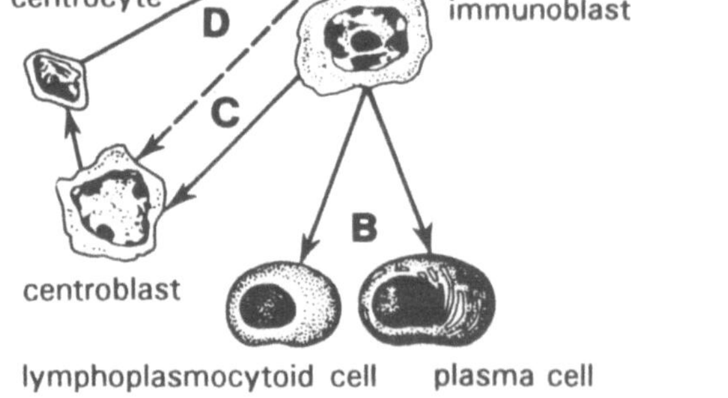


Fig. I/2

stimulate synthesis of different immunoglobulins. The details of this process, however, need further investigation. Genes coding for the chains of antibody polypeptides are found on three chromosome pairs. The chromosome 2 encodes the light chain κ, the chromosome 14 - the heavy chain and the chromosome 22 - the light chain λ. Therefore, every B lymphocyte has a gene coding for a heavy chain and another one coding for the light chains. These genes are rearranged on the chromosome in the course of maturation and differentiation of the B cells. Thus a clone with a certain light chain isotype and fixed structure of the variable region is formed. These immunoglobulins, however, can develop different isotypes of their heavy chains (for example co-expression of IgM and IgD). The bone marrow origin of the cells carrying surface immunoglobulin markers is of no doubt. These cells, however, can not proliferate, so fast changes in their number are maintained by precursor (pre-B) cells. Null-cell acute lymphoblast leukemias-neoplastic diseases affecting different stages of pre-B cell maturation are used as models for these studies.

Bone marrow is proved to meet the requirements of the organism providing it with a considerable amount of lymphocytes. The life span of the peripheral b lymphocytes lasts from several days to several weeks. The newly formed B cells are activated following an antigen stimulus and migrate between various organs as mentioned above. Hence, their adequate supply is of vital importance. On the other hand, if the peripheral B cell pool is emptied, for example due to cytostatic drug therapy, the cells can "home" in the secondary lymphoid organs even without antigen stimuli. Antigen challenge of the B cells might rearrange its genes to increase antibody diversity if necessary.

Analysis of the secondary lymphoid organs indicates that the small lymphocytes in the follicles are the circulating cells carrying IgM and IgD markers. Though the circulatory system is the main input of these lymphocytes, some cells are transported by the afferent lymphatics. Lymphocytes mainly enter the lymph nodes via the high endothelial venules (HEV). Lymphocyte receptors can interact with the specific binding places assumed on the HEV. Of all the secondary lymphoid organs only the spleen is free from HEV, though the significance of this observation is not yet clear. The lymphocytes usually "home" in the T zones. B cells activated by the antigen - T lymphocyte cooperation are also present as B immunoblasts. Interdigital dendritic cells carrying MHC II receptors are also found in these zones. Interdigital cells maintain the process of antigen presentation, for morphological studies show lymphocytes bound to dendritic cells. Cells of similar phenotype are also found in the afferent lymphatics and skin (Langerhans cells). The above data suggest that lymphatics might be a regular route of migration of different antigen-carrying cells into the T zones. The activated immunoblasts stay for at least 48 hours in the areas of their primary activation. Consideration of the transferrin receptor expression gives a better insight into the events occurring in the secondary lymphatics. Following the first 48 hours the cells migrate into the medulla of the nodes becoming plasma cells. Similar process occurs in the red pulp of the spleen. Antigen persistence is the basic requirement for the induction of a normal secondary immune response.

As it has already been mentioned the interdigital dendritic cells guarantee the antigen persistence. Inasmuch as the activated B cells from the secondary lymphoid organs are the main source of the bone marrow plasma cells, the stimulating effect of the antigen-containing dendritic cells might be of importance in the pathomechanism of some gammapathies.

*B cell follicles* have complex structure, they mainly contain small lymphocytes carrying IgM and IgD markers, as well as the follicular dendritic cells. In the secondary lymphoid organs these follicles form germinative centers. After a while these centers replace their cells. Follicular dendritic cells start to play the APC part, covering the whole colonies of the centroblasts. It is noteworthy that such cells are not present around the T zones and the areas of small lymphocytes. 6-8 days following the antigen stimulus, small, actively proliferating cells - centrocytes - appear in the enlarged swollen germinal centers distally to the T zones. The relationship between the centrocytes and centroblasts inducing the enlargement of germinative centers, is not yet elucidated; however close relationship between these cells can be observed in the centroblast-centrocyte neoplastic form of non-Hodgkin lymphoma. The centrocyte-enriched area of the follicles is determined as a lucid zone. It also contains a considerable amount of macrophages, which have nuclear debris inside. They might remove the quickly proliferating and decaying centrocytes. The centroblast-containing area is called a dark zone, or germinal center. These cells are derived from B lymphocytes activated by the T cells. In the course of their differentiation memory cells and small lymphocytes carrying surface immunoglobulin markers are formed. The latter give rise to plasma cells, which produce immunoglobulins. Small lymphocytes with immunoglobulin markers as precursors of memory and plasma cells, are derived not directly from the centroblasts, but rather from intermediate cells morphologically defined as centrocytes. Differentiation into plasma cells occurs in the lymph nodes or bone marrow (Fig. I/3).

Presentation of the morphology, localization and migration of the cells is followed by the analysis of the immunoglobulin secretion.

Several days after a primary antigen stimulus, but prior to the appearance of the specific

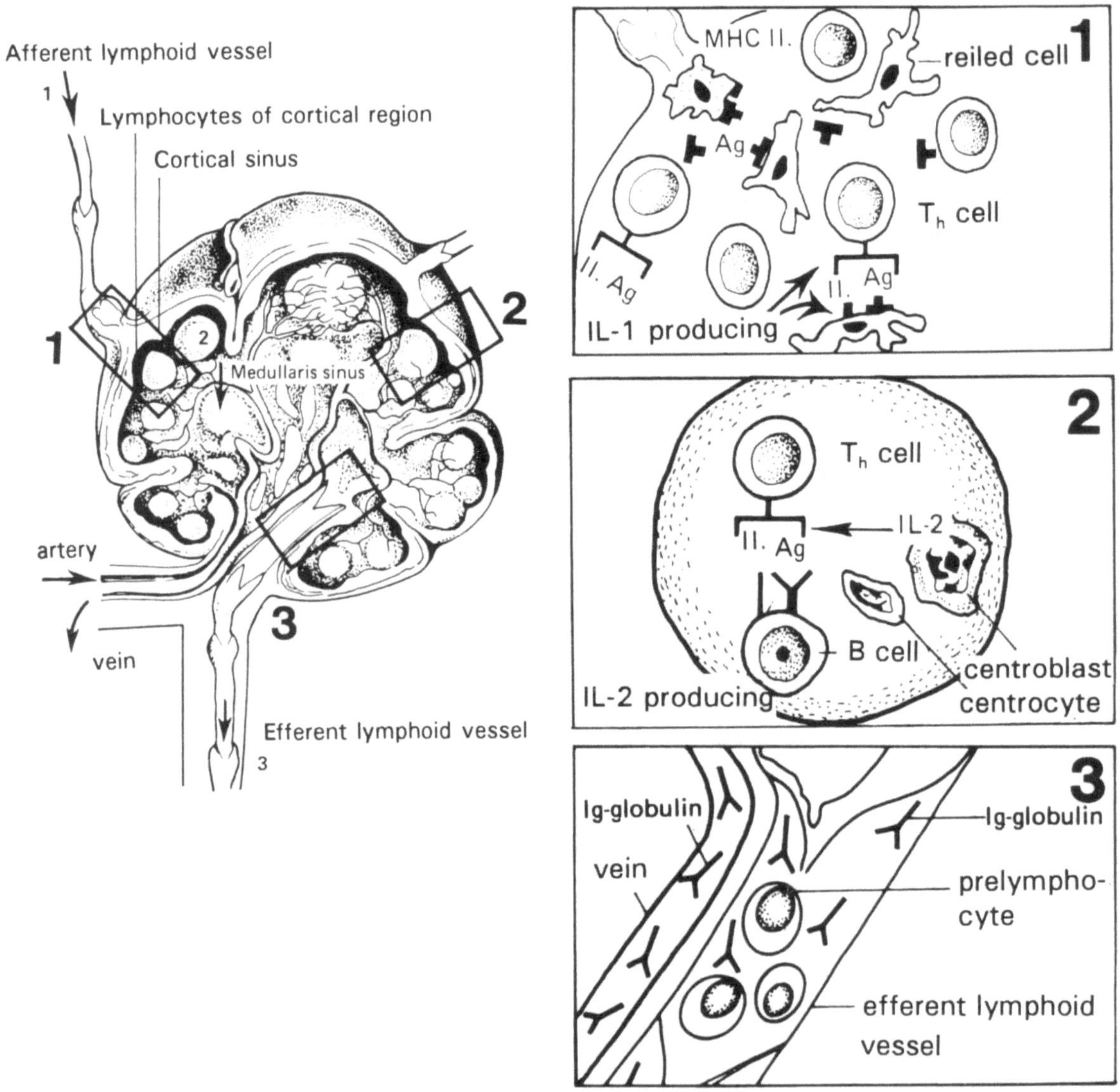


Fig. I/3

PFC in the lymphoid tissue, the surface Ig-producing RNA must switch to production of their secreted forms. Such genetically encoded switching is induced by a trigger stimulus. Following stimulation, these cells will synthesize uniform antibodies.

It is important to stress that it is the DNA which encodes the protein structure; environmental factors or intracellular/intramolecular cooperations do not affect the final structure of Ig. The modern theory of antibody production explains the three main features of the immune response:

1. The DNA of the organism a priori encodes the structures of antibodies against all possible antigens even before such encounter occurs. As the molecular size of the epitopes is very small and their structure is only determined by the presence of the H, C, N, O, P and S atoms, the total number of all possible epitopes is limited, possibly comprising only 1-2% of mammalian genome.
2. The problem of self - non-self discrimination. The self-reacting clones are either eliminated during ontogenesis, or suppressed due to some unknown mechanism.
3. The immune memory. A special form of clonal expansion, which triggers the antibody production after a primary response, when a considerable amount of cells carrying complementary receptors has already been produced. Under these conditions a repeated antigen stimulus will initiate a more elective and pronounced secondary immune response. The type of the antigen and the individual characteristics determine the period of immune memory. In these terms, tolerance means negative immune memory.

The following intracellular reactions lead to the formation of the immunoglobulins: the ribosomes aggregate in the cytoplasm, stimulating the energy producing system of the cell required for DNA synthesis and mitotic activity, i.e. cell proliferation. Activation of the cells takes about 24 hours. The cycle period is 12-15 hours, however there are no available data on the number of mitotic cycles in the activated cells. The amount of the extracellular antibodies depends on the character and intensity of the stimulus and the factors inhibiting proliferation. However the nature and action mechanism of the latter is unknown. The maximum antibody level in the course of a primary immune response develops by the 5th-7th day. The presence and persistence of the antigen predetermines the specificity of the immune reactions. Antigens can be eliminated via a dual mechanism - direct degradation or by forming complexes with performated antibodies, which is filtered by RES. After an antigen stimulus either the AFC (Antibody Forming Cell), or the memory cells are formed. Recent data indicate that while plasma cells are the final forms of differentiation, the memory cells provide a possibility of further specific proliferation following an antigen stimulus. The surface antigens of the memory cells represent antigen-specific immunoglobulins.

In the early phase of primary immune response the lymph nodes and the serum contain only IgM type antibodies. Later IgG and IgA classes can also be observed. The level of IgM antibodies is very low in the secondary immune response, while the IgG and IgA classes become predominantl. The low level of IgM antibodies might be due to a new, memory cell-independent B cell response to a repeated antigen stimulus. In this phase the cells previously producing IgM antibodies are switched to IgG or IgA synthesis; hence, all the IgG secreting cells are derived from an IgM producing precursor. Such switching is a physiological process involving the T cells. This process is accompanied by a related formation of the memory cells. This statement is consistent with experimental data showing that the clones incapable of switching fail to produce memory cells. However some special antigens can induce only IgM antibody production.

Affinity maturation does not actually mean maturation, but rather alteration: only the clones producing antibodies with maximum affinity to antigens prevail. Therefore, the changes do not affect the intracellular (intra-AFC) processes, but cause rearrangement within the whole cell population.

For an accurate definition of the already mentioned term "affinity", the mechanisms of antigen-antibody binding are analyzed.

Antigen-antibody binding means formation

of non-covalent bonds. The reaction follows the laws of thermodynamics. The antigen-antibody binding resembles to the interaction between enzymes and substrates; the only difference between the enzymes and the antibodies is that the latter never alter the structure of antigens. The antigen-antibody interaction requires formation of the non-covalent, otherwise similar chemical bonds, such as:

- contact of the apolar groups of the antigen determinants with the variable domains of the antibodies. This interaction is minimized by the water molecules present in the system. The concomitant decrease of entropy can not be considered as a real obstacle, but might only postpone the mentioned interactions. The above mentioned process is termed as "solvation". In the next step the liberated apolar groups form hydrophobic bonds. It is well known that if a hydrogen atom covalently binds to an electronegative one, it can also interact with a free unpaired electron of another electronegative atom, "stabilizing" the energy of the whole system. O-H-O, N-H-N and O-H-N variations are the most important hydrogen bonds in the antigen-antibody reactions.
- Ionic interactions. They are due to attraction of the oppositely charged groups. In terms of thermodynamics, however, this Coulomb's effect is insignificant, but important. Antibodies with different molecular structures can combine with more than one antigen. The number of the binding sites shows whether the antibody molecule is mono- or polyvalent. The binding capacity implies possible heterogeneity. The above information is important for comprehending the terms "avidity" and "affinity". If one binding fragment of a bivalent IgG molecule binds a certain structure, it would greatly "promote" the binding reaction of its other binding site. The binding capability of a polyvalent antibody to bind a multideterminant antigen is termed avidity to differentiate it from the strength of binding defined as affinity.

*Antibody specificity.* It has already been mentioned that according to one of the basic statements of immunology, genetic information encoding the structure of the antigen-binding sites of an antibody is antigen-independent (pre-synthetized). It suggests that antigens have no instructive effect, so all the antibody-producing lymphocytes already have the full spectrum of surface antigen determinants prior to the contact with antigens. This theory suggests that the probability of formation of absolutely specific antibodies to a certain antigen reciprocally correlates with the whole antigen spectrum ever existed. Antibodies reacting with antigens are classified as agglutinins, opsonins, antitoxins, cytotoxins and precipitins. These immunoglobulins can play both physiological and pathological role, which can not be always evaluated in quantitative terms (in contrast to monoclonal antibodies). It should be stressed, however, that the increased levels of monoclonal antibodies, as well as the elevation of auto-antibody level can reflect a polyclonal antibody reaction. The following presentation of the molecular structure of the immunoglobulins is based on the analysis of the most recent data. Two basic properties define the structure and function of the immunoglobulins:

1. all the immunoglobulins exhibit certain *structural identity*, having a common $H_2L_2$ structure;
2. selective binding to the antigen which had elicited their synthesis is a basic functional feature of immunoglobulins. Structural identity, however, implies a vast heterogeneity of the antigen-binding fragments within the same species and individuals. Genetic homogeneity and similarity of all the Ig classes, subclasses and antigen-binding fragments as gene products is typical of all the representatives of the species. Such similar epitopes are termed isotopes.

### *Immunoglobulins*

First observations concerning the existence of immunoglobulins date back to 1847. A special fraction of serum proteins with minimum anode electrophoretic mobility contains immunoglobulins. This protein fraction exhibits heterogeneity. Heremans was the first to introduce the concept of immunoglobulins, which was based on the marked structural and functional similarity and typical reactions with antigens. The uniform terminology has been introduced by the WHO Committee of experts. The classes and the most important features of immunoglobulins are summarized in Table I/I.

## IgM

IgM is a pentamer globulin of the primary immune response and it is the first immunoglobulin synthesized by a newborn (Fig. I/4).

Thiol reagents split the IgM molecule into 5 basic fragments with molecular weight of 180,000 and sedimentation constant of 6.5 s. The subunits are bound with four disulfide bonds and an additional J chain. The latter is produced by the B cells following the antigen stimulus. Evidence exists that the synthesis of J chains is the rate-limiting step of the IgM production. The molecular weight of the heavy chain μ is about 70,000. Electronmicroscopic studies clearly visualize these immunoglobulins (see Figure). A heavy chain consists of five domains: one variable (VH) and four constant ($C\mu_1$, $C\mu_2$, $C\mu_3$, $C\mu_4$). The VH and $C\mu_1$ comprise the Fd-, while the $C\mu_2$, $C\mu_3$, and $C\mu_4$ - the Fc domain. Three laboratories published data on the structure of the μ chain subclasses ($\mu_1$, $\mu_2$, $\mu_3$) however, evidence on genetic level still needs further confirmation.

The 8s subunit - IgMs - is frequently detected under pathological conditions, such as lupus erythematosus, cirrhosis, macroglobulin-emia, and other lymphoprolipherative diseases. Recent data indicate that IgMs is not a degradation product, but an unlinked subunit.

According to the data of the literature 76% of the pentamer structure is found in the vascular bed. Our studies failed to support this view, indicating that similar to other low

| | IgM | IgG | AgA | IgD | IgE |
|---|---|---|---|---|---|
| Antigenicity of heavy chain | $\mu_1$ $\mu_2$ $\mu_3$ | $\gamma_1$ $\gamma_2$ $\gamma_3$ $\gamma_4$ | $\alpha_1$ $\alpha_2$ | δ | ε |
| Light chain | κλ | κλ | κλ | κλ | κλ |
| Form of molecule | $(\mu_2\ \kappa_2)^5$ $(\mu_2\ \lambda_2)^5$ | $\gamma_2\ \kappa_2$ $\gamma_2\ \lambda_2$ | $\alpha_2\ \kappa_2$* $\alpha_2\ \lambda_2$* $(\alpha_2\ \kappa_2)^2$ $(\alpha_2\ \lambda_2)^2$ $(\alpha_2\ \kappa_2)^3$ $(\alpha_2\ \lambda_2)^3$ | $\delta_2\ \kappa_2$ $\delta_2\ \lambda_2$ | $\varepsilon_2\ \kappa_2$ $\varepsilon_2\ \lambda_2$ |
| Molecular weight | 900,000 | 150,000 | 170,000 300,000 | 180,000 | 200,000 |
| Sedimentation coefficient | 19 S | 6,7 S | 7 S 15 S | 7 S | 8 S |
| Carbohydrate content % | 9,8 | 2,6 | 5 10 | 10–12 | |
| T 1/2 days | 5,1 | 21 | 5,8 | 2,8 | 2,3 |
| Serum concentration mg/ml | 1 | 11,4 | 1,8 | 0,03 | 0,0003 |
| Immunoglobulin percent | 5 | 74 | 21 | 0,2 | 0,002 |
| Percentage of intra-vasal volume | 76 (50)! | 45 | 42 | 75 | 51 |
| Daily catabolism of intravasal volumen % | 18 | 6,7 | 25 | 37 | 89 |
| Rate of synthesis mg/kg/day | 6,7 | 25,4 | 24 | 0,4 | 0,02 |
| Complement fixation | yes | yes | alternative way | no | no |
| Placental transfer | no | yes | no | no | no |

Table I/I

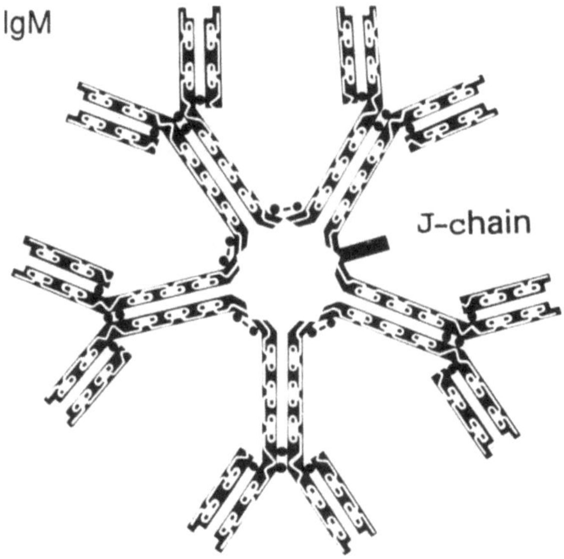


Fig. I/4

molecular weight immunoglobulins, this fragment shows a 50-50% distribution between intra- and extravasal space.

Functionally IgM is a cold agglutinin, isoagglutinin, rheumatoid factor, heterophilic antibody, immunoglobulin of Wasserman's reaction and a general antibody of Gram-negative infections. It also exhibits marked complement-binding activity.

IgM also occurs as a membrane-bound form, whose heavy chain is larger, compared to the secreted molecule. An amino acid residue links to the C-terminal of its $C\mu_4$ domain. This amino acid binds to the submembranous cytoskeletal elements. The presence of such hinge makes the whole molecule flexible, permitting rotation around its axis. In one of the Waldenström-type macroglobulinemias such "enlarged" molecules are also detected in plasma.

### IgG

This monomer antibody occurs at highest concentrations in human plasma. Table I/II presents the physico-chemical and physiological properties of IgG. This molecule consists of the two heavy and two light chains (Fig. I/5).

The heavy chain, which has four domains is subdivided into four classes. A polypeptide chain consists of 440-450 amino acids. Light chains of molecular weight 22,500 contain 214 amino acids. Fragments produced after cleavage have different physiological func-

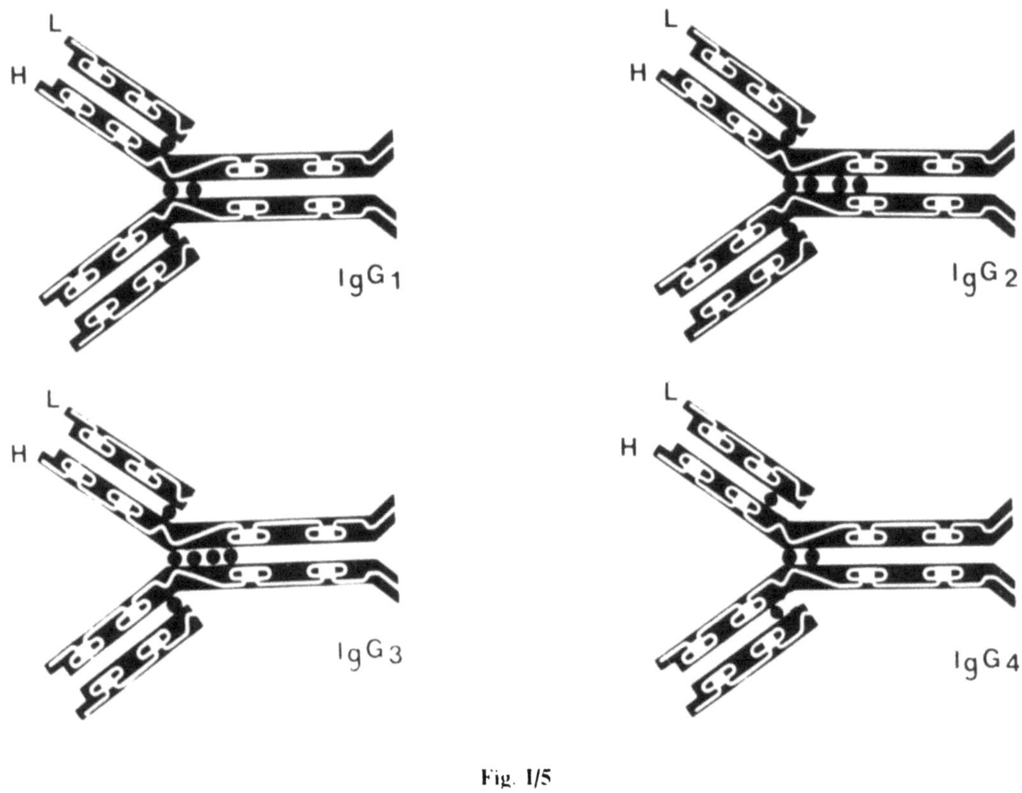


Fig. I/5

| | $G_1$ | $G_2$ | $G_3$ | $G_4$ |
|---|---|---|---|---|
| Amount in serum g/l | 6.63 | 3.22 | 0.58 | 0.46 |
| Percent rate between G immunoglobulins | 60.8 | 29.5 | 5.4 | 4.3 |
| Molecular weight | 54,200 | 54,200 | 60,950 | 54,200 |
| T 1/2 days | 21 | 21 | 8 | 21 |
| Daily catabolisation % | 6.9–9.0 | 6.9–9.0 | 16.8 | 6.9–9.0 |
| Synthesis mg/kg/day | 25.4 | 25.4(?) | 3.4 | 25.4(?) |
| Electrophoretic mobility | cathodal | anodal | cathodal | anodal |
| Isoelectrofocus | 8.3–9.5 | 7.0–7.3 | 8.4–8.9 | ? |
| Papain sensitivity | sensitive | resistent | very sensitive | resistent |
| Pepsin sensitivity | very resistent | resistent | very sensitive | sensitive |
| Number of disulfide bridges between heavy chains | 2 | 4 | 5 | 2 |
| Position of disulfide bridges between heavy chains on aminoacid chain | 226, 229 | 226, 229, 219, 220 | 226, 229, 210, 214, 220 | 226, 229 |
| Position of disulfide bridges between heavy and light chains | 220 | 131 | 131 | 131 |
| Complement fixation | yes | yes | yes | yes |
| PCA | yes | no | yes | yes |
| Binding to monocyte | yes | no | yes | yes |
| Binding to Staph. A protein | yes | yes | no | no |
| Placental transfer | yes | no | yes | yes |
| Antibody against Diphteria | yes | several | several | rarely |
| Antibody against Tetanus | yes | yes | yes | yes |
| Antibody against Cholera | no | yes (against polysacch. part) | no | yes (against toxin) |
| Antibody against Typhoid | yes | yes | no | no |
| Antibody against Rubella | yes | no | no | no |
| Antibody against dextran | no | yes | no | no |
| Antibody against Rh factor | yes | no | yes | rarely |
| Antibody against thyreoglobulin | yes | yes | yes | rarely |
| ANA | yes | yes | yes | yes |
| Antibody against VIII factor | no | no | rarely | yes |
| Antibody against IgG (AF, RF) | yes | yes | yes | yes |
| Antibody against GBM | yes | yes | yes | yes |
| Antibody against surface antigen of spermium | yes | yes | no | no |

**Table I/II**

tions (Fig. I/6). The functions of different subclasses of the molecule are summarized in Table I/II.

Apart from the four subclasses, $IgG_{4a}$, $IgG_{4b}$ and $IgG_5$ were also identified, the latter being a variant of $IgG_2$.

$IgG_1$ and $IgG_3$ mediate the ADCC reactions. Both heavy and light chains carry genetic surface markers (Table I/III)

IgG is transported into the fetal organism with the active transport.

It is noteworthy that at low plasma levels of immunoglobulins their half-life period increases to 35 days, while the catabolism is

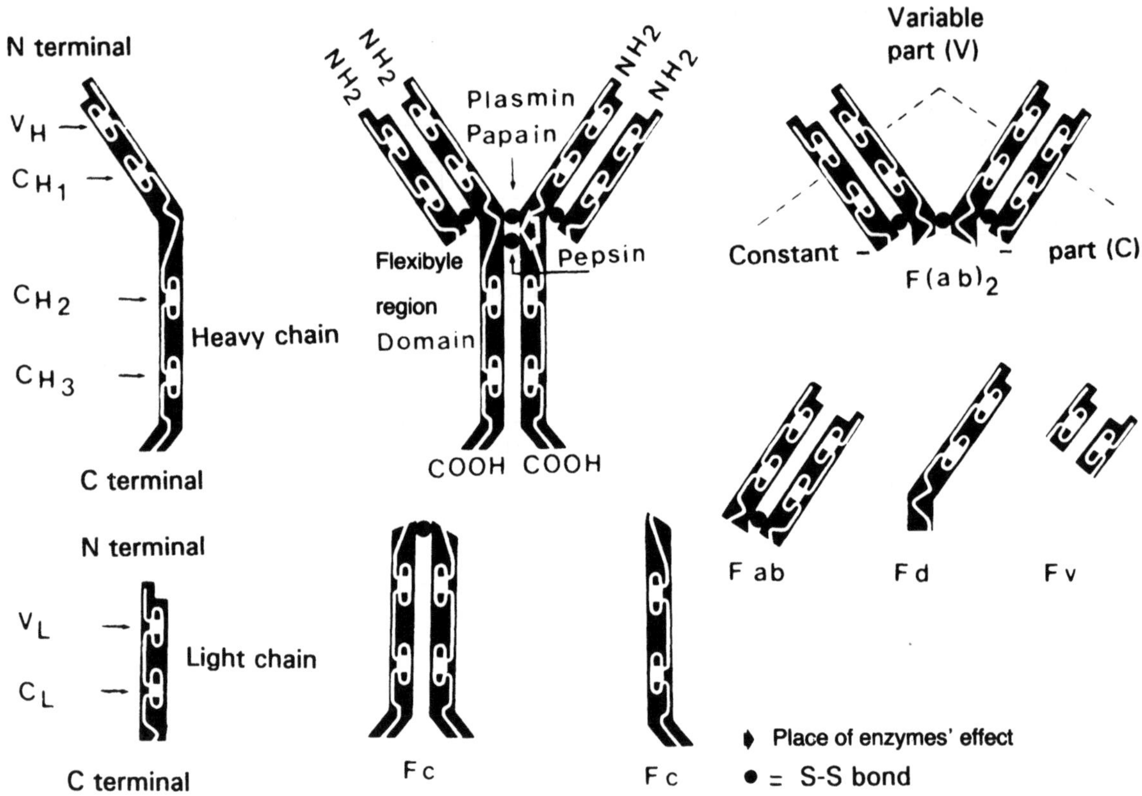


Fig. I/6

decreased by 1-2%. Reverse situation occurs in hyperglobulinemia.

**IgA**

Heremans was the first to analyze the IgA molecule. Similarly to IgG, it consists of the two heavy and two light chains; Kappa or Lambda-type light chains can be linked to a pair of heavy chains of a given class (Table I/I). It is worth mentioning that the light chains to the alpha$_2$-subclass heavy chains with non-covalent bonds; hence, $IgA_1$ has normal, while $IgA_2$ - deficient antigenity. $IgA_1$ comprises about 90-94%, while $IgA_2$ only 6-10% of the alpha-immunoglobulins with antigenity. The possibility of IgA existence was first suggested in the mid seventies, However the IgA in question proved to be an immune defficient $IgA_1$ whose heavy chain was 100 amino acids shorter than normal. IgA also occur in a dimeric form linked with a J fragment, which consists of 106 amino acids. The latter, though produced by plasma cells, is not an immunoglobulin and has a molecular

| | $\gamma_1$ | $\gamma_2$ | $\gamma_3$ | $\gamma_4$ | $\kappa$ | $\lambda$ |
|---|---|---|---|---|---|---|
| Gm(1–23) | 1, 2, 8, 9, 22 (Fc) | 3, 4, 8, 9, 17, 18, 23 (Fd) | 5, 6, 10–16, 21 | 7, 19, 20 | | |
| Inv | | | | | 1, 2, 3 | |
| Oz | | | | | | yes |
| Kern | | | | | | yes |

Table I/III

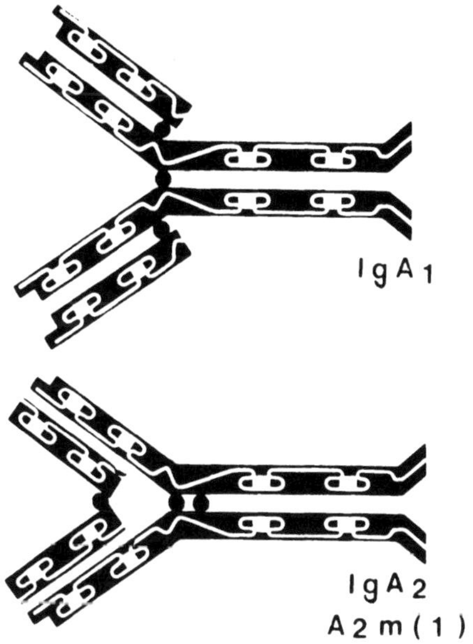


Fig. I/7

weight of 15,000. The J chains covalently bind to the Fc fragment of the heavy chain. It is never bound to monomer immunoglobulins. In different polymer molecules (dimers, trimers or pentamers) J chains can bind to all the basic molecules forming a giant structure (Fig. I/7).

The amino acid sequence of the J chains in the IgM and IgA molecules is identical and never resembles to that of the light chains.

Some researchers suggest that the catabolic rate of $IgA_1$ and $IgA_2$ is similar, while others showed a higher catabolic rate for $IgA_2$.

These molecules are of importance in antibacterial and antiviral immunity. Due to their aggregation capability, they can induce an alternative activation of the complement. Secretory IgA (SIgA) is a form performing special tasks. It mainly occurs in colostrum, tears and mucus secreted by the lung and gut. The molecular weight of this product is 390,000 and its sedimentation constant is 11S. J chain links two IgA monomers leading to a formation of dimer molecules in the plasma cells. Abluminal surface of the epithelial cells possess IgA-binding receptors. The IgA molecules are taken up with endocytosis and are transported to the luminal surface becoming for a while constituents of the cell membrane. Later, after the expression of IgA, the dimer and a part of its receptor linked with a J fragment are detached from the cell. This complex is termed Secretory Piece (SP) or Secretory Component (SC) (Fig. I/8). Such free fragments can appear on the mucosal surfaces. There is no difference between the bound forms occurring in serum and mucosal surfaces and the free form. This accessory protein protects the IgA against the action of proteolytic enzymes.

The secretory IgA possesses antibacterial and antiviral activity. It inhibits the adherence of streptococci to the epithelial cells, thus protecting the mucosal surfaces.

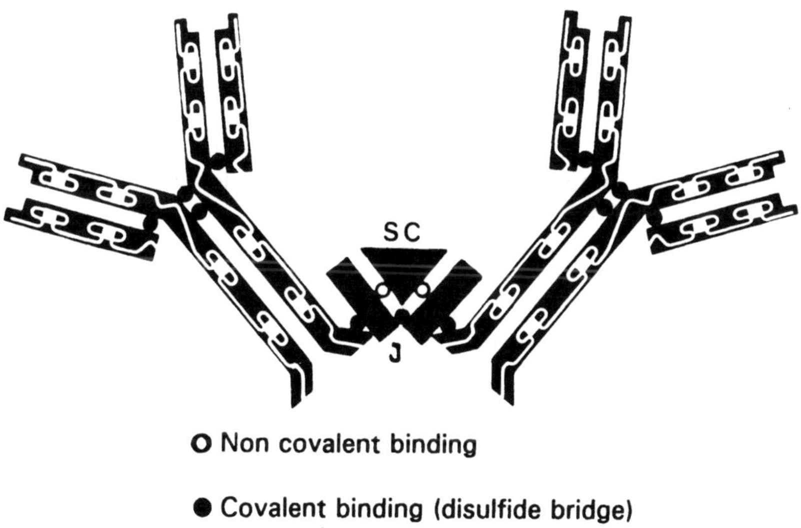


Fig. I/8

### IgD

Rowe and Fahey have identified this Ig while studying a myeloma protein with unknown antigenity. This Ig has an extremely high catabolic rate (see Table I). This Ig is considered as a surface immunoglobulin, since lymphocytes usually carry it as a surface marker. The function of this Ig is still unknown. No sufficient information is available on the heterogeneity of its heavy chain, though different types of antigenity are already reported in the literature. Its antibody properties are verified.

### IgE

Ishizaka was the first to identify IgE from the serum of an allergic patient. Since a ragweed E antigen has been used in that study, the new immunoglobulin was called IgE. The molecular weight of the protein is 200,000 and its sedimentation constant is 8 S (see Table I). The serum concentration of IgE is very low, hence it can not be identified with routine immunochemical methods. This antibody is a reagin of allergic reactions and can fix to the Fc fragments of basophils and mast cells. Bound to allergen, the five-domain molecule delivers a signal stimulating a cAMP and $Ca^{++}$-dependent degranulation with subsequent release of histamine. It is interesting to mention, that while the IgE level decreases in B cell disease, it is elevated in T cell-mediated malignomas. IgE is normally present in mucosal secretes (nose, eye, bronchi, intestines).

IgE is secreted in the mucosal tissue and gets into the circulation via the lymphatics. Sensitive double-antibody radioimmunoassay tests failed to support the old view that IgE occurs in breast milk and urine. Mechanisms others than those involved in the IgA or IgG transport, bring the IgE molecules to the mucosal surface. IgE has no hinge region! IgE plays a crucial role in the ADCC type immune response to parasite infection acompanying eosinophilia and immediate hypersensitivity. Serum IgE concentration shows a marked elevation in allergic and atopic diseases.

### IgU

A series of antigen-deficient microglobulins, though exhibiting immunoglobulin antigenity occur in the serum and urine. It is a mixture of Kappa and Lambda chains, their fragments and polymers. Their molecular weight varies between 10.000 and 150.000. Although the exact nature of these proteins needs further investigation, it is assumed that they are synthesized de novo. The wide spectrum of molecular weights is due to the presence of both monomer and polymer forms. If the light chains can be identified by gel electrophoresis or immunoelectrophoresis in urine or serum, this state is known as a light chain disease.

*Allotypes and idiotypes*

Antigen determinants determined by the primary amino acid sequence define the classes of immunoglobulins. These determinants, or isotypes, occur in all the representatives of a species. The DNA, however, contains at least one separate gene for all the isotypes of immunoglobulins; therefore, all genes encoding the heavy and the light chains can occur in allelic forms. This os a genetic basis of polymorphism and development of allotypes. In humans the allelic variance refers to the kappa chains $\mu$ 1-2, $\gamma$ 1-4, $\alpha$ 1-2. Amino acid sequences determining the allotypes in humans are localized in the H and V domains of the constant region. Allotypes are autosomally inherited according to simple Mendelian laws. In heterozygous individuals the inheritance is codominant, though in the individual cells, only this or that allele is expressed (allele exclusion).

- The first 120 amino acids of an H chain contain all the variants for the Ig subclasses, while the rest part of the chain is constant, with the only exception of allotypic variations.
- Idiotypes are the variations of antigen determinants on the variable domain, they can be identical with the specific antigen-binding structures. Antibodies produced against these antigen determinants are termed anti-idiotypes. Analysis of anti-idiotype antibodies showed that the $V_L$-$V_H$ conformation itself can bind anti-idiotype antibodies, so the real internal antibody-binding epitope structure is unnecessary. These data suggest that the $V_L$-$V_H$ conformation can behave as an epitope. It is one, but not the sole mechanism, since the studies of the idiotype - anti-idiotypes sys-

tems in immunization show that the replacement of an original - probably dangerous - antigen, which elicits the 1st idiotype production by an anti-idiotype proved to be efficient: these organisms were capable of antibody production against the original pathogenic antigen. The $V_L$-$V_H$ combination also implies optimum conformation. These studies revealed that the most important epitopes occur at the $V_H$ domain, the less important ones - at the $V_L$. It is important to emphasize that these molecules possessing pronounced antigenic characteristics are immunoglobulins. Their constant domains are the most immunogenic, while the variable areas exhibit less immunogenity; therefore, formation of the anti-variable domain antibodies is a difficult task. Antibodies with similar features simultaneously occur in all the classes and subclasses of immunoglobulins, indicating that the V and the C domains are encoded by different genes. Immunoglobulins synthesized by the same clone possess identical idiotypes even if an IgM-IgG switching had occurred in the course of the response. The lymphocytes of the clone which produces antibodies carry surface immunoglobulins of the same idiotype.

*The fine structure of the antibodies*

Peptide chains constituting immunoglobulin molecules are joined with disulfide bonds forming loops within a chain. These loops comprising 60 amino acids are termed domains. This term has already been mentioned, but not yet defined. According to the existing view, the domain structure of immunoglobulins is very functional. The $V_L$-$V_H$ domains bind the antigens, whereas the $C_L$-$C_H$ structure rather stabilizes the molecule, making its interaction with antigens more efficient. The $C_{H3}$ (Fc fragment) binds to the surface of the cells, changing the molecule into a secretory form. It also promotes its passage through the placenta. The $CH_3$ domain possesses allotypic determinants.

- The hinge area. The distance between $C_{H1}$ and $C_{H2}$ in IgG, IgA and IgD molecules is significantly (by 12-60 amino acids) larger, than that between the other domains. IgE and IgM do not possess hinge area, however, one more domain is included into their heavy chain structure. The hinge region maintains flexibility between the $F(ab)_2$ and the Fc fragments. The flexibility is differently pronounced in the different classes of immunoglobulins. Immunoglobulins are compact globular structures. It is interesting to mention that the Fab, $F(ab)_2$, Fc and even Fb fragments do possess similar structure. Characteristics of their molecular structure guarantee the conformational independence and the integrity of immunoglobulins. Alpha-helical structure is not typical of the Ig molecules, the domains rather exhibit the beta-folded structure.
- IgG are very inhomogenous as far as their amino acid sequence is concerned. They are usually enriched in proline and hydrophilic residues, making them both flexible and available for *proteases*. These regions are preferentially affected by pepsin or papain cleavage

*Fv fragments*

They are comprised of $V_L$-$V_H$ domains, bearing full antigen-binding capacity. These domains are bound to each other with non-covalent bonds. Segregated chains are capable of spontaneous reaggregation. This observation suggests that the chains have sufficient information on their three-dimensional tertiary and even quaternary structure, which determines the antigen-binding properties. V domains have special, so-called hypervariable regions, which contain about 10 amino acids. These regions are responsible for complementary properties, that is for the antigen binding.

Apart from the hypervariable regions, amino acid sequences show very little heterogeneity within the V region. Therefore, the V region has a stable "framework region" and the variable structures (complementarity-determining region, CDR) are localized within this invariant framework. CDR comprises only 15-20% of the amino acids of the V region. In this area the antigen is bound to the tyrosine, histidine and lysine residues of the variable regions by non-covalent bonds.

## References


1. Abdul-Cader C., Richardson P., Valsh, L., Ling, N. R., MacLennan, I. C. M., Jones, E. L., Leyland, M.
The incidence of B cell leukaemia and lymphopenia in B cell neoplasia in adults. A study using the Kiel classification of non-Hodgkin's lymphoma.
British Journal of Cancer 48, 185, 1983.
2. Alexander, J.
Some intracellular aspects of life and disease.
Protoplasma 14, 296, 1931.
3. Amzel, L. M., Poljak, R., Saul, F., Varga, J., Richards, F.
The three-dimensional structure of a combining region-ligand complex of immunoglobulin NEW at 3.5-Á resolution.
Proc. Natl. Acad. Sci. USA 71, 1427, 1974.
4. Anderson, K. C., Bates, M. P., Slaughenhoupt, B., Schlossman, G. F., Nadler, L. M.
A monoclonal antibody with reactivity restricted to normal and neoplastic plasma cells.
Journal of Immunology 132, 3172, 1984.
5. Anderson, K. C., Park, E. K., Bates, M. P., Leonard, R. C. F., Hardy, R., Schlossman, S. F., Nadler, L. M.
Antigen on human plasma cells indentified by monoclonal antibodies.
Journal of Immunology 130, 1132, 1983.
6. Armerding, D., Sachs, D. H., Katz, D. H. Activation of T and B lymphocytes in vitro. III. Presence of Ia determinants on allogeneic effect factor.
J. Exp. Med. 140, 1717, 1974.
7. Asano, Y., Singer, A., Hodes, R.
Role of the major histocompatibility complex in T cell activation of B cell subpopulations, MHC-restricted and -unrestricted B cell responses are mediated by distinct B cell subpopulations.
J. Exp. Med. 154, 1100, 1981.
8. Ashwell, J., de Franco, A., Paul, W., Schwartz, R.
Antigen Presentation by Resting B cells.
J. Exp. Med. 159, 881, 1984.
9. Baroni, C. D.
The human thymus in aging; Histologic involution paralleled by increased mitogen response and by enrichment of OKT3$^{+}$ lymphocytes.
Immunology 50, 519, 1983.
10. Bazin, H., Platteau, B., MacLennan, I. C. M., Johnson, G. D.
B cell production and differentation in adult rats.
Immunology 54, 79, 1985.
11. Benner, R., Hijmans, W., Haaijman, J. J.
The bone marrow: the major source of serum immunoglobulins, but still a neglected site of antibody formation.
Clinical and Experimental Immunology 46, 1, 1981.
12. Berek, C., Griffiths, G. M., Milstein, C.
Molecular events during maturation of the mirurine response to oxazalone.
Nature 316, 412, 1985.
13. Bessis, M.
Living Blood Cells and Their Ultrastructure,
trans. by R. I. Weed.
Springer-Verlag, Berlin, Heidelberg, New York 1973.
14. Bienenstock, J., Befus, A. D.
Mucosal immunology.
Immunology 41, 249, 1980.
15. Billingham, R. E., Brent, L., Medawar, P.
Quantitative studies on tissue transplantation immunity. III. Actively scquired tolerance.
R. Soc. (London) Phil. Trans, Series B, 239, 357, 1956.
16. Bodmer, J., Bodmer, W.
Histocompartibility.
Immunology Today 5, 251, 1984.
17. Botzenhardt, V., Klein, J., Ziff, M.
Primary in vitro cell mediated lympholysis reaction of NZB mice against unmodified targets syngeneic at the major histocompatibility complex.
J. Exp. Med. 147, 1435, 1978.
18. Boudart, D., Guimbretiere, L., Guimbretiere, J.
Role of IgD in „in" protective system.
Pathologie Biologie 185, 1977.
19. Brambell, F. W. R.
The transmission of immunity from mother to young and the catabolism of immunoglobulins.
Lancet 2, 564, 1968.
20. Brandtzaeg, P., Prydz, H.
Direct evidence for an integrated function of J chain and secretory component in epithelial transport.
Nature 311, 71, 1984.
21. Breinl, F., Haurowitz, F.
Chemische Untersuchung des Präzipitäts aus Hämoglobin und anti-Hämoglobin-Serum, und Bemerkungen über die Natur der Antikörper.
Z. Physiol, Chem, 192, 45, 1930.
22. Brichier, J., Schmitt, D., Yonish-Rouach, E., Cordier, F., Viac, J.
Use of tissue distribution studies to determine the specificity of monoclonal anti-lymphocyte antibodies.
In: Bernard, A., Boumsell, L., Dansset, J., Milstein, C., Schlossman, S. F. (eds) Leukocyte typing. Springer-Verlag, New York 465, 1984.
23. Brodeur, P., Riblet, R.
The IgH chain V region locus in mice. 100 V Genes comprise seven families of homolugous genes.
Eur. J. Immunol. 14, 922, 1984.
24. Burnet, F. M.
A modification of Jern's theory of antibody production using the concept of clonal selection.
Australian J. Sci. 20, 67, 1957.
25. Burnet, F. M.
The Clonal Selection Theory of Acquired Immunity.
Cambridge University Press, London 1959.
26. Bush, S. T., Swedlung, H. A., Gleich, G. J.
Low molecular weight IgM in human sera.
J. Lab. Clin. Med. 73, 194, 1969.
27. Butcher, E. R., Scolley, S. Weissman, I.
Organ specificity of lymphocyte migration; Mediation by highly selective lymphocyte interaction with organ-specific determinants on high endothelial venules.
Eur. J. Immunol. 10, 556, 1980.
28. Butterfield, J. H., Maddox, D., Gleich, G.
The eosinophilic leucocyte; Maturation and function.
Clinical Immunology Reviews 2, 187, 1983.
29. Capra, J. D., Kehoe, J. M.
Hypervariable regions, idiotypy, and the antibody-combining site.
Adv. Immunol. 20, 1, 1975.
30. Capron, M., Bazin, H., Joseph, M., Capron, A.
Evidence for IgE-dependent cytotoxicity by rat eosinophils.
J. Immunol. 126, 1764, 1981.
31. Capron, A., Dessaint, J. P.
IgE: A molecule in search of a function.
Annales D 'Immunologie 132, 3, 1981.
32. Carbonara, A. O., Heremans, J. F.
Subunits of Normal and Pathological $\gamma_1$A-Globulins ($\beta_2$A-Globulins)
Arch. Biochem. Biophys. 102, 137, 1963.
33. Chesebro, B., Bloth, B., Svehag, S. E.
The ultra structure of normal and pathological IgM immunoglobulins
J. Exp. Med. 127, 399, 1968.
34. Chiller, J., Habicht, G., Wigle, W.
Kinetics differences in unresponsiveness of thymus and bone marrow cells.
Science 171, 813, 1971.
35. Chin, Y. H., Carey, G. D., Woodruff, J. J.
Lymphocyte recognition of lymph node high endothelium. I. Inhibition of in vitro binding by a component of thoracic duct lymph.
J. Immunol. 125, 1764, 1980.
36. Cline, M. J.

The White Cell.
Harward University Press, Cambridge 1975.
37. Coico, R. F., Bhogal, B. S., Thorbecke, G. J.
Relationship of germinal centres in lymphoid tissue to immunologic memory. VI. Transfer of B cell memory with lymph node cells fractionated according to their receptors for peanut agglutinin.
Journal of Immunology 131, 2254, 1983.
38. Cooper, M. D.
Pre-B cells: normal and abnormal development.
Journal of Clinical Immunology 1, 81, 1981.
39. Cunningham-Rundles, C., Lamm, M. E., Franklin, E. C.
Primary structure of human secretory component (abstract)
Fed. Proc. 33, 747, 1974.
40. De Cotean, W. E.
The role of secretory IgA in defense of the distol lung.
Ann. NY. Acad. Sci. 221, 214, 1974.
41. Dekruyff, R. H., Clayberger, C., Cantor, H.
Hapten reactive inducer T cells. II. Evidence that a secreted form of the T cell receptor induces antibody production.
J. Exp. Med. 158, 1881, 1983.
42. Dorf, M. E. (ed.)
The Role of the Major Histocompatibility Complex in Immunobiology.
Garland STPM Press, New York 1981.
43. Drayson, M., Smith, M., Ford, W.
The sequences of changes in blood flow and lymphocyte influx to stimulated rat lymph nodes.
Immunology 44, 125, 1981.
44. Dresser, D. V.
Specific inhibition of antibody production. I. Protein overloading paralysis.
Immunology 5, 161, 1962.
45. Edelman, G. M.
Antibody structure and molecular immunology.
Science 180, 830, 1973.
46. Eerola, E., Jalkanen, S., Granfors, K., Toivanen, A.
Immune capacity of the chicken bursectomized at 60 hours of incubation: Mitogen-induced cell proliferation and Ig secretion.
J. Immunol. 131, 120, 1983.
47. Ehrlich, P.
On immunity with special reference to cell life.
Proc. Roy, Soc. London (Biol.) 66, 424, 1900.
48. Erb, P., Meier, B., Matsunaga, T., Feldmann, M.
Nature of T-cell macrophage interaction in helpercell induction in vitro. II. Two stages of T-helper–cell differentation analyzed in irradiation and allogenic chimeras.
J. Exp. Med. 149, 686, 1979.
49. Feldmann, M., Basten, A.
Cell interactions in the immune response. Specific collaboration across a cell-impermeable membrane.
J. Exp. Med. 136, 49, 1972.
50. Feldmann, M.
Cell interactions in the immune response in vitro. V. Specific collaboration via complexes of antigen and thymus-derived cell immunoglobulin.
J. Exp. Med. 136, 737, 1972.
51. Felton, Lloyd, D.
The significance of antigen in animal tissues.
J. Immunology 61, 107, 1949.
52. Fossum, S.
The architecture of rat lymph nodes. II. Lymph node compartments.
Scand, J. Immunol. 12, 411, 1980.
53. Franklin, E. C., Frangione, B.
Structural differences between macroglobulins belonging to two serologically distinguishable subclones.
Biochemistry 7, 4203, 1968.
54. Fu, S. M., Winchester, R. J., Kunkel, H. G.
Similar idiotypic specificity for the membrane IgD and IgH of human B lymphocytes.
Journal of Immunology 114, 250, 1975.
55. Gagnon, R. F. MacLennan, I. C. M.
Regulation of secondary antibody. II. The requirement for continued antigen supply to maintain IgG production.
Immunology 45, 283, 1982.
56. Gallatin, W. I., Weissman, Butcher, E.
A cell surface molecule involved in organ-specific homing of lymphocytes.
Nature 304, 30, 1983.
57. Gathings, E. W., Lawton, A. R., Cooper, M. D.
Immunofluorescent studies of the development of pre-B cells, B lymphocytes and immunoglobulin isotype diversity in humans.
European Journal of Immunology 7, 804, 1977.
58. Germain, R. N., Theze, J., Kapp, J. A., Benacerraf, B.
Antigen-specific T-cell-mediated suppression. I. Induction of L-glutamic acid$^{60}$-L-alanine$^{30}$-L-tyrosine$^{10}$-specific suppressor T cells in vitro requires both antigen-specific T-cell-suppressor factor and antigen.
J. Exp. Med. 147, 123, 1978.
59. Gershon, R. K., Kondo, K.
Infectious immunological tolerance.
Immunology 21, 903, 1971.
60. Gershon, R. K.
T cell Control of antibody production.
Contemp. Topics Immunobiol 3, 1, 1974.
61. Gillis, S., Inman, F. P. (eds.)
The interleukins.
Contemp, Topics in Molec. Immunol. 10, 1985.
62. Givol, D.
Structural analysis of the antibody combining site
Contemp. Topics Mol. Immunol 2, 27, 1973.
63. Glick, B., Chang, T., Jaap, R.
The bursa of Fabricius and antibody production.
Poultry Science 35, 224, 1956.
64. Gobbi, M., Caligaris-Cappio, F., Janossy, G.
Normal equivalent cell of B cell malignancies: analysis with monoclonal antibodies.
British Journal of Haematology 54, 393, 1983.
65. Gordon, J., Nathan, P., Ling, N. R.
In preparation Gray, D., Chassoux, D., MacLennan, I. C. M., Bazin, H.
Selective depression of thymus-independent anti-DNP antibody responses induced by adult but not neonatal splenectomy.
Clinical and Experimental Immunology 60, 78, 1985.
66. Gowans, J. L., Knight, E. J.
The route off recirculation of lymphocytes in the rat.
Proc. r. Soc. London 159, 257, 1964.
67. Gray, D.. MacLennan, I. C. M., Bazin, H., Khan, M.
Migrant $\mu^+$ and static $\mu^+$ -B lymphocyte subsets.
European Journal of Immunology 12, 564, 1982.
68. Greaves, M. F., Hariri, G., Newman, R. A., Sutherland, D. R., Ritter, M. A., Ritz, J.
Selective expression of the common acute lymphoblastic leukaemia (gp100) antigens on immature lymphoid cells and their malignant counterparts.
Blood 61, 628, 1983.
69. Grey, H. M., Abel, C. A., Yount, W. J., Kunkel, H. G.
A Subclass of human $\gamma$A-globulins ($\gamma A_2$) wich lacks disulfide bonds linking heavy and light chains.
J. Exp. Med. 128, 1223, 1968.
70. Griffiths, G. M., Berek, C., Kaartinen, M., Milstein, C.
Somatic mutation of the immune response to Z-phenyl oxazalone.
Nature 304, 320, 1984.
71. Guyton, A. C.
Interstitial fluid pressure and dynamics of lymph formation.
Fed. Proc. 35, 1861, 1976.
72. Halpern, M. S., Koshland, M. E.
Novel submit in secretory IgA.
Nature 228, 1276, 1970.
73. Hanna Jr. M. G., Szakal, A. K.
Localisation of $125_I$ labelled antigen in germinal centres of mouse spleen: histologic and ultrastructural autoradiographic studies of the secondary immune reaction.
Journal of Immunology 101, 949, 1968.

74. Harboe, M., Deverill, J., Godal, H. C.
Antigenic heterogeneity of Waldenström type γM globulins.
Scand, J. Haematol. 2, 137. 1965.
75. Hay, J. B., Hobbs, B. B.
The flow of blood to lymph nodes and its relation to lymphocyte trafic and the immune response.
J. Exp. Med. 145, 31, 1977.
76. Heidelberger, M., Avery, O. T.
The soluble specific substance of pneumococcus.
J. Exp. Med. 38, 73, 1923.
77. Heidelberger, M., Kendall, F.
A quantitative study of the precipitin reaction between type III pneumococcus polysaccharide and purified homologous antibody.
J. Exp. Med. 50. 809, 1929.
78. Heremans, J. F., Heremans, M. T., Schultze, H. E.
Isolation and description of a few properties of the $\beta_{2A}$-globulin of human serum.
Clin. Chim. Acta 4, 96, 1959.
79. Heremans, J. F.
Immunochemical studies on protein pathology: the immunoglobulin concept.
Clin. Chim Acta 4, 639, 1959.
80. Hijmans, W., Schuit, H. R. E., Hulsing-Hesselink, E.
An immunofluorescence study on intracellular immunoglobulins in human bone marrow cells.
Annals of the New York Academy of Science 177, 290, 1971.
81. Hitzig, W. H., Jákó, J.
Fraction of the plasma proteins.
Cit. Clinical biochemistry, Ed. Curtius, H. Ch., Roth, M. Walter de Gruyter. Berlin–New York 1974.
82. Honjo, T.
Immunoglobulin genes.
Annual Review of Immunology 1, 499, 1984.
83. Howard, M., Mizel, S. B., Lachman, L., Ansel, J., Johnson, B., Paul, W. E.
The role of interleukin l in anti-immunoglobulin-induced B cell proliferation.
Journal of Experimental Medicine 157, 1529, 1983.
84. Howard, M., Paul, W. E.
Regulation of B cell growth and differentiation by soluble factors.
Annual Review of Immunology 1, 307, 1983.
85. Humphrey, J. H., Grennan, D.
Different macrophage populations distinguished by means of fluorescent polysaccharides: recognition and properties of marginal zone macrophages.
European Journal of Immunology II. 221, 1981.
86. Humphrey, J. H., Grennan, D., Sundaram, V.
The origin of folicular dendritic cells in the mouse and the mechanisms of trapping of immune complexes on them.
European Journal of Immunology 14, 859, 1984.
87. Hünig, Th., Schimpl. A., Wecker, E.
Mechanism of T-cell help in the immune response to soluble antigens. I. Evidence for in situ generation and action of T-cell-replacing factor during the anamnestic response to dinitrophenyl keyhole limpet hemacyanin in vitro.
J. Exp. Med. 145, 1216, 1977.
88. Inaba, K., Witmer, M. D., Steinman, R. M.
Clustering of dendritic cells helper T lymphocytes and histocompatible B cells during primary antibody responses in vitro.
Journal of Experimental Medicine 160, 858, 1984.
89. Ishizaka, S., Geneste, C., Pillot, J.
Molecular state of secretory component in human sera.
Immunology Letters 1, 217, 1980.
90. Ishizaka, K., Ishizaka, T., Horn Brook, M. M.
Physicochemical properties of reaginic antibody. V. Correlation of reaginic activity with γE globulin antibody.
J. Immunol. 97, 840, 1966.
91. Ishizaka, K., Ishizaka, T.
Identification of γE antibodies as a carrier of reaginic activity.
J. Immunol. 99, 1187, 1967.
92. Ishizaka, K., Ishizaka, T.
Mechanism of reaginic hypersensitivity.
Immunol. Rev. 41, 109, 1978.
93. Janossy, G. J., Bradstock, K. F., McMichael, A., Rapson, N., Greaves, M. F.
Terminal transferase positive human bone marrow cells exhibit the antigenic phenotype of non-T, non-B acute lymphoblastic leukaemia.
Journal of Immunology 123, 1525, 1979.
94. Jerne, N. K.
The natural selection theory of antibody formation.
Proc. Natl. Acad. Sci USA 41, 849, 1955.
95. Johnson, G. D., Ling, N. R., Narhan, P. D., Hardie, D. L.
Use of monoclonal antibodies reactive with secretory epithelial cells for the immunocytochemical identification of plasma cells.
Immunological letters, 1986.
96. Jones, B. H.
On a new substance occuring in the urine of a patient with „mollities ossium".
Phil. Tr. Royal. Soc. London 55–62, 1848.
97. Kaartinen, M., Griffiths, G. M., Markham, A. F., Milstein, C.
mRNA sequences define an unusually restricted IgG response to 2-phenyloxazalone and its early diversification.
Nature 304, 320, 1983.
98. Kabat, E. A., Padlan, E. A., Davis, D. R.
Evolutionary and structural influences on light chain constant ($C_L$) region of human and mouse immunoglobulins.
Proc. Natl. Acad. Sci. USA 72, 2785, 1975.
99. Kabat, E. A.
Structural Concepts in Immunology and Immunochemistry, Second Edition. Holt, Rinehart and Winston, New York 1978.
100. Kaji, H., Parkhorse, R. M. E.
Control of J chain biosynthesis in relation to heavy and light chain synthesis, polymerization and secretion.
J. Immunol. 114, 1218, 1975.
101. Kapadis, S. B.
Multiple myeloma: a clinicopathologic study of 62 consecutive autopsied cases.
Medicine 59, 380, 1980.
102. Kappler, J. W.
Antigen-inducible, H-2 restricted, IL-2 producing T cell hybridomas, Lack of independent antigen and H-2 recognition.
J. Exp. Med. 153, 1198, 1981.
103. Katz, D. H., Dorf, M. E., Benacerraf, B.
Cell interactions between histoincompatible T and B lymphocytes. VI. Cooperative responses between lymphocytes derived from mouse donor strains differing at genes in the S and D regions of the H-2 complex.
J. Exp. Med. 140, 290, 1974.
104. Kawanishi, H., Saltzman, L., Strober, W.
Mechanisms regulating IgA class-specific Ig production is murine gut-associated lympoid tissues. II. Terminal differentation of postswitch sIgA-bearing Peyer's patch B cells.
J. Exp. Med. 158, 649, 1983.
105. Keller, R.
Tissue mast cells in immune reactions.
Monographs in Allergy, Vol 2, Karger, Basel 1966.
106. Kelly, R. H., Balfour, B. M., Armstrong, J. A., Griffiths, S.
Functional anatomy of lymph nodes. II. Peripheral lymphborne mononuclear cells.
Anatomical Record 190, 5, 1978.
107. Kindred, B., Shreffler, D. H-2 dependence of cooperation between T and B cells in vivo.
J. Immunol. 109, 940, 1972.
108. Kishimoto, T.
Factors affecting B cell growth and differentation.
Annuel Review Immunol. 3, 133, 1985.
109. Kitamura, K., Nakauchi, H., Koyasu, S., Yahara, I., Okumura, K., Tada, T.
Characterization of an antigen-specific suppressive factor derived from a cloned suppressor effector T cell line.
J. Immunol. 133, 1371, 1984.

110. Kitamura, Y., Yokoyama, M., Matsuda, H., Ohno, T., Mori, K. J.
Spleen colony-forming cell ascommon precursor for tissue mast cells and granulocytes.
Nature 291, 159, 1981.
111. Klaus, G. G. B.
The follicular dendritic cell: Its role in antigen presentation in the generation of immunological memory.
Immunol. Rev. 53, 3, 1980.
112. Klaus, G. G. B., Kunkl, A.
The role of germinal centres in the generation of immunological memory,
In: Microenvironments and haemopoietic and lymphoid differentation.
Ciba Foundation Symposium 84. Pitman Medical, London 265, 1981.
113. Klinkert, W., LaBadie, J., Bowers, W.
Accessory and stimulating properties of dendritic cells and macrophages isolated from various rat tissues.
J. Exp. Med. 156, 1, 1982.
114. Koch, G., Lok, B. D., van Oudenaren, A., Benner, R.
The capacity and mechanism of bone marrow antibody formation by thymus-independent antigens.
Journal of Immunology 128, 1497, 1982.
115. Koch, G., Osmond, D. G., Julius, M. H., Benner, R.
The mechanism of thymus-dependent antibody formation in bone marrow.
Journal of Immunology 126, 1447, 1981.
116. Kolmer, J. A.
A Practical Textbook of Infection, Immunity, and Specific Therapy.
Saunders, Philadelphia 1917.
117. Kontiainen, S., Feldmann, M.
Suppressor-cell induction in vitro. IV. Target of antigenspecific suppressor factor and its genetic relationships.
J. Exp. Med. 147, 110, 1978.
118. Korsmeyer, S. J., Arnold, A., Bakhshi, A., Ravetch, J. V., Siebencist, V., Hieter, P. A., Sharrow, S. O., LeBien, T. W., Kersey, J. H., Poplack, D. G., Leder, P., Waldmann, T. A.
Immunoglobulin gene rearrangement and cell surface antigen expression in acute lymphoblastic leukaemias of T cell and B cell precursor origins.
Journal of Clinical Investigation 71, 301, 1983.
119. Korsmeyer, S. J., Hieter, R. A., Ravetch, J. V., Poplack, D. G., Waldmann, T. A., Leder, P.
Development hierarchy of immunoglobulin gene rearrangements in human leukaemia pre B cells.
Proceedings of the National Academy of Science USA 78, 7096, 1981.
120. Koshland, M. E.
Structure and function of the J chain.
Adv. Immunol. 20, 41, 1975.
121. Kuan, T.
Three monoclonal immunoglobulins; An $IgG_2$, an IgM and an IgM/A hybrid, in one patient. II. Sharing of common variable regions.
Immunology 44, 265, 1981.
122. Kubagawa, H., Vogler, L. B., Lawton, A. R., Cooper, M. D.
The extent of clonal involvement in multiple myeloma.
In: Potter M (ed) Progress in myeloma. Elsevier/North Holland, New York 195, 1980.
123. Kumararatne, D. S., MacLennan, I. C. M.
Cells of the marginal zones of the spleen are lymphocytes derived from recirculating precursors.
European Journal of Immunology 3, 865, 1981.
124. Kunkel, H. G., Joslin, F. G., Penn, G. M., Natvig, J. B.
Genetic variants of $\gamma G_4$ glogulin: unique relationship to other clones of $\gamma$G globulin.
J. Exp. Med. 132, 508, 1970.
125. Kyle, A. R., Bayard, E. D.
The monoclonal gammopathies.
Charles C. Thomas Publisher Springfield Illinois USA 1976.
126. Lamm, M. E.
Cellular aspects of immunoglobulin A.
Adv. Immunol. 22, 233, 1976.
127. Landsteiner, K.
The Specificity of Serological Reactions.
Thomas, Springfield Ill, 1936.
128. Langevoort, H. L.
The histophysiology of the antibody response. I. Histogenesis of the plasma cell reaction in rabbit spleen.
Laboratory Investigations 12, 106, 1963.
129. Lebman, D., Trucio, M., Bo Hero, L.
A monoclonal antibody, the detects expression of transferrin receptor on human erythroid precursor cells.
Blood 59, 671, 1982.
130. Lennox, E. S.
A search for biosynthetic subunits of light and heavy chains of immunoglobulins.
Cold Spring Harbor Symp. Quant. Biol. 32, 249, 1967.
131. Leonard, R. C. F., MacLennan, I. C. M., Smart, Y., Vanhegan, R. I., Cuzick, J.
Light chain isotype-associated suppression of normal plasma cell numbers in patients with multiple myeloma.
International Journal of Cancer 24, 385, 1979.
132. Lipscomb, M. F.
Antigen presentation by guinea pig alveolar macrophages.
J. Immunol. 126, 286, 1981.
133. Mackenzic, M. R., Warner, N., Linscott, D., Fundenberg, H. H.
Differentiation of human IgM subclass by the ability or inability to interact with C'l (abstract).
J. Lab. Clin. Med. 72, 993, 1968.
134. MacLennan, I. C. M., Gray, D., Kumararatne, D. S., MacLennan, I. C. M.
The lymphocytes of splenic marginal zones: a distinct B cell lineage.
Immunology Today 3, 305, 1982.
135. Malkovsky, M., Medawar, P.
Is immunological tolerance a consequence of IL-2 deficiency during the recognition of antigen.
Immunology Today 5, 340, 1984.
136. Mandel, T. E.
The follicular dendritic cell: Long term antigen retention during immunity.
Immunol. Rev. 53, 29, 1980.
137. Marrack, P.
Nonspecific factors in B cell responses.
Immunol. Rev. 63, 33, 1982.
138. Mayrhoffer, G., Fisher, R.
Mast cells in severely T cell depleted rats, and the response to infestation with N. brasiliensis.
Immunology 37, 145, 1979.
139. McGregor, D. D., Gowas, J. L.
The antibody response of rats depleted of lymphocytes by chronic drainage from the thoracic duct.
J. Exp. Med. 117, 303, 1963.
140. Meinke, G. C.
Antigenic studies of J. chain (abstract).
Fed. Proc. 32, 968, 1973.
141. Melchers, F., Braun, V., Galanos, C.
The lipoprotein of the outer membrane of Escherichia coli a B-lymphocytes mitogen.
J. Exp. Med. 142, 473, 1975.
142. Mellstedt, H., Petterson, D., Holm, G.
Idiotype-bearing lymphoid cells in plasma cell neoplasia. Clinical Haematology 11, 65, 1982.
143. Merler, E., Rosen, F. S.
The structure and synthesis of the immunoglobuline.
New Engl. J. Med. 275, 480, 1966.
144. Metchnikoff, E.
Immunity in Infections Disease,
trans. by F. Binnie, Cambridge University Press, London 1905.
145. Metzger, H.
The IgE-mast cell system as a paradigm for the study of antibody mechanisms.

Immunol. Rev. 41, 186, 1978.
146. Mihaesco, C., Seligmann, M.
In „Protides of Biological Fluids".
H. Peeters (ed) Pergamon Press, Oxford 16, 227, 1969.
147. Miller, J. F. A. P.
Immunological Memory.
Contemp. Topics Immunobiol. 2, 151, 1973.
148. Mole, J. E., Benett, J. C.
J-chain: amino acid sequence of the c-terminal cyanogen bromide fragment (abstract).
Fed. Proc. 33, 747, 1974.
149. Mole, J. E., Bhown, A. S., Benett, J. C.
The primary structure of J-chain: sequence analyses of the major proteolytic fragments (abstract).
Fed. Proc. 34, 953, 1975.
150. Morell, A., Skvaril, F., Hijmans, N., Scherz, R.
Cytoplasmic immunofluorescence of bone marrow plasma cells producing immunoglobulins of the four IgG subclasses.
Journal of Immunology 115, 579, 1975.
151. Morell, A., Skvaril, F., Van Loghem, E., Kleemola, H.
Human IgG subclasses in matural and fetal serum.
Vox Sang. 21, 481, 1971.
152. Morell, A., Skvaril, F., Noseda, G., Barandun, S.
Metabolic properties of human IgA subclasses.
Clin. Exp. Immunol. 13, 521, 1973.
153. Mosier, D. E., Subbara, B.
Thymus-independent antigens: complexity of B-lymphocyte activation revealed.
In: Inglis J. R. (ed) B lymphocytes today Elsevier biomedical. Amsterdam 70, 1982.
154. Mostov, K. E., Kraehenbuhl, J. P., Blobel, G.
Receptor-mediated trancellular transport of immunoglobulin: Synthesis of secretory components as multiple and larger transmembrane forms.
Proc. Natl. Acad, Sci. USA 77, 7257, 1980.
155. Möller, G.
Immunoglobulin D: Structure, synthesis, membrane representation and function.
Immunol. Rev. 37. 1977.
156. Möller, G.
Immunoglobulin E.
Immunol. Rev. 41, 1978.
157. Mudd, S.
A hypothetical mechanism of antibody formation.
J. Immunol. 23, 423, 1932.
158. Munro, A. J.
Antigen-specific T cell factor in cell cooperation: Physical properties and mapping in the left-hand (K) half of H-2.
J. Exp. Med. 140, 1579, 1979.
159. Mitchison, N. A.
Induction of immunological paralysis in two zones of dosage.
Proc. Royal Society (B) 161, 275, 1964.
160. Natvig, J. B., Kunkel, H. G.
Human immunoglobulins: Classes, subclasses, genetic variants and idiotypes.
Adv. Immunol. 16, 1, 1973.
161. Nadler, L. M., Anderson, K. C., Marti, G., Bates, M. P., Parke, E. K., Daley, J. F., Schlossman, S. F.
B4 a human B lymphocyte-associated antigen expressed on normal, mitogen-activated and malignant B lymphocytes.
Journal of Immunology 131, 244, 1983.
162. Nadler, L. M., Anderson, K. C., Fishe, D., Slaughenhoupt, B., Boyd, A. N., Schlossman, S. F.
Immunologic classification of B-cell leukaemias and lymphomas.
In: Ford, R. J., Fuller, L. M., Hagmeister, F. B (eds) U. T. M. D. Anderson Clinical Conference on Cancer,
Raven Press, New York, 225, 1984.
163. Nieuwnhuis, P., Ford, W. L.
Comparative migration of B and T lymphocytes in rat spleen and lymph nodes.
Cellular Immunology 23, 254, 1976.
164. Nieuwenhuis, P., Nouhuijs, C. E., van Eggens, J. H., Keuning, F. J.
Germinal centres and the origin of the B cell system. 1. Germinal centres in rabbit appendix.
Immunology 26, 497, 1974.
165. Nisonoff, A. J., Hopper, E., Spring, S. B.
The Antibody Molecule.
Academic Press, New York 1975.
166. Nomenclature for Human Immunolglobulins.
Bull. WHO 30, 447, 1964.
167. Nomenclature Committee of the IUIS: Notation for human immunoglobulin subclasses.
Bull. WHO 35, 953, 1966.
168. Nossal, G. J. V., Abbot, A., Mitchell, J., Lummus, Z.
Antigens in immunity. XV. Ultrastructural features of antigen capture in primary and secondary lymphoid follicles.
Journal of Experimental medicine 127, 277, 1968.
169. Notation for genetic factors of human immunoglobulins.
Bull. WHO 33, 721, 1965.
170. Novogrodsky, N., Ravid, A., Glaser, T., Rubin, A. L., Stenzel, K. H.
Role of iron in transferrin-dependent lymphocyte mitogenesis in serum free medium.
Experimental Cell Research 139, 419, 1982.
171. Nussenzweig, M. C., Steinman, R. M., Unkeless, J. C., Witmer, M. D., Gutchinov, B., Cohn, Z. A.
Studies of the cell surface of mouse dendritic cells and other leukocytes.
Journal of Experimental Medicine 154, 168, 1981.
172. Nutall, G.
Experimente über die bakterienfeindlichen Einflüsse des tierischen Körpers.
Z. Hyg. Infekt. Krankh. 4, 353, 1888.
173. Opstelten, D., Osmond, D. G.
Pre-B cells in mouse bone marrow: immunofluorescence statonokinetic studies of the proliferation of cytoplasmic $\mu$ chain-bearing cells in normal mice.
Journal of Immunology 131, 2635, 1983.
174. Origins of lymphocyte diversity.
Cold Spring Harbor Symp. Quant. Biol. Vol 41/part 2, 627 et seq. 1976.
175. Osmond, D. G., Nossal, G. J. V.
Differentiation of lymphocytes in mouse bone marrow. II. Kinetics of maturation in renewal of antiglobulinbilding cells studied by double labelling.
Cellular Immunology 13, 132, 1974.
176. Ottaway, C. A., Parrott, D. M. V.
Regional blood flow and its relationship to lymphocyte and lymphoblast traffic during a primary immune reaction.
J. Exp. Med. 150, 218, 1979.
177. Owen, R. D.
Immunogenetic consequences of vascular anastomoses between bovine twins.
Science 102, 400, 1945.
178. Palm, W.
Zur Raumstruktur der Antikörper. Ein Modell, basierend auf Ergebnissen der Untersuchungen am humanen Bence Jones Protein Rei und am humanen Kryoglobulin Kol.
Österreichische Gesellschaft für Hämatologie und Onkologie, Graz 25, 1981.
179. Pasteur L.
De l'attenuation du virus du cholera des poules.
C. R. Acad. Sci Paris 91, 673, 1880.
180. Parish, H. J.
A History of Immunization,
E. S. Livingstone, Ltd. Edinburgh 1965.
181. Parrot, D., Ferguson, A.
Selective Migration of lymphoblasts with in the small intestine.
Immunology 26, 571, 1974.
182. Pauling, L.
A theory of the structure and process of formation of antibodies.
J. Am. Chem. Soc. 62, 2643, 1940.
183. Penn, G. M., Cawley, L. P., O'Grady, J.
Antigenic differences between monoclonal IgA proteins: IgA subclass or deletion.

Am. J. Clin. Pathol. 62, 143, 1974.
184. Petterson, D., Mellstedt, H., Holm, G.
Monoclonal B lymphocytes in multiple myeloma.
Scandinavian Journal of Immunology 12, 375, 1980.
185. Platts-Mills, T. A. E.
Local production of IgG, IgA and IgE antibodies in grass pollen hay fever.
J. Immunol. 122, 2218, 1979.
186. Pollock, R. R., Mescher, M. F.
Murine cell surface immunoglobulin. Two native IgD structures.
J. Immunol. 124, 1668, 1980.
187. Poortmans, J. R., Jeanloz, R. W., Schmid, K.
3 $S\gamma_1$-globulin levels of normal human serum and urine.
Biochim. Biophys. Acta 133, 363, 1967.
188. Porter, R. R.
The hydrolysis of rabbit $\gamma$-globulin and antibodies with crystalline papain.
Biochem. J. 73, 119, 1959.
189. Porter, R. R.
Structural studies of immunoglobulins.
Science 180, 713, 1973.
190. Potter, M.
Antigen-binding myeloma proteins of mice.
Adv. Immunol. 25, 141, 1977.
191. Putnam, F. W., Easley, C. W.
Structural studies of the immunoglobulins. I. The tryptic peptides of the Bence Jones proteins.
J. biol. Chem. 240, 1626, 1965.
192. Putnam, F. W., Takahashi, N., Tetaert, D.,
Li-C, Lin, Debuire, B.
The last of the immunoglobulins: Complete amino acid sequence of human IgD.
In Immunoglobulin D: Structure and Function. J. Thorbecke and G. Leslie, eds.
Annals New York Academy of Sciences 399, 41, 1982.
193. Reinherz, E. L., Haynes, B. F., Nadler, L. M., Bersten, I. D.
Leukocyte typing II.
Springer-Verlag, New York 1985.
194. Reinherz, E. L., Kung, P. C., Goldstein, G.
Discrete stages of human intra thymic differentation: analysis normal thymocytes and leukaemic lymphoblasts of the T cell lineage.
Proceedings of the National Academy of Science 77, 1588, 1980.
195. Richards, F., Konigsberg, W., Rosenstein, R., Varga, J.
On the specificity of antibodies.
Science 189, 130, 1975.
196. Riva, C., Ropartz, C., Rowe, D. S.
Antigenic heterogeneity of human IgD immunoglobulins
Nature (New Biol.) 231, 279, 1971.
197. Rogoff, T. M., Lipsky, P. E.
Antigen presentation by isolated guinea pig Kupffer cells.
J. Immunol. 124, 1740, 1980.
198. Rosenberg, Y. J., Parish, C. R.
Ontogeny of antibody forming cells in mice. IV. Appearance of cells bearing FC receptors, complement receptors and surface immunoglobulin.
Journal of Immunology 118, 612, 1977.
199. Rowe, D. S., Fahey, J. L.
A new class of human immunoglobulin I. A unique myeloma protein.
J. Exp. Med. 121, 171, 1965.
200. Ruiz-Arguelles, G. J., Katzmann, J. A., Greipp, P. R., Gonchoroff, N. J., Garton, J. P., Kyle, R. A.
Multiple myeloma: circulating lymphocytes that express plasma cell antigens.
Blood 64, 352, 1984.
201. Schrader, J. W., Schrader, S.
In vitro studies on lymphocyte differentation. I. Long term in vitro culture of cells giving rise to functional lymphocytes in irradiated mice.
Journal of Experimental medicine 148, 823, 1978.
202. Schultze, H. E., Heremans, J. F.
Molecular Biology of human proteins; nature and metabolism of extracellular proteins.
Elsevier Publ. Co. Amsterdam–London–New York 1, 1966.
203. Schur, P. H.
Human gamma-G subclasses.
Prog. Clin. Immunol. 1, 71, 1972.
204. Seidman, J. G., Leder, A., Neu, M., Norman, B., Leder, P.
Antibody diversity.
Science 202, 11, 1978.
205. Sell, S.
Immunoglobulins of the germfree guinea pig.
J. Immunol. 93, 122, 1964.
206. Sercarz, E., Coons, A.
Specific inhibition of antibody formation during immunological paralysis and unresponsiveness.
Nature 184, 1080, 1959.
207. Silberberg-Sinakin, I., Baer, R. L., Thorbecke, G. J.
Langerhans cells.
Progress in Allergy 24, 268, 1978.
208. Silverstein, A. M.
History of immunology. Cellular versus humoral immunity: Determinants and consequences of an epic 19th century battle.
Cell. Immunol. 48, 208, 1979.
209. Singer, A.
Role of the major histocompatibility complex in T cell activation of B cell subpopulations. Lyb-5 and Lyb-5 subpopulations differ in their requirement for MHC-restricted T cell recognitions.
J. Exp. Med. 154, 501, 1981.
210. Singer, A., Hathcock, K., Hodes, R.
Cellular and genetic control of antibody responses VIII. MHC restricted recognition of accessory cells, but not B cells, by parent-specific subpopulations of normal $F_1$, T. helper cells.
J. Immunol, 124, 1079, 1980.
211. Singer, A., Hathcock, K., Hodes, R.
Self recognition in allogeneic radiation bone marrow chimeras.
J. Exp. Med. 153, 1286, 1981.
212. Skvaril, F., Morell, A.
Distribution of IgA subclasses in sera and bone marrow plasma cells of 21 normal individuals.
International Symposium on the Immunoglobulin A System. In Meslecky, J. and Lawton, A. R. (ed).
Advences in Experimental Medicine and Biology New York, Plenum Pr. Plenum Pub. 45, 433, 1974.
213. Socken, D. J., Jeejeebhoy, K. N., Bazin, H., Underdown, B. J.
Identification of secretory component as an IgA receptor on rat hepatocytes.
J. Exp. Med. 150, 1538, 1979.
214. Spielgelberg, H. L.
Biological activities of immunoglobulins of different classes and subclasses.
Adv. Immunol. 19, 259, 1974.
215. Sprent, J.
Antigen-induced selective sequestration of T lymphocytes: role of the major histocompatibility complex.
Monographs in Allergy 16, 233, 1980.
216. Sprent, J., Basten, A.
Circulating T and B lymphocytes of the mouse. II. Lifespan.
Cellular Immonology 7, 40, 1973.
217. Stashenko, P., Nadler, L. M., Hardy, R., Schlossman, S. F.
Characterisation of human B lymphocytes-specific antigen.
Journal of Immunology 125, 1678, 1980.
218. Stein, H., Gerdes, J., Mason, D. Y.
The normal of malignant germinal centres.
Clinics in Haematology II. 531, 1982.
219. Steinman, R. M., Lustig, D. S., Cohn, Z. A.
Identification of a novel cell type in peripheral lymphoid organs of mice. III. Functional properties in vivo
J. Exp. Med. 139, 1431, 1974.
220. Stokes, C. R., Hosking, C. S., Turner, M. W., Johansson, S. G. O.

Urinary IgE: a reappraisal.
Europ. J. Immunol. 3, 241, 1973.

221. Szakál, A.
Isolated follicular dentritic cells: Cytochemical antigen localization. Nomarski, SEM and TEM morphology.
J. Immunol. 134, 1349, 1985.

222. Tada, T., Ishizaka, K.
Distribution of γE-forming cells in lymphoid tissues of the human and monkey.
J. Immunol. 104, 377, 1970.

223. Taniguchi, M., Tada, T., Tokuhisa, T.
Properties of antigen-specific suppressive T cell factor in the regulation on antibody responses in mouse. III. Dual gene Control of T cell-mediated suppression of the antibody response.
J. Exp. Med. 144, 20. 1976.

224. Taussig, M. J.
T cell factor which can replace T cells in vivo.
Nature 248, 234, 1974.

225. Taussig, M. J.
Antigen-specific T cell factor in cell cooperation. Mapping with in the 1 region of H-2 and the ability to cooperate across allogeneic barriers.
J. Exp. Med. 142, 694, 1975.

226. Terry, W. D., Fahey, J. L., Steinberg, A. G.
Gm and Inv Factors in subclasses of human IgG.
J. Exp. Med. 122, 1087, 1965.

227. Tew, J., Thorbecke, G., Steinman, R.
Dendritic cells in the immune response: Characteristics and recommended nomenclature
J. Reticuloendothelial Soc. 31, 371, 1982.

228. Thorbecke, G. J., Gordon, H. A., Wostnam, B., Wagner, M., Rayniers, J. A.
Lymphoid tissue and serum gammaglobulin in young germ free chickens.
Journal of Infection Diseases 101, 237, 1957.

229. Thorbecke, G. J., Keuning, F. J.
Antibody formation in vitro by haemapoietic organs after subcutaneous and intravenous immunisation.
Journal of Immunology 70, 129, 1953.

230. Thorbecke, G. J., Keuning, F. J.
Antibody and gammaglobulin formation in vitro in haemopoietic organs.
Journal of Infections Diseases 98, 157, 1956.

231. Thorbecke, G. J., Leslie, G.
Immonoglobulin D: Structure and function.
Ann. N. Y. Acad. Sci. 399, 1982.

232. Tiselius, A., Kabat, E. A.
An electrophoretic study of immune sera and purified antibody preparations. J. Exp. med. 69, 119, 1939.

233. Tonegawa, S.
Somatic generation of antibody diversity
Nature 302, 575, 1983.

234. Turner, M. W., McClelland, D. B. L., Medlan, A. R., Stokes, C. R.
IgE in human urine and milk.
Scand. J. Immunol. 6, 343, 1977.

235. Underdown, B. J., Dorrington, K. J.
Studies on the structural and conformational basis for the relative resistence of serum and secretory immunoglobulin A to proteolysis.
J. Immunol. 112, 949, 1974.

236. Underdown, B. J., Knight, A., Papsin, F. R.
The relative paucity of IgE in human milk.
J. Immunol. 116, 1415, 1976.

237. Van Boxel, J. A., Paul, W. E., Terry, W. D., Green, I.
IgD-bearing human lymphocytes.
J. Immunol. 109, 648, 1972.

238. Veldman, J. E. Histophysiology and electronmicroscopy of the immune response.
N. V. Boekdonkley Dijkstra Niemeyer, Groningen. 1970.

239. Waldmann, Th. A., Strober, W.
Kinetic studies of Immunoglobulin metabolism in immunologic deficiency.
In: „Immunologic Deficiency Diseases in Man." D. Bergman and R. A. Good (ed) Birth Defects IV. 1, 338, 1968.

240. Waldmann, Th. A., Bull, J. M., Bruce, R. M., Broder, S., Jost, M. C., Balestra, S. T., Suer, M. E.
Serum immunoglobulin E levels in patients with neoplastic disease.
J. Immunol. 113, 379, 1974.

241. Wall, R., Kuehl, M.
Biosynthesis and regulation of immunoglobulins.
Ann. Rev. Immunol. 1, 343, 1983.

242. Weaver, E.
Secretion of IgA by human milk leukocytes initiated by surface membrane stimuli.
J. Immunol. 132, 684, 1984.

243. Weigle, W. O.
Immunological unresponsiveness.
Adv. Immunol. 16, 61, 1973.

244. Wigzell, H., Anderson, B.
Cell separation on antigen coated columns. Elimination of antibody forming cells and memory cells.
J. Exp. Med. 129, 23, 1969.

245. White, R. G.
Immunological functions of lymporeticular cells.
In: Gell P. G. H., Coombs R. R. A., Lachmann P. J. (ed) Clinical aspects of immunology 3 rd edn. Blackwells Scienfitic. Oxford 432, 1975.

246. Williams, R. C., Pinnel, S. R., Bratt, G. T., Sheldon, R.
Low molecular weight L-chain components related to Bence Jones proteins.
J. Lab. Clin. Med. 86, 81, 1966.

247. Wright, A. E., Douglas, S. R.
An experimental investigation of the role of the blood fluids in connection with phagocytosis.
Proc. R. Soc. Lond. (Biol.) 72, 357, 1903.

248. Wu, T. T, Kabat, E. A.
An analysis of the sequences of the variable regions of Bence Jones proteins and myeloma light chains and their implications of antibody complementarity.
J. Exp. Medicine 132, 211, 1970.

249. Zajicek, J. Studies on the histogenesis of blood platelets and megakaryocytes.
Acta Physiol. Scand. 40/Supplement 138, 1957.

250. Zinkernagel, R. M., Doherty, P. C.
H-2 compatibility requirement for T-cell-mediated lysis of target cells infected with lymphocytic choriomeningitis virus.
J. Exp. Med. 141, 1427, 1975.

251. Yodoi, J., Ishizaka, K.
Lymphocyte bearing receptors for IgE. – III. Transition of FcyR( + ) cells to $Fc_{g}R$( + ) cells by IgE.
J. Immunol. 123, 2004, 1979.

# II. Immunology of the gastrointestinal tract

The digested metabolized substrates provide energy required for the locomotion and the maintenance of the structure. Digestion is a sequel of the concomitant physical and chemical processes, due to which the nutrients are split and their antigenic properties are lost. Different mechanisms would transport these substrates promoting their absorption. The total skin surface of a normal adult is about 1.5-1.9 $m^2$, while the surface of the gastrointestinal tract (GI), which is in contact with the "internal" milieu comprises about 140 $m^2$. A system termed by Burnet as GALT regulates and controls the defense mechanisms of this enormous surface.

The Waldeyer's ring, the Payer's patches and the plasma cells of the colon mucosa all having similar immunological functions, are identified as a T-independent B-dependent system. Similar to the well-known functions of the thymus, peripheral lymph nodes or bone marrow, a considerable amount of data exists concerning the activity and information exchange within the above mentioned B-dependent system.

The cells, morphological entities and molecules comprising the morphological basis of this system will be specially emphasized.

**Epithelial cells** - aside from absorption - together with the other active systems - they maintain about 50% of the mucoperistaltic function (111). The epithelial cells also play a unique role in the production of Secretory Component (SC) responsible for the transport of the IgA molecules into the lumen (9).

**Mucus-secreting Goblet cells** - are one of the final forms of differentiation of the epithelial cells. The rate of mucus secretion affects the intraluminal processes.

**Glandular epithelium** - though it plays a minor role in the absorption of the nutrients as antigens, its major purpose is the formation of antigen-reacting islets (9, 5, 123), which comprise an outer "core" of the Peyer's patches.

**"M"- microfold cells** - specialized epithelial cells responsible for a special form of the antigen uptake and presentation. Other organs fail to exhibit similar mechanisms (9, 95, 96).

Compared to the other epithelial cells, these elements contain very few microvilli and, therefore, can be easily identified. Their number depends on the size of the follicle, however, their ratio to the other absorbing glandular epithelial cells (M:glandular epithelial cells) is always constant - 1:12. Raster electronoptical studies indicate that the M cells promote absorption of the minor amount of both natural substances (for example seeds, 40 μ in diameter) and artificial particles (latex, 0.5-1 μ in diameter). Hence, these molecules rather participate in the information transfer, than in the energy supply processes.

**Lymphocytes** - T lymphocytes comprise about 50-90% of the whole lymphocyte population and directly or indirectly control the differentiation of the mucosal cells. They can cause hyperplasia of the mucus-secreting cells and increase the deepth of the crypts. Concomitantly atrophy of the villous cells occurs. This interesting morphological reaction has been first attributed to infection, however, transfusion of the lymphocytes separated from the lymph nodes of the infected animal to a healthy syngenic animal elicited mucosal alterations similar to those caused by the infection. This finding proves the regulatory role of the lymphocytes. The infection-induced changes of gastrointestinal cells, which develop by the 14th day post-infection would alter the normal structure of the mucosa. The donor lymphocytes from the thoracic duct of the infected animal transfused to the recipient with a simultaneous infection transfer induce similar changes a week earlier - by the 8th day post-infection (86).

The above experiment fails to fully reveal the individual characteristics and the functions of the intraepithelial (IEL) lymphocytes.

The available data of the literature indicate that this field needs further investigation. The IEL lymphocytes undoubtedly possess specific individual properties, compared to the other lymphocyte pools, including the LP.

50% of these cells are derived from the thymus (25, 54, 75, 80, 98, 101, 112) and follow the same course of migration as the T-cells or the precursors of the IgA-producing plasma cells of the LP (20, 54, 84, 102, 116). These cells, starting from the Peyer's patches perform special functions. Following an intraluminal antigen stimulus, the IEL-precursor starting from the Peyer's patches traverses the mesenteric lymphatic nodes (MLN) and get into the blood stream via the thoracic duct. In the circulation these lymphocytes become large proliferating cells; they return into the intestinal mucosa, where they are transformed into quiescent cells (54). With regard to their functions and membrane properties, these cells can be assessed into two large subgroups:

A - Cytotoxic suppressor phenotype (CD $8^{+}$) (25, 65, 75, 80, 98, 101, 112), or MHC-I restricted cytotoxic cells (37, 64).

B - population carrying no pan-T antigens, comprising about 50% of the IEL (22, 25, 65, 80, 98,101). These cells might origin from the bone marrow (80). It is noteworthy that not all the cells belonging to this population can be identified with the CD 8 serum (53). Furthermore, there is no evidence concerning the function, differentiation and the relationship of these cells with the thymus-derived lymphocytes. Only one team succeeded to reveal the so-called HML-1 antigen on their membranes (23) using monoclonal antibodies. 97-99% of the IEL are HML-1 positive, whereas only 0.1-0.7% of the systemic lymphocytes obtained from the venous blood of the same individual possess these surface antigens. 68-74% of the CD $8^{+}$ intraepithelial cells carry this marker, while it occurs only in 20-38% of the circulating CD $8^{+}$ cells and in 11-20% of the mucosal CD $4^{+}$ lymphocytes. 42-62% of the systemic CD $4^{+}$ cells also exhibit this membrane antigen. Immune electromicroscopic studies of the HML-1 positive cells from the epithelial region or lamina propria indicate characteristic intracytoplasmatic granules of these cells, their other characteristics resemble those of the small and medium lymphocytes. It must be stressed that the macrophages, granulocytes, mast- and plasma cells are HML-1 negative. The bone marrow contains about 1% of the positive cells; 0.7% occur in the systemic circulation and only 1-2 cells can be found in the medullar zone of the thymus. Cortical areas of the spleen and the T-zones of the Peyer's patches sometimes have HML-1 positive cells. There are relatively more HML-1 positive cells (2-5%) in the tonsils and the peripheral lymph nodes.

In summary, the HML-1 antigens act as the markers of the gastrointestinal tract, including the T and the IgA-producing B cells of the mesenteric lymph nodes (22).

This antibody-marked antigen is a membrane receptor responsible for the specific localization of the lymphocytes, termed "intestinal homing". However despite the establishment of a specific relationship between the mucosa and the IgA-producing lymphocytes (81, 82), only few HML-1 receptors are localized on the B cells, hence, their special task is not yet fully understood.

The presence of the identical differentiation antigen markers in all the IEL subpopulations suggests similar differentiation of all the IEL cells in the intestinal micro-environment.

**The LGL cells.** About 50% of the intraepithelial lymphocytes contain large granules, and are, therefore, called "large granular lymphocytes" (LGL) (122). A growing body of evidence identify them as NK cells. Some researchers. however, find very low count of these cells. The LGL cells are derived from the bone marrow and their differentiation is independent of the thymus (40, 92).

4-6% of the LGL cells are responsible for the non-specific cytotoxicity (19, 101, 122). 25% of these CD $8^{+}$ T cells contain sulphatated polysaccharides in their granules (21, 24, 35, 51, 114, 115). Interesting enough, the early phase of development of the sporozoites of Eimeria tenella, E.necatrix and pichinde virus occurs inside these cells (5). Since the further migration of these microorganisms might result in a generalized infection, our current knowledge about the mechanisms of gastrointestinal infections should be revised.

**Mastocytes.** According to recent findings, mastocytes represent a special form of differentiated lymphocytes (2.10). They can be cultured from the lymph nodes, thymus and the spleen. It has been believed that the mast cell precursors origin from the thymus. However, evidence exists that these hemopoietic precursors are rather derived from the bone marrow, since the IL-3 signal transforms them into mastocyte-like cells (59, 121). In vivo these precursors commonly found in the connective tissue, skin, around the vessels or in subepithelial spaces, differentiate into mature mastocytes. Mature forms having a short lifespan, exit the circulatory bed but can move within the LP. Their granules contain histamine, heparin, PAF, SRS-A, ECF-A, NCF and chymase and their membranes carry $FC_{\varepsilon}$ receptor. The morphology and probably the function of the epithelial mastocytes differ from that of the subcutaneous forms. In parasite infections their count dramatically increases. The molecular weight of the "heparin" (chondroitin sulphate) molecules found in the mucosal cells ranges between 5,000 and 30,000 Daltons (4, 46).

**Macrophages.** Classically this term means phagocytes. Beside this well known function, macrophages carrying MHC II receptors can present antigens. They also possess MHC I, FCγ, complement binding surface markers, etc. The cytoplasm of macrophages contains heparin. This cell population plays an extremely important role in the course of inflammation and in the control of the lymphocyte function (58,125).

**Peyer's patches.**

Peyer's patches comprise a special organ within the gastrointestinal immune system. Their number increases from the duodenum to the coecum. Similar to the lymph nodes, Peyer's patches contain T and B lymphocytes, antigen-presenting and accessory cells, thus forming a special unit of the local immune response. Peyer's patches slightly protrude into the intestinal lumen and their epithelial membrane contains special "presenting" M (microfold) cells. The cytoplasm processes enclose lymphocytes. Due to the endocytosis of the M cells, the abluminal antigens come to contact with the lymphocytes (95) (Fig. II/1), which would migrate into the subepithelial follicles of the Peyer's patches to form germinative centers.

Macrophages as antigen-presenting cells interact with the T and B lymphocytes in the peripheral germinative centers. The former, acting as accessory cells, elicit aimed lymphocyte proliferation with subsequent development of lymphoblasts. 85% of the B lymphocytes would proliferate into IgA-producing plasma cells (29, 88). Leaving the site of their primary interaction with antigens, the stimulated B lymphoblasts migrate (54) from the Peyer's patches into systemic circulation via the regional mesenteric lymph nodes; they pass the spleen finally arriving to the lamina propria (Fig. II/2).

Transfer experiments validate the fact of migration: lymph cells from the Peyer's patches were transferred to a lethally irradiated syngenic recipient, which scarcely had its own lymphatic cells. The donor cells migrated to the mucosal lamina propria (30). In the experiments with donor lymphocytes originating from mesenteric lymph nodes and the thoracic duct the cells exhibited similar course of the migration.

Mesenteric lymph nodes of the newborn lambs were removed and the Peyer's patch lymphocytes were labeled with radionuclides. Following this procedure a pronounced cell exchange was observed between the lymphocytes of the remote lymph nodes, spleen, thymus and the bone marrow. It is suggested that the alimentary antigens are spread all over the organism following this route of migration subsequently activating the effector mechanisms (12).

After surgical removal of the Peyer's patches in the adult rats, the animals were orally infected with the strains of E. coli 055 and B5 and the immune response of the mesenteric nodes, spleen and lamina propria was compared to that of the identically infected healthy animals.

The control group developed pronounced local IgA and IgM responses, whereas the

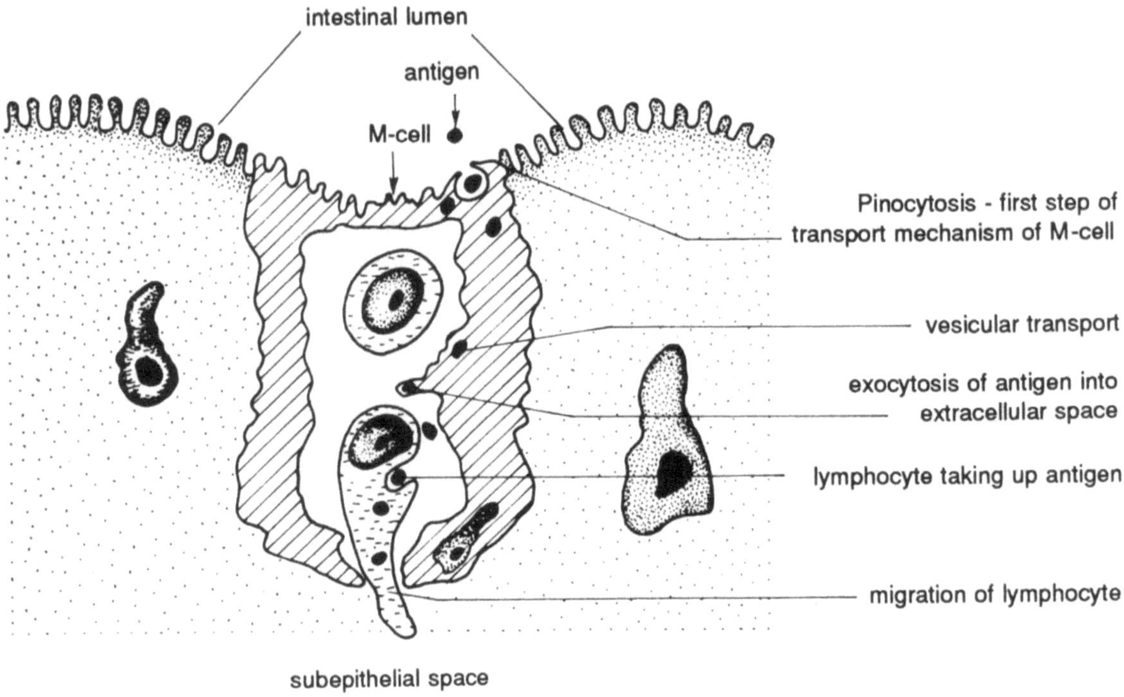


**Fig. II/1**

operated rats responded with a suppressed IgA and atypical IgG antibody production.

Despite these experimental results the fine mechanisms of the above mentioned immune response are not yet fully understood. A special receptor-ligand interaction between the lymphocytes and the endothelial cells is suggested. Similar interaction was observed in the postcapillary venules during lymphocyte migration between the primary and the secondary organs (28, 117). Further investigations of the above discussed HML-1 membrane antigens (23), certain chemotactic factors and the specific regulatory role of the continuous antigen presentation (88, 97) could answer this question.

In summary, after a specific antigenic stimulus the differentiated lymphoblasts leave the Peyer's patches and enter the lamina propria via the systemic circulation. Lamina propria is the site of their final differentiation into IgA-producing plasma cells (Fig. II/3).

**Mucosal lamina propria** is localized beneath the mucosal layer of the epithelial cells. Lamina propria is not a homogeneous layer of specific cells, but contains different cells also occurring in other organs. The latter are embedded in the matrix of fibroblasts and mesenchimal cells. Lymphocytes, mastocytes, neutrophil and eosinophil leukocytes are the most typical cellular elements of the lamina propria. The ratio between these constantly migrating cells dynamically alters meeting the requirements of the organism. The old cells are destroyed in the lamina propria. This "independent" activity is only apparent, for complex relationships between the cells control the mucosal function. The anatomical (I), functional (II) and biochemical (III) changes following an antigen stimulus support the functional entity of the mucosal lamina propria and the epithelial cells.

**I.** The evident modification of the mesenchimal cells and the pericryptal fibroblasts is of less importance, compared to that observed during differentiation and proliferation of the epithelial cells (74). These events are parts of the inflammatory process, although no exact definition of inflammation or the sequence of its events still exists. It is well known, however, that the activation of the complement, coagulation system, kinins and fibrinolytic mechanism contribute to the migration of these cells leading to mucosal damage. Nevertheless, there has been doubt, whether it is the inflammation or the mucosal lamina propria,

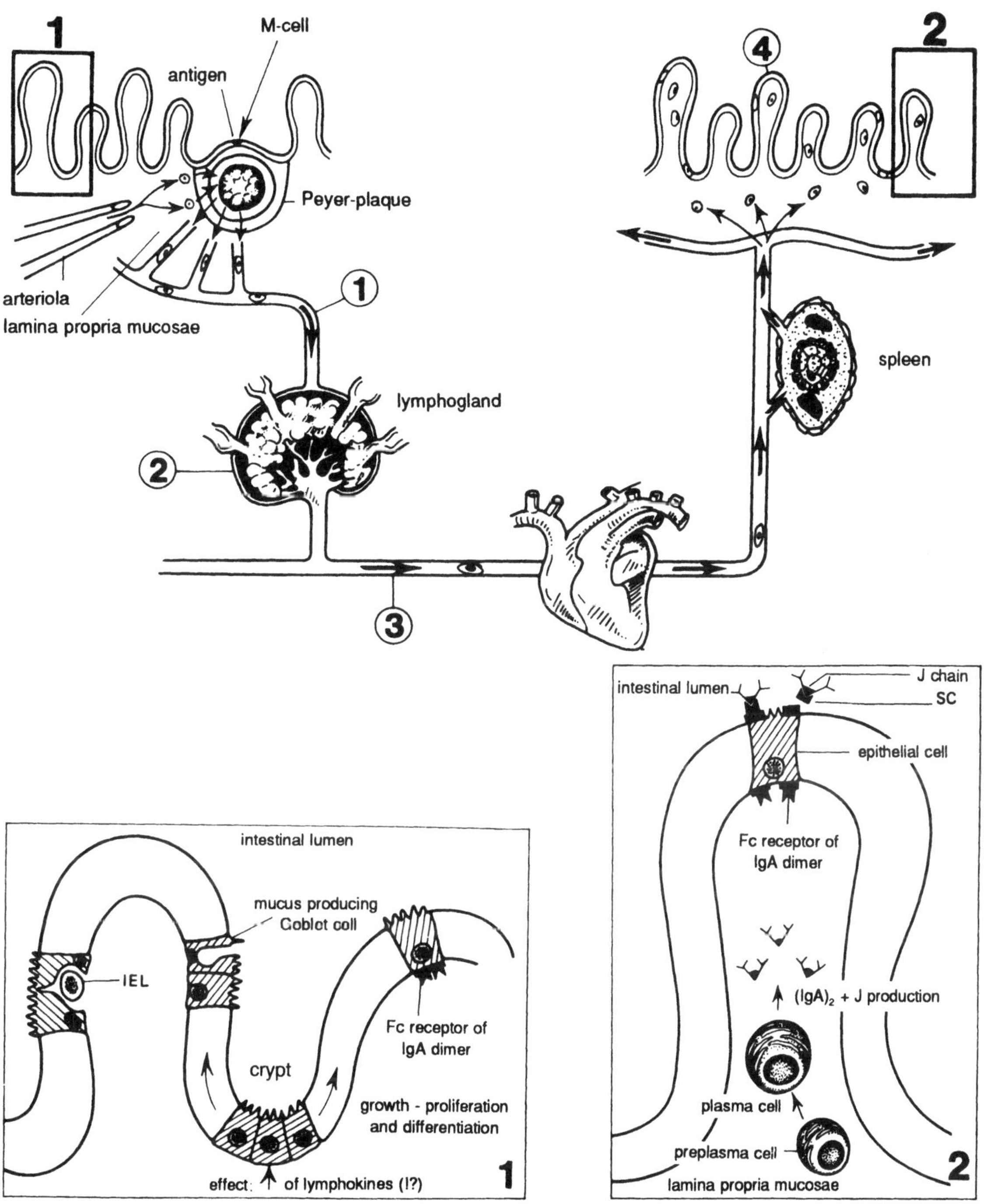


Fig. II/2

which play the primary role in these pathological processes.

**II.** As the number of IEL grows, the epithelial cell-mediated sIgA transport into the lumen is promoted (5). Both cellular and molecular changes of the mucosa are the consequences of the active function of the lamina propria - epithelial cells can only accept or transport different molecules, but are incapable of their synthesis. The phospholipase activity of the eosiniphils (70) and the myeloperoxydase activity of the myeloid cells (44) are elevated. Activation of both enzymes, as well as a subsequent decrease of the

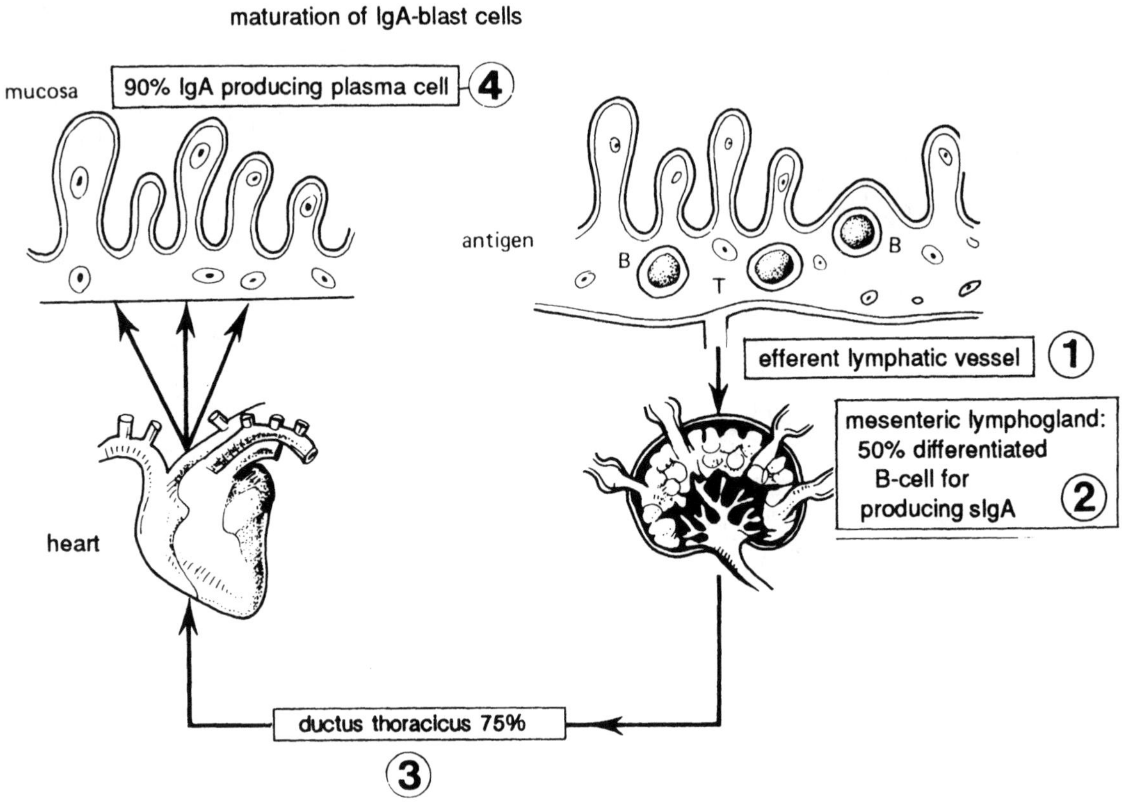


Fig. II/3

enzyme activity in the epithelial brush border indicates inflammatory process. The characteristic cytoplasm of the mastocytes plays a complex role in the inflammatory reactions: it triggers the IgE-mediated histamine release; releases serotonin in its environment without involvement of any accessory immune mechanisms; and finally accumulates 5-hydroxytriptophan and histamine (90). The mucosal lamina propria has a dual effect on the GI function. One is the histamine-mediated smooth muscle contraction, observed for example in the Schultz-Dale reaction (126). The other alters the mucosal function, however, its complex mechanism needs further investigation. It is generally assumed that though histamine triggers the pathological process (127), the final result is rather due to an interaction of the released biologically active substances.

The watery jejunal and ileal secrete is released into the lumen (71, 124) reducing absorption, including that of $Ca^{++}$. The latter is required by the mastocytes for the release of serotonin into the extracellular space and for VIP secretion (11, 32, 110). VIP induces a cAMP-mediated secretion of NaCl in the biliary, pancreatic and intestinal epithelium (39, 42). This secretion mechanism resembles that of the cholera toxin, prostaglandins (39) and alpha-interferon. Serotonin mechanism is not the sole trigger of the above reactions, for it also has secretogogue properties (33, 63). VIP, however, has another action mechanism: being released as a neurotransmitter (45, 131) it activates the cholinergic cells with subsequent release of ACh. The latter stimulates secretion of fluid, electrolytes (56, 130) and mucus (1). The secreted VIP is inactivated in situ by the kinases of the mastocytes (60) (Fig. II/4).

The remote effects include modification of the pancreatic exocrine function and altering the secretin and cholecystokinin levels.

**III.** The synthesis of glycoproteins derived from the cell membrane increases, consequently their concentration in the microenvironment grows (41). Glycoproteins inhibit the mediator interactions at the level of the membrane-bound glycoproteins, receptors or lectins. This inhibition indirectly prevents the alteration of the cell "behavior". This protection is, however, only partial, since the mucus secretion is still enhanced.

### The IgA immune response

Though the mucosal IgA production also termed as secretory immune response, is well documented (69), recent findings have urged the researchers to review this problem (111). Thus the formation of sIgA-mediated immune complexes seems to be a basic factor of tolerance. Without the knowledge of immunoregulatory mechanisms its effector function can not be understood. The experiments with double Thiry-Vella fistulas suggest that the secretion and regulation of sIgA is a local response.

Proximal and distal fistulas were made on the abdominal wall in rabbits. Peyer's patches were removed from the proximal fistula and left in the distal one. At the 7th post-operation day the immune response to an antigen applied before surgery was evaluated. The antibody titer found in the fistula was considered as control. Then the antigen was repeatedly administered into the fistula lumen. Administration into the patch-free fistula resulted in a non-significant increase of the specific sIgA antibodies in both fistulas, whereas same administration into the intact lumen revealed a ten-fold elevation of the antibody level in the intact, and a 4,5-fold increase in the patch-free fistulas (62).

LPS treatment of the B-cells cultured from the Peyer's patches and the spleen resulted in a parallel elevation of the IgG, IgM, and IgA production. Addition of the spleen T cells stimulated with Concanavalin A resulted in the inhibition of the IgG, IgM, and IgA production induced by the suppressor T cells. Replacement of spleen T lymphocytes by the stimulated T cells from the Peyer's patches resulted in similar depression of IgG and IgM synthesis, while the production of IgA antibodies was paradoxically enhanced (36). These findings are consistent with the assumption that the Ts lymphocytes derived from the Peyer's patches activate the sIgA-producing plasma cells and inhibit those synthesizing IgG and IgM molecules.

Other experimental findings support the above theory: T cells from the Peyer's patches of mice immunized by enteral administration of ovalbumin were transfused into syngenic animals, which were simultaneously treated with ovalbumin. Similar to the above mentioned experiment, the IgG and IgM-producing plasma cells were inhibited, while the synthesis of the ovalbumin-specific IgA antibodies was enhanced (119). T lymphocytes from the Peyer's patches of non-immunized animals also stimulated the IgG production.

### The gastrointestinal tract as immune system

The overview of the morphological and functional features of the gastrointestinal tract, is followed by the discussion of its protective role against the abluminal antigens. Both specific and non-specific immune response act as a synergist that forms the first and the second lines of defense against foreign agents. Specific response mainly induces tolerance (A), while the non-specific one enhances mucus production and peristalsis, moving forward the contents of the lumen (B) (85, 113).

**A.** Local immune response and induction of systemic immune tolerance are the main tasks of the gut-associated immune system (GALT). The first experiments in this field indicated that following intragastric administration of cow milk (6) or corn (129), the repeated parenteral administration of these substances evoked no reaction. Recent data indicate that guinea-pigs fed with dinitrochlorobenzene exhibited contact cutaneous sensitization (26), while animals fed with oxasolon or picrilchloride, both capable of inducing contact allergy, failed to induce dermatitis in mice (3). The role of T and B suppressor cells and suppressor substances produced by these lymphocytes might explain the above experimental data. The following information is, however, needed to get a deeper insight in these pathomechanisms:

**1. The role of sIgA in the specific immune response and tolerance.** This antibody

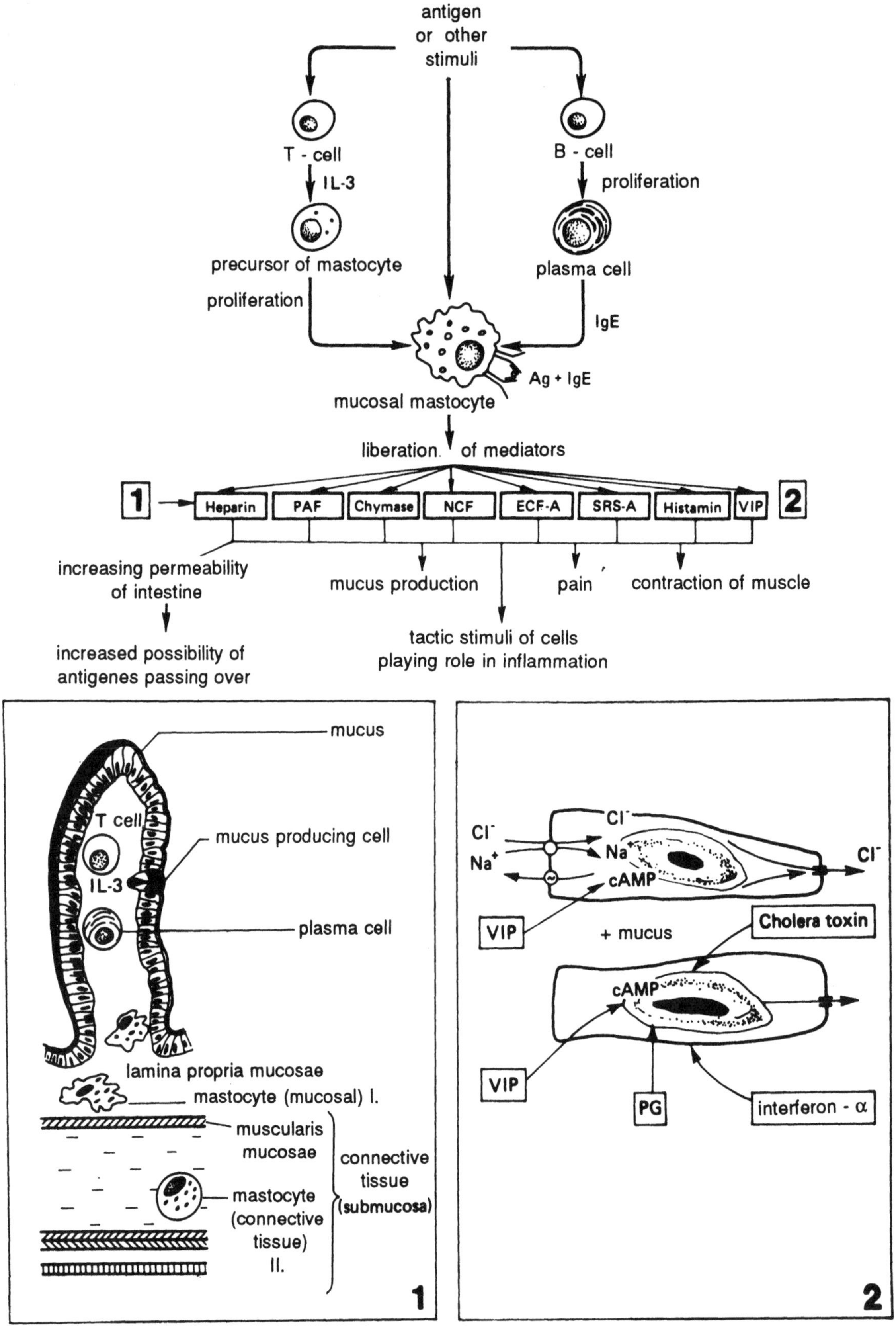

antigen or other stimuli
T - cell
IL-3
precursor of mastocyte
proliferation
B - cell
proliferation
plasma cell
IgE
Ag + IgE
mucosal mastocyte
liberation of mediators
1
Heparin
PAF
Chymase
NCF
ECF-A
SRS-A
Histamin
VIP
2
increasing permeability of intestine
increased possibility of antigenes passing over
mucus production
tactic stimuli of cells playing role in inflammation
pain
contraction of muscle
mucus
T cell
mucus producing cell
IL-3
plasma cell
lamina propria mucosae
mastocyte (mucosal) I.
muscularis mucosae
connective tissue (submucosa)
mastocyte (connective tissue) II.
1
Cl⁻
Na⁺
Cl⁻
Na⁺
cAMP
Cl⁻
VIP
+ mucus
Cholera toxin
cAMP
VIP
PG
interferon - α
2


Fig. II/4

with molecular weight of 400,000 Daltons is made up of two IgA molecules (160,000 Daltons each) connected with a J chain (15,000 Daltons) and an SC molecule (70,000 Daltons) (10, 91, 103). Epithelial cells transport this molecule into the lumen: at the site of mucosal lamina propria $F_c$ receptors bind the dimeric IgA molecules and transport them into the intracellular space via pynocytosis. The free IgA is finally released into the lumen, and a part of $F_c$ receptor (SC) still remains attached to the molecule. Thus altered IgA molecule is unavailable to the proteolytic enzymes (Fig. II/5).

The dimeric IgA molecules produced by the GALT and released into the systemic circulation are uptaken by the liver. Binding to similar receptors, they migrate from the Disse spaces into the primary bile canalicules and the intestinal lumen (72, 94). In contrast with the data of literature, our experiments show that the medullar non-dimeric IgA-s fail to follow this course of migration (8). It is interesting to mention that in gammapathies, the anti-idiotype antibodies could identify independent clone products in the saliva. These molecules of different size occurred at very low concentrations.

The analysis of the absolute values in plasma and saliva, as well as the ratios between the similar, but polyclonal immunoglobulins to other immunoglobulins indicates that the salivary idiotype positive monoclonal gamma globulin was passively transported. Evidence exists that the IgA producing plasma cells of the bone marrow markedly differ from those of the lamina propria even in cases of myeloma (67).

Though the regulatory aspects of the luminal IgA transport remain uncertain, its existence is extremely important. The immune complexes are either digested by the proteolytic enzymes, or are excreted with feces. Furthermore, it is assumed that the complexes can be absorbed by the intestinal wall being split or not split (54) by the complement. Due to the latter mechanism the complex is recycled in the enterohepatic circulation (100), so its hepatic or luminal decomposition would be delayed. By this transport mechanism the local immune response neutralizes the antigens and prevents the development of a systemic immune response with non-desired consequences (enteral allergy). Besides the above mechanisms, a systemic immune tolerance can also develop, since oral or intragastric antigen exposition induces immune tolerance in the non-GALT immune system, independ-

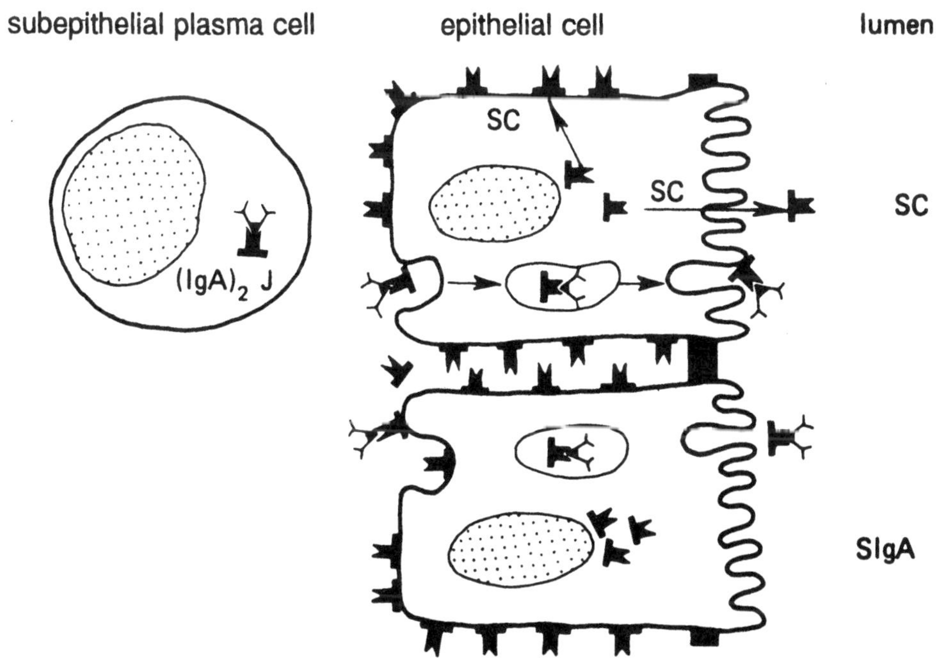


Fig. II/5

ently of the cellular or humoral character of the immune response (26, 36, 78).

Though the significance of these mechanisms is still uncertain and the conclusions are rather of theoretical character, still tolerance occurs only if the experimental animals had not received systemic immunization prior to the antigen exposition (38). Antrum is the site of the primary contact of alimentary antigens with the host cells. This contact is sufficient to increase the level of the plasma cells and mastocytes in the mucosal lamina propria, and to increase the MHC II restricted antigen presenting cells. These data undoubtedly indicate that the antigen recognition occurs not only in the antrum. The enterally acquired antigens can also reduce the level of the circulating antibodies capable of interaction. Immune complexes are formed at high rates and the organism is finally desensitized. These data indicate that the nutritional antigens not only affect the systemic immune response, but also regulate their own absorption.

Experimental data show that following enteral sensitization, the animals uptake only about 50% of the original antigen.

**2. Enhanced activation of the IgG specific suppressor cells.** The basic mechanisms of tolerance were assessed in transfer experiments. The cells isolated from the Peyer's patches, mesenteric lymph nodes and spleen of the enterally immunized animals were transfused into syngenic animals. Following a successful cell transfer, the same antigen was parenterally administered to the host. The controls received no enteral antigens; other untreated controls were only administered cell suspensions from the above mentioned organs. The cells of the immunized animals induced tolerance in the host. Membrane marker analysis proved them to be T lymphocytes (3, 57). After treatment of these cells with an anti-T antiserum the suspension lost its immune suppressive properties and failed to elicit tolerance.

B cells can also induce systemic tolerance, though their capability is less pronounced: antibodies produced by B lymphocytes can form immune complexes with alimentary antigens, thus the antigens can not further transform information to the appropriate specialized cells (128).

Following an oral immunization of mice with sheep erythrocytes a humoral suppressor factor appeared in their plasma. Administration of this factor in vivo or in cell cultures induced antigen-specific suppressor effect (61). Subsequent studies have identified this factor as IgA and $IgG_1$, the latter occurring at lower levels (2). Therefore, the immune suppressive effect of the antigen-specific antibodies is mediated by the formation of immune complexes (133). Not only the antigens, but also the antigen receptors of the cells bound to the anti-idiotype antibodies can participate in the immune complex formation. This mechanism can act as a negative feedback regulation. B lymphoblasts proliferating after an antigen stimulus, can activate the T inducer cells, which would stimulate the T suppressors in a chain reaction fashion (Fig. II/6).

**In summary** the tolerance is primarily mediated via the T suppressor cells, involving both the IgG and IgE antibody production and the delayed hypersensitivity (87, 93). Uncertainty exists about the nature of the non-cellular, non Ig-mediated suppressor factors. Anyway, low molecular weight substances are identified in supernatants obtained from the cultures of specific suppressor cells. Following treatment of the suppressor cells with anti lymphocyte serum, these substances are absent from the supernatant (79).

The synergism of the specific and non-specific mechanisms is also worth mentioning. The latter comprises arachidonic acid, biogenic amines, and the metabolites of the complement (76). Development of tolerance controlling the complex process of the nutrient absorption is based on the two immune mechanisms: T helper cells regulating the local restricted immune response and the concomitantly activated T suppressor cells mediating the paradox systemic antigen-specific tolerance. These findings, as well as the above presented data indicate that the T helper cells initiate a dimeric IgA response in the lamina propria with a subsequent local formation of immune complexes. The importance of the latter can not be overestimated. Evidence exists that at decreased or selectively inhibited IgA production free alimentary antigens or their complexes with the IgG antibodies occur

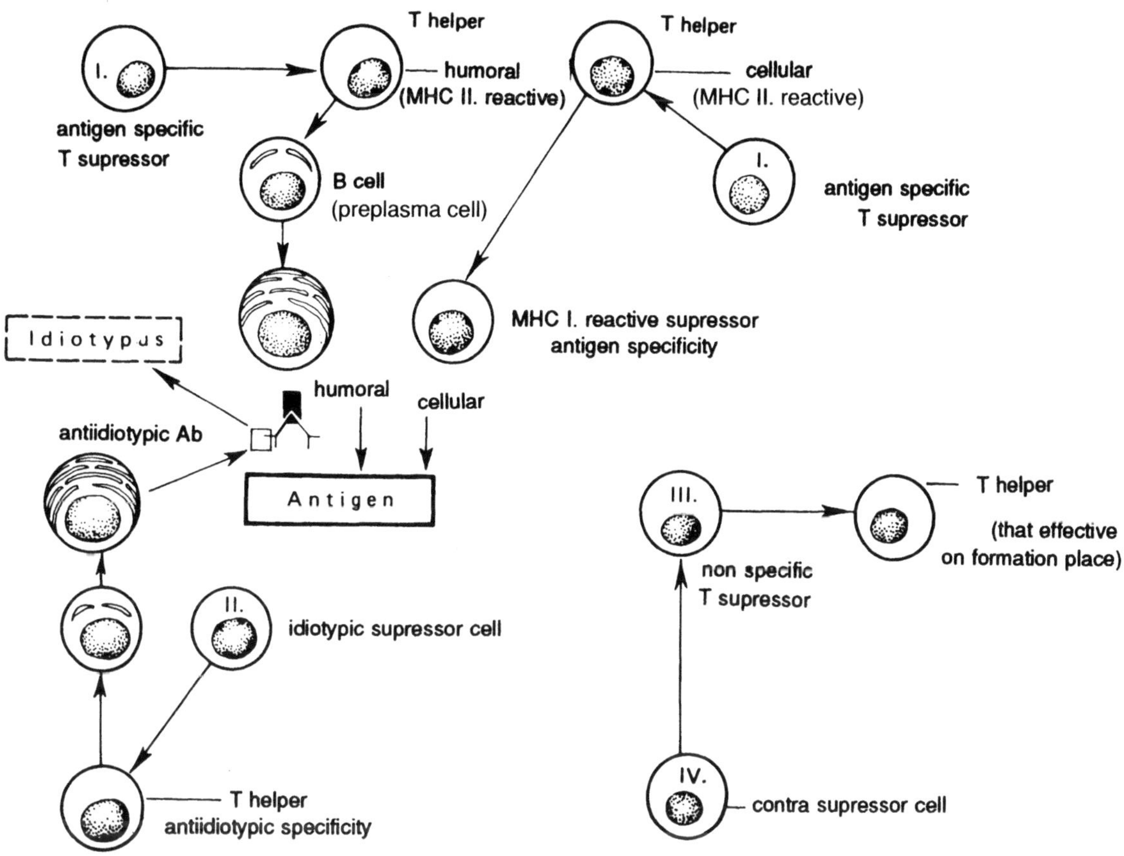


Fig. II/6

in systemic circulation (66) thus leading to tolerance. Systemic tolerance is mediated via the T helper cells activating the T suppressor lymphocytes (15, 16, 34, 47, 83, 89, 120).

Contrasuppression is another factor contributing to these complex regulatory process. In this case the contrasuppressor-inducers (T cell subpopulation), which has been identified in animal models, make the T suppressor lymphocytes to alter their markers turning them into helper cells (48, 49, 50,107, 132). This subpopulation is formed together with the suppressor lymphocytes.

Finally a very sensitive dynamic balance between different lymphocyte subpopulations develops: the non-Peyer's plaque suppressors play a role of general dampers; cells originating from the Peyer's patches might equally behave as selective activators or inhibitors, while the contrasuppressors modulate the activity of both pools. Therefore, due to intracellular cooperations, even the results of the most accurate in vitro cell preparations aimed can easily mislead the researcher in his attempts to evaluate there significance in vivo.

Protective mucosal immunoglobulins have other function aside of formation of the immune complexes. They are capable, for example, of blocking the entrance and absorption of pathogenic organisms and toxic macromolecules into the intestinal mucosa (43, 56).

ADCC reactions also play in important in the mucosal immune reactions. Beside lymphocytes, which were believed to be the only effectors of the ADCC reactions in other organs, monocytes, macrophages, eosinophils and even mastocytes actively participate in these interactions. Mastocytes, however, play a specific role in the ADCC reactions. Positive experimental evidence has been obtained only with the peritoneal, but not abdominal mastocytes. This finding might be of importance, since the mast cells of the lamina propria have different morphology, histochemistry and physiology (5).

sIgA (109) (and other molecules) recognize the antigen receptors in the ADCC reactions. In protozoon or helminth infections eosinophils function as killers; this ADCC function requires IgG or IgE membrane receptors (13, 14, 17, 18). IgM and sIgA fail to activate eosinophils, furthermore, they act as blocking antibodies (52, 77, 118). A paradox stimulation of the IgG-mediated cellular responses by the sIgA antibodies is also documented (5, 7).

**B.** The following process, though simple and clearly "visible", involves many complex, sometimes independently acting mechanisms.

The human small intestine is 640 cm long and weighs about 1-2 kg. 100-150 new cells are daily formed in a crypt, totally yielding $7x10^{10}$ cells or 70 g. In 1000 days the epithelial loss (apoptosis), due to regeneration processes, equals to an average body weight (73, 105). This loss, however, does not mean any real loss of body weight (55). The acidophilic gland cells, Paneth cells and the mucin secreting (Goblet) cells are the main constituents of the Lieberkuhn crypts. The base of the crypt consistes of a ring of stem cells that replace the losses due to apoptosis (106). These specially differentiated Paneth cells give rise to the so-called enteroendocrine cells. Lymphocytes also migrate to this region (27, 104, 108). The T lymphocytes, as well as the cells originating from the LP secrete lymphokines thus stimulating the stem cells. This **direct** effect of the immune system on the epithelial cells enhances mucus secretion. Experimental evidence in favor of this view is the enhancement of mucus secretion in the repeatedly infected animals, due to hyperplasia of the **thymus-dependent** Goblet cells (5). The actual mechanism is, however, more complex, and also involves **indirect** mechanisms, for example a reagin (IgE)-mediated histamine release. IgE is always present because the lamina propria contains IgE-producing plasma cells (99). Furthermore, in 20% of sterile animals the mononuclear cells of the Peyer's patches carry IgE, while these markers scarcely occur on the cell membranes of the animals kept under normal conditions (66). "Normal conditions" imply continuous sensitization with a subsequent IgE response. Under these conditions IgE molecules bind to the mastocytes promoting histamine release after a repeated antigen stimulus. The final result is the enhanced secretion of mucus (5, 7).

In summary, the complex interactions mentioned in this chapter support the idea of Goethe:

*Achtzig Prozent des Glücks eines Menschen liegt in seinen Därmen.*

### References


1. Allan, A.
   Structure and function of gastrointestinal mucus.
   In: Physiology of the Gastrointestinal Tract.
   Ed.: by Johnson, L. R.
   Raven Press, New York 21, 617. 1981.
2. André, C., Hereman, J. F., Vaerman, J. P., Cambiaso, C. L.
   A mechanism for the induction of immunological tolerance by antigen feeding: antigen-antibody|complexes.
   J. Exp. Med. 142, 1509. 1975.
3. Asherson, G. L., Zembala, M., Perara, M., Mayhew, B., Thomas, W. R.
   Production of immunity and unresponsiveness in the mouse by feeding contact sensitizing agents and the role of suppressor cells in the Peyer's patches, mesenteric lymph nodes and other lymphoid tissues.
   Cell. Immunol. 33, 145. 1977.
4. Barrett K. E., Metcalfe D. D.
   Mucosal mast cells and IgE.
   In: Immunology of the Gastrointestinal Tract and Liver.
   Ed. by Heywroth, M. F. and Jones, A. L.
   Raven Press, New York. 65. 1988.
5. Befus, D., Bienenstock, J.
   Factors involved in symbiosis and host resistance at the mucosa-parasite interface.
   Progr. Allergy. 31, 76. 1982.
6. Besredka, A.
   De l'anaphylaxie sixième mémoire de l'anaphylaxie lactique.
   An. Inst. Pasteur Gille. 23, 166. 1909.
7. Bienenstock, J., Befus, A. D.
   Mucosal immunology.
   Immunology. 41, 249. 1980.
8. Birbeck, M. S. C., Carrwright, P., Hall, J. G., Orlans, E., Pappard, J.
   The transport by hepatocytes of immunoglobulin A from blood to bile visualized by autoradigraphy and electron microscopy.
   Immunology. 37, 477. 1979.
9. Brandtzaeg, P.
   Transport models for secretory IgA and secretory IgM.
   Clin. Exp. Immunol. 44, 221. 1981.
10. Brown, W. R.
    Relationships between immunoglobulins and the intestinal epithelium.
    Gastroenterology. 75, 129. 1978.
11. Buffa, R., Capella, C., Solcia, E., Frigerio, B., Said, S. I.
    Vasoactive intestinal peptide (VIP) cells in the pancreas and gastrointestinal mucosa: an immunochemical and ultrastructural study.
    Histochemistry. 50, 217. 1977.
12. Butcher, E. C.
    Lymphocyte migration and mucosal immunity.
    In: Immunology of the Gastrointestinal Tract and Liver.
    Ed. by Heyworth, M. F. and Jones, A. L.
    Raven Press, New York. 93. 1988.
13. Butterworth, A. E., Remold, G. H., Houba, V., David, J. R., Franks, D., David, P. H., Sturrock, R. F.
    Antibody-dependent eosinophil-mediated damage to $^{51}Cr$-labeled schistosomula of Schistosoma mansoni: Mediation

by IgG, and inhibition by antigen-antibody complexes.
J. Immunol. 6, 2230. 1977.

14. Butterworth, A. E., Taylor, D. W., Veith, M. C., Vadas, M. A., Dessein, A., Sturrock, R. F., Wells, E.
Studies on the mechanisms of immunity in human schistosomiasis.
Immunol. Rev. 61, 5. 1982.
15. Cantor, H., Boyse, E. A.
Regulation of cellular and humoral immune responses by T-cell subclasses.
Cold Spring Harbor Symp. Quant. Biol. 41, 23. 1977.
16. Cantor, H., Gershon, R. K.
Immunological circuits: cellular compositions.
Fed. Proc. 38, 2051. 1979.
17. Capron, A., Dessaint, J. P., Capron, M., Joseph, M., Gorpier, G.
Effector mechanisms of immunity to schistosomes and their regulation.
Immunol. Rev. 61, 41. 1982.
18. Capron, M., Spiegelberg, H. L., Prin, L., Bennich, H., Butterworth, A. E., Pierce, R. J., Quaissi, M. A., Capron, A.
Role of IgE receptors in effector function of human eosinophils.
J. Immunol. 132, 462. 1984.
19. Carman, P. S., Ernst, P. B., Rosenthal, K. L., Clark, D. A., Befus, A. D., Bienenstock, J.:
Intraepithelial leukocytes contain a unique subpopulation of NK-like cytotoxic cells active in the defense of GUT epithelium to enteric murine coronavirus.
J. Immunol. 136, 1548. 1986.
20. Cebra, J. J., Gearthart, P. J., Kamat, R., Robertson, S. M., Tseng, J.
Origin and differentation of lymphocytes involved in the secretory IgA responses.
Cold Spring Harbor Symp. Quant. Biol. 41, 201. 1977.
21. Cerf-Bensussan, N., Guy-Grand, D., Griscelli, C.
Intraepithelial lymphocytes of human gut:
isolation, characterisation and study of natural killer activity.
Gut. 26, 81. 1985.
22. Cerf-Bensussan, N., Guy-Grand, D., Lisowska-Grospierre, B., Griscelli, C., Bhan, A. K.
A monoclonal antibody specific for rat intestinal lymphocytes.
J. Immunol. 136, 76. 1986.
23. Cerf-Bensussan, N., Jarry, A., Brousse, N., Lisowska-Grospierre, B., Guy-Grand, D., Griscelli, C.
A monoclonal antibody (HML-1) defining a novel membrane molecule present on human intestinal lymphocytes.
Eur. J. Immunol. 17, 1279. 1987.
24. Cerf-Bensussan N., Schneeberger, E. E., Ghan, A. K.
Immunohistologic and immunoelectron microscopic characterisation of the mucosal lymphocytes of human small intestine by the use of monoclonal antibodies.
J. Immunol. 130, 2615. 1983.
25. Cerf-Bensussan, N., Quaroni, A., Kurnick, J. T., Bhan, A. K.
Intraepithelial lymphocites modulate Ia expression by intestinal intraepithelial cells.
J. Immunol. 132, 2244. 1984.
26. Chase, M. W.
Inhibition of experimental drug allergy by prior feeding of sensitized agent.
Proc. Soc. Exp. Biol. Med. 61, 257. 1946.
27. Cheng, H., Leblond, C. P.
Origin, differentiation and renewal of the four main epithelial cell types in the mouse small intestine V. Unitarian theory of the origin of the four epithelial cell types.
Am. J. Anat. 141, 537. 1974.
28. Chin, Y. H., Garey, G. D., Woodruff, J. J.
Lymphocyte recognition of lymph node high endothelium. 1. Inhibition of in vitro binding by a component of thoracic duct lymph.
J. Immunol. 125, 1764. 1980.
29. Crago, S. S., Tomasi, T. B.
Immunoglobulin circulation and secretion.
In: Immunology of the Gastrointestinal Tract and Liver.
Ed.: by Heyworth, M. F. Jones, A. L.
Raven Press, New York. 105. 1988.
30. Craig, S. W., Cebra, J. J.
Peyer's patches, an entiched source of precursors for IgA producing immunocytes in the rabbit.
J. exp. Med. 134, 188. 1971.
31. Cutz, E., Chan, W., Track, N. S., Goth, A., Said, S. I.
Release of vasoactive intestinal polypeptide in mast cells by histamine liberators.
Nature. 275, 661. 1978.
32. Donowitz, M., Asarkof, N., Pike, G.
Calcium dependence of serotonin-induced changes in rabbit ileal electrolyte transport.
J. Clin. Invest. 66, 341. 1980.
33. Donowitz, M., Charney, A. N., Heffernan, J. M.
Effect of serotonin treatment on intestinal transport in the rabbit.
Am. J. Physiol. 232/Endocrinol Metab. Gastrointest. Physiol. 1./: E 85.– E 94. 1977.
34. Eardley, D. D., Murphy, D. B., Kemp, J. D., Shen, F. W., Cantor, H., Gershon, R. K.
Ly-1 inducer and Ly 1,2 acceptor T-cells in the feedback suppression, circuit bear an I-J subregion controlled determinant.
Immunogenetics. 11, 549. 1980.
35. Ebert, E. C., Roberts, A. K., Brolin, R. E., Raska, K.
Examination of the low proliferative capacity of human jejunal intraepithelial lymphocytes.
Clin. Exp. Immunol. 65, 148. 1986.
36. Elson, C. O., Heck, J. A., Strober, W.
T-cell regulation of murine IgA synthesis.
J. Exp. Med. 149, 632. 1979.
37. Ernst, P. B., Clark, D. A., Rosenthal, K. L., Befus, A. D., Bienenstock, J.
Detection and characterisation of cytotoxic T lymphocyte precursor in the murine intestinal intraepithelial leukocyte population.
J. Immunol. 136, 2121. 1986.
38. Ernst, P. B., Scicchitano, R., Underdown, B. J., Bienestock, J.
Oral immunization and tolerance.
Immunology of the Gastrointestinal Tract and Liver.
Ed.: by Heyworth, M. F. Jones A. L.
Raven Press, New York. 125. 1988.
39. Field, M.
Regulation of small intestinal ion transport by cyclic nucleotides and calcium.
In Secretory Diarrhea.
Ed.: by Fiel, M., Fordtran, J. S. Schultz, S. G. Bethesda, MD.
Am. Physiol. Soc. 3, 21. 1980.
40. Flexman, J. P., Shellam, G. R., Mayrhofer, G.
Natural cytotoxicity, responsiveness to interferon and morphology of intra-epithelial lymphocytes from the small intestine of the rat.
Immunology. 48, 733. 1983.
41. Fox, R. A.
Membrane glycoproteins in defense of the cells of the gastrointestinal tract.
Med. Hypotheses. 5, 669. 1979.
42. Frizzell, R. A., Heintze, K., Stewart, C. P.
Mechanism of intestinal chloride secretion.
In: Secretory Diarrhea.
Ed.: by Fiels, M., Fordtran, J. S., Schultz, S. G. Bethesda MD.
Am. Physiol. Soc. 2, 11. 1980.
43. Fubura, E. S., Freter, R.
Protection against enteric bacterial diarrheas.
N. Engl. J. Med. 285, 831. 1973.
44. Furth, R. van, Hirsh, J. G., Fedorko, M. E.
Morphology and peroxidase cytochemistry of mouse promonocytes, monocytes and macrophages.
J. Exp. Med. 132, 794. 1970.
45. Gaginella, T. A., O'Dorisio, T. M.

Vasoactive intestinal polypeptide: neuromodulator of intestinal secretion?
In: Mechanisms of Intestinal Secretion.
Ed.: by Binder, H. J.
New York: Liss. 18, 231. 1979.

46. Galli, S. J., Dvorak, A. M., Dvorak, H. F.
Basophils and mast cells: Morphologic insights into their biology, secretory patterns, and function.
Prog. Allergy. 34, 1. 1984.

47. Germain, R., Benacerraf, B.
Helper and suppressor T-cell factors.
Springer-Sem. Immunopathol. 3, 92. 1980.

48. Gershon, R. K., Eardley, D. D., Durum, S., Green, D. R., Shen, F. W., Yamauchi, K., Cantor, H., Murphy, D. B.
Contra suppression. A novel immunoregulatory activity.
J. Exp. Med. 153, 1533. 1981.

49. Green, D. R., Eardly, D. D., Kimura, A., Murphy, D. B., Yamauchi, K., Gershon, R. K.
Immunoregulatory circuits which modulate responsiveness to suppressor cell signals: characterization of an effector cell.
Eur. J. Immunol. 11, 973. 1981.

50. Green, D. R., Gold, J. St., Martin, S., Gershon, R., Gershon, R. K.
Microenvironmental immunoregulation: possible role of contrasuppressor cells in maintaining immune responses in gut-associated lymphoid tissue.
Proc. Natl. Acad. Sci. 79, 889. 1982.

51. Greenwood, J. H., Austin, L. L., Dobbins, W. O.
In vitro characterisation of human intestinal intraepithelial lymphocytes.
Gastroenterology. 85, 1023. 1983.

52. Grzych, J. M., Capron, M., Dissous, C., Capron, A.
Blocking activity of rat monoclonal antibodies in experimental schistosomiasis.
J. Immunol. 133, 998. 1984.

53. Guy-Grand, D., Cerf-Bensussan, N., Griscelli, C., Vassalli, P.
International symposium on the immunology of the Gastrointestinal Tract and the Liver.
Jerusalem 1987.

54. Guy-Grand, D., Griscelli, C., Vassalli, P.
The GUT associated lymphoid system: nature and properties of the large dividing cells.
Europ. J. Immunol. 4, 435. 1974.

55. Hanson, W. R., Henninger, D. L., Fry, R. J. F., Sallese, A. R.
In: Cellproliteration in the Gastrointestinal Tract.
Ed.: by Appleton, D. R., Sunter, J. P., Watson, A. J.
Tunbridge Wells Pitman Medical. 198, 1980.

56. Hardcastle, P. T., Eggerton, J.
The effect of acetylcholine on the electrical activity of intestinal epithelial cells.
Biochim. Biophys. Acta. 298, 95. 1973.

57. Heyworth, M. F.
T Cells and other non-B lymphocytes.
In: Immunology of the Gastrointestinal Tract and Liver.
Ed.: by Heyworth, M. F., Jones, A. L.
Raven Press, New York. 1. 1988.

58. Hume D.A., William, F. D.
Role of Macrophages in Cellular Defense.
In: Immunology of the Gastrointestinal Tract and Liver.
Ed.: by Heyworth, M. F. and Jones, A. L.
Raven Press, New York. 23. 1988.

59. Ishizaka, T., Dvorak, A. M., Conrad, D. H., Niebyl, J. R., Marquette, J. P. and Ishizaka, K.
Morphologic and immunologic characterization of human basophils developed in cultures of cord blood mononuclear cells.
J. Immunol. 134, 532. 1985.

60. Jahnke, G. D., Lazarus, L. H., DiAugustine, P. P., Soldato, C. M., Erisman, M. D.
Peptide hormone degradation by a rat mast cell chymase-heparin complex.
Life Sci. 29, 397. 1981.

61. Kagnoff, M. F.
Effects of antigenfeeding on intestinal and systemic immune response. III. Antigen-specific serum mediated suppression of humoral antibody responses after antigen feeding.
Cell. Immunol. 40, 186. 1978.

62. Keren, D. F., Elliott, H. L., Brown, C. D., Jardley, J. H.
Atrophy of villi hypertrophy and hyperplasia of Paneth cells in isolated (Thiry-Vella) ileal loops in rabits. Light microscopic studies.
Gastroenterology. 68, 83. 1975.

63. Kisloff, B., Moore E. W.
Effect of serotonin on water and electrolyte transport in the in vivo rabbit small intestine.
Gastroenterology. 71, 1033. 1976.

64. Klein, J. R., Lefrancois, L., Kagnoff, M. F.
A murine cytotoxic T lymphocyte clone from the intestinal mucosa that is antigen specific for proliferation and displays broadly reactive inducible cytotoxic activity.
J. Immunol. 135, 3697. 1985.

65. Klein, J. R.
Ontogeny of the Thy-1, Lyt-2 + murine intestinal intraepithelial lymphocytes. Characterisation of an unique population of thymus independent cytotoxic effector cells in the intestinal mucosa.
J. Exp. Med. 164, 309. 1986.

66. Kohler, P. F., Brown, W. R.
Immunologic aspects of hepatic and gastrointestinal tract disease.
JAMA, 248, 2704, 1982.

67. Kubagawa, H., Bertoli, L. F., Barton, J. C., Koopman, W. J., Mestecky, J., Cooper, M. D.
Analysis of paraprotein transport into the saliva by using anidiotype antibodies.
J. Immunol. 138, 435. 1987.

68. L'Age-Stehr, J., Teichmann, H., Gershon, R. K., Cantor, H.
Stimulation of regulatory T-cell circuits by immunoglobulin dependent structures on activated B-cells.
Eur. J. Immunol. 10, 21. 1980.

69. Lamm, M. E.
Cellular aspects of immunoglobulin A.
In: Advances in Immunology.
Ed.: by Dixon, F. J., Kunkel, H. G.
Academic Press, New York–London–Toronto–Sidney–San Francisco. 22, 223. 1976.

70. Larsh, J. E., Ottolenghi, J. R. A., Weatherly, N. F.
Trichinella spiralis: phospholipase in challenged mice and rats.
Exp. Parasitol. 36, 299. 1974.

71. Lee, J. S., Silverberg, J. W.
Effect of histamine on intestinal fluid secretion in the dog.
Am. J. Physiol. 231, 793. 1976.

72. Lemaitre-Coelho, J., Jackson, G. D. F., Vaerman, J. P.
Rat bile as a convenuent source of secretory IgA and free secretory component.
Eur. J. Immunol. 8, 588. 1977.

73. Lipkin, M.
Proliferation and differentiation of gastrointestinal cells in normal and disease states.
In: Physiology of the Gastrointestinal Tract.
Ed.: by Johnson, L. R.
Raven Press, New York, 4, 146, 1981.

74. Lipkin, M.
Proliferation and differentiation of gastrointestinal cells in normal and disease states.
In: Physiology of the Gastrointestinal Tract.
Ed.: by Johnson, L. R.
Raven Press, New York, 4, 145. 1981

75. Lyscom, N., Brueton, M. J.
Intraepithelial lamina propria and Peyer's patch lymphocytes of the rat small intestine: isolation and characterisation in terms of immune globuline markers and receptors for monoclonal antibodies.
J. Immunology. 45, 775. 1982.

76. MacDonald, T. T., Dillon, B. S.
Chemica mediators of Cellular Communication.
In: Immunology of the Gastrointestinal Tract and Liver.

Ed.: by Heyworth, M. F. and Jones, A. I.
Raven Press, New York. 47. 1988.

77. Mathew, G. D., Qualtierre, L. F., Neel, H. B., Pearson, G. R.
IgA antibody, antibody-dependent cellular cytotoxicity and prognosis in patients with nasopharyngeal carcinoma.
Int. J. Cancer. 27, 175. 1980.

78. Mattingly, J. A., Waksman, B. H.
Immunologic suppression after oral administration of antigen. I. Specific supressor cells formed in rat Peyer's patches after oral administration of sheep erythrocytes and their systemic migration.
J. Immunol. 121, 1878. 1978.

79. Mattingly, J. A., Waksman, B. H.
Immunologic suppression after oral administration of antigen. II. Antigen specific helper and suppressor factors produced by spleen cells of rats fed sheep erythrocytes.
J. Immunol. 125, 1044. 1980.

80. Mayrhofer, G., Whately, R. J.
Granular intraepithelial lymphocytes of the rat small intestine. I. Isolation, presence in T-lymphocyte-deficient rats and bone marrow origine.
Int. Arch. Allergy Appl. Immunol. 71, 317. 1983.

81. McDermott, M. R., Bienenstock, J.
Evidence for a common mucosal immunologic system. I. Migration of B immunoblasts into intestinal, respiratory and genital tissues.
J. Immunol. 122, 1892. 1979.

82. McDermott, M. R., Clark, D. A., Bienenstock, J.
Evidence for a common mucosal immunologic system. II. Influence of the estrous cycle on B immunoblast migration into genital and intestinal tissues.
J. Immunol. 124, 2536. 1980.

83. McDougal, J. S., Shen, F. W., Cort, S. A., Bard, J.
Feedback suppression: phenotypes of T-cell subsets involved in the Ly-1 T-cell-induced immunoregulatory circuit.
J. Immunol. 125, 1157. 1980.

84. McWilliams, M., Philips-Quagliata, J. M., Lamm, M. E.
Mesenteric lymphnode B lymphoblasts which home to the small intestine are precommitted to IgA synthesis.
J. Exp. Med. 145, 866. 1977.

85. Miller, H. R. P., Huntley, J. F., Wallace, G.
Immune exclusion and mucus-trapping during the rapid expulsion of Nippostrongylus brasiliensis from primed rats.
Immunology. 44, 419. 1981.

86. Miller, H. R. P., Nawa, Y.
Nippostrongylus brasiliensis: Intestinal goblet-cell response in adoptively immunized rats.
Exp. Parasitol. 47, 81. 1979.

87. Miller, S. D., Hanson, D. G.
Inhibition of specific immune responses by feeding protein antigens. IV. Evidence for tolerance and specific active suppression of cell mediated immune responses to ovalbumin.
J. Immunol. 123, 2344. 1979.

88. Moore, A. R., Hall, J. G.
Evidence for a primary association between immunoblasts and small gut.
Nature. 239, 161. 1972.

89. Murphy, D. B., Herzenberg, L. A., Okumura, K., McDevitt, H. O.
A new I subregion (I-J) marked by a locus (Ia-4) controlling surface determinants on suppressor T-lymphocytes.
J. Exp. Med. 144, 699. 1976.

90. Murray, M., Miller, H. R. P., Sanford, J. Jarrett, W. F. H.
5-Hydroxytryptamine in intestinal immunological reactions.
Int. Arch. Allergy Appl. Immunol. 40, 236. 1971.

91. Nagwa, H., Nakane, P. K., Brown, W. R.
Translocation of dimeric IgA throug neoplastic colon cells in vitro.
J. Immunol. 123, 2359. 1979.

92. Nauss, K. M., Pavlina, T. M., Kumar, V., Newberne, P. M.
Functional characteristics of lymphocytes isolated from the rat large intestine.
Response to T-cell mitogens and natural killer cell activity.
Gastroenterology. 86, 468. 1984.

93. Ngan, J., Kind, L. S.
Suppressor T-cells for IgE and IgG in Peyer's patches of mice made tolerant by the administration of ovalbumin.
J. Immunol. 120, 861. 1978.

94. Orlans, E., Pappard, J., Reynolds, J., Hall, G. J.
Rapid active transport of immunoglobulin A from blood to bile.
J. exp. Med. 147, 588. 1978.

95. Owen, R. L.
Sequential uptake of horse radish peroxidase by lymphoid follicle epithelium of Peyer's patches in the normal, unobstructed mouse intestine. An ultrastuctural study.
Gastroenterology. 72, 440. 1977.

96. Owen, R. L., Nemanic, P.
Antigen processing structures of the mammalian intestinal tract: an SEM study of lymphoepithelial organs.
Scanning Electron Microsc. 11, 367. 1978.

97. Parrott, D. M. V., Ferguson, A.
Selective migration of lymphocytes within the mouse small intestine.
Immunology. 26, 571. 1974.

98. Parrott, D. M. V., Tait, C., MacKensie, S., McI. Mowat, A., Davies, M. D. J., Micklem, H. S.
Analysis of the effective functions of different populations of mucosal lymphocytes.
Ann. N. Y. Acad. Sci. 409, 307. 1983.

99. Patterson, S., Roebuck, Pamela, Plats-Mill, T. A. E., Shiner, Margot, Kingston, D., Pearson, Joyr.
IgE plasma cells in human jejunum demonstrated by immune electron microscopy.
Clin. exp. Immunol. 46, 301. 1981.

100. Peppard, J., Orlans, E., Payne, W. R., Andrew, E.
The elimination of circulating complexes containing polymeric IgA excretion in the bile.
Immunology. 42, 83. 1981.

101. Petit, A., Ernst, P. B., Befus, A. D., Clark, D. A., Rosenthal, K. R., Ishizaka, T., Bienenstock, J.
Murine intestinal intraepithelial lymphocytes I. Relationship of a novel tThy-1, Lyt-1, Lyt-2 +, granulated subpopulation to natural killer cells and mast cells.
Europ. J. Immunol. 15, 211. 1985.

102. Pierce, N. F., Gowans, J. L.:
Cellular kinetics of the intestinal immun response to cholera toxoid in rats.
J. Exp. Med. 142, 1550. 1975.

103. Porter, P., Noakes, D. E., Allen, D. W.
Intestinal secretion of immunoglobulins in the preruminant calf.
Immunology. 23, 299. 1972.

104. Potten, C. S., Chwalinski, S., Swindell, R., Palmer, M.
The spatial organization of the hierarchical proliferative cells of the crypts of the small intestine into clusters of „synchronized" cells.
Cell and Tissue Kinet. 15, 351. 1982.

105. Potten, C. S., Hendry, J. H., Moore, J. V., Chwalinski, S.
Cytotoxic Insult to Tissues: Effects on Cell Lineages.
Ed.: by Potten, C. S.
Churchill-Livingstone, Edinburg, 105, 1983.

106. Potten, C. S., Hendry, J. H.
Stem Cells: Identification and Characterisation
Ed.: by Potten, C. S.
Churchill-Livingstone, Edinburg, 155, 1983.

107. Ptak, W., Green, D. R., Durum, S. K., Kimura, A., Murphy, D. B., Gershon, R. K.
Immunoregulatory circuits which modulate responsiveness to suppressor cell signals: contrasuppressor cells can convert an in vivo tolerance signal into an immunogenic one.
Eur. J. Immunol. 11, 980. 1981.

108. Quastler, H., Sherman, F. G.
Cell population, kinetics in the intestinal epithelium of the mouse.
Exp. Cell. Res. 17, 420. 1959.

109. Richman, L., Graeff, A. S., Strober, W.

Antigen presentation by macrophage-enriched cells from the mouse Peyer's patch.
Cell Immunol. 62, 110. 1981.
110. Said, S. I., Mutt, V.
Polypeptide with broad biological activity isolated from small intestine.
Science. 169, 1217. 1970.
111. Schmiegel, W. H., Hamann, A., Thiele, H. G.
Das Darm-Assoziierte-Immunsystem. Lokale Immun-antvort-systemische Toleranz.
DMW. 107, 67. 1982.
112. Schrader, J. W., Scollay, R., Battye, F. J.
Intramucosal lymphocytes of the GUT: Lyt-2 and thy-1 phenotype of the granulated cells and evidence for the presence of both Talls and mast cell precursors.
J. Immunol. 130, 558. 1983.
113. Schrank, G. D., Verwey, W. F.
Distribution of cholera organisms in exerimental Vibrio cholerae infections: proposed mechanisms of pathogenesis and antibacterial immunity.
Infect. Immun. 13, 195. 1976.
114. Selby, W. S., Jánossy, G., Bofill, M., Jewell, D. P. Lymphocyte subpopulations in the human small intestine. The findings in normal mucosa and the mucosa of patients with adult coeliac disease.
Clin. Exp. Immunol. 52, 219. 1983.
115. Selby, W. S., Jánossy, G., Goldstein, G., Jewell, D. P.
T lymphocyte subsets in human intestinal mucosa: the distribution and relationship to MHC-derived antigens.
Clin. Exp. Immunol. 44, 453. 1981.
116. Sprent, J.
Fate of H-2 activited T lymphocytes in syngenyc hosts. I. Fate in Lymphoid tissues and intestines traced with $^3$H-thymidine, 1251-deoxyuridin and SI chromium.
Cell. Immunol. 21, 278. 1976.
117. Stamper, H. B., Woodruff, J. J.
Lymphocyte homing into lymph nodes. In vitro demonstration of the selective affinity of recirculating lymphocytes for highendothelial venules.
J. exp. Med. 144, 828. 1976.
118. Steele, J. R., Amkerst, J., Sjögren, H. O., Vang, H., Lannerstad, O.
Absorption of blocking activity from human tumor-bearer sera by staphylococcus aureus cowan I.
Int. J. Cancer. 15, 180. 1975.
119. Strober, W., Richman, L. K., Elson, C. O.
The regulation of gastrointestinal immune responses.
Immunology today. 14, 156. 1981.
120. Tada, T., Taniguchi, M., Okumura, K.
Multiple MHC loci controlling lymphocyte interactions.
J. Supramol. Struct. Cell. Biochem. (Suppl.) 3, 236. 1979.
121. Tadokoro, K., Stadler, B. M., De Weck, A. L.
Factordependent in vitro growth of human normal bone marrow-derived basophil-like cells.
J. Exp. Med. 158, 857. 1983.
122. Tagliabue, A., Befus, A. D., Clark, D. A.,
Bienenstock, J. Characteristics of natural killer cells in the murine intestinal epithelium and lamina propria.
J. Exp. Med. 155, 1785. 1982.
123. Toner, P. G., Carr, K. E., Wyburn, G. M.
Intraepithelial cells in gastrointestinal mucosa.
In: The Digestive System. An Ultrastructural Atlas and Riview.
Appleton-Century-Crofts, New York, 13, 166. 1971.
124. Tutton, P. J. M.
The influence of serotonin on crypt cell proliferation in the jejunum of the rat.
J. Anat. 118, 389. 1974.
125. Unanue, E. R.
Macrophages as regulators of lymphocyte function,
In: The biology of immunologic disease
Ed.: by Dixon, F. J., Fisher, D. W.
Sinauer Associates, Inc. 87, 1983.
126. Uvnäs, B.
Mechanism of histamine release in mast cells.
Ann. NY Acad. Sci. 103, 278. 1963.
127. Uvnäs, B.
The mechanism of histamine release from mast cells.
In: Histamine II and Antihistaminics; Chemistry, Metabolism and Physiological and Pharmacological Actions. Handbook of Experimental Pharmacology.
Ed.: by Roche Silva, M.
Springer-Verlag, New York, 18, 75. 1978.
128. Weigle, W. O.
Cyclical production of antibody as a regulatory mechanism in the immune response.
Adv. Immunol. 21, 87. 1975.
129. Wells, H. G., Osborne, T. B.
The biological reactions of the vegetable proteins.
J. Infect. Dis. 8, 66. 1911.
130. Wright, R. D., Jenning, M. A., Florey H. W., Lium, R. The influence of nerves and drugs on secretion by the small intestine and an investigation of the enzymes in the intestinal juices.
Quant. J. Exp. Physiol. 30, 73. 1940.
131. Woods, J. D.
Physiology of the enteric nervous system.
In: Physiology of the Gastrointestinal Tract.
Ed.: by Johnson, L. R.
Raven Press, New York, 1, 1. 1981.
132. Yamauchi, K., Green, D. R., Eardley, D. D., Murphy, D. B., Gershon, R. K.
Immunoregulatory circuits that modulate responsiveness to suppressor cell signals.
J. Exp. Med. 153, 1547. 1981.
133. Zubler, R. H., Benacerraf, B., Germain, B.
Feedback suppression of the immune response in vitro: II. IgVH-restricted antibody dependent suppression.
J. Exp. Med. 151, 681. 1980.

# III. Immunology of the liver

Liver, being the largest "gland" of the gastrointestinal tract performs a wide variety of known and unknown functions. To fully comprehend the immunological role of liver, the present chapter gives an overview of the liver functions with respect to A) protein metabolism; B) detoxication; and C) relationship with the immune system. Protein metabolism is a balance between synthetic and catabolic reaction.

The structure and the turnover of the proteins differ from organ to organ, from very slow, muscle-type pattern to fast one typical of the liver. The liver can replace about one half of its protein content in 10 days (26). The protein turnover shows no marked differences under normal and pathological conditions - active catabolism occurs even at negative nitrogen balance, whereas synthetic activity continues during fasting. Exogenous proteins are fully digested by the gastrointestinal tract and only about 3-5% is excreted with feces (103, 107, 117). Digestion means hydrolyzation of proteins carrying antigens into amino acids or oligopeptides. However there is evidence that some protein molecules can escape digestion and be absorbed in their native form. The subsequently activated "janitor" immunoglobulins will form immune complexes with the intact proteins. The latter is absorbed in a "second step" before the antigens could reveal their antigenic properties. This "second step" is mediated via the complement, which is the most potent general protein degrading system of the organism. Whilst the mechanism of the protein synthesis is well known, few data exist concerning the mechanisms and localization of the catabolic reactions. Proteins might be degraded in various organs due to the protease activity of the final product of activated complement cascade system. The protease selectively damages the membranes of certain cells without affecting other structures in their environment. Despite this double functional barrier, protein molecules can pass through the liver sensitizing the host and eliciting antibody response. Allergy is not always revealed as urticaria or diarrhoea, but can also damage the liver as an atopic organ (46). Injection of I125 labelled latex particles of 2300 Å in diameter shows the distribution of the large molecules. RHS eliminates the large particles from the circulation in 30-50 minutes. Similar time course was detected for smaller particles of 50-200 Å in diameter. 1g liver tissue exhibits the highest activity, compared to the other organs: the elimination activity of the spleen and the lung makes 50% and 12.5% respectively, while the activity of all the other tissues is only 1% (104). Later the relative ration of radioactivity in the organs and excreted with feces shows an increasing tendency, due to the diminished activity of the liver. These data suggest that the liver excretes the foreign particles via the bile ducts, whereas other organs lacking this natural route are incapable of that. Therefore, the labelled latex particles are excreted by the liver, with the assistance of the circulatory system and the lymphatics (1, 82). The size of the latex particles significantly exceeds that of the macromolecules occurring in the organism, including large molecular complexes.

The following part deals with the fate of the products of degradation. Since liver plays a central role in this process, the molecules transported from different sites by the blood stream and lymphatics would interact with the liver cells. This interaction, however, is extremely complex, for the molecules are moving across inhomogenous gels of poorly defined properties. There is evidence that for homologous substances molecular size is proportional tothemolecular distribution.

$$K = \exp. \times (\text{const.} \times M)$$

Brönsted showed that the high molecular weight substances can belong to either phase of distribution, whereas low molecular weight molecules show a more homogenous pattern of distribution (12). The RHS efficiently clears the circulatory system from the foreign particles, so only few of them are absorbed; in the course of their transport most of the particles are excreted or metabolized. Therefore, excretion is not a sequellae of passive distribution, but is rather determined by the active cellular processes.

The following experiment was designed to assess the relationship between the molecular size and the excretion rate (47, 48) (Fig. III/1)

The common bile duct of the rats was exposed and the bile samples were collected in plastic tubes every 10 min. After the experiment the liver, the thyroid, the kidney, and the lung tissue were studied with histological* and autoradiographic** methods.

Group I: $I^{125}$ labelled human fibrinogen (O.S.S.K.I. $I^{125}$ human fibrinogen Fg-II-71 122309 $L_{11}$/8) was i.v. administered following the bile-fistula surgery. The pooled bile fractions, urine and plasma (0.1 ml Heparin + 1.9 ml blood) samples obtained by the end of the experiment were analysed with electro-immunodiffusion using antifibrinogen serum (Human L-113 N). Activity was detected only in the blood; in the course of a 2.5-hour experiment the labelled substances were not excreted into the urine or bile (Fig. III/2).

Results of electroimmunodiffusion and autoradiographic histological studies indicate that 2.5 hours following administration, human FBG is not distributed between the intra- and extravascular spaces and all the activity is detected intravascularly (Fig. III/3).

Group II. This group was administered degradation products of $I^{125}$ labelled human fibrinogen in vitro treated with plasmin.

The course of the experiment did not differ from that of the first group, however FDP, D, and E fragments appeared in some bile fraction of these animals. Since the activity of the 80- and the 90 min. samples showed no marked increase, it was assumed that the liver had cleared the blood from all (or at least from the most) the low-molecule labelled substances (Fig. III/4).

* Histological studies were carried out by János Sebők, MD. (Dept. of Pathology) .

** Radiographic studies were performed by József Kiss, MD. (2nd Dept. of Histology).

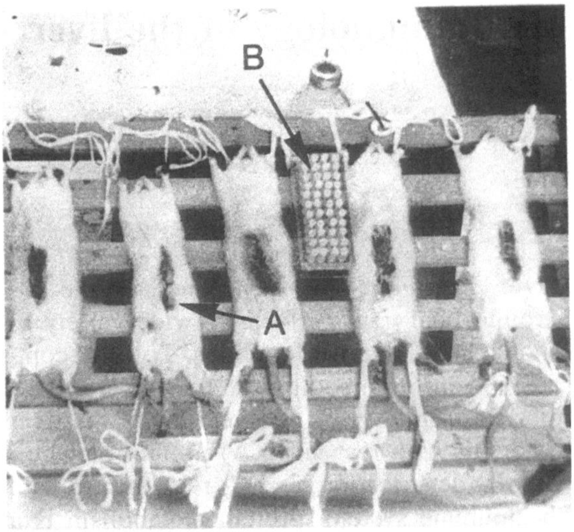


**Rats were fixed and tilted "head up" at 45o. Bile from the common bile duct was collected (A) and every 10 minutes the pooled samples were placed into the test tubes.**

**Fig. III/1**

Our calculations strongly support the above experimental data. After assessment of the data presented in Table III/I, the activity of the total excreted fraction was:

$I_{bile}$ = 69.19 μCi (Fig. III/4).

The values of the diffusion coefficients indicate that certain activity of plasma after 2.5 hours was attributed to fibrinogen-like molecules or fragments, which were not digested by the plasmin. Calculations based on the activity and plasma volumes (Table III/1) yield a value of the total plasma activity I plasma = 3.21 μCi.

The excreted activity is 69.19 μCi, that of the plasma 3.21 μCi, total detected activity is 72.4 μCi, while the total activity at administration was 93 μCi. The difference, or at least

| | |
|---|---|
| PLASMA (ml) | 11.016 |
| ACTIVITY (in 0.003 ml) | $8.75 \times 10^{-4}$ |
| TOTAL PLASMA ACTIVITY (μCi) | 3.21 |

**Table III/1**

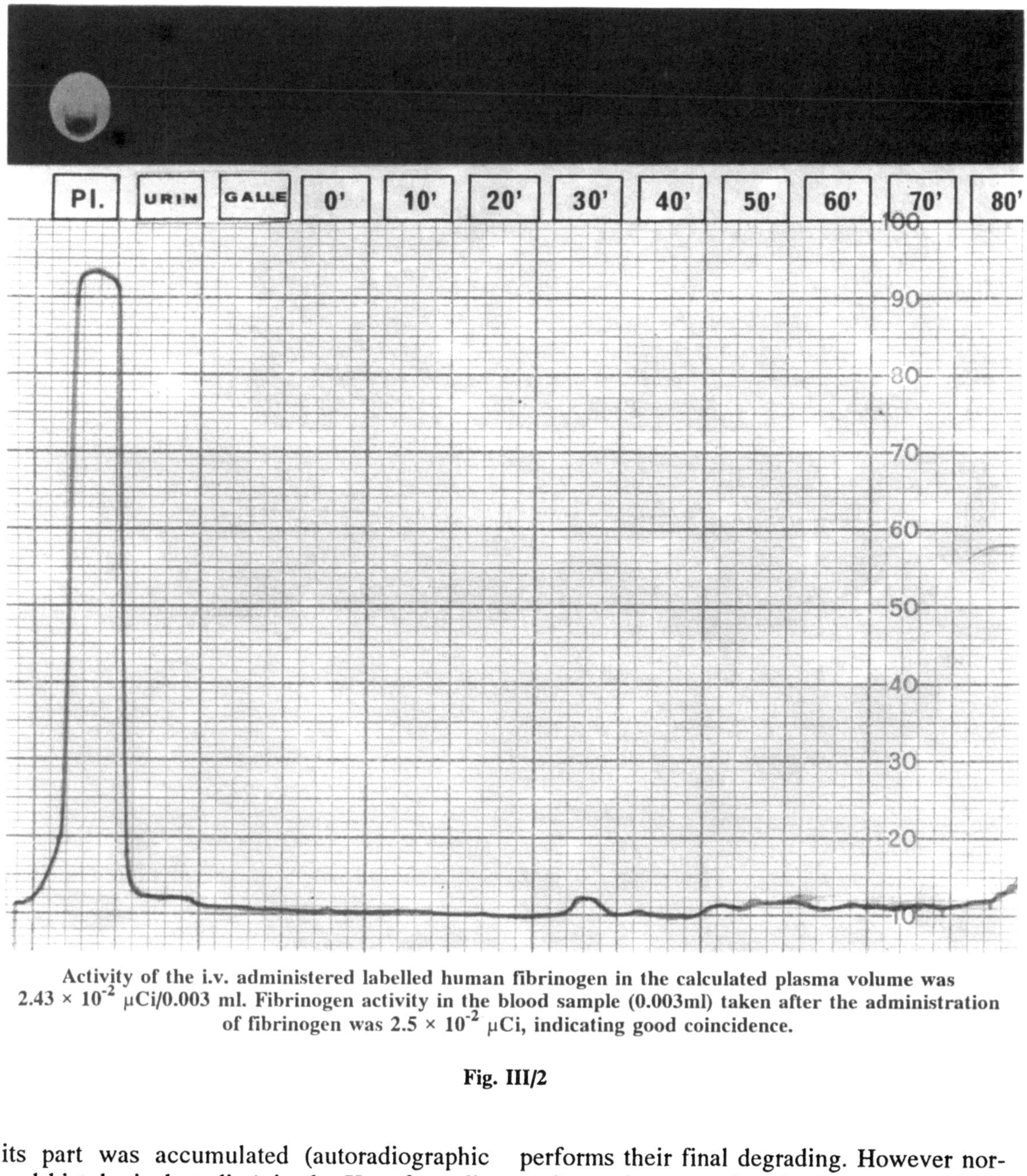


**Activity of the i.v. administered labelled human fibrinogen in the calculated plasma volume was $2.43 \times 10^{-2}$ μCi/0.003 ml. Fibrinogen activity in the blood sample (0.003ml) taken after the administration of fibrinogen was $2.5 \times 10^{-2}$ μCi, indicating good coincidence.**

**Fig. III/2**

its part was accumulated (autoradiographic and histological studies) in the Kuppfer cells (Fig. III/5).

The analysis of the results obtained in groups I and II strongly indicates that the liver is an efficient filter, its activity depending on the molecular weight of the particles.

Catabolism of the endogenous (own) proteins is at least as important as degrading the extrinsic molecules of alimentary origin. A lesser part of the endogenous proteins consists of the "rejections" removed by the immune system from the site of their synthesis. Liver performs their final degrading. However normal proteins comprise the larger part of the endogenous pool catabolized by the liver. These old molecules are replaced by the newly synthesized ones. The average half-life period of the proteins alters in different organs, similar to their already mentioned anabolic rate. For liver proteins the half-life period is about 3.5 days, however, that of the ornitin decarboxilase is 19 days (55, 61, 70, 73). In summary, biological and pathological factors both affect the rate of protein metabolism.

The liver and muscle are the largest protein

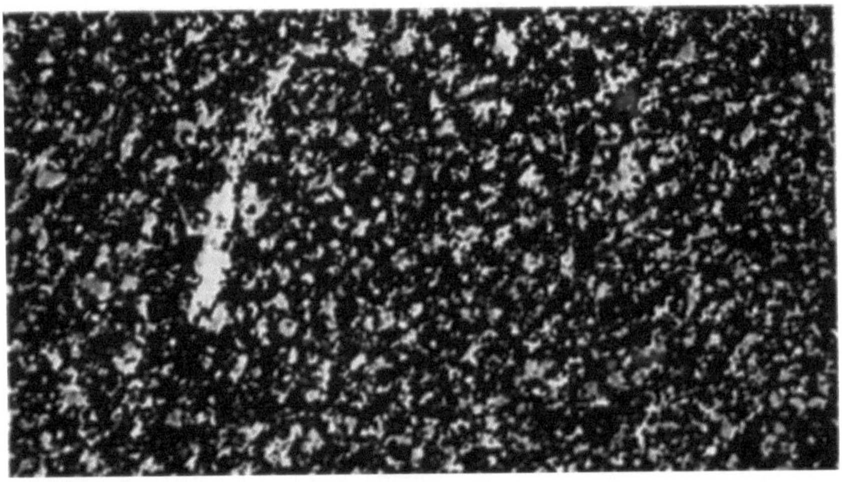

**Autoradiographic study of the liver of a sacrificed rat: after an i.v. injection the tissue is "filled" with the labelled fibrinogen.**

**Fig. III/3**

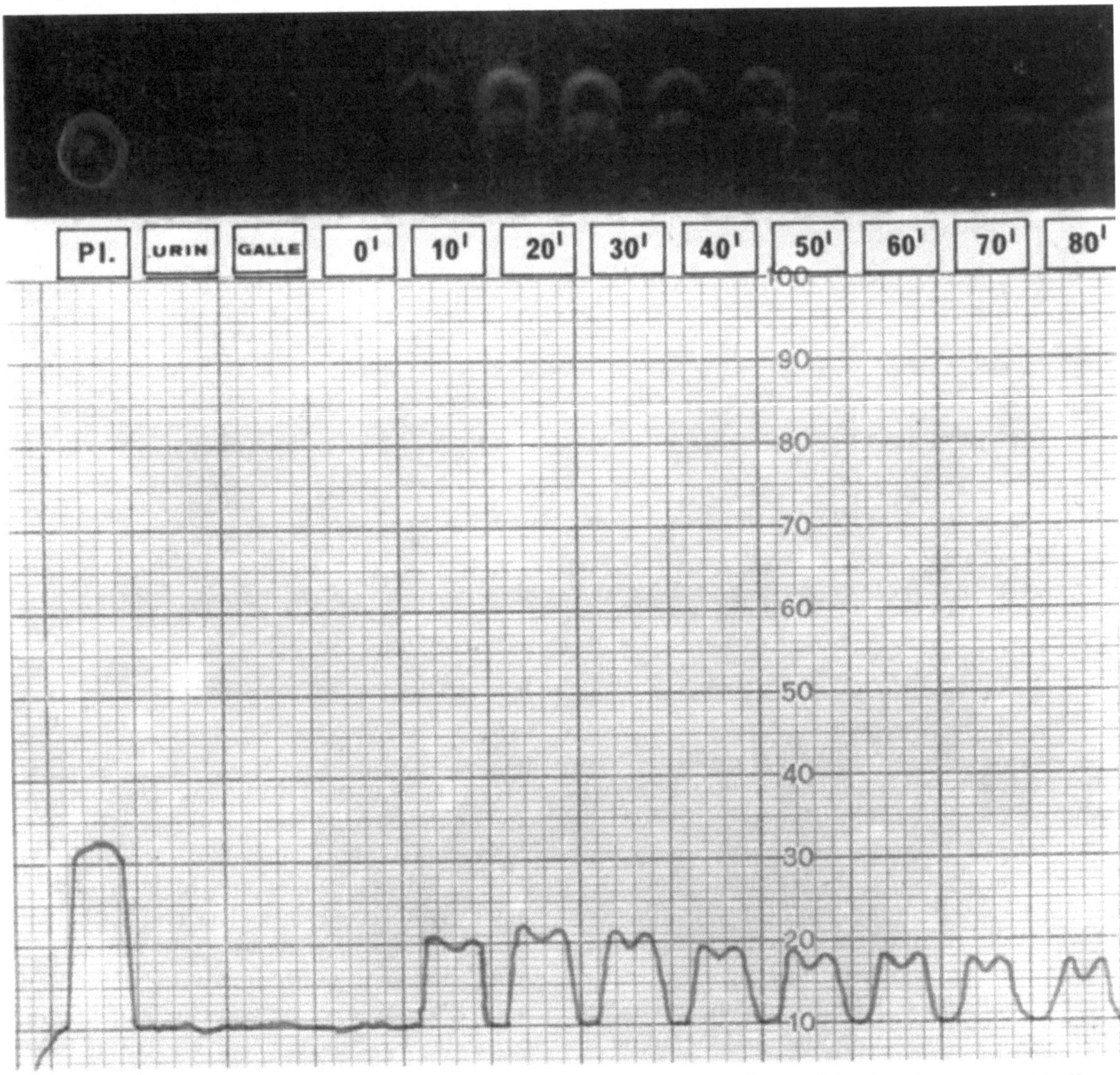


**Fast hepatic clearence of blood from the E and XY degradation products of the in vitro enzymatically treated human fibrinogen i.v. administered to the rats.**

**Fig. III/4**

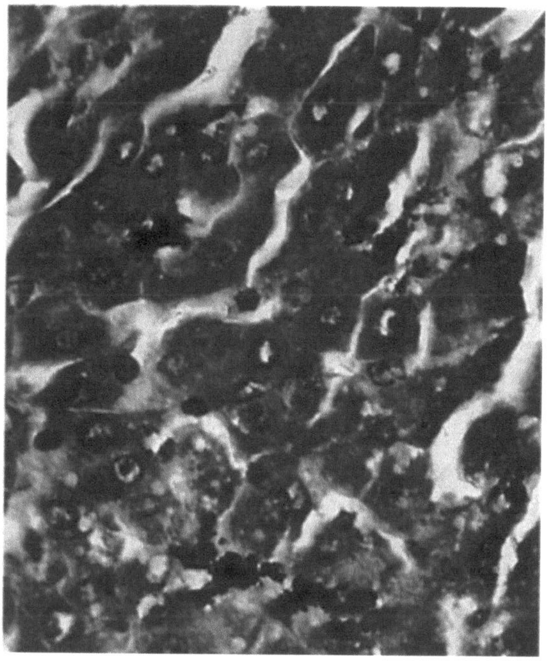

Methods presented in Fig. III/3 detect activity in some Kuppfer cells.

Fig. III/5

reserves of the organism. Amino acids released from one organs are uptaken by the other. Liver plays a central role in this adaptive regulatory process: it is well known that liver is the site of the most intense intracellular proteolytic reactions (for example, hemoglobin metabolism) (66, 90).

Liver synthesizes a wide variety of proteins, however, recently the interest of the researchers has been focused on a group of the so-called adhesive proteins (for details see later). The following general remarks provide a more deep insight in the role of adhesive proteins regulating cellular interactions.

Generally, normal differentiation, growth and survival of the cells necessitate their binding to the cellular matrix. Extracellular matrix, however, is at least as important, for it facilitates migration of the particles (cells) within the tissue. Thus, life begins with the migration of embryonic cells in the course of their morphogenesis, and is terminated when the tumor cells are propagated amidst the normal tissue. Leaving the extremal situations alone, the reversible binding to the extracellular matrix is the basis of the adaptational reserve of the cells, determining whether it starts migrating or remains motionless. The borders of these two possibilities are obscured by the continuous proliferation and regeneration of the cells with subsequent alterations of their morphology and size: proliferating cells must occupy more space, so they spread on the extracellular matrix. The interaction between the cells and the extracellular matrix is mediated via the adhesive proteins, which bind with one end to the cellular membrane receptors and adhere to the matrix with the other. The extracellular matrix is an insoluble complex consisting of collagen, elastin, glycoproteins, proteoglycans and hyaluronic acid (94). Glycoproteins generally contain fibronectin (56, 95), laminin (111, 114), vitronectin (35, 41), thrombospondin (67, 115), von-Willebrand's factor (29, 97) and tenascin. Proteoglycans maintain the above mentioned balance by inhibiting adhesion (13, 59, 91). They "shade" the the cell receptor-binding sites of the adhesion proteins (18, 60, 64).

Following the above mentioned adhesion, hepatocytes start to proliferate or regenerate (10, 14, 42). An additional specific factor, though not identified, specifically promotes growth of the liver cells if liver matrix is included in the cell culture system. It is also worth pointing out that the adhesion proteins are synthesized by the liver (106, 110), hence, apart from its above mentioned functions, fibronectin, which is always present in plasma also behaves as a non specific opsonin. It binds to the tissue debris and antigen molecules of alimentary origin promoting their absorption by the reticuloendothelial system and the liver cells with the subsequent activation of the complemet components and final elimination of the antigens (96). Vitronectin also has similar action mechanism (44, 49).

In summary: the liver accumulates proteins, peptides and amino acids from the gut and promotes their catabolism. Concomitantly new proteins meeting the needs of the organism and necessary for the liver regeneration are synthesized. Therefore, the liver protects the organism against the foreign protein molecules preventing the initiation of the immune response.

Apart from the nutritients, different toxic substances are absorbed from the gut. Most of them are lipid-soluble and can be easily uptaken by hepatocytes. Some metabolites formed in the smooth endoplasmatic reticulum, mitochondria or the nuclear membrane can interact with different macromolecules, whilst others inhibit such binding. The toxic substances are generally inactivated by oxidation and hydrolysis. The products of degradation, though non-toxic, can have carcinogenic properties. The next step is conjugation and hydrolysis. For some substances no preliminary oxydation occurs before the hydrolysis or conjugation take place. The exact character of these reactions depends on genetic polymorphism, observed, for example, in the inhomogenuous activity of the drug- metabolizing enzymes (8).

The binding to the above mentioned proteins triggers a series of unpredictable hypersensitive metabolic and immune reactions. Abnormal metabolic background of these hypersensitive reactions may develop either following a previous exposition with subsequent impairment of the cytochrome P 450, or as a sequellae of genetical polymorphism which is revealed in altered oxidation and acethylation. It is worth mentioning that under physiological conditions the immune system of the gut is permanently exposed to the endotoxins (lipopolysaccharides) of the intestinal bacteria. Constant synthesis of the antibodies (IgM) is the first normal reaction to the antigen challenge (98). The next step is the liver peroxidase-mediated elimination of the antigens preventing them from getting into the systemic circulation (for details see Chapter IV). Crohn's disease and surgery reducing the intestinal length performed to decrease body weight are usually accompanied by the development of cryoglobulinemia, IgM being precipitated. A five-fold elevation of the IgA levels excreted with bile is reported after the operations, which markedly reduce the length of the bowel and consequently the number of the Payer's plaques (Fig. III/6).

In immature (newborns) liver (126) or in cases when the detoxication function is impaired (patients) endotoxins elicit inflammation, central zone necrosis, microvascular lipid accumulation (3, 33, 40, 68, 127). Leucocytes might play an indirect role in the development of these tissue alterations. The uptake of endotoxins in the sinuses leads to microthrombosis and impairment of circulation with a subsequent hepatocellular necrosis of the central zones and more rarely in pericentral and periportal regions (3,4). At ultrastructural level endotoxins directly damage liver mitochondria (120).

In summary, the character and the capacity of liver detoxication not only controls its barrier function preventing the unpredictable reactions caused by the toxic agents, but also regulates their elimination (119).

In the embryonic period the progenitor cells of the yolk sac form homogenous islets amidst the parenchimal cells contributing to the formation of hemopeietic tissue in the liver and spleen. Apart from the well known migration of the T and B lymphocytes between the thymus, GALT and the liver, a special local migration may be induced by an antigen stimulus.

Certain progenitors give rise to the Kuppfer cells prior to the development of the bone marrow (81), though in the later phases the circulating monocytes can act as progenitors. Kuppfer cells eliminate non-desirable substance even as large as a bacterial cell. These molecules are excreted into the intestinal lumen with bile. They produce different cytokins, for example IL-1 and TNF, and carry Fc and complement receptors. Their antigen presenting role must also be emphasized.

Under physiological conditions liver contains many lymphocytes, though their role and the full scope of their functions are not yet understood.

A natural heterogenous population of the natural cytotoxic lymphocytes eliminates the poorly functioning (35, 74) and the virus-infected cells. In contrast to the cytotoxic T lymphocytes requiring MHC I antigens (92, 93), the natural cytotoxic cells elicit their functions without sensitization or memory and their action is not restricted by MHC. The NK lymphocytes are the classical effectors of the natural cellular cytotoxicity (37, 54). The subtypes of such cells were identified in erythroleukemia. These lymphocytes primarily protect the splanchnic organs. The antigen receptors and target specificity of the Natural Cytotoxic cells (NC) present in the splanchnic organs, bone marrow and the thymus alter from that of the NK cells. α-interferon and

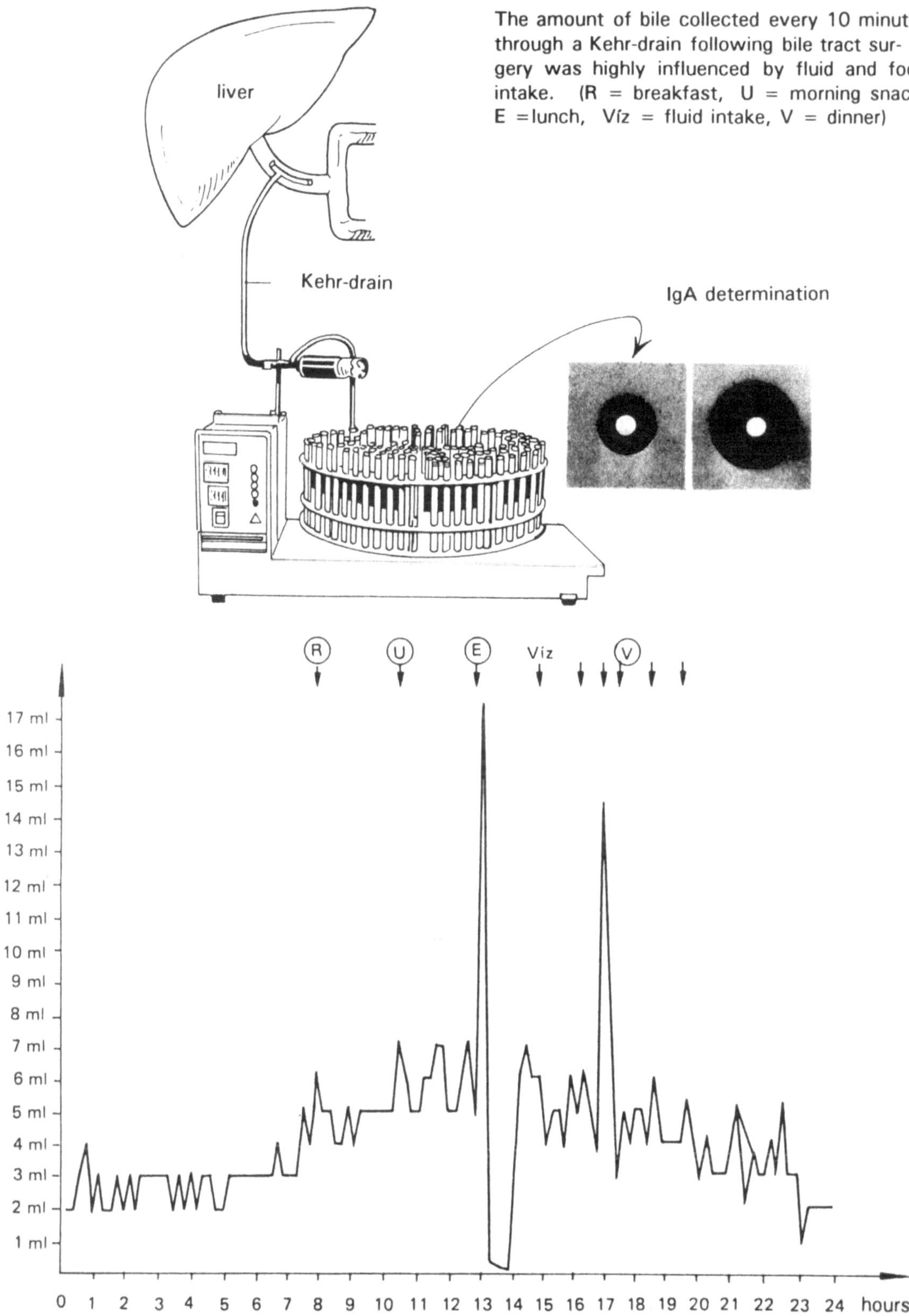


**A five-fold increase of the bile IgA concentration in patients undergone gallbladder surgery and small intestine resection, compared to those without resection.**

**Fig. III/6**

IL-2, but not IL-3 stimulate the NK activity, while the NC cells are activated by IL-2 and IL-3, and α-interferon fails to exhibit stimulatory effect (105, 123, 124).

The tissue NK and NC cells are termed as LGL cells (88, 89, 113). An other type of the so-called pit-cells has been identified in the liver by Wisse et al. Pseudopodiums and cytolplasmic granules (9, 53) are typical of these cytolytic cells (20, 52). Lymphocytes comprise about 30-40% of the extractable, non-parenchimal cells of the rat liver (122). Of these 3-9% are identified as LGL (69, 122), which can be stimulated by different inductors (maleic anhydride divynil eter), interferon and viral infections. In hepatitis B and C the content of the $T_8^+$ lymphocytes in the portal and lobular infiltration is markedly elevated, suggesting their involvement in the pathogenesis of hepatocellular necrosis (31, 34, 58, 85). In the A-virus induced hepatitis, however, only the count of the NK cells is increased (62). In contrast to the above mechanisms, dendritic $T_4^+$ and B lymphocytes are the cellular elements observed in the areas of piecemeal necrosis observed in CAH (6, 25, 74). Their presense unambiguously indicates the prevalence of the humoral reactions (7, 71, 98, 116, 121) with a special emphasise on ADCC (112).

Normal liver contains very few cytotoxic CTL ($T_8^+$) lymphocytes. In PBC $T_4^+$ lymphocytes comprise about 1-2% of the extractable non-parenchimal cells; in Stage I the lymphocyte infiltration is mainly localized around the bile ducts (5, 17, 25, 83). Few $T_8^+$ lymphocytes are always present too (125). The activated $T_4^+$ lymphocytes carry MHC II and IL-2 receptors (85). Under pathological conditions the epithelial cells of the bile ducts express MHC II antigens (5, 17), whereas physiologically they can carry only MHC I.

Liver plasmoblasts, as precursors of the plasma cells, should rather be termed as lymphoblasts, centroblasts and centrocytes. Real immunoglobulin producing cells seldom occur in the areas not specific for the plasma cells (bone marrow, lymph nodes, mucosal lamina propria and areas of inflammation). In hepatitis, as a special form of inflammation, plasma cells appear in the liver (13). They participate in the so-called secondary tissue reactions, however, their function is more complicated: on one hand, at the ligand-receptor level, they participate in the non-specific homing; on the other hand, considering the paracrinic effect of the interleukins, specific homing might also occur (57). The altered plasma cells migrating into the liver might participate in the "secondary" immune reactions, for example, producing antibodies against the virus-modified tissue autoantigens. The blast cells and hepatocytes develop typical desmosomal contacts never occurring under physiological conditions (45, 86) (Fig. III/7).

Furthermore, T-cell derived plasmocytoid elements were isolated from the liver extracts (84, 101).

In summary, the current knowledge of the liver immune system still has gaps and controversies, for example, only few immunoglobulin-secreting and containing cells are found in the portal tract and around the lobular sinu-

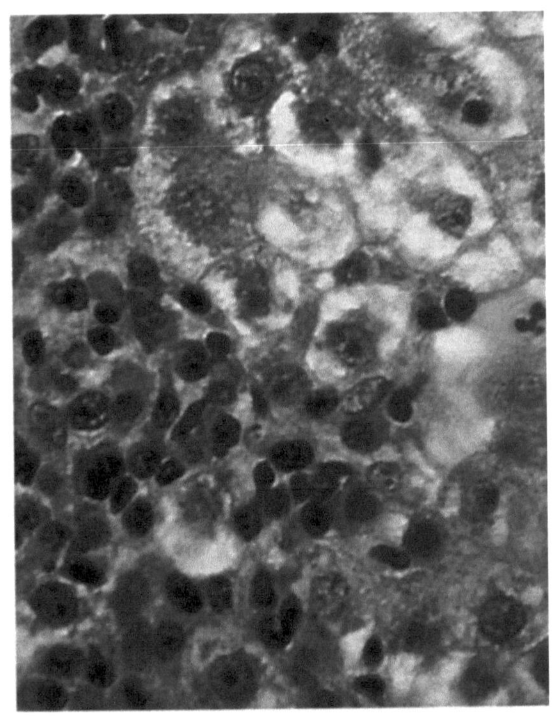

**Lympho-plasmocyte infiltration in the portal area develops amidst the hepatocytes (moth-hole necrosis, piecemeal necrosis). Hepatocyte membranes exhibit contacts with the inflammatory cells. Chronic active hepatitis. H.E. 800 × (data of Dr. Gábor Kendrey, M.D.).**

**Fig. III/7**

soids of the liver (16, 79), while about 50% of the IgA detected in the liver is locally produced by this organ (22, 23, 24). Plasma cells of the accessory glands of the intra- and extrahepatic bile ducts might be responsible for the IgA and the J-chain synthesis (80). The SC-mediated transport from blood performed by the epithelial cells of the bile ducts also contributes to the high IgA levels (22, 79, 80). Immunoglobulins of other classes also directly or indirectly detected in the bile (Ig-producing cell transfer) are transported from the spleen via the portal vein. The splenic origin of these cells, otherwise active in the liver, is shown in the experiments with splenectomy, After surgery no antibodies were secreted into the bile! This observation holds true despite the important part of the liver in the transport of immunoglobulins into the gastrointestinal tract (63, 79). Dual mechanism for IgA transport has been revealed.

The first, so-called direct transport of IgA immunoglobulins from the circulation into the bile ducts, involves pinocytosis and IgA receptors (50, 76). The receptor which plays a central role, is synthesized in the rough endoplasmatic reticulum and is expressed on the liver cell membranes by the vesicles of the Golgi apparatus (76, 77, 78, 108). The IgA receptor interacts with the dimeric IgA molecule. Pinocytotic vesicles transport the complex to the site of the bile ducts. The enzymes split the dimeric IgA leaving an attached part of its receptor, the latter is termed SC (75).

The second indirect or lysosomal mechanism involves fusion of the endocytotic vacuoles (pinocytosis) with the Golgi apparatus containing lysosomal enzymes. Thus the IgA molecule is splitted intracellularly. The cytoskeleton of the liver controls both mechanisms (32). The rate of transport decreases with age, and is suppressed by the toxic agents. Production of tubulin which forms a microtubular network of the cytoskeleton might also control the interaction of the liver with the immune system, therefore, the behaviour of the transplanted liver markedly differs from that of the other transplanted organs.

The largest IgA-producing cell population of the liver is observed in the A-type hepatitis supporting the gastrointestinal character of the infection. In hepatitis type B or C the count of the immunoglobulin-producing plasma cells in liver is very low (51).

The lymphoid system participates not only in the immune responses, but also plays an important part in the homeostasis regulation (2, 27). As a trophocyte it promotes the organ growth; regulates the chronic inflammatory reactions, wound healing, tissue regeneration, vascularization; moreover, it stimulates compensatory hyperplasia, enhances the fibroblast proliferation, including collagen formation (118). Though all the above mentioned effects have been discovered at least 20 years ago (15, 21, 28, 30, 87, 100, 102), they are still of importance, for the discovery of the limphokins provided a unique new insight in their mechanisms (36, 65, 72). IL-6, being the most potent growth factor, is especially worth pointing out, because it affects not only the immune cells, but also others carrying IL-6 receptors. Liver cells possess IL-6 receptors (19, 109), so the paracrin role of immunocytes (11, 39, 43, 57) in hepatitis greatly affects the regeneration of the liver cells. As the immune cells also promote the growth of the hepatic matrix, they might markedly contribute to the development of cirrhotic degeneration (38).

## References


1. Adlersberg, L., Singer, J. M., Ende, E.
   Redistribution and elimination of intravenously injected latex particles in mice
   J. Reticuloendoth. Soc. 6, 536. 1969.
2. Ashwell, J. D.
   Lymphocytes: Propertier and strategies.
   In: The liver: biology and pathobiology.
   Ed.: by Arias, I. M., Jakoby, W. B., Popper, H., Schachter, D., Shafritz, D. A.
   Raven Press. N. Y. 49, 867. 1988.
3. Balis, J. U., Rappaport, E. S., Gerber, L., Fareed, J., Buddingh, F., Messmore, H. L.
   A primate model for prolonged endotoxin shock.
   Lab. Invest. 38, 511. 1978.
4. Balis, J. U., Paterson, J. F., Shelley, S. A., Larson, C. H., Faréed, J., Gerber, L. I.
   Glucocorticoid and antibiotic effects on hepatic microcirculation and associated host responses in lethal gram-negative bacteremia.
   Lab. Invest. 40, 55. 1979.
5. Ballardini, G., Bianchi, F. B., Doniach, D., Mirakian, R., Pisi, E., Bottazzo, G. F.
   Aberrant expression of HLA-DR antigens on bile duct epithelium in primary biliary cirrhosis: Relevance to pathogenesis.
   Lancet. ii, 1009. 1984.
6. Bardadin, K. A., Desmet, V. J.
   Interdigitating and dendritic reticulum cells in chronic active hepatitis.
   Histopathology. 8, 657. 1984.

7. Barnada, V., Levrero, M., Franco A.
Characterization of effector cells in lymphocyto-toxicity to autologous hepatocytes in HBsAg-positive and autoimmune chronic active hepatitis (CAH).
Liver. 6, 45. 1986.
8. Biesel, K. W., Ehrinpreis, M. N., Bhathal, P. S., Mackey, I. R., Rose, N. R.
Genetics of carbon tetrachloride-induced liver injury in mice. II. Multigenic regulation.
Brit. J. Pathol. 65, 125. 1984.
9. Bioulac-Sage, P., Latry, P., Dubroca, J., Quinton, A., Balabaud, C.
Pit cells in human liver.
In: Cells of the Hepatic Sinusoid.
Ed: by Kirn, A., Knook, D. L., Wisse, E.
The Kupffer Cell Foundation, Rijswijk, The Netherlands 1, 415. 1986.
10. Bissell, D. M., Stamatoglou, S. C., Nermut, M. V., Hughes, R. C.
Interactions of rat hepatocytes with type IV collagen, fibronectin and laminin matrices. Distinct matrix-controlled modes of attachment and spreading.
Eur. J. Cell Biol. 40, 72. 1986.
11. Bowcock, A. M, Kidd, J. R., Lathrop, M., Danshvar, L., May, L. T., Ray, A., Sehgal, P. B., Kidd, K. K., Cavallisforza, L. L.
The human „beta-2" interferon (hepatocyte stimulating factor) „interleukin-6" gene: DNA polymorphism studies and localization to chromosome 7p21.
Genomics. 3, 8. 1988.
12. Brönsted, J.
In: An introduction to gel chromatography.
Ed. by Fischer, L.
North-Holland Publishing Co. Amsterdam-London. 1969.
13. Brennan, M. J., Oldberg, A., Hayman, E. G., Ruoslahti, E.
Effect of a proteoglycan produced by rat tumor cells on their adhesion to fibronectin-collagen substrata.
Cancer Res. 43, 4302. 1983.
14. Carlsson, R. N. K., Engvall, E., Freeman, A., Ruoslahti, E.
Laminin and fibronectin in cell adhesion: Enhanced adhesion of cells from regenerating liver to laminin.
Proc. Natl. Acad. Sci. USA. 78, 2403. 1981.
15. Carrel, A.
Growth promoting function of leukocytes.
J. Exp. Med. 36, 385. 1922.
16. Cohen, S., Ohta, G., Singer, E. J., Popper, H.
Immunocytochemical study of gamma-globulin in liver in hepatitis and postnecrotic cirrhosis.
J. Exp. Med. 111, 285. 1960.
17. Colucci, G., Schaffner, F., Paronetto, F.
Lymphocyte subpopulations and distribution of HLA antigens in liver biopsies of patients with primary biliary cirrhosis (PBC). (Abstr.).
Hepatology. 4, 1087. 1984.
18. Couchman, J. R., Woods, A., Höök, M., Christner, J. E.
Characterization of a dermatan sulfate proteoglycan synthesized by murine parietal yolk sac (PYS-2) cells.
J. Biol. Chem. 260, 13755. 1985.
19. Coulie, P. G., Vanhecke, A., Van Damme, J., Cayphas, S., Poupart, P., De Wit, L., Content, J.
High affinity sites for human 26 kDa protein (interleukin 6, B-cell stimulatory factor-2, human hybridoma plasmacytoma growth factor, interferon-beta 2), different from those of type-1 interferon (alpha, beta) on lymphoblastoid cells.
Eur. J. Immunol. 17, 1435. 1987.
20. Dan, C., Kaneda, K., Wake, K.
Pit cells and liver-associated natural killer activity following admistration of OK-432.
In: Cells of the Hepatic Sinusoid.
Ed: by Kirn, A., Knook, D. L., Wisse, E.
The Kupffer Cell Foundation, Rijswijk, The Netherlands. 1, 427. 1986.
21. David, J. R.
Macrophage activation by lymphocyte mediators.
Fed. Proc. 34, 1730. 1975.
22. Delacroix, D. L., Hodgson, H. J. F., McPherson, A., Dive, C., Vaerman, J. P.
Selective transport of polymeric immunoglobulin A in bile. Quantitative relationships of monomeric and polymeric immunoglobulin A, immunoglobulin M, and other proteins in serum, bile, and saliva.
J. Clin. Invest. 70, 230. 1982.
23. Dive, Ch., Nadalini, R. A., Vaerman, J. P., Heremans, J. F.
Origin and nature of the proteins of bile. II. A comparative analysis of serum, hepatic lymph and bile proteins in the dog.
Eur. J. Clin. Invest. 4, 241. 1974.
24. Dive, Ch., Heremans, J. F.
Nature and origin of the proteins of bile. I. A comparative analysis of serum and bile proteins in man.
Eur. J. Clin. Invest. 4, 235. 1974.
25. Eggink, H. F., Houthoff, H. J., Huitema, S., Gips, C. H., Poppema, S.
Cellular and humoral immune reactions in chronic active liver disease. I. Lymphocyte subsets in liver biopsies of patients with untreated idiopathic autoimmune hepatitis. chronic active hepatitis B and primary biliary cirrhosis.
Clin. Exp. Immunol. 50, 17. 1982.
26. Fever, G., Iglesia, F. A.
Molecular biochemistry of human disease.
CRC Press. Inc. 1, 80. 1985.
27. Fidler, I. J.
Lymphocytes are not only immunocytes.
Biomedicine. 32, 1. 1980.
28. Fox, M.
Etiology of the compensatory response: observations on the role of the lymphoid system.
Proc. roy. Soc. Med. 17, 62. 1969.
29. Fretto, L. J., Fowler, W. E., McCaslin, D. R., Erickson, H. P., McKee, P. A.
Substructure of human von Willebrand factor. Proteolysis by V8 and characterization of two functional domains.
J. Biol. Chem. 261, 15679. 1986.
30. Gersten, D. M., Fidler, I. J.
Effect of syngeneic lymphocytes on the vascularity, growth and induced metastasis of the B16 melanoma.
In: The Chicago Symposium.
Ed.: by Crispen, V. S. R. 1979.
31. Gocke, D. J., Ackley, A. M., Kim, H. C., Benson, G. D.
Alterations in helper and suppressor T lymphocytes in hepatitis (Abstr.).
Hepatology. 1, 511. 1981.
32. Goldman, I. S., Jones, A. L., Hradek, G. T., Huling, S.
Hepatocyte handling of immunoglobulin A in the rat: The role of microtubules.
Gastroenterology. 85, 130. 1983.
33. Goodman, M. L., Way, B. A., Irwin, J. W.
The inflammatory response to endotoxin.
J. Pathol. 128, 7. 1979.
34. Govindarajan, S., Uchida, T., Peters, R. L.
Identification of T lymphocytes and subsets in liver biopsy cores of acute viral hepatitis.
Liver. 3, 13. 1983.
35. Hayman, E. G., Pierschbacher, M. D., Öhgren, Y., Ruoslahti, E.
Serum spreading factor (vitronectin) is present at the cell surface in tissues.
Proc. Natl. Acad. Sci. USA: 80, 4003. 1983.
36. Henney, C. S., Kuribayashi, K., Kern, D. E., Gillis, S.
Interleukin-2 augments natural killer cell activity.
Nature. 291, 335. 1981.
37. Herberman, R. B., Holden, H. T.
Natural cell-mediated immunity.
Adv. Cancer Res. 27, 305. 1978.
38. Hirano, T., Kishimoto, T.
Interleukin 6 and plasma cell neoplasias.
Progress in Growth Factor Research. 1, 133. 1989.
39. Hirano, T., Kishimoto, T.

Interleukin 6. Peptide growth factors and their receptors.
In: Handbook of Experimental Pharmacology.
Ed.: by Sporn, M. B., Roberts, A. B.
Springer-Verlag, Berlin. 1, 95. 1989.
40. Hoffman, E. O., DiLuzio, N. R., Holper, K., Brettschneider, L., Coover, J.
Ultrastructural changes in the liver of baboons following lead and endotoxin administration.
Lab. Invest. 30, 311. 1974.
41. Holmes, R.
Preparation from human serum of an alpha-one protein which induces the immediate growth of unadapted cells in vitro.
J. Cell Biol. 32, 297. 1967.
42. Höök, M., Rubin, K., Oldberg, Á., Öbrink, B., Vaheri, A.
Cold-insoluble globulin mediates the adhesion of rat liver cells to plastic petri dishes.
Biochem. Biophys. Res. Commun. 79, 726. 1977.
43. Ikebuchi, K., Wong, G. G., Clark, S. C., Ihle, J. N., Hirai, Y., Ogawa, M.
Interleukin-6 enhancement of interleukin-3-dependent proliferation of multipotential hemopoietic progenitors.
Proc. Natl. Acad. Sci. USA. 84, 9035. 1987.
44. Ill, C. R., Ruoslahti, E.
Association of thrombin-antithrombin III complex with vitronectin in serum.
J. Biol. Chem. 100, 1948. 1985.
45. Imai, Y., Kasajima, T., Terashima, K., Dobashi, M., Matsuda, M.
Non Hodgkin's malignant lymphoma with special reference to desmosome-like junctions.
Acta Pathol. Jpn. 30, 195. 1980.
46. Jancsó, N.
Speicherung. Stoffanreicherung in Retikuloendothel und in der Niere.
Akadémiai Kiadó. Budapest. 1955.
47. Jákó, J.
Fehérjeanyagcsere-vizsgálatok májbetegségekben.
Kandidátusi értekezés I–II. 1973.
48. Jákó, J., Sebők, J.
Experimentelle Angaben über den Eliminationsweg der Degradationsprodukte von Fibrinogen/Fibrin.
Forschungsergebnisse der Transfusions-medizin und Immunhaematologie. 6, 225. 1979.
49. Jenne, D., Hugo, F., Bhakdi, S.
Interaction of complement S-protein with thrombin-antithrombin complexes: A role for the S-protein in haemostasis.
Thromb. Res. 38, 401. 1985.
50. Jones, A. L., Renston, R. H., Burwen, S. J.
Uptake and intracellular disposition of plasma-derived proteins and apoproteins by hepatocytes.
In: Progress in Liver Diseases.
Ed.: by Popper, H., Schaffner, F.
Grune & Stratton, N. Y. 7, 51. 1982.
51. Jones, A. L., Altorter, J.
Immunological Functions of the Liver.
In.: Immunology of the Gastrointestinal Tract and Liver.
Ed.: by Heyworth M. F., Jones, A. L.
Raven Press. N. Y. 1988.
52. Kaneda, K., Dan, C., Wake, K.
Pit cells as natural-killer cells.
Biomed. Res. 4, 567. 1983.
53. Kaneda, K., Kurioka, N., Seki, S., Wake, K., Yamamoto, S.
Pit cell-hepatocyte contact in autoimmune hepatitis.
Hepatology. 4, 955. 1984.
54. Kiessling, R., Haller, O.
Natural killer cells in the mouse: An alternative immune surveillance mechanism?
Contemp. Top. Immunobiol. 8, 171. 1978.
55. Kirsch, E., Ihme, A., Müller, P., Krieg, T.
Molecular defects in from disorders of collagen metabolism.
Enzyme. 27, 239. 1982.
56. Klebe, R. J.
Isolation of a collagen-dependent cell attachment factor.
Nature. 250, 248. 1974.
57. Klein, B., Zhang, X.-G., Jourdan, M., Content, J., Houssiau, F., Aarden, L., Piechaczyk, M., Bataille, R.
Paracrine rather than autocrine regulation of myeloma-cell growth and differentiation by interleukin-6.
Blood. 73, 517. 1989.
58. Klingenstein, R. J., Savarese, A. M., Dienstag, J. L., Rubin, R. H., Bhan, A. K.
Immunoregulatory T cell subsets in acute and chronic hepatitis (Abstr.).
Hepatology. 1, 523. 1981.
59. Knox, P., Wells, P.
Cell adhesion and proteoglycans.
J. Cell Sci. 40, 77. 1979.
60. Koda, J. E., Bernfield, M.
Heparan sulfate proteoglycans from mouse mammary epithelial cells. Basal extracellular proteoglycan binds specifically to native type I collagen fibrils.
J. Biol. Chem. 259, 11763. 1984.
61. Kolodny, E. H., Cable, W. J.
In born errors of metabolism.
Ann. Neurol. 11, 221. 1982.
62. Kurane, I., Binn, L. N., Bancroft, W. H., Ennis, F. A.
Human lymphocyte responses to hepatitis A virus-infected cells:
Interferon production and lysis of infected cells.
J. Immunol. 135, 2140. 1985.
63. Kutteh, W. H., Prince , S. J., Phillips, J. O., Spenney, J. G., Mestecky, J.
Properties of immunoglobulin A in serum of individuals with liver diseases and in hepatic bile.
Gastroenterology. 82, 184. 1982.
64. Laterra, J., Silbert, J. E. Culp, L. A.
Cell surface heparan sulfate mediates some adhesive responses to glycosaminoglycan-binding matrices, including fibronectin.
J. Cell Biol. 96, 112. 1983.
65. Lattime, E. C., Pecoraro, G. A., Stutman, O.
The activity of natural cytotoxic cells is augmented by interleukin 2 and interleukin 3.
J. Exp. Med. 157, 1070. 1983.
66. Lavrell, C. B., Jeppsson, J. O.
Protease inhibitors in plasma.
In: The Plasma proteins.
Ed.: by Putnam, F. W.
Academic Press. N. Y. 1, 229. 1975.
67. Lawler, J.
Review: The structural and functional properties of thrombospondin.
Blood. 67, 1197. 1986.
68. Liehr, H., Grun, M., Thiel, H., Brunswig, D., Rasenack, U.
Endotoxin induced liver necrosis and intravascular coagulation in rats enhanced by portacaval collateral circulation.
Gut. 16, 429. 1975.
69. McIntyre, K. W., Welsh, R. M.
Accumulation of natural killer and cytotoxic T large granular lymphocytes in the liver during virus infection.
J. Exp. Med., 164, 1667. 1986.
70. Marden, R. J., Rent, R., Choi, E. Y. C., Gewurz, H.
C1q deficiency associated with urticarial-like lesions and cutaneous vasculitis.
Am. J. Med. 61, 560. 1976.
71. Meyer zum Büschenfelde, K.-H., Manns, M.
Mechanisms of autoimmune liver disease.
Semin. Liver. Dis. 4, 26. 1984.
72. Mietkiewski, J. M., Scheuer, P. J.
Immunoglobulin-containing plasma cells in acute hepatitis.
Liver. 5, 84. 1985.
73. Mittman, C.
Pulmonary emphysema and proteolysis.
Academic Press. N. Y. 1972.
74. Montano, L., Aranguibel, F., Boffill, M., Goodall, A. H., Janossy, G. Thomas, H. C.

An anlysis of the composition of the inflammatory infiltrate in autoimmune and hepatitis B virus-induced chronic liver disease.
Hepatology. 3, 292. 1983.

75. Mostov, K. E., Friedlander, M., Blobel, G.
The receptor for transepithelial transport of IgA and IgM contains multiple immunoglobulin-like domains.
Nature. 308, 37. 1984.

76. Mostov, K. E., Kraehenbuhl, J.-P., Blobel, G.
Receptor-mediated transcellular transport of immunoglobulin: Synthesis of secretory component as multiple and larger transmembrane forms.
Proc. Natl. Acad. Sci. USA. 77, 7257. 1980.

77. Mullock, B. M., Hinton, R. H., Dobrota, M., Peppard, J., Orlans, E.
Distribution of secretory component in hepatocytes and its mode of transfer into bile.
Biochem. J. 190, 819. 1980.

78. Mullock, B. M., Jones, R. S., Hinton, R. H.
Movement of endocytic shuttle vesicles from the sinusoidal to the bile canalicular face of hepatocytes does not depend on occupation of receptor sites.
FEBS Lett. 113, 201. 1980.

79. Nagura, H., Smith, P. D., Nakane, P. K., Brown, W. R.
IgA in human bile and liver.
J. Immunol. 126, 587. 1981.

80. Nagura, H., Tsutsumi, Y., Hasegawa, H., Watanabe, K., Nakane, P. K., Brown, W. R.
IgA plasma cells in biliary mucosa: A likely source of locally synthesized IgA im human hepatic bile.
Clin. Exp. Immunol. 54, 671. 1983.

81. Naito, M., Wisse, E.
Observations on the fine structure and cytochemistry of sinusoidal cells in fetal and neonatal rat liver.
In: Kupffer Cells and Other Liver Sinusoidal Cells.
Ed.: by Wisse, E., Knook, D. L.
Elsevier/North-Holland Biomedical Press. Amsterdam. 497. 1977.

82. Olarte, J. A., Sanchetto, J. R., Vilarino, J., Rubio, H. H. Jr., Fernandez, F. J.
Fecal excretion of PVP-$I^{131}$ in liver cirrhosis.
Gastroenterologia. 101, 137. 1964.

83. van den Oord, J. J., Fevery, J., De Groote, J., Desmet, V. J.
Immunohistochemical characterization of inflammatory infiltrates in primary biliary cirrhosis.
Liver. 4, 264. 1984.

84. Pape, G. R., Rieber, E. P., Eisenburg, J., Hoffmann, R., Balch, C. M., Paumgartner, G., Riethmüller, G.
Involvement of the cytotoxic/suppressor T cell subset in liver tissue injury of patients with acute and chronic liver diseases.
Gastroenterology. 85, 657. 1983.

85. Paronetto, F., Colucci, G., Colombo, M.
Lymphocytes in liver diseases.
In: Progress in Liver Disease.
Ed.: by Popper, H., Schaffner F.
Grune & Stratton, N. Y. 8, 191. 1986.

86. Perez-Atayde, A. R., Chang, Y. C., Seiler, M. W.
Retroperitoneal lymphoma with ribosome-lamellae complexes, cytoplasmic projections and desmosome-like junctions: an ultrastructural study.
Ultrastruct. Pathol. 3, 43. 1982.

87. Pliskin, M. E., Prehn, R. T.
Stimulation of liver regeneration and compensatory kidney hyperplasia by passive transfer of spleen cells.
J. Reticuloendothel. Soc. 17, 290. 1975.

88. Podack, E. R., Konigsberg, P. J.
Cytolytic T cell granules. Isolation, structural, biochemical, and functional characterization.
J. Exp. Med. 160, 695. 1984.

89. Reynolds, C. W., Timonen, T., Herberman, R. B.
Natural killer (NK) cell activity in the rat. I. Isolation and characterization of the effector cells.
J. Immunol. 127, 282. 1981.

90. Ratnoff, O. D.
Disordered homeostasis in liver disease.
In: Diseases of the Liver.
Ed.: by Schiff, L.
Lippincott, Philadelphia. 184, 1975.

91. Rich, A. M., Pearlstein, E., Weissman, G., Hoffstein, S. T.
Cartilage proteoglycans inhibit firbonectin-mediated adhesion.
Nature. 293, 224. 1981.

92. Rola-Pleszczynski, M., Lieu, H.
Human natural cytotoxic lymphocytes: Definition by a monoclonal antibody of a subset which kills an anchorage-dependent target cell line but not the K-562 cell line.
Cell. Immunol. 82, 326. 1983.

93. Rola-Pleszczynski, M., Lieu, H., Sullivan, A. K., Girard, M.
Membrane markers, target cell specificity, and sensitivity to biological response modifiers distinguish human natural cytotoxic from human natural killer cells.
J. Clin. Invest. 76, 1927. 1985.

94. Rubin, K., Höök, M., Öbrink, B., Timpl. R.
Substrate adhesion of rat hepatocytes: Mechanism of attachment to collagen substrates.
Cell. 24, 463. 1981.

95. Ruoslahti, E., Hayman, E. G., Pierschbacher, M. D.
Extracellular matrices and cell adhesion.
Arteriosclerosis. 5, 581. 1985.

96. Saba, T. M., Jaffe, E.
Plasma fibronectin (opsonic glycoprotein): Its synthesis by vascular endothelial cells and role in cardiopulmonary integrity after trauma as related to reticuloendothelial function.
Am. J. Med. 68, 577. 1980.

97. Santoro, S. A., Cowan, J. F.
Collagen Relat. Res. 2, 31. 1982.

98. Schaffner, F.
Autoimmune chronic active hepatitis: Three decades of progress.
In: Progress in Liver Diseases.
Ed.: by Popper, H., Schaffner, F.
Grune & Stratton, N. Y. 8, 485. 1986.

99. Seifert, J., Schröder, D.
Immunologische Vorgänge bei der Verdaung.
Z. Gastroenterologie. 22, 713. 1984.

100. Shearer, W. T.
Stimulation of cells by antibody.
Science. 182, 1357. 1973.

101. Shibata, M., Kojima, T., Aoyama, K., Kouda, T., Inoue, K., Sasaki, H., Imoto, S.
Morphological studies on the mechanism of hepatocellular injury in type A acute viral hepatitis. Ultrastructural investigation and identification of T cell subsets using peroxidase antibody technique.
Acta Hepatol. Jpn. 25, 1. 1984.

102. Sidkey, Y. A., Auerback, R.
Lymphocyte induced angiogenesis: a quantitative and sensitive assay of the graft vs. host reaction.
J. exp. Med. 141, 1084. 1975.

103. Simpson, L. O.
A hypothesis proposing increased viscosity as a cause of proteinuria and increased vascular permeability.
Nephron. 31, 89. 1982.

104. Singer, J. M., Adlersberg, L., Hoenig, E. M., Ende, E., Tchorsch, Y.
Radiolabelled latex particles in the investigation of phagocytosis in vivo; clearance curves and histological observations.
J. reticuloendothelial Soc. 6, 561. 1969.

105. Stankova, J., Rola-Pleszczynski, M.
Human natural cytotoxic cells: Distinctive membrane markers and patterns of reactivity.
Immunol. Res. 5, 33. 1986.

106. Suzuki, S., Oldberg, Á., Hayman, E. G., Pierschbacher, M. D., Ruoslahti, E.
Complete amino acid sequence of human vitronectin deduced from cDNA. Similarity of cell attachment sites in vitronectin and fibronectin.

EMBO J. 4, 2519. 1985.
107. Sveger, T., Ekelund, H.
Variations of protease inhibitors in foetuses, newborn infants and in some neonatal disorders.
Acta Paediatr. Scand. 64, 763. 1975.
108. Sztul, E. S., Howell, K. E., Palade, G. E.
Intracellular and transcellular transport of secretory component and albumin in rat hepatocytes.
J. Cell. Biol. 97, 1582. 1983.
109. Taga, T., Kawanishi, K., Hardy, R. R., Hirano, T., Kishimoto, T.
Receptors for B cell stimulatory factor 2 (BSF–2): Quantitation, specificity, distribution and regulation of the expression.
J. Exp. Med. 166, 967. 1987.
110. Tamkun, J. W., Hynes, R. O.
Plasma fibronectin is synthesized and secreted by hepatocytes.
J. Biol. Chem. 254, 4641. 1983.
111. Terranova, V. P., Rohrbach, D. H., Martin, G. R.
Role of laminin in the attachment of PAM 212 (epithelial) cells to basement membrane collagen.
Cell. 22, 719. 1983.
112. Timonen, T., Ranki, A., Saksela, E., Häyry, P.
Human natural cell-mediated cytotoxicity against fetal fibroblasts. III. Morphological and functional characterization of the effector cells.
Cell. Immunol. 48, 121. 1979.
113. Timonen, T., Ortaldo, J. R., Herberman, R. B.
Characteristics of human large granular lymphocytes and relationship to natural killer and K cells.
J. Exp. Med. 153, 569. 1981.
114. Timpl, R., Engel, J., Martin, G. R.
Laminin–A multifunctional protein of basement membranes.
Trends Biochem. Sci. 8, 207. 1983.
115. Varani, J., Dixit, V. M., Fligiel, S. E. G., McKeever, P. E., Carey, T. E.
Thrombospondin-induced attachment and spreading of human squamous carcinoma cells.
Exp. Cell Res. 167, 376. 1986.
116. Vento, S., Nouri–Aria, K. T., Eddleston, A. L. W. F.
Immune mechanisms in autoimmune chronic active hepatitis.
Scand. J. Gastroenterol. 20, (Suppl. 114), 91, 1985.
117. Viberti, G. C., Hill, R. D., Jarett, R. J., Mahmud, V., Keen, H.
Microalbuminuria as a predictor of clinical nephropathy in insulin-dependent diabetes mellitus.
Lancet. i, 1430. 1982.
118. Wahl, S. M., Wahl, L. M., McCarthy, J. B.
Lymphocyte-mediated activation of fibroblast proliferation and collagen production.
J. Immunol. 121, 942. 1978.
119. Wellmann, W., Fink, P. C., Benner, F., Schmidt, F. W.
Endotoxaemia in active Crohn's disease. Treatment with whole gut irrigation and 5-aminosalicylic acid.
Gut. 27, 814. 1986.
120. White, R. R., Mela, L., Bacalzo, V., Olofsson, K., Miller, L. D.
Hepatic ultrastructure in endotoxemia, hemorrhage, and hypoxia.
Surgery. 73, 525. 1973.
121. Wiedmann, K. H., Bartholomew, T. C., Brown, D. J. C., Thomas, H. C.
Liver membrane antibodies detected by immunoradiometric assay in acute and chronic virus-induced and autoimmune liver disease.
Hepatology. 4, 199. 1984.
122. Wiltrout, R. H., Mathieson, B. J., Talmadge, J. E.
Augmentation of organ-associated natural killer activity by biological response modifiers. Isolation and characterization of large granular lymphocytes from the liver
J. Exp. Med. 160, 1431. 1984.
123. Zinkernagel, R. M., Doherty, P. C.
Restriction of in vitro T cell-mediated cytotoxicity in lymphocytic choriomeningitis within a syngeneic or semiallogeneic system.
Nature. 248, 701. 1974.
124. Zinkernagel, R. M., Welsh, R. M.
H–2 compatibility requirement for virus-specific T cell-mediated effector functions in vivo. I. Specificity of T cells conferring antiviral protection against lymphocytic choriomeningitis virus is associated with H–2K and H–2D.
J. Immunol. 117, 1495. 1976.
125. Yamada, G., Hyodo, I., Tobe, K.
Ultrastructural immunocytochemical analysis of lymphocytes infiltrating bile duct epithelia in primary biliary cirrhosis.
Hepatology. 6, 385. 1986.
126. Young, R. S. K., Woods Cynthia, Towfighi, J.
Hepatic damage in neonatal rat due to E. Coli endotoxin.
Digestive Diseases and Sciences. 31, 651. 1986.
127. Young, R. S. K., Yagel, S. K., Towfighi, J.
Systemic and neuropathologic effects of E. coli endotoxin in neonatal dogs.
Pediatr. Res. 17, 349. 1983.

# IV. Polyclonal gammopathies

Gammopathies are defined as pathological changes of the quantity, disproportions between different fractions, and the alteration of the physico-chemical and functional properties of immunoglobulins. Pathomorphological changes are always involved in this pathology even if it is difficult to detect them. Waldenstrôm was the first to introduce the term gammopathia (137).

Polyclonal gammopathies usually occur in acute and chronic infections (bacterial, viral, fungal), tumors (56), connective tissue diseases (31) and acute and chronic liver diseases (3). The wide spectrum of pathological states inducing gammopathies suggests that in most cases the B response is not specific (75). A question arises whether the above changes reflect the direct activation of the B cells (short-cut activation), or they are rather activated via a physiological route (antigen stimulus, activation of the helper cells etc.). At present our knowledge is insufficient to answer this question, though accurate systematization of the available data should be helpful.

Under physiological conditions polyclonal immunoglobulins are produced following an antigen stimulus (26), since the antigens carry different epitopes. However some antigens also exhibit mitogenic properties, which activate the cells in a non-specific way. Mitogens rather bind to the membrane receptors of the cells promoting their differentiation into plasma cells. LPS (71, 83), LPP (4), PPD (123) and the mitogens exhibit these properties (51, 53). B cell proliferation can be stimulated by binding of specific antibodies to the surface immunoglobulins (79) or with certain activators used in clinical practice (7) (Fig. IV/1)

The decreased activity of the suppressor T cells (6, 139) plays a crucial role in the stimulation of polyclonal antibody-synthesis elicited by the stimuli presented in Fig. IV/1. Genetic defects might also play an important role in the above mechanisms (42, 43, 44, 45). In Epstein-Barr viral infection genetic defects can contribute (46) or directly elicit functional inhibition of the suppressor T cells (109, 128). Endogenous $PGE_2$ can also inhibit the activity of the CD8(+) suppressor T cells (18). $PGE_2$ is produced by the monocytes and it probably acts as a physiological feed-back mechanism (17, 33, 54). Polyclonal B cell-activator (PBA), or rather a series of lipopolysaccharides (53) is (are) a second factor of the autoregulation (27, 41, 58, 81, 104). Evidence exists that α-2 macroglobulin also acts as PBA (125). It can as well be a signal, which binds to specific structures or receptors (24, 59) of the suppressor cells, thus stimulating the synthesis of polyclonal antibodies (19, 88).

As far as the antigen-induced antibody synthesis is concerned, the antigen presentation can trigger the production of polyclonal immunoglobulins. It is assumed that some antigens do not get into the intracellular space of the APS (lyzosomal processing etc.) before being presented, but can directly interact with the MHC II receptors. The latter stimulates a prolonged, usually undesirable antigen stimulus. Similarly, the interdigital dendritic cells can also store and continuously present the antigen.

Primary alterations of the T cells might also contribute to the activation of the antibody production.

The thymus-dependent T cells are derived from the progenitors of the liver, bone marrow and the yolk sac. In the course of their maturation in the thymus the T cells are differentiated into helper or suppressor T cells depending on the exact character of their interactions with the B lymphocytes (10) (Fig. IV/2).

Spontaneous increase of the helper cell activity independent of other factors - the so-called sensitized T helpers - are extremely

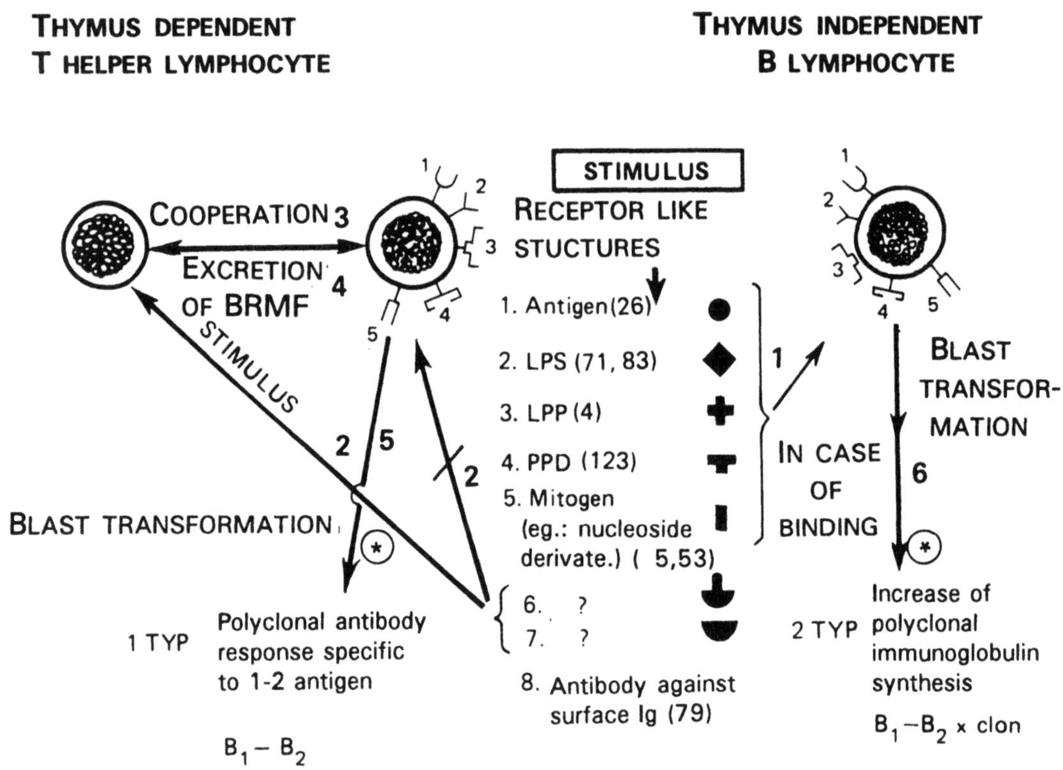


6,7 unknown factor cannot be able to bind receptor

BRMF = B cell replication and maturation producing factor

(*) role of activators realised in this stage, but yet there is a therapeutic trial (7)

anti-$\mu$ antibodies binding to surface IgM markers have stimulatory effect (79)

**Fig. IV/1 (107, 108)**

sensitive to the antigen presentation. The combination of both mechanisms is termed "absolute increase of the response" (28, 97, 105). Mitogenic effect of either auto- or heteroantibodies is a potent stimulus promoting the helper cells activity. Some drugs, for example, procainamide can directly stimulate the helper cells (85).

A relative increase of the helper cells caused by three possible mechanisms is frequently observed:

a) self-maintaining synthesis of the specific anti-suppressor cell antibodies due to the primary defect of B cells;
b) lack of the suppressor cells due to a genetic defect;
c) progenitor defect also affecting differentiation of all the three cell types, including the macrophages (112), which leads to $T_H$ predominance.

The first possibility implies that the increased B cell activity is responsible for the decreased function of the suppressor cells; while in the two other cases the B cell activation is a final result. It is also worth noting that all the three reactions are not antigen specific or antigen dependent, similar to the helper - T cells interaction, when the helper cell "presents" to a responder T cells. The latter is called autopresentation.

In the course of the other - specific - T cell cooperation one of the antigen epitopes specifically (?!) alters the helper or suppressor cell activity (117). The critical epitopes are se-

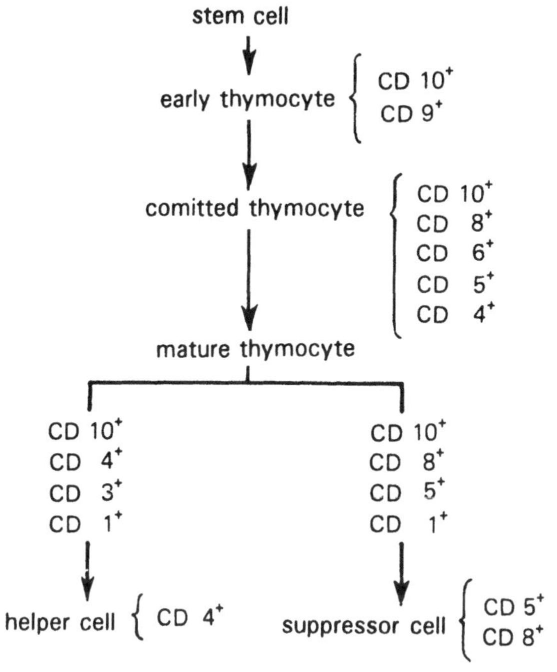


**Fig. IV/2**

lected by the macrophages (110), which thus contribute to the final result of the T-B interaction (118). The selected epitope would shift the existing physiological balance triggering undesirable clinical sequellae (61, 91, 135).

Of the widely used drugs Cimetidin directly inhibits the suppressor cells (49, 95). Monocytes can both stimulate or inhibit the production of immunoglobulins (55, 74, 80). They can also enhance the suppressor activity and affect the helper cells increasing or inhibiting their functions. In contrast to the T cells, the regulation of the B cell activity might be antigen-dependent.

The $F(ab)_2$ fragments of the surface Ig (sIg) of the B cells can bind antigens thus triggering the synthesis of specific antibodies (89, 124). A wide variety of the polyclonal B cell activators stimulate the production of clonal immunoglobulins. Apart from the above mentioned cellular interactions, some "independent" unstimulated B cells can produce immunoglobulins at a higher rate.

The synthesized immunoglobulin level is also of importance, since it adjusts the feedback regulation (90, 111). Moreover, a passive transfer of immunoglobulins, though decreases the antigen-producing cell count, fails to affect the rate of proliferation in vivo. These reactions have been studied in considerable detail. The results show that the transferred antibodies can bind only the existing antigens, hence they elicit their effect only in the macrophage-dependent phase of antigen processing, but fail to exhibit direct effects (102). However a direct effect of these antibodies is reported: the Fc fragments of the antibodies bind to the T helper cells, preventing the T-B cellular interaction (2, 120).

Interaction between different immunoglobin antibodies, can regulate *the rate of their synthesis.* This concept has been first introduced by Jerne. According to this theory the variable part of the synthesized antibody - an idiotype - acts as an antigenic determinant, eliciting antibody production against the newly formed immunoglobulins. The following step is directed against the anti-idiotypes etc.: antibodies will be always synthesized against the new immunoglobulins... until an idiotype-anti-idiotype network is formed (73). That means that the antigen-specific parts of immunoglobulins not only recognize the antigen determinants, but also participate in a self-recognition process. The main steps of this process are:

1. epitopes of a foreign substance acts as stimulators;
2. specific antibodies are synthesized, however, their idiotype of the antibodies is again recognized as an "epitope";
3. a second generation of the so-called anti-idiotype antibodies is produced, the latter showing great *similarities* (1) to the original molecules (Fig. IV/3).

In a more intensive process more and more clones are involved and, finally, the synthesis of the polyclonal antibodies is stimulated. The idiotype recognition occurring within this network can either depress or stimulate the idiotype-specific B cell reaction, the final result depending on whether suppressor or helper T lymphocytes responsible for clonal restriction are *predominantly* activated. In this phase the whole network acts as self-regulating, independently of the primary antigen stimulus.

Of the diseases accompanied with the enhancement of immunoglobulin synthesis, liver pathology plays a central role and will be discussed in more details (67, 69, 96).

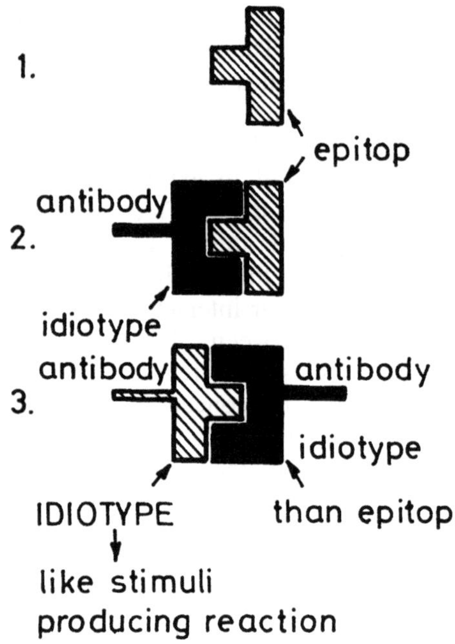


Fig. IV/3

The relationship between the damage of liver parenchyma and the activation of B lymphocytes has not yet been proved and the overall significance of such interactions is obscure. However, data of the literature suggest some basic relationships:

- the decrease of the oncotic pressure as a signal: the lower extracellular pressure stimulates the intracellular protein synthesis, this response being diminished in liver diseases. As a compensatory mechanism, B cells, which are also capable of producing plasma proteins, are activated, therefore, immunoglobulins replace plasma proteins (86, 141).
- Accumulation (release) of biologically active substances promoting protein synthesis (9). There is experimental evidence showing accumulation of an α-1 type glycoprotein in rats following semihepatectomy. Administration of this substance into normal animals markedly stimulated mitotic activity of the liver cells (15, 39, 66). If this agent is not selective, it can also stimulate the B line cells of the bone marrow, thus inducing protein synthesis.
- Role of immunoglobulins in membrane stabilization of the affected liver cells.

Experiments of Najjar support the dual role of immunoglobulins, since their Fab fragments can bind not only antigens, but also to the membranes of the living cells (78) stabilizing the latter and increasing their resistance to physical stimuli, hence, a quick fall in the intracellular ionic concentration is prevented. The importance of the transport processes is illustrated in a well known effect that the bi-concave normal RBC turn into spherocytes with multiple protrusions (echynocytes) in hypotonic 0.15 M saline. At addition of normal plasma they shape is immediately restored (47, 131, 132). Najjar has identified erythrophilic and leucophilic gamma-globulins (35, 36, 37), moreover, he has shown that immunoglobulins stabilize membranes carrying any organ-specific antigens. In his opinion, the lymphoid tissue plays a *strategic* role producing adequate immunoglobulin molecules, which can bind to the membranes of different organs increasing their viability and functional activity (34, 92).

The spleen as a largest lymphoid organ, drains its blood into the portal circulation, therefore, the synthesized immunoglobulins are first exposed to the surface of the liver cells and the proteins (alimentary?) of the portal blood. Similar to gammaglobulins binding to RBC, it can be assumed that the excess of the γ-globulins is required for stabilization of the deranged liver cell membranes.

- Mononuclear cells "infiltrating" the liver might in situ produce high amounts of polyclonal immunoglobulins (22, 98, 99, 101). It was also assumed that reticulohistiocytes (RHS) are capable of immunoglobulin synthesis in liver cirrhosis and hepatitis. The DNA content of the RHS cells is very high suggesting the role of these cells as additional sources of protein (16, 20, 21, 23). It is, however, well known that immunoglobulins can only be synthesized by lymphoplasmocytes, the RHS cells fail to exhibit this function (113, 142) and, therefore, cannot synthesize immunoglobulins.
- Antibodies against intestinal bacteria and the products of their metabolism contribute to the elevated titer of immunoglobulins (13, 77, 116).

In patients with hepatitis the maximum titer of the antibodies against E. Coli is observed two months post-infection; later the titer declines to the normal value. At the same time the level of the antibodies against the haemophilus influensae remains unchanged throughout the disease. These observation indicate that the pathologically affected liver fails to detoxicate products of bacterial metabolism (toxins) that are transported with the portal blood. The latter get into the systemic circulation and antibodies are produced; hyperglobulinemia is typical of the liver diseases (134). It was also found that the intestinal bacterial flora of the patients with cirrhosis is more developed, compared to the healthy individuals (40). The predominance of either mechanism must be thoroughly considered. It seems plausible that several reactions might occur simultaneously, but the role of each mechanism is different.

In our opinion endotoxins of the intestinal bacteria physiologically stimulate the activity and the sensitivity of the B cell-mediated system. The extent of the stimulus is regulated by the portal system and the liver (70).

Coexistence of human organism with E. Coli is not a simple "symbiosis", the presence of the antigens affect the early period of the extrauterine life.

A normal newborn has only maternal immunoglobulins (IgG) (84). The development of his own antibodies is triggered by the appearance of intestinal bacteria (E. Coli). It is well known that the "germ free" animals develop "agammaglobulinemia" (51, 57, 126, 127).

The above observations indicate that the immunological maturity of the fetus at term is not yet developed. Environmental antigenic stimuli are responsible for the normal development of the immune system. The first circulating antibodies produced by a newborn are of class IgM; 30-40 days later IgG immunoglobulins also appear in the circulation. Similarly, in adults the primary immunization is followed by the synthesis of IgM-type antibodies and 3-4 days late IgG antibodies appear (38, 136).

Liver plays a special role regulating the stimulatory effect of endotoxin (30, 76). Endotoxins, being absorbed from gastrointestinal tract, are metabolized in the liver due to its lipoperoxydase activity (39, 133). However if the liver function decreases or the portal blood is directly shunted into the systemic circulation, the lasting antigen stimulus would activate the B cell-mediated system (Fig. IV/4).

Additional evidence in favour of this view is obtained in rat experiments: following the

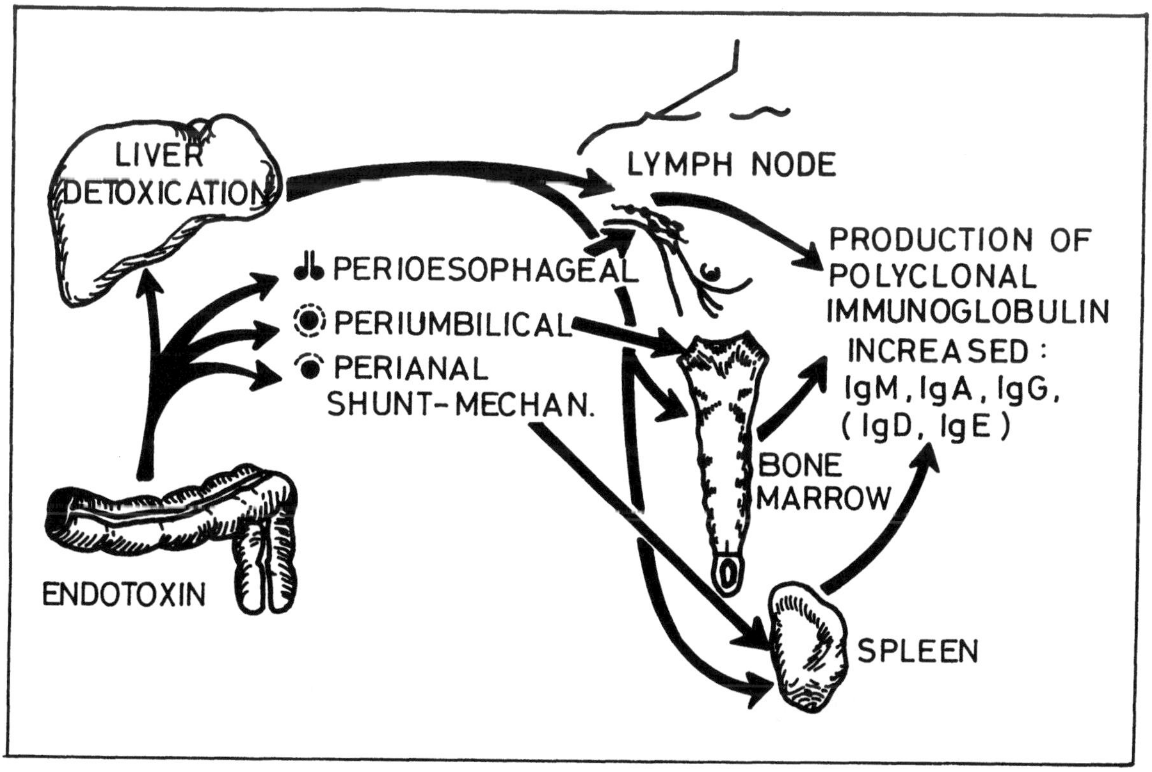


Fig. IV/4

porto-caval shunt surgery the γ-globulin titer increases and the albumin concentration decreases (129).

The stimulating effect of endotoxins on the B cell-mediated system was assessed in the in vitro lymphoblast transformation experiments (94). The stimulating effect of phytohaemagglutinin (PHA) and endotoxin on the blast transformation reaction was compared and their ratio calculated. The blast transformation effect of the antigens was taken from the literature (Fig. IV/5).

The results indicate that though endotoxin is a significantly less potent stimulating agent, which elicits rather mitogenic than antigenic effect, its stimulatory role can not be precluded. Data of the literature confirm our assumption that the thymus-independent, B-dependent cells undergo blast transformation following endotoxin, but not PHA stimulus (25, 121). The percent of blast transformation in these experiments is, however, lower than that observed in PHA experiments (48, 100, 170). These results confirm the non-specific stimulatory role of endotoxins in the development of the polyclonal gammopathies in vivo. The colloid lability tests in patients with final phase of cirrhosis give negative results and immunoglobulin levels return to the normal values. It is assumed that the bone marrow in these patients is exposed to higher concentrations of endotoxin, which rather inhibits than

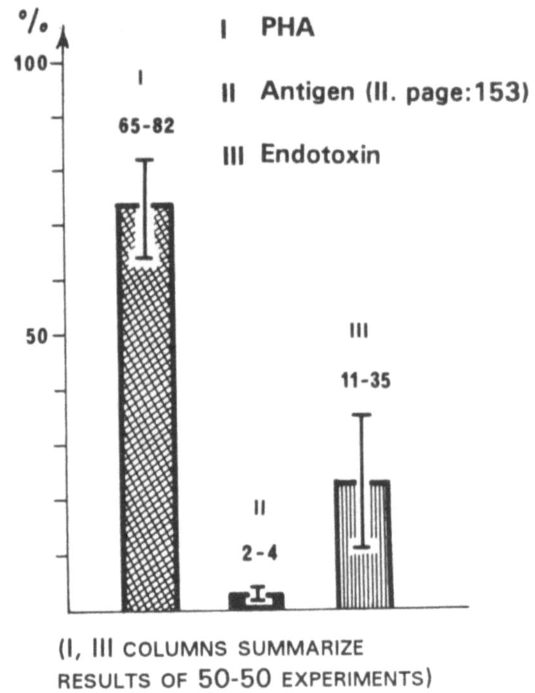


Fig. IV/5

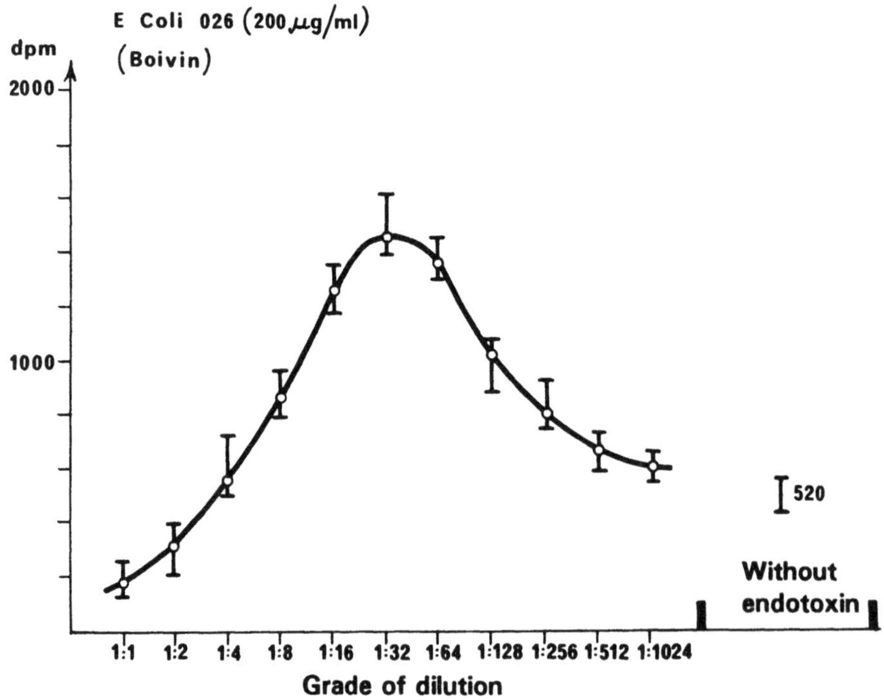


Fig. IV/6

stimulates its function (11, 103). In vitro experiments confirm this observation: concentrations of endotoxins under optimum are ineffective, while those exceeding it inhibit mitosis (Fig. IV/6).

No significant difference was observed between the Westphal and the Boivin-produced endotoxins (93). In the next series of experiments biological activity of the sterile supernatant of homogenized fetal liver alone or in combination with endotoxin has been studied in lymphocyte culture. In 5 experiments we failed to detect any signs of biological activity in homogenized liver cells. Not only the $H^3$-thymidin incorporation, but also morphological changes in the smears (% of blasts) were evaluated. In contrast to the PHA group completely lacking eosinophils significant eosinophilia (8-14%) was observed in the endotoxin-treated samples.

At present we can give no explanation of the above findings. Eosinophilia can not be caused by the presence of Dextran in vivo, otherwise the PHA-cultures should have developed similar increase of the eosinophil count (140).

Furthermore, if the lymphocytes would be accepted as possible precursors of eosinophils (114), still the effect of the two transformation-inducing substance could not differ so greatly. Therefore, it is the endotoxin, which is responsible for the above differences, and the incorporation of $T^3$-thymidin into both lymphocytes and eosinophils (114).

Plasmocellular reaction of the bone marrow underlies the cytological changes developing in polyclonal gammopathies in liver pathology. In 1963 Schneiderbauer and Narbeschuber have studied this problem in considerable detail, specially emphasizing the differences between the enhanced monoclonal and polyclonal immunoglobulin synthesis (115).

Many authors agree that polyclonal gammopathies can be transformed into monoclonal ones (8, 143). Hitzig (63) mentioned that the so-called "M" component is typical of the liver diseases (cirrhosis). In contrast to his view, Waldenström precludes the possibility of the polyclonal - monoclonal transfer (138). On the basis of our experience we rather support the latter theory.

The bone marrow cytology analysed in 115 sternal punctates of patients with cirrhosis markedly altered from that observed in multiple myeloma. In normal individuals the ratio of the plasma cells to the total WBC count is always constant (about 1-2%). Therefore, plasma cell content over 2% was evaluated as abnormal. 100% of these plasma cells are normally mature, plasmoblasts or proplasmocytes are never identified in normal smears. Our results are summarized in Fig. IV/7.

The following typical morphological changes of the bone marrow plasma cells were found: though most of the cells were unchanged, the morphology of some plasmocytes resembled the myeloid cells. The similarity was so striking that if the diagnosis should be based only on the microscopic findings of several visual fields, a subclinical myeloma would be surely diagnosed. We found high incidence of endomitotic activity: the nuclei were of different size and density. The nucleus/plasma ratio in the mononuclear cells showed typical alterations: large centrally localized nuclei were surrounded with a

*Distribution of the cases studied according to the bone marrow plasma cell content. (In the myelogram 1000 WBC were evaluated)*

| | | | |
|---|---|---|---|
| 1-2% | normal level | 25 cases | 21.73% |
| 3-4% | slightly elevated | 30 cases | 26.08% |
| 5-6% | elevated | 25 cases | 21.73% |
| 7-10% | markedly elevated | 25 cases | 21.73% |
| > 10% | significantly elevated | 10 cases | 8.69% |

**Fig. IV/7**

narrow strip of cytoplasm without vacuoles. In three cases the cells contained "bunch of grapes", being filled with protein globules. These cells differed from the well known Mott cells ("blackberry" cells) filled with the Russel's bodies. They were first described in Hungarian literature by Hollán and Solti (65); supporting the data of Stich, the authors regard these cells as typical of myeloma (122) (Fig.IV/8).

Morphological signs indicating active secretory activity of the plasma cells were also identified: marked vacuolization, formation of pseudopodium-like processes, honey-comb like cloudy cytoplasm with lumpy inclusions.

- We never observed monoclonal gammopathies in our patients with liver pathology;
- In contrast to the findings of Jarrold and Vilter (72), we failed to observe quantitative relationship between the cytomorphology of the bone marrow and the immunoglobulin spectrum.
- Apart from some similarities reflecting the enhanced protein synthesis, the bone marrow findings in cirrhosis greatly differ from those in myeloma. Still the above mentioned similarities can sometimes be misleading.

Immunoglobulin spectra of the patients with liver pathology were studied with TIEF and gel-diffusion methods. Though these techniques are widely used for already 25 years, still no concepts exist for a uniform evaluation of the results (14, 50, 52, 60, 62, 106).

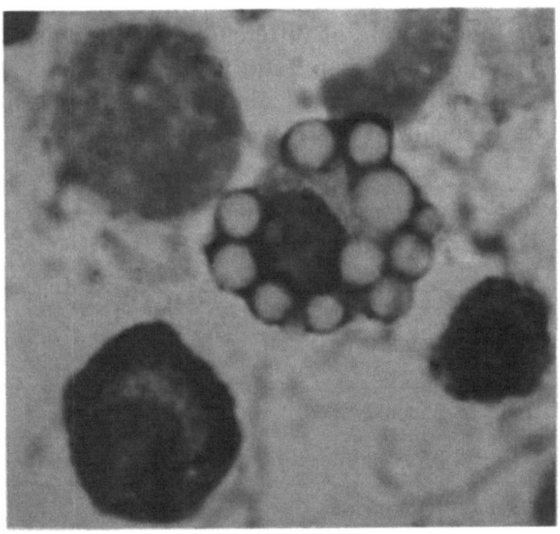

Fig. IV/8

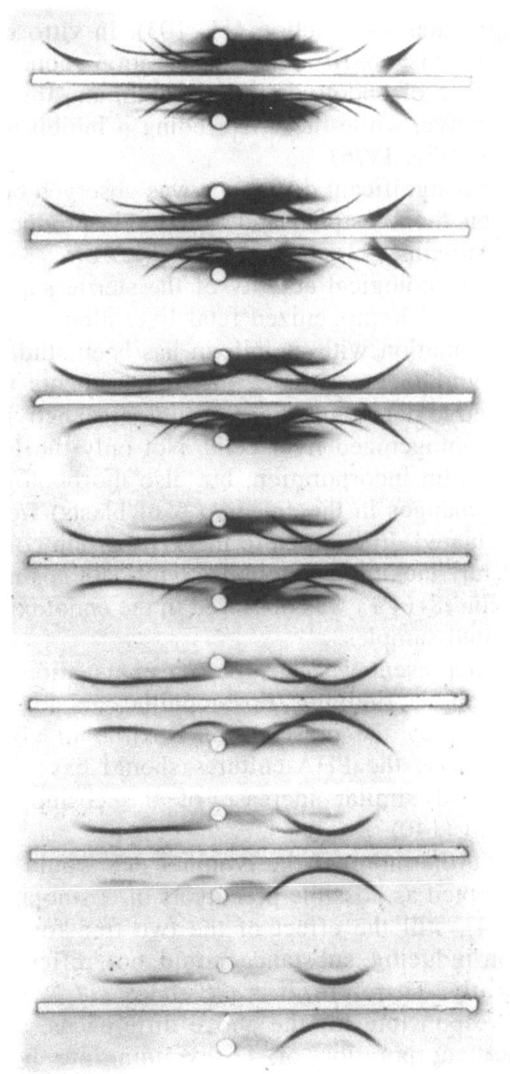

Fig. IV/9

Beside quantitative changes, a typical split of the slowest IgG fraction appears on the immunoelectrophoretic tracings (double-tailed IgG precipitation arc) in the severe cases (68) (Fig. IV/9).

Before a detailed evaluation of our quantitative findings, a Table of the most important diagnostic values in liver diseases based on the data of the literature is presented (Table IV/I).

The most relevant data are specially emphasized:

- After hepatitis the immunoglobulin levels return to their normal values; elevated levels indicate chronic hepatitis or development of cirrhosis (130).

| | IgA mg/100 ml | IgD mg/100 ml | IgG mg/100 ml | IgM mg/100 ml | IgE mg/100 ml | |
|---|---|---|---|---|---|---|
| Hepatitis (A, B) | 130 300 | 280 552 | 690 2185 | – | – | |
| | 1, 2, 3, 4, 5, 6, 7 | 1, 2, 3, 4, 5, 6, 7 | 1, 2, 3, 4, 5, 6, 7 | | | |
| Hepatitis chr. | 154 302 | 278 616 | 1750 6080 | – | – | |
| | 1, 2, 5, 8 | 1, 2, 5, 8 | 1, 2, 5, 8 | | | |
| Hepatitis chr. Aggressiva | 250 320 | 600 950 | 3000 4000 | 10 30 | – | |
| | 2, 8, 9 | 2, 8, 9 | 2, 8, 9 | 9 | | |
| Cirrhosis sec. Laënnec | 432 | 999 | 2866 | 21 | 0,35 | Average values |
| | 10, 11, 12, 13, 14, 15, 16 | 11, 12, 13, 14, 15, 16 | 11, 12, 13, 14, 15, 16 | 10 | 10 | |
| Cirrhosis Alcoholica | 120 396 | 652 2000 | 1160 4400 | 10 30 | – | |
| | 1, 2, 9, 20 | 1, 2, 9, 20 | 1, 2, 9, 20 | 9 | | |
| Cirrhosis Cryptogenica | 380 | 700 | 2500 3000 | 10 30 | – | Average values |
| | 3, 9, 14 | 3, 9, 14 | 3, 9, 14 | 3, 9, 14 | | |
| Cirrhosis Biliaris sec. | 92 350 | 400 900 | 1400 3000 | 10 30 | | |
| | 1, 8, 9, 17, 18, 19 | 1, 8, 9, 17, 18, 19 | 1, 8, 9, 17, 18, 19 | 9 | | |
| Cirrhosis Biliaris sec. | 180 | 400 | 1500 2000 | 10 30 | | Average values |
| | 9, 21 | 9, 21 | 9, 21 | 9 | | |
| Normal values | 39 117 | 110 180 | 900 1500 | 0,3 20 | 0,03 | |
| | 22, 23, 24 | 22, 23, 25 | 8 | 15 | 10 | |

**Table IV/I**


1. Roser, H.
Quantitative Immunoglobulin-Bestimmungen bei Lebererkrankungen.
Münch. med. Wschr. 112, 1392. 1970.
2. Akdamar, K., Epps, A. C., Maumus, L. T., Sparks, R. D.
Immunoglobulin changes in liver disease.
Ann. N. Y. Acad. Sci. 197, 101. 1972.
3. Hirayama, S., Tominaga, D., Koga, S., Toda, T., Murai, J.
Proporties of anti-liver antibody in sera from patients with and without liver disease.
Clin. Chim. Acta. 19, 37. 1968.
4. Moghadam, F., Lamers, A., Reisenburg, J., Kendel, M. I.
Significance of immunoglobulins to the diagnosis of liver and assesment of evolution of liver diseases.
Klin. Wschr. 47, 129. 1969.
5., 13. Tomasi, T. B., Tisdale, W. A.
Serum gamma-globulins in acute chronic liver disease.
Nature. 201, 834. 1964.
6. Gleichman, E. Deicher, H.
Differenzierung der Hypergammaglobulinämie bei entzündlichen Lebererkrankungen.
Klin. Wschr. 45, 684. 1967.
7., 16. Gleichmann, E., Deicher, H.
Quantitative Immunoglobulin-Bestimmungen im Serum bei entzündlichen Lebererkrankungen.
I. Normalwerte und Untersuchungen im Verlauf der akuten Hepatitis.
Klin. Wschr. 46, 171. 1968.
II. Chronisch entzündlichen Leberkrankheiten.
Leberkrankheiten.
Klin. Wschr. 46, 793. 1968.
Acta hepato-splenol. 14, 152. 1967/b.
8. Schumacher, K., Gross, R.
Das quantitative Verhalten der Immunoglobuline bei chronisch entzündlichen Lebererkrankungen.
Klin. Wschr. 47, 123. 1969.
9. Feizi, T.
Immunoglobulins in chronic liver disease.
Gut. 9, 193. 1968.
10. Heiner, D. C., Rose, B.
Elevated levels of IgE in conditons other then classical allergy.
J. Allergy. 45, 30. 1970.
11., 12. Bonomo, L., Gillardi, U.
Broad-band hypergammaglobulinemia and chronic liver disease.
I. Diagnostic and prognostic meaning of the elektrophoretic patterns.
Acta hepato-splenol. 14, 142. 1967/a.
13. Bonomo, L., Gillardi, U.
Broad-band hypergammaglobulinemia and chronic liver disease.
II. Immunoglobulin levels in liver cirrhosis.
Acta hepato-splenol 14, 152. 1967/b.
14., 18. McKelvey, E. M., Fahey, J. L.
Immunoglobulin changes in disease: quantitation on the basis of heavy polypeptide chains, IgG ($\gamma$G), IgA ($\gamma$A), and IgM ($\gamma$M), and of light polypeptide chains, type $\kappa$ (I) and type $\lambda$ (II).
J. clin. Invest. 44, 1778. 1965.
15., 26. Rowe, D. S., Fahey, J. L.
A new class of human immunoglobulins.
I. A unique myeloma protein.
J. exp. Med. 121, 171. 1965/a.
16. Rowe, D. S., Fahey, J. L.
A new class of immunoglobins.
II. Normal serum IgD.
J. exp. Med. 121, 185. 1965/b.
17. Paronetto, F., Schaffner, F., Popper, H.
Immunocytochemical and serologic observations in primary biliary cirrhosis.
New Eng. J. Med. 271, 1123. 1964.
19. Hobbs, J. R.
Primary biliary cirrhosis: positive antibody tests associated with increased immunoglobulin IgM.
Proc. Roy. Med. 59, 568. 1966.
20. Lee, F. I.
Immunoglobulins in viral hepatitis and active alcoholic liver disease.
Lancet. ii. 1043. 1965.
21. Hobbs, J. R.
Serum proteins in liver disease.
Proc. Roy. Soc. Med. 60, 1250. 1967.
22. Chodirker, W. B., Tomasi, R. B.
Gamma-globulins: quantitative relationship in human serum and non-vascular fluids.
Science. 142, 1080. 1963.

23. Fahey, J. L., Lawrence, M. E.
Quantative determination of 6,6 S gamma-globulin, ß$_2$A-globulins and $\gamma_1$-macroglobulins in human serum.
J. Immunol. 91, 597. 1963.
24. Walton, K. W., Rowe, D. S., Soothill, J., Stanworth, D. R.
An investigation of methods of isolation of ß$_2$M globulin (syn: Iota protein, 19 S $\gamma$-globulin, $\gamma_1$ macroglobulin, ß$_2$M globulin) and its association with iso-agglutinin activity, together with preliminary observations on other macroglobulins of slow elektrophoretic mobility in normal human serum.
Immunology. 6, 305. 1963.
25. Heremans, J. F.
Les globulines sériques du système gamma.
Arscia, Brussels and Masson. Paris. 1960.

- High immunoglobulin levels do not reflect increased activity of the immune system (12).
- No correlation exists between the immunoglobulin concentration and the extent of hepatic derangement (blood level of ALT, AST etc.) (87)
- Bile duct obstruction is doubtful if the symptoms of chronic cholelithiasis are accompanied with increased levels of IgM (119).
- Marked elevation of IgA is typical of alcoholic cirrhosis.
- The highest IgG concentrations are measured in active forms of chr. hepatitis (lupoid hepatitis) (32).

Evaluation of our quantitative analysis has lead to formation of new diagnostic groups based on the further histological classification of cirrhosis.

WHO standards have been used for the quantitative assessment of the IgM, IgA and IgG classes.

Standard serum contains the following amounts of immunoglobulins:

*IgM = 60 mg%
IgA = 100 mg%
IgG = 2940 mg%

Since the sera studied contained immunoglobulins at 5-9-6-fold higher concentrations, compared to the WHO standards, the control sera could be only indirectly used for the construction of the calibration curve.

In cases of Laennec cirrhosis the IgM concentration is significantly elevated, in general, macroglobulin titers in all the patients studied markedly exceeded those of the normal individuals (Fig. IV/10).

Similar to the IgM levels, the IgA concentration also exceeded the normal values supporting the data of the literature. The most pronounced difference occurred in alcoholic cirrhosis (Fig. IV/11).

High titers of IgG also indicated the stimulation of polyclonal immunoglobulin synthesis. The highest IgG values were obtained in cases of active chronic hepatitis (Fig. IV/12).

No definite results could be obtained with the IgD levels. On one hand their values remained within the normal range presented in literature, on the other hand the precipitin rings were too small to be accurately evaluated (Fig. IV/13).

*In summary,* neither of the mentioned changes can be considered as specific. The levels of immunoglobulins and the ratios between the different classes provide no adequate information on the activity or the progress of the liver diseases. Moreover, as other diseases might also interfere with liver pathology, the final evaluation should be based only on the complex analysis of all findings. These conclusions are in agreement with the data of Hobbs and Sherlock, that the quantitative assessment of immunoglobulins alone, though *helpful* in the diagnostics of liver pathology (64), is insufficient for an accurate diagnosis (119).

Our results suggest that TIEF analysis is more informative, compared to the quantitative assessment of 3 or 4 immunoglobulins: the former method simultaneously shows the amount of the liver and the B cell-derived proteins, indicating the activity (stage) of the disease. Therefore, clinically TIEF analysis is a more informative method for accurate evaluation of certain proteins.

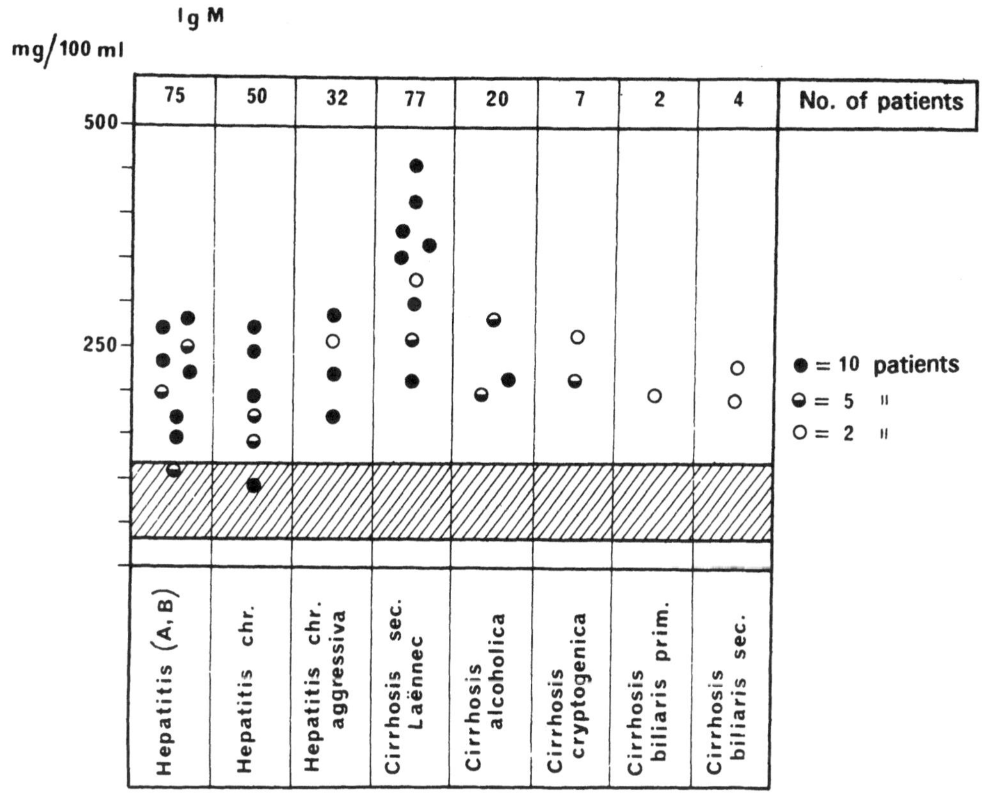


Fig. IV/10

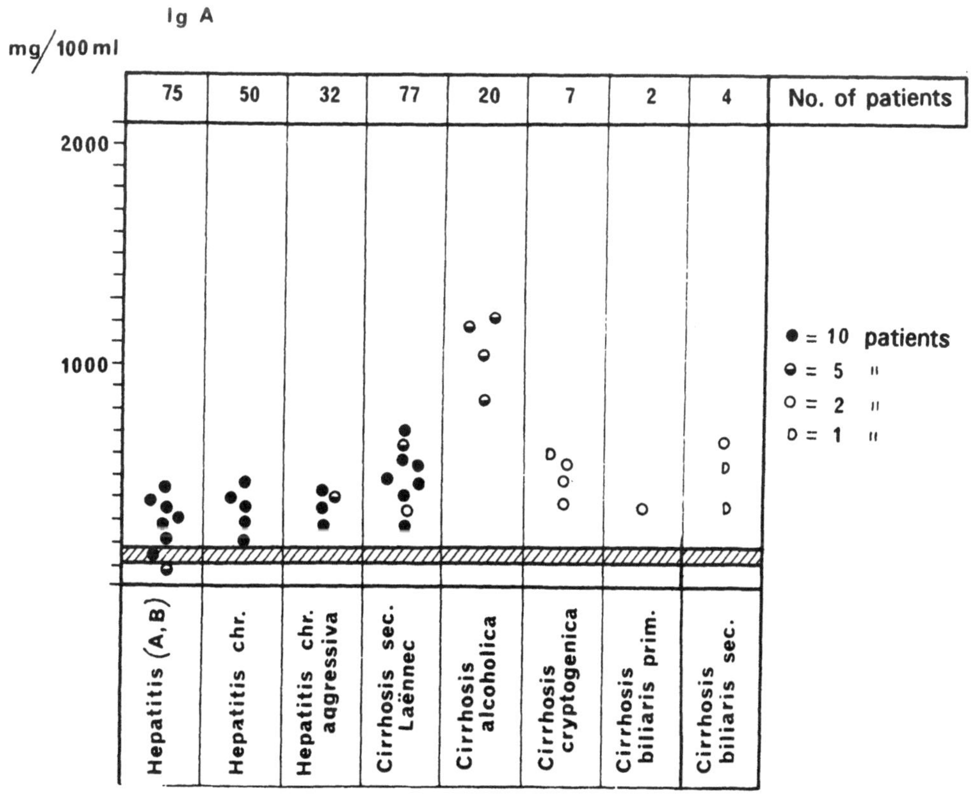


Fig. IV/11

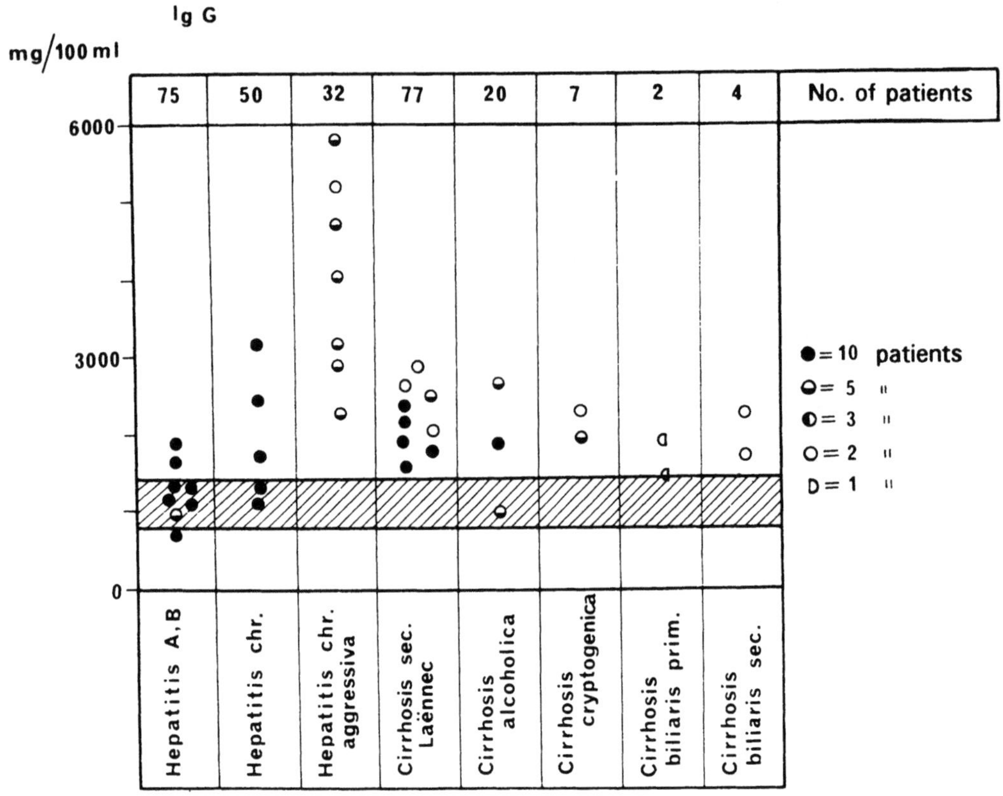


Fig. IV/12

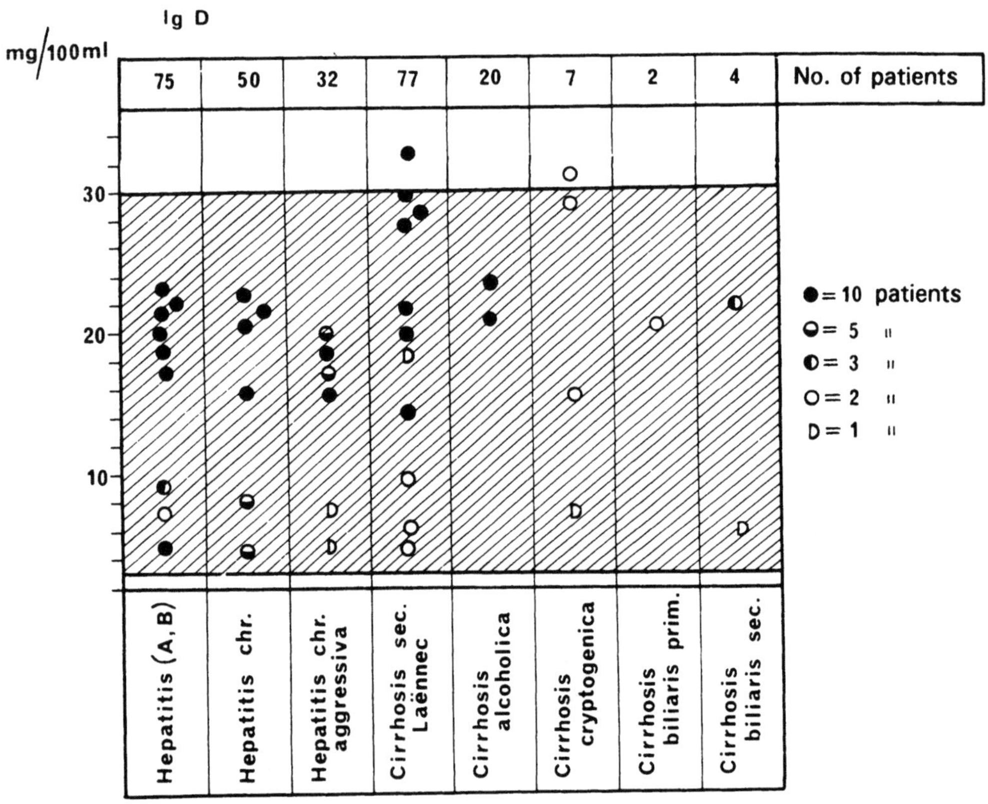


Fig. IV/13

## References


1. Adorini, L.
Basic strategies of the immune system in the regulation of antibody response.
La Ricerca. 10, 2. 1980.
2. Adorini, L., Eipert, E. F., Couderc, J.
Immunoadsorbent removal of antibodies produced during in vitro culture relieves inhibition of the response.
Cell. Immunol. 47, 356. 1979.
3. Akdamar, K., Epps, A. C., Maumus, L. T., Sparks, R. D.
Immunoglobulin changes in liver disease.
Ann. N. Y. Acad. Sci. 197, 101. 1972.
4. Andersson, J., Sjöberg, O., Möller, G.
Induction of immunoglobulin and antibody synthesis in vitro by lipopolysaccharide.
Eur. J. Immunol. 2, 349. 1972.
5. Andersson, J., Sjöberg, O., Möller, G.
Mitogens as probes for immunocyte activation and cellular cooperation.
Transplant. Rev. 11, 131. 1972.
6. Argyris, B. F.
Role of suppressor cells in immunological maturation.
Transplantation. 31, 334. 1981.
7. Arrigoni – Martelli, E.
Developments in drugs enhancing the immune responser.
Meth. and Find. Exptl. Clin. Pharmacol. 3, 247. 1981.
8. Bach, G.
Ein Fall von subakuter Hepatitis mit Übergang in Cirrhose, Gamma-G-Paraproteinaemia und positiver Rheumaserologie.
Wien. med. Wschr. 120, 726. 1970.
9. Bauer, H., Paronetto, F., Porro, R. F., Einheber, A.
Gamma-globulin elevation in chronic liver injury of germfree and conventionalised rats.
Fed. Proc. 25, 667. 1966.
10. Bellanti, J. A.
Immunology.
JAMA 247, 2957. 1982.
11. Bergman, S., Weibull, C.
Effects of endotoxin on tissue culture cells.
Acta Path. Microbiol. Scand. 77, 698. 1969.
12. Bjorneboe, M., Jensen, K. B., Scheibel, I., Thomsen, A. Chr., Bentzon, M. W.
Tetanus antitoxin production and gamma-globulin levels in patients with cirrhosis of the liver.
Acta. Med. Scand. 188, 541. 1970.
13. Bjorneboe, M., Proytz, H., Örekov, F.
Antibodies to intestinal microbes in serum of patients with cirrhosis of the liver.
Lancet. i. 58. 1972.
14. Boulet, P., Mirouze, P., Barjon, M., Robinet, M.
L'immunoélectrophorèse sérique des cirrhoses hépatiques.
Sem. Ho. Paris. 37, 415. 1961.
15. Bucher, N. L. R., Schott, J. F., Aub, J. C.
Regeneration of the liver in parabiotic rats.
Cancer Res. 10, 207. 1950.
16. Bufano, M.
Über die Pathogenese der Dysproteinämien.
Münch. med. Wschr. 32, 1448. 1970.
17. Ceuppens, J., Goodwin, J. S.
Prostaglandins and the Immune Response to Cancer.
Anticancer Research. 1, 71. 1981.
18. Ceuppens, J. L., Goodwin, J. S.
Endogenous prostaglandin $E_2$ enhances polyclonal immunoglobulin production by tonically inhibiting T suppressor cell activity.
Cellular immunology. 70, 41. 1982.
19. Chiorazzi, N., Fu, S. M., Kundel, H. G.
Induction of polyclonal antibody synthesis by human allogeneic and autologous helper factors.
J. Exp. med. 149, 1543. 1979.
20. Coburg, A. J., Seifert, E., Deicker, H.
Zur Differenzierung immunoglobulinhaltiger Zellen in der Leber bei entzündlichen Lebererkrankungen.
Acta hepato-splenol. 18, 1. 1971.
21. Cohen, S., Ohta, G., Singer, E. J., Rosenfield, R. E., Permann, E., Popper, H.
The demonstration of gamma-globulin in livers with postnecrotic cirrhosis and fulminating viral hepatitis, using the flourescent antibody technique.
Amer. J. Path. 35, 685. 1959.
22. Cohen, S., Ohta, G., Singer, E. J., Popper, H.
Immunochemical study of gamma-globulin in liver in hepatitis and postnecrotic cirrhosis.
J. exp. Med. 111, 285. 1960.
23. Coons, A. H., Kaplan, M. H.
Localisation of antigen in tissue cells.
J. exp. med. 91, 1. 1950.
24. Coutinho, A., Möller, G.
Immune activation of B cells evidence for „one nonspecific triggering signal" not delivered by the Ig receptors.
Scand. J. Immunol. 3, 133. 1974.
25. Davies, A. J., Carter, R. L., Leuchars, E., Wallis, V., Dietrich, F. M.
The morphology of immune reactions in normal thymectomised and reconstitued mice.
III. Response to bacterial antigens: salmonellar flagellar and pneumococcal polysaccharide.
Immunology. 19, 945. 1970.
26. Dutton, R. W., Eady, J. D.
An in vitro system for the study of the mechanism of antigenic stimulation in the secondary response.
Immunology. 7, 40. 1964.
27. Eisbach, M., Roubinian, J. R., Talal, N.
Lipopolysaccharide induction of IgM antibodies to polyadenylic acid in normal mice.
J. Immunol. 120, 1856. 1978.
28. Falkoff, R. M., Scavulli, J. F., Dutton, R. W.
Independent analysis of T helper and T killer cell function in young and old NZB mice.
J. Immunol. 121, 897. 1978.
29. Farrar, W. R., Corwin, L. M.
The essential role of the liver in detoxification of endotoxin.
Ann. N. Y. Acad. Sci. 133, 668. 1966.
30. Farrar, W. E.
In „La structure et les effects biologiques des products bacteriens provevant de germes Gramnegativs".
Cent. Mat. Rech. Sci. 455. Paris 1969.
31. Fauci, A. S., Steinberg, A. D., Haynes, B. F., Whalen, G.
Immunoregulatory abberations in systemic lupus erythematosus.
J. Immunol. 121, 1473. 1978.
32. Feizi, T.
Immunoglobulins in chronic liver disease.
Gut. 9, 193. 1968.
33. Ferraris, V. A., DeRubertis, F. R.
Release of prostaglandin by mitogen – and antigen – stimulated leukocytes in culture.
J. Clin. Invest. 54, 378. 1974.
34. Fidalgo, B. V., Najjar, V. A., Zukoski, C. F., Katayama, Y.
The physiological role of the lymphoid system.
II. Erythrophilic gamma-globulin and the survival of the erythrocyte.
Nat. Acad. Sci. 57, 665. 1967.
35. Fidalgo, B. V., Najjar, V. A.
The physiological role of the lymphoid system.
III. Leucophilic globulin and the phagocytic activity of the polymorphonuclear leucocyte.
Proc. Nat. Acad. Sci. 57, 957. 1967/a.
36. Fidalgo, B. V., Najjar, V. A.
The physiological role of the lymphoid system.
IV. The stimulatory effect of leucophilic globulin (leucokinen) an the phagocytic activity of human polymorphonuclear leucocyte.
Biochemistry. 6, 3386. 1967/b.
37. Fidalgo, B. V., Katayama, Y., Najjar, V. A.

The physiological role of the lymphoid system.
V. The binding of autologous (erythrophilic) gamma-globulin to human red blood cells.
Biochemistry. 6, 3378. 1967/c.
38. Fink, C. W., Miller, W. E., Dorward, B., Lospalutto, J.
The formation of macroglobulin antibodies. II. Studies on neonatal infants and older children.
J. Clin. Invest. 41, 1422. 1962.
39. Fischer, A., Kerekes E., Bona, A.
Májregenerációt serkentő humoralis tényező kimutatása.
Orvosi Hetilap. 102, 66. 1961.
40. Floch, M. H., Katz, J., Conn, H. O.
Qualitative and quantitative relationsips of the fecal flora in cirrhotic patients wirt portal systemic encephalopathy and following portacaval anastomosis.
Gastroenterol. 59, 70. 1970.
41. Fournié, G. J., Lambert, P. H., Miescher, P. A.
Release of DNA in circulating blood and induction of anti-DNA antibodies after injection of bacterial lipopolysaccharides.
J. exp. Med. 140, 1189. 1974.
42. Fudenberg, H. H.
Are autoimmune diseases immunologic deficiency states?
Hosp. Pract. 3, 43. 1968.
43. Fudenberg, H. H.
Genetically determined immune deficiency as the predisposing cause of „autoimmunity" and lymphoid neoplasia.
Amer. J. Med. 51, 295. 1971.
44. Fudenberg, H. H.
Are autoimmune diseases and malignancy due to selective T-cell deficiencies? Critical factors.
Cancer. Immunol. 10, 179. 1975.
45. Fudenberg, H. H., Wells, J. V.
The paradox of immunosuppression: T-cell deficiency as the cause of autoimmunity.
In: Infection and immunology in rheuma diseases.
Ed.: by Dumonde, D. C.
Blackwell Scientific Publications, Oxford 549. 1976.
46. Fudenberg, H. H.
Molecular theology, immunophilosophy, and autoimmune disease.
Scand. J. Immunol. 12, 459. 1980.
47. Furchgott, R. E., Ponder, E.
Disc-sphere transformation in mammalian red cells.
II. The nature of the antisphering factor.
J. Exptl. Biol. 17, 117. 1940.
48. Gerry, I., Krüger, J., Spiesel, S. Z.
Stimulation of B-lymphocytes by endotoxin. Reactions of thymus-deprived mice and karyotipic analysis of dividing cells in mice bearing $T_6$ thymus grafts.
J. Immunol. 108, 1088. 1972.
49. Gifford, R. R. M., Ferguson, R. N., Voss, B. V.
Cimetidine reduction of tumor formation in mice.
Lancet, i, 638. 1981.
50. Gleichmann, E., Deicher, H.
Differenzierung der Hypergammaglobulinämie bei entzündlichen Lebererkrankungen.
Klin. Wschr. 45, 684. 1967.
51. Gleichmann, E., Deicher, H.
Quantitative Immunoglobulin-Bestimmungen im Serum bei entzündlichen Lebererkrankungen.
I. Normalwerte und Untersuchungen im Verlauf der akuten Hepatitis.
Klin. Wschr. 46, 171. 1968.
52. Gleichmann, E., Deicher, H.
Quantitative immunoglobulin-Bestimmungen im Serum bei entzündlichen Leberkrankheiten.
II. Chronisch entzündlichen Leberkrankheiten.
Klin. Wschr. 46, 793. 1968.
53. Goodman, M. G., Weigle, W. O.
Induction of immunoglobulin secretion by a simple nucleoside derivative.
J. of Immunol. 128, 2399. 1982.
54. Goodwin, J. S., Messner, R. P., Bankhurst, A. D., Peake, G. T., Saiki, J., Williams, R. C. Jr.
Prostaglandinproducing suppressor cells in Hodgkin's Disease.
N. Engl. J. Med. 297, 963. 1977.
55. Goodwin, J. S., DcHoratius, R., Israel, H., Peake, G. T., Messner, R. P.
Suppressor cell function is sarcoidosis.
Ann. Intern. Med. 90, 169. 1979.
56. Guardia, J., Pedreira, J. D., Vidal, M. T., Roca, A.
Malignant lymphoma with plasmocytoid differentation and polyclonal gammopathy.
Acta Haematol 55, 346. 1976.
57. Gustaffson, B. E., Laurell, C. B.
Gamma-globulin production in germfree rats after bacterial contamination.
J. Exp. Med. 110, 675. 1959.
58. Hammarström, L., Smith, E., Primi, D., Möller, G.
Induction of autoantibodies to red blood cells by polyclonal B-cell activators.
Nature. 263, 60. 1976
59. Hammarström, L., Bird, A. G., Smith, C. I. E.
Mitogenic activation of human lymphocytes: a protein A plaque assay evaluation of polyclonal B-cell activators.
Scand. J. Immunol. 11, 1. 1980.
60. Hartmann, G., Burtin, P., Grabar, P., Fauvert, P.
L'analyse immuno-electrophoretique des serum de malades atteints d'affections hepatiques.
C. R. Acad. Sci. (Paris) 243, 1937. 1956.
61. Hashim, G. A.
Myelin basic protein: structure, function and antigenic determinants.
Immunol. Rev. 39, 60. 1978.
62. Heremans, J. F.
Quelques données de l'immunoélectrophorèse en pathologie hépatique.
Acta. Gastroent. Belg. 22, 582, 1959.
63. Hitzig, W. H.
Clinical effects of anomalous serum globulins.
Clin. Prot. Chem. Basel. 1, 72. 1968.
64. Hobbs, J. R.
Serum proteins in liver disease.
Proc. Roy. Soc. Med. 60, 1250. 1967.
65. Hollán, Zs., Solti, V.
A serum fehérjekép dynamikus változásával és lipaemiával járó „Szőlőfürt-sejtes" plasmocytoma.
Haemat. Hung. 3, 157. 1963.
66. Horváth. É., Biliczky, F., Jákó, J., Hámori, É.
Some problems of humoral regulation of liver regeneration.
Acta Biol. Acad. Sci. Hung. 5, 47. 1963.
67. Jákó, J., Szirtes, M., Szigeti, I.
Latex vizsgálatok különböző betegségekben, s ezek positivitásának összefüggése az immunoglobulinokkal.
Rheum. Baln. Allerg. 2, 99. 1966.
68. Jákó, J., Tóth, B., Szigeti, I.
A titratios immunelectrophoresis klinikai alkalmazása.
Magyar Belorv. Arch. 20, 120. 1967.
69. Jákó, J., Sas, G.
Vérfehérje változások chronicus májbetegségekben: a „közvetlen" és „közvetett" májfunctios próbák.
Orv. Hetil. 109, 1526. 1968.
70. Jákó, J., Sas, G.
The effect of endotoxin on blood proteins.
Ann. Immunol. Hung. 12, 79. 1968.
71. Jákó, J.
Fehérjeanyagcsere vizsgálatok májbetegségekben.
Kandidátusi dissertatio 1973.
72. Jarrold, T., Vilter, R. W.
Hematologic observation in patients with chronic hepatic insufficiency. Sternal marrow morphology and bone marrow plasmacytosis.
J. Clin. Invest. 28, 286. 1949.
73. Jerne, N. K.

Towards a network theory of the immune system.
Ann. Immunol. (Inst. Pasteur) 1250, 373. 1974.
74. Katz, P., Fauci, A. S.
Inhibition of polyclonal B-cell activation by suppressor monocytes in patients with sarcoidosis.
Clin. Exp. Immunol. 32, 554. 1978.
75. Kelly, R. H., Hardy, T. J.
Multiple clonally-restricted immunoglobulins in human sera: disease associations.
Inl. Arch. Allergy Appl. Immun. 69, 56. 1982.
76. Kennan, A. L.
Changes in plasma proteins after portocaval shunt in the rat.
Proc. Soc. exp. Biol. 116, 453. 1964.
77. Kim, Y. B., Bradley, S. G., Watson, D. W.
IV. the role of antigen elimination in the true primary immune response in germfree, colostrum-deprimed piglets.
J. Immunol. 99, 320. 1967.
78. Lahiri, A. K.
Dissertation in partial fulfilment of the requirement for the degree of Doctor of Phyilosophy, June, 1968. Vanderbilt University, Mashville, Tenn.
Cit: Najjar, V. A.: The physiolocical role of the lymphoid system.
Leb. Med. J. 22, 401. 1969.
79. Manning, D. D.
In vivo immune stimulation of mice by anti-$\mu$ antibodies: generation of high serum levels of a low molecular weight IgM product.
Cell. Immunol. 50, 348. 1980.
80. McCombs, C., Hom, J., Talai, N., Michel, R.
Functional deficiency of splenic adherent cells in NZ black mice.
J. Immunol. Rev. 115, 1695. 1975.
81. McHough, Y. E., Bonavida, B.
Autoreactiv antibody-forming cells directed against thymocytes and thymus-derived lymphocytes.
J. Immunol. 121, 1090. 1978.
82. Melchers, F., Andersson, J., Lernhardt, W., Schreier, M. H.
Role of surface-bound immunoglobulin molecular in regulating the replication an maturation to immunoglobulin secretion of B lymphocytes.
Immunol. Rev. 52, 89. 1980.
83. Metcalf, D.
Role of mercaptoethanol and endotoxin in stimulating B-lymphocyte colony formation in vitro.
J. Immunol. 166, 635. 1976.
84. Miller, J. A. F. P.
Immunity in the foetus and the newborn.
Brit. med. Bulletin. 22, 21. 1966.
85. Miller, K. B., Salem, D.
Immune regulatory abnormalities produced by procainamid.
Amer. J. Med. 73, 487. 1982.
86. Miller, L. L., Bly, C. G., Bale, W. F.
Plasma and tissue proteins produced by non hepatic rat organs as studied with lysine E-$C^{14}$. Gamma-globulins the chief plasma protein fraction produced by non-hepatic tissues.
J. Exp. Med. 99, 133. 1954.
87. Moghadam, F., Lamers, A., Reisenburg, J., Kendel, M. I.
Significance of immunoglobulins to the diagnosis of liver and assesment of evolution of liver diseases.
Klin. Wschr. 47, 129. 1969.
88. Morito, T., Bankhurst, A. D., Williams, R. C.
Studies on the pharmacologic manipulation of suppressor cells associated with impaired immunoglobulin production.
J. Lab. Clin. Med. 96, 232. 1980.
89. Möller, G.
Autoimmunity and self-nonself discrimination.
Transplant. Rev. 31, 1. 1976.
90. Möller, G.
Activation of antibody synthesis in human B lymphocytes.
Immunol. Rev. 45, 1. 1979.
91. Muckerheide, A., Pesce, A. J., Michael, G. J.
Immunosuppressive properties of a peptic fragment of BSA.
J. Immunol. 199, 1340. 1977.
92. Najjar, V. A.
The physiological role of the lymphoid system.
Lab. Med. J. 22, 4. 1969.
93. Nowotny, A.
Recent problems of endotoxin research.
An. Immunol. Hung. 12, 15. 1968.
94. Oppenheim, J. J., Perry, S.
Effects of endotoxins on cultured leukocytes.
Proc. Sec. exp. Biol. Med. 118, 1014. 1965.
95. Osband, M. E., Jinshen, Y., Shlesinger, M., Brown, A., Hamilton, D., Cohen, E., Lavin, P. McAffrey, R.
Successful tumor immunotherapy with Cimetidine in mice.
Lancet. i. 636. 1981.
96. Ossermann, E. F., Takatsuki, K.
The plasma proteins in liver disease.
Med. clin. N. Amer. 47, 679. 1963.
97. Palmer, D. W., Dauphinee, M. J., Hurphy, E., Talal, N.
Hyperactive T cell function in young NZB mice: Increased proliferative responses to allogenetic cells.
Clin. Exp. Immunol. 23, 578. 1976.
98. Paronetto, E., Rubin, E., Popper, H.
Local formation of $\gamma$-globulin in the diseased liver and its relation to hepatic necrosis.
Lab. Invest. 11, 150. 1962.
99. Paronetto, E., Schaffner, F., Popper, H.
Immunocytochemical and serologic observations in primary biliary cirrhosis.
New Eng. J. Med. 271, 1123. 1964.
100. Peavy, D. L., Adler, W. H., Smith, R. T.
The mitogenic effects of endotoxin and staphylococcal enterotoxin B on mouse spleen cells and human peripherial lymphocytes.
J. Immunol. 105, 1453. 1970.
101. Peterman, M. L.
Alterations in plasma proteins patterns in disease.
In: The plasma proteins.
Ed.: Putnam, R. W.
Academic press New York. 2, 310. 1960.
102. Pierce, C. W.
Immune responses in vitro. II. Suppression of the immune response in vitro by specific antibody.
J. exp. Med. 143, 672. 1976.
103. Plaut, E. M., Goldman, J. K.
Inhibition of substrate oxidation by endotoxin in vitro.
Proc. Soc. exp. Biol. Med. 133, 433. 1970.
104. Primi, D., Hammarström, L., Smith, C. I. E., Möller, G.
Characterization of selfreactive B cells by polyclonal B-cell activators.
J. exp. Med. 145, 21. 1977.
105. Raeder, R., Preimer, E. H., Pansky, B., Senitzer, D.
Response of peripheral blood lymphocytes from the (NZ B) (NZ W) $F_1$ mouse to phytohemagglutinin.
Clin. Immunol. Immunopathol. 6, 174. 1976.
106. Rebeyrotte, P., Baujat, J. P., Labbe, J. P., Koutsoukos, A.
Application de la technique d'immunoelectrophorese bi-dimensionelle au probleme des cirrhoses.
Pathol. biol. 18, 653. 1970.
107. Reinherz, E. L., Schlossman, S. F.
The differentiation and function of human T lymphocytes.
Cell. 19, 821. 1980.
108. Reinherz, E. L., Schlossman, S. F.
The characterization and function of human immunoregulatory T lymphocyte subsets.
Immunol. Today. 2, 69. 1981.
109. Rosen, A., Gergely, P., Jondal, M., Klein, G., Britton, S.
Polyclonal Ig production after Epstein-Barr virus infection of human lymphocytes in vitro.
Nature. 267, 52. 1977.
110. Rosenthal, A. S., Barcinski, M. A., Blake, J. T.
Determinant selection is a macrophage dependent immune response gene function.
Nature. 267, 156. 1977.
111. Rowley, D. A., Fitch, F. W., Axelrad, M. A., Pierce, C. W.
The immune response suppressed by specific antibody.

Immunology. 16, 549. 1969.
112. Sagawa, A., Abdov, N. I.
Suppressor cell dysfunction in systemic lupus erythematosus. Cells involved and in vitro correlation.
J. Clin. Invest. 62, 789. 1978.
113. Schenkein, I., Uhr, J. W.
Immunoglobulin synthesis and secretion.
I. Biosynthetic studies of the addition of the carbohydrate moieties.
J. Cell. Biol. 46, 42. 1970.
114. Schneiberg, K., Jonecko, A., Bartnikova, W., Szymaniec, S.
Lymphocyte as a myeloid cell precursor.
Japan, J. Exp. Med. 39, 331. 1969.
115. Schneiderbaur, A., Narbeshuber, H.
The significance of plasmocytosis in the sternal marrow in liver cirrhoses.
Wien. med. Wschr. 113, 370. 1963.
116. Sell, S., Fahey, J.
Relationship between gamma-globulin metabolism low serum gamma-globulin in germfree mice.
J. Immun. 93, 81. 1964.
117. Sercarz, E. E., Yowell, R. L., Turkin, D., Miller, A., Araneo, B. A., Adorini, L.
Different funtcional specificity repertoires for suppressor and helper T cells.
Immunol. Rev. 39, 108. 1978.
118. Shearer, G. M., Rehn, T. G., Garbarino, C. A.
Cell mediated lympholysis of trinitrophenylmodified autologous lymphocytes. Effector cell specificity to modified cell surface components controlled by the H-2D and H-2K serological regions of the major histocompatibility gene products.
J. exp. Med. 141, 1348. 1975.
119. Scherlock, S.
The immunology of liver disease.
Amer. J. Med. 49, 693. 1970.
120. Sinclair, N. R.
Immunoregulation by antibody and antigen-antibody complexes.
Transplant. Proc. 10, 349. 1978.
121. Sjöberg, O.
Antigen-binding cells in mice immune or tolerant to Escherichia Coli polysaccharide.
J. exp. Med. 133, 1015. 1971.
122. Stich, M. H., Swiller, A. I., Morrison, M.
The grape cells of multiple myeloma.
Amer. J. Clin. Path. 25, 601. 1955.
123. Sultzer, B. M., Nilsson, B. S.
PPD tuberculin – a B cell mitogen.
Nature New Biol. 240, 198. 1972.
124. Talal, N.
Autoimmunity: genetic, immunologic, virologic and clinical aspects.
New York. Academic Press. Inc. 1977.
125. Teodorescu, M.
Characterization and role in autoimmune diseases of the polyclonal B-cell activator produced by T-cells – The helper factor.
Immunological Rev. 55, 155. 1981.
126. Thorbecke, G. J.
Some histological and functional aspects of lymphoid tissue in germfree animals.
I. Morpological studies.
Ann. N. Y. Acad. Sci. 78, 237. 1959/a.
127. Thorbecke, G. J.
Some histological and functional aspects of lymphoid tissue in germfree animals.
II. Studies on phagocytosis in vivo.
Ann. N. Y. Acad. Sci. 78, 247. 1959/b.
128. Tholey-Lawson, D. A.
The suppression of Epstein-Barr virus infection in vitro occurs after infection but before transformation of the cell.
J. Immunol. 124, 745. 1980.
129. Tiselius, A., Kabat, E. A.
An electrophoretic study of immune sera and purified antibody preparations.
J. exp. Med. 69, 119. 1939.
130. Tomasi, T. B., Tisdale, W. A.
Serum gamma-globulins in acute chronic liver disease.
Nature. 201, 834. 1964.
131. Tompkins, E. H.
The effects of washing and of serum albumin upon erythrocyte.
I. Upon morphology as seen with phase microscopy and with polarized light.
J. Lab. Clin. Med. 43, 518. 1954/a.
132. Tompkins. E. H.
The effects of washing and of serum albumin upon erythrocytes.
II. Upon the internal movement of erythrocytes.
J. Lab. Clin. Med. 43, 527. 1954/b.
133. Trejo, R. A., Di Luzio, N. R.
Influence of reticuloendothelial system (RES) functional modification on endotoxin detoxication by liver and spleen.
J. Reticuloendth. Soc. 10, 515. 1971.
134. Triger, D. R., Alp, M. H., Wright, R.
Bacterial and dietary antibodies in liver disease.
Lancet. i. 60. 1972.
135. Turkin, D., Sercarz, E. E.
Key antigenic determinants in regulation of the immune response.
Proc. nat. Acad. Sci. 74, 3984. 1977.
136. Uhr, J. W., Dancis, J., Franklin, E. C., Finkelstein, M. S., Lewis, E. W.
The antibody response to bacteriophage ∅X 174 in newborn premature infants.
J. Clin. Invest. 41, 1509. 1962.
137. Waldenström, J.
Studies on conditions associated with disturbed gamma-globulin formation (gammapathies).
Harvey Lect. 56, 211. 1960.
138. Waldenström, J.
The different types of disturbed gamma-globulin formation.
In: Structure and control biosynthesis. Nobel Symposium 3.
Ed.: by Kilander, J.
Interscience Publications. New York. 527, 1967.
139. Waldmann, T. A., Broder, S.
Suppressor cells in the regulation of the immune response.
Prog. Clin. Immunol. 3, 155. 1977.
140. Walls, R. S., Beeson, P. B.
Mechanism of eosinophilia.
IX. Induction of eosinophilia in rats by certain forms of dextran.
Exp. Biol. Med. 140, 689. 1970.
141. Wuhrmann, F., Wunderly, Ch.
The human blood proteins.
Grune-Stratton New York and London. 1960.
142. Zagury, D., Uhr, J. W., Jamieson, J. D., Palade, G. E.
Immunoglobulin synthesis and secretion.
II. Radioautographic studies of sites of addition of carbohydrated moities and intracelluar transport.
J. Cell. Biol. 46, 52. 1970.
143. Zawadzky, Z. A., Edwards, G. A.
Dysimmunoglobulinemia associated with hepatobiliary disorders.
Amer. J. Med. 48, 196. 1970.

# V. Monoclonal gammopathies

This chapter summarizes the experience of the last 25 years in the field the monoclonal gammopathies. This malignant proliferation of plasma cells is accompanied by high plasma (and fluid) levels of a homogeneous protein with normal structure (at least the existing available methods fail to reveal any pathological changes). This protein being a marker of the malignant plasma cells is a monoclonal immunoglobulin. At present the available literature in this field comprises thousands of volumes. Not only individual researchers, but large scientific research centers are studying monoclonal gammopathies. Compared to this extensive basic research, patients with monoclonal gammopathies are followed up and treated at only few clinics; therefore, the case records and the patients' data acquisition is insufficient.

Since monoclonal gammopathy is a rare condition, one might agree with those who claim that extensive research is unnecessary. However the following statistical data strongly oppose the above view: in Hungary there are 25,250 young patients under 35 years suffering from monoclonal gammopathy; of these patients 3,100 have a malignant form of the disease, which requires special care, treatment and nursing.

Apart from their sociological importance, these statistical data also show that only 2-3 patients with gammopathies are generally admitted to the hospitals every year, hence the staff can obtain no adequate experience in the diagnostics and treatment of this disease; furthermore, the data processing in these rare cases can not be used for further scientific purposes.

Centralization of these patients (or at least most of them) at specialized institutes would markedly simplify the adequate organization of the work (76).

The new drugs and medication schemes can only prolong the life of the patients, but not cure the disease. However the quality of life of the patients in the final stage of their disease is significantly improved, compared to that of a decade ago. At present we can only reduce the sufferings of these patients before their death and it is noteworthy.

## Differential diagnostics of monoclonal gammopathies

Despite a significant development of laboratory methods during the last three decades, differential diagnostics of monoclonal gammopathies (MG) still remains a problem. Classification of the disease can be considered as a first step forward. At present the MG-s are assessed into three groups: clinical idiopathic (IMG), facultative (FMG) and obligatory (OMG) (89).

*I. Idiopathic (benign?) monoclonal gammopathies (IMG)*

*II. Facultative monoclonal gammopathies (FMG)*

A) 1. Non-Hodgkin lymphomas (Kiel nomenclature) (low degree of malignity).
   2. Non-Hodgkin lymphomas (high degree of malignity).
   3. Hodgkin's disease.

B) Symptomatic or simultaneous gammopathies
   1. Autoimmune diseases
      Periarthritis nodosa
      LED
      PCP
      Cold agglutinin disease
      Myxoedematous lychen
   2. Chronic inflammations
   3. Liver diseases (?) (chronic hepatitis, cirrhosis)

4. Metabolic diseases (tessaurosis)
5. Neoplastic diseases (carcinomas of the lung, kidney, colon etc.).

*III. Obligatory monoclonal gammopathies*

A) Lymphoplasmocyte type (Kiel) (104).
   1. Diffuse plasmocytoma (30)
   2. Multiplex myeloma (170)
      a) Light-chain disease (Bence-Jones myeloma) (187)
      b) Non-secretory myeloma
   3. Plasmoblastoma
      a) Secretory
      b) Non-secretory
      1. and 2. can both turn into plasmacellular leukemia: plasma cells leave the bone marrow and appear in the systemic circulation.

B) Lymphoplasmocytoid or mixed-cellular type (Kiel)
   1. Waldenström's disease
      a) Immunoblastoma

C) Heavy chain disease (?)

In patients with IMG only MG are present in the circulation without any other clinical or laboratory signs (58, 90, 146, 188). Literature gives a variety of synonyms for the IMG: benign, essential, asymptomatic, non-myelomal, kryptogenic, rudimentary, essential homogeneous immunoglobulinemia (EHI) etc. (108, 174).

MG-s induced by other diseases then neoplastic proliferation of the thymus-independent B lymphocytes are termed facultative monoclonal gammopathies. Under these conditions the paraproteins are produced by the non-tumor cells.

MG-s included by neoplastic proliferation of B-dependent cells are referred to as obligatory monoclonal gammopathies (OMG). The term "obligatory" does not necessarily mean that the paraproteins are detectable in the serum; it implies that the proliferating cells both produce, and release these proteins (132, 152), furthermore, the mass of the protein-producing cells should be increased.

In normal population the MG incidence in the aged individuals increases. In the age groups between 40-60 years MG occurs in about 1% of the population (0.1-0.9%) (7, 36), while in the individuals of the age groups between 70-90 years the incidence of the disease increases to 2.5 - 3.2% and 4.8 - 5% respectively (59, 75, 100). The correlation between the incidence of the IMG and aging has been proved in mice experiments (144).

The familial occurrence of MG suggests genetic preredisposition and might partially explain the low rate of incidence (6, 180). Considering the HL-A gene incidence, the IMG for genes A1, A11, B5, B7, B12, B18, W15, W18 and DR occurs significantly more frequently, but never occurs in OMG (129, 174). It is worth pointing out that the incidence of MG in the families of patients with myeloma is markedly lower than that in the whole population (49).

Screening of healthy (?!) individuals of different age groups revealed MG in 0.3-0.9%. Of 51.124 individuals studied MG was found in 0.48% (Table V/I) (7, 22, 51, 64, 174).

Further studies of the revealed protein pathology indicated that 64-92% of observed changes belonged to IMG group. It is noteworthy that apart from the laboratory differential diagnostic methods (69, 90, 108) the disease can be diagnosed retrospectively, as for example, in the IMG cases diagnosed on the basis of the 15, 16 18 and 24 year-old data (96, 131). Clinical practice indicates that a 5-year period is sufficient for differential diagnosis of an MG as IMG (162). There is, however, debate (?) about whether the patients should or should not be treated during the latent phase of the disease, since not all the IMG-s are malignant. The time factor is very important and probably all the MG-s would sooner or later develop into malignant forms. These patients, however, usually die of other diseases prior to manifest malignancy. The depressed cellular and humoral immune response and, therefore, the insufficient regulatory control might be in the background of the transformation of IMG into OMG (179).

Data of the literature evaluating 4,537 cases conclude that the IMG-s are only "innocent" protein changes occurring in 3-53% (average 24.2%) of the patients and therefore need no treatment (?) (Table V/II) (1, 2, 41, 77, 89, 146, 169).

| DISTRIBUTION OF MONOCLONAL GAMMOPATHIES | | | | |
|---|---|---|---|---|
| Author | # of cases | % of obligatory and facultative | % of IMG | Literature |
| DUGUE, M., ET AL. | 559 | 83,5 | 16,5 | CLIN. CHIM. ACTA 33,75, 1971. |
| ISOBE, T., OSSERMAN, E. F. | 806 | 70,0 | 30,0 | ANN. N. Y. ACAD. SCI 190, 507, 1971. |
| STREIFF, F., ET AL. | 394 | 47,0 | 53,0 | NOUV. REV. FR. HEMATOL. 11, 636, 1971. |
| KINDLER, M. | 170 | 92,4 | 7,6 | DTSCH, MED. WSCHR. 97, 646, 1972. |
| ABRAMSON, N. S., AHATTIL, J. | 59 | 71,2 | 28,8 | JAMA 223, 156, 1973. 3. |
| RITZMANN, S. E., ET AL. | 673 | 70,0 | 30,0 | ARCH. INTERN. MED. 135, 95, 1975. |
| AMEIS, A., ET AL. | 1242 | 75,8 | 24,2 | CMA J. 144, 889, 1976. |
| JÁKÓ, J., ET AL. | 684 | 97 | 3 | IN 1988. |
| Total | 4587 | Average % 75,8 | Average % 24,2 | |

**Table V/I**

The clinician is fully *responsible* for the correct differential diagnosis in these cases; the latter is an extremely difficult task with numerous laboratory and clinical pitfalls. Moreover, in certain cases the differentiation between IMG, OMG and FMG is impossible (14, 160).

The following guidelines help in the differential diagnosis:

I. protein-chemical analysis;
II. quantitative and qualitative evaluation of the bone marrow cells (percent distribution, cytomorphological changes etc.);

| SCREENING OF HEALTHY INDIVIDUALS | | | | | | |
|---|---|---|---|---|---|---|
| Author | # of cases | MG | % incidence of the MG | IMG | % incidence of the IMG within the MG diseases | Literature |
| HOBBS, J. R. | 7200 | 57 | 0,8 | | | SCI. BASIS MED. ANNU. REV. pp. 106, 1966 |
| AXELSSON, M. | 6995 | 64 | 0,9 | 57 | 89 | ACTA MED. SCAND. 179, 235, 1966. |
| CARREL, R. W. ET AL. | 2192 | 11 | 0,5 | 7 | 64 | AUST. N. ZEAL. J. OF. MED! 1, 398, 1971. |
| FINE, J. M. | 13400 | 20 | 0,15 | 16 | 80 | REV. FR. TRANSF. 15, 47, 1972. |
| van CAMP, B. G. K. ET AL. | 21337 | 83 | 0,3 | 76 | 92 | CLIN. IMMUNOL. IMMUNPATH. 7, 315, 1977. |
| Total | 51124 | | | | | |
| Total MG | | 235 | | | | |
| Average incidence of the MG in the cases studied | | | 0,48 | | | |
| Total IMG: (of 178 cases) | | | | 156 | | |
| Average IMG as % ratio to 178 cases | | | | | 88 | |

**Table V/II**

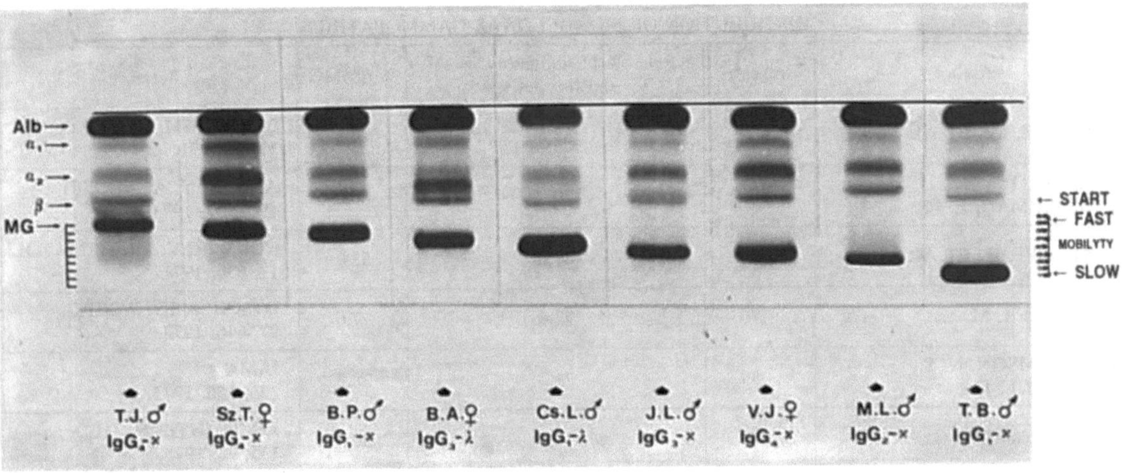


Fig. V/1

III. X-ray and scintigraphic studies of the bones.

Other clinical symptoms should also be considered before the final diagnosis is established.

## I. protein-chemical analysis

1. Though the *total protein level* rarely indicates the above mentioned diseases, it is a necessary diagnostic step, which is probably more informative in the diagnostics of the hyperviscosity syndromes. High total protein levels are usually typical of the malignant forms of the disease.
2. *Gel* (or foil)-*electrophoresis* for identification of the "M" fraction (82). That is the first tool informing about the presence or absence of the "M" fraction. However this method is not always sensitive if the M-concentration is low. Electrophoresis also detects the mobility and the percent distribution of the pathological protein fractions (Fig. V/1 and V/2)
3. *Immunoelectrophoresis.* It is recommended to put the patient's (diluted 1:10), and not the control serum against the well filled with concentrated serum. This modification would immediately reveal protein anomalies. In MG-s with secondary antibodies and high levels of the "M" component a 10x dilution usually proves to be insufficient to visualize the precipitin arc of paraproteins. In such cases titration immunoelectrophoresis (TIEF) is a method of choice (80). With this method - even if interference ocurs - there will always be a dilution yielding visible precipitation solely with monoclonal gammaglobulins (Fig. V/3).

The patient's serum is studied in a so-called analytic system (Fig. V/4), which reveals the type of the light and the heavy

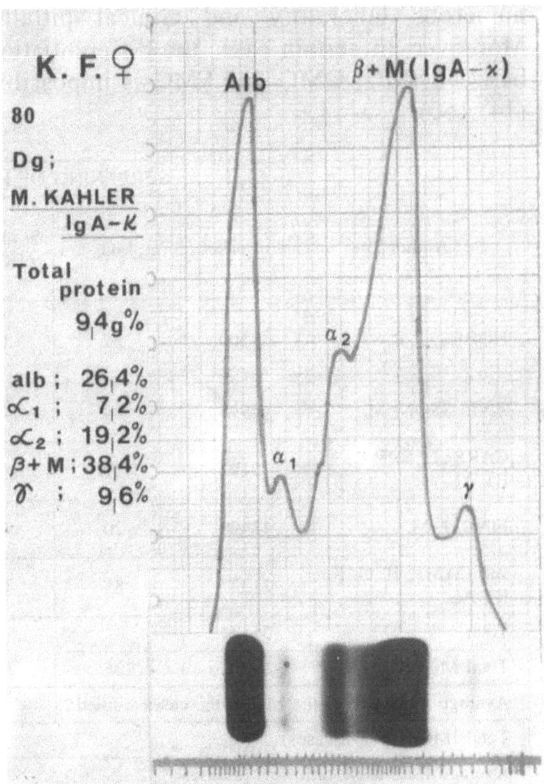


Fig. V/2

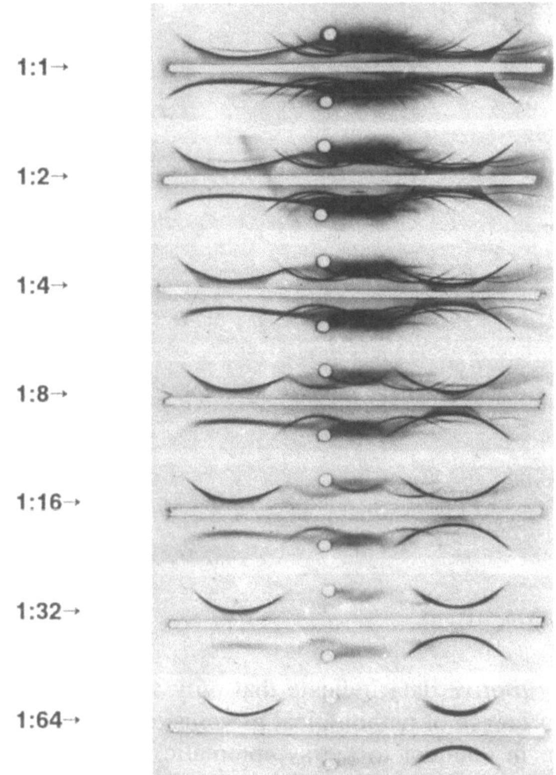


Fig. V/3

Dilution 1:16

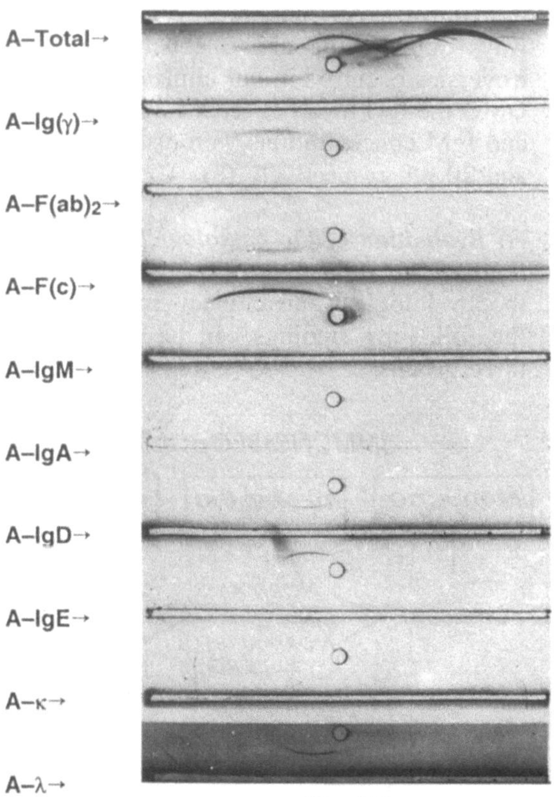


Fig. V/4

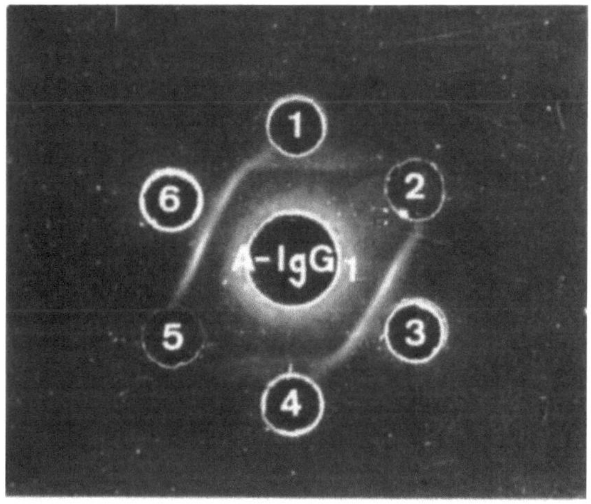


1, 3, 4, 6 POS. 2, 5 NEG.

Fig. V/5

chain that contains the "M" component. For example: Dg.: IgD-λ monoclonal gammopathy.

4. *The subclasses of the heavy chains can be determined* with specific immune sera using the gel diffusion rosette technique. Determination of the γ1, γ2, γ3, γ4 subgroups of the IgG immunoglobulins is a routine task nowadays (Fig.V/5). The α, μ, δ and the κ and λ subtypes are identified only at specialized laboratories, for the antisera are not widely available.
5. *Quantitative determination of the monoclonal gammaglobulin* is usually performed with Mancini's radial gel diffusion technique (39, 83, 116); however, due to some technical reasons (monoclonal gammaglobulin can not be accurately detected with polyclonal anti-immunoglobulin antiserum) only approximate evaluation (Fig. V/6) is possible.

It is noteworthy that the immunoglobulin producing capacity of the plasma cells is heterogeneous. Data of the literature mention the maximal synthesis rate as 100,000 molecules/cell/min (132, 151). In our opinion, the secretion rate ranges between the non-secretory and the hypersecretory plasma cells, and it also depends on the individual characteristics and maturity of the plasma cell. These data imply existance of hypersecretory *plasma cell clones* in the benign idiopathic MG-s forms and low-secretory plasma cells in OMG; in such situations the quantitative dif-

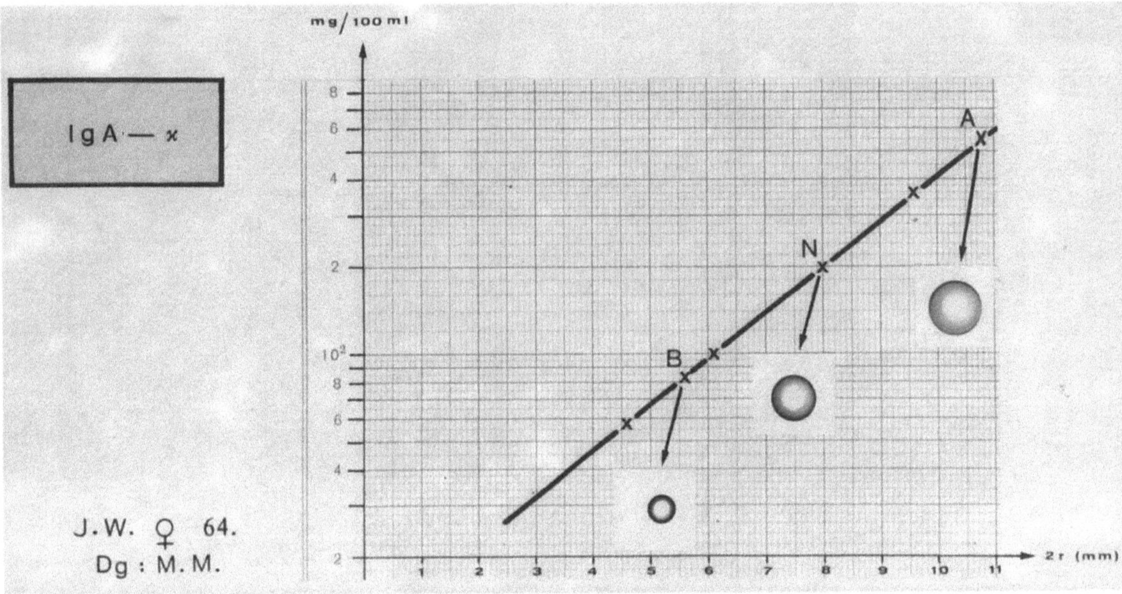


Fig. V/6

ferentiation between these diseases based on the concentration of paraproteins is impossible.

Lack of correlation between the elevated monoclonal immunoglobulins and the bone marrow-derived plasma cells supports the above view (Fig. V/7) (81).

*Evaluation of immunochemical data*

A) Different types of *heavy chains* imply different prognosis. For example, an obligatory monoclonal increase of IgM indicates Waldenström's macroglobulinemia, while their facultative elevation suggests a benign process, such as chronic lymphoid leukemia. These findings are IgM-specific and are not typical of the α, γ, δ and ε -type heavy chains. *Quantitative* data indicate that only a sustained increase of pathological proteins occurs in IMG, in contrast to the symptomatic forms (OMG, FMG). This finding is strongly supported by the data of the literature (21, 58, 125, 175).

Clinical experience, however, does not always support the above conclusions. Controversies occurred in our clinical material: in OMG the IgG levels exceeded 20 g/L, the IgA and IgM concentrations were usually over 10 and 20 g/L respectively (8).

B) *Evaluation of the pathological role of the light chains is also extremely important.* Pathophysiological and clinical implications of the following findings can be assessed into three groups:

*Classification of the experimental results according to the values of the variables*

| g(%) / Cells(%) | -1,2 | 1,3-2,0 | 2,1-3,0 | 3,1- |
|---|---|---|---|---|
| 6-10 | 1 | 3 | 0 | 0 |
| 11-25 | 25 | 37 | 10 | 2 |
| 26-93 | 33 | 38 | 19 | 4 |

*Corellation coefficient = 0.203281*

Fig. V/7

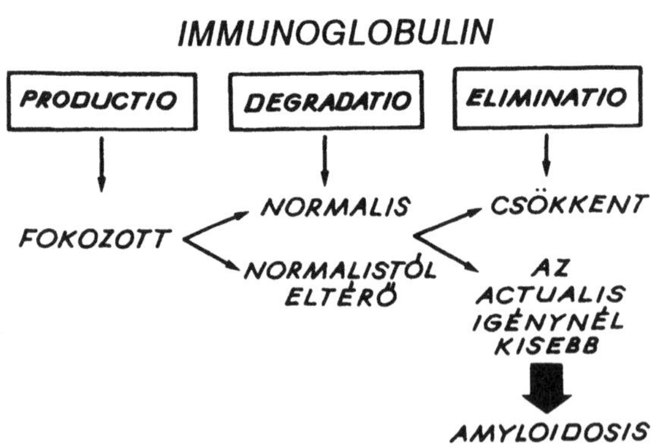


Fig. V/8

a) Which of the light-chain types occur in monoclonal immunoglobulins?
b) Is this light chain excreted with the urine?
c) Can the "free light chains" be detected in the serum?

a) Data of statistical analysis indicate that the κ-λ ratio in IMG is 0.4:1, while in symptomatic MG it is shifted to 2:1 (33, 125). Individual assessment of the k-λ ratio, however, is insufficient for a differential diagnosis. At the same time high levels of the λ light chain are potecionally dangerous and the benign course of the disease is doubtful. The λ chains are directly nephrotoxic (27, 84, 85, 95), and promote the development of amyloidosis, as well as other metabolic alterations (56, 68, 70, 106, 109) (Fig. V/8).
Thus, for example, the above mentioned disease is just a rare symptom of IMG and is not an individual entity (67, 96, 168). The presence of monoclonal λ-type gamma-globulins in OMG indicates worse prognosis.

b) Excretion of the light chains in the urine - the Bence-Jones type proteinuria (BJ) - is an accepted *sign* of malignity (175); if its concentration exceeds 10 mg/L, than it is a *proof* of malignity (65).
Thermostability of the light chains is another useful feature used for the differential diagnostics.
The typical temperature test - precipitation with subsequent dissolving of the molecules - is only positive for the dimeric structures containing two light chains in a molecule. Hence this structure is termed pyroglobulin (50). Sometimes the opposite reaction occurs, when the cold-sensitive kryoglobuline-type light chains are excreted in urine (61).
In summary, the excretion of either pyro-, or kryoglobulin-type light chains indicates malignity, the higher is their excreted level, the more severe is the malignity.
In clinical practice the BJ proteinuria is subdivided into three groups.

c) Free light chains and monoclonal Ig-s appear in the serum
In one patient, however, we failed to detect light chains even after the concentration of the urine sample. At the same time the urinal excretion of the light chains occurs more frequently.

1. A concomitant sign of the light chain disease. In this case only the light chain-producing clones show enhanced proliferation. Despite the low molecular weight (37) and the fast excretion of these molecules by the kidney (156, 183), the light chains are produced at such a high rate and concentrations that the process is usually accompanied by BJ proteinemia and is, therefore, defined as a BJ myeloma (sometimes an "IgU myeloma" is also used) (9).
2. Idiopathic or benign BJ proteinuria. Some patients excrete light chains for 7 or more years without developing any signs of the disease (102). It is worth pointing out that in all these cases only the *κ-type light chains* were excreted.
3. Light chains (BJ proteins) can appear in the urine in all types of MG (IgM, IgG, IgA, IgD and IgE). However no uniform view exists on the origin of these proteins. *Sometimes* the type of the excreted light chain coincides with the monoclonal-light chain, whereas *in other cases* the excreted and the monoclonal light chains differ. It is assumed that *in the first group* the excess of the light chains is produced by the pathologically altered plasma cells (4), which secrete these proteins together with the normally assembled immunoglobulins. These cases can be defined to a certain extent as biclonal gammopathies. Such "switch" mechanism has been proved for some cases of biclonal gammopathy, (67). It is worth pointing out that these light chains are not the degradation products of paraproteins (21, 124). *In the second group* the light chain type excreted with urine, does not coincide with that of the light chains of MG. These are the unambiguous cases of biclonal gammopathy.

C) *The importance of the presence or absence of the secondary antibody deficiency syndrome.*

It is generally accepted that in IMG no secondary antibody-deficiency syndrome occurs: the monoclonal production of proteins does not affect the normal synthesis of other immunoglobulins (58), though exceptions are documented in clinical practice: in obligatory - that is malignant - MG the antibody-deficiency syndrome important for the differential diagnosis sometimes fails to develop. Moreover, in IMG the titer of polyclonal immunoglobulins sometimes decreases.

It was generally accepted that in the secondary antibody-deficient syndrome the normal clones of the bone marrow are displaced by pathological ones. Recent studies have somewhat modified our views about the role of neoplastic clones in the process of desintegration of the immune response. Available data suggest several mechanisms, some of which have been studied in considerable detail:

1. Chalon's effect *
   pathological plasma cells produced at enormous concentrations release biologically active substances, which do not affect the pathological forms, but inhibit the function of the normal clones finally eliminating them from the bone marrow (128).
2. The RNA-containing factor **
   Humoral immunosuppressive factor is released by the plasma cells or their precursors into the systemic circulation. Chemically this factor contains proteins with poly-A sequencies with sedimentation constant value = 14-18W (62, 88). They elicit their effect via the following mechanisms:
   a) lymphocytes exposed to this factor start synthesizing surface immunoglobulins of the same idiotype (idiotype conversion) as that of the malignant cells, thus inactivating the lymphocytes within the network of the immune reaction*** (15, 26)
   b) stimulation, activation of the suppressor macrophages, which inhibit the normal polyclonal antibody synthesis either via a direct intercellular interaction (177), or by releasing a low molecular weight immunesuppressive factor (94) (Fig. V/9).
   It is suggested that the factor inhibiting normal clones might be a monoclonal

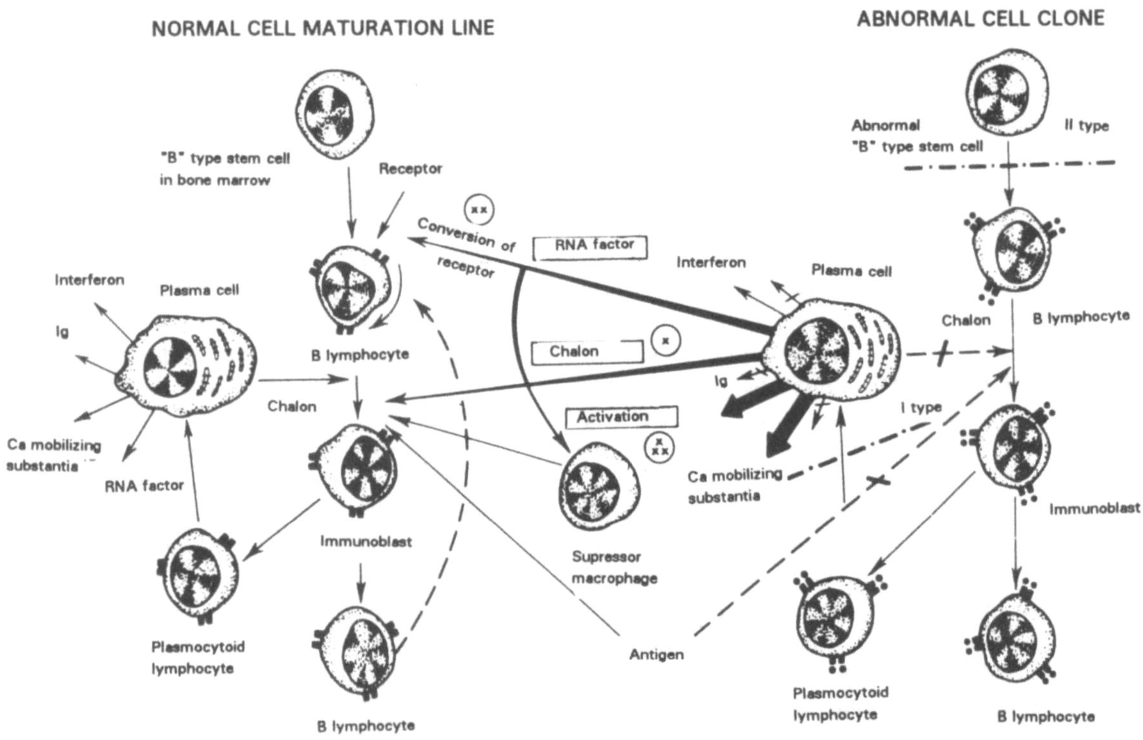


The effect of a pathological clone on the function of the normal clones

Fig. V/9

immunoglobulin. Data of the literature, however, rather stress the role proteins (178): high levels of immunoglobulins stimulate non-selective catabolism, so the polyclonal immunoglobulins will be destroyed. Such mechanism can not be assumed in the secondary antibody-deficiency syndrome, which develops in non-secretory myeloma (45). The high β-2 microglobulin titer is an accepted finding in the differential diagnostics of IMG, OMG and FMG, though this method is not yet routinely used.

In summary the immunochemical analysis of the proteins is insufficient for the differential diagnosis of the benign (requiring no treatment) and malignant forms (OMG, FMG requiring therapy). Our experience and data of the literature suggest that IMG can be the first *laboratory* sign of OMG (MM) (19); typical alterations of the bone marrow and osteolysis would develop later (66, 99, 167). Hence an otherwise "harmless" IMG might be the first phase of a facultative (sometimes) or obligatory (nearly always) MG.

## II. The studies of the bone marrow

Immunoglobulines are produced in the plasmocytes of the bone marrow, lymph nodes, *gastrointestinal tract* and other areas (120). Therefore, MG diagnostics requires qualitative and quantitative assessment of the plasma cell count.

A) Light microscope studies

a) *The marrow aspirate* is of diagnostic value in 90% of the multiple myeloma cases. In one-two percents of the cases the aspirate rather contains lymphoid elements, suggesting non-Hodgkin's lymphoma (for example Waldenström's macroglobulinemia). In 7-8% of the cases even the repeated aspirate samples provides no sufficient information for the diagnosis of OMG. In 1% of the cases the aspirates contained no cellular elements (98). The latter two types of findings are the major obstacles in the differential diagnosis between IMG and OMG. The bone marrow findings can not always differentiate between the reactive plasma cell increase (cirrhosis, neoplasma) or MM. Furthermore, the bone marrow analysis can not supply ample information for the Kahler disease diagnosis (184).

b) Though the percent distribution of the plasma cells is of great importance, it is not the only diagnostic finding. The analysis of 326 cases indicates that in 84% of the patients the plasma cell ratio exceeded 10% reaching the value of 36%. The values ranged between 5-100% (101). Our data showed the plasma cell ratio of 32.7% ( a study of 172 cases).

Values below 10% rather indicated reactive plasmocytosis or IMG. We have observed numerous cases when the marrow plasma cell ratio remained below 5%. The puncture biopsies of the X-ray - verified osteolytic areas exhibited a uniform distribution of the plasma cells (Fig. V/10). In these case dissemination of soliter plasmocytoma or a real MM can be equally assumed (Fig,V/11 and V/12). We could also observe patients with a typical protein profile and signs of the X-ray verified osteoporosis in whom the plasma cell count was under 5%. 1-2 years later the latter value had reached the diagnostically significant level. Such type of distribution is typical of the diffuse plasmocytoma.

In summary, the examples presented above indicate that in cases of diffuse osteoporosis without bone marrow lesions the % ratio is insufficient for bringing either positive, or negative decisions. It is, however, worth pointing out that values of about 5-10% must not be accepted as the differential diagnostic ones.

c) Cytomorphological evaluation is of greater importance than the quantitative analysis of the *plasma cells* (185).

1. The first step is the assessment of the nuclear to cytoplasmic (N/C) ratio (162). In OMG the cytoplasm is somewhat narrower. In pathological cases the cytoplasm can be greatly enlarged, the cells sometimes resemble the Gaucher cells.
2. Following the analysis of the N/C ratio, the nucleus itself is evaluated. It has an

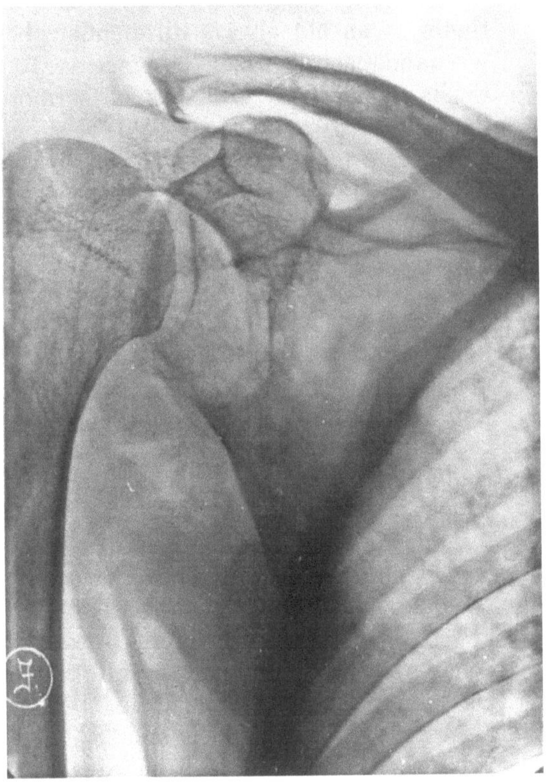

Fig. V/10

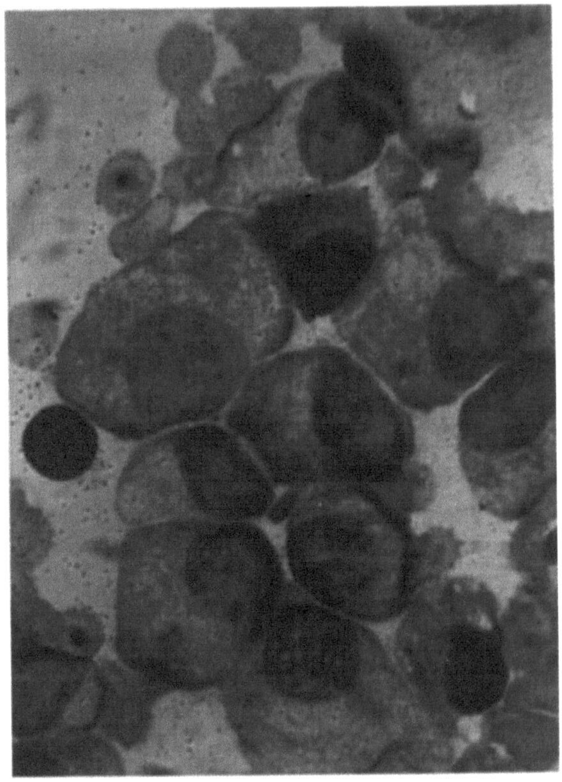

Fig. V/11

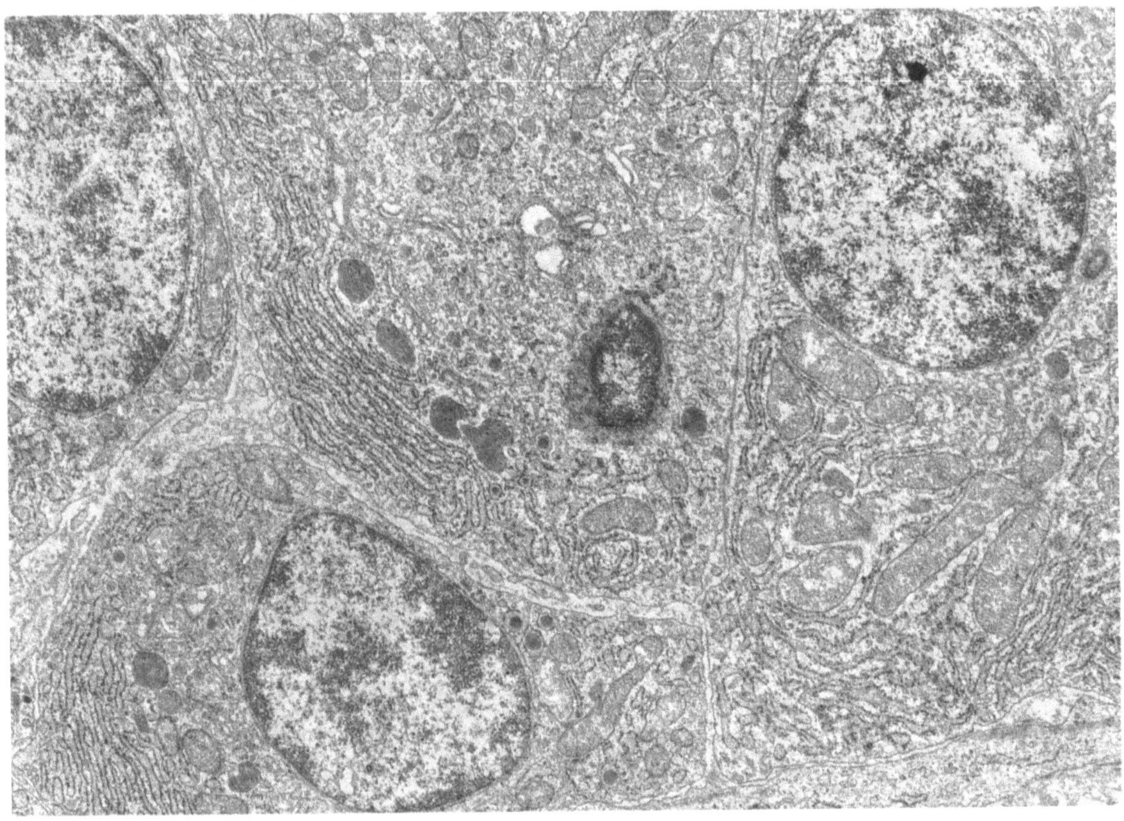

Fig. V/12

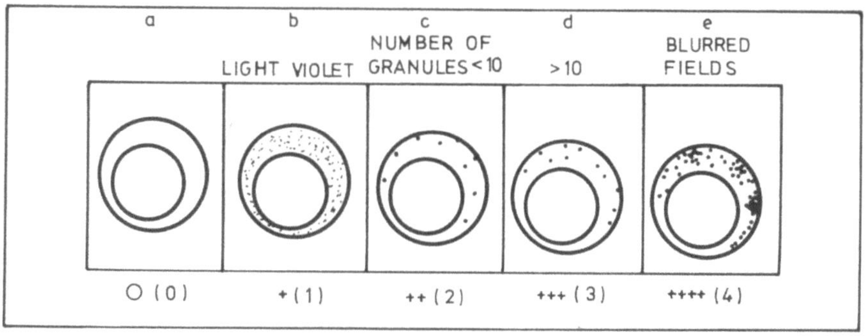


PAP (PLASMACELL ACID PHOSPHATASE) SCORE

$$\text{SCORE} = \frac{a \times 0 + b \times 1 + c \times 2 + d \times 3 + e \times 4}{\text{COUNTED CELLS} \times 0{,}01}$$

**Fig. V/13**

excentric localization and its chromatin resembles that observed in the lymphoid or reticular cells. The presence of polynuclear cells, though typical, but is not pathognostic. For example polynuclear cells generally occur in reactive bone marrow, however the increased mitotic activity and the nest-like localization of these cells indicate MM (Fig. V/12). The well visualized nucleolus, if enlarged and widened, suggests malignity (172). Other inclusions of cytoplasmatic origin rarely occur; according to the accepted view, protein is stored in these particles (53). If the cytoplasmic inclusions do not occur in parallel with the nuclear ones, this micromorphological feature is explained by the nuclear protein synthesis (172). Lipid inclusions are also documented (135), though their significance needs further investigation.

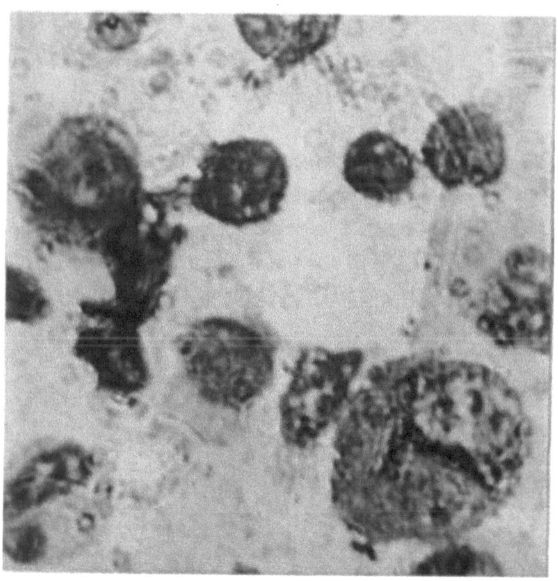

**Fig. V/14**

3. Evaluation of the cytoplasm also gives important information. If the cytoplasm surrounding the nucleus forms a light ring, it is a sign of malignity. Sometimes the cytoplasm surrounding the nucleus forms flame-tongues, these cells are called "burning" plasma cells. Their edges exhibit typical eosiniphilia. Such cells are characteristic of IgA myeloma, though they also occur in the normal bone marrow of healthy individuals (40, 138).
   Inclusions, vacuolization or Russel bodies occurring in the cytoplasm indicate malignity. Inclusions storing protein are usually PAS-positive (53). If substantial amounts of protein are stored in the cytoplasm, the retineal substance can transform into needle-like crystals (87) with extracysternal localization (165).
4. The mentioned studies were all performed with May-Grünwald staining. It is extremely important, however, to use other staining methods and cytochemical studies if the diagnosis is still uncertain. Thus, the differential diagnosis is facilitated by specific staining for gly-

| OMG | | | | | | | | | | | | | | | |
|---|---|---|---|---|---|---|---|---|---|---|---|---|---|---|---|
| IgG | | | IgA | | | IgD | | | LIGHT CHAIN | | | IMG | | CONTROL | |
| | STA-DIUM | SCO-RE | | STA-DIUM | SCO-RE | | STA-DIUM | SCO-RE | | STA-DIUM | SCO-RE | | SCO-RE | | SCO-RE |
| IgG$_2$-λ | III A | 160 | IgA-λ | II A | 180 | Ig□-λ/λ | II B | 190 | κ | III A | 310 | IgG$_1$-κ (5 y) | *210* | I | 120 |
| IgG$_1$-κ | II A | 210 | IgA-λ/Ig-A-κ | III A | 200 | | | | | | | IgG$_1$-κ (10 y) | 100 | II | 40 |
| IgG$_1$-κ | I A | 300 | IgA-l | II A | 180 | | | | | | | IgG$_1$-κ (7 y) | 75 | III | 90 |
| IgG$_2$-κ | I A | *90* | IgA-λ | III A | *100* | | | | | | | | | IV | 20 |
| IgG$_1$-κ | I A | 190 | IgA-l/IgG$_1$-λ | III B | 210 | | | | | | | | | V | 100 |
| IgG$_2$-κ | II A | 240 | | | | | | | | | | | | VI | 95 |
| IgG$_1$-κ | III A | 290 | | | | | | | | | | | | | |

**Fig. V/15**

curonidase, β-acetylglycoseaminidase, specific esterase, adenosin-triphosphatase and acid phosphatase. Positivity of these findings unambiguously indicates malignity (111, 112, 162). Data of the literature consider acid phosphatase as a discriminative criterion of IMG and OMG (24), but our findings doubt this statement. About 52% of the normal plasma cells are acid phosphatase-positive, though the deviation of the values is high enough, ranging between 20% - 90%. In myeloma the positivity is 95% and the deviation ranges between 73% - 100%. A plasma cell can be either negative, showing no activity; + positive with direct light-pink staining; ++ positive having less than 10 areas of pink granulation; +++ positive exhibiting more than 10 pink granules. In ++++ positive cells the granules are fused into homogenuous areas of pink staining. This method of evaluation

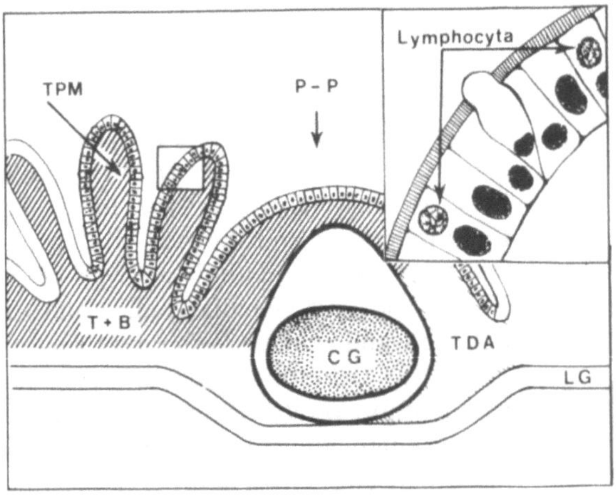


**Fig. V/16**

| Authors: Cif: Brandtzaeg v. Baklien (1976) | IgA% | IgM% | IgG% |
|---|---|---|---|
| Crabbe v. Heremans 1966 | 83,5 | 12,4 | 3,8 |
| Soltoft 1969 | 58,6 | 35,4 | 6,0 |
| Soltoft v. Soeberg 1972 | 66,5 | 22,9 | 10,6 |
| Savilakti 1972 | 72,8 | 21,0 | 6,2 |
| Jos et al 1972 | 86,7 | 8,0 | 5,2 |
| Gasbarrini et al 1974 | 58,9 | 23,4 | 17,7 |
| Brandtzaeg v. Baklien 1976 | 80,8 | 16,5 | 2,6 |
| Jákó et al 1978 | 60,4 | 24,6 | 15,0 |

**Table V/III**

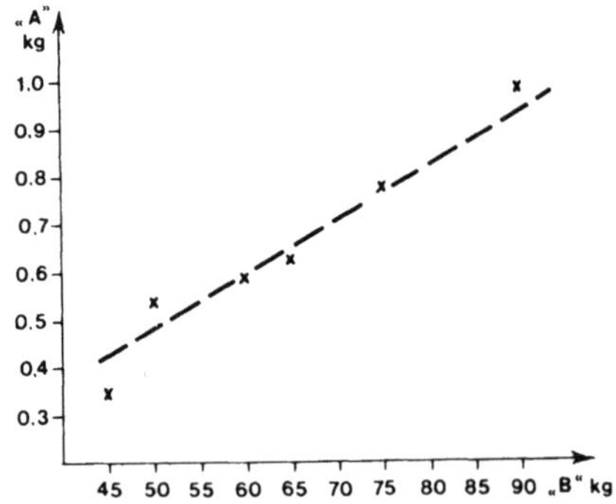


**Fig. V/17**

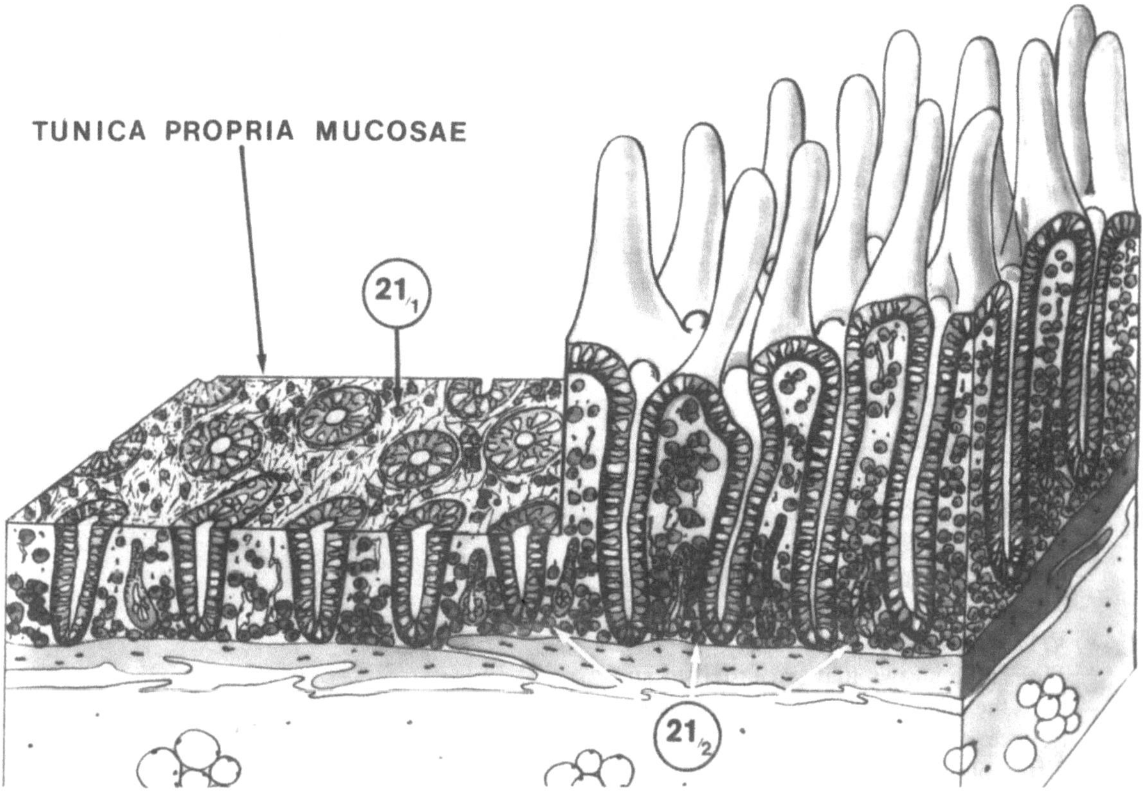


Fig. V/18

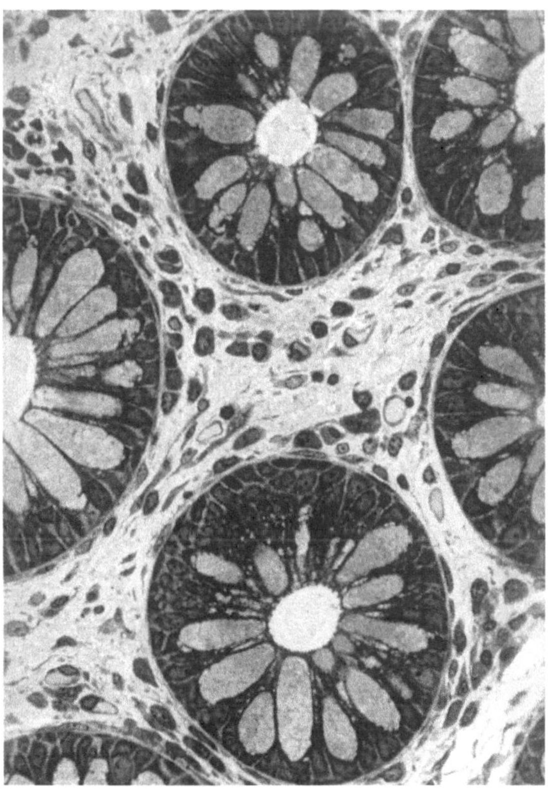

Fig. V/19 (21/1)

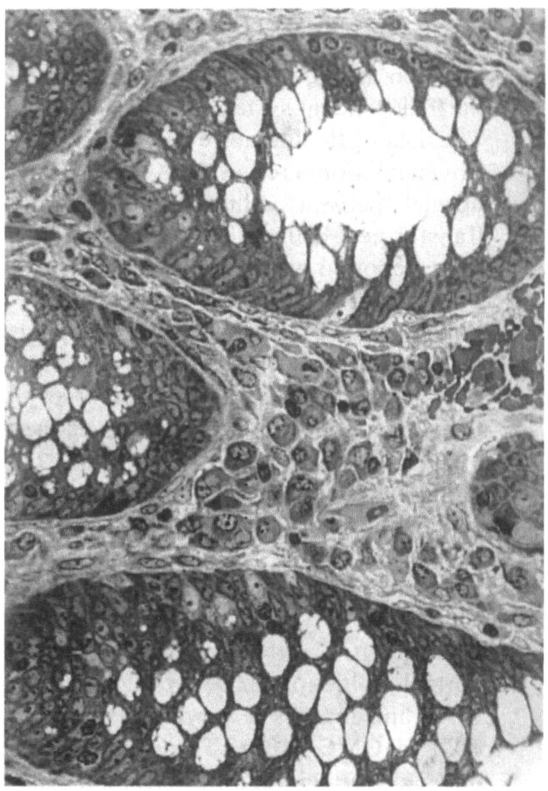

Fig. V/20 (21/2)

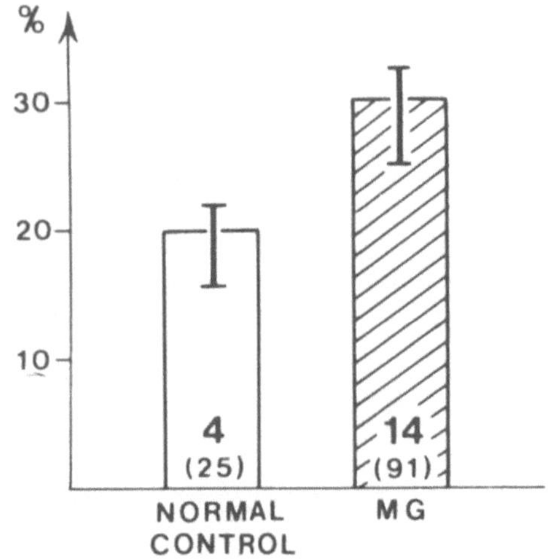


Fig. V/21

yields a so-called PAP score (*P*lasma cell *A*cid *P*hosphatase *s*core) (24) (Fig. V/13, V/14, V/15).
The Table shows that we failed to support the absolute discriminative significance of the acid phosphatase in the differential diagnosis of OMG and IMG.

5. Pathology of T-independent, B-dependent cells lies in the background of monoclonal gammopathies. Hence a question arises, whether the function of the T-dependent system is affected (71). The data of the literature are rather controversial. Some researchers have found normal percent values of the blast-transformed cells (121), others, however, reported very low values of the blast transformation used as a functional test (179). It is generally accepted that regardless of the benign or malignant nature of a pathological process, the depressed immune response shows inverse correlation with the titer of monoclonal immunoglobulins: at high concentrations the T cell activity is low (146).
In monoclonal gammopathies the percent of the B lymphocytes can markedly differ. It slightly decreases from the normal value of 22.9% to 21.4% in IMG, and reaches a very low value of 10% in OMG (107). These data refer only to the circulating lymphocytes. The marrow B cell count comprising both the lymphoid elements and the plasma cells can be assessed with immunofluorescent technique. This count is usually expressed as a percent of the total nuclear cells of the marrow (86, 100, 107).
Of the immunoglobulin-producing cells the ratio of the paraprotein-synthesizing paraproteins can be evaluated with the similar methods (126), though some shortcomings discussed below can occur.

6. The marrow is the main source of the plasma cells studied. We would also suggest to expand the studies by evaluation of the immunologically competent

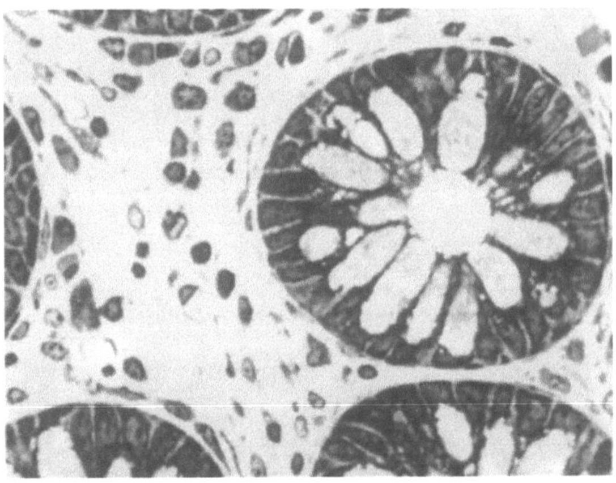

TANGENTIAL SECTION OF THE RECTAL MUCOSA; MAGNIFICATION 560x.

TAGO A–IgG(γ) T. J. ♂ $IgG_1$–λ M. G.
FITC 2190

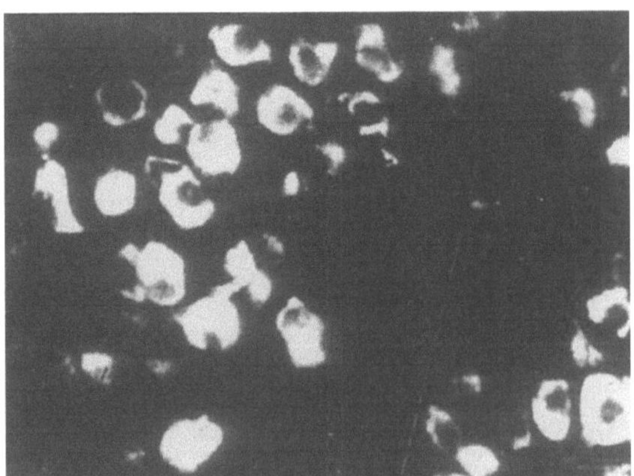

TAGO A–IgG(γ) T. J. ♂ $IgG_1$–λ M. G.
FITC 2190

Fig. V/22

cellular system of the gastrointestinal tract (18, 155). With respect to their immune functions, the small and the large intestines are rather similar (118, 139). The only difference is observed in the duodenal Payer's patches and interepithelial migrating cells, which were later proved to be lymphocytes (29, 134). 63.5 migrating lymphocytes termed "theliocytes" occur in every 1000 of the epithelial cells (Fig. V/16).

Though the 0 lymphocytes are predominant, some T and more B cells are also found between the epithelial cells (48). The lymphocytes of the lamina propria are called propriocytes. The immunoglobulin-producing cells are localized in this area. The mucosal lamina propria can be regarded as a diffuse lymphoid organ (141). Table V/III compares the pattern of cell distribution and the classes of produced immunoglobulins obtained in our studies and presented in the literature.

The tunica propria of the large intestine contains mature plasma cells, histiocytes, eosinophils and some lymphocytes (34, 35, 54). No transient forms between the plasma cells and the lymphocytes exist (43). Typical lymphoblasts do not occur in the tunica propria, but are observed in the well developed lymphoid follicules, which resemble germinative centres. Therefore, the mucosal tunica propria of the large intestine is a continuation of the diffuse lymphoid organ existing in the small intestine (42, 54).

The above findings made us to study quantitative and qualitative changes of the plasma cells in rectal biopsies in

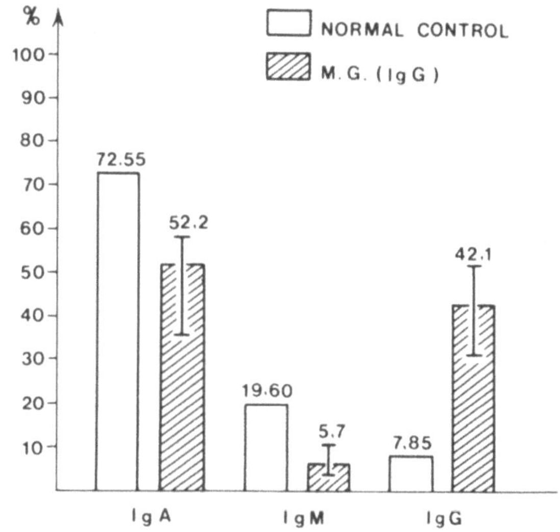


Fig. V/23

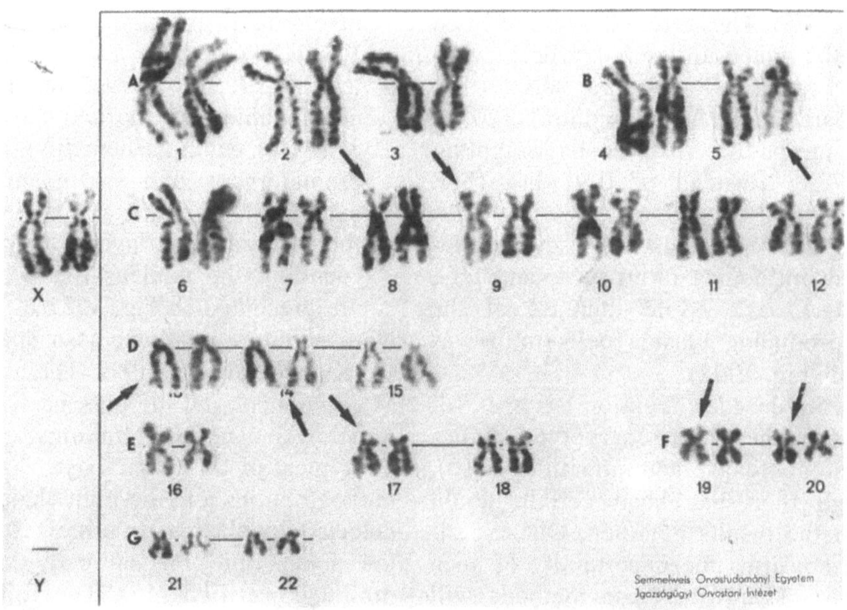


J. Juhász ♀ 65 y. Diagnosis: IgD myeloma

Fig. V/24

IMG and OMG. First the weight of the cadaver large intestine samples were determined and found to comprise about 1% of the body weight. The weight of mucosa makes about 1/5 of that of the whole large intestine. Plasma cell count in the mucosal tunica propria was determined in semi-thin histological sections. It yields about 20% of the total nuclear cell count (Fig. V/17, V/18, V/19, V/20). The significance of the mentioned studies is shown in the following example: the large intestine weight of an average person weighing 70 kg is 0.7 kg; 1/5 of this value - 0.15 kg - is the weight of the mucosa. If a 20% plasma cell ratio is assumed for the interglandular cells in the tunica propria, than their weight is 0.01 kg. Evidence exists that the weight of $10^9$ cells corresponds to 0.001 kg (26). Hence, 0.01 kg contains $10^{10}$ cells. These results show that the weight of the ICS cells in the small and large intestine is similar, resembling that of the cells in the bone marrow, lymph nodes and spleen.

At least two rectal mucosal biopsies from different areas were studied in patients with monoclonal gammopathies, who had the marrow plasma cell ratios under 5%. The serial study indicated that the mucosal lamina propria contained much more plasma cells (30%), compared to the healthy controls (20%). 30% means $1.5 \times 10^{10}$ cells weighting 0.015 kg (instead of 0.01 kg) (Fig. V/21).

Using the fluorescein-isothiocyanate-labelled antibodies (Anti-IgM, anti-IgG, anti-IgA) a 6-7-fold increase of the IgG-producing plasma cells ratio was found (Fig. V/22).

7. In recent decades attention has been focused on the marrow caryotype studies in monoclonal gammopathies (10), though their has been certain doubt about the results obtained. Our experience confirms the uncertainity of such studies. Therefore these methods will play no important role in the differential diagnostics of IMG, OMG and FMG (Fig. V/24). The so-called MG chromosome occurs very rarely and this abnormality is not congenital, but rather acquired (78).

**B) Electronmicroscopic** studies (EM) of the immuno-globulin-producing cells might also be of importance. Normal plasma cells have typical nuclear structure, nucleolus, dense endoplasmatic reticulum (ER), large mitochondria and a well developed Golgi area (91). Though these structures also occur in the malignant cells, certain quantitative and qualitative differences, are observed. The latter can be accepted as having differential diagnostic value only in some special cases (55) (Fig. V/25).

EM can primarily detect the so-called asynchronous plasma cells, known as Bernier-Graham disparity, when the nuclear and the cytoplasmic maturation is disproportional. Furthermore, the non-aggregated chromatin, prominent nucleoli and distended ER are also typical (56, 136). The endoplasmatic reticulum can exhibit three main types of informative alternations:

a) the cysterns of the distended ER contain electrondense and non-electrondense inclusions. The former are egg-shaped black rings of different sizes included in the otherwise normal cytoplasm. The latter represent the typical structures of the BJ myeloma (light chain disease) (28) (Fig. V/26).
b) tubular ER is observed in many diseases accompanied by the activation of plasma cells (for example hepatitis), as well as in premalignant and malignant states, for example MM (5).
c) the onion-shaped layers of cytoplasm surrounding the nucleus like a "whirl" (85) are presented in Fig. V/27.

Non-secretory myeloma is a special case of the Kahler's disease (92). The absence of the M component and the presence of the characteristic bone marrow findings and osteolysis are typical of this pathology. Monoclonal immunoglobulins and the light chains can not be detected in plasma or urine respectively. At the same time the antibody-deficient syndrome is verified (45). The following three possibilities arise from the analysis of the typical myeloma plasma cells:

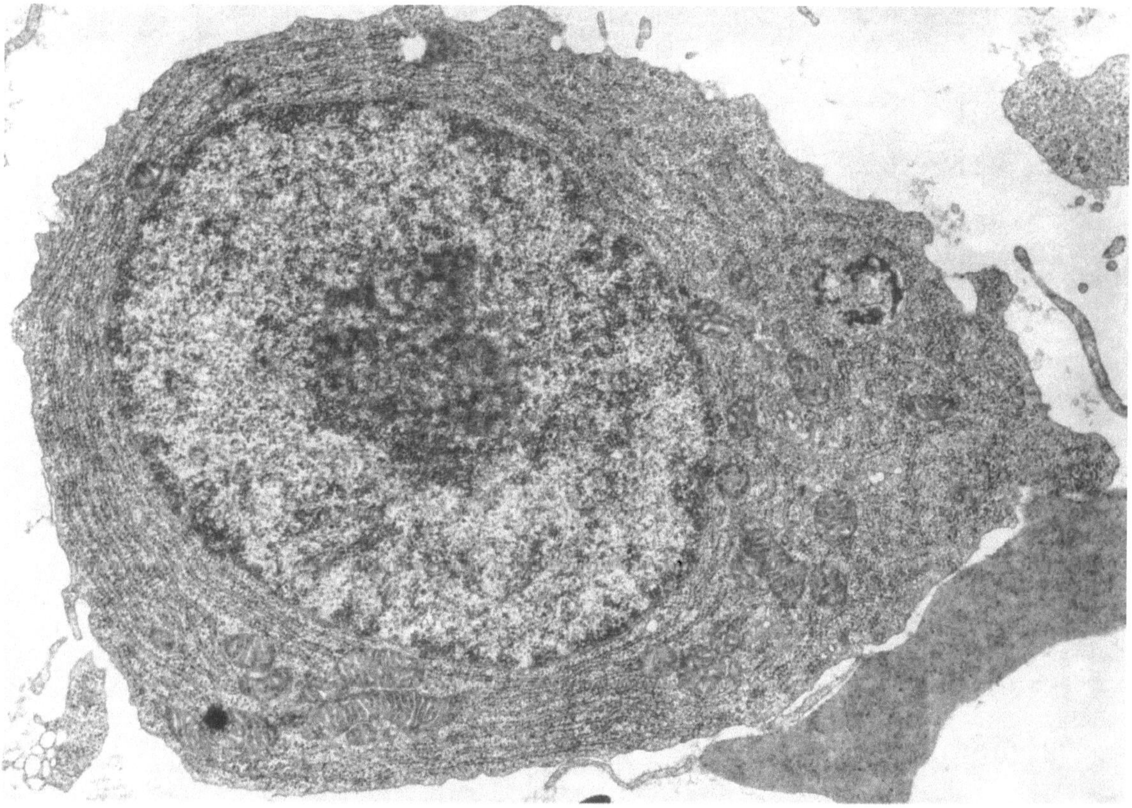

Fig. V/25

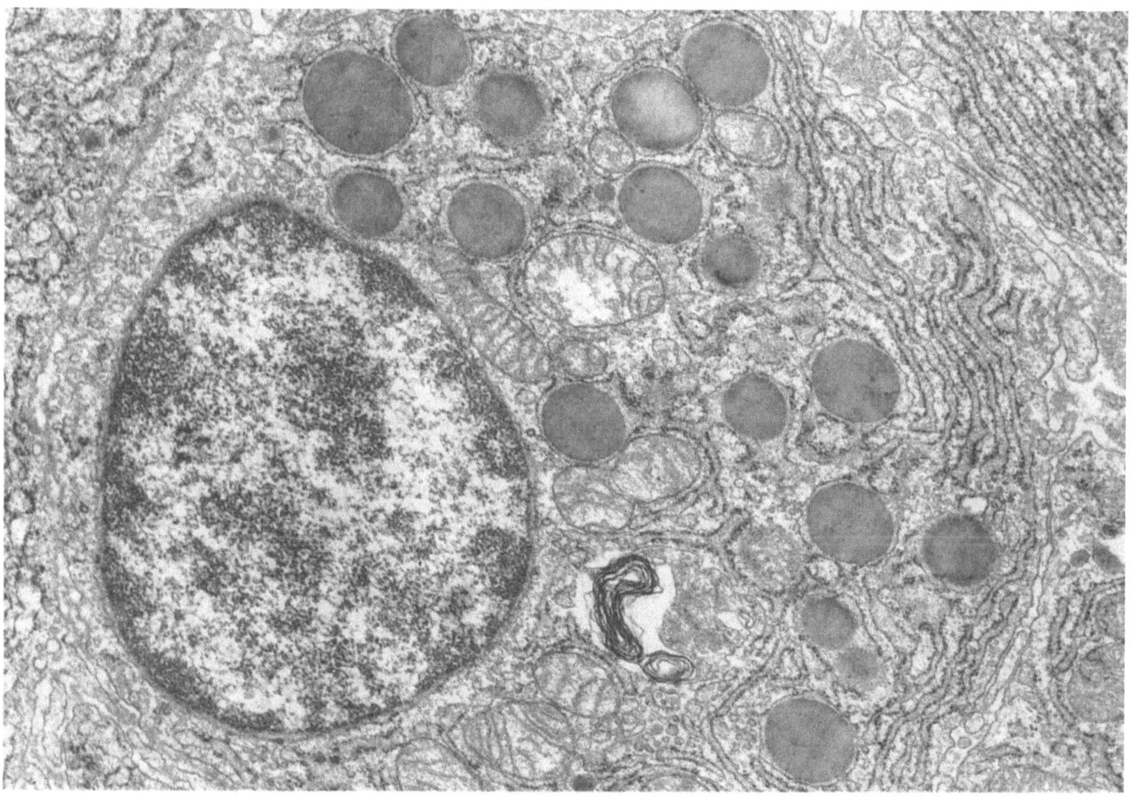

Fig. V/26

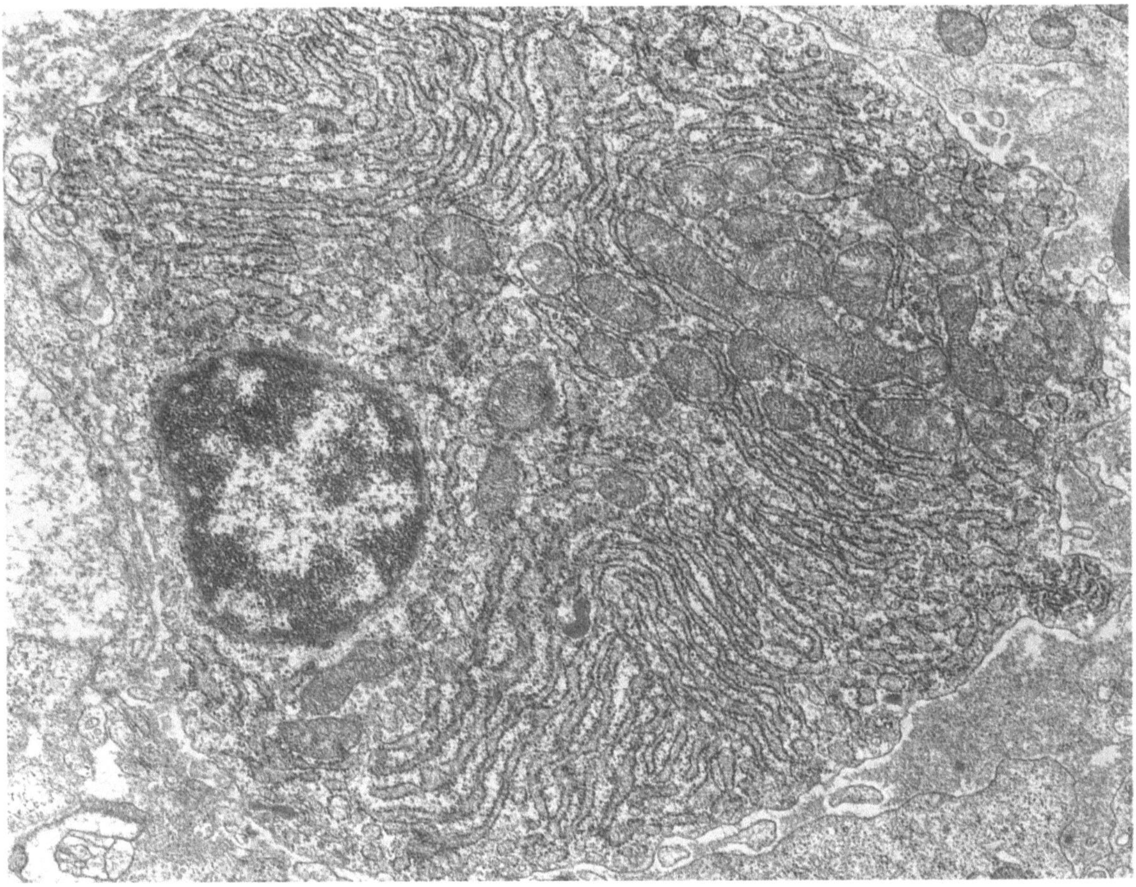

Fig. V/27

1. Decreased activity of the immunoglobulin-*producing* cells. Due to that, monoclonal immunoglobulins or light chains can be periodically detected in the serum or urine. Thereby the secretion is of intermittant character, so this pathology can not be defined as a "non-secretory" myeloma. Immunofluorescence methods always identify intracellular immunoglobulins or light chains (123). It might be a cause of the so-called transitory MG.
2. Monoclonal immunoglobulins produced by the cells are not complete molecules, but only fragments or minor parts of the molecules. Such "erroneous" molecules are real paraproteins (105, 163, 164). They are very quickly synthesized and as quickly degraded (143). The loci of degradation, similar to other immunoglobulins, are still unknown. The intracellular immunoglobulins are detected with heavy- or light-chain specific labelled antisera (3).
3. In some myelomas the present methods fail to detect immunoglobulin production in the proliferating tumor plasma cells. The "molecular machinery" of these cells (DNA transcription, messenger- or ribosomal RNA translation etc) is quiescent; but *"the devil is never asleep"* so production of heavy chains of unknown antigenity, which, therefore, can not be identified, is possible (166).

*In summary,* extreme fluctuations of the plasma cell count, heterogeneity of its morphology including all possible forms between a reactive cell to a highly proliferating one, prevent accurate differential diagnosis, especially in the atypical cases. Therefore, the bone marrow analysis alone is insufficient. Ad absurdum even if the bone marrow punctate gives repeated normal results a possibility of MM diagnosis still exists. A combination of the bone marrow studies with chemical analysis of proteins yields a more valuable informa-

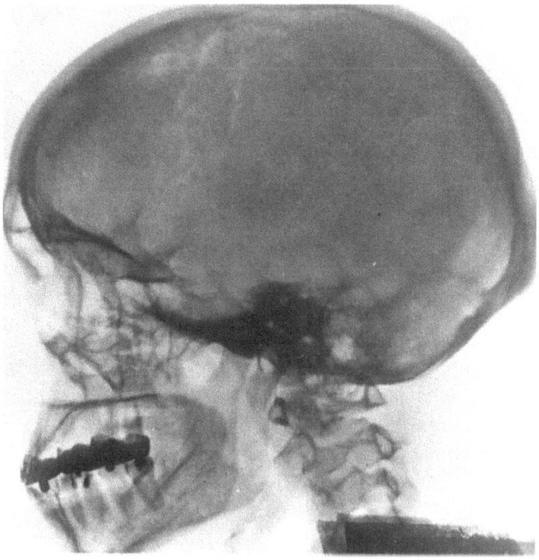

Multiple myeloma $IgG_3$-λ

Fig. V/28

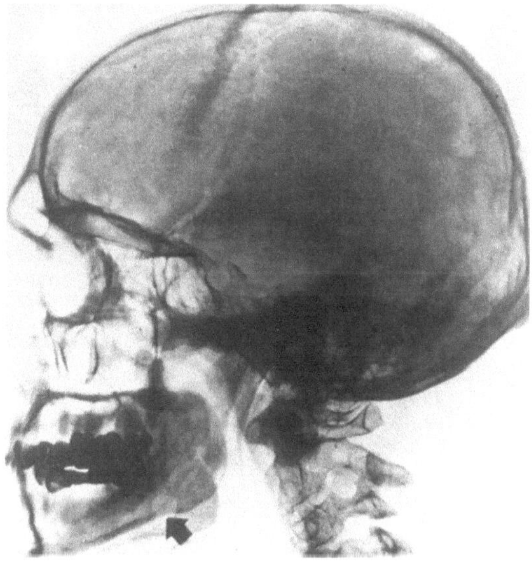

Hypernephroma. Round-shaped osteoporosis in the lawer jaw $IgG_1$-κ

Fig. V/29

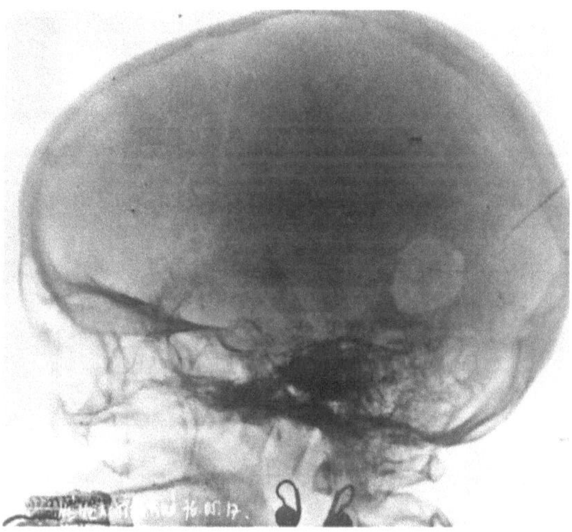

Cc. mammae. M component is not observed

Fig. V/30

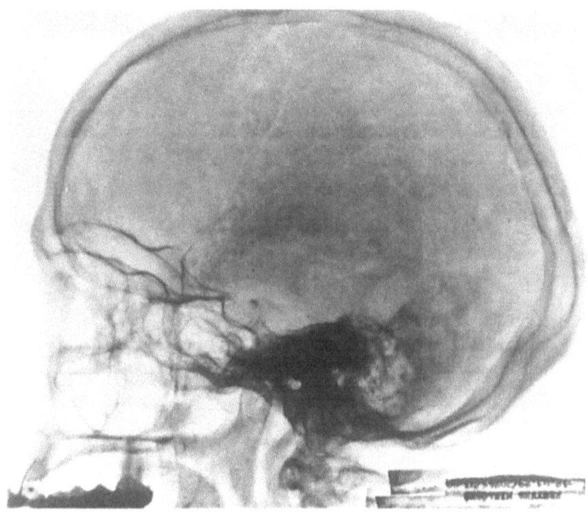

Histiocytosis "x". $IgG_1$-κ

Fig. V/31

tion. Considering the above mentioned, a complex analysis which can be mathematically evaluated, has been introduced:

1. The immunoglobulin-producing plasma cells and the lymphoid elements of the bone marrow are detected with direct immunofluorescent technique. Their count is expressed as a ratio of a total nuclear cell count and is termed $S_1$.
2. The same analysis is repeated with specific labeled anti-paraprotein immune sera and the paraprotein-producing cell count is calculated. The ratio of the latter to the total count of the immunoglobulin-secreting cells is expressed as $S_2$.
3. $S_3$ is the total serum protein content in g%.
4. The percent ratio of paraproteins to the total protein content is termed $S_4$.

Discriminative analysis has been used for the mathematical evaluation of the data. The

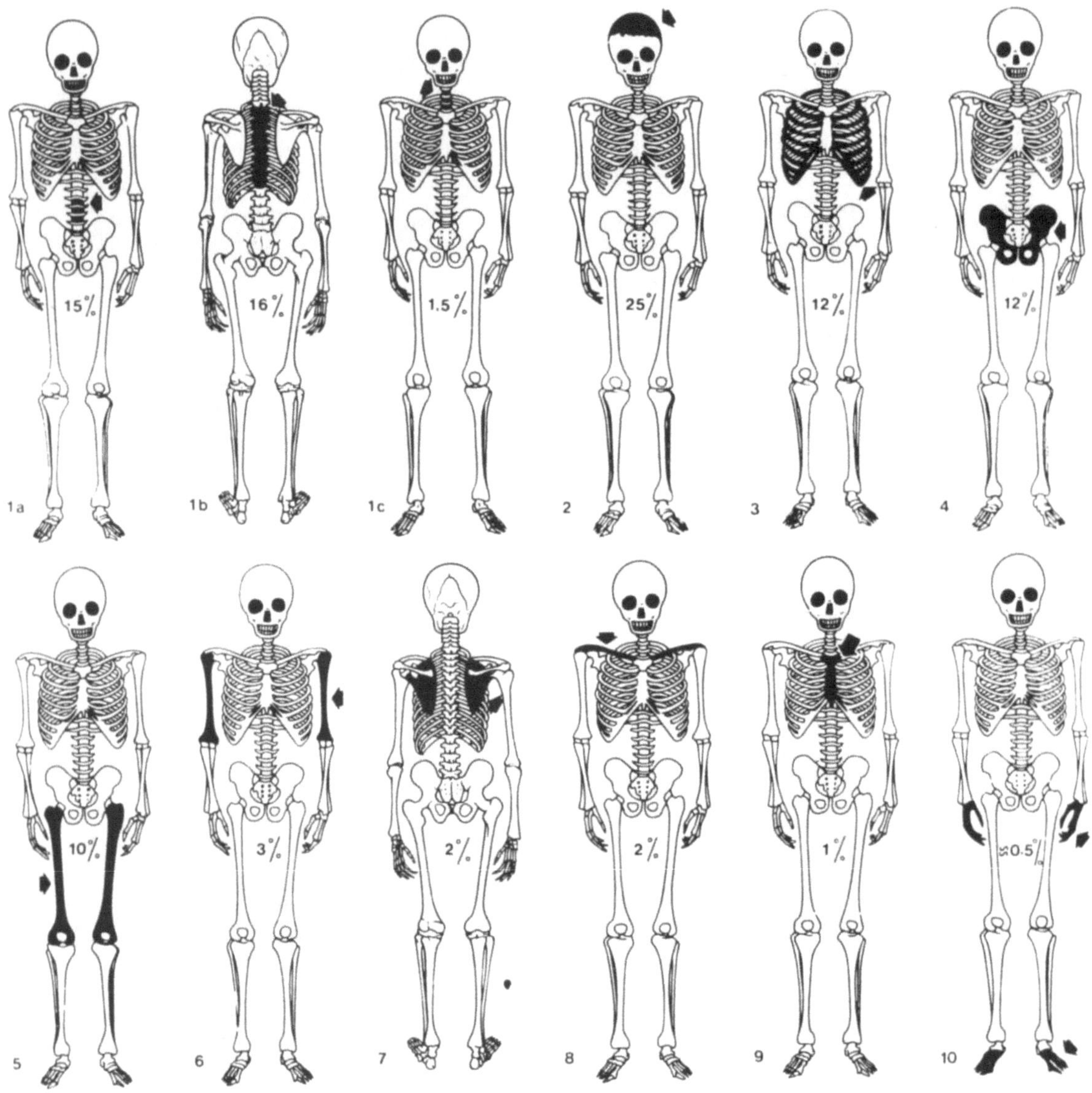


Fig. V/32

log10 values of $S_1$, $S_3$, $S_4$ and the discounted value of the $S_2$ yield values between 0 and 1. The value is 1, if the cell ratio is 98% or more and is 0 or about 0, if the ratio is less than 98%. $S_3$ can be omitted, but $S_4$ is important for the calculations.

$D(D_1, S_2, S_4) =$
$5.99 - 1.39 \log S_1 - 1.37 S_2D - 2.42 \log S_4.$

If the D value exceeds 0.25, the process is benign, if it is under 0.25 than it is malignant (129). Though this method is a secure tool for differentiation between IMG and OMG, it is not introduced for routine clinical use in Europe.

## III. X-ray and scintigraphic study of the bones

Interpretation of the scintigraphic and the X-ray data for the differentiatiation between IMG, OMG and FMG is extremely difficult. In aged patients natural osteoporosis obscure the diagnostic role of this pathology. Moreover, in 20-25% of OMG diffuse plasmocytoma is not accompanied by osteolysis.

Osteolysis, however, is an important third criterion and sometimes the only visible alteration, which helps in the diagnostics of multiple myeloma. Typically the skull is affected. The following figures show that osteolysis is not a typical feature and can cause numerous problems (Fig. V/28, V/29, V/30, V/31).

Statistical analysis indicates that osteolytic changes have different spinal localization in about 32.5% of the cases (lumbal - 15%; dorsal - 16%; cervical - 1.5%). The so-called "map-skull" occurs in 25% of the patients. Hip-bone osteolysis occurs in 12%, the femur and the ribs are affected in 10% and 12% respectively. The latter usually leads to rib fractures. In the other areas osteolysis incidence is as follows: humerus - 3%; clavicula - 2%; scapula - 2%; sternum - 1%. Sternal osteolysis should be specially emphasized, for it can be visulized only with a special X-ray technique and, thereby, can be frequently left unnoticed. If a 1% incidence of sternum osteolysis is accepted, *sternal* punction should be replaced by that in the areas, where osteolysis occurs more frequently (the cell proliferation is undisturbed and the aspiratum would be more informative). The rest 3.5% are distributed among the other bones, for example affecting the phalanxes of the fingers (Fig. V/32).

In rare cases osteolysis occurs in Waldenström's macroglobulinemia (103, 150).

It is worth pointing out that osteolysis would not necessarily be the third component of the MM-triad; osteosclerotic changes and their combination with osteolysis are documented. Totally 60-65 cases are reported in the literature (110, 117, 127. 147, 148).

It was first assumed that the plasma cells might synthesize reticulin, however, no evidence exists supporting this view. Transplantation of the murine myeloma cells into the bone marrow induced osteoporosis. If the cells were transplanted under the periosteum, osteosclerosis developed (93), probably indicating the irritation of the periosteum by the myeloma cells (?) with subsequent sclerotization.

Diagnosis of the so-called "solitary plasmocytoma" (182) is extremely difficult. No X-ray criteria exist for the differentiation between the tumor metastasis and the plasmocytoma (186).

Solitary plasmocytomas are sometimes, but not always, accompanied by MG, the latter is more typical. Thus only histological findings can help in the correct diagnosis. The presence of the MG and a solitary osteolysis does not necessarily indicate myeloma, for the MG can be of an FMG type (for example bone lesions due to hypernephroma metastasis).

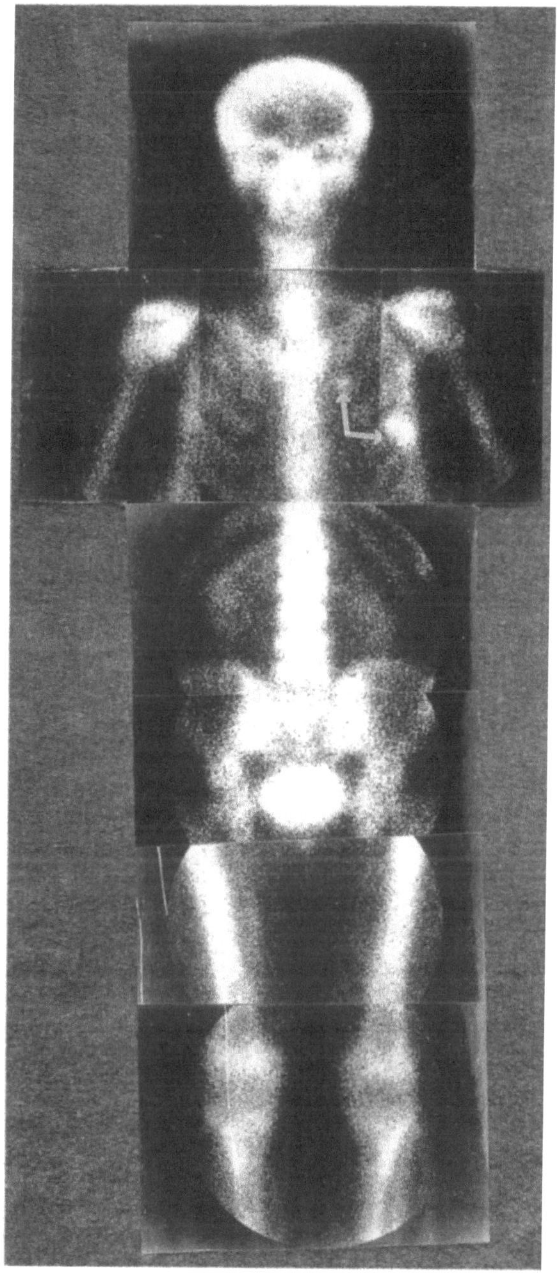

Fig. V/33

The X-ray determination of osteolytic and osteosclerotic bone changes, as well as the multiple myeloma diagnosis are not as complicated as the above mentioned cases of diffuse osteoporosis (145). Osteoporosis is the disease of the aged, similarly, myeloma can equally develop at the age of 50-60-70 years.

In our opinion, diffuse osteoporosis with typical marrow aspirates and MG indicate myeloma. If, however, the marrow finding is negative, than diffuse osteoporosis and the presence of MM are insufficient for the diagnosis. A facultative, idiopatic or, very rarely, pseudomyeloma, should be suggested (20, 115).

Since the X-ray findings are sometimes insufficient for the assessment of the bone lesions especially in the early phases of MM, bone scintigraphy should be a method of choice (Fig. V/33). $^{67}$Ga-citrate ($^{67}$Ga), $^{99m}$Tc-diphosphonate ($^{99m}$Tc-DP), $^{99m}$Tc-polyphosphate ($^{99m}$Tc-PP), $^{99m}$Tc-sulphur colloid ($^{99m}$Tc-SC), $^{171}$ER-, $^{157}$Dy-, $^{167}$Tm-citrate alone or as two-component combinations can give good results. However clinical experience shows that bone scintigraphy can be used as a valuable tool only within certain limits. It is widely accepted that compared to the classic myeloma loci, extramedullar plasmocytomas might develop from the lymphoid tissue (79). Since they are well vascularized (130), $^{67}$Ga can be used as a good tracer, though not 100% of the cases studied have yielded positive results! Revascularization is the first step in the recalcification of the bone lesions following successful (?) therapy. In these cases $^{99m}$Tc-PP gives the best results (72). Amyloid accumulating in the tissues prior to the development of any clinical symptoms, readily uptakes the $^{99m}$Tc-PP (173).

*In summary,* bone scintigraphy is a method of restricted value (though can be informative in special cases) in the detection of the MM bone lesions and its value does not exceed that of the X-ray studies. Systematic X-ray studies and those directed by the "pain" are of utmost importance and can not be omitted.

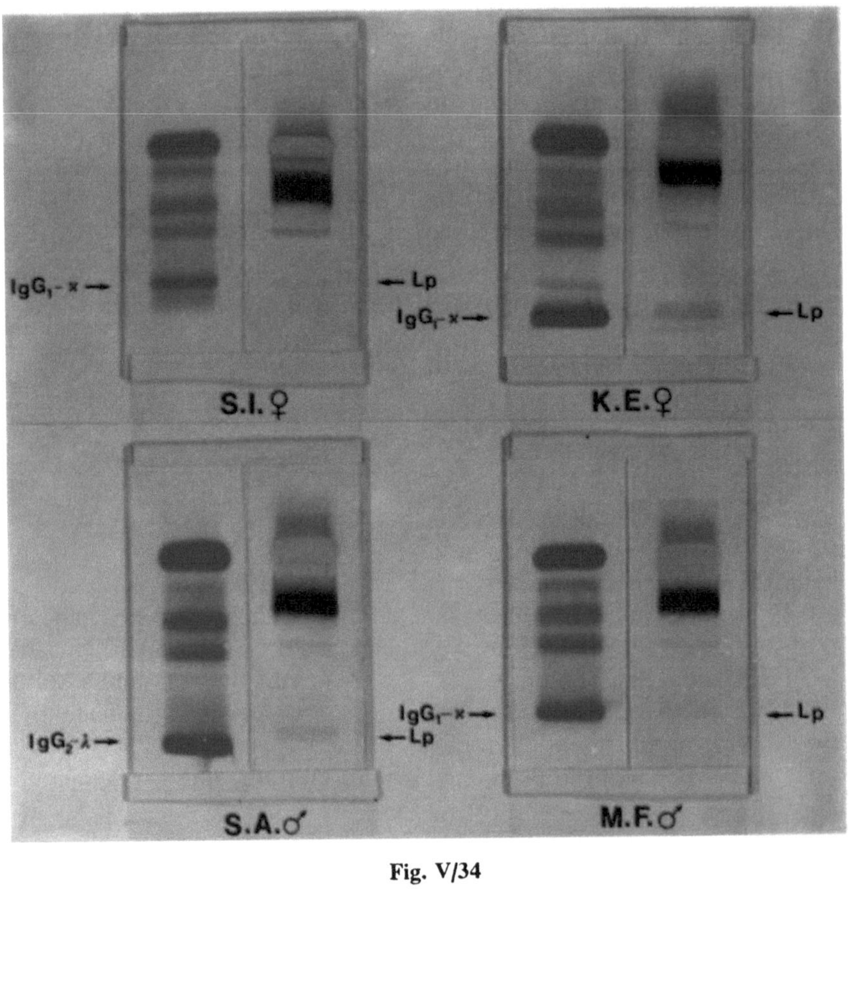


Fig. V/34

Clinically IMG and OMG can scarcely be differentiated. It is well documented that a real diffuse plasmocytoma is not a common form of the MG. Clinical experience indicates that MG is generally revealed as OMG because these patients have complaints and certain clinical and laboratory findings. So this cohort of patients is hospitalized and diagnosed, while patients with IMG stay complaint-free, "healthy" and, therefore, are not examined. Inpatients with IMG never have complaints or other signs (history) suggesting paraproteinemia. Though the monoclonal gammopathia indicating hospital admission of the OMG patients is generally atypical (137), this paraneoplastic syndrome must always be regarded as typical.

*Infections*

Infection is the most frequent diagnosis. Usually purulent pathogenic agents causing furunculosis, pneumonia or even septicemia occur (47).

Tuberculosis occurs rather rarely and is usually due to the endogenous reinfection caused by the decreased resistance of the organism (?!) and by the so-called calcium-mobilizing substance.

Compared to the high incidence of the bacterial infections, viral infections occur rather rarely, because plasma cells produce substantial amounts of interferon. The latter (α-INF) is produced together with immunoglobulins independently of the normal or pathological function of the latter (153). We observed herpes zoster in 37 cases, this observation being in apparent controversy with the interferon theory. The analysis of the case histories, however, show that no such contradiction exists: of 291 OMG patients 94 were admitted primarily due to bacterial infections; 7 cases of herpes zoster developed following cystostatic treatment.

*Impairment of the renal function*

It is well known that the growing tumors scarcely elicit clinical symptoms, the diagnostically important signs are usually caused by the concomitant complications. Thus renal failure is a frequent symptom in patients with MG, and myeloma is identified later. Correlation between MM and the impairment of the renal function is well documented (52, 101, 137). Both tubular and glomerular failure develop, usually in combination. Data of the literature indicate that apart from the infections, renal failure is the most frequent cause of death (12, 23).

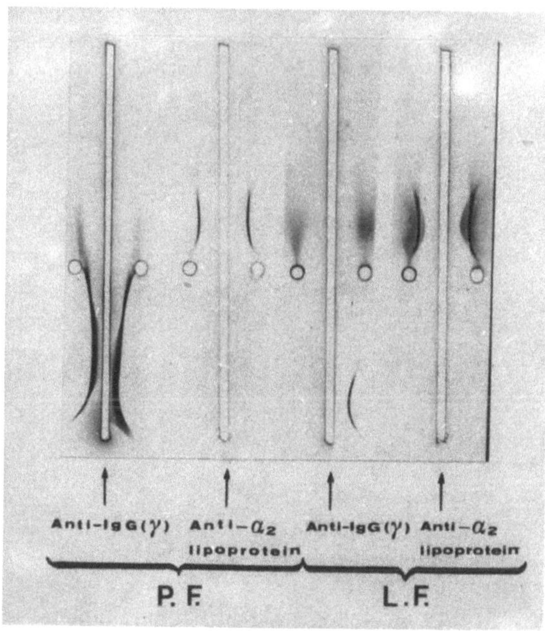


Fig. V/35

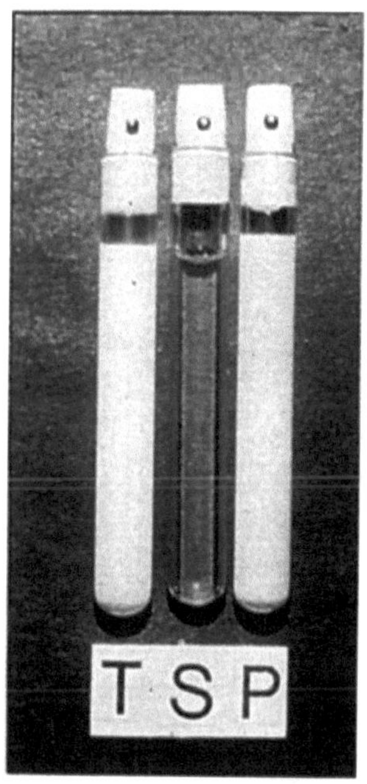


T = milk
S = NHS
p = patient's serum

Fig. V/36

*Ostalgia*

Osteoporosis-induced ostalgia, osteolysis or osteosclerosis is the third dominant symptom frequently occurring in this pathology (13). Primary pathological rib, vertebral or tubular bone fracture is a frequent cause of admission.

*Paraneoplastic syndrome*

The signs of this syndrome can be assessed into three groups:

A) Pathology, due to the "M"-component antibody activity:
- Cold agglutinine disease. The anti-I antibodies are produced against the RBC membranes, bacterial cells and proteins. Though IgM is a predominant form, IgA can also occur (44, 157, 158).
- Autoimmune hemolytic anemia is always mediated via IgG antibodies.
- Hemophilia induced by the so-called inhibitory bodies. Formation of the anti-AHG antibodies is attributed to the primary pathomechanism of this disease. Both IgM and IgA (antibodies against the VIII clotting factor) can occur (57, 140).
- Hyperlipemia. Anti-lipoprotein IgG or IgA antibodies generally occur (142). We observed 4 patients with this pathology (Fig, V/34, V/35, V/36).

Hyperlipemia can, however, develop without anti-lipoprotein antibodies.
- Mixed kryoglobulinemia is usually mediated via the anti-IgG-type monoclonal IgM antibodies. Other classes, for example IgA-type anti IgG; IgM-type anti IgA and IgG-type anti IgG are also reported (31, 60, 122, 181).

The relationship between FMG and the paraneoplastic syndrome is worth pointing out. It is assumed and experimentally verified that the observed "M" component is a tumor-specific antibody (25): normal and tumor tissues of the patients were treated with the anti "M" compound-antibodies produced in animals and labeled with FITC. Fluorescence was detected only in the tumor tissue, probably showing a "protective" reaction (176). However the presence of the complement was not detected (153).

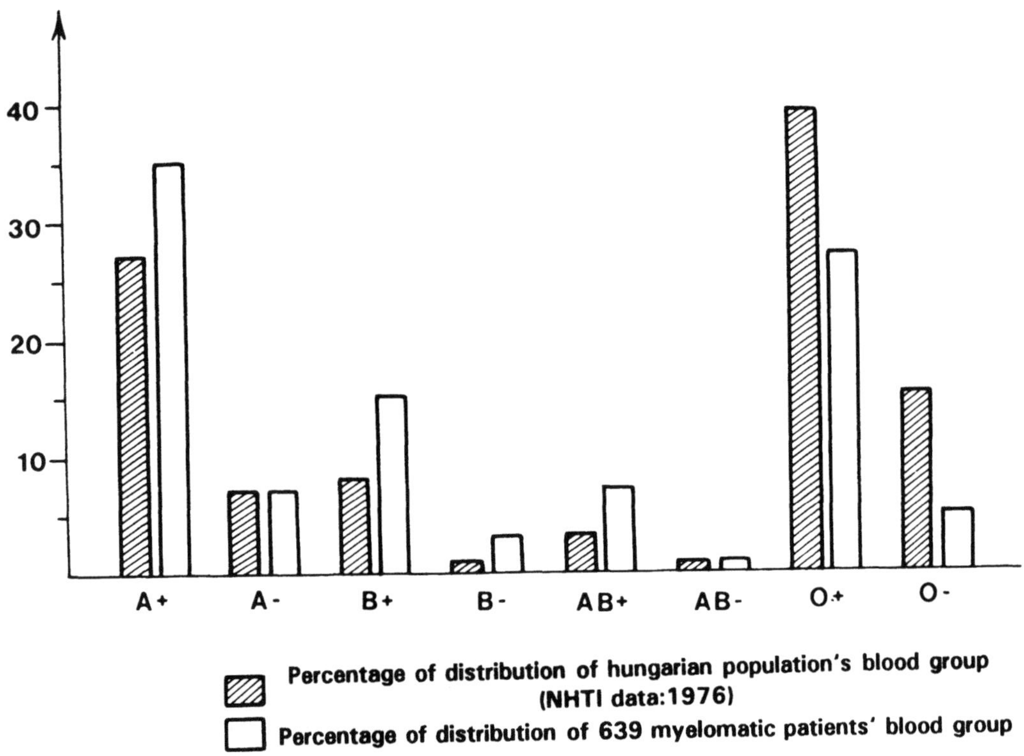


Fig. V/37

The above findings are explained as follows:

a) if these antibodies can really detect the K cells, it might explain the longer survival rate of the FMG patients (?!).

b) the enhancement reaction with the subsequent inhibition of the natural immune response can not be precluded (16).

At present the significance of these mechanisms still remains uncertain.

B) Pathological changes due to the physico-chemical properties of the M component:

- hyperviscosity syndrome. IgM (and other Ig capable of aggregation) frequently causes microcirculatory damages (113).
- Primary amyloidosis. In MG-s amyloid fragments (?) are formed from the light chains of immunoglobulins. The insolubility of the amyloid molecules contributes to the pathological changes. In some cases of amyloidosis occurring in the non-secretory myelomas (153) the stored substance was not derived from immunoglobulins (153).
- Kryoglobulinemia. At low temperatures the M components aggregate without complex-formation (119).
- Pyroglobulinemia. Thermal aggregation of the free kappa-chains in the serum. After cooling the chains are re-solubilized, this process being the opposite of kryoglobulinemia.

C) Clinical sequellae of the non-immunoglobulin secretion (?) occurring in the M-producing cells.

- Secondary antibody-deficient syndrome (45)
- Hematological changes (11, 189)
  Anemia and leukopenia are usually secondary or develop in the course of therapy or at the terminal stages of the disease. Thereby they can not be considered as the early signs of pathology. An interesting relationship between the blood types and the Kahler disease is observed by us, however, its clinical use meets obstacles (51, 74) (Fig. V/37).
- Multifocal leukoencephalopathy or demyelinization.
  Myeloma patients are rarely admitted to neurological wards due to peripheral neuropathy (38, 63). Two forms of neurological changes exist:
  a) mild chronic symmetric forms usually accompanied with paresthesia;
  b) acute- subacute sensomotoric neuropathy (32) with frequently developing palsies (137). These symptoms usually occur together with osteosclerosis (117, 127), but show no correlations with bone lesions, amyloidosis and the size of tumor. Histologically neuronal degeneration and demyelinization occur. Though amyloid is never detected in the neuronal fibres, the damage still develop. Lymph nodes are palpable in these patients. Their high serum alkaline phosphatase and increased levels of immunoglobulins (MG) in the CSF are of diagnostic value (147).

The carpal tunnel syndrome should be regarded as a special case. Hypesthesia, hypoalgesia, paresthesia and/or atrophy of the opponent adductor muscles are worth pointing out. Furthermore, excretion of the lambda light-chain and primary amyloidosis are also typical of these cases (97).

In summary, clinical and laboratory sudies provide possibilities for the differential diagnosis of the most IMG, OMG and FMG cases; however, in some patients the correct diagnosis mainly depends on time, on the accurate constant follow-up of the patients (154).

**References**


1. Abramson, N., Shattils, S. J.
   M-components.
   J. Amer. med. Ass. 223, 156, 1973.
2. Ameis, A., Ko, H. S., Pruzanski, W.
   M-components – a review of 1242 cases.
   Canad. med. Ass. J. 114, 889, 1976.
3. Arend, W. P., Adamson, J. W.
   Nonsecretory myeloma: immunofluorescent demonstration of paraprotein within bone marrow plasma cells.
   Cancer 33, 721. 1974.
4. Asconas, B. A., Fahey, J. L.
   Formation of Bence Jones protein and myeloma protein in vitro by the plasmacell tumor MPC-2.
   Biochem. J. 80, 261. 1961.
5. Avertack, P.
   A human tubular array plasma cell.
   Virchows Arch. path. Anat. 337, 17. 1977.
6. Axelsson, U., Hällén, J.
   Familial occurence of pathological serum proteins of different gamma-globulin groups.
   Lancet. ii. 369. 1965.

7. Axelsson, U., Bachmann, R., Hällén, J.
Frequency of pathological proteins (M-components) in 6.995 sera from an adult population.
Acta med. scand. 179, 235. 1966.
8. Bachmann, R.
The diagnostic significance of the serum concentration of pathological proteins (M-components).
Acta med. scand. 178, 801. 1965.
9. Backhausz, R.
Immunodiffusion und Immunoelektrophorese.
Akadémiai Kiadó Budapest. 454, 1967.
10. Balogh, E., Kaszás, I., Krizsa, F., Fleischmann, T.
Chromosoma vizsgálatok myeloma multiplexben.
Orvosi Hetilap. 123, 35, 1982.
11. Balogh, E., Krizsa, F., Varga, Gy., Cserháti, I., Husz, S.
Myeloma multiplexben szerzett tapasztalataink 25 beteg észlelése során.
Orvosi Hetilap. 121, 2505, 1980.
12. Bayrd, E. D., Heck, F. J.
Multiple myeloma: Review of eighty-three proved cases.
J. Amer. med. Ass. 133, 147. 1947.
13. Begemann, H.
Klinische Hämatologie.
Thieme: Stuttgart. 652, 1970.
14. Berrebi, A., Estrov, Z.
Twenty years follow-up in a patient with multiple myeloma.
Acta haemat. 66, 269. 1981.
15. Bhoopalam, N., Chen, Y., Yakuls, V., Heller, P.
The effect of RNA-rich extract from mouse plasmacytoma MOPC 104 E on the immune response.
Clin. exp. Immunol. 24, 357. 1976.
16. Billingham, R. E., Brent, L., Medewar, P. B.
„Enhancement" in normal homografts, with a note on its possible mechanism.
Transplant, Bull. 3, 84. 1956.
17. Boga, M., Jákó, J., Domán, J., Magyar, É., Kónyár, É.
Familial myeloma.
Folia haemat. 100, 201. 1973.
18. Brandtzaeg, P., Baklien, K.
Immunoglobulin producing cells in the intestine in health and disease.
Clinics in Gastroenterology. 5, 251. 1976.
19. Brücher, H.
Das beginnende Plasmozitom.
Blut. 28, 136. 1974.
20. Buonocore, E., Solomon, A., Kerley, H. E.
Pseudomyeloma.
Radiology. 95, 41. 1970.
21. Caluvel, J. P., Damon, F., Seligmann, M.
Immunoglobulines monoclonales decelees en l'absence de myelome an de macroglobulinemia de Waldenström.
Nouv. Rev. Franc. Hémat. 11, 677. 1971.
22. Carrel, R. W., Colls, B. M., Murray, J. T.
The significance of monoclonal gammopathy in a normal population.
Aust. N. Z. J. med. 1, 398. 1971.
23. Carson, C. P., Ackerman, L. V., Maltby, J. D.
Plasmacell myeloma: clinical, pathologic and roentgenologic review of 90 cases.
Amer. J. clin. Path. 25, 849. 1955.
24. Cassuto, J. P., Hammov, J. C., Pastrolle, E., Dujardin, P., Masseyeff, R.
Plasmacell acid phosphatase, a discriminative test for benign and malignant monoclonal gammopathies.
Biomedicine. 27, 197. 1977.
25. Causey, J. Q.
IgG paraproteinaemia associated with bronchogenic carcinoma.
Arch. intern. med. 119, 407. 1967.
26. Chen, Y., Bhoopalam, N., Yakulis, V., Heller, P.
Change in lymphocyte immunoglobulins in myeloma and the effect of an RNA-containing plasma factor.
Ann. intern. med. 83, 625. 1975.
27. Clyn, D. H., Brendstrup, L., First, M. P., Pesce, A. J., Finkel, P. N., Pollak, V. E., Pirani, C. L.
Renal effects of intraperitoneal kappa chain injection. Induction of crystals in renal tubular cells.
Lab. Inves. 31, 131. 1974.
28. Cohen, H. J., Lefere, L. G.
Intranuclear inclusions in Bence Jones lambda plasma cell myeloma.
Blood. 45, 131. 1975.
29. Collan, Y.
Characteristics of non epithelial cells in the epithelium of normal rat ileum.
Scand. J. Gastroenterol. 7, 18. 1972.
30. Conklin, R., Alexanian, R.
Clinical classification of plasma cell myeloma.
Arch. intern. med. 135, 139. 1975.
31. Cooper, A. G., Hobbs, J. R.
Immunoglobulins in chronic cold haemogglutinin disease.
Brit. J. Haemat. 19, 383. 1970.
32. Croft, P. B., Urich, H., Wilkinson, M.
Peripheral neuropathy of sensorimotor type associated with malignant disease.
Brain. 90, 31. 1967.
33. Danon, F., Clauvel, J. P., Seligmann, M.
The IgG and IgA paraproteines in non myelomatous disseases.
Rev. franc. Étud. clin. biol. 12, 681. 1967.
34. Deane, H. W.
Some electron microscopic observations of the lamina propria of the gut, with comments on the close association of macrophages, plasma cells and eosinophils.
Anat. Rec. 149, 453. 1964.
35. Dennellan, W. L.
The structure of the colonic mucosa. The epithelium and subepithelial reticulohistiocytic complex.
Gastroenterology. 49, 496. 1965.
36. Derycke, C., Fine, J. M., Boffa, G. A.
Dysglobulinémies essentielles chez les sujets ages.
Nouv. Rev. franc. Hémat. 5, 729. 1965.
37. Deutsch, H. F.
Relationships of two light chain immunoglobulins isolated from the same human source (low molecular weight immunoglobulins).
Immunoschemistry. 2, 207. 1965.
38. Doll, D. C., Weiss, R. B.
Unusual presentations of multiple myeloma.
Postgrad. med. J. 61, 116. 1977.
39. Domán, J., Jákó, J., Szárits, Á.
Adatok az antitest radiál-immundiffúziós módszer kivitelezéséhez.
Orvosi Hetilap. 116, 1639. 1975.
40. Drivsholm, A., Clausen, J.
The relationship between the cytology and the immunoelektrophoretic pattern in 105 cases of myelomatosis.
Acta med. scand. 175, 609. 1964.
41. Dugue, M., Rousselet, F., Kahn, M. F., et al.
Etudes biologiques sur 559 cas de paraproteinemies.
Clin. chim. Acta. 33, 75. 1971.
42. Eidelman, S., Davis, S. D.
Immunoglobulin content of intestinal mucosal plasma-cells in ataxia teleangiectasia.
Lancet. i. 884. 1968.
43. Eidelman, S., Lagunoff, D.
The morphology of the normal human rectal biopsy.
Human Pathology. 3, 389. 1972.
44. Eisen, H. N., Michaelaelides, M. C., Underdown, B. J., Schulenburg, E. P., Simms, E. S.
Myeloma proteins with anti-hapten antibody activity.
Fed. Proc. 29, 78. 1970.
45. Fahey, J. L., Scoggins, R., Utz, J. P.
Infection, antibody response and gammaglobulin components in multiple myeloma and marcroglobulinemia.
Amer. J. med. 35, 698. 1963.

46. Farman, J., Degnan, T. J.
Multiple myeloma with small-bone involvement.
N. Y. St. J. Med. 76, 990. 1976.
47. Fateh-Moghadam, A., Lamerz, R., Knedel, M., Baner, B.
Quantitative immunologische Bestimmung von Serumproteinen bei Paraproteinaemien.
Dtsch. med. Wschr. 98, 309. 1973.
48. Ferguson, A.
Lymphocytes in coeliac disease.
In: Coeliac Disease. Proceedings of the second international coeliacs symposium.
Ed.: by Hekkens W. Th. J. M., Pena, A. S., Stefert, H. E., Kroese, B. V.
Leiden. 1974.
49. Feslen, J. J. M., Marrink, J., dr. Waard-Kuiper, E. H., Mandema, E.
Immunoglobulins in families of myeloma patients.
Scand. J. Immunol. 6, 887. 1977.
50. Fine, J. M., Dormont, J., Mery, J. P., Creyseel, R., Grovlade, J., Debray-Sachs, M.
Études sur la nature d'une pyroglobuline sérique observée an cours du myelome.
Rev. franc. Étud. clin. biol. 6, 884. 1961.
51. Fine, J. M., Lambin, P., Leroux, P.
Frequency of monoclonal gammopathy (M-components) in 13 400 sera from blood donors.
Vox Sang. 23, 336. 1972.
52. Fine, J., Luke, R. G., Rees, E. D.
Multiple myeloma and renal impairment.
Lancet. ii, 1205. 1973.
53. Fischer, E. R., Zawadzki, Z. A.
Ultrastructural features of plasma cells in patients with paraproteinemias.
Amer. J. clin. Path. 54, 779. 1970.
54. Grabbe, P. H., Heremans, J. F.
The distribution of immunoglobulin-containing cells along the human gastrointestinal tract.
Gastroenterology. 51, 305. 1966.
55. Graham, R. C., Bernier, G. M.
The bone marrow in multiple myeloma: correlation of plasma cell ultrastructure an clinical state.
Medicine. 54, 225. 1975.
56. Glenner, G. G., Ein, D., Terry, W. D.
The immunoglobulin origin of amyloid.
Amer. J. Med. 52, 141. 1972.
57. Glveck, H. I., Hong, R. A.
Circulating anticoagulant in IgA multiple myeloma its modification by penicillin.
J. clin. Invest. 44, 1866. 1965.
58. Hallén, J.
Discrete gamma-globulin M-components in serum. Clinical study of 150 subjects without myelomatosis.
Acta med. scand. 180, 462. 1966.
59. Hallén, J.
Frequency of „abnormal" serum globulins (M-components) in the aged.
Acta med. scand. 173, 737. 1963.
60. Harboe, M., Torsvik, H.
Protein abnormalities in the cold haemagglutinin syndrome.
Scand. J. Haematol. 6, 416. 1969.
61. Harris, R. I., Kohn, J.
An urinary cryo-Bence Jones protein gelling at room temperature.
Clin. chim. Acta. 53, 233. 1974.
62. Heller, P., Yakulis, V., Bhoopalam, N., Costea, N., Cabana, V., Nathan, R. D.
Surface immunoglobulins on circulating lymphocytes in mouse and human plasmacytoma.
Transact. Assoc. Amer. Physicians. 85, 192. 1972.
63. Hesselvik, M.
Neuropathological studies on myelomatosis.
Acta neurol. scand. 45, 95. 1969.
64. Hobbs, J. R.
Disturbances of the immunoglobulins.
Sci. Basis Med. 106, 27. 1966.
65. Hobbs, J. R.
Paraproteins, benign or malignant?
Brit. med. J. 3, 699. 1967.
66. Hobbs, J. R.
Monitoring myelomatosis.
Arch. Intern. Med. 135, 125. 1975.
67. Hopper, J. E., Cera, L.
The structure of human immunoglobulins.
Ann. clin. Lab. Sci. 8, 201. 1978.
68. Husby, G., Sletten, K., Michaelsen, T. E.
Amyloid fibril protein subunit, „protein AS" distribution in tissue and serum in different clinical types of amyloidosis including that associated with myelomatosis and Waldenströms macroglobulinemia.
Scand. J. Immunol. 2, 395. 1973.
69. Husz, S., Krizsa, F., Dobozy, A., Hunyadi, J., Kenderessi, Sz. A., Simon, M.
Egy újabb lehetőség a benignus monoklonalis gammopathia és a myeloma multiplex elkülönítő kórismézésében.
Magyar Belorvosi Arch. 33, 162. 1980.
70. Husz, S., Török, L., Kovács, L., Szabó, É., Simon, M.
Ein Fall von mit IgU ($\alpha$)-Myeloma multiplex komplizierter Amyloidose.
Z. Hautkr. 54. 585. 1979.
71. Husz, S., Krizsa, F., Dobozy, A., Hunyadi, J., Simon, M.
B and T Cell markers of Bone Marrow and peripheral Blood Lymphoid Cells in Patients with Paraproteinaemia.
Acta haemat. 57, 321. 1977.
72. Hübner, K. F., Andrews, G. A., Hayes, R. L., Poggenburg, J. K., Solomon, A.
The use of rare-earth radionuclides and other boneseekers in the evaluation of bone lesions in patients with multiple myeloma or solitary plasmocytoma.
Radiology. 125, 171. 1977.
73. Hyun, B. H., Kwa, D., Gabaldon, Herminia, Ashton, J. K.
Reactive plasmacytic lesions of the bone marrow.
Amer. J. clin. Path. 65, 921. 1976.
74. István, L.
Vércsoportok és betegségek.
Orvosi Hetilap. 102, 202. 1961.
75. István, L.
Plasmocytome im Alter.
In: Haematologie im Alter.
Ed.: by Böhnel, J., Heinz, R., Sacher, A.
Urban-Schwarzenberg. Wien–München–Baltimore.
167. 1982.
76. István, L., Hindy, I., Eckhardt, S.
Adatok a plasmocytoma klinikumához, kezeléséhez és gondozásához.
Orvosi Hetilap. 117, 11. 1976.
77. Isobe, T., Osserman, E. F.
Pathologic conditions associated with plasma cell dyscrasia. A study of 806 cases.
Ann. N. Y. Acad. Sci. 190, 507. 1971.
78. Itani, S., Hoshino, T., Kawasaki, S., Nakayama, S.
Chromosome abnormality and its significance in human multiple myeloma.
Acta haemat. Jap. 33, 54. 1970.
79. Jaeger, E.
Das extramedulläre Plasmocytom.
Z. Krebsforsch. 52, 349. 1942.
80. Jákó, J.
Die Titrationsimmunoelektrophorese als ein Form der quantitativen Immunoelektrophorese.
Hung. Sci. Instruments Sonderausgabe Achema. 55. 1967.
81. Jákó, J.
Ist die Immunoglobulinsynthese der wuchernden Plasmacellen immer die gleiche?
Österreichische Gessellschaft für Hämatologie und Okologie Graz, 25–26. Sept. 1981.
82. Jákó, J., Hitzig, W. H.

In „Clinical Biochemistry" Principles and Method
Ed.: by **Curtius H. Ch., Roth, M.**
Walter de Gruyter. Berlin-NY. 1974.
83. Jákó, J., Pádár, J., Domán, J.
Quantitative measurement of proteins by immunodiffusion methods.
Hung. Sci. Instruments. 30, 33. 1974.
84. Jákó, J., Virágh, Sz.
Nierenschädigung durch Lambda-Leichtketten.
**1. Donausymposium für Nephrologie.**
Linz 252. 1976.
85. Jákó J., Virágh, Sz., Boga, M., Brooser, G., Dóbiás, Gy., Domán, J., Ottó, Sz., Riskó, Z., Szemere, P.
A case of IgD-lambda myeloma.
Haemat. hung. 9, 261. 1975.
86. Jones, S. V., McFarlone, H.
T and B cells in myelomatosis.
Brit. J. Haemat. 31, 545. 1976.
87. Kalderon, A. E., Bogaars, H. A., Diamond, J. C., Cummings, F. J., Kaplan, S. R., Calabresi, P.
Ultrastructure of myeloma cells in a case with of crystal cryo globulinemia.
88. Katzmann, J., Giacomoni, D., Yakulis, V., Heller, P.
Characterization of two plasmacytoma fractions and their RNA capable of changing surface immunoglobulin (cell convertion).
Cell. Immunol. 18, 98. 1975.
89. Kindler, V.
Zur Differenzierung der Paraproteinaemieformen.
Dtsch. med. Wschr. 97, 646. 1972.
90. Kitamura, M., Yamaguchi, H., Murakawa, K., Murao, T., Tizuka, Y.
Screening for multiple myeloma using routine laboratory test results.
Clin, Biochem. 15, 17. 1982.
91. Kitani, T., Yonezawa, T., Imanaka, T., Hiraoka, A., Nasu, T.
Ultrastructural analysis of membrane-bound polysomes in human myeloma cells.
Blut. 44, 51. 1982.
92. Klinghofer, L., Boroviczeny, K. G.
Ein atypisches Plasmocytom.
Schweiz. med. Wschr. 91, 1193. 1961.
93. Kobayashi, H., Potter, M., Dunn, T. B.
Bone lesions produced by transplanted plasma-cell tumors in Balb/c mice.
J. nat. Cancer Instr. 28, 649. 1962.
94. Krakauer, R. S., Strober, W., Waldmann, T. A.
Hypogammaglobulinemia in experimental myeloma: The role of suppressor factors from mononuclear phagocytes.
J. Immunol. 118, 1385. 1977.
95. Krull, P., Kühn, K., Zobl., H., Sterzel, R. B.
Akutes Nierenversagen bei Bence Jones-Plasmozytom.
Dtsch. med. Wschr. 89, 318. 1973.
96. Kyle, R. A., Bayrd, E. D.
„Benign" monoclonal gammopathy: a potentially malignant condition?
Amer. J. Med. 40. 426. 1966.
97. Kyle, R. A., Bayrd, E. D.
Amyloidosis: Review of 236 cases.
Medicine. 54, 271. 1975.
98. Kyle, R. A., Bayrd, E. D.
The Monoclonal Gammopathies.
Charles C. Thomas Publischer.
Springfield Illinois USA. 133. 1976.
99. Kyle, R. A., Elveback. L. R.
Management and prognosis of multiple myeloma.
Mayo Clin. Proc. 51, 751. 1975.
100. Kyle, R. A., Finkelstein, S., Elveback, L. R., Kurland, L. T.
Incidence of monoclonal proteins in a Minnesota Community with a cluster of multiple myeloma.
Blood 40, 719. 1972.
101. Kyle, R. A., Greipp. P. R.
Smouldering multiple myeloma.
N. Engl. J. Med. 302, 1347. 1980.
102. Kyle, R. A., Maldonado, J. E., Bayrd, E. D.
Idiopathic Bence Jones proteinuria. A distinct entity?
Amer. J. Med. 55, 222. 1973.
103. Leb, L., Grimes, E. T., Balogh, K., Meritt, J. A.
Monoclonal macroglobulinemia with osteolytic lesions.
Cancer. 39, 227. 1977.
104. Lennert, K.
Malignant lymphomas other than Hodgkin's disease.
Springer-Verlag Berlin–Heidelberg–New York. 1978.
105. Lieberman, R., Mushinski, J. F., Potter, M.
Two chain immunoglobulin. A molecules: abnormal or normal intermediates in synthesis.
Science. 159, 1355. 1968.
106. Limas, C.
Amyloidosis and multiple myeloma.
Amer, J. Med. 54, 116. 1973.
107. Lindström, F. D., Hardy, W. R., Eberle, B. J., Williams, R. C.
Multiple myeloma and benign monoclonal gammopathy: differentiation by immunfluorescence of limphocytes.
Ann. Intern. Med. 78, 837. 1973.
108. Lindström, F. D., Dahlström, U.
Multiple myeloma or benign monoclonal gammopathy? A study of differential diagnostic criteria in 44 cases.
Clin. Immunol. Immunopathol. 10, 168. 1978.
109. Linke, R. P., Sipe, J. D., Pollock, P. S.
Isolation of low-molecularweight serum component antigenically related to an amyloid fibril protein of unknown origin.
Proc. nat. Acad. Sci. 72, 1473. 1975.
110. Lowbeer. L.
Occurence of osteosclerosis in multiple myeloma.
Lab. Med. Bull. Pathol. 10, 397. 1969.
111. Löffler, H., Schubert, J. C. F.
Cytochemische Unterschiede zwischen Plasmazellen und Myelomazellen.
Klin. Wschr. 41, 484. 1963.
112. Löffler, H.
Unspezifische Esterase und säure Phosphatase bei Plasmocytomen.
Blut. 15, 330. 1967.
113. MacKenzie, M. R., Fundenberg, H. H., O'Reilly, R. A.
The hyperviscosity syndrome. I. In IgG myeloma. The role of protein concentration and molecular shape.
J. Clin. Invest. 49, 15. 1970.
114. Maldonado, J. E., Kyle, R. A.
Familial myeloma. Report of eight families and a study of serum proteins in their relationes.
Amer. J. Med. 57, 875. 1974.
115. Maldonado, J. E., Riggs, B. L., Bayrd, E. D.
Pseudomyeloma: is association of severe osteoporosis with serum monoclonal gammopathy an entity or a coincedence?
Arch. Intern. Med. 135, 267. 1975.
116. Mancini, G. A., Carbonara, D., Heremans, J. F.
Immunochemical quantitation of antigens by single radial immunodiffusion.
Immunochemistry. 2, 235. 1965.
117. Mangalih, A., Veliath, A. J.
Osteosclerotic myeloma and peripherial neuropathy: A case report.
Cancer. 28, 1040. 1971.
118. March, M. N.
Studies of intestinal lymphoid tissue.
Gut. 16, 665. 1975.
119. Meinke, C. G., Spegelberg, H. L.
Amino acid sequence of the first hypervariable region of 2 kappa and a lambda Bence Jones cryoglobulin.
Immunochemistry 13, 915. 1976.
120. Mellors, R. C., Korngold, L.
The cellular origin of human immunoglobulins.
J. exp. Med. 118, 387. 1963.
121. Mellstedt, H., Holm, G.
In vitro studies of lymphocytes from patients with plasmacell myeloma. I. Stimulation by mitogens and cytotoxic activities.

Clin. exp. Immunol. 15, 309. 1973.
122. Meltzer, H. M., Franklin, E. C., Elias, K., Mc Luskey, F. T., Cooper, N.
Cryoglobulinemia. A clinical and laboratory study.
Amer. J. Med. 40, 837. 1966.
123. Menkes, C. J., Hereman, G., Preud'homme, J. L., Dodeau, P., Delbarre, F.
Non secretory plasma cell myeloma. Two new cases.
Nouv. Pres. Med. 29, 309. 1972.
124. Meyer. F.
In vitro incoporation of $C_{14}$-lysine into Bence-Jones protein.
Proc. Soc. Exp. Biol. 110, 106. 1962.
125. Michaux, J. L., Heremans, J. F.
Thirty cases of monoclonal immunoglobulin disorders other than myeloma or macroglobulinemia.
Amer. J. Med. 46, 562. 1969.
126. Morell, A., Skvaril, F., Spengler, G., Barandun, S.
Unterscheidung benigner und maligner monoklonaler Gammopathien aufgrund von Knochenmark- und Serumuntersuchungen.
Schweiz. med. Wschr. 107, 1463. 1977.
127. Morley, J. B., Schwieger, A. C.
The relation between chronic polyneuropathy and osteosclerotic myeloma.
J. Neurol. Neurosurg. psychiat. 30, 432. 1967.
128. Mundy, G. R., Raisz, L. G., Cooper, R. A.
Evidence for the secretion of on osteoclast stimulating factor in myeloma.
New Engl. J. Med. 291, 1041. 1974.
129. Muylle, L., Smet, D. De., Cole, J., Peetermans, M. E.
HLA-DR and monoclonal gammopathy.
Tissue Antigens. 20, 397. 1982.
130. Nagaoka, M.
On the histological structure of extramedullary lesion in multiple myeloma and of extramedullare plasma-cell tumor.
J. Kyushu Hamatol. Soc. 6, 88, 1956.
131. Norgaard, o.
Three cases of multiple myeloma in which the preclinical asymptomatic phases persisted throughout 15 to 24 years.
Brit. J. Cancer. 25, 417, 1971.
132. Nossal, F. J. V., Mäkela, O.
Elaboration of antibodies by single cells.
Ann. Rev. Microbiol. 16, 53. 1962.
133. Otto, H. F., Walke, A.
Über lympho-epitheliale Bezichungen bei Enteropathien.
Virchows Arch. path. Anat. 355, 85. 1972.
134. Otto, H. F.
The interepithelial lymphocytes of the intestinum. Morphological observation and immunological aspects of intestinal enteropathy.
Curr. Top. path. 57, 81. 1973.
135. Ottó, Sz., Balogh, I., Jákó, J.
Elektron microscopic study of plasma cells in immunoproliferative diseases.
Acta morph. Acad. Sci. hung. 22, 331. 1974.
136. Ottó, Sz., Balogh, I., Jákó, J., Kellner, R.
Monoclonal IgA-kappa type protein of alpha-2 mobility and various plasma cell alterations in plasmocytoma.
Acta morph. Acad. Sci hung. 22, 337. 1974.
137. Paine, C. J., Richardson, J. V., Eichner, E. R.
Diverse clinical expression of multiple myeloma-atypical presentations.
Amer. J. med. Sci. 267, 99. 1974.
138. Paraskevas, F., Heremans, J., Waldenström, J.
Cytology and electrophoretic pattern $gamma_1A$ ($beta_2A$) myeloma.
Acta med. scand. 170, 575. 1961.
139. Parrot, D. M. V.
The gut as a lymphoid organ.
Gastroenterology. 5, 211. 1976.
140. Perkins, H. A., MacKenzie, M. R., Fudenberg, H. H.
Hemostatic defects in dysproteinemias.
Blood. 35, 695. 1970.
141. Plaut, A. G., Keonil, P.
Immunoglobulins in human small intestinal fluid.
Gastroenterology. 56, 522. 1969.
142. Potter, M.
Myeloma proteins (M-components) with antibodylike activity.
New Engl. J. Med. 284, 831. 1971.
143. Preud'homme, J. L., Hurez, D., Danon Francoise, Brovet, J. C., Seligmann, M.
Intracytoplasmic and surface-bound immunoglobulins in „nonsecretory" and Bence Jones myeloma.
Clin. exp. Immunol. 25, 428. 1976.
144. Rádl, J., Hollander, G. F.
Homogeneous immunoglobulins in sera of mice during aging.
J. Immunol. 112, 2271. 1974.
145. Ringe, J. D., Mertelsmann, R.
Fehldiagnose „Osteoporose" beim diffusen Plasmocytom.
Dtsch. med. Wschr. 102, 928. 1977.
146. Ritzmann, S. E., Loukas, D., Sakai, H., Daniels, J. C., Levin, W. C.
Idiopathic (asymptomatic) monoclonal gammotpathies.
Arch. intern. Med. 135, 95. 1975.
147. Rodriquez, A. R., Lutcher, Ch. L., Coleman, F. W.
Osteosclerotic myeloma.
J. Amer. med. Ass. 236, 1872. 1976.
148. Rogers, J. S., Spahr, J., Judge, D. M., Varano, L. A., Eyster, M. E.
IgE myeloma with osteoblastic lesions.
Blood. 49, 295. 1977.
149. Roques, C. F., Fournie, A., Ruffie, R.
La maladie de Kahler.
Sem. Hop. Paris. 53, 1343. 1977.
150. Rywlin, A. M., Civantos, I., Ortega, R. S., Dominquez, C. J.
Bone marrow histology in monoclonal macroglobulinaemia.
Amer. J. clin. Path. 63, 769. 1975.
151. Salmon, S. E., Smith, B. A.
Immunoglobulin synthesis and total body tumor cell number in IgG multiple myeloma.
J. clin. Invest. 49, 1114. 1970.
152. Salmon, S. E., Smith, B. A.
Sandwich solid phase radioimmunoassays for the characterization of human immunoglobulins synthesized in vitro.
J. Immunol. 104, 665. 1970.
153. Salmon, S. E.
„Paraneoplastic" syndromes associated with monoclonal lymphocyte and plasma cell proliferation.
Amer. N. Y. Acad. Sci. 230, 228. 1974.
154. Salmon, S. E.
Myeloma and related disorders.
Clinics in haematology. 11. 1982.
155. Schmiegel, W. H., Hamann, A., Thiele, H. G.
Darmassoziiertes Immunsystem.
Lokale Immunantwort – systemische Toleranz.
Dtsch. med. Wschr. 107, 267. 1982.
156. Schneiderbaur, A., Rettenbacher, F.
Die Paraproteinurie beim Plasmocytom.
Wien. med. Wschr. 112, 801. 1962.
157. Seligmann, M., Danon, F., Basch, A., Bernard, J.
IgA myeloma-cryoglobulin with antistreptolysin activity.
Nature. 220, 711. 1968.
158. Seligmann, M., Sassy, C., Chevalier, A.
A human IgG myeloma protein with anti-$alfa_2$-macroglobulin antibody activity.
J. Immunol. 110, 85. 1973.
159. Shimizu, K., Murate, T., Kunii, A.
Circulating immunoglobulin-secreting cells in patients with plasma cell dyscrasia.
Blood. 55, 590. 1980.
160. Shimizu, K., Ohinishi, K., Kunii, A.
Differentiation of benign monoclonal gammopathy and smouldering multiple myeloma from frank myeloma.
Clin. exp. Immunol. 50, 596. 1982.
161. Shuster, J., Gold, P., Povlik, M. D.
$Beta_2$ microglobulin levels on cancerous and other disease

162. Smetana, K., Györkey, F., Györkey, P., Busch, H.
Ultrastructural studies on human myeloma plasmocytes.
Cancer Res. 33, 2300. 1973.
163. Spiegelberg, H. L., Heath, V. C., Lang. J. E.
Human nyeloma IgG half-molecules. Structural and antigenic analyses.
Biochemistry. 14, 2157. 1975.
164. Spiegelberg, H. L., Fishkin, B. G.
Human myeloma IgA half-molecules.
J. clin. Invest. 58, 1259. 1976.
165. Staven, P., Vandvik, B., Skrede, S., Hovig, T.
Nedle-like crystals in plasma cells in a patient with a plasma cell proliferative disorder.
Scand. J. Haematol. 14, 24. 1975.
166. Staven, P., Froland, S. S., Haugen, H. F., Lislerud, A.
Nonsecretory myelomatosis without intracellular immunoglobulin.
Scand. J. Haematol. 17, 89. 1976.
167. Stocker, W. G.
Monoklonale Gammopathie.
Med. Welt. 28, 1286. 1977.
168. Stone, M. J., Frendel, E. P.
The clinical spectrum of light chain myeloma: A study of 35 patients with special reference to the occurence of amyloidosis.
Amer. J. Med. 58, 601. 1975.
169. Streiff, F., Humberl, J. C., Bertrand, F., et. al.
Fréquence des gammapathies monoclonales isolées. Confrontation bioclinique á propos de 394 cas d' immunoglobulines pathologiques.
Nouv. Rev. franc. Hémat. 11, 636. 1971.
170. Togawa, A., Inoue, N., Miyamoto, K., Hyodo, H., Namba, M.
Establishment and characterization of a human myeloma cell line (KKM–1).
Int. J. Cancer. 29, 495. 1982.
171. Toner, P. G., Fergusson, A.
Intraepithelial cells in the human intestinal mucosa.
J. Ultrastruct. Res. 34, 329. 1971.
172. Turesson, I.
Nucleolar size in benign and malignant plasma cell proliferation.
Acta med. scand. 197, 7. 1975.
173. Van Antwerp. J. D., O'Mara, R. E., Pitt, M. J.
Technetium – $^{99m}$diphosphonate accumulation in amyloid.
J. nucl. med. 16, 238. 1975.
174. Van Camp, B. G. K., Cole, J., Peetermans, M. E.
HLA antigens and homogeneus immunologlobulins.
Clin. Immunol. Immunpath. 7, 315. 1977.
175. Waldenström, J.
Diagnosis and treatment of multiple myeloma.
Grune and Stratton, New York and London, 1970.
176. Waldenström, J.
Clinical diagnosis and biochemical findings in material of 296 sera with M-type, Narrow gamma globulins.
Acta med. scand. 367 (suppl.) 110. 1961.
177. Waldmann, T. A., Blaese, R. M., Broder, S., Krakauer, R. S.
Disorders of suppressor immunoregulatory cells in the pathogenesis of immunodeficiency and autoimmunity.
Ann. intern. Med. 88, 226. 1978.
178. Waldmann, T. A., Strober, W.
Metabolism of immunoglobulins.
Progr. Allergy. 13, 1. 1979.
179. Weits, J., de Gast. G. C., The, T. H., Esselink, M. T., Mandema, R.
Cellular immunocompetence in asymptomatic paraproteinemia.
Acta med. scand. 202, 17. 1977.
180. Williams, R. C. Jr., Erickson, J. L., Polerky, H. F., Swaim, W. R.
Studies of monoclonal immunoglobulins (M-components) in various kindreds.
Ann. intern. Med. 67, 309. 1967.
181. Williams, R. C.
Cold agglutinins: studies of primary structure serologic activity and antigenic uniqueness.
Ann. N. Y. Acad. Sci. 190, 330. 1971.
182. Wiltshaw, E.
The natural history of extramedullary plasmacytoma and its relation to solitary myeloma of bone and myelomatosis.
Medicine. 55, 217. 1976.
183. Wochner, R. D., Strober, W., Waldmann, T. A.
The role of kidney in the catabolism of Bence Jones proteins and immunoglobulin fragments.
J. exp. Med. 126, 207. 1967.
184. Wutke, K., Rüdiger, K. D., Kelényi, G., István, L., Herold, M., Anger, G.
Das multiple-Myelom-erste Erfahrungen mit einer kooperativen Studie.
Deutsche Gesundheitswesen. 35, 268. 1980.
185. Wutke, K., Várbíró, M., Rüdiger, K. D., Kelényi G.
Citological and Histological Classification of Multiple Myeloma.
Haematologia. 14, 315. 1981.
186. Yentis, I.
Radiological aspects of myelomatosis.
Clin. Radiol. 12, 1. 1961.
187. Yonezawa, T., Kitani, R., Tamaki, R., Kanayama, Y., Hiraoka, A., Tarui, S.
Immunoglobulin production and secretion in Bence Jones protein myeloma and „nonsecretory" myeloma.
Ultrastructural and immunofluorescence study.
Blut. 45, 121. 1982.
188. Zawadzki, Z. A., Edwards, G. A.
Nonmyelomatous monoclonal immunoglobulinemia.
Prog. clin. Immunol. 1, 105. 1972.
189. Zielinski, Ch., Lanzer, G., Ludwig, H. P.
Defects of leukocyte locomotion in multiple myeloma.
J. Clin. Lab. Immunol. 7, 111. 1982.

# VI. The oligoclonal gammopathies

## Evaluation and laboratory diagnostics

In bi- or oligoclonal gammopathies two or more (?) homogeneous populations of monoclonal immunoglobulins can be simultaneously identified in the serum of the patients (39).

This pathology occurs in 1-3% of the patients having the "M"-component (8, 80, 103, 111); other statistical data show a 5-10% incidence (4, 32, 47, 49).

We observed biclonal gammopathy in 1.9% of the patients (Table VI/I).

While in the MG the sex distribution is near equal (51.7% males - 48.3% females), it is altered in biclonal gammopathies (61.2% - 38.8%) (105). In our studies the sex distribution was 36.7% for males and 63.3% for the female patients.

There is no uniform view on the origin of the proteins. Some researchers assume that they are derived from the same cell, while others rather support the theory of the different clones. Tissue culture studies (78), experiments with fluorescent technique (8, 20, 23, 40, 57, 64) and the analysis of the cellular morphology (31, 85) seem to support the hypothesis "one cell - one immunoglobulin".

Additional evidence in favour of this view is obtained in clinical studies. For example, after removal of a solitary protein-producing plasma-cell tumor in a patient with diffuse plasmocytoma biclonal gammopathy turned into a monoclonal one (59).

Data of the literature indicate that in some cases of paraprtoteinemia, one was verified as IMG-, while the other was undoubtedly (?) of the OMG type. In the course of the disease the latter was maintained, while IMG paraproteinemia regressed (evaluated as a transitory type). This observation supports the theory one cell - one MG? (6, 94).

Researchers supporting the theory one cell, two paraproteins also use fluorescent label technique (9, 10, 11, 13, 14, 16, 33, 34, 44, 45, 71, 83, 84, 95, 103) and tissue culture methods (100) to prove their view. In the course of these studies a possibility of hybridization as a cytologic tool that could prove the synthesis of two different proteins by the same cell has been discussed (70, 72). Despite the still existing debate about this theory, the existence of in vivo hybrids producing two types of proteins: $IgG_4$-$IgG_2$ is documented in the literature (66). Moreover, some lymphoid cells with dual staining are observed in the normal bone marrow (42, 96).

Furthermore, in normal individuals or immunized animals plasma cells capable of the synthesis of *two* different heavy chains were identified. This finding ab ovo precludes malignity (37, 38, 58). Following the stimulation of human "B" lymphocytes by the Pokeweed mitogen, the terminally differentiated plasma cells produced double heavy chains: 15% - IgM + IgG and IgM + IgA; and 2% of IgG + IgA (60).

Recent studies regard the bi- and triclonal reactions as virtual, for the pathological immunoglobulines have the same idiotype (86, 92) at the light chain level, suggesting a common precursor (62). The different origin of the heavy chains within the same clone is due to the class-switch mechanism (29, 61): it is documented that some plasma cells can synthesize two heavy chains, because in tumors or other pathologies the coupling phase is affected or prevented (69). The class-switch effect is a part of a normal differentiation program of the B cells, suggesting complex or sequential rearrangements of the DNA (22, 56).

The presented controversial data indicate that no uniform view still exists and only genetic experiments seem to be an adequate tool to solve this problem.

Studies of the patients with two or three M

| | Serum | | | Urine | | Biclonal | |
|---|---|---|---|---|---|---|---|
| | ♂ | ♀ | | ♂ | ♀ | | |
| IgM-κ | 53 | 72 | | | | IgM-k - $IgG_1$-κ | 1 |
| IgM-λ | 22 | 53 | | | | IgM-κ - $IgG_2$-κ | 1 |
| $IgG_1$-κ | 165 | 201 | | 7 | 5 | IgM-κ - IgG polyclonal | 4 |
| $IgG_1$-κ | 103 | 160 | | 2 | 7 | IgM-κ - $IgG_1$-λ | 1 |
| $IgG_2$-κ | 55 | 58 | | | | IgM-κ - λ | 1 |
| $IgG_2$-λ | 19 | 50 | | | | IgM-λ - IgG polyclonal | 2 |
| $IgG_3$-κ | 10 | 12 | | 2 | 5 | $IgA_1$-κ - $IgG_1$-κ | 5 |
| $IgG_3$-λ | 24 | 26 | | | | $IgA_1$-κ - $IgG_3$-κ | 1 |
| $IgG_4$-κ | 14 | 12 | | | | $IgA_1$-κ - IgG polyclonal | 1 |
| $IgG_4$-λ | 7 | 12 | | | | $IgA_1$-κ - λ | 1 |
| $IgA_1$-κ | 76 | 81 | | | | $IgA_1$-λ - λ | 2 |
| $IgA_1$-λ | 45 | 69 | | | 1 | $IgG_1$-κ - λ | 2 |
| $IgA_2$-κ | 12 | 10 | | | | $IgG_3$-κ - κ | 1 |
| $IgA_2$-λ | 5 | 5 | | | | $IgG_1$-λ - λ | 3 |
| IgD-κ | - | - | | | | $IgG_2$-λ - Fab | 1 |
| IgD-λ | 5 | 12 | | | | $IgG_2$-λ - urine l-Fab | 1 |
| IgE | - | - | | 1 polyclonal | | | |
| κ | 2 | 5 | | | 7 | | |
| λ | 7 | 5 | | 2 | 5 | | |
| Summarized | 624 | 843 | | 13 | 30 | ♂: 15 ♀: 13 | 28 |

**Table VI/I**

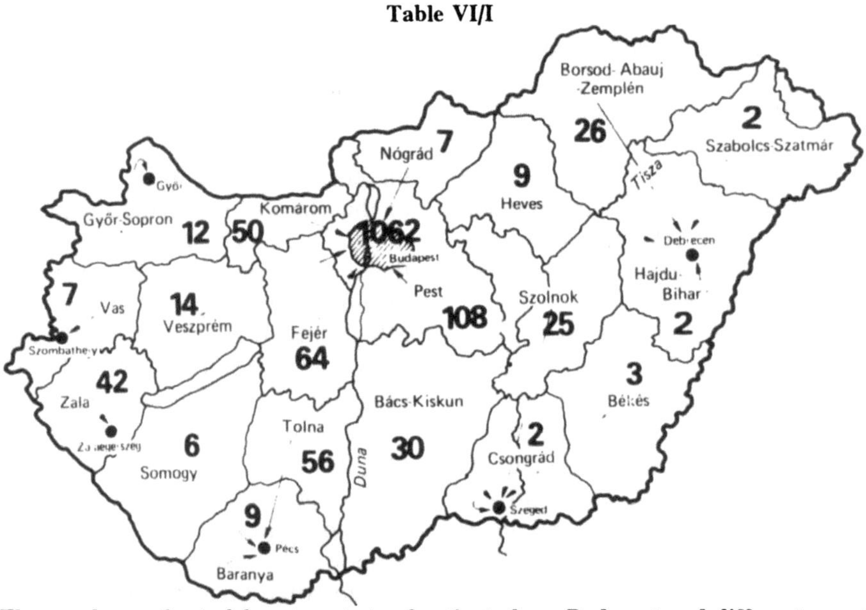


**The numbers refer to laboratory tests of patients from Budapest and different counties.**

components provide a deeper insight in the mechanisms of the immunoglobulin biosyntesis (19). The well documented great diversity of the primary structure of immunoglobulins as "gene products" is a unique tool for genetic studies, which is not available with other molecules. Based on the available data, a two genes - one polypeptide concept (24, 67), the allel exclusion theory and the class "switch" mechanism for the heavy chains (15, 55, 107) have been defined.

The V and C regions of the light and heavy chains are genetically controlled (108). Though high variability (A) and conservation (B) of structure is typical of the V regions, while the maintenance of the constant structure is characteristic of the C regions (C).

A. Antigen stimulus elicits specific changes of the V-region synthesis, which reflect the specificity of the antigen determinants. The appropriate genetic mutation preceding the V-region changes (12, 36, 93) is not accidental, but is predetermined by the genetic constitution of an individual.
B. The following step is the maintenance of the changes, which is realized via the inhibition of somatic mutations or recombinations of the V-genes (both $V_L$ and $V_H$) in the cells belonging to the *same clone*. Among numerous cells of this clone some will "remember" the type of the V-region synthesis behaving as *memory cells.*
C. C region is a constant one, however certain changes are responsible for the properties of specific antibodies: the C regions *do not alter,* but are rather exchanged. It occurs when the IgM synthesis is switched to IgG or IgA (83). Therefore, *one clone* is capable of a simultaneous or subsequent synthesis of even three immunoglobulines (46). For example the shift between the primary and the secondary immune response might be explained by the switch mechanism (17, 65, 68). It is assumed that the rivalization between the two gene generations, or rather their subsequent activation might underlie the above events (89).

Data of the literature show that at least 10-15% of the IgM-producing clones can switch to the IgG-synthesis (68, 97), indicating a physiological mechanism. The switch mechanism is experimentally proved as follows: administration of anti-IgM serum not only inhibits the synthesis of IgM, but also abolishes the production of other immunoglobulins (68, 73, 77).

The switch mechanism, therefore, is a characteristic feature of the cells belonging to the same clone. At present the mechanism of the information transfer, its biochemical and genetic background is unknown. It is assumed that the B lymphocytes have certain genetically predetermined structures responsible for the synthesis of the different immunoglobulin classes (104), however the experimental evidence supporting this view is insufficient. Still it is suggested that an erroneous information transfer or a switching error might lead to the synthesis of two different immunoglobulins in the same cell (16). This theory would consider all double paraproteinemies produced by the same cell as pathological, though in terms of a clone product they would not necessarily indicate malignity. Moreover, since some of the committed B cells develop multiple specificity in the course of their development, the final biclonal gammopathy might be due to a disease of the early, but not the late state of the plasma cells.

The available data of the literature suggest the following five versions of the possible cytological background of double paraproteinemias (Fig. VI/1).

Classification of oligoclonal gammopathies and the most important features of the biclonal gammopathies with respect to the secreted proteins are presented.

## Classification of oligoclonal gammopathies

quinclonal (1 case) (81),
quatroclonal (3 cases) (39, 53, 98),
triclonal (10 cases) (39, 51, 72, 76, 81,84),
biclonal (>300 cases) (2, 18, 82, 99).

## Classification of biclonal gammopathies

*Not true biclonal gammopathies*

1. Free IgG light chain in the serum (same type of the light chains);

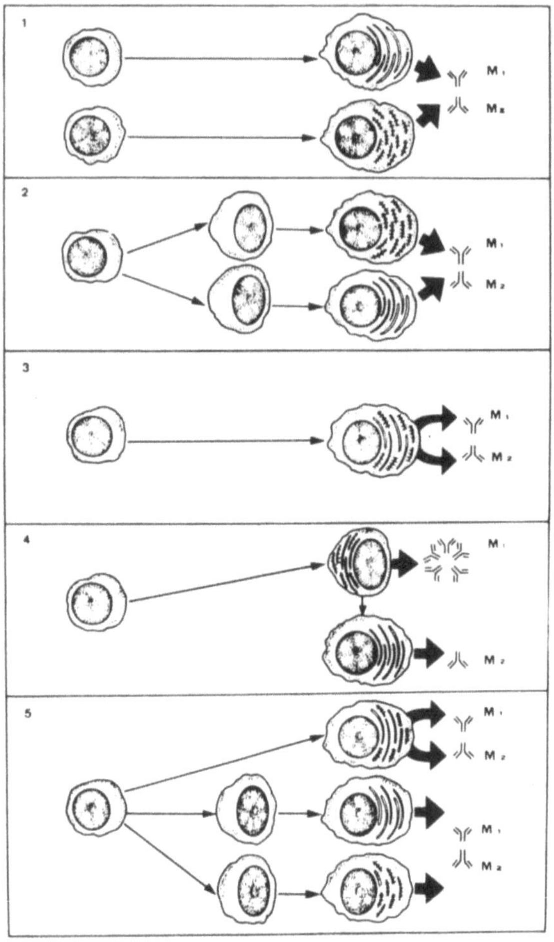

Fig. VI/1

2. Polymer heterogeneity (19, 26)
   - IgA and polymeric IgA together;
   - IgM and 7S IgM (non-degradation product);
   - BJ monomers and dimers (or trimers).
3. Polymorphic heterogeneity (3, 26, 74)
   A) as sequellae of degradation:
      a) in vivo
      b) in vitro, but due to an enzyme with in vivo activity (50).
   B) Some of the cells produce incomplete immunoglobulin molecules - cells are synthesizing double products (88).

Although this classification is clear enough, there has been doubt about its 1st item: in our opinion the free light chains detected in the serum *can not be* the products of Ig degradation, for the presence of the free heavy chains was never detected. Therefore, these molecules must be synthesized either by the same cells, or by the other clones. Additional evidence in favour of this view is the physiological production of extremely small amounts of the light chains. Pathological proliferation of this clone leads to malignity and light-chain myeloma. Therefore, the occurrence of free light chains together with full Ig molecules might either be a consequence of a true biclonal gammopathy, or is due to a switch mechanism indicating proliferation activity.

The 3.A.b. item is worth pointing out. It is well known that the storage of Ig in vitro undergo degradation, due to the proteolytic enzyme activity, sometimes leading to an erroneous diagnosis of paraproteinemia or biclonal gammopathy (50). Hence such findings must be critically revised especially in pediatric patients. Of 118 mother-infant pairs observed by us, in vitro degradation has been detected in 44 cases. High incidence of positive results proves that these findings can not be due to a pathological process (Fig. VI/2, VI/3, VI/4).

Without a thorough analysis of the case, the detection of the degradation alone might lead to a false diagnosis.

Considering the 3.B. item it is worth pointing out that biclonal gammopathy can be precluded only with difficulties, for the cells of the clone, which synthesize immunoglobulin fragments are more immature, thereby two forms of maturity and two different subpopulations would exist within the same clone, despite the common progenitor.

| | |
|---|---|
| M | Maternal serum |
| S | Fetal serum |
| A | Anti-human serum |
| B | Anti-IgG |
| C | Anti- IgHCh |
| D | Anti-κ (not monovalent, but κ-specific) |
| E | Anti-λ (not monovalent, but λ-specific) |
| F | Anti-Fab |
| G | Anti-Fc |
| H | Anti-IgD |
| I | Anti-IgA |
| J | Anti-IgM |

Fig. VI/2

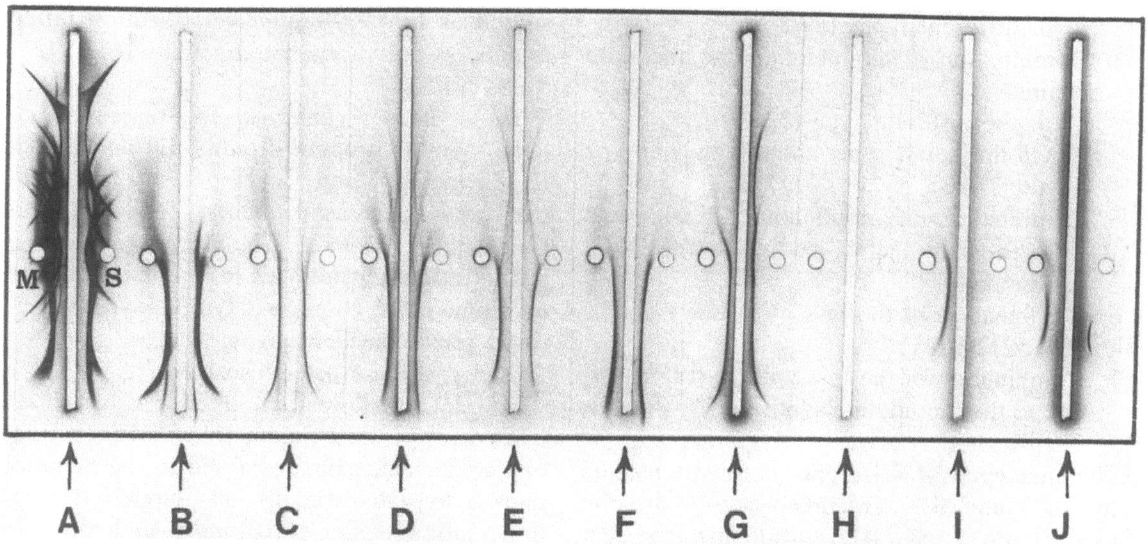


Fig. VI/3

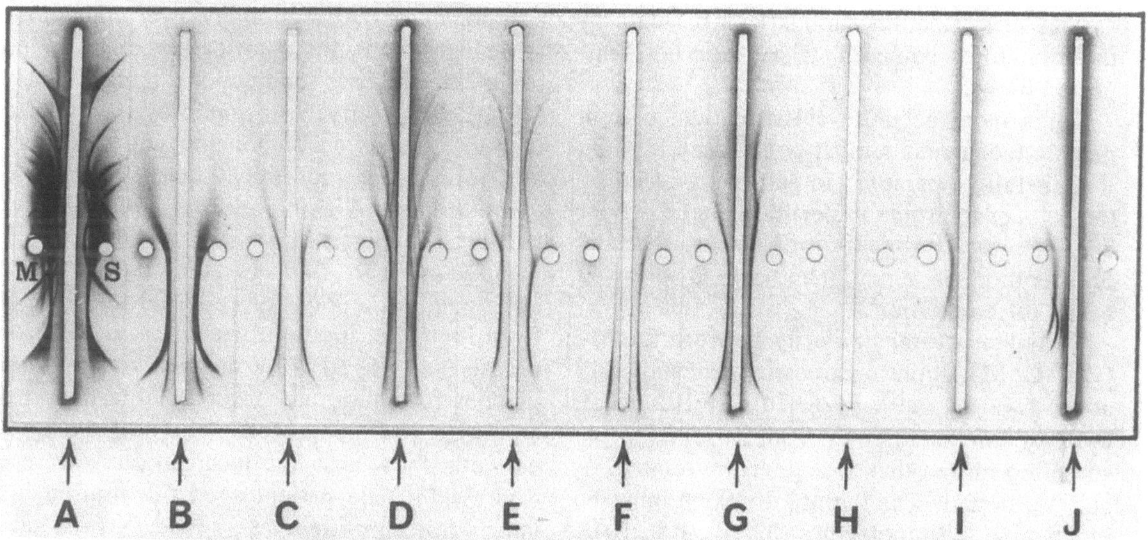


Fig. VI/4

The following classification summarizes the true gammopathies and gives some special possibilities and criteria for their diagnosis.

*True biclonal gammopathies*

I. Homotypical (the Ig-s belong to the same class);
II. Heterotypical (the Ig-s belong to the different classes)
   1. concordant (same type of the light chains)
   2. discordant (different types of the light chains) (10, 11).

*Special biclonal gammopathies*

1. One of the components is not extracted from the cells but its existence is verified. It is a combination of secretory and non-secretory clones (7, 95).
2. One of the components is IgD (1, 75,81).
3. The light chains of both M components are:
   a) of the same type (30)

b) of different types (87).
4. Combination of monoclonal Ig-s and light chains:
   a) of the different type (5, 89)
   b) Of the same type, though its idiotype alters (63).
5. Combination of monoclonal Ig with two different light chains, leading to a triclonal gammopathy (21, 25, 72, 105).
6. Combination of the Ig-s with heavy chains (19, 52, 54, 91).
7. Combination of the Ig-s with Ig-fragments, due to the incomplete synthesis (?) (48, 51, 90) by another clone.

The incidence of different combinations, the IgA-IgG and the IgM-IgG versions are the most frequent (84, 109), while the IgM-IgA occurs rarely (41, 43, 101). A IgD-IgG pair is a very rare combination (27).

This classification summarizes the data of the literature, some of which are not supported by us.

For a more accurate classification, certain genetic mechanisms must be stressed.

Generally a paraprotein pair is produced by the cells of the *same* molecular genetic origin, and the *same* progenitor. It means that the proliferation of a malignant cell is followed by its differentiation.

Partial structural similarity between the M-G, A-G, M-A combinations indicates identity at the level of the $V_L$ (28, 76, 79, 102, 110), or $V_H$ domains (35, 45, 106, 107). This assumption means that the either the related $V_L$ and $V_H$ domains or the $V_H$ domains must be identical in both molecules (33, 42, 100, 101).

It is also assumed that only the constant regions of the heavy (33, 78, 83, 87, 99, 101, 103) and light (65) chains alter.

In 25 years we have identified 30 cases of double paraproteinemia, presents the first 12. (Fig. VI/5) Typical case histories are presented herein:

*1. T.D.,* female patient aged 70; complaints - loss of body weight and fatigue. RBC sedimentation - 107 mm/h. Gel electrophoresis revealed two pale strips indicating two M components at low concentrations (Fig. VI/6). Apart from the polyclonal (Pč) IgM, immunoelectrophoresis identified two additional precipitation arcs corresponding to the M components. These components were assumed as IgM-λ, though gel filtration failed to support polymeric heterogeneity (Fig.VI/7., VI/8., VI/9).

*Bone marrow findings:* the marrow contains medium amount of cells, fat and blood. Erythropoiesis and myelopoiesis are both markedly decreased. Only few megacaryocytes are observed. The morphology of the proliferating plasmocytes resemble that of the myeloma cells. Numerous lymphocytes sometimes form confluent areas.

X-ray: diffuse osteoporosis.

*2. F.S.,* male patient aged 64 has been treated for years at rheumatological outpatient clinic for back pain. Admitted to the hospital due to hypertension disease. High RBC sedimentation values (110 mm/h) and multiple spinal osteolytic changes suggested MM. Gel electrophoresis showed a well detectable M component and s fraction at lower concentration (Fig. VI/10). Immunoelectrophoresis failed to reveal the two components (Fig. VI/11, VI/12). Preparative electrophoresis was used to purify the two components with different electrophoretic mobilities that have been identified as IgA-κ (higher titer and mobility) and $IgG_1$-λ (Fig. VI/13).

*Bone marrow findings:* the marrow contains medium amount of cells, fat and blood. Erythropoiesis is slightly decreased, myelopoiesis unchanged. Few megacaryocytes and platelet formation are observed. Plasmocytes comprise 26% of the cells. The uniform typical cells show nest-like localization.

*3. M.J.,* male patient aged 60. injured his back while carrying a heavy load. X-ray study revealed that the I and III lumbal vertebra are only half as thick as the rest. RBC sedimentation 85 mm/h. Gel electrophoresis showed two neighbouring, intensively staining M component arcs (Fig. VI/14).

Immunoelectrophoresis detected a *single* abnormal arc (Fig. VI/15, VI/16). Using preparative gel electrophoresis the two proteins have been isolated and identified (Fig. VI/17). The faster paraprotein corresponded to the IgA-κ, while the slower one - to the IgA-λ.

*Bone marrow findings:* the marrow contains medium amount of cells. Erythropoiesis and myelopoiesis are slightly decreased. Platelet formation is observed in the megaca-

| Number | Name (sex) | | Age | Immunoglobulins | Urine | Diagnosis |
|---|---|---|---|---|---|---|
| 1 | T. D. | ♀ | 70 | IgM-λ, -IgM-λ | - | W. M. |
| 2 | F. S. | ♂ | 64 | IgA-κ, $-IgG_1$-λ | - | M. M. |
| 3 | M. J. | ♂ | 60 | IgA-κ, -IgG-λ | - | M. M. |
| 4 | J. S. | ♂ | 60 | IgA-κ, $-IgG_1$-λ | - | M. M. |
| 5 | Gy. N. | ♀ | 70 | IgA-κ, $-IgG_1$-λ | 1 | M. M. |
| 6 | I. H. | ♀ | 60 | IgA-κ, $-IgG_3$-λ | - | M. M. |
| 7 | J. N. | ♂ | 62 | $IgG_1$-κ, $-IgG_3$-λ | - | M. M. |
| 8 | dr. E. N. | ♀ | 64 | $IgG_1$-λ, λ | λ | M. M. |
| 9 | F. F. | ♀ | 84 | $IgG_3$-κ, $-IgG_1$-κ | - | M. M. |
| 10 | dr. J. M. | ♂ | 70 | $IgG_2$-κ, $-IgG_1$-κ | - | M. M. ? |
| 11 | J. Sz. | ♀ | 57 | IgD-λ, λ | λ | M. M. |
| 12 | L. J. | ♂ | 78 | IgA-κ, $-IgG_1$-λ | - | I. M. G. |
| 13 | D. A. | ♀ | 57 | $IgG_1$-λ, λ | 1 | M. M. |
| 14 | M. P. | ♂ | 60 | IgA-κ, $-IgG_2$-λ | - | M. M. |
| 15 | Sz. F. | ♂ | 62 | IgA-κ, $-IgG_1$-κ | - | M. M. |
| 16 | K. P. | ♀ | 67 | $IgG_1$-κ, $-IgG_3$-κ | - | M. M. |
| 17 | G. Gy. | ♀ | 71 | $IgG_1$-λ, $-IgG_1$-κ | 1 | M. M. |
| 18 | B. F. | ♀ | 58 | IgA-κ, -IgE-κ | - | M. M. |
| 19 | T. F. | ♀ | 65 | IgM-λ, $-IgG_1$-κ, κ | κ | M. M. (TRI) |
| 20 | T. P. | ♂ | 64 | IgA-κ, $-IgG_1$-λ | - | M. M. |
| 21 | H. J. | ♀ | 63 | IgA-λ, $-IgG_1$-λ | - | M. M. |
| 22 | K. P. | ♀ | 53 | $IgG_1$-κ, $-IgG_3$-κ | - | M. M. |
| 23 | M. J. | ♀ | 57 | $IgG_1$-λ, $-IgG_1$-λ, $-IgG_3$-λ | - | M. M. (TRI) |
| 24 | S. Gy. | ♀ | 61 | $IgG_1$-λ, κ | k | M. M. |
| 25 | S. M. | ♂ | 70 | IgA-κ, $-IgG_1$-κ, $-IgG_1$-λ | - | M. M. (TRI) |
| 26 | L. M. | ♀ | 71 | IgM-κ, $-IgG_1$-λ | - | M. M. |
| 27 | P. Gy. | ♂ | 68 | $IgG_2$-κ, λ | 1 | M. M. |
| 28 | T. D. | ♀ | 63 | IgM-κ, λ | 1 | M. M. |
| 29 | T. M. | ♀ | 43 | IgM-κ, $-IgG_1$-κ | - | ? |
| 30 | M. J. | ♀ | 70 | IgM-κ, -IgM-κ | - | W. M. |

**Fig. VI/5**

ryocytes. Polymorphic plasmocytes and plasmoblasts comprise 64% of the cells.

*4. J.S.*, male patient aged 60. Admitted to the Department of laryngology due to the repeated tonsilar abscesses. Since the patient was febrile and resistant to therapy, hematologic disorder was suspected and the patient has been transferred to our Clinic. Antibody deficiency as a possible cause of the repeated infections was assumed and gel- and immu-

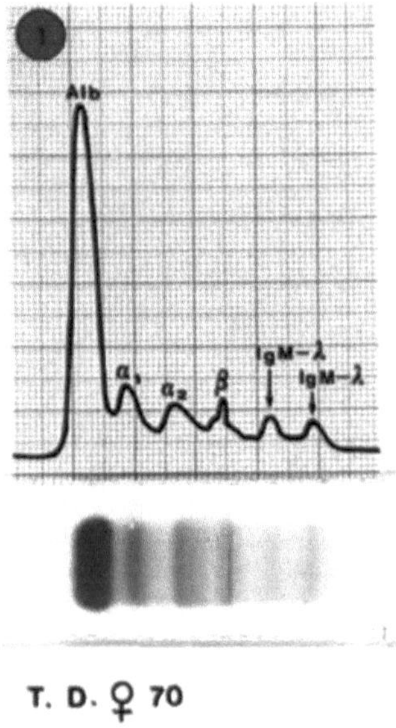


Fig. VI/6

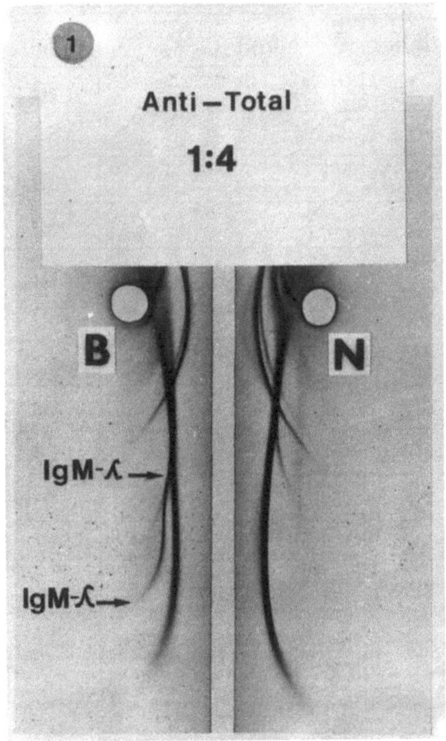


Fig. VI/8

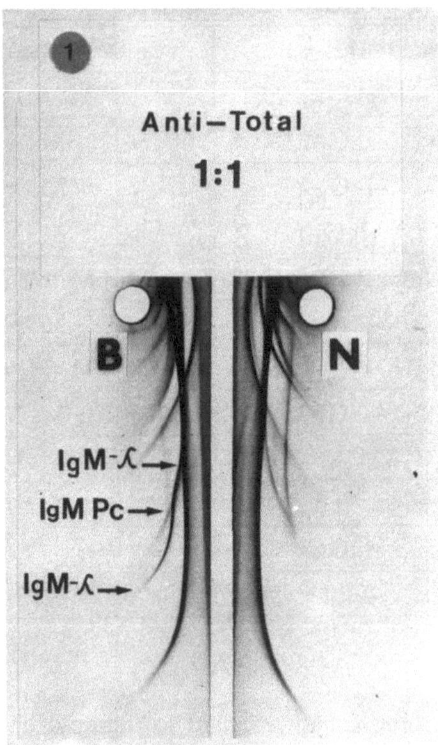


Fig. VI/7

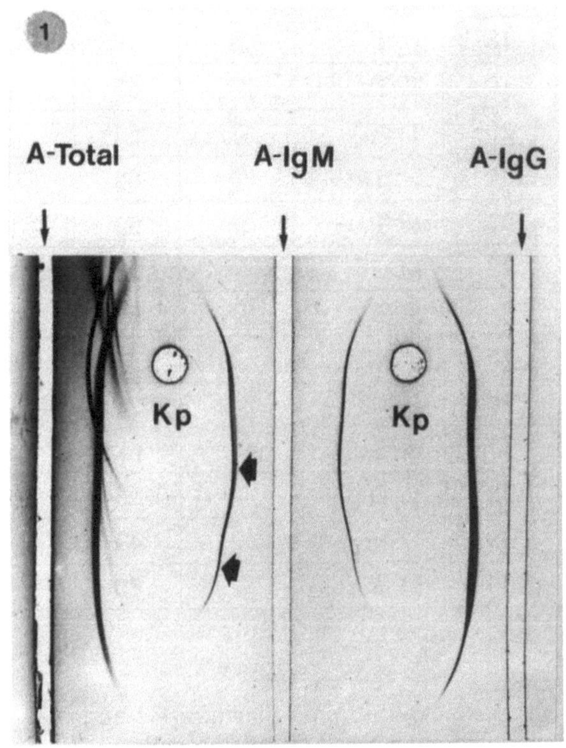


Fig. VI/9

noelctrophoresis performed (Fig. VI/18, VI/19, VI/20). The former has verified double paraproteinemia, whereas the latter revealed a single pathological precipitin arc indicating a secondary antibody-deficient syndrome. This disease explained the recurrent bacterial infections and the decreased resistance of the organism. Preparative electrophoresis revealed two components - Ig-κ and $IgG_1$-λ.

*Bone marrow study:* low cell count, decreased erythropoiesis. Myelopoiesis shifted to the right. Platelet formation is seen, though the megacaryocyte count is low. Marked plasma cell proliferation. The cells are diffuse (52%), showing no nest-like localization.

X-ray: sustained diffuse osteoporosis. Osteolytic changes of about 2 cm in diameter in the ileal bone.

*5. Gy.N.* female patient aged 70. Observed at our Clinic for a long period due to liver cirrhosis and monoclonal gammopathy. Anemia could only be transiently controlled by repeated transfusions and ascites was treated every 3-4 months with abdominal puncture. Between the punctures the patient took combined diuretics. In the last years of her life the patient had developed biclonal gammopathy (Fig. VI/22); immunoelectrophoresis visualized both components (Fig.VI/23, VI/24, VI/25).

Its M-components separated and purified by preparative gel electrophoresis proved to belong antigenically to the IgA and $IgG_1$ class (Fig. VI/26).

By that time the lambda-light chains could be detected in the urine. Post-mortem histological study revealed both glomerular and tubular damages.

*Bone marrow analysis:* the marrow rich in cells. Markedly depressed erythropoiesis and slightly decreased myelopoiesis. Only several promyelocytes and mature myeloid cells detected. No megacaryocytes. Most of the cells are polymorphic myeloma cells.

X-ray: pronounced diffuse osteoporosis. Bone scintigraphy of the whole body revealed no local alterations.

*6. I.H.,* female patient aged 60. Admitted to ophthalmologic clinic due to fundic hemorrhages. RBC sedimentation of 78 mm/h and the fundus examination suggested paraproteinemia and the patient was transferred to our Clinic. Gel electrophoresis revealed a well

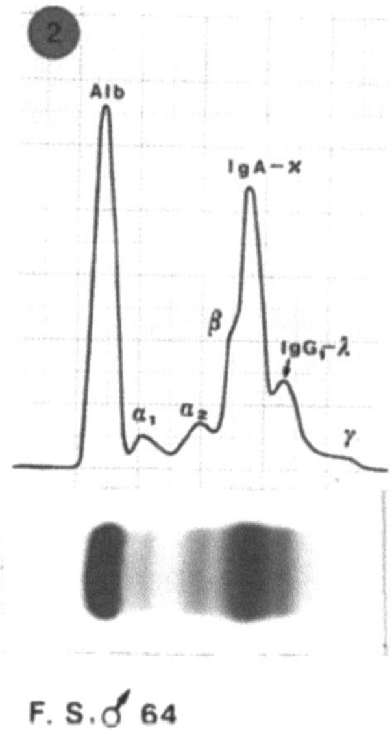


Fig. VI/10

stained slow arc and a scarcely visible slower M component (VI/27). Immunoelectrophoresis showed two typical precipitin arcs (Fig. VI/28, VI/29, VI/30).

Fundus angiography (papillo-macular region): edematous, stripped, white-greyish re-

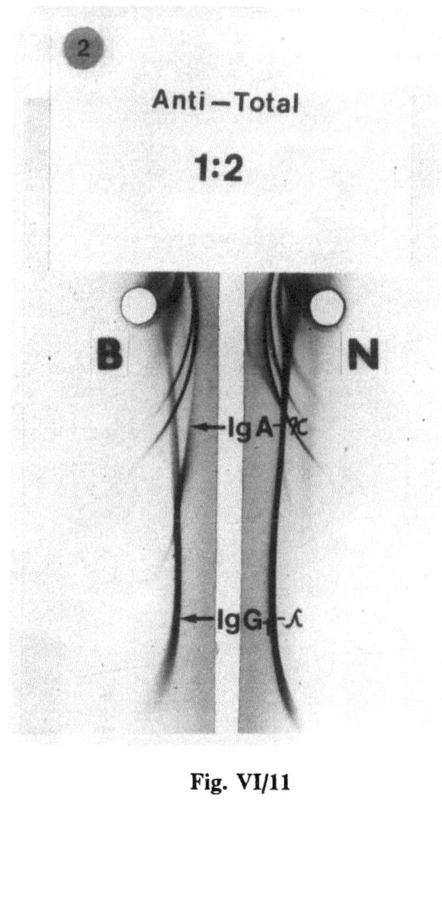


Fig. VI/11

tina. Tortuous, irregular veins; dilated, irregular capillaries (Fig. VI/31).

*Bone marrow analysis:* decreased cell count in the marrow. Depressed erythropoiesis and myelopoiesis. Signs of platelet formation, though the megacaryocyte count is low. Abundant typical myeloma cells (Fig. VI/32).

X-ray: osteolytic changes in the epiphysis of the humerus.

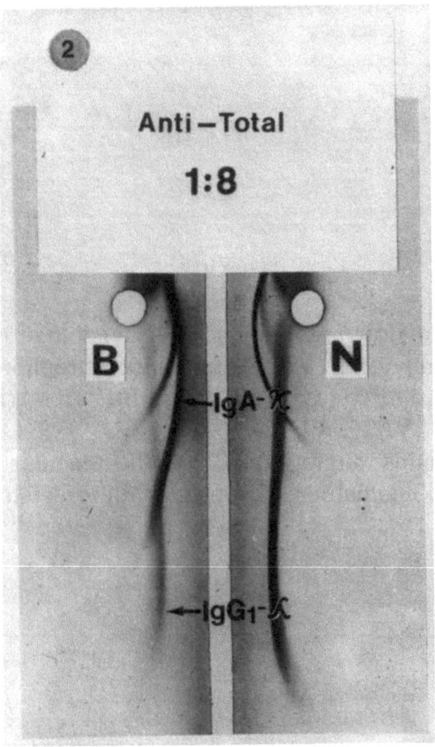


Fig. VI/12

*7. J.N.*, male patient age 62. Consulted the doctor due to the pain in the right shoulder during car driving. The X-ray examination revealed multiple lesions in the right shoulder joint. Since an MM was suspected, sternum punctures were twice performed. The aspirate, however, did not support the MM diagnosis. Gel elctrophoresis revealed two neighbouring M components of the same intensity (Fig. VI/33). Immunoelectrophoresis showed an elongated asymmetric precipitin arc (Fig. VI/34, VI/35, VI/36).

Preparative electrophoresis revealed pure paraproteins (Fig.VI/37) and their types could be identified (Fig. VI/38).

The punctures of the left humerus epiphysis and the scapula (Prof. T.Riskó), as well as the electronmictroscopic studies of the removed tissues revealed a very high content of the plasma cells (Fig.VI/39). The dilated rough endoplasmatic reticulum contained homogeneous material (Fig.VI/40) probably identical with the non-secreted proteins.

8. Dr.E.N., female patient, aged 64. Admitted due to drug allergy, loss of body weight and fatigue. Marked anemia and anisopoikilocytosis in the peripheral blood indicated bone marrow aspiration. Bone marrow analysis alone was sufficient for the MM diagnosis. Gel electrophoresis indicated three staining arcs corresponding to the three M components (Fig. VI/41). Immunoelectrophoresis could only reveal two arcs of precipitation (Fig. VI/42, VI/43). Later the third

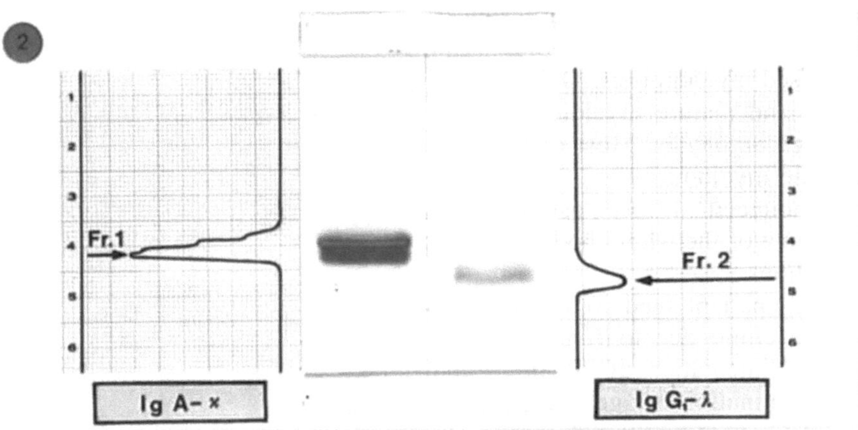


Fig. VI/13

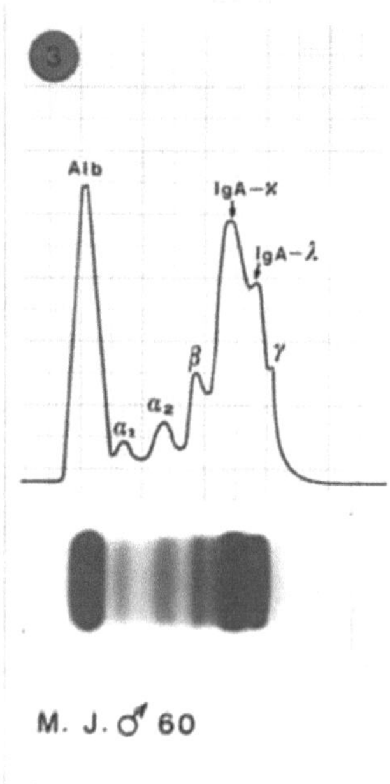


Fig. VI/14

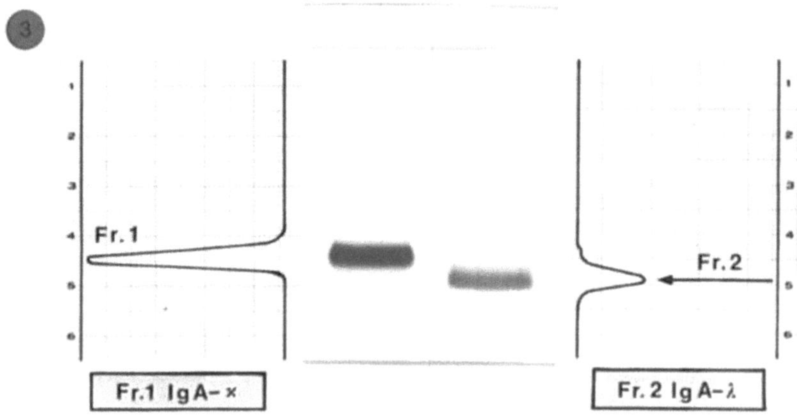


Fig. VI/17

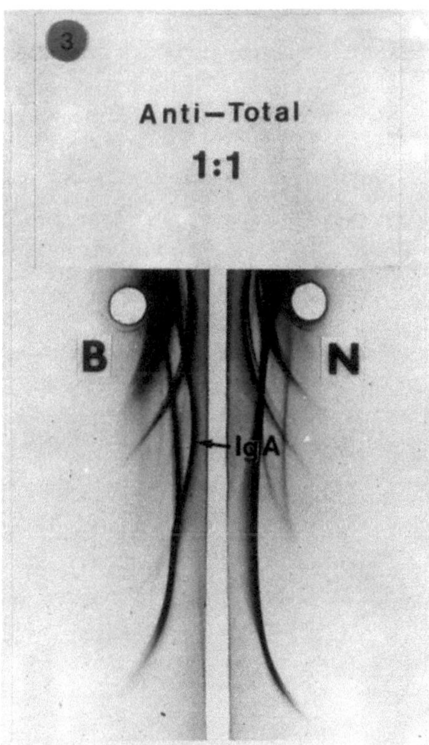


Fig. VI/15

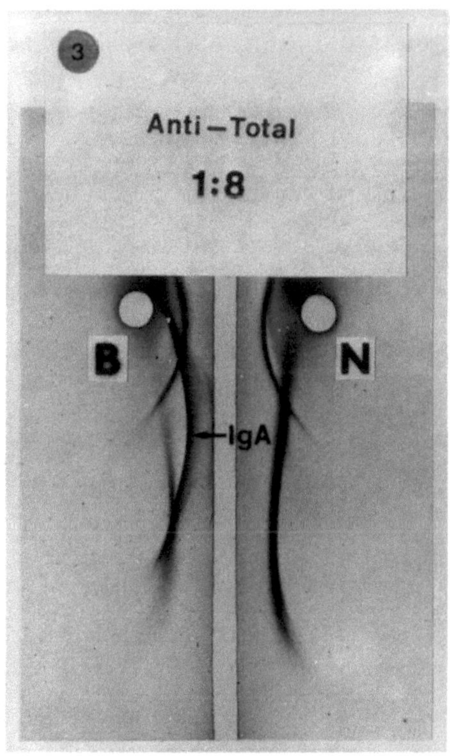


Fig. VI/16

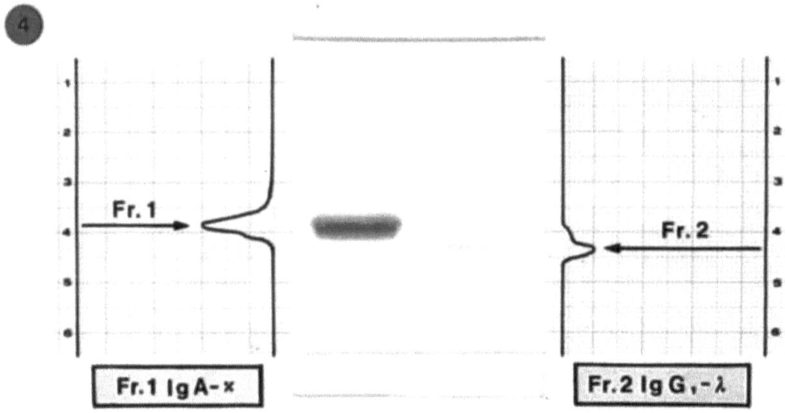


Fig. VI/18

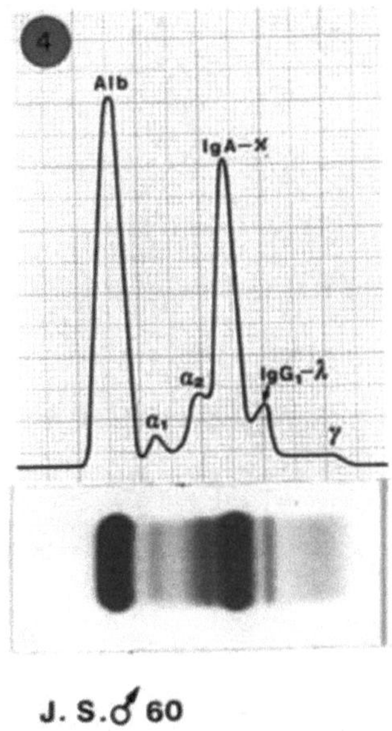


Fig. VI/21

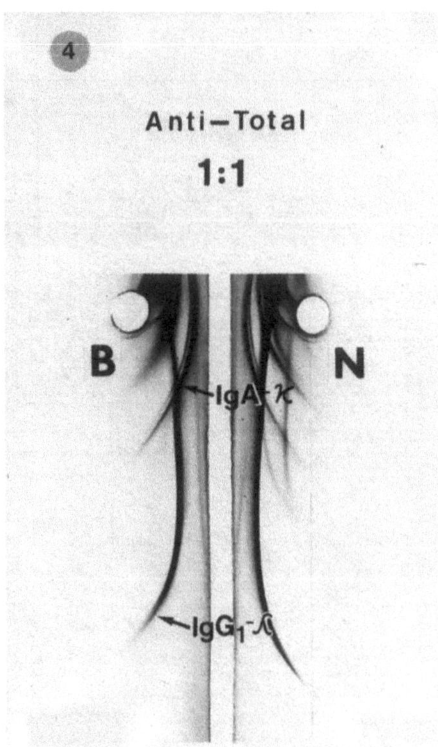


Fig. VI/19

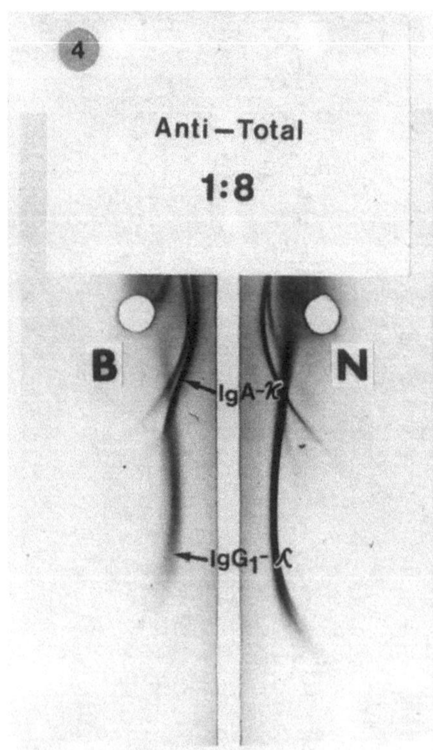


Fig. VI/20

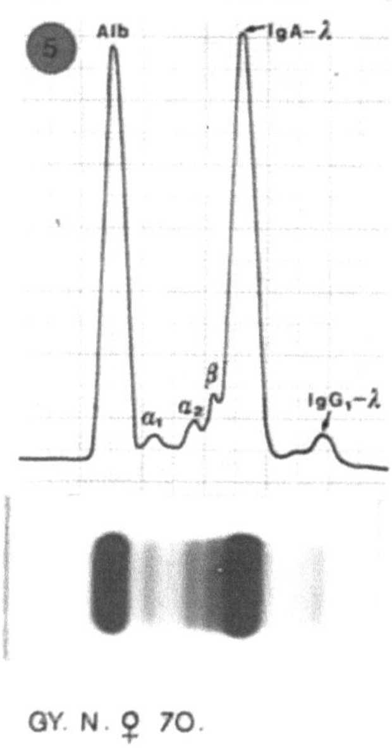


Fig. VI/22

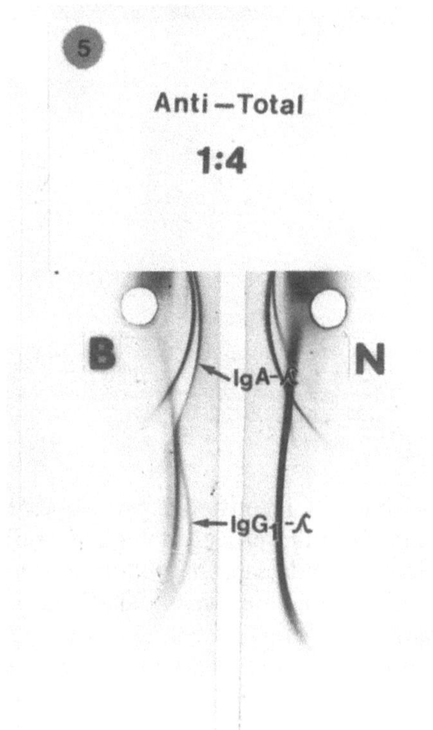


Fig. VI/24

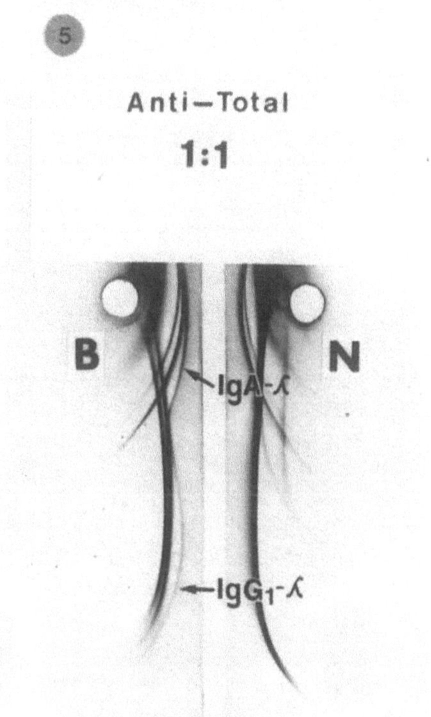


Fig. VI/23

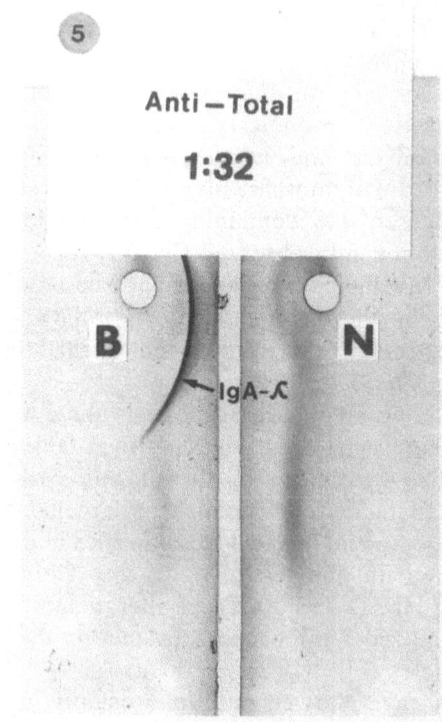


Fig. VI/25

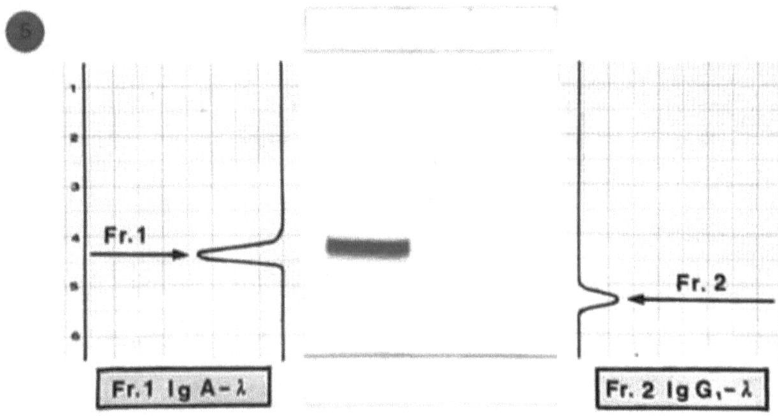


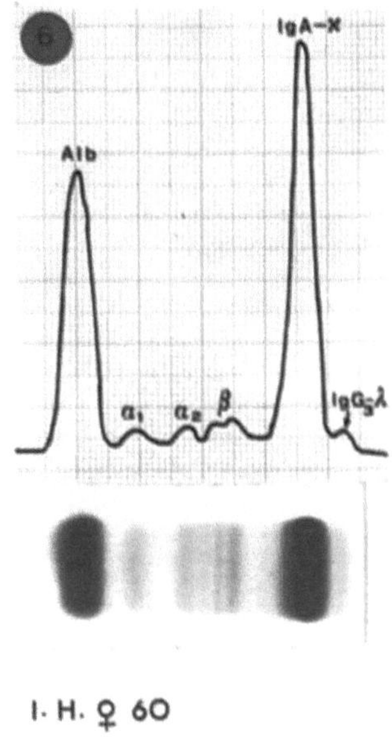


Fig. VI/27

component was identified as haemopexin. Laboratory findings proved a typical antibody-deficiency syndrome, in which the normal antibody production was visualized as the remnants of (Pc)IgG.

Paraproteins separated with preparative gel elctrophoresis showed $IgG_1$ λ-light chains (Fig.VI/44).

Column chromatographic analysis was used to detect the molecular weight of the immunochemically pure light chain occurring at high concentration and corresponding to the M component (Fig.VI/45). Further analysis of the fractions indicated the presence of pure λ light chains in the 4th fraction (Fig.VI/46).

*Bone marrow studies:* high cell count; depressed hemopoiesis. Few normoblasts, myelocytes and megacaryocytes were visualized. Polymorphic myeloma cells and those of lesser size containing large nuclei were predominantly observed (Fig.VI/47).

EM studies revealed two basic types of cells having different size (Fig. VI/48 - magnification 7,000x) and pathological structure (Fig.VI/49).

Shows autophage vacuoles in the cytoplasm of the plasma cells (Fig.VI/50) (38,000x).

The cytoplasm of the plasma cells contained increased amount of filaments. Autophage vacuoles and lysosome-like bodies occurred in the vicinity of the Golgi area (Fig.VI/51) (13,000x). Higher magnification visualized semiopaque substance in the Golgi vesicles (Fig.VI/52) (30,000x).

X-ray: showed multiple osteolytic alterations along the whole spine.

Lethal outcome due to the sepsis after a purulent bronchitis. At autopsy histological studies revealed plasma cell infiltration of the spleen and the myeloma-kidney with tubular damages and less affected glomeruli. Plasma cell infiltration was not observed in the kidney.

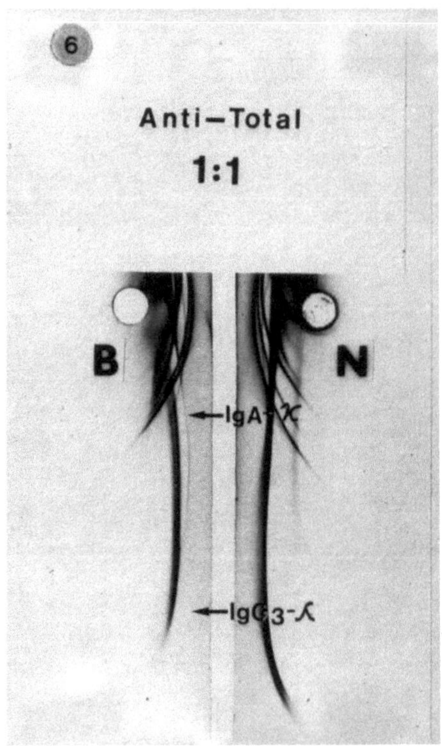


Fig. VI/28

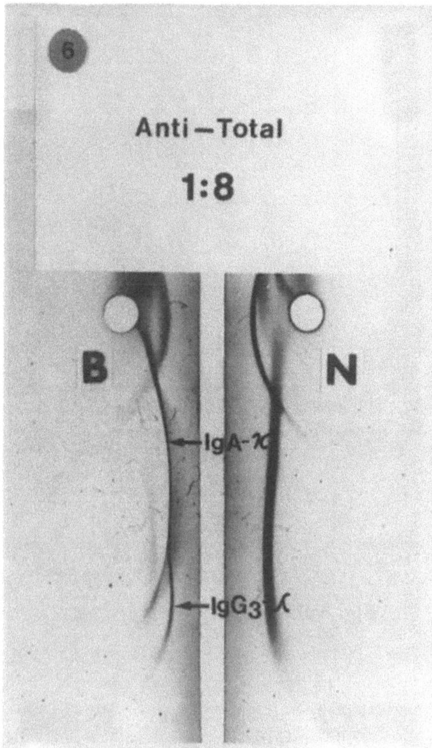


Fig. VI/29

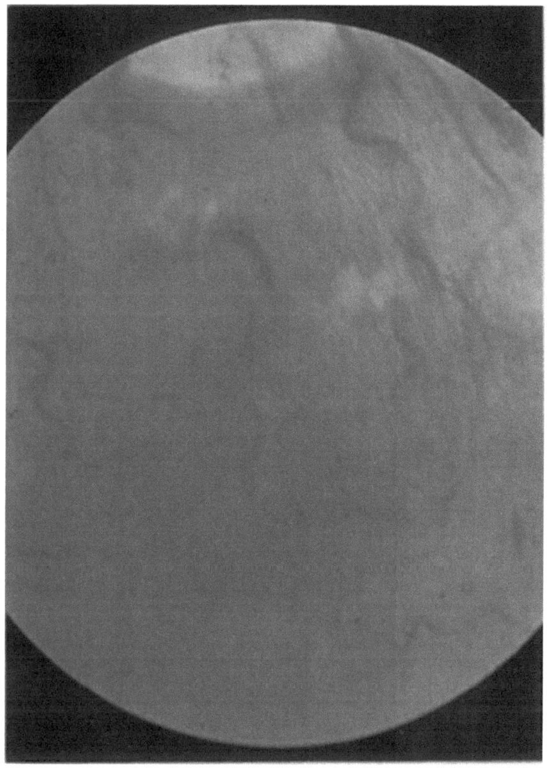

Fig. VI/31

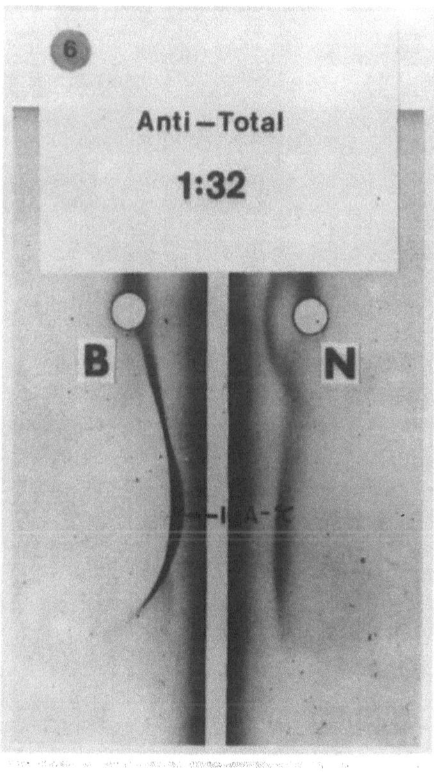


Fig. VI/30

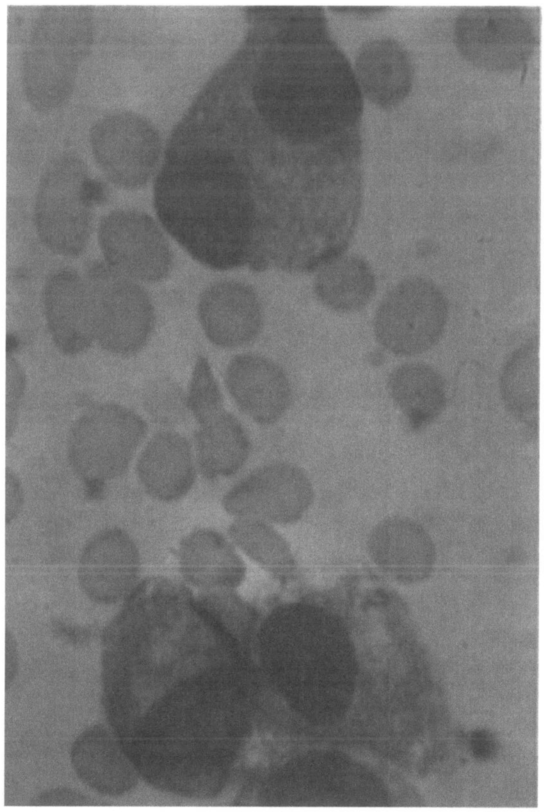

Fig. VI/32

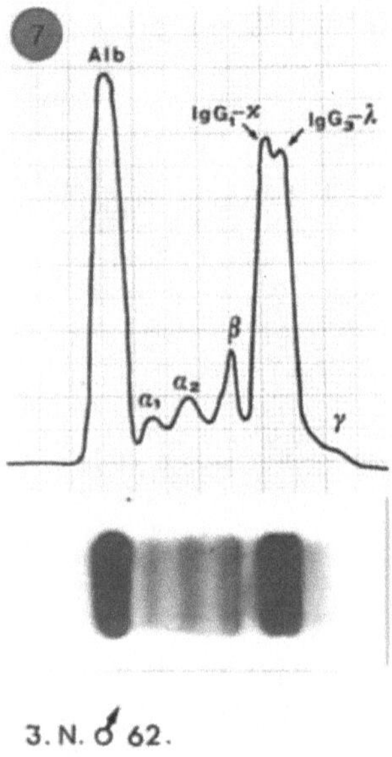


Fig. VI/33

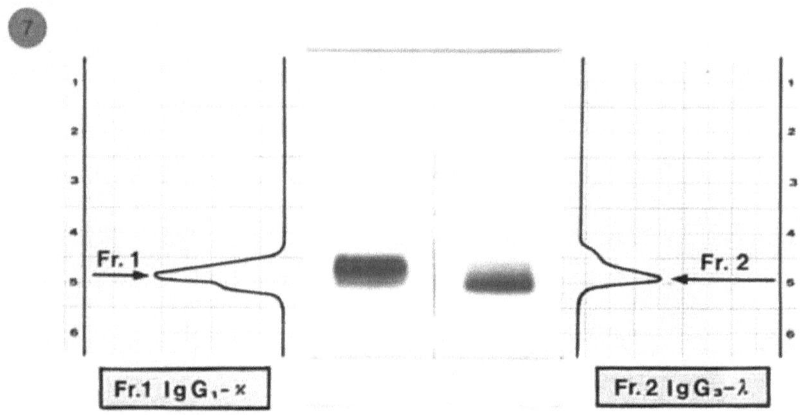


Fig. VI/37

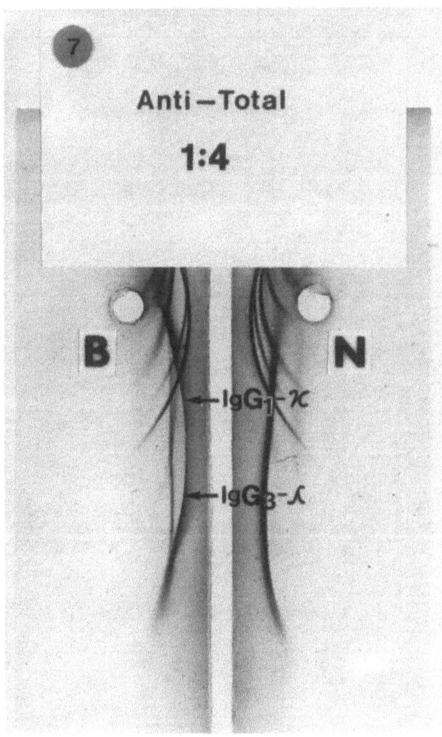


Fig. VI/34

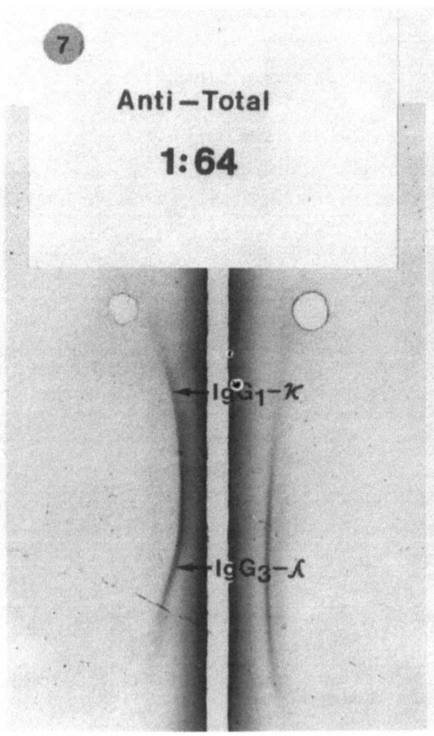


Fig. VI/35

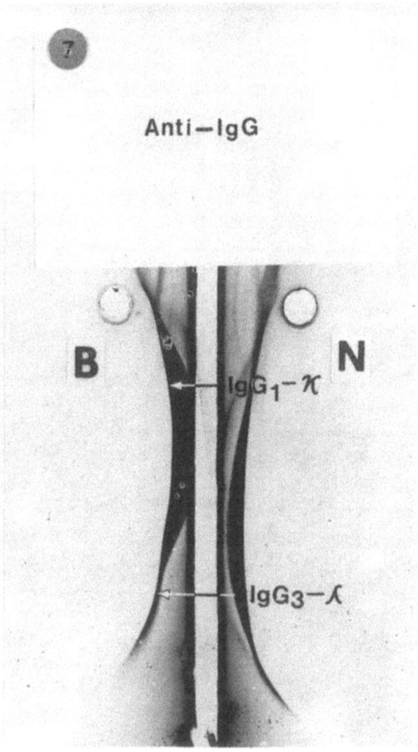


Fig. VI/36

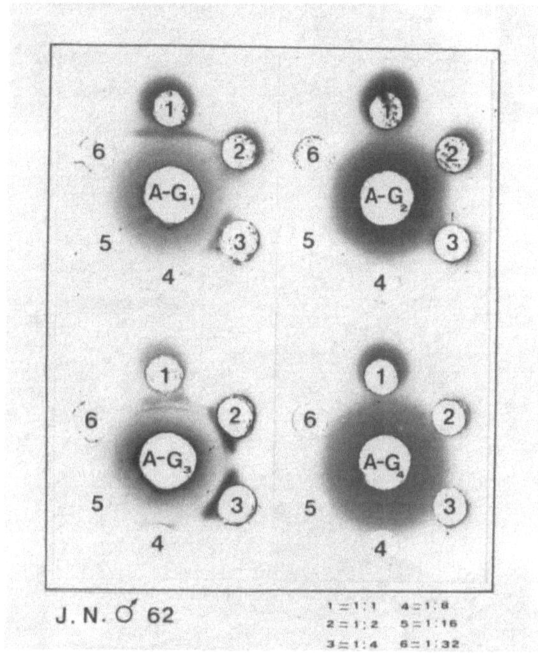


Fig. VI/38

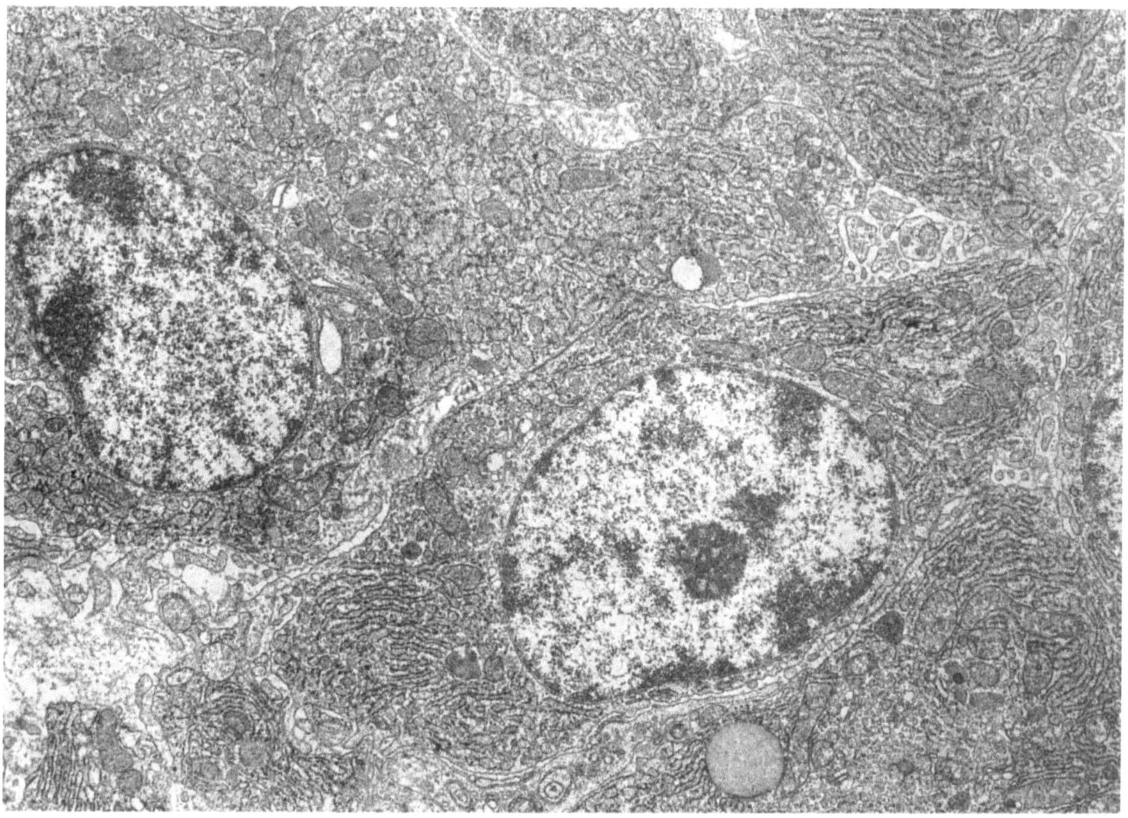

Fig. VI/39

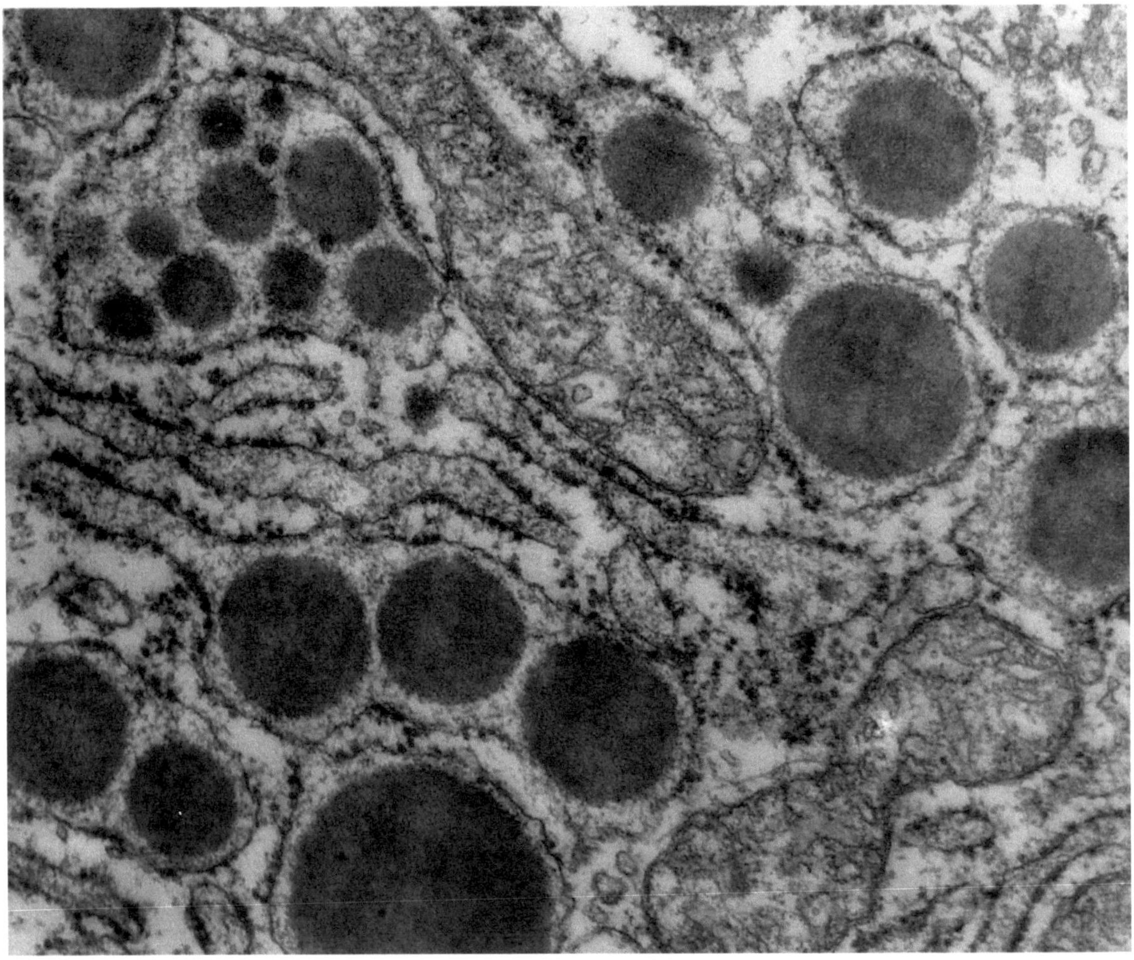

Fig. VI/40

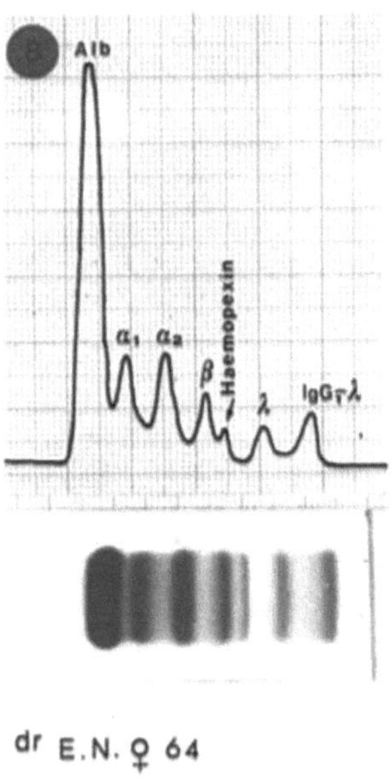


Fig. VI/41

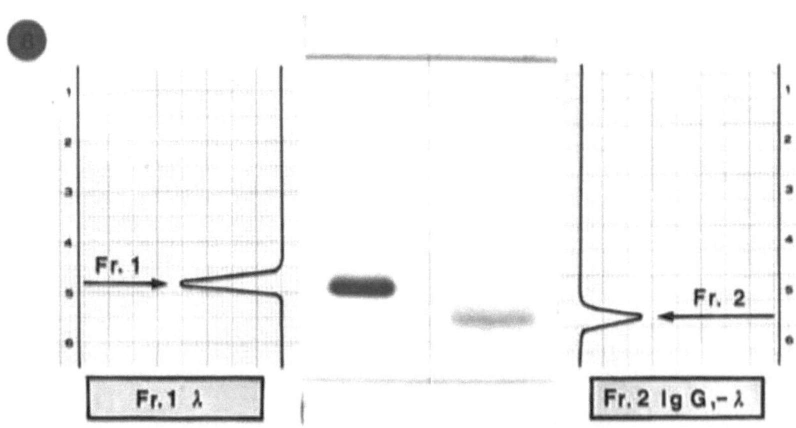


Fig. VI/44

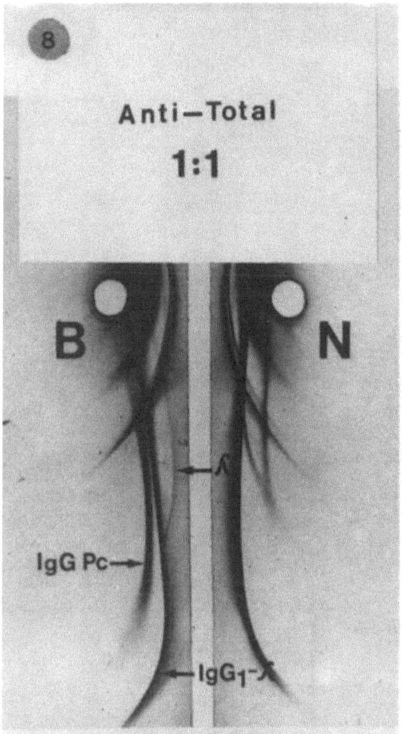


Fig. VI/42

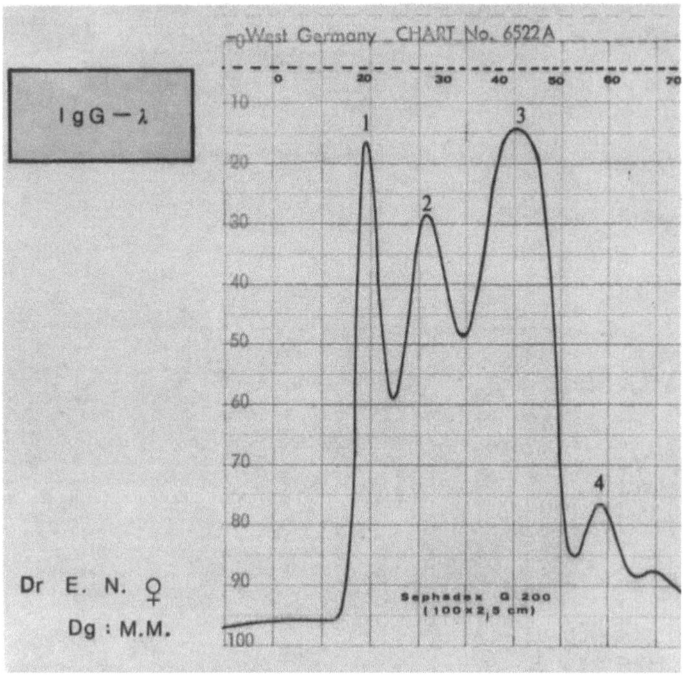


Fig. VI/45

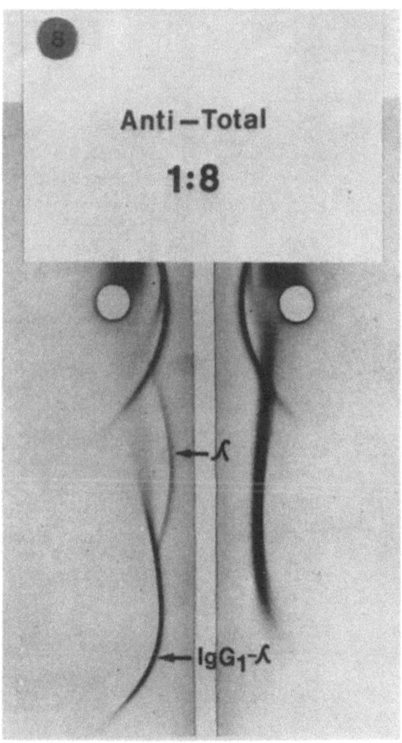


Fig. VI/43

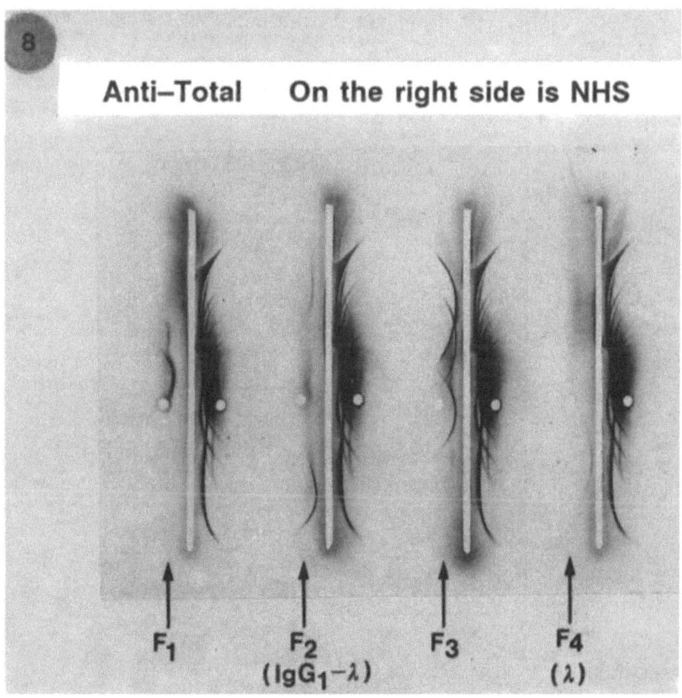


Fig. VI/46

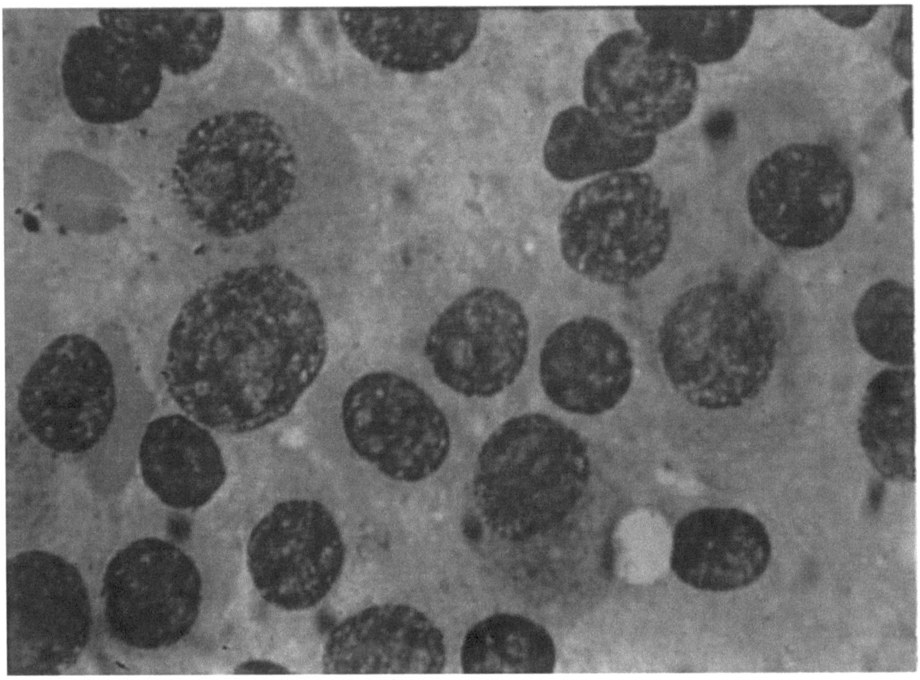

Fig. VI/47

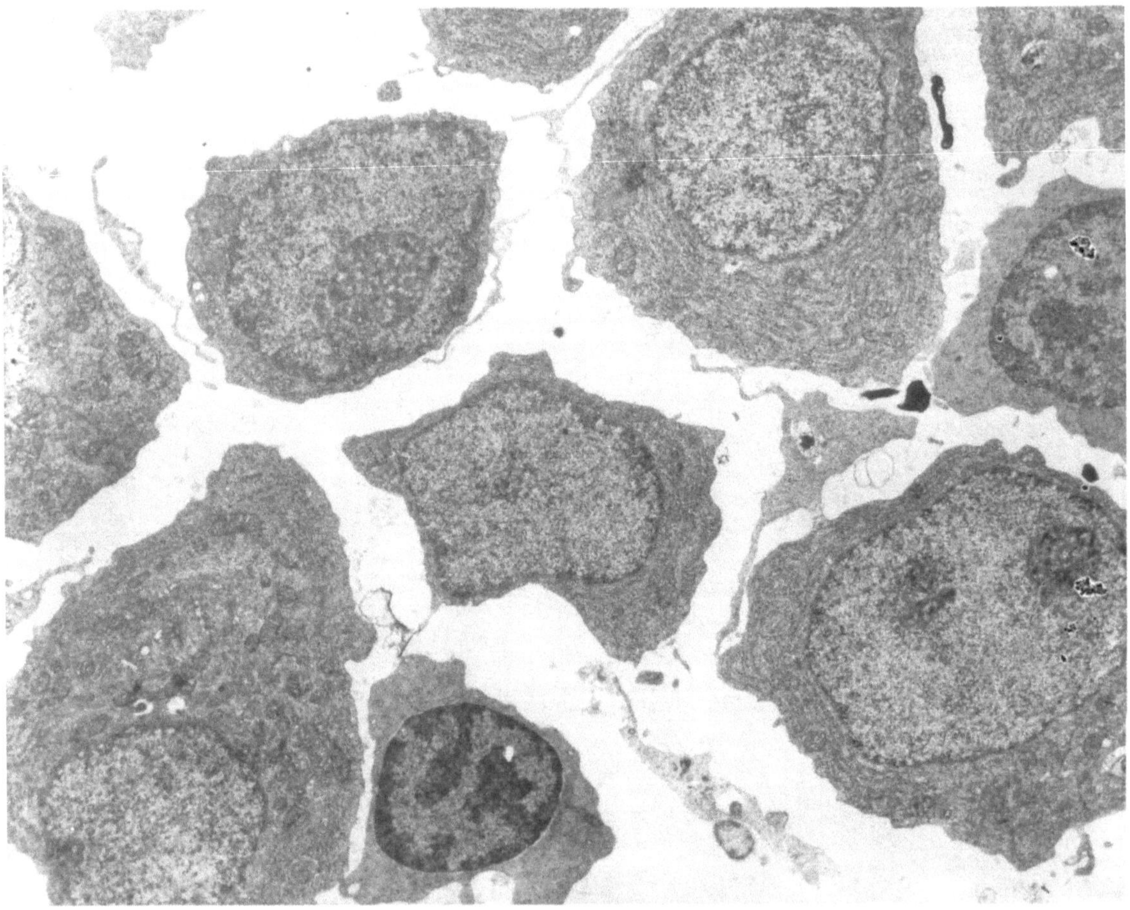

Fig. VI/48

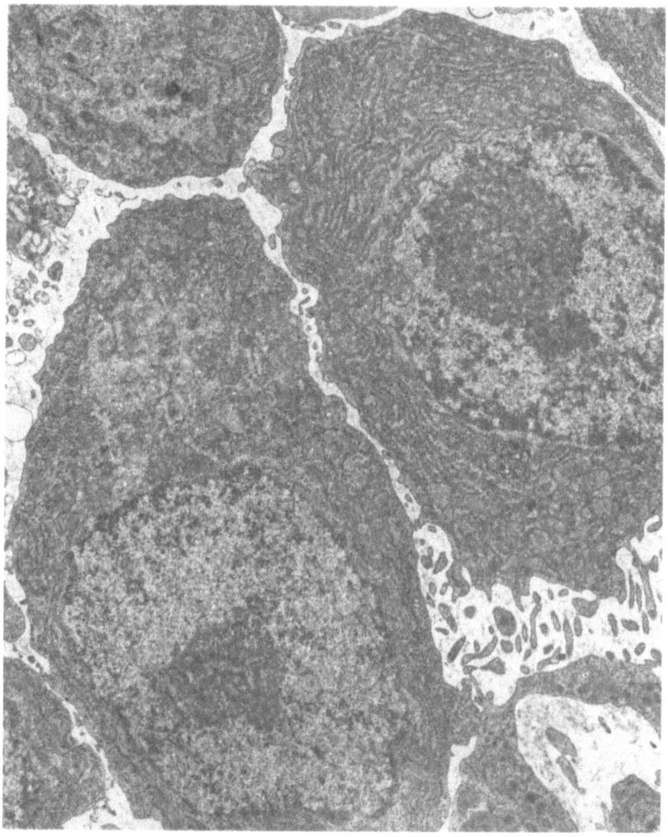

Fig. VI/49

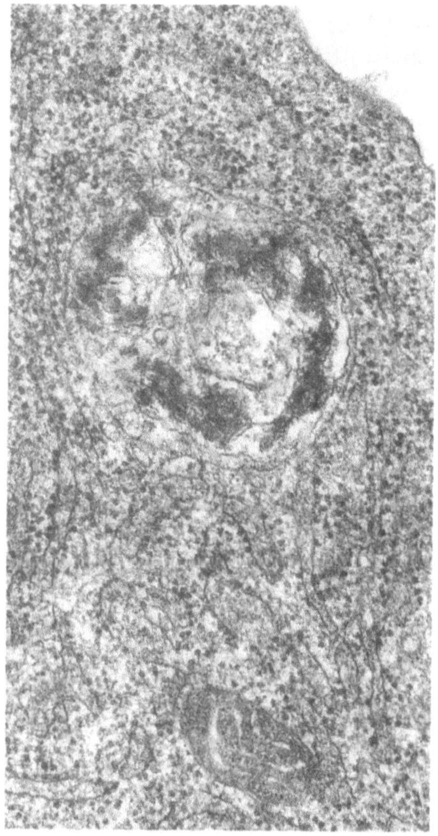

Fig. VI/50

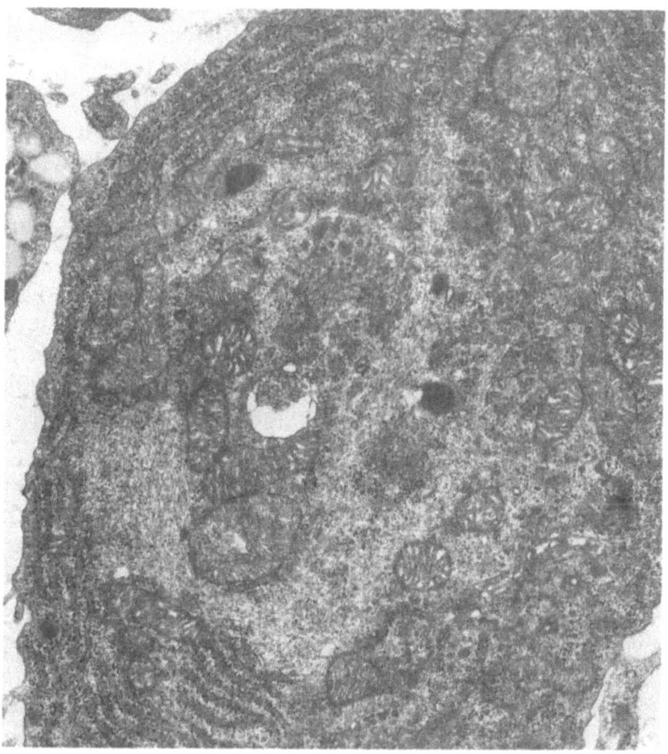

Fig. VI/51

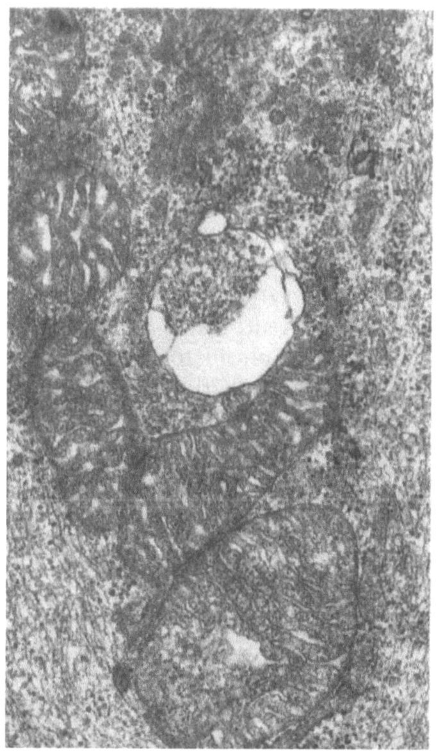

Fig. VI/52

*9. F.F.*, female patient, aged 84. Admitted due to the typical symptoms of bronchopneumonia and high RBC sedimentation (85 mm/h). Anemia and resistance to therapy indicated further examinations. Gel elctrophoresis revealed an intense wide M component which did not suggest paraproteinemia (Fig.VI/53).

Immunochemical analysis identified two well separated precipitin arcs (Fig. VI/54, VI/55, VI/56).

This case is worth pointing out because in contrast to the previous cases, where gel electrophoresis had always revealed double arcs and immunoelectrophoresis failed to support it, in this patient it was the immunoelectrophoresis, which supported biclonal pathology while gel phoresis was not informative.

*Bone marrow analysis:* yielded extraordinary findings. Apart from few large plasmocytes with vacuolized cytoplasm, the lymphoid and lymphoplasmocytoid elements were predominant (Fig. VI/57). Erythro- and myelopoiesis both markedly depressed. Normal platelet formation, though the megacaryocyte count was very low.

X-ray of the skull: unambiguously supports the MM diagnosis. Osteolytic changes in other regions were not observed (Fig. VI/58), which is a rare finding.

*10. Dr.J.M.*, male patient aged 70 with a long history of Waldenström macroglobulinemia. For more than 10 years (!) the RBC sedimentation value had been 80-100 mm/h. The patient was admitted to the Department of Urology and undergone surgery due to the retention of urine. Several days later the patient died of circulatory insufficiency. Clinical and laboratory findings supported double paraproteinemia (Fig. VI/59) identified immunochemically as not IgM, but as IgG. At the same time the immunoelectrophoretic arcs indicated IgM monoclonal gammopathy (Fig. VI/60, VI/61, VI/62, VI/63). Monovalent immune serum identified the paraprotein as IgG-κ (Fig. VI/64). All these findings suggested formation of the IgG complexes. To answer this question column chromatography with AcA 44 was performed and monomeric immunoglobulins identified (Fig. VI/65, VI/66).

*Bone marrow histology:* uniform malignant proliferation of lymphoplasmocytes with high rate of protein synthesis (Fig. VI/67, VI/68).

Despite clear laboratory findings, the clinical diagnosis was uncertain, rather suggesting the WM.

X-ray: revealed only diffuse osteoporosis.

*11. J.Sz.*, female patient aged 57. Admitted to the regional Dept. of internal medicine with MM diagnosis for further examination and therapy. The patient had severe anemia and her blood serum showed milky hyperlipemia. Gel electrophoresis showed two M components (Fig.VI/69). Immunoelectrophoresis also revealed two abnormal scarcely visible precipitin arcs. Monovalent antiserum verified the proteins as IgD-λ and λ light chain (Fig. VI/70, VI/71, VI/72). The serum was further purified with gel filtration and the fractions analysed with immunochemical methods, supporting the above mentioned results (Fig. VI/73, VI/74). High concentrations of the lambda light chains were detected even in the non-concentrated urine samples.

Bone marrow analysis: few cells in the marrow smears. Few normoblasts and more cells of the myeloid line. No megacaryocytes. Polymorphic plasma cells comprise 90% of the nuclear elements.

The analysis of the bones indicates MM with signs of multiplex rib fractures.

The patient was followed-up for two years. After a chest herpes zoster the patient died in the stage III/B. Kidney histology indicated tubular and diffuse glomerular damages, proving that the excreted light chains are nephrotoxic.

*12. L.J.*, male patient aged 78. Was admitted with biclonal gammopathy after a screening test (Fig. VI/75). This aged patient was complaint-free. Immunoelectrophoresis showed IgA-κ and $IgG_1$-λ components (Fig. VI/76).

*Bone marrow analysis:* highly enhanced erythropoiesis; normal myelopoiesis. Increased eosinophil and lymphocyte count. Megacaryocytes show normal platelet formation activity. Sustained increase in plasma cell formation; the cells exhibit pronounced acid phosphatase activity.

X-ray: osteoporosis corresponding to the patient's age; spondylarthrosis. No other pathological signs.

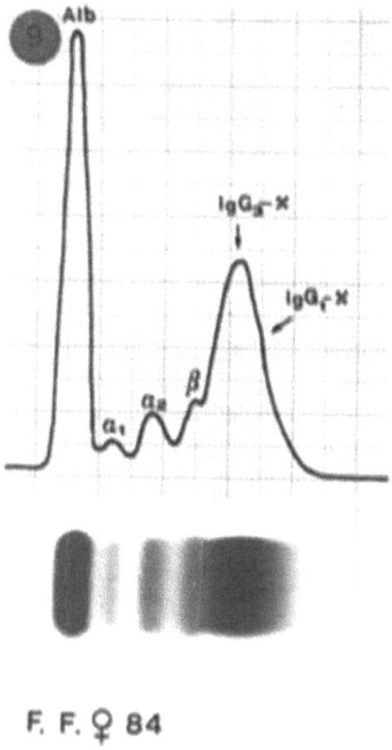


Fig. VI/53

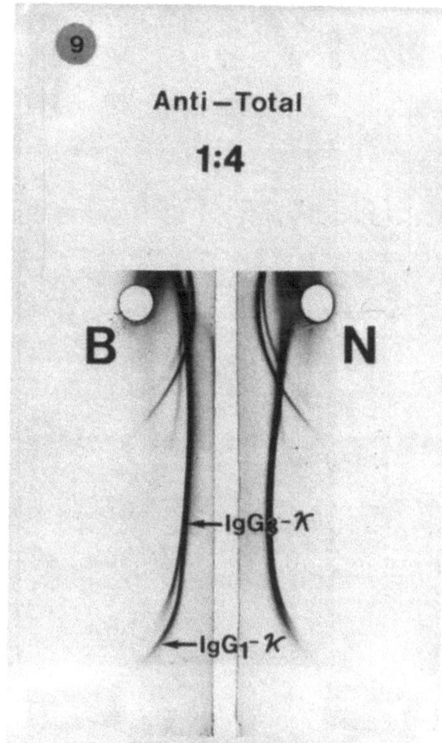


Fig. VI/55

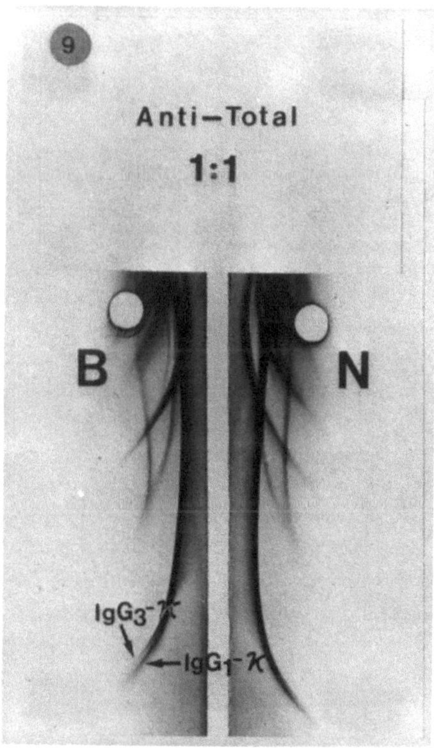


Fig. VI/54

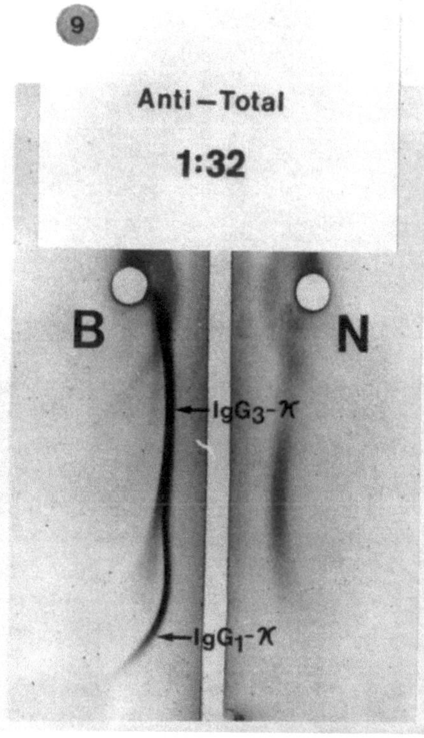


Fig. VI/56

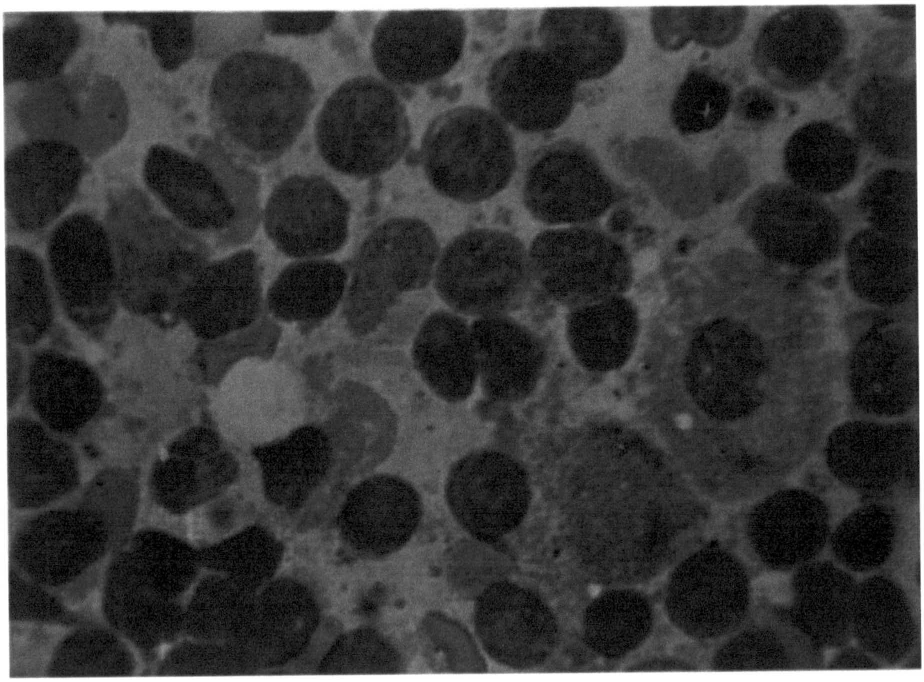

Fig. VI/57

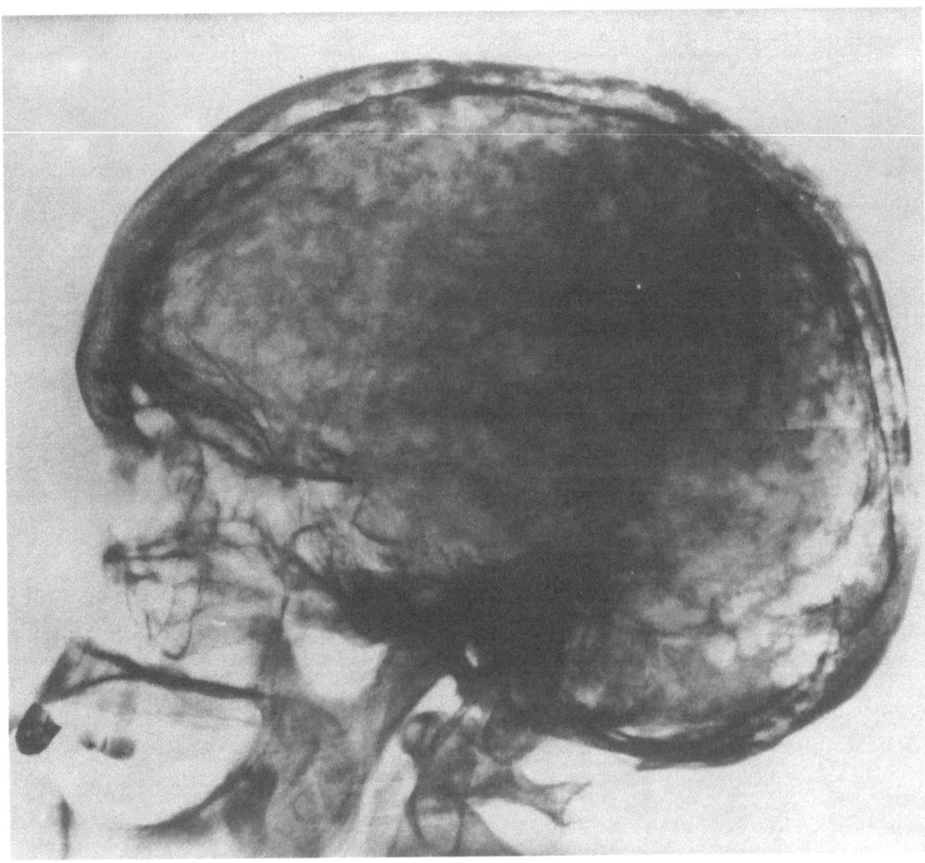

Fig. VI/58

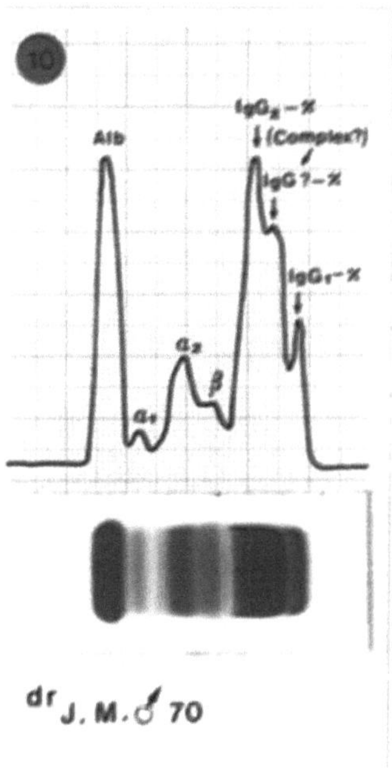


Fig. VI/59

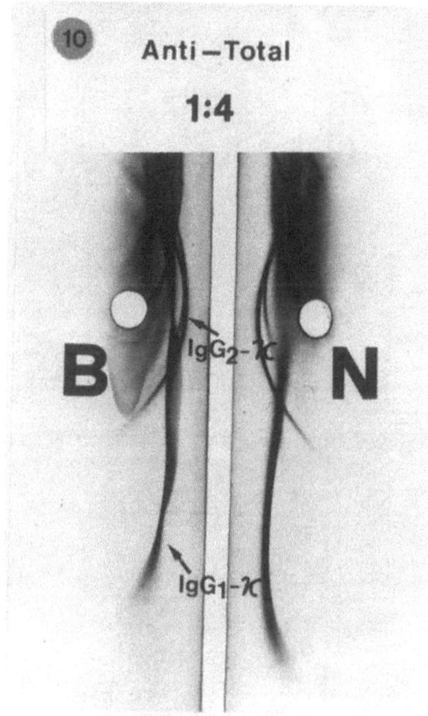


Fig. VI/61

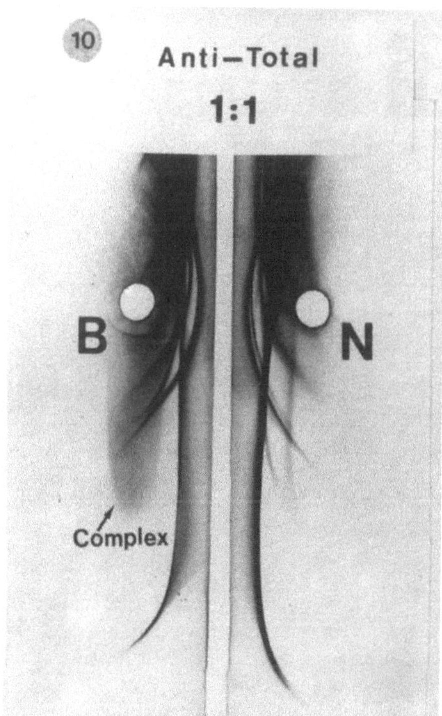


Fig. VI/60

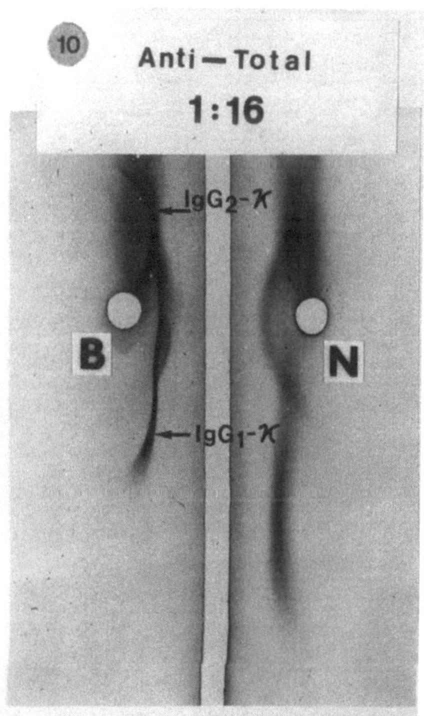


Fig. VI/62

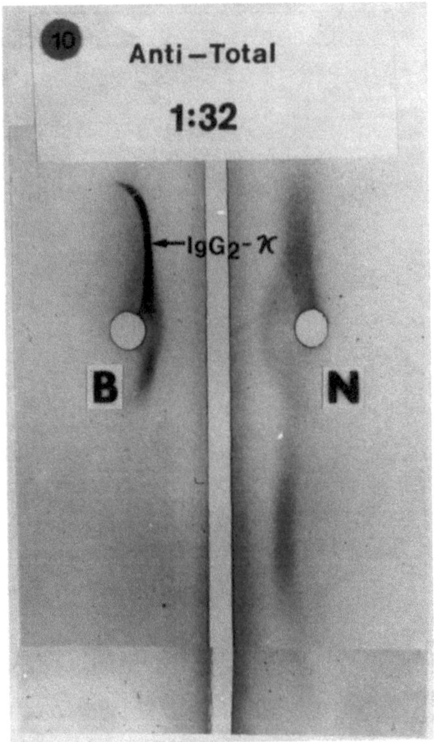


Fig. VI/63

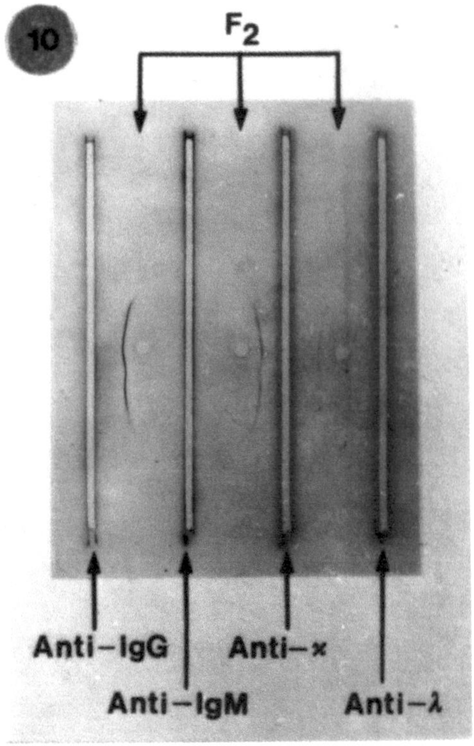


Fig. VI/64

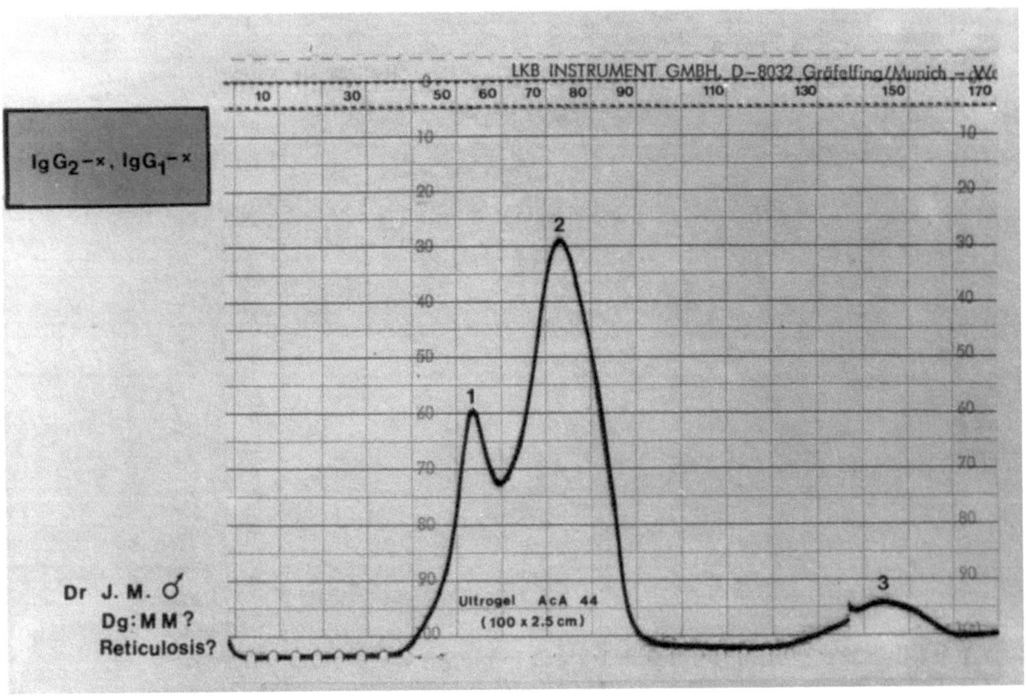


Fig. VI/65

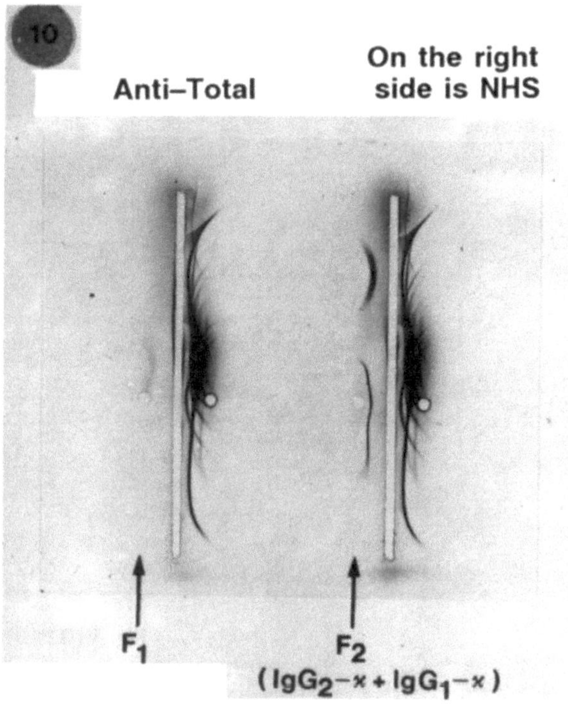


Fig. VI/66

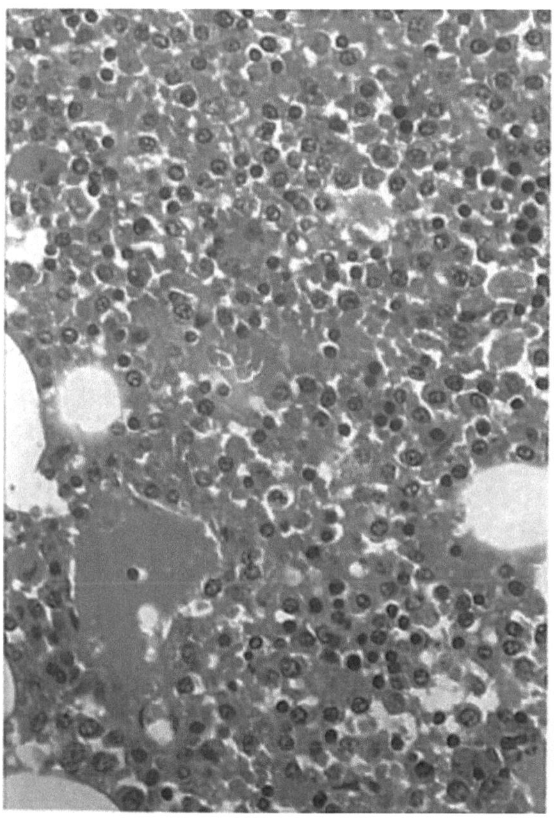

Fig. VI/67

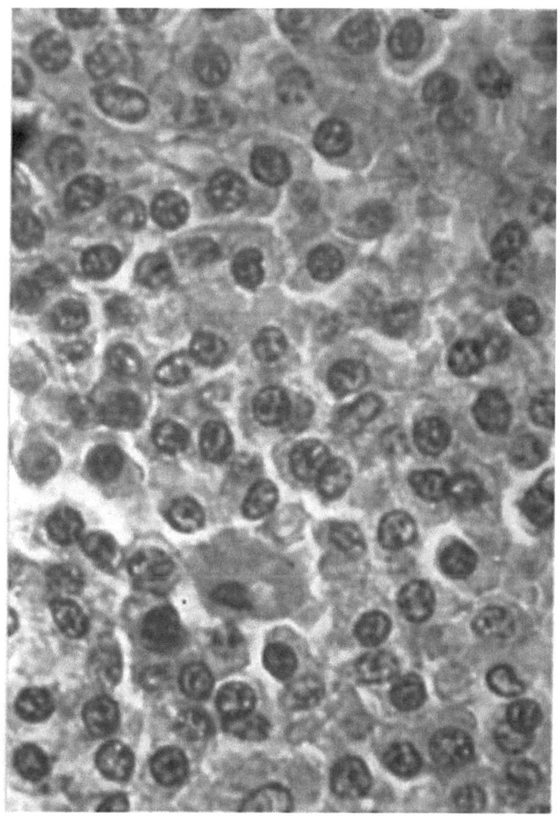

Fig. VI/68

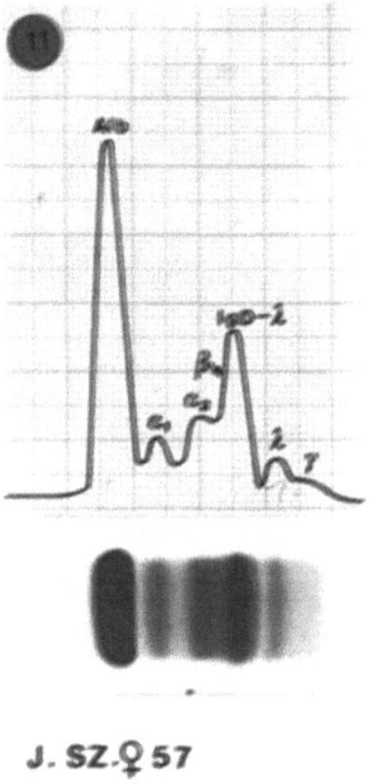


Fig. VI/69

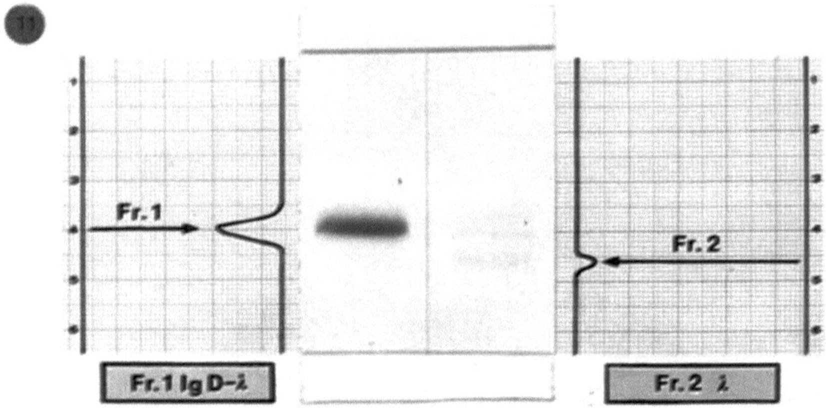


Fig. VI/72

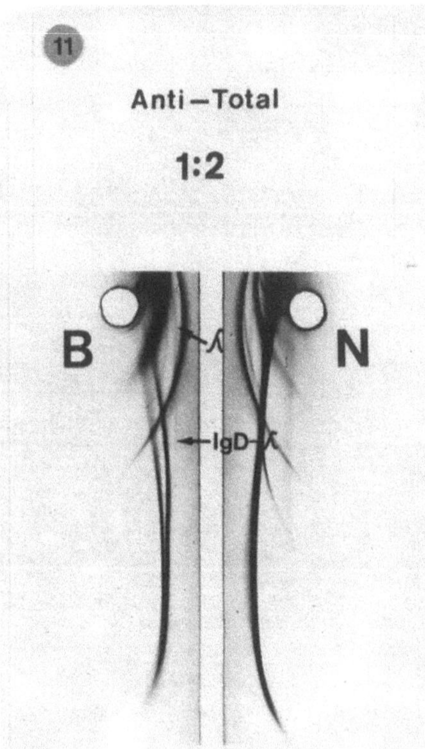


Fig. VI/70

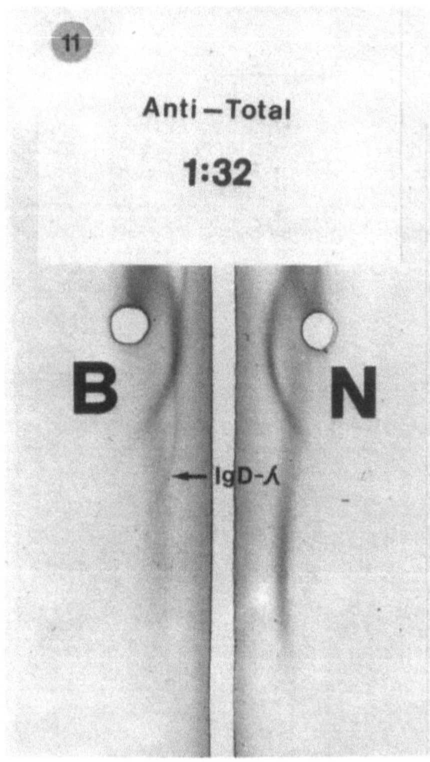


Fig. VI/71

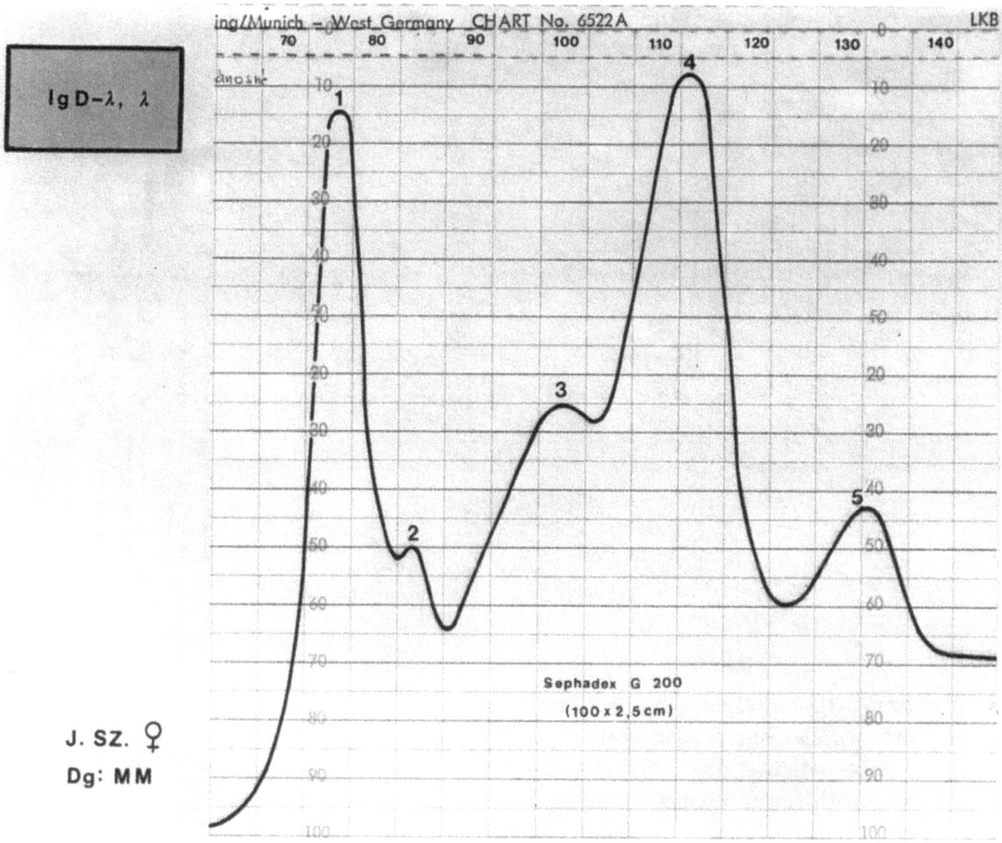


Fig. VI/73

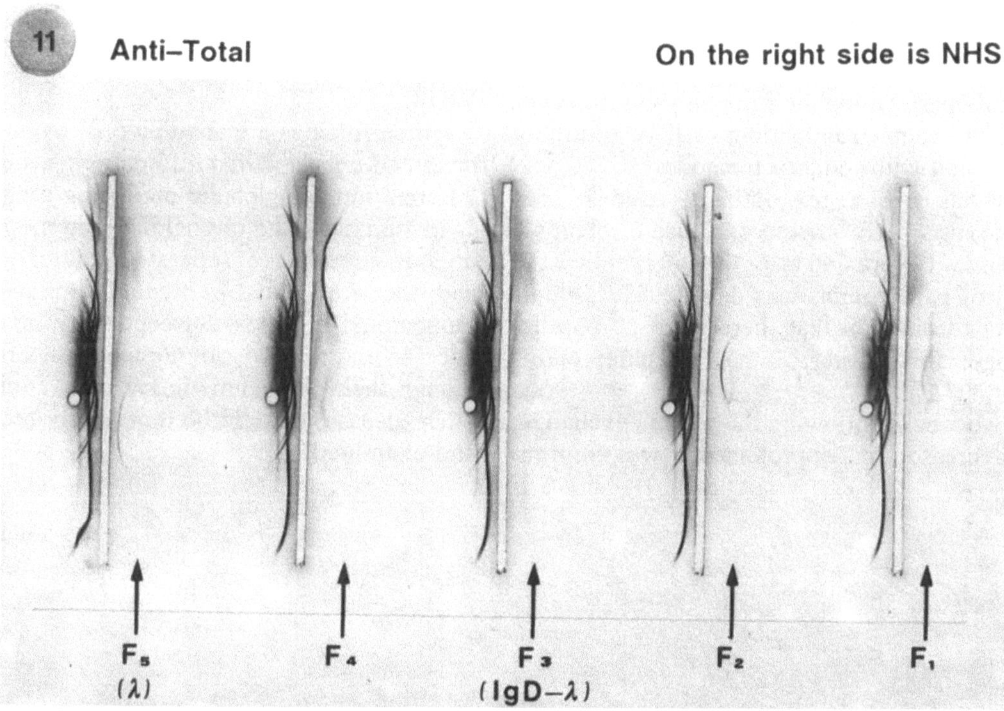


Fig. VI/74

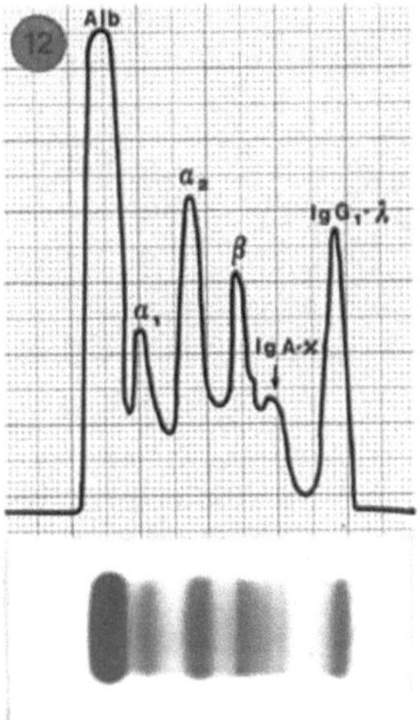


Fig. VI/75

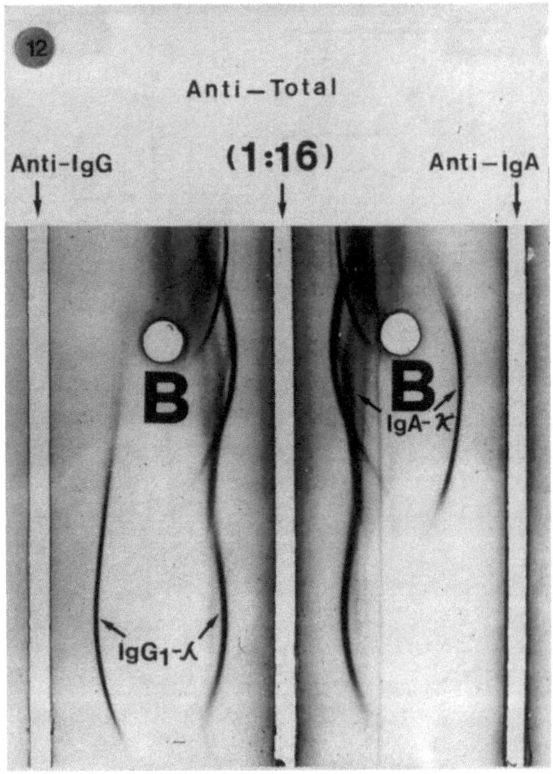


Fig. VI/76

In summary, the analysis of the twelve cases *was incomplete,* because in neither case we could detect, whether the cells originate from one or two different clones.

The main reason is that such studies are extremely time-consuming: for fluorescent and other labeling techniques fresh marrow samples or deep-freezed marrow must be available. These technical obstacles prevented us from performing the full analysis. However in future such examinations will be routinely performed with modern methods.

Plasma exchange routinely used in our Clinic, was performed in one case of paraproteinemia. Before and after the intervention the level of paraproteins was detected. It is interesting to mention that there were no parallel changes in the titer of the two paraproteins (Fig. VI/77).

Two weeks following the plasma exchange, the decreased IgA-κ protein titer was continuously increasing, reaching a 50%, of its initial level; while the IgG$_1$-λ could increase to the same value only by the seventh week (Fig. VI/78).

These findings are an additional evidence in favour of our view that different clones exhibit different immunoglobulin-producing capacity.

In summary, the oligoclonal gammopathies do not comprise a separated clinical group and, therefore, must be treated similar to the monoclonal diseases. Susceptibility or resistance to therapy do not indicate differences, though their characteristic features, similar to other diseases, should be thoroughly recorded and examined.

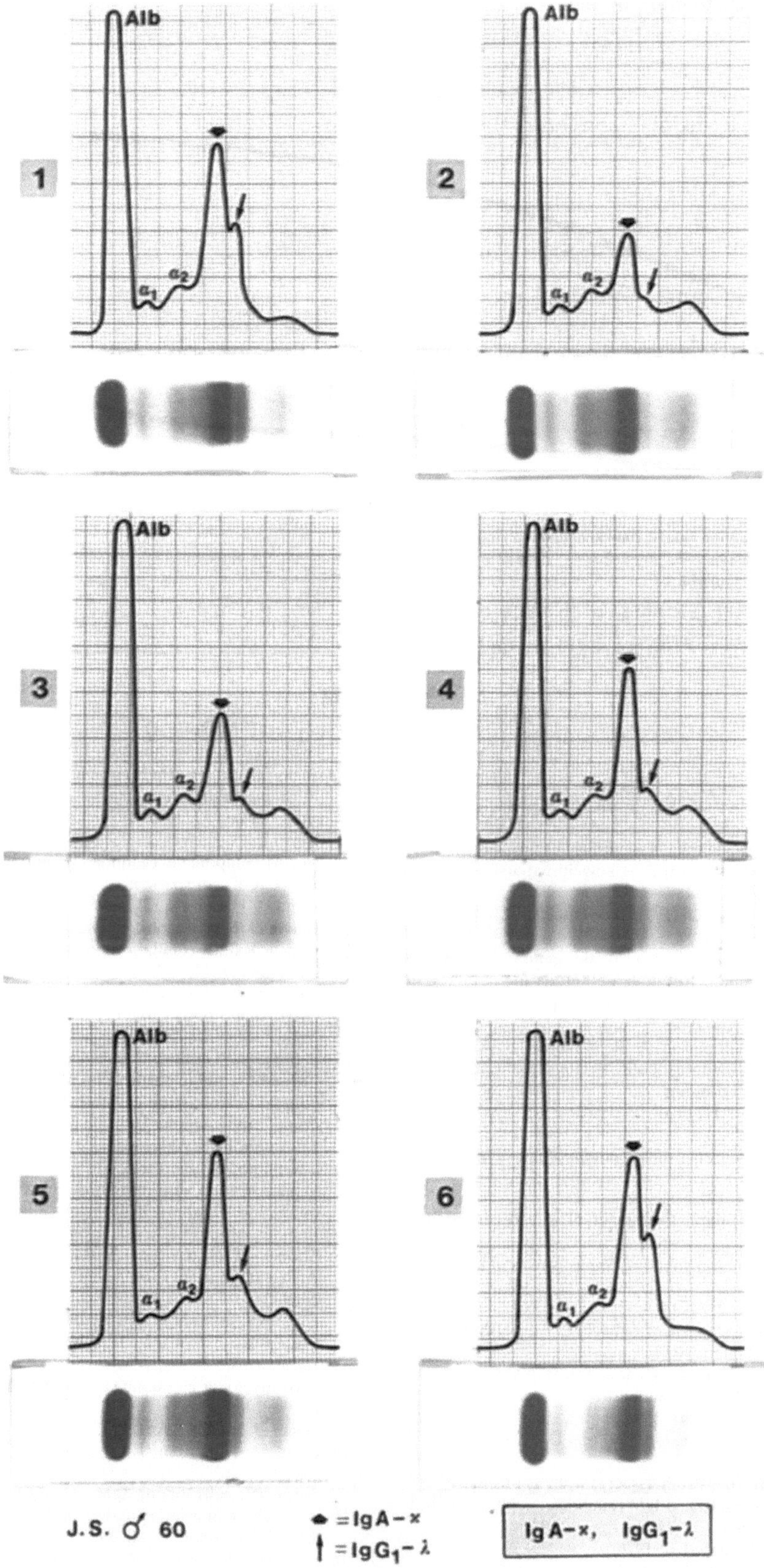


Fig. VI/77

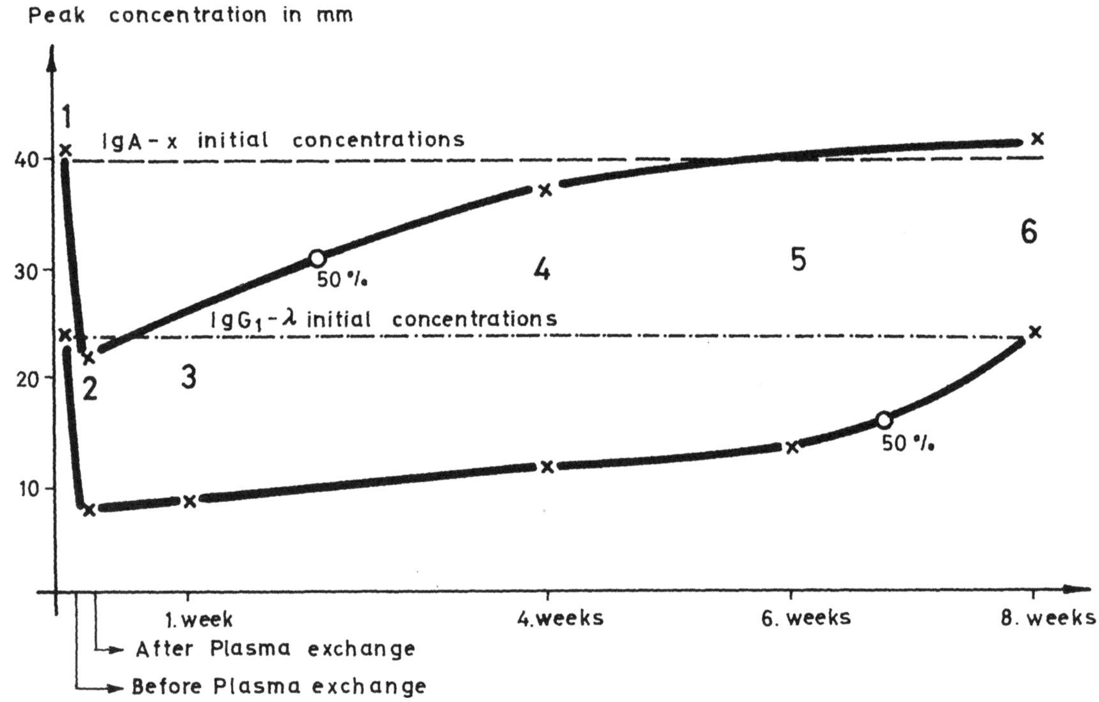


Fig. VI/78

## References


1. Alcalay, M., Reboux, J. F., Vivant, R., Rauvet, J. M., Bontoux, D.
Polyarthrite séronégative, vascularité, et double gammapathie monoclonale IgA et IgD non myélomateuse.
Rev. Rheum. 41, 693. 1974.
2. Arnaud, Ph., Courpron, P., Dechavanne, M., Meunier, P., Vignon, G., Creysell, R.
Deux cas de gammapathies biclonales gamma G – lambda–gamma A – lambda. Résultats de l'étude métabolique des composants monoclonaux chez l'un des patients.
Lyon méd. 236, 129. 1976.
3. Awdeh, Z. L., Williamson, A. R., Askonas, B. A.
One cell – one immunoglobulin, origin of limited heterogeneity of myeloma proteins.
Biochem. J. 116, 241. 1970.
4. Bachman, R.
Simultaneous occurence of two immunologically different M components in serum.
Acta med. scand. 177, 593. 1965.
5. Ballieux, R. E., Imhof, J. W., Mül, N. A., Zegers, B. J. M. M., Stoop, J. W.
Pathology of immunoglobulins. Some aspects of monoclonal gammapathies.
Clin. chim. Acta. 22, 7. 1968.
6. Basl, E. J. E. G., Slaper-Cortenbach, L. F., Verdonck, L. F., Samson, W., Ballieux, R. E.
Transient expression of a second monoclonal component in two forms of biclonal gammopathy.
Brit. J. Haematol. 60, 91. 1985.
7. Becq-Giraudon, B., Yver, L., Hurez, D., Gombert, L., Alcalay, D.
Maladie de Waldenström associée à un syndrome de Sjögren. Double population d'IgM (lambda et kappa). Evolution sous Chlorambucil.
Nouv. Presse Méd. 5, 137. 1976.
8. Bihrer, R., Flury, R., Morell, A.
Biklonale Paraproteinämie.
Schweiz. med. Wschr. 104, 39, 1974.
9. Bouvet, J. P., Buffe, D., Oriol, R., Liacopoulos, P.
Two myeloma globulins $IgG_1$-kappa and $IgG_1$-lambda from a single patient: II. Their common cellular origin as revealed by immunofluorescence studies.
Immunology. 27, 1095. 1974.
10. Bjerrum, O. J., Weeke, B.
Two M components (gamma GK and gamma ML) in different cells of same patient.
Scand. J. Haemat. 5, 215. 1968.
11. Bouvet, J. P., Feingold, J., Oriol, R., Liacopoulos, P.
Statistical study cellular origin of both myeloma globulins.
Biomedicine. 22, 517. 1975.
12. Brenner, S., Milstein, C.
Origin of antibody variation.
Nature. 211, 242. 1966.
13. Cebra, J. J., Goldstein, G.
Chromatographic purification of tetramethylrhodamine-immunoglobulin conjugates and their use in cellular localization of rabbit gamma-globulin polypeptide chains.
J. Immunol. 95, 230. 1965.
14. Chiappino, G., Pernis, B.
Demonstration with immunofluorescence of 19S macroglobulins and 7S gamma in different cells of human spleen.
Path. et. Mocrobiol. 27, 8. 1964.
15. Cooper, M. D., Lawton, A. R., Kincade, P. W.
A two stage model for development of antibodyproducing cells.
Clin. exp. Immunol. 11, 143. 1972.
16. Costea, N., Yakulis, V. J., Libnoch, J. A., Pilz, C. G., Heller, P.
Two myeloma globulins (IgG and IgA) in one subject and one cell-line.
Amer. J. Med. 42, 630. 1967.
17. Couderc, J., Birrien, J. L., Oriol, R., Bleux, C., Liacopoulos, P.
Bispecific cells among IgM and IgG producers during the early phase of primary and secondary responses.
Eur. J. Immunol. 5, 140. 1975.
18. Creyssel, R., Gibaud, A., Arnaud, Ph., Cordier, J. F. Gammapathies biclonales et oligoclonales.
Lyon med. 236, 101. 1976.

19. Creyssel, R., Brizard, C. P., Gibaud, A., Cordier, J. F., Gibaud, H.
Maladie des chaines lourdes gamma d'apparition secondaire au cours de l'évolution d'une macroglobulinémie.
Lyon med. 237, 101. 1977.
20. Curtain, C. C.
Immunocytochemical localization of two abnormal serum globulins in one bone-marrow smear.
Austr. Ann. Med. 13, 136. 1964.
21. Dammacco, F., Trizio, D., Bonomo, L.
A case of IgA-kappa myelomatosis with two urinary Bence Jones proteins (BJ-kappa and BJ-lambda) and multiple chromosomal abnormalities.
Acta haemat. 41, 309. 1969.
22. Davis, M. M., Kim, S. K., Hood, L.
Immunoglobulin class switching: developmentally regulated DNA rearrangements during differentiation.
Cell. 22, 1. 1980.
23. Dittmar, K., Kochwa, S., Zucker-Franklin, D., Wasserman, L. R.
Coexistence of polycythaemia vera and biclonal gammopathy (gamma GK and gamma AL) with two Bence Jones proteins (BJK and BJL).
Blood. 31, 81. 1968.
24. Dreyerm, W. J., Benett, J. C.
The molecular basis of antibody formation: a paradox.
Proc. Nat. Acad. Sci. 54, 864. 1965.
25. Engle, R. L., Nachman, R. L.
Two Bence Jones proteins of different immunologic types in the same patient with multiple myeloma.
Blood. 27, 74. 1966.
26. Fahey, J. L.
Heterogenety of myeloma proteins.
J. Clin. Invest. 42, 111. 1963.
27. Fahey, J. L., Carbone, P. P., Rowe, D. S., Bachmann, R.
Plasma cell myeloma with D-myeloma protein (IgD myeloma).
Amer. J. Med. 45, 373. 1968.
28. Fair, D. S., Krueger, R. G., Gleich, G. J., Kyle, R. A.
Studies on IgA and IgG monoclonal proteins derived from a single patient. Evidence of shared individually specific antigenic determinants.
J. Immun. 112, 201. 1974.
29. Fair, D. S., Krueger, R. G.
Analysis of biclonal immunoglobulins and their contributions to understanding the developmental aspects of the antibody response.
Contem. Top. mol. Immunol. 7, 51. 1978.
30. Fateh-Moghadam, A., Würz, H., Oeser, R. M., Knedel, M.
Doppelparaproteinämie.
Dtsch. med. Wschr. 36, 1965. 1968.
31. Fateh-Moghadam, A., Beil, E., Borchers, H., Asamer, M., Raab, Ch.
Plasmozytom, Makroglobulinämie Waldenström und Morbus Paget bei einem Patienten mit IgA-kappa + IgM-kappa Doppelparaproteinämie und Bence Jones Protein (Typ kappa).
Blut. 21, 146. 1970.
32. Fine, J. M., Zakin, M. M., Faure, A., Ropertz, G., Rivat, C., Rivat, L., Boffa, G. A.
Étude immunochemique de 5 cas dysglobulinémies „biclonales".
Rev. Franc. de Transfusion. 11, 223. 1968.
33. Finegold, I., Fahey, J. L., Dutcher, T. F.
Immunofluorescent studies of immunoglobulins in human lymphoid cells in continuous culture.
J. Immunol. 101, 366. 1968.
34. Fu, S. M., Winchester, R. J., Feizi, T., Walzer, P. D., Kunkel, H. G.
Occurence of surface IgM IgD and free light chains of human lymphocytes.
J. exp. Med. 139, 451. 1974.
35. Fudenberg, H. H., Wang, A. C., Ping, J. R. L., Levin, A. S.
Studies of an unusual gammapathy: implications with regard to genetic control of normal immunoglobulin synthesis.
Amer. N. Y. Acad. Sci. 190, 501. 1971.
36. Gally, J. A., Edelman, G. M.
The genetic control of immunoglobulin synthesis.
Ann. Rev. Genet. 6, 1. 1972.
37. Gearhart, P. J., Sigal, N. H., Klinman, N. R.
Production of antibodies of identical idiotype but diverse immunoglobulin classes by cells derived from a single stimulated B cell.
Proc. nat. Acad. Sci. 72, 1707. 1975.
38. Gearhart, P. J., Hurwitz, J. L., Cebra, J. J.
Successive switching of antibody isotypes expressed within the lines of a B-cell clone.
Proc. nat. Acad. Sci. 77, 5424. 1980.
39. Harboe, M., Hannestad, K., Sletten, K.
Oligoclonal macroglobulinaemia.
Scand. J. Immun. 1, 13. 1972.
40. Hijmans, W., Schuit, H. R. E., Klein, R.
An immunfluorescence procedure for the detection of intracellular immunoglobulins.
Clin. exp. Immunol. 4, 457. 1969.
41. Hijmans, W., Schuit, H. R. E., Hulsing-Hesselink, E.
An immunfluorescence study an intracellular immunoglobulins.
Ann. N. Y. Acad. Sci. 177, 290. 1971.
42. Hijmans, W., Schuit, H. R. E.
Immunfluorescence studies an immunoglobulins in the lymphoid cells of human peripherial blood
Clin. exp. Immunol. 11, 483. 1972.
43. Hijmans, W.
Waldenström's disease with an IgA paraprotein.
Acta med. scand. 198, 519. 1975.
44. Hopper, J. E.
Immunofluorescent evidence for a common clonal origin of IgM-lambda and IgG-kappa paraproteins.
Clin. Res. 21, 557. 1973.
45. Hopper, J. E.
Comparative studies on monotypic IgM-lambda and IgG-kappa from an individual patient.
Evidence for shared $V_H$ idiotypic determinants.
J. Immun. 115, 1101. 1975.
46. Hopper, J. E., Cera, L.
The structure of human immunoglobulins.
Ann. Clin. Lab. Sci. 8, 201. 1978.
47. Imhof, J. W., Ballieux, R. E., Mul, N. A. J., Poen, H.
Monoclonal and diclonal gammopathies.
Acta med. scand. 179, (suppl. 445.) 102. 1966.
48. Isobe, T., Osserman, E. F.
Plasma-cell dyscrasia associated with the production of incomplete deleted IgG-lambda molecules, gamma heavy-chains and free lambda-chains containing carbohydrates, description of the first case.
Blood. 43, 505. 1974.
49. István, L., Hindy, I., Eckhardt, S.
Adatok a plasmocytoma klinikumához, kezeléséhez és gondozásához
Orv. Hetil. 117, 11. 1976.
50. Jákó, J.
Fehérjeanyagcsere-vizsgálatok májbetegségekben I., Kandidátusi értekezés, 1973.
51. Jensen, K. Jensen, K. B., Olsen, H.
Three M components in serum from an apparently healthy person.
Scand. J. Haemat. 4, 485. 1967.
52. Josephson, A. S., Nicastri, A., Price, E., Bíró, L.
H $\mu$-chain and monoclonal IgA in an hyperproliferative disorder.
Amer. J. Med. 54, 127. 1973.
53. Kaboth, U., Kattermann, R., Arnold, R., Hauswaldt, C.
Immunoelektrophoretische Untersuchungen an einer pluriklonalen Gammopathie.
Verh. Dtsch. Ges. inn. Med. 74, 469. 1968.

54. Keller, H., Spengler, G. A., Skvaril, F., Flury, W., Noseda, G., Riva, G.
Zur Frage der Heavy-Chain Disease. Ein Fall von IgG-heavy-chain Fragment und IgM Typ K Paraproteinämie mit Plasmazellenleukämie.
Schweiz. med. Wschr. 100, 1012. 1970.
55. Keshgegian, A. A.
Supression of one monoclonal immunoglobulin in the presence of another in multiple myeloma: evidence for benign B cell neoplasia.
Cancer. 51, 1097. 1983.
56. Kubugawa, H., Vogler, C. B., Capra, J. D., Conrad, M. E., Lawton, A. R., Cooper, M. D.
Studies on the clonal origin of multiple myeloma.
J. Exp. Med. 150, 792. 1979.
57. Levin, A. S., Rundenberg, H. H., Hopper, J. E., Wilson, S. K.
Immunoflourescent evidence for cellular control of synthesis of variable regions of light and heavy chains of immunoglobulins G and M by the same gene.
Proc. nat. Acad. Sci. 68, 169. 1971.
58. Liacopoulos, P., Amstutz, H., Gille, F.
Early antibody-forming cells of double specificity.
Immunology. 20, 57. 1971.
59. Lievre, J. A., Camus, J. P., Darcy, M., Crouzet, J., Guillien, P.
Myélome avec tumeur iliaque (résection, reconstruction au ciment plastique, effect sur les paraprotéines).
Ann. Med. Int. 122, 663. 1971.
60. Lucivero, G., Lawton, A. R., Cooper, M. D.
Pokeweed-mitogen induced differentiation of human peripperal blood B-lymphocytes.
II. Suppression of plasma cell differentiation by heavy-chain-specific antibodies and development of immunoglobulin class restriction.
Hum. Lymph. Diff. 1, 27. 1981.
61. Lucivero, G., Miglietta, A., Dell'Osso, A., Theodossio, G., Bonomo, L.
Double (IgAK + IgGK) paraproteinaemia in a single patient: immunofluorescence evidence for a common plasma cell clone „frozen" at the switch phase.
Acta Haemat. 75, 224. 1986.
62. Metz, K., Leder, L. D.
A case of multiple myeloma with two subclones producing separately kappa and defective gamma chains.
Cancer. 56, 1392. 1985.
63. Maldonato, J. E., Williams, R. C., Silverstein, M. N., Harrison, E. G.
Lymphosarcoma with Bence Jones proteinuria (type kappa) and IgG serum M component (type lambda): a bitypic gammopathy.
Proc. Mayo Clin. 45, 413. 1970.
64. McNutt, D. R., Fudenberg, H. H.
IgG myeloma and Waldenström macroglobulinemia.
Coexistence and clinical manifestations in one patient.
Arch. Intern. Med. 131, 731. 1973.
65. Mellors, R. C., Korngold, L.
The cellular origin of human immunoglobulins (gamma-2, gamma-1M, gamma-1A).
J. exp. Med. 118, 387. 1963.
66. Natvig, J. B., Kunkel, H. G.
A hybrid IgG4–IgG2 immunoglobulin.
J. Immunol. 112, 1277. 1974.
67. Nisonoff, A., Hopper, J. E., Spring, S. B.
The antibody molecule.
Academic Press, N. Y., 1, 542. 1975.
68. Nossal, G. J. V., Szenberg, A., Ada, G. L., Austin, C. M.
Single cell studies on 19S antibody production.
J. exp. Med. 119, 485. 1964.
69. Nowell, P. C.
Genetic instability in cancer cells: relationship to tumor cell heterogeneity.
In: Tumor cell heterogeneity: origins and implications.
Ed. by Owens, A. H., Coffey, D. S., Baylin, J. B.
New York Academic Press. 351. 1982.
70. Oriol, R., Barvier, Y., Biozzi, G.
Hybrid 7S immunoglobulin molecules formed in vivo by association of heavy and light chains of differenct parental origin.
European J. Immunol. 2, 105. 1972.
71. Oriol, R., Huerta, J., Bouvet, J. P., Liacopoulos, P.
Two myeloma globulins, IgG1–K and IgG1–L, from a single patient (Im) I. Purification and immunochemical characterization.
Immunology. 27, 1081. 1974.
72. Otto, Sz., Puskás, E., Medgyesi, G. A., Gergely, J.
Diclonal and multiple gammapathies.
Haematologia. 6, 471. 1972.
73. Oudin, J., Michel, M.
Idiotype of rabbit antibodies. II. Comparison of idiotype of various kinds of antibodies formed in the same rabbits against Salmonella Typhi.
J. exp. Med. 130, 619. 1969.
74. Owen, J. A., Got, C., Silberman, H. J.
Heterogeneity of serum myeloma globulins.
Clin. chim. Acta. 3, 605. 1958.
75. Oxelius, V. A.
Alternating appearance of IgD and IgG myeloma protein during treatment.
Scand. J. Haemat. 8, 439. 1971.
76. Penn, G. M., Kunkel, H. G., Grey, H. M.
Sharing of individual antigenic determinants between a gamma G and a gamma M proteins in the same myeloma serum.
Proc. Soc. exp. Biol. Med. 35, 66. 1970.
77. Pernis, B., Forni, L., Amantel, L.
Immunoglobulins as cell-receptors. Conference on immunoglobulins.
Ann. N. Y. Acad. Sci. 190, 420. 1971.
78. Potter, M.
Plasma-cell neoplasia in single host: a mosaic of different protein-producing cell types.
J. exp. Med. 115, 339. 1962.
79. Prendergast, R. A., Grey, H. M., Kunkel, H. G.
Recombination of heavy and light chains of human gamma A myeloma proteins: formation of hybrid molecules and configurational specificity.
J. exp. Med. 124, 185. 1966.
80. Pruzanski, V., Underdown, B., Silver, E. H., Katz, A.
Macroglobulinemia – myeloma double gammopathy.
A study of four cases and a review of the literature.
Amer. J. Med. 57, 259. 1974.
81. Radl, J., Berg, P. V. D.
Transitory appearance of homogenous immunoglobulins – „paraproteins" – in children with severe combined immunodeficiency, before and after transplantation treatment.
In: Protides of the Biological Fluids 21st Colloquium.
Ed: by Peeters H.
Pergamon Press. 263. 1973.
82. Rosen, B. J., Smith, T. W., Bloch, K. J.
Multiple myeloma associated with two serum M components, gamma G type K and gamma A type L.
Engl. J. Med. 17, 2. 1967.
83. Rudders, R. A., Heller, P.
Double myeloma production of both IgG type lambda and IgA type lambda myeloma proteins by a single plasma cell line.
Amer. J. Med. 55, 215. 1973.
84. Sanders, J. H., Fahey, J. L., Finegold. I., Ein, D., Reisfeld, R.
Multiple anomalous immunoglobulins. Clinical, structural and cellular studies in the patients.
Amer. J. Med. 47, 43. 1969.
85. Scolari, L., Masal, C., Di Guglielmo, R.
Il metodo dell'immunoperassidase Applicacioni allo studio delle cellule produttrici di immunglobuline nel midolio asseo di pazienti affetti da gammopatie monoclonali.
Boll. Soc. ital. Biol. sper. 50, 549. 1974.

86. Scolari, L., Vaerman, J. P., Castigli, E., Voliani, D., Salvano, F., Masala, C., DiGuglielmo, R.
Late appearance of an IgA-kappa monoclonal protein in a patient with IgG-kappa multiple myleoma: sharing of idiotypic specificities between the two serum proteins.
Scand J. Immunol. 8, 201. 1978.
87. Sega, E., Isabella, R., Citro, G.
Two Bence Jones proteins of different immunological types light-chains (kappa and lambda) in a case of micromolecular myeloma.
Boll. Ist. Sieroter. Milanese. 52, 436. 1973.
88. Seidel, K., Kühn, R. A., Kunstmann, F. W., Winnefeld, K.
Mehrfache Proteinanomalien mit inkompletter Synthese und Spontanpräzipitation beim Plasmozytom.
2. Ges. inn. Med. 25, 704. 1970.
89. Silverman, A. Y., Yagi, Y., Pressman, D., Ellison, R. R., Tormey, D.
Monoclonal IgA and IgM in the of a single patient III. Immunofluorescent identification of cells.
J. Immunol. 110, 350. 1973.
90. Singurowicz, J., Murawski, K. Kraj, M., Jozwiak, W., Zupranska, B.
Defective (incomplete) synthesis of immunoglobulin heavy polypeptide chains as the cause of a myeloma-like.
Clin. chim. Acta. 22, 187. 1968.
91. Skvaril, F., Juricic, D., Spengler, G. A., Morell, A.
The IgG subclass distribution in dobule M component sera.
In: Protides of the Biological Fluids 21st Colloquium.
Ed. by Peeters, H.
Pergamon Press. 273. 1973.
92. Sledge, C., Rair, D. S., Black, D., Krueger, R. G., Hood, L.
Antibody differentation apparent sequence indentity between variable regions shared by IgA and IgG immunoglobulins.
Proc. nat. Acad. Sci. 73, 923. 1976.
93. Smithies, O.
Antibody variability.
Science. 157, 267. 1967.
94. Spengler, G. A., Steinberg, A. G., Skvaril, F.
Development of a second monoclonal immunoglobulin G in a patient with late manifestation of myeloma.
Acta med. scand. 192, 309. 1972.
95. Stein, H., Kaiserling, E.
Myeloma producing non secretory IgM and secretory IgG.
Scand. J. Haemat. 12, 274. 1974.
96. Sterzl, J., Nordin, A.
Demonstration of the cammon precursor for production of IgM and IgG antibodies.
Folia microbiol. 16, 1. 1971.
97. Sterzl, J., Nordin, A.
The common cell precursor for cells producing different immunoglobulins.
In: Cell interactions and receptor antibodies in immune responses.
Ed.: by Mäkelä O., Cross, Anne, Kosunen, T. V.
Academic Press, Inc., New York 213. 1971.
98. Stoop, J. W., Ballieux, R. E., Weyers, H. A.
Paraproteinemia with secondary immunoglobulin deficiency in an infant.
Pediatrics. 29. 97. 1962.
99. Szemere, P., Jákó, J., Szilágyi, G., Virágh, Sz.
Oligoclonalis gammopathiák klinikai jelentősége.
A Magyar Allergológiai és Klinikai Immunológiai Társaság VIII. Vándorgyűlése, Budapest, 1978.
100. Takahashi, M., Tanigaki, N., Yagi, Y., Moore, G. E., Pressman, D.
Presence of two different immunoglobulin heavy-chains in individual cells of established human hematopoietic celllines.
J. Immun. 100, 1176. 1968.
101. Tormey, D. C., Ellison, R. R., Hossfeld, D.
Concurrent monoclonal IgM and IgA proteins in lymphocytic lymphoma.
Cancer. 36, 1321. 1975.
102. Yagi, Y., Pressman, D.
Monoclonal IgA and IgM in the serum of a single patient. Sharing of individually specific determinants between IgA and IgM.
J. Immunol. 110, 335. 1973.
103. Villiaumey, J., Bouvet, J. P., Di Menza, C., Boccara, M.
Myélome plasmocytaire „monoclonal" à deux paraprotéines, IgG kappa et IgG lambda.
Nouv. Presse Méd. 3, 2707. 1974.
104. Vitetta, E. S., Uhr, J. W.
Immunoglobulin-receptors, revisited.
Science. 189, 964. 1975.
105. Von Oberdörfer, A., Schnauffer, K., Lange, H. J., Neiss, A.
Zur Verteilung von Paraproteinämien nach Geschlecht und Alter der Patienten, Paraprotein-Klassen-Subklassen und Leichketten-Typen.
Z. klin. Chem. 8, 51. 1973.
106. Wang, A. C., Wang, I. Y. F., McCormick, J. N., Fudenberg, H. H.
The identity of light-chains of monoclonal IgG and monoclonal IgM in one patient.
Immunochemistry. 6, 451. 1969.
107. Wang, A. C., Wilson, S. K., Hopper, J. E., Fudenberg, H. H., Nisonoff, A.
Evidence for control of synthesis of the regions of the heavy chains of immunoglobulins G and M by same gene.
Proc. nat. Acad. Sci. 66, 337. 1970.
108. Wang, A. C., Fudenberg, H. H., Goldrosen, M. H., Freedman, M. H.
Chemical studies of heavy chains of two $IgG_1$-lambda myeloma proteins from a single patient.
Immunochemistry. 9, 173. 1972.
109. Wang, A. C., Gergely, J., Fudenberg, H. H.
Amino acid sequences at constant and variable regions of heavy chains of monotypic immunoglobulins G and M of a single patient.
Biochemistry. 12, 528. 1973.
110. Wolfenstein-Todel, C., Franklin, E. C., Rudders, R. A.
Similarities of the light-chains and the variable regions of the heavy-chains of the $IgG_2$-lambda and $IgA_1$-lambda myeloma proteins from a single patient.
J. Immunol. 112, 871. 1974.
111. Zawadzki, Z. A., Edwards, G. A.
M components in immunoproliferative disorders.
Amer. J. Clin. Path. 48, 418. 1967.

# VII. The heavy chain disease

The heavy chain disease is a "true" form of paraproteinemia. The existance of the μ, γ, and α heavy chain disease is well documented. The δ-type heavy chain seems to be a special form of a IgD-myeloma (20, 102). The immunochemical differences within the - μ, γ, α - are accompanied with specific clinical symptoms. While no clinical peculiarities exist within the IgM, IgG, IgA, IgD and IgE-type myelomas, the well-defined three forms of the heavy chain pathology can be assessed into a separate group of B-cell malignant diseases.

## μ-Heavy chain disease

Generally occurs in the age group over 40, but patients aged 15 were also described. Lymphoproliferative symptoms: lymphocyte infiltration of the marrow, vacuolization of the plasma cells, chronic lymphoid leukemia, enlarged peripheral lymph nodes, splenomegalia, osteal pain, osteolysis, anemia, amyloidosis, bilateral carpal tunnel syndrome and the kappa chains in the urine are typical of this pathology. Gel electrophoresis does not reveal M components, however, immunoelectrophoresis visualizes a μ-heavy chain precipitin arc. Usually hypogammaglobulinemia is also typical (4, 32, 37).

Available literature presents only 15 cases. Furthermore, some patients did not show CLL and did not excrete κ-light chains in the urine. Gel electrophoresis showed M component, but their bone marrow plasma cells had no vacuolization (7, 10).

The peptide structure of the heavy chain was normal, but only fragments of the J chain were linked to it (34). Besides the μ-heavy chains, the bone marrow cells contained kappa light chains, thereby it was assumed that the interchain linkage mechanisms in the plasma cells are impaired, so the light- and the heavy chains are secreted separately (109). In most patients, with very few exceptions the kappa light chain is excreted with the urine.

Peripheral lymphocytes are less prone to the stimulation by PHA, Co-nA or PWM (106).

## γ-Heavy chain disease

No heavy chain disease-specific clinical symptoms exist, hence the diagnosis is based on immunochemical findings. In most cases the M component is a $F_c$-fragment-containing heavy chain identified from the serum (rarely from the urine) with gel electrophoresis. The excretion rate ranges between 1–20 g daily. Immunoelectrophoresis reveals a well defined precipitin arc with β-mobility (35, 62). Females over 40 comprise two-thirds of the patients. Loss of body weight and appetite, as well as the enlargement of the lymph nodes are the leading symptoms. Typically the cervical lymph nodes are enlarged, but in some cases generalized forms occur. About 50% of the patients have fever without any signs of the infection.

Hepatosplenomegalia, though typical, but not compulsory. The obstruction of the lymphatics in the Waldeyer's ring causes uvular and palatial edema as a clinical symptom.

Normocyte-normochromic anemia and leukopenia with relative lymphocytosis are always present. Peripheral smears show atypical lymphocytes and plasma cells; in one extreme case the γ-heavy chain disease was accompanied by plasmacellular leukemia. Thrombocytopenia and eosinophilia are also the typical features of the disease. Increased plasma cell count, lymphocytosis and eosinophilia in the bone marrow were also detected. X-ray show no changes in the bone structure. Osteolysis was found only in one patient (8) of 50 (15).

Lymph node biopsy greatly resembles the Hodgkin's disease, for the giant cells, resembling the Sternberg-Reed elements occur in the smears (29). Rheumatoid arthritis complications are also documented in the literature (1, 44).

The survival period following the diagnosis and adequate therapy ranges between 5 months and 5 years. Etiology of this disease is unknown.

Immunochemically the detectable heavy chain is a dimeric molecule with sedimentation constant value of 3.5-4 and molecular weight between 45,000-55,000. The heavy chains generally originate from $IgG_1$, but sometimes $IgG_2$ or $IgG_3$-derived molecules also occur (15). Heavy chains never display normal structure, being always incomplete (96). Four heavy chain types are defined (33):

Type 1 partially disrupted Fd fragments with repeated synthesis at position 216.
Type 2 deletion of the hinge region.
Type 3 partially disrupted hinge region and Fd fragment.
Type 4 enzymatic postsynthetic formation of the heavy chains as degradation products.

Other heavy chain types (41) and hybrid molecules ($IgG_3$-$IgG_1$) (2) are also described in the literature. Available data indicate occurrence of the heavy chains belonging to the genetic group Gm4, 10 and 23 (19). Some researchers are strongly convinced that the Type 4 does not exist, because these pathological proteins are not degradation products, but are newly synthesized (35).

In a documented case of a γ-heavy chain disease a benign form, similar to monoclonal gammopathies was assumed (104).

### α-heavy chain disease (case report)

The diagnosis of α-heavy chain disease is also based on the immunochemical findings (83). Other diagnostic means might reveal only the involvement of the lymph nodes of the gastrointestinal and (rarely) respiratory tracts (30, 80, 100) suggesting changes of the B-dependent, T-independent systems; these methods can be used for the diagnosis of non-Hodgkin lymphoma.

Considering the exceptions (31, 83), clinical and pathological evidence rather supports the view that the α-heavy chain disease is a disfunction of the gastrointestinal IgA-secretory plasma cells (72, 83).

Histologically a marked plasma cell infiltration of the small intestine is a typical feature of the disease (24).

Of 25 cases of the primary intestinal lymphoma the α-heavy chain could be identified in the serum in four patients (16%) (93). This finding suggests that it might be rather a version of intestinal lymphoma and not an individual entity (45). Diarrhea with all the typical signs of the malabsorption syndrome (loss of body weight, steatorrhea (?), hypocalcemia with subsequent tetany, marked salt and fluid loss (69) is a predominant symptom (28, 73). Such conditions are usually erroneously diagnosed as sprue, however, the patients are resistant to the gluten-free diet. According to the literature in four cases of sprue IgA deficiency has been detected. Genetic studies indicated deletion of the 18 chromosome (21). The above mentioned data suggest that the following three mechanisms lead to the loss of the mucosal immune functions with a subsequent wide range of the malabsorption symptoms:

1. Lack(?) or loss of the secretory capacity (IgA-deficient sprue) of the IgA-producing plasma cells;
2. mucosal cells carrying non-functioning pathological molecules (α-heavy chain disease);
3. IgA exhaustion due to allergic reactions (true sprue, gluten enteropathy).

Though most of the patients have abdominal pain, abdominal lymph nodes are seldom revealed by manual examination. Hepatosplenomegaly is not typical, its development indicates the final stage of the disease. The peripheral lymph nodes have normal histological and morphological structure. Lymphoangiography shows involvement of the lymphatic system (6, 48), which is otherwise not typical of this disease. We do not recommend to perform this method due to its low diagnostic value.

The researchers never reported osteolytic changes. Seligmann was the first to describe this syndrome in 1968 (83); he pointed out the preferential occurrence of this disease in the

Mediterranean region (5, 9, 11, 17, 22, 25, 27, 39, 42, 65, 70, 71, 74, 76, 81,82, 84, 85, 86, 89, 90, 91, 92, 94, 95), therefore it is also termed Mediterranean lymphoma. WHO suggests the term *i*mmuno*p*roliferative *s*mall *i*ntestinal *d*isease (IPSID) (107). Since 1988 more than 120 cases have been reported. The early assumption of the prevalence of this pathology among the Israeli Arabs and the non-european Jews is reconsidered (28, 73). Still most of the patients belong to the non-caucasian type (103) (Fig. VII/1, VII/2).

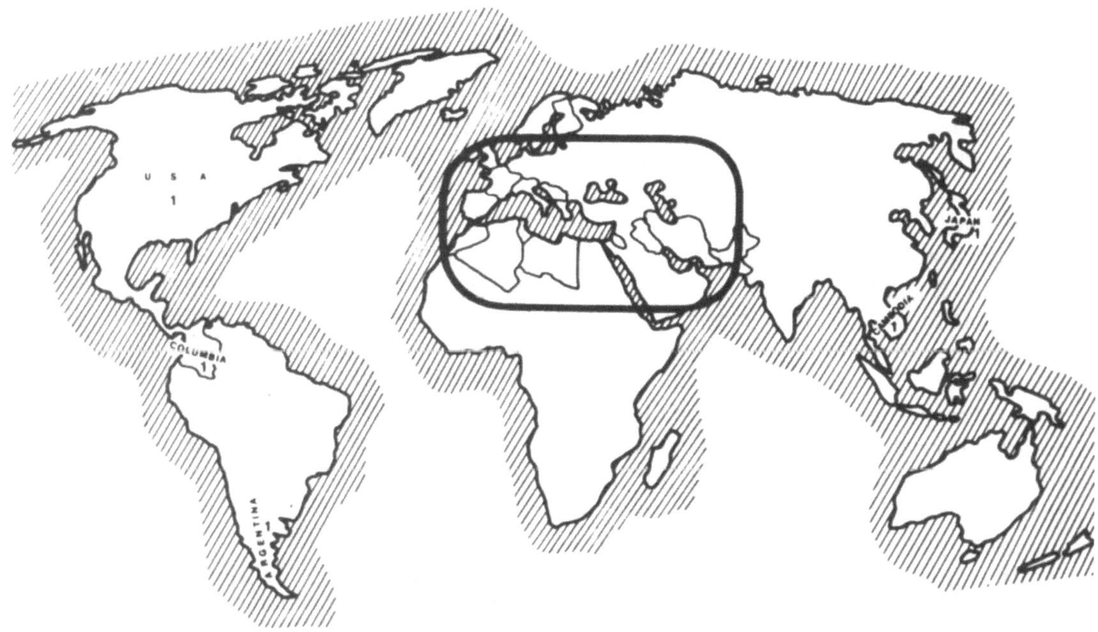


Fig. VII/1

Fig. VII/2

The significance of the geographic localization of the disease is still obscure, though a genetic predisposition or regional infections are assumed. The latter is supported by the fact that antibiotic therapy elicits long remissions in the course of this otherwise lethal progressive disease (57, 78 79, 105). Furthermore, one patient presented by British researchers had spent several years in the "infected" region (103). In other cases Prednisolon and Cyclophosphamide combination was found as effective (48, 87, 108). Some patients were treated by a combination of antibiotics, Prednisolon and Cyclophosphamide. Even one year following the cessation of the therapy the patients were complaint-free. All these findings support the theory of a benign, infectious (?) syndrome later turning into a malignant disease.

Controversy exists between the age distribution observed in the α-heavy chain disease and OMG, the former occurring in the second- third decades. A sustained prevalence of the male patients is also observed (83).

After a general clinical and epidemiologic introduction, the typical features of the IgA immunoglobulins are overviewed. According to the IPSID definition:

*The α-heavy chain disease is a proliferative pathology of the B-dependent IgA-secreting lymphocytes, when the $F_c$-like fragments containing no light chains are excreted* (83).

Lymphocytes and lymphoblasts of the affected lymph nodes are the precursor elements of the α heavy chain-producing plasmocytes (16, 66, 75).

The basic IgA structure is similar to that of the other immunoglobulins. The molecule contains two light (κ or λ) and two heavy chains. Heavy chains with $a_1$ or $a_2$ antigenity equally occur. The $\alpha_1$-heavy chain has two oligosaccharide components (49, 50, 51, 64, 67), whilst the $\alpha_2$ - 4 or 5 giving rise to the allotypes A 2m*(1)* (4 oligosaccharides) and A 2m*(2)* (5 oligosaccharides) (101). The usual disulphide bonds link the light chains to the $\alpha_1$ heavy chain in the $IgA_1$ molecules. The chains of the $IgA_2$ immunoglobulins are linked not with the disulphide- but rather with the covalent bonds (58). Both IgA-s occur in human serum at the following concentrations:

$IgA_1$ 2.1 ± 0.9 mg/ml (82 ± 6%)
$IgA_2$ 0.5 ± 0.3 mg/ml (18 ± 6%).

Consequently, the ratio of the plasmocytes producing both types of IgA is 76–94% and 6–24% respectively (97). Apart from the antigenic differences in the heavy chain structure, the IgA immunoglobulins are further subdivided as those of the serum and the gastrointestinal tract. The former is the marrow-derived monomer form, whereas the latter is a di- or trimeric secretory molecule.

The J-chain links two or three IgA molecules (40). Three J-chains with different structures are identified: J-A, J-B, J-C with molecular weights of 25,620–22,710 and 19,800 D respectively. Their percent incidence is 30–55–15% (3, 56). The J-chain concentration in plasma, milk and saliva is 1.74 ± 0.65 μM, 1.94 ± 1.21 μM, and 0.48 ± 0.26 μM respectively (12, 13, 14, 54).

Both T and B lymphoid cells producing *IgG*, *IgA* or IgM molecules have J chains in their intracellular space, not carrying them on their membranes (54).

This laboratory finding indicates that: a) these cells are probably not the sites of the synthesis, b) the J chains play no "active" part in binding, but are rather passively linked to the available C-terminal sites of the molecules via the covalent disulphide bonds (55).

The dimeric molecules are bound to the IgA receptors of the mucosal cells carrying a part of this receptor, which is termed SC (46, 47, 53, 68). Their molecular weight varies between 50,000-60,000 D (60). It is assumed that such structure is a prerequisite of the maintenance of IgA on the mucosal surface! This statement is, however, opposed by the observation that most of the intestinal mucosal IgA molecules have no SC (77, 98).

The antigenity of the free and the bound SC is slightly different (59).

The mucosal distribution of the $IgA_1$ and $IgA_2$ differs from that in plasma, $IgA_2$ being a predominant secretory form (58). In contrast to $IgA_1$, the above mentioned peculiarities of the $IgA_2$ structure probably make this molecule more resistant to proteolytic enzymes. Moreover, $IgA_1$ can not be split as easily as an IgG. Many researchers have failed to isolate $F_c$ fragments from $IgA_1$. At present this fragment can be obtained only with bacterial proteolysis (77).

A more detailed analysis of the a heavy chain disease, compared to the μ and γ pathology, is explained by the utmost importance of the former in clinical practice. A detailed case report of a patient with such pathology is presented herein:

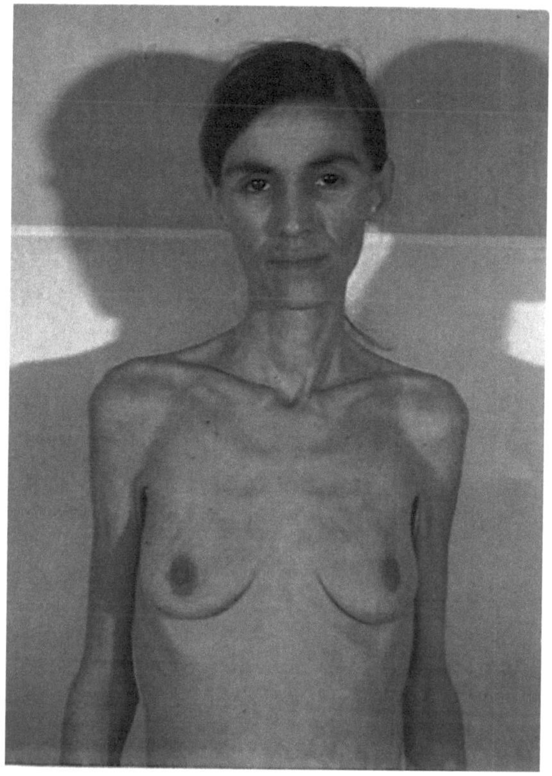

Fig. VII/3

**Case report:**

T.B. female patient, Albanian citizen, born in 1952 in Korca, was admitted to our Clinic in August, 1979. Both parents have thyroid disease. In childhood the patient had also been treated for hypothyroidism. Lambliasis was diagnosed in 1974. Other diseases not found. The complaints before admittance had started in 1976. The patient had diarrhea with tenesms and epigastric pain 4-6 times daily. The stool was stenchy and contained indigested food. In two years the patient lost 23 kg. After irrigoscopy sigma tumour was assumed, however, the results of colonocscopy unambiguously precluded this diagnosis.

*X-ray of gastrointestinal tract:* normal esophagus and cardia. Hook-shaped stomach with increased tone. Cascade formation in its upper portion throughout the whole period of study. Rough plication, peristaltic waves moving along the visualized structures. No secretion. Normal pylorus. Spontaneous filling of bulbus indicates normal contours. The lumen of the horse-shoe shaped duodenum is normal with very rough plication. At the lower horizontal segment the passage of the contrast substance was sometimes interrupted, however no strictures were observed. In three hours the contrast appeared in the coecum. The small intestine is atonic, with rough plication, without strictures.

*Jejunoscopy:* Kerkring-plication of the horizontal superior and inferior parts of the duodenum and jejune is knob-shaped, edematous, with pseudopolypoid structures amidst the permanent plication.

*Colonoscopy:* poor haustration, no other alterations.

*Histology of the small intestine biopsy:* detectable mucosal structure containing very few glands. Intercellular signs of chronic inflammation. Atypical cells with distorted giant nuclei. This process breaks through the muscular mucosa penetrating into submucosa.

*Conclusion:* Stage II sprue or pseudolymphoma.

Physical examination: cachexia (Fig. VII/3), meteorism and sustained hypotonia. During the hospital stay the patient had several fits of tetany due to the low $Ca^{++}$ level.

The results of the clinical examination and the histology of the small intestine suggested involvement of the immune system (similar to abdominal lymphoma) and indicated immunochemical studies. Immunoelectrophoresis indicated a well detectable pathological precipitin arc between the α-β regions (Fig. VII/4). Analytical system was used for identification of the antigenity of this pathological precipitin arc (Fig. VII/5).

Data of the literature strongly emphasize the problems while working with anti-light chain sera: difficulties occur in the reaction with the IgA light chains and the formation of precipitin arc (36, 63, 88).

Two antisera showing best reactions with the light chains of monoclonal IgA-s from the myelomal sera have been used for the test. In our opinion the expected negative result would be of great diagnostic value (26, 38). In contrast to the results of Seligmann (88), gel electrophoresis revealed M component (Fig. VII/6).

Lipid gel electrophoresis also yielded surprising results, indicating a lipid-positive M component between the β-zone and the start line (Fig. VII/7).

It is well known that IgA myeloma proteins can sometimes form complexes with the serum components, for example β-lipoproteins (61). Specific anti β and anti α-lipoproteins did not precipitate with either part of the α-heavy chain arc, suggesting that not the lipoptorein, but the free lipid binds to the α-heavy chain.

All diagnosed cases had $\alpha_1$-antigenity (88). Similarly, we could identify $\alpha_1$ heavy chain in our patient (Fig.VII/8).

In accordance with the data of the literature, the quantitative analysis of the main types of immunoglobulins indicated secondary antibody deficiency syndrome (IgG 690 mg/dl, IgM 24 mg/dl, IgA 44.5 mg/dl). At the same time the $\alpha_1$-heavy chain concentration reached an extremely high level of 1424 mg/dl ((Fig.VII/9., VII/10., VII/11).

Since the normal IgA was present in the serum, and the results of gel electrophoresis indicated pathological

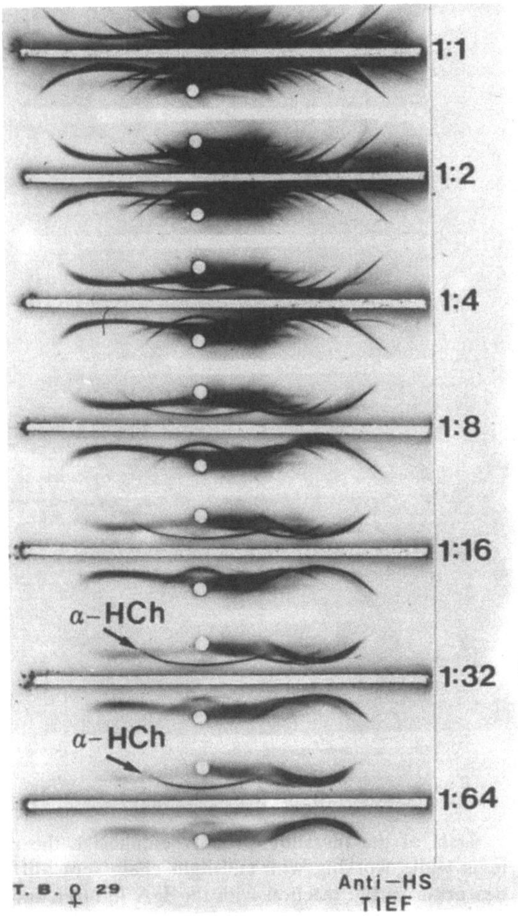


Fig. VII/4

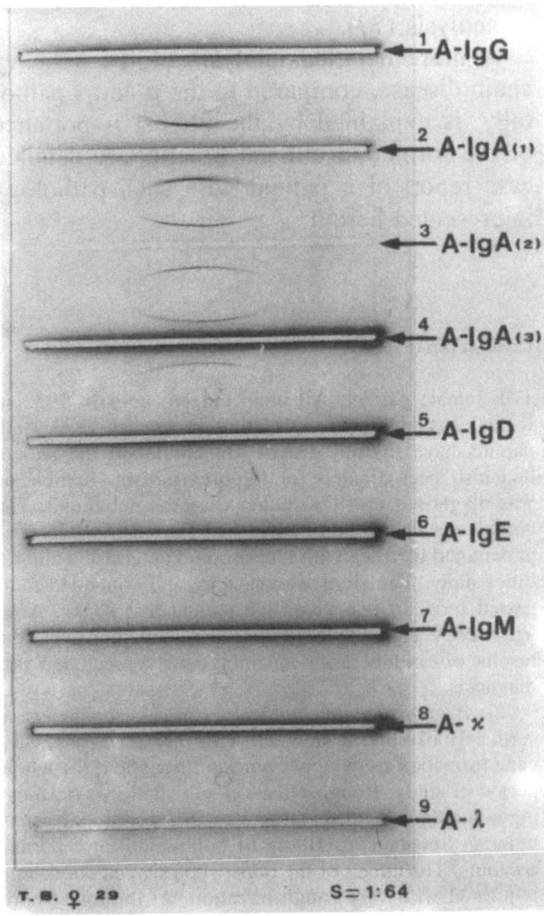


Fig. VII/5

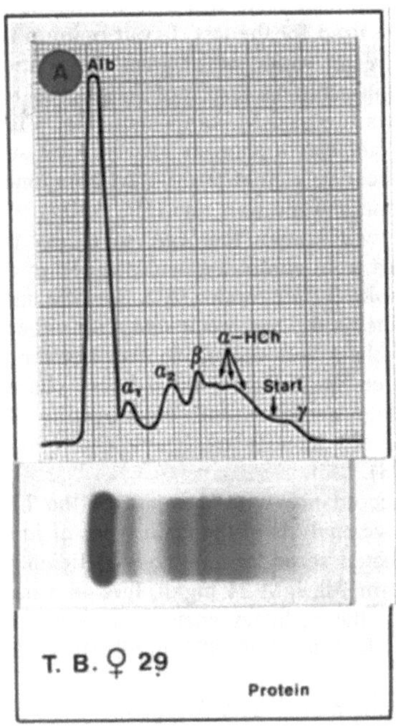


Fig. VII/6

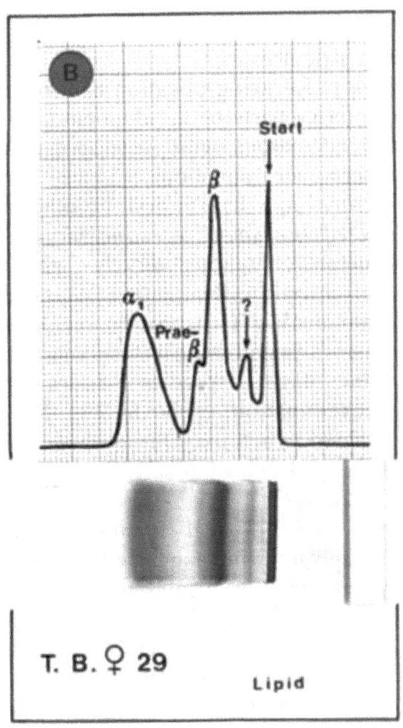


Fig. VII/7

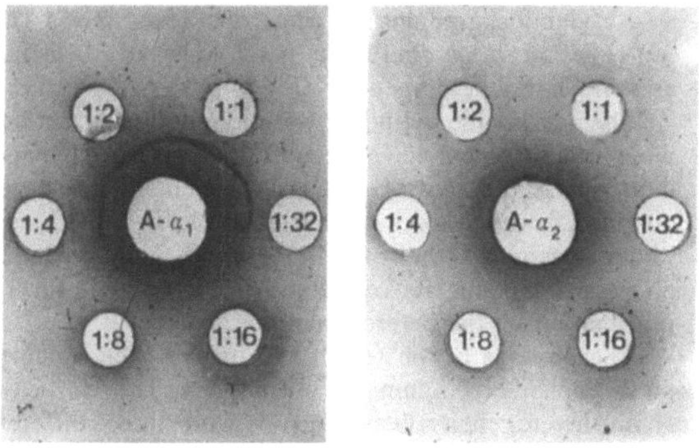


**Anti-$\alpha_1$ heavy chain monovalent serum**
**T. B. ♀ 29**

**Anti-$\alpha_2$ heavy chain monovalent serum**
**Diagnosis: $\alpha_1$-HChD**

**Fig. VII/8**

protein with α-heavy chain antigenity, a heterogeneous protein population was assumed and further studies performed (Fig.VII/12).

Immunological analysis of the peaks unambiguously indicated the presence of polymeric (probably trimeric) secretory IgA in the serum. Another heterogeneous population of $\alpha_1$-heavy chain polymers with different molecular weight was also identified (Fig.VII/13, VII/14).

Heterogeneity of the monoclonal proteins occurring in the heavy chain disease depends on the following three factors: *1.* proneness to polymerization; *2.* high carbohydrate content; *3.* heterogeneity of the N-terminal (88). The monomer units have molecular weight of 29,000-34,000 (88) or 39,000-41,000 (other sources). The α-chain + J-chain complex yields a molecular weight of 67,000-69,000, while that of the α-chain dimer + J-chain complex is 96,000-98,000 Dalton No

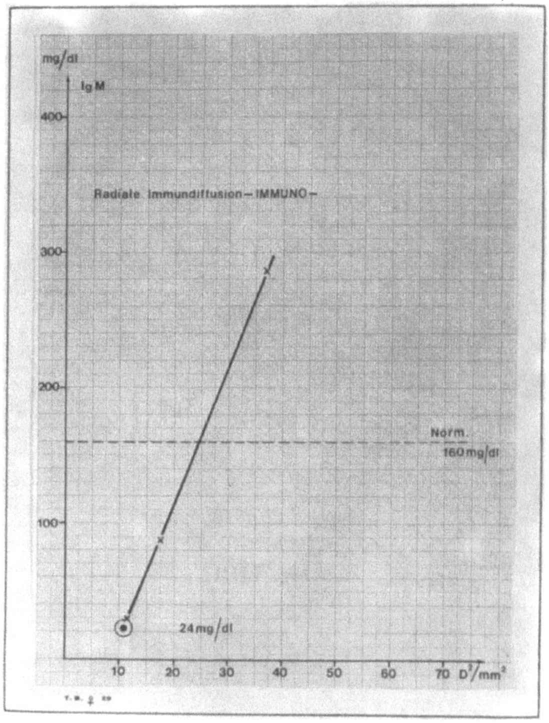


**Fig. VII/10**

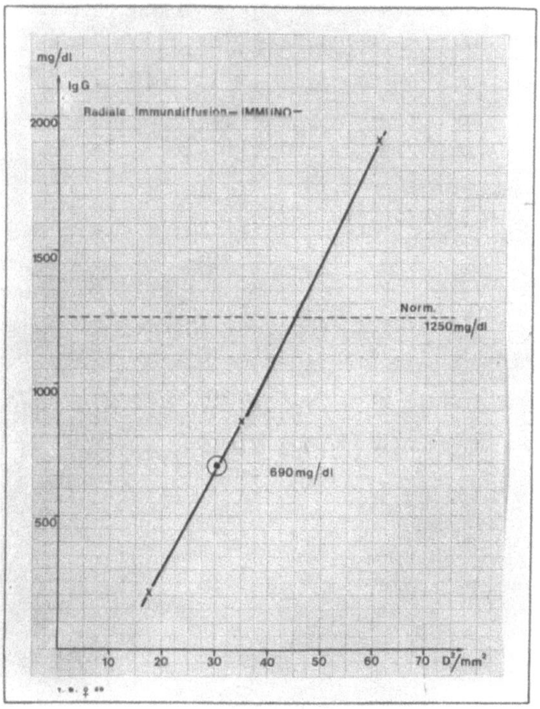


**Fig. VII/9**

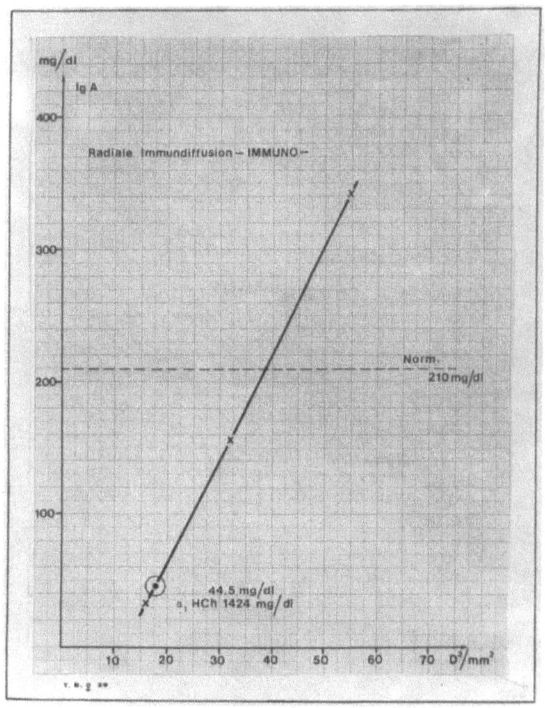


**Fig. VII/11**

uniform view on the structure of the α-heavy chains exists. It is assumed that they are longer than a half of a full chain, but shorter than its three-quarters. Thus it seems correct enough to regard these chains as $F_c$ fragments of the IgA molecules (23).

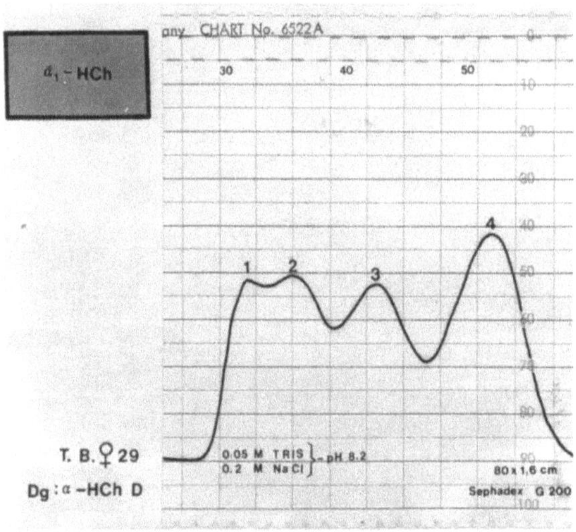


Fig. VII/12

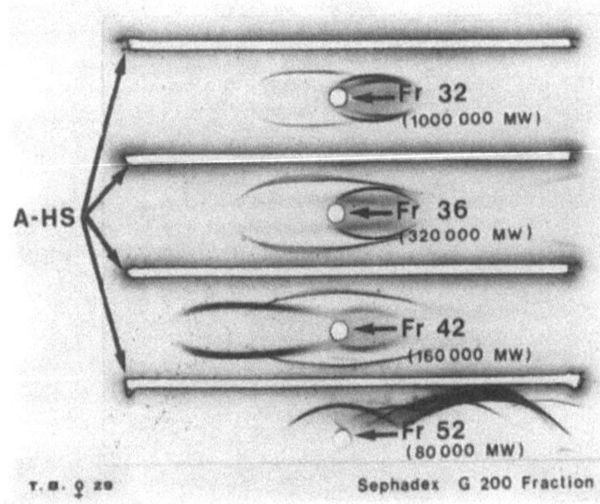


Fig. VII/13

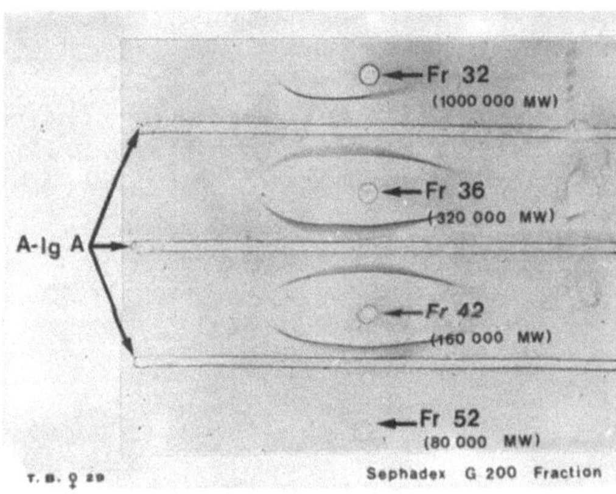


Fig. VII/14

Acrylamide gradient gel electrophoresis (SDS) (18, 99) reveals a large confluent "patch" instead of distinct arcs indicating heterogeneity (Fig. VII/15).

The acrylamide cylinders were longitudinally cut and the protein content have been further separated in gel that contained serum against α heavy chain. The peaks obtained from the discs had no clear profiles, the image rather resembled the two-dimensional immunoelectro phoresis. Thus, the above methods failed to yield further information on the heterogeneity.

The proteins of the serum and its 3rd peak obtained on a Sephadex column were analysed. The $\alpha_1$ heavy chain heterogeneity was proved with three different antisera in an Ouchterlony system, since two precipitin arcs were formed at the third well. One of them might probably be the lipid - $\alpha_1$ heavy chain complex. In contrast to the available data of literature, we failed to identify SC and J-chain structures, otherwise considered as obligatory (88) (Fig. VII/16).

The urine and saliva samples were also studied. A 100x concentrated urine contains all the proteins of the

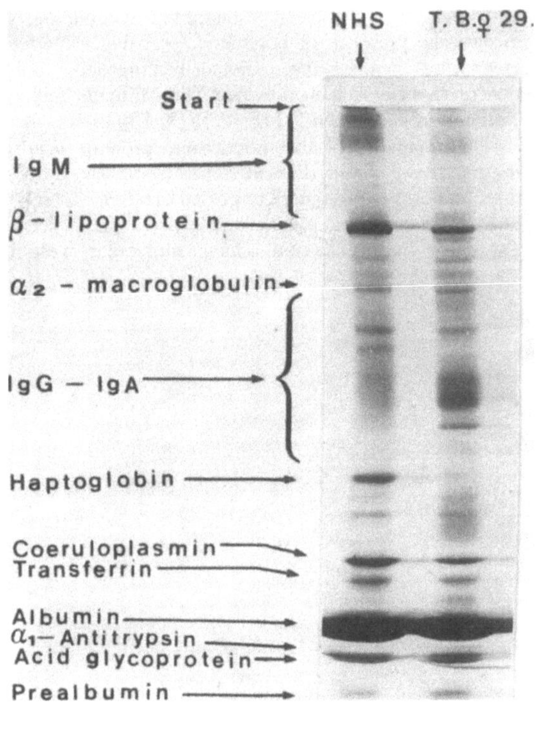


Fig. VII/15

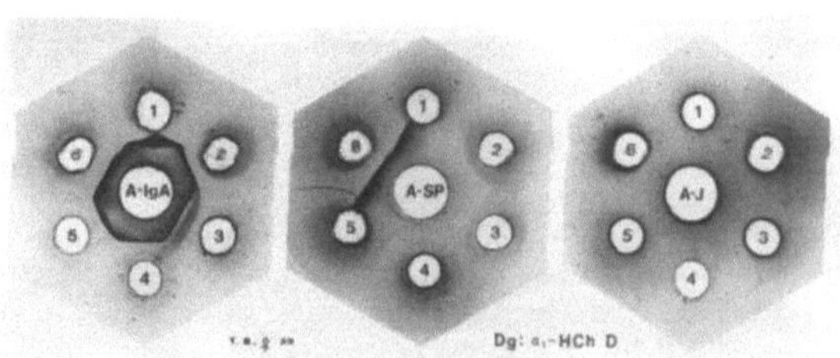


Fig. VII/16

serum, except the macromolecules (1,1). Neither the concentrated urine, nor the saliva (20x concentration) exhibited precipitin arcs with anti-J chain (2,3). Colostrum known to contain high amounts of SC develops a definite precipitin arc with anti-SC (A-SP), which corresponded to both IgA and SC with a slower mobility (4). The serum contains no SC (5). The saliva and colostrum of the patient revealed both precipitin arcs (6), whereas the urine sample gives precipitation corresponding to the free SC (7). Anti-IgA in the saliva and urine proves the presence of a protein with α-antigenity (8,9) (Fig.VII/17)

In the last step of the immunochemical analysis the presence of the heavy chains in the serum was identified with the rocket technique using a light chain trap (23) (Fig. VII/18).

The Figure clearly demonstrates that neither the NHS IgA, nor the monoclonal IgA (MM) can pass the trap. Only the patient's α-heavy chain containing serum can give precipitin arcs in the anti-IgA area. Histological studies were performed on three occasions; the samples were obtained from the inferior horizontal part of the duodenum, from the sternum punctate, and the rectoscope biopsy. The duodenal sample was fixated with buffered formalin (pH 7.2), that from the bone marrow with 5% glutaric aldehyde, and the colon biopsy material - with 96% ethanol at 4 °C and 2% glutaric aldehyde. The samples fixated with formalin and ethanol were directly embedded in Araldyte, while those placed in glutaric aldehyde were first post-fixated with $OsO_4$. The duodenal tissue samples were retrospectively studied with immunohistochemical methods. The α, γ and μ chains were identified with the reagents (1:20 and 1:40 dilutions) produced in the rabbit. For the light chain analysis the

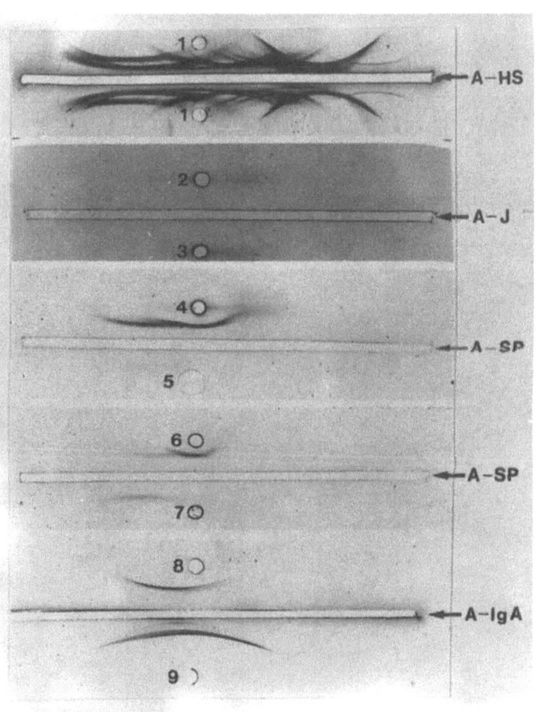


Fig. VII/17

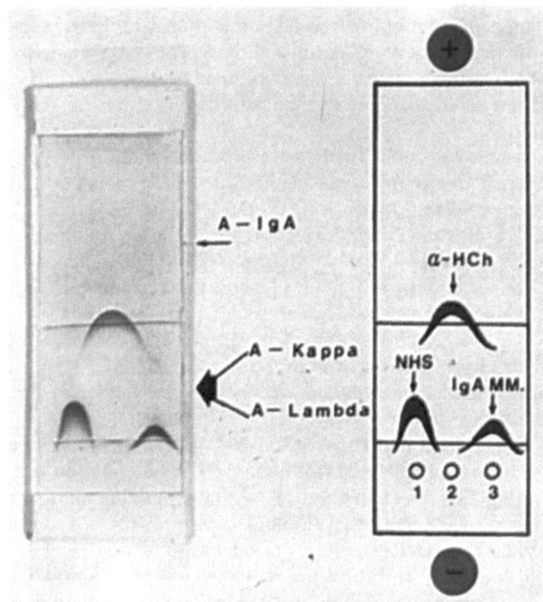


Fig. VII/18

R-78 rabbit serum (obtained from P. Brandtzaeg) with protein concentration of 1 mg/ml. The binding of the rabbit sera was visualized with the peroxidase-labelled anti-rabbit serum (Miles) at 1:20 dilution (Fig. VII/19).

The duodenal mucosa was infiltrated with the lymphocyte-derived mature plasmocytes and immunoblasts. Immunocytochemically most of the cells had α-chains, which could be identified on the apical site of the epithelial-, and the luminal site of the glandular cells. These cells did not contain light chains, the latter occurred extremely rarely. Similarly, cells containing γ and μ chains could be seldom visualized. Cells localized amidst the rectal glands contained light chains, as well as the γ and the α chains. Normal hemopoietic cells were identified in the bone marrow sample. The rectal and marrow plasma cells had normal structure.

Considering the possible genetic predisposition described in the literature, chromosome analysis was also performed.

The chromosomes were obtained from blood follow-

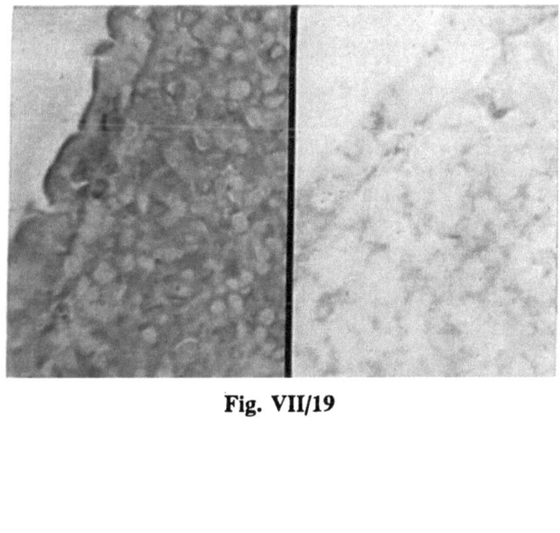

Fig. VII/19

ing a 72-hour culturing and stimulation with PHA. Chromosomes showing mitotic activity could be visualized with a direct method. The chromosome strips were identified with different methods following culturing (G-Q-strips).

Mitosis count (metaphase chromosomes) in the sternum karyogram (direct method) was 70 evaluated 36, in the peripheral blood - 30 (Fig. VII/20-VII/25).

Following tripsin digestion (cryst. SERVA), the samples were stained with Atebrin (G. T. GURR) using the Sunner's method for the G-, and the Casperson's technique for the Q strips analysis. (Table VII/I).

Of other chromosomal deviations 1,16 p occurred in one chromosome group in the sternum sample; 1g,6 g in one mitosis group blood. Minits were visualized in three groups.

The high incidence of the 2nd and 12th chromosome aberrations revealed following PHA stimulation even using the direct method is worth pointing out. Evaluation of the abnormal findings is very difficult, for individual aberrations can be considered as pathological only if compared to the results obtained in other individuals with the same pathology. Available literature provides no data on the karyotype studies in the α-heavy chain disease. Therefore, only other cases of IgA pathology were used as the reference. In two clinically healthy but IgA-deficient siblings the deletion of the structural gene was considered as the cause of pathology (21). Cytogenetic studies attribute the IgA synthesis and secretion to the 18th chromosome: the 18th chromosome deletion is found in most of the IgA-deficient cases, including our patient. Though the IgA deficiency naturally differs from the α-heavy chain disease, it seems plausible to assume that this genetic error might underlie the development of the heavy chain disease after "infections" or "antigen stimulus".

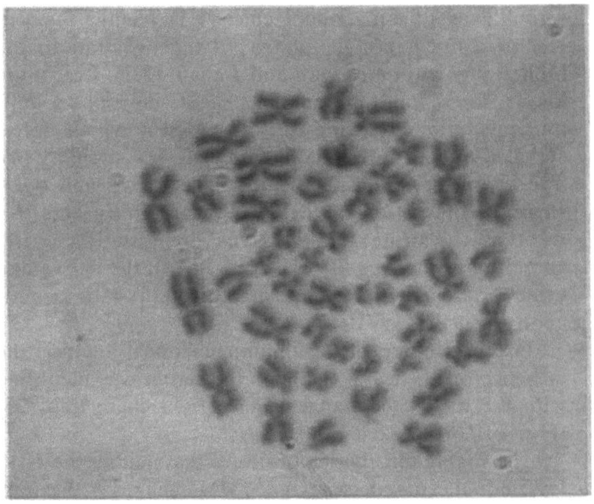

Fig. VII/20

In summary, though we had a doubtless case of heavy

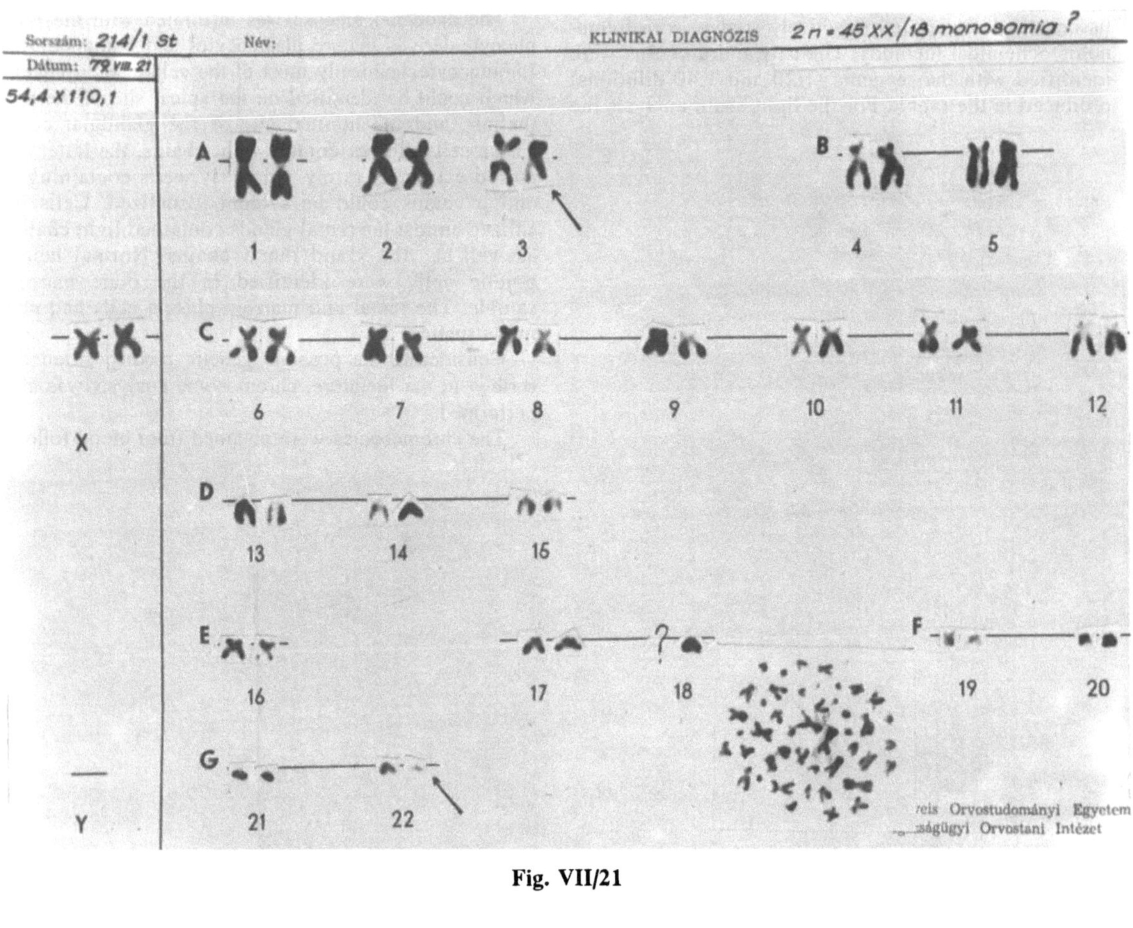


Fig. VII/21

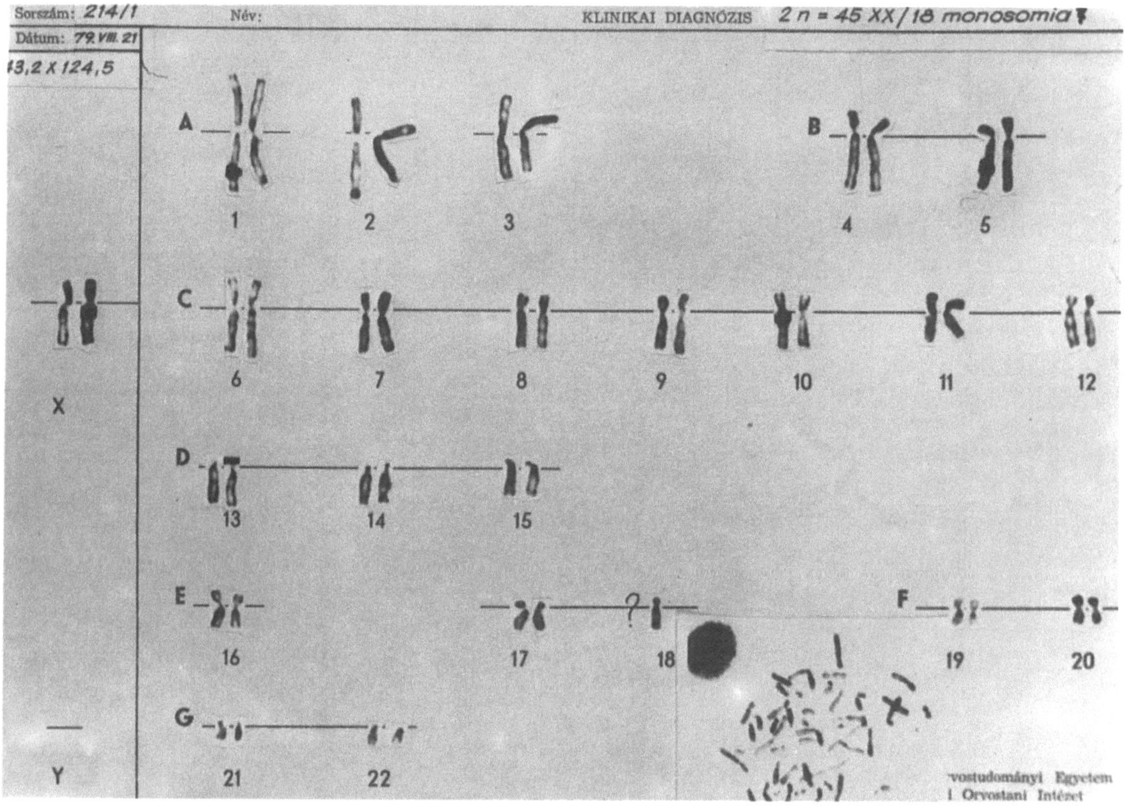


Fig. VII/22

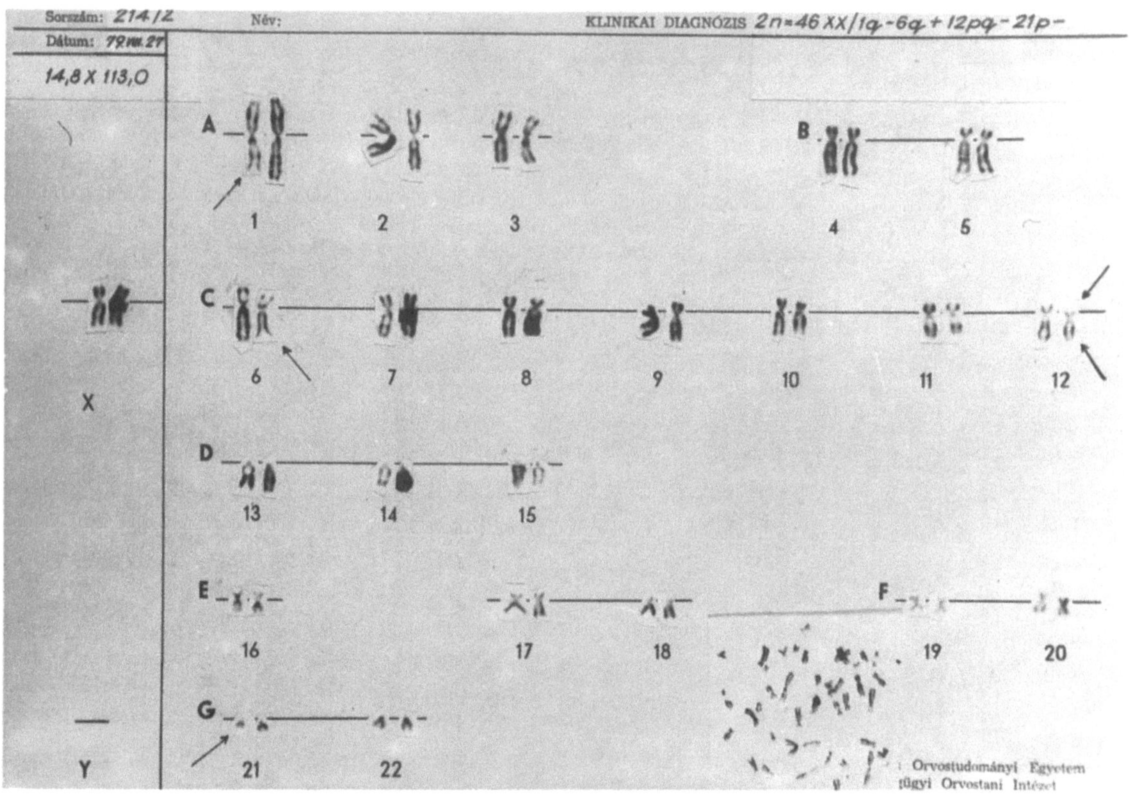


Fig. VII/23

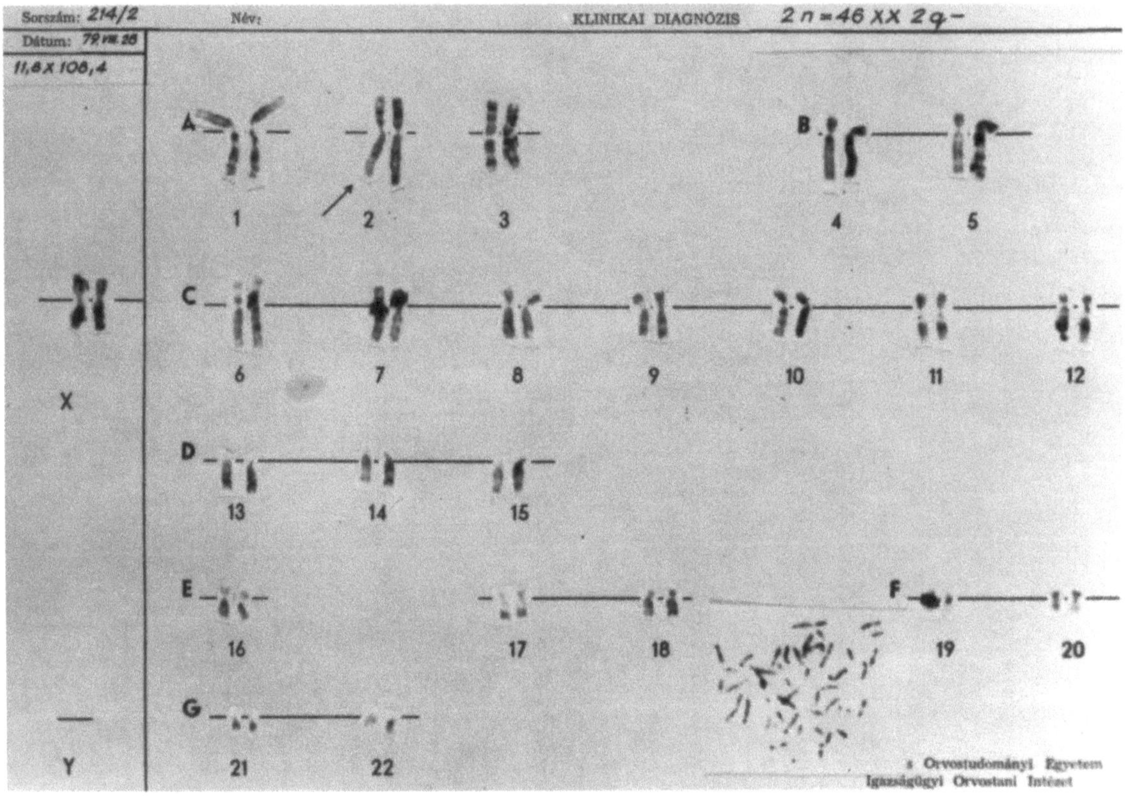


Fig. VII/24

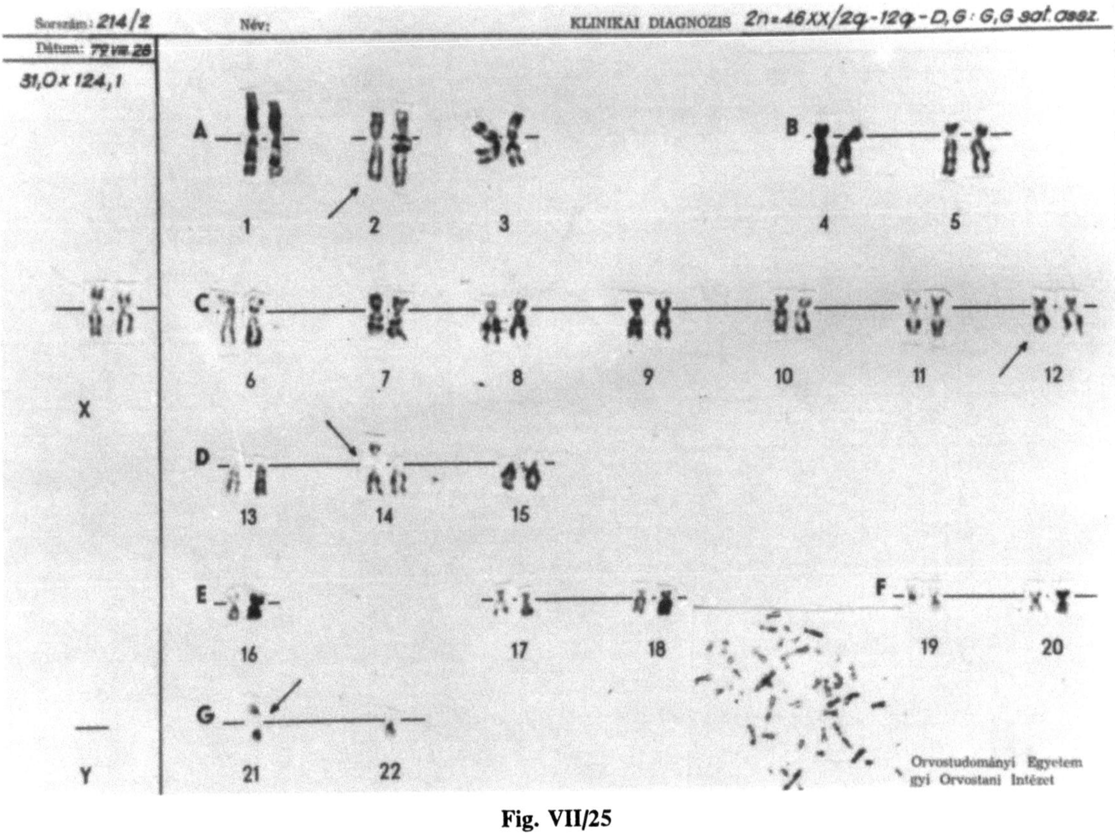


Fig. VII/25

chain disease, numerous findings altered from those mentioned in the literature:

a) in the previous studies J chains were found to be always linked to the α-heavy chains - we failed to confirm it in our patient (88).
b) Similarly to our patient, in other reported cases the α-heavy chains have been identified in the saliva only once (24).
c/ In contrast to other publications, the colon and the bone marrow were not affected in our patient.
d/ We have verified the deletion of the 18th chromosome.

## Disorders

| |
|---|
| 1 q terminal (—) |
| 2 q terminal (—) |
| 12 q terminal (—) |
| 12 p, q terminal (—) |
| 18 monosomia (—) |
| 1, 3 c spiraled lesion |
| minit's |
| Marker |

**Table VII/I**

### References


1. Albutt, E. C., Hawker, P. C., Hine, K. R. , Northam, B. E. Diagnosis of gamma heavy-chain disease. Ann. Clin. biochem. 18, 207. 1981.
2. Arnaud, Ph., Wang, An-Ch., Gianazza, Elisabetta, Wang, Irene, Lasne, Y., Greyssel, R., Fudenberg, H. H. Gamma heavy chain disease protein CHA: immunological and structural studies. Molecular Immunology. 18, 379. 1981.
3. Baezinger, J. V. Structure of the oligosaccharide of human J chain. J. Biol. Chem. 254, 4063. 1979.
4. Ballard, H. S., Hamilton, L. M., Marcus, A. J., Illés, G. H. A new variant of heavy-chain disease ($\mu$ chain disease). N. Engl. J. Med. 282, 1060. 1970.
5. Berenguer, J. Maladie des chaines alpha: A propos d'un cas, abstracted. Arch. Fr. Mal. App. Dig. 61, 950. 1972.
6. Bernadou, A. La maladie des chaines lourdes alpha: A propos d'une observatio Nouv. Rev. Fr. Hematol. 12, 333. 1972.
7. Biserte, G., Lebreton, J. P., Ropartz, G., Tijov, G., Morty, B., Rodat, G., Maillet, J. Y., Letournov, A., Guimbretiere, J., Bray, B., Boudart, D. Un cas de maladie des chaines lourdes $\mu$ (letter to the editor). Novo Presse Med. 2, 1997. 1973.
8. Bloch, K. J., Lee, L., Mills, J. A., Haber, E. Gamma heavy chain disease: an expanding clinical and laboratory spectrum. Am. J. Med. 55, 61. 1973.
9. Bognel, J. C., Rambaud, J. C., Modigliani, R., Matuchansky, C., Bognell. C., Bernier. J. J., Scotto, J., Hautefeuille, P., Mihaesco, E., Hurez, D., Preud'homme, J. L., Seligmann, M. Etude clinique, anatomo-pathologique et immunochimique d'un nouveau cas de maladie des chaines alpha suivi pendant cinq ans. Rev. Europ. Et. Clin. Biol. 17, 362. 1972.
10. Bonhomme, J., Seligmann, M., Mihaesco, C., Clauvel, J. P., Danon, F., Brovet, J. C., Bouvry, P., Martine, J., Clerc , M. M$\mu$-chain disease in an African patient. Blood. 43, 485. 1974.
11. Bonomo, L., Dammacco, F., Marano, R., Bonomo, G. M. Abdominal lymphoma and alpha chain disease. Amer. J. Med. 52, 73. 1972.
12. Brandtzaeg, P. Presence of J chain in human immunocytes containing various immunoglobulin classes. Nature. 252, 418. 1974.
13. Brandtzaeg, P. Immunochemical studies on free and bound J chain of human IgA and IgM. Scand. J. Immunol. 4, 439. 1975.
14. Brandtzaeg, P. Complex formation between secretory component and human immunoglobulins and human immunoglobulins related to their content of J chain. Scand. J. Immunol. 5, 411. 1976.
15. Briät, E., van Nieuwkoop, J. A., Jol-v. d. Zijde, C. M., den Ottolander, G. J., Rádl, J. Chronic lymphocytic leukaemia and IgG heavy chain disease. Scand. J. Haematol. 25, 244. 1980.
16. Brouet, J. C., Mason, D. Y., Danon, F., Preud'homme, J. L., Seligmann, M., Reyes, F., Navab, F., Galian, A., Rene, E., Rambaud, J. C. Alpha-chain disease: evidence for a common clonal origin of intestinal immunoblastic lymphoma and plasmacytic proliferation. Lancet. i, 861. 1977.
17. Buxbaum, J. N., Preud'homme, J. L. Alpha- and gamma-heavy chain diseases in man: Intracellular origin of the aberrant polypeptides. J. Immunol. 109, 1131. 1972.
18. Calvanico, N. J., Tomasi, Th. B. High temperature trypsinolysis of human IgA: isolation of the Fc-alpha fragment. Immunochemistry. 13, 203. 1976.
19. Carbonara, A. O., Paolino, W., Malavasi, F., Levis, A., Stramignoni, A., de Lange, G., van Loghem, E., Franklin, E. C. A new case of gamma heavy chain disease: clinical, immunochemical and structural characterisation. Vox Sanquinis. 37, 89. 1979.
20. Cejka, J., Kithier, K. IgD myeloma protein with „unreactive" light chain determinants. Clin. Chem. 25, 1495. 1979.
21. Crabbé, H., Ponce, R. G., Heremans, J. F. The syndrome of IgA-deficient sprue and its genetic aspects. Europ. Soc. clin. Invest, 3rd Ann. Meet. Scheveningen 1969.
22. Dammacco, F., Bonomo, L. Abdominal lymphoma and $\alpha$-chain disease: report of three cases. In: Protides of Biological Fluids. Ed.: by Peeters, H. Pergamon Press. 18, 273. 1970.
23. Dammacco, F., Antonaci, S., Miglietta, A. Two dimensional immunoelectrophoresis as a diagnostic tool for heavy chain disease. Boll. Ist. Steroter Milanese. 54, 460. 1975.
24. Doe, W. F., Henry, K., Hobbs, J. R., Avery Jones, F., Dent, C. E., Booth, C. C. Five cases of alpha chain disease. Gut. 13, 947. 1972.

25. Doe, W. F.
Alpha chain disease. Clinocopathological features and relationship to so-called Mediterranean lymphoma.
Brit, J. Cancer. 31. Suppl. 2, 350. 1975.
26. Doe, W. F., Danon, Francoise, Seligmann, M.
Immunodiagnosis of alpha chain disease.
Clin, exp. Immunol. 36, 189. 1979.
27. Dorrington, K. F., Mihaesco, E., Seligmann, M.
The molecular size of three alpha-chain disease proteins.
Biochim. Biphys. Acta. 221, 647. 1970.
28. Eidelman, S., Parkins, R. A., Rubin, C.
Abdominal lymphoma presenting as malabsorption: A clinicopathologic study of nine cases in Israel and a review of the literature.
Medicine 45, 111. 1966.
29. Ellman, L. L., Bloch, K. J.
Heavy-chain disease: report of a seventh case.
N. Engl. J. Med. 278, 1195. 1968.
30. Faux, J. A.
An alpha-heavy chain abnormality in a child with hypogammaglobulinemia.
Clin. Immunol. Immunopathol. 1, 282. 1973.
31. Florin-Christensen, A., Doniach. D., Newcomb, P. B.
Alpha chain disease with pulmonary manifestations.
Br. med. J. 2, 413. 1974.
32. Forte, F. A., Prelli, F., Yount, W. J., Jerry, L. M., Kochwa, I., Franklin, E. C., Kunkel, H. G.
Heavy chain disease of the $\mu$ (JM) type: report of the first case.
Blood. 36, 137. 1970.
33. Frangione, B., Franklin, E. C.
Heavy chain disease; clinical features and molecular significance of the disordered immunoglobulin structure.
Semin. Hematol. 10, 53. 1973.
34. Frangione, B., Franklin, E. C., Prelli, F.
$\mu$-Heavy-chain disease – a defect in immunoglobulin assembly. Structural studies of the X chain.
Scand. J. Immunol. 5, 623. 1976.
35. Franklin, E. C., Lowenstein, J., Bigelow, B., Meltzer, M.:
Heavy chain disease – a new disorder of serum $\gamma$-globulins. Report of the first case.
Am. J. med. 37, 332. 1964.
36. Franklin, E. C., Feinstein, D., Fudenberg, H. H.
The problem of untypable M-protein.
Acta Medica Scand, 179. suppl. 445, 80. 1966.
37. Franklin, E. C.
$\mu$-Chain disease.
Arch. Intern. Med. 135, 71. 1975.
38. Gale, D. S. J., Versey, J. M. B., Hobbs, J. R.
Rocket immunoselection for detection of heavy chain disease.
Clin. Chem. 20, 1292. 1974.
39. Galian, A., Lecestre, M. J., Scotto, J., Bognel, C., Matushansky, C. Rambaud, J. C.
Pathological study of alpha-chain disease, with special emphasis on evolution.
Cancer. 39, 2081. 1976.
40. Grubb, A. O.
Quantitation of J chain in human biological fluids by a simple immunochemical procedure.
Acta med. Scand. 204, 453. 1978.
41. Isobe, T., Osserman, E. F.
Plasma cell dyscrasia associated with the production of incomplete (? deleted) $IgG_h$ molekules gamma heavy chains and free lambda chains containing carbohydrate: description of the first case.
Blood. 43, 505. 1974.
42. Irunberry, J.
Trois cas de maladie des chaines alpha observés en Algerie.
Nouv. Rev. Fr. Hematol. 10, 609. 1970.
43. Kobayashi, K., Vaerman. J. P., Bazin, H., Lebacq-Verheden, A. M., Heremans, J. F.
Identification of J chain in polymeric immunoglobulins from a variety of species by cross-reaction with rabbit antisera to human J chain.
J. Immunol. 111. 1590. 1973.
44. Kretschmer, R. R., Pizzuto, J., Gouzales, J., Lopes-Osuna, M.
Heavy chain disease, rheumatoid arthritis, and cryoglobulinemia.
Clin. Immunol. Immunopathol. 2, 195. 1974.
45. Kyle, R. A., Bayrd, E. D.
Monoclonal gammopathies.
Charles C. Thomas Publisher.
Springfield, Illinois USA. pp: 226. 1976.
46. Lai A. Fat, R. F. M., McClelland, D. B. L., van Furth, R.
The synthesis of secretory component by lymphoid tissues in vitro.
J. Immunol. 113, 1199. 1974.
47. Lai A. Fat, R.F.M., McClelland, D.B.L., van Furth, R.
In vitro synthesis of immunoglobulins, secretory component, complement and lysozyme by human ganstrointestinal tissues, I. Normal tissues.
Clin. exp. Immunol. 23, 9. 1976.
48. Laroche, C.
Nouvelle observation d'une maladie des chaines lourdes alpha au cours d'un lympho-sarcome abdominal de type „méditerranéen" avec tuberculose isolée des ganglions mésentériques et pelvispondylite.
Presse Med. 78, 55. 1970.
49. Liu, Y-Sh. V., Low, Teresa L. K., Putnam, F. W.
Primary structure of a human $IgA_1$ immunoglobulin III. Isolation, composition, and amino acid sequence of the thermolysin peptides.
J. Biol, Chem. 254, 2859. 1979.
50. Liu, Y-Sh. V., Putnam, F. W.
Primary structure of a human $IgA_1$ immunoglobulin, I. Isolation, composition, and amino acid sequence of the chymotryptic peptides.
J. Biol. Chem. 254, 2839. 1979.
51. Low, Teresa, L. K., Liu, Y-Sh. V., Putnam, F. W.
Primary structure of a human $IgA_1$ immunoglobulin, II. Isolation, composition, and amino acid sequence of the tryptic peptides of the whole alpha-1 chain and its cyanogen bromide fragments.
J. Biol. Chem. 254, 2850. 1979.
52. Manousos, O. N., Economidou, J. C., Georgiadou, D. E., Pratsika-Ougourloglou, K. G., Hadziyannis, S. J., Merikas, G. E., Henry, K., Doe, W. F.
Alpha chain disease with clinical, immunological and histological recovery.
British Medical Journal. 2, 409. 1974.
53. McClelland, D.B.L., Shearman, D.J.C., van Furth, R.
Synthesis of immunoglobulin and secretory component by gastrointestinal mucosa in patients with hypogammaglobulinaemia or IgA deficiency.
Clin, exp. Immunol. 25, 103. 1976.
54. Mestecky, J., Winchester, R. J., Hoffman, T., Kunkel, H. G.
Parallel synthesis of immunoglobulins and J chain in pokeweed mitogen-stimulated normal cells and in lymphoblastoid cell lines.
J. Exp. Med. 145. 760. 1977.
55. Mole, J. E., Bhown, A. S., Bennett, J. C.
Sequence analysis of human J chain. Amino terminal location of a disulfide bound linking the immunoglobulin heavy chain.
Biochemical and biophysical research communications. 73, 92. 1976.
56. Mole, J. E., Bhown, A. S., Bennett, J. C. Primary structure of human J chain: Alignment of peptides from chemical and enzymatic hydrolyses.
Biochemistry. 16, 3507. 1977.
57. Monges, H.
Un nouveau cas de maladie des chaines lourdes alpha à forme intestinale traité par antibiothérapie: Rémission clini-

que, histologique et immunologique.
Arch. Fr. Mal. App. 66, 591. 1977.

58. Mote, Gabriela.
Relationship between structure and properties of human IgA subpopulations.
Rev. roum. Biochim. 15, 137. 1978.

59. van Munster, P. J. J., Stoelinga, G.B.A., Poels-Zanders, S.
Evidence of immunochemical differences between free and bound secretory piece.
Immunology, 17, 165. 1969.

60. van Munster, P.J.J., Stoelinga, G.B.A., Poels-Zanders, S.
Isolation of free secretory component (SC) from human milk, determination of its molecular weight.
Immunochemistry, 8, 471. 1971.

61. Musiani, P., Lauriola, L., Piantelli, M.
Inhibitory activity of alpha-1-antriripsin bound to human IgA.
Clin, Shim. Acta. 85, 61. 1978.

62. Osserman, E. F., Takatsuki, K.
Clinical and immunochemical studies of four cases of heavy (H $\gamma$-2) chain disease.
Am. J. Med. 37, 351. 1964.

63. Osterland, K. C., Chaplin, H.
Atypical antigenic properties of $\gamma$ A myeloma protein.
Fed. Proc. 24, 502. 1965.

64. Perez Pena, F.
Enfermedad de cadenas alfa.
Rev Clin Esp. 123, 237. 1971.

65. Pittman, F. E.
Alpha-Heavy chain disease: The first proven case detected in the Western hemisphere, abstracted.
Gastroenterology. 60, 792. 1971.

66. Preud'homme, J. L., Brouet, J. C., Seligmann, M. Cellular immunoglobulins in human $\gamma$ and $\alpha$-heavy chain disease.
Clin. exp. Immunol. 37, 283. 1979.

67. Putnam, F. W., Liu, Y-Sh. V., Low, Teresa, L. K.
Primary structure of a human $IgA_1$ immunoglobulin, IV. Streptococcal $IgA_1$ protease digestion, Fab and Fc fragments, and the complete amino acid sequence of the alpha-1 heavy chain.
J. Biol. Chem. 254, 2865. 1979.

68. Radl, J., Berg, P.V.D., De Bruyn, A.M., Hijmans, W.
Binding of secretory component to polymeric IgA and IgM Paraproteins in vitro.
Immunology. 20, 843. 1971.

69. Rambaud, J. C., Bognel. C., Prost, A., Bernier, I. J., Le Quintec, Y., Lambling, A., Danon, F., Hurez, D., Seligmann, M.
Clinico-pathological study of a patient with „Mediterranean" type of abdominal lymphoma and a new type of IgA abnormality (Alpha chain disease).
Digestion. 1, 321. 1968.

70. Rambaud, J. C. et al.
Nouveau cas de maladie des chaines alpha chez un Eurasien.
An. Med. Intern. 121, 135. 1970.

71. Rambaud, J. C., Matuchansky, C.
Alpha-chain disease: pathogenesis and relation to Mediterranean lymphoma.
Lancet. i. 1430. 1973.

72. Rambaud, J. C., Seligmann, M.
Alpha chain disease.
Clinics in Gastroenterology. 5, 341. 1976.

73. Ramot, B., Shanin, N., Bubif, J. J.
Malabsorption syndrome in lymphoma of small intestine: A study of 13 cases.
Isr. J. Med. Sci. 1, 221. 1965.

74. Ramot, B., Hulu, N.
Primary intestinal lymphoma and its relation to alpha heavy chain disease.
British Journal of Cancer. 31 (Suppl. 2.) 343. 1975.

75. Ramot, B., Levanon, M., Hahn, Y., Lahat, N., Moroz, C.
The mutual clonal origin of the lymphoplasmocytic and lymphoma cell in alpha-heavy chain disease.
Clin. exp. Immunol. 27, 440. 1977.

76. Rappaport, H., Ramot, B., Hulu, N., Park. J. K.
The pathology of so-called Mediterranean abdominal lymphoma with malabsorption.
Cancer. 29, 1502. 1972.

77. Rivat, C., Bourguignon, Jeannette, Fontaine, M., Ropartz, C.
Enzymatic fragmentation of human IgA F (ab)'$_2$: A new peptic fragment.
Immunochemistry. 14, 69. 1977.

78. Rogé, J., Druet, P., Marche, C.
Lymphome méditerranéen avec maladie des chaines alpha: Triple rémission clinique, anatomique et immunologique.
Pathol, Biol. 18, 851. 1970.

79. Rogé, J., Druet, P., Marche, C., Silvereano D. R., Roissard, F., Camilleri, J. P.
Alpha-chain disease cured with antibiotics.
Br. Med. J. 4, 225. 1975.

80. Rosner, F., Vallejo, Violetta, Khan, A. F., Vessely, Z., Grünwald, H. W., Calas, C.
Hypogammaglobulinemia and selective immunoglobulin – A deficiency (double consanguinity in family).
New York State Journal of Medicine. 78, 1459. 1978.

81. Savilahti, E., Brandtzaeg, P.
An atypical case of alpha-chain disease, abstracted.
Scand. J. Immunol. 2, 322. 1973.

82. Scotto, J., Stralin, H., Caroli, J.
Ultrastructural study of two cases of alpha-chain disease.
Gut. 11, 782. 1970.

83. Seligmann, M., Danon, F., Hurez, D., Mihaesco, E.
Preud'homme, J. L.
Alpha chain disease: a new immunoglobulin abnormality.
Science. 162, 1396. 1968.

84. Seligmann, M., Mihaesco, E., Hurez, D., Mihaesco, C., Preud'homme, J. L. Rambaud, J. C.
Immunochemical studies in four cases of alpha chain disease.
J. clin. Invest. 48, 2374. 1969.

85. Seligmann, M., Rambaud, J. C.
IgA abnormalities in abdominal lymphoma (alpha-chain disease).
Isr. J. Med. Sci 5, 151. 1969.

86. Seligmann, M.
La maladie des chaines alpha.
La Presse Medicale. 78, 51. 1970.

87. Seligmann, M., Mihaesco, E., Frangione, B.
Studies on alpha chain disease.
Ann. N. Y. Acad. Sci. 190, 487. 1971.

88. Seligmann, M.
Immunological, clinical and pathological features of alpha-chain disease.
Archives of Internal Medicine, 135, 78. 1975/a.

89. Seligmann, M.
Alpha heavy chain disease.
Blut. 31, 1. 1975/b.

90. Seligmann, M.
Alpha chain disease: immunoglobulin abnormalities, pathogenesis and current concepts.
British Journal of Cancer. 31 (Suppl. 2.), 356. 1975/c.

91. Seligmann, M.
Immunobiology and pathogenesis of alpha chain disease, In: Immunology of the Gut.
Ed: Ciba Foundation, Symposium 46 (New Series).
Elsevier, Amsterdam. 263. 1977.

92. Seligmann, M., Rambaud, J. C.
Alpha chain disease, a possible model for the pathogenesis of human lymphomas.
In: Immunopathology of lymphoreticular neoplasma.
Ed.: by Good, R. A., Twomey, J. J.
Plenum Press, New York, p 425. 1978.

93. Shadid, M. J., Alami, S. Y., Nassar, V. H., Balikian, J. B., Salem, A. A.
Primary intestinal lymphoma with paraproteinemia.
Cancer. 35, 848. 1975.

94. Shulman, G., Lai York, E., Grieve, S.

Alpha-chain disease in a non-mediterranean climate.
South African Medical Journal. 49, 2188. 1975.

95. Skinner, J. M., Manousos, O. N., Economidov, V., Nicolaou, A., Merikas, G.
Alpha chain disease with localized plasmocytoma of the intestine. Immunoperoxidase study.
Clin. exp. Immunol. 25, 112. 1976.

96. Skvaril, F., Morell. A., Barandun, S.
The IgG subclass distribution in 659 myeloma sera.
Vox Sang. 23, 546. 1972.

97. Skvaril, F., Morell, A.
Distribution of IgA subclasses in sera and bone marrow plasma cells of 21 normal individuals,
In: The immunoglobulin A system.
Ed.: by Mestecky, J., Lawton, A. R.
Plenum Publishing Corporation N. Y. 433. 1974.

98. Spohn, Martha, McColl, I.
Studies on human gastric mucosal immunoglobulin-A.
Biochim. Biohys. Acta. 576, 1. 1979.

99. Spohn, Martha, McColl, I.
Studies on human gastric mucosal immunoglobulin-A,
II. Further evidence for the absence of the secretory component from the predominant immunoglobulin-A of human gastric mucus.
Biochim. Biophys. Acta. 576, 9. 1979.

100. Stoop, J. W.
Alpha-Chain disease with involvement of the respiratory tract in a Dutch child.
Clin. Exp. Immunol. 8, 625. 1971.

101. Torano, A., Tsuzukida, Y., Liu, Y-S. V., Putnam, F. W.
Location and structural significance of the oligosaccharides in human $IgA_1$ and $IgA_2$ immunoglobulins.
Proc. Nath. Acad. Sci. 74, 2301. 1977.

102. Vladutiv, A. O.
IgD myeloma or $\delta$ heavy-chain disease?
Clin. Chem. 26. 353. 1980.

103. Wang, S. W. S.
Alpha-Chain disease in an Englishman.
Postgraduate Medical Journal 52, 798. 1976.

104. Westin, J., Eyrich, R., Falsen, E., Lindholm, L., Lundin, P., Lonuroth, I., Weinfeld, A.
Gamma heavy chain disease: reports of three patients.
Acta Med. Scand. 192, 281. 1972.

105. Wetter, O.
Alpha-chain disease cured with antibiotics.
Br. Med. J. 1, 835. 1976.

106. Wetter, O., Schmidt, G. G., Linder, K. H., Leene, W.
H – Kettenkrankheit: Humorale und cellulare Befunde bei sechs Fällen vom $\lambda$-Kettentyp.
J. Cancer. Res. Clin. Oncol. 94, 207. 1979.

107. WHO Meeting Report
Alpha chain disease and related small intestinal lymphoma.
Arch. Fr. Mal. App. Dig. 65, 591. 1976.

108. Zlotnick, A., Levy, M.
Alpha-Heavy chain disease: A variant of Mediterranean lymphoma.
Arch. Intern. Med. 128, 432. 1971.

109. Zucker-Franklin, D., Franklin, E. C.
Ultrastructural and immunofluorescence studies of the cells associated with $\mu$-chain disease.
Blood. 37, 257. 1971.

# VIII. Forms and differentiation of the plasma cell tumors

Autonomic irreversible proliferation of the cells that produce immunoglobulins of different specificity is termed plasma cell tumor. Such tumors either infiltrate the bone marrow, or form isolated islets; combined forms also occur.

Histologically the fiber spectrum produced by these cells is characteristic. The nuclei are homogeneous and have typical structure already mentioned in Chapter V. Inflammatory reaction does not develop around the tumor (2).

Typical radiological findings indicating osteolysis are documented (34). The extramedullar character of proliferation is also characteristic (6) - plasmacytomas rather "ask" for more room, than elicit tissue damage.

If morphologically identical cells are found, the tumors are usually considered as identical; however, a question arises whether the observed cells represent different forms of the same tumor (19, 40, 50, 100), or do they rather reflect similar morphological forms of different pathological processes (29, 47)?

The following groups with different clinical manifestation occur:

I. Myelomatosis, having two different forms:
   a) multiple myeloma (MM)
   b) diffuse myeloma (DM)
II. Solitary bone myeloma (SM)
III. Extramedullar plasmocytoma (EMP)
IV. Plasma cell leukemia (PL)
V. Premyeloma (PM), lanthanic, or smouldering myeloma (71, 137).

This classification is based clinical experience. Definition of each category is not difficult but, as we will see later, their understanding is more problematic (20).

ad I. The diagnosis is based on the verification of the well-known triad. Differentiation between the MM and DM is mainly based on the bone marrow findings: while the nesting of the plasma cells is typical of the former, diffuse proliferation with clearly visible plasma cells rather occurs in the latter form of tumor. These cells probably do not produce Ca-dissolving substances or lack the cell contact properties (137). These two features might explain the differences in clinical pictures of the disease. In all forms of myelomatosis plasma cells proliferate only in the active - red, but not yellow bone marrow. In the course of the disease the yellow marrow is transformed into the red one, becoming *"suitable"* for the plasma cell proliferation. Therefore, osteolysis in the distal parts of the peripheral bones of the extremities is a sign of tumor growth and generalization of the disease: as it has already been mentioned, transformation of the yellow bone marrow into the red one is a basic prerequisite for proliferation; the decrease of the red marrow mass might trigger this reaction.

Apart from the marrow of the tubular bones, plasma cells occur in the "progenitor" zones - the liver, the spleen and to a lesser extent (29) in the lymph nodes (93). In some cases only infiltration in the spleen sinuses or the portal or sinuses of the liver indicates pathological changes. More rarely individual plasma cell colonies can be visualized. Three forms of the liver cell infiltration exist:

- purely sinusoid
- purely portal
- mixed form.

The first form occurs very rarely, however, plasma cell leukemia induces simusoid changes. Still plasmocyte infiltration of the liver does not mean terminal state (99, 127). A case of myelomal infiltration of the thyroid cartilage has been discovered because the surrounding cartilage tissue has been replaced by hemopoietic cells (137).

ad II. Normal bone marrow, characteristic osteolytic changes and histological signs of plasma cell tumor are typical of solitary bone myeloma. Monoclonal gammopathy occurs very rarely. If, however, paraproteins are present, dissemination is suspected. Sometimes after the removal of the tumor protein anomalies are normalized. The latter finding proves that paraproteinemia does not necessarily mean dissemination.

Radiological signs of SM generally differ from those of MM. Two main forms of pathology - the so-called "soap-bubble" (or trabecular) form and osteolysis with definite borders. The latter usually developes in the spine penetrating into the bone tissue from the medulla (92). We emphasize these findings, since the SM is mainly diagnosed with radiological methods (22, 32, 135).

It is next to impossible to decide whether all of the plasma cells occurring at the normal ratio in the "negative" "healthy" bone marrow belong to the normal clones, or are there some malignantly proliferating forms too (136)? As a rule solitary changes generalize within several months. Thus the diagnosis of a solitary form should be clinically and radiologically confirmed at least for three years (28).

Similarly to mammal carcinoma, the question arises, whether the SM originates from *one site,* proliferates and generalizes, or does it rather simultaneously affects the whole organism and only several sites can be visualized (120) The latter should then correspond to the early state of the disease. This assumption is supported by the incidence of 42% of SM in the individuals aged under 50, while only 16.7% of MM occur in this age group (137).

ad III. Extramedullary plasmocytoma (EMP) never involves the bone marrow, primarily occurring in the pharynx, upper respiratory tract (8), lamina propria of the gastrointestinal tract and the lymph nodes (48). It always spreads at the surrounding soft tissues (107) and is regarded as the cause of the so-called local amyloidosis.

The relationship of this tumor with bones and bone marrow should be specially emphasized. It elicits bone lesions only if the primary tumor tissue can infiltrate the bones; the bone marrow is affected due to dissemination. These bone and bone marrow changes are reported in myeloma and are required for monoclonal Ig sunthesis (27).

EMP cells do not require active bone marrow; the participation of the plasma cells in the processes affecting bone marrow does never exceed 10% even in the dissemination stage.

This plasma cell tumor develops not only in the aged individuals, but, in contrast with MM or SM, is sometimes observed in children. Histological diagnosis is usually erroneous, because plasma cells are considered as participants of an inflammatory process (7). This is a very rare disease that deserves more attention. It is well known that the plasma cells occur in the bone marrow, spleen, lymph nodes, respiratory and gastrointestinal tracts (35). It can be, therefore, assumed that the incidence of plasmocyte tumors corresponds to the above pattern. Clinical evidence opposes this view, since EMP comprises only 2-5% of all the plasmocyte tumors. It usually originates from subepithelial tissue of the airways (74%). Its occurrence in the lymph nodes and gastrointestinal tract is 5.5% and 3.6% respectively. The rest 16.9% are distributed between the vagina, mamma, pancreas, parotis, thyroid gland, and testes (137). Hypercalcemia, renal or bone marrow insufficiency have never been reported.

ad IV. Plasma cell leukemia is not an independent entity (68). WBC count can exceed 50,000. Monoclonal gammopathia is obligatory and bone lesions occur in nearly 100% of the pathients (94), but sometimes only smear analysis revealed pathological changes. These symptoms unambiguously reflect dissemination of the plasmocyte tumor. This assumption is supported by the reported coincidence of EMP (6) with true paraproteins (halved Ig: 1 heavy chain + 1 light chain) (121), heavy chain disease (42), and Kaposi-sarcoma (75). It occurs in 1-2% of typical MM. Coincidence of PL with IgD myeloma is about 5%, for IgE myeloma the value is even higher (70).

ad V. Premyeloma (PM) is difficult to define, since MG is its only symptom. This pathology comprises a certain (not yet identified) part of the IMG cases.

Presentation of the five types of the disease is not complete without the outline of some basic relationships. The assumption that in contrast to MM or SM, the EMP-plasmocytes do not require hemopoietic background also suggests the existennce of two large cell populations with all the subpopulations (19, 50). Thereby plasmocytes of the respiratory and gastrointestinal tracts should be separated from the bone marrow plasma cells.

The finding that EMP spreads via the lymphatics, while MM uses the blood stream does not indicate difference between these cells, but only stresses the role of the lymphatics (morphology and function) in the development of this pathology. "Centrifugal" and "centripetal" spread of the MM and EMP is assumed (5, 131).

The above data indicate that plasma cell tumors originate from the bone marrow or extramedullar precursors (Fig. VIII/1).

The Figure summarizes all the possibilities of the type-transition. Transitions exist not only between the plasma cell tumors, but also frequently occur in the malignant diseases of the lymphocytes. Block in the certain stages of the lymphocyte differentiation process leads to proliferation. These cells either maintain their differentiation (but erroneous) capability, or aquire it spontaneously or due to some external stimulus. Therefore, clinicians should be aware that some histopathological entities are not always identical and not always imply the final stage of the development (?!!). It is generally recognized that while the clinical picture of the lymphoid pathology is rather constant, its pathohistological findings usually undergo changes (41). The latter might be due to de-differentiation trend in some cell types (105).

MG as an accompanying symptom or obligatory form occurs in the following cases:

1. Immunoblastic lymphadenopathy + kryoglobulinemia (116)
2. Angio-immunoblastic lymphadenopathy + MG (43)
3. Lymphoma + MG (3, 66, 91)
4. Lymphoma + CLL + MG (9)
5. Lymphoma + IgM MG (85, 97)

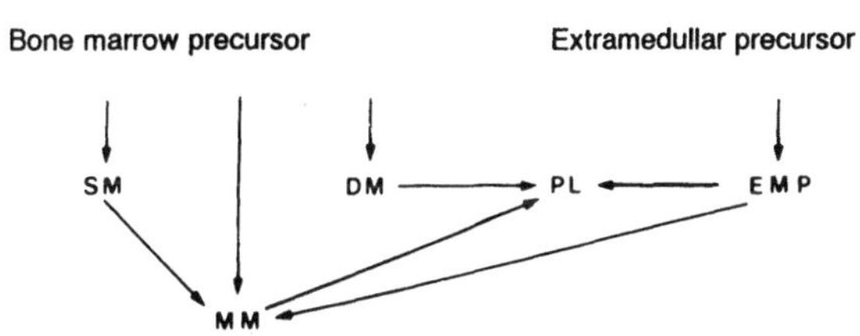


Fig. VIII/1

6. Leukemic lymphosarcoma (LLS) + IgM MG (123)

7. CLL + MG (30, 106)

I. WM + lymphosarcoma + CLL (33)
II. IgG MM + CLL (86, 98, 122, 129)
III. IgG MM + CLL + lymphosarcoma (87)
IV. IgG MM + CML + lymphosarcoma (57)
V. IgG MM + AML (46, 67, 72)
VI. MM + LGR (109)

a) CLL + Lymphosarcoma (117)
b) EMP + terminal RCS (23)

Increased synthesis of polyclonal immunoglobulins is typical of malignant diseases and hematologic systemic pathology.

The occurrence of MG can be induced by the following three mechanisms:

A) MG-s are synthesized by the tumor tissue;
B) The tumor as an antigen-stimulus is responsible for the reduction of monoclonal Ig;
C) Antigen stimulus independent of the disease (?) causes rare reactive changes in the non-healthy organism (60).

Two morphologically different processes can simultaneously occur *in the same tissue of the same organism.*

I. Co-development of two independent diseases (45, 134).
II. Lymphoid leukemoid reaction (in CLL) occurring with lymphoma or RCS. However variabilities at the onset of the two diseases are in controversy with the above assumption (79).
III. The nature of the disease is different, but the cause is the same (62, 78).
IV. In accordance with the generally accepted view, histogenetical relationship between the diseases could explain the transition between different forms (31, 115, 126).
V. Treatment of the first recognized pathology triggers the second disease (Fig.VIII/2) (106).

Basically supporting the primary importance of the IV. item, we regard item V. as the most

| Carcionegenesis<br>(E. Hecker; Gann, 67, 471-481, 1976) | The Committee on International Collaborative Activities (CICA) of the International Union Against Cancer (UICC) | | International Agency for Research on Cancer (IARC)<br>World Health Organization (WHO) |
|---|---|---|---|
| | Solitary cytostatics | | Radiation |
| | Physical | UV., X-ray etc... | |
| | Chemical | Alkylating cytostatics | Immunosuppressive drugs |
| Exogenous<br>↙ Unifactorial ↘ Multifactorial | Viruses | Diethylstilbestrol<br>DDT<br>Aflatoxin | Alcohol<br>Terpenes<br>Iron-oxyde<br>Phenols<br>Detergents<br>Papilloma-, fibroma-/virus |
| Endogenous<br>↙ Unifactorial ↘ Multifactorial | Chemical | Anthranyl | Estrogenes<br>Androgenes<br>Bile acids |
| | Viral | Latent viruses | Helper virus (?) |

**Fig. VIII/2**

probable mechanism, which might develop on the basis of the mechanism IV. (24, 64, 69, 90).

Considering the above possibilities, 2 of the 7 verified cases of MM are presented (triad: MG + secondary antibody-deficiency syndrome; bone marrow - osteolytic areas (61) (Fig.VIII/3-VIII/11).

The patients were treated with a combination of Cyclophosphamide - Prednisolone. In the course of the intermittent cytostatic therapy hazel-nut, walnut- and apple-sized tumors developed throughout the whole body in a period of 1 - 1.5 years. The tumors were accompanied by pathological fractures (Fig. VIII/12-VIII/15).

Marked X-ray differences can be observed between the myelomal osteolytic changes and the bone lesions elicited by the plasmoblasts ("inflated" bones with visible cortical zone in the former and the lack of these zones in the latter) (Fig.VIII/16, VIII/17).

The tissue sample from the palpable tumor was studied under light- and electronmicroscope. The former identified RCS (80, 104), whilst the latter revealed plasmoblastoma (Fig.VIII/18, VIII/19).

Besides the I-V mechanisms that explain the possible development of different structures within the same tissue of the same organ, another version has been suggested.

B lymphocytes carrying surface markers are the direct precursors of the plasma cells (63, 96, 102, 103, 132). Therefore, in cases of MG the precursor cells should also carry surface markers corresponding to the monoclonal gammaglobulin. These markers comprise specific antigens of the tumor cells (83). Some researchers have succeeded to prove the above statement (49, 58, 77, 82, 101). At present it is generally accepted that the lymphocyte surface marker distribution in MM can remain normal (65, 125). This view precludes the involvement of the cells at the early phase of differentiation, otherwise the distribution of the surface immunoglobulin markers would reflect that of the paraproteins (1, 51, 52).

These observations suggest that the MM is a neoplastic disease of the differentiated cells localized in the bone marrow and capable of Ig secretion. In contrast to this statement we succeeded to identify two different types of myeloma. The first develops due independent

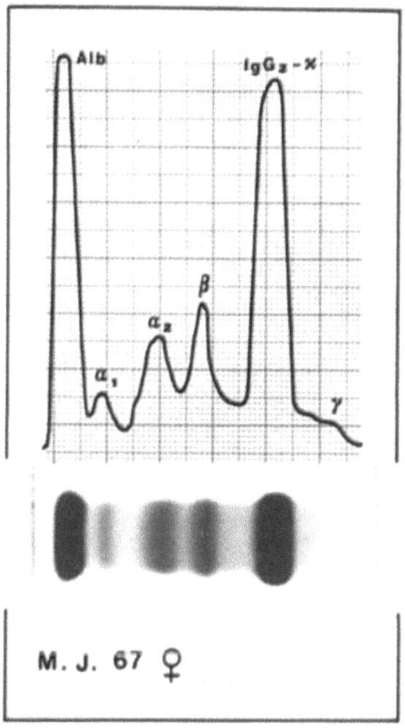


Fig. VIII/3

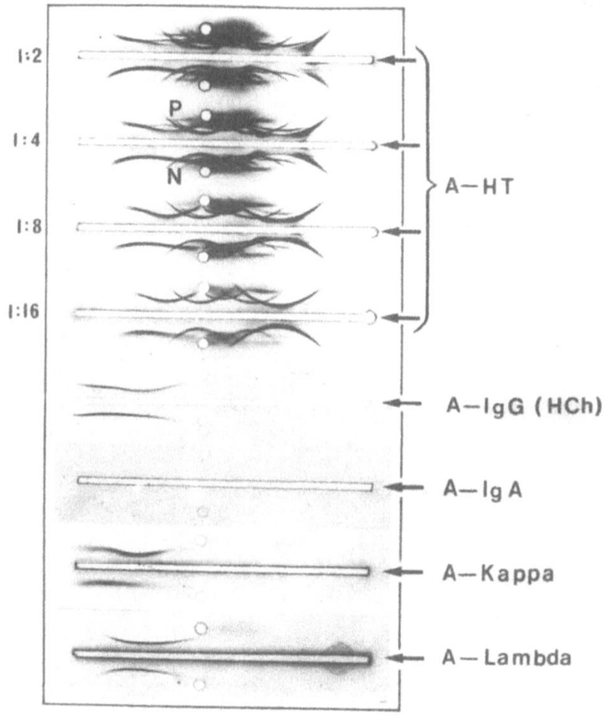


P = Patient
N = Control

Fig. VIII/4

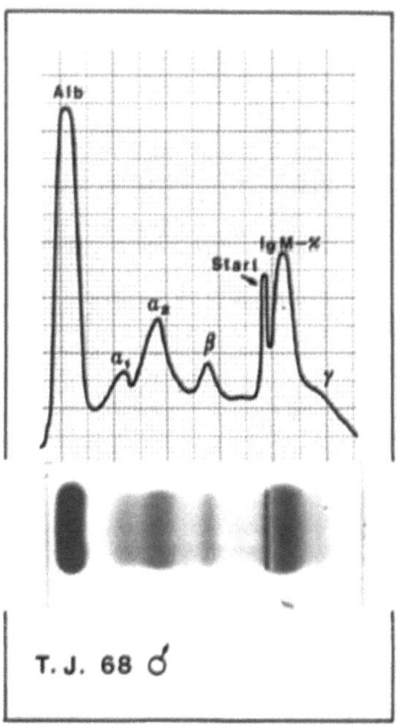


Fig. VIII/5

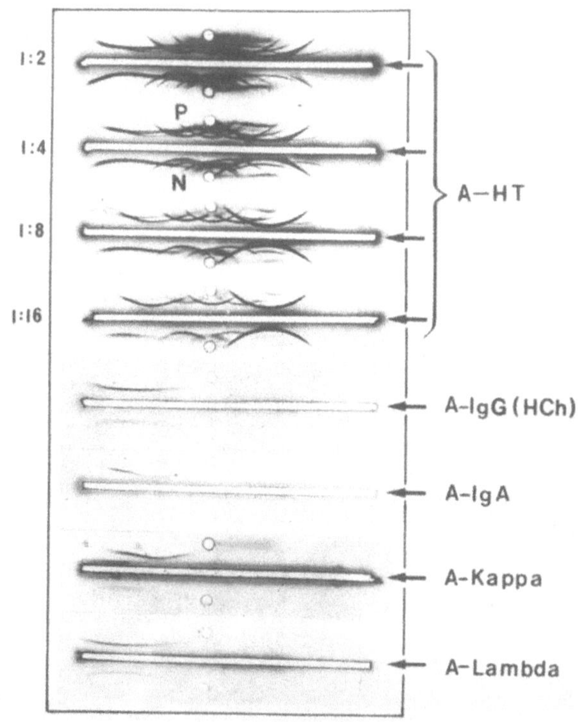


P = Patient
N = Control

Fig. VIII/6

| | IgG | IgA | IgM | IgD | IgE |
|---|---|---|---|---|---|
| M. J. ♀ | ≈ 2800 mg% ($IgG_1$) | 25 mg% | 170 mg% | >0,2 mg% | ≈ o |
| T. J. ♂ | 450 mg% | ≈ 1250 mg% | >10 mg% | >0,4 mg% | ≈ 0,04 |

**Fig. VIII/7**

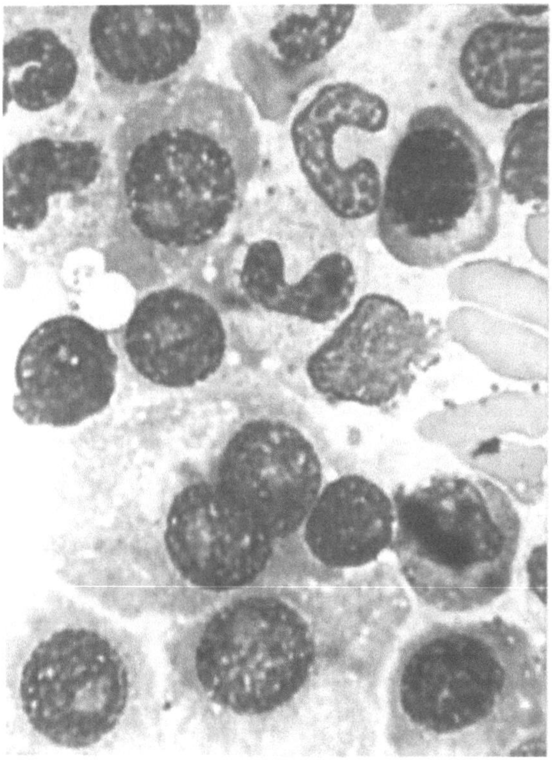

**Fig. VIII/8**

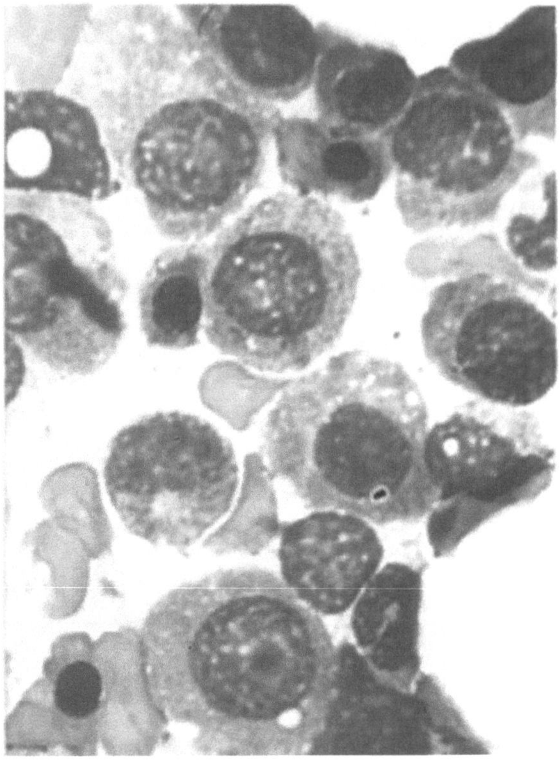

**Fig. VIII/9**

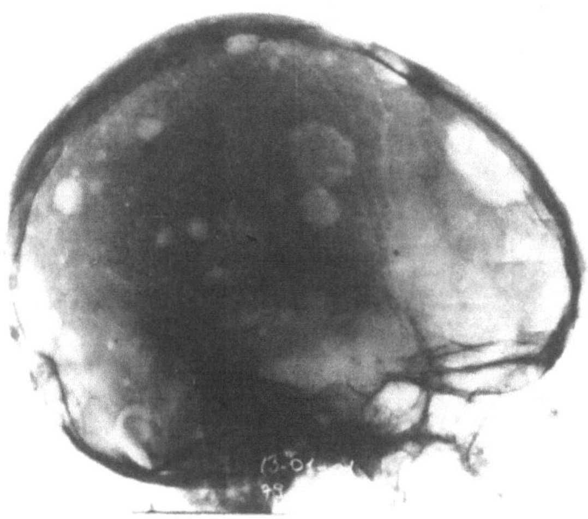

**Fig. VIII/10**

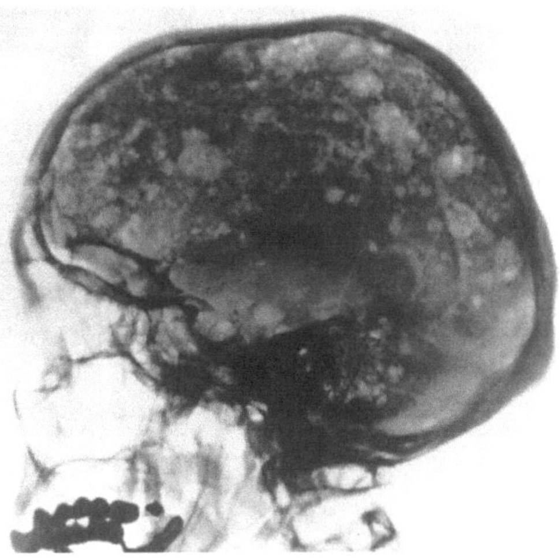

**Fig. VIII/11**

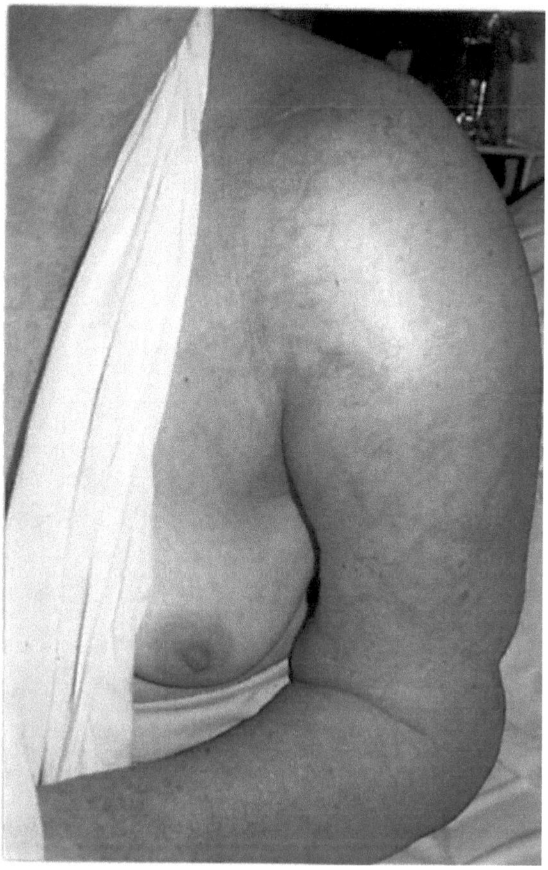

Fig. VIII/12

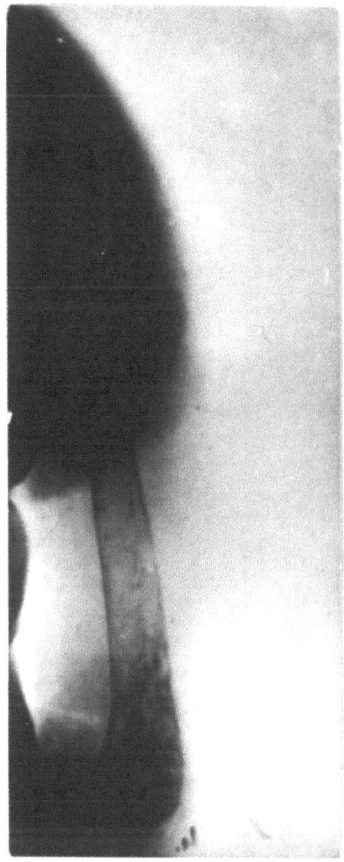

Fig. VIII/13

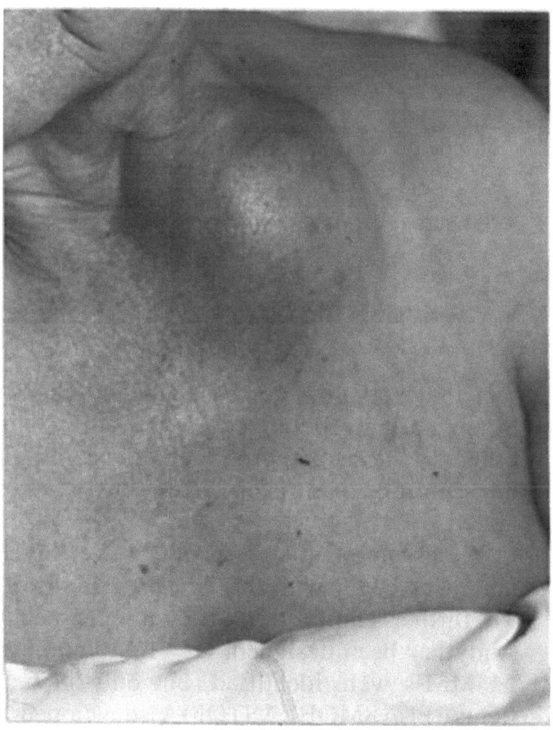

Fig. VIII/14

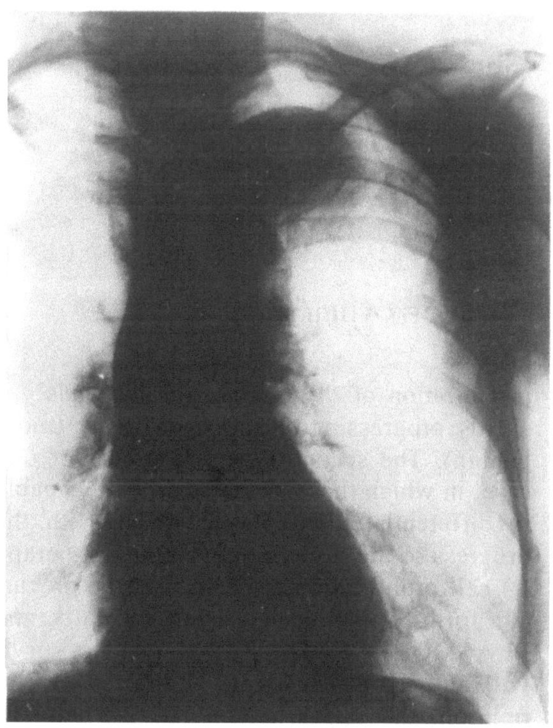

Fig. VIII/15

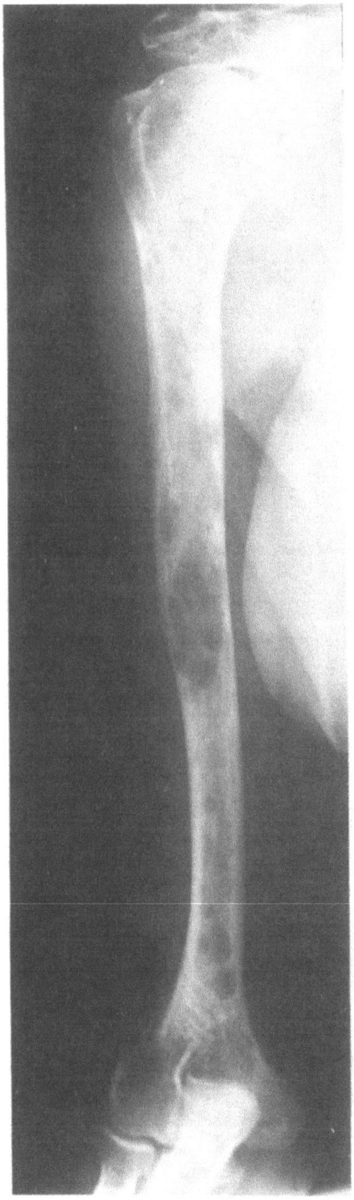

Fig. VIII/16

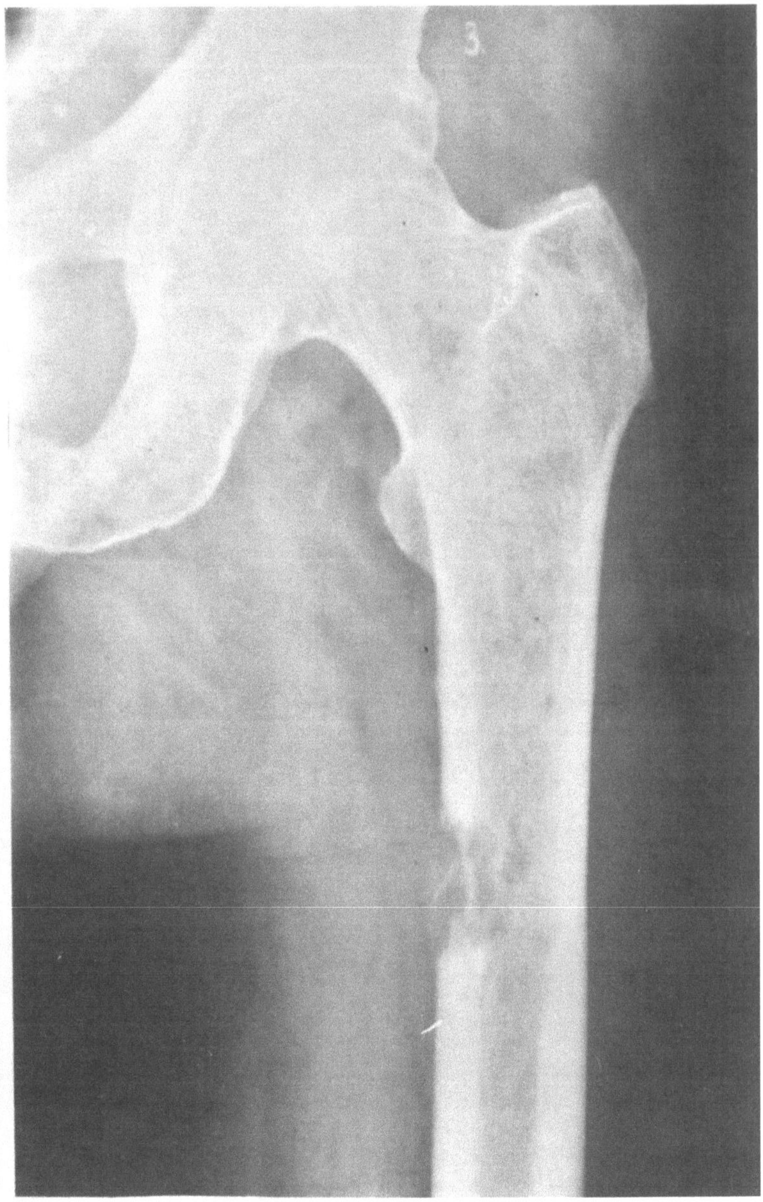

Fig. VIII/17

proliferation of the marrow plasma cells - a slowly progressing plasma cell disease (more benign). The second type is a precursor disease, in which these elements are still capable to differentiate into plasma cells. With the progress of the disease or even due to therapy (!) cell proliferation is blocked at the precursor forms, first locally, later generally and PLASMOBLASTOMA develops. The survival rate of these patients is shorter. In our opinion, the apparent controversy between the data of the literature is only due to the fact that very few cases had been analysed. Some patients belonged to the first, while others to the second group, and these individual findings had been erroneously generalized.

We assume that significantly fewer myeloma cases begin with the "precursor" disease. To further clarify this ratio, a retrospective study has been designed. Of 11,130 autopsies 51 MM-s were identified, but only in 6 patients PLASMOBLASTOMA developed in the final phase of the disease (108). Hence,

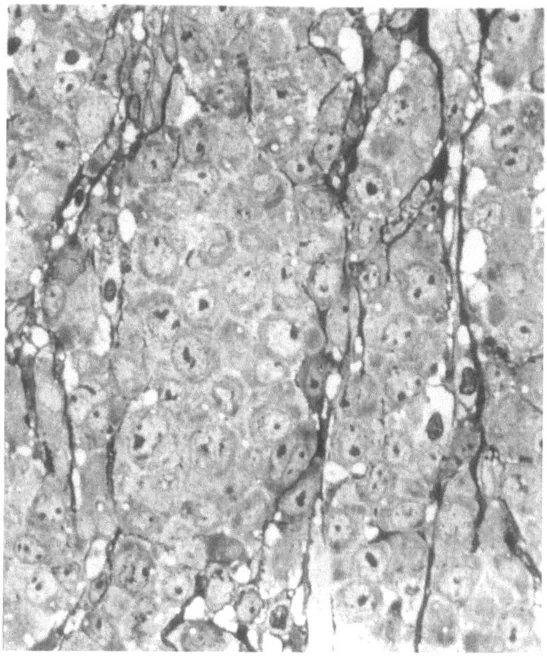

Fig. VIII/18

about 10% of myeloma cases are assessed into the second group (Fig. VIII/20).

To verify our conclusions, a reverse study had been carried out: since 1970 in our Institute all the biopsies diagnosed as RCS are stained with gallocyanin to reveal RNA and DNA as markers of the stimulated protein synthesis.This method is a useful tool for differentiation between immunoblastoma and plasmoblastoma (88). In 17 cases studied gallocyanin staining indicated increased protein synthesis. Three randomly selected cases were compared with two gallocyanin-negative findings. A small part of the paraffin-embedded sample was deparaffinated, fixated with glutaric aldehyde and $OsO_4$ and embedded in araldyte for electronmicroscopic studies. In cases *exhibiting no plasma cell differentiation* the ER could scarcely be visualized (Fig.VIII/21), whereas in tissues where *dif-*

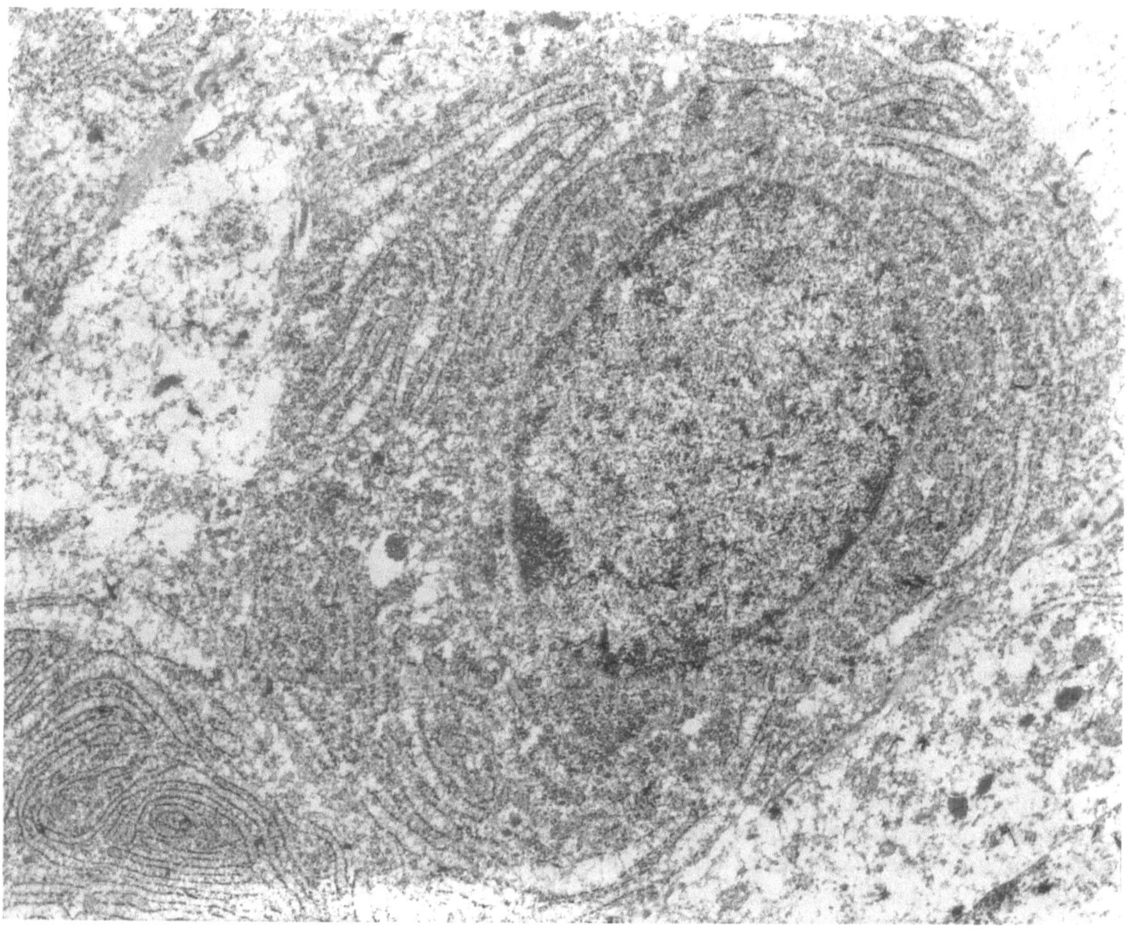

Fig. VIII/19

| | | | | | | | |
|---|---|---|---|---|---|---|---|
| | | | | | RCS | 6 | 11,7% |
| | | Myeloma multiplex | 51 | 0,43% | | | |
| Number of autopsied patients | 11,130 | | | | | | |
| | | Waldenström's macroglobulinemia | 5 | 0,04% | | | |
| | | | | | RCS | 2 | — |

**Fig. VIII/20**

*ferentiation was identified,* polyribosomes, developed ER and sometimes even mature plasma cells could be visualized (Fig. VIII/22).

5 plasmoblastomas diagnosed in 17 cases studied indicate that plasmoblastomas are the frequent forms of RCS, compared to their lower occurrence rate in MM "complicated" with plasmoblastoma. The limited material prevents us from calculating exact ratios. Further data access is required for the recognition of the complex relationships within this system.

The recent special marker study, though stresses the complexity of the above mechanisms, rather supports the possibility of the malignant proliferation of the plasma cells. Mitosis could be rarely observed in the final stage in some plasmocytoma cases suggesting that these cells have no proliferative capability. It is also well known that in diffuse plasmocytoma (DM) plasma cells exhibit homogeneous distribution and appear in the circulation in the terminal phase of the disease. Therefore, these plasma cells can not function as a so-called "self-maintaining" mitotic pool, hence special "clonogenic" precursors are assumed for the replenishment of the psalmocyte population. Morphologically these cells are of lymphoid character, but can not be further separated on the basis of their idiotype antigen properties. The existence of the standardized antibodies (MABs) against the surface antigens of the lymphoid elements has been recently proved (21) and used for the separation of the lymphoid elements. Their differentiation into plasma cells was triggered with phorbol esters (12-0-tetradecanoyl-phorbol-13-acetate) (TPA).

The CALLA (*C*ommon *A*cute *L*ymphoblastic *L*eukemia *A*ntigen) (CD10) reagent (a version of MABs) is a new tool for detection of the CALLA+ plasma cells and circulating lymphocytes (in 6 patients of 10) (Fig. VIII/23, VIII/24). This positivity showed no correlation with the age and sex of the pa-

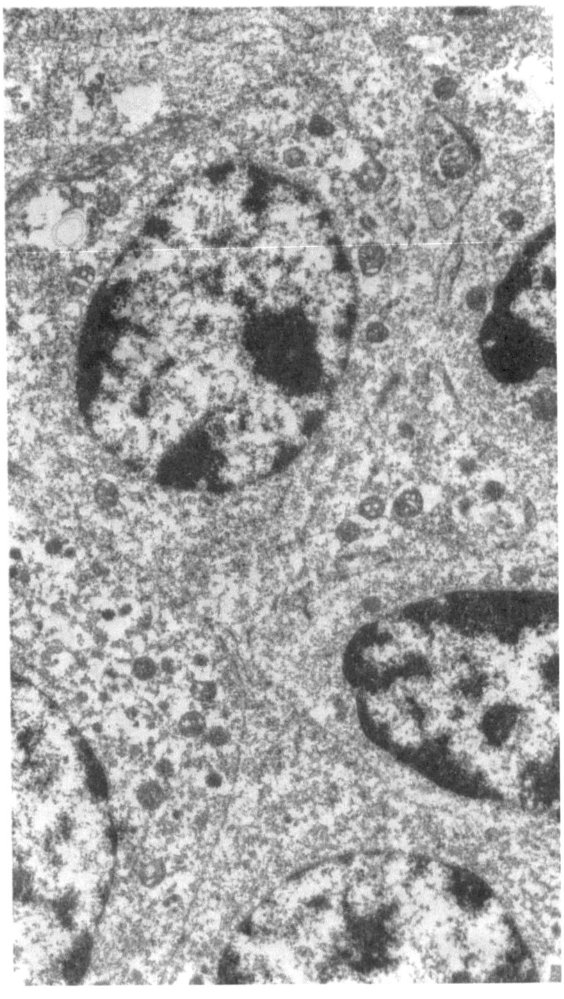

**Fig. VIII/21**

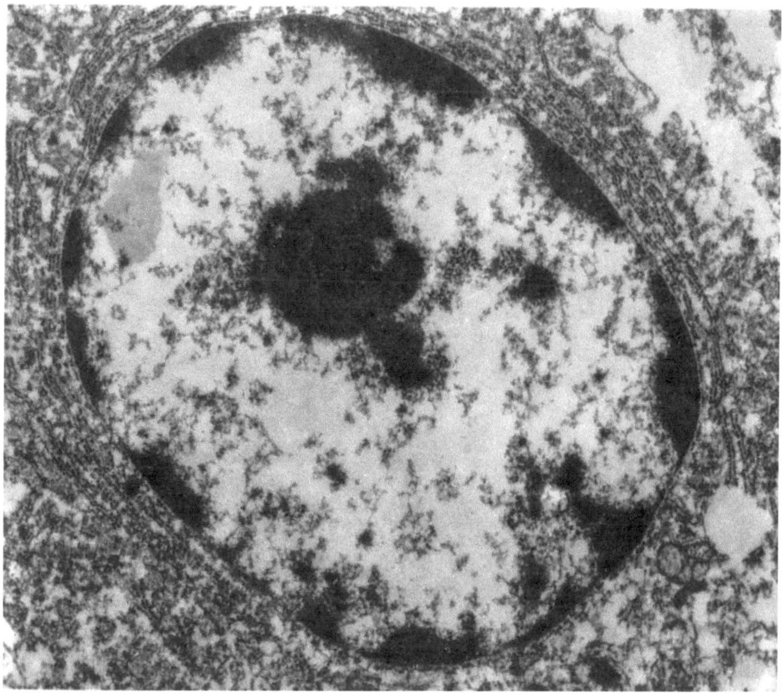

**Fig. VIII/22**

tients and did not depend on the character of the MG.

In CALLA+ (CD10) plasmocytoma the count of the lymphocytes carrying membrane antigens CD9, CD19, CD20, CD21, CD22, CD24 is extremely low (under 2%). Only these lymphocytes carry surface sIgM immunoglobulin with a normal kappa/lambda ratio. The CALLA-positive cells carry only HAN-$PC_1$ and HLA-DR markers but not surface immunoglobulins. However in the later stages of their plasmocytoid maturation they would also carry heavy and light chains typical of the monoclonal gammaglobulins that develop in the course of the disease. It should be stressed again that this study has been designed to reveal, whether plasmocytoma can be considered as a tumor differentiating from precursor elements? The answer is yes! (25). Clinically it is important to stress *1.* the lack of the CALLA+ cells in IMG; *2.* cell count exceeding 5% indicates rapid progression; *3.* positive peripheral lymphoid cells in stages I and II indicate fast fatal outcome (38). Thereby this study is of diagnostic value. Reconsidering our theory presented above, a certain balance between the CALLA+ and CALLA- cells is assumed in plasmocytoma: the CALLA- cells are more prone to cytostatic treatment. If their count is aggressively reduced due to therapy, the relative, and later the absolute count of the CALLA+ cells will grow, reducing the survival rate of the patients. The series of bone marrow studies of the plasmocytoma patients have revealed that a tendency towards immaturity (blast transformation) reflects progression (39, 59, 88).

The possible existence of the types of myeloma is supported by the studies, in which the tumor mass has been evaluated. Data of bivariant correlation- and multivariant regression analysis indicate that the mass of the myeloma cells can be predicted, provided the following parameters are known:

*a)* the spread of the bone lesion; *b)* the hemoglobin value; *c)* serum calcium level; *d)* serum and urine concentration of the M component (36). The following calculations are required:

1. *FCC* (*F*ractional *C*reatinine *C*learance):
   FCC= Creatinine Clearance ml/ (min × 60) plasma volume (Liters × 1000).
2. *FMR* (*F*ractional *M*etabolic *R*ate)
   FMR = 0.081 × FCC + 0.028
3. *FPR* (*F*ractional *P*roteinuria *R*ate)
   FPR = FCC × (0.0253-0.0078)

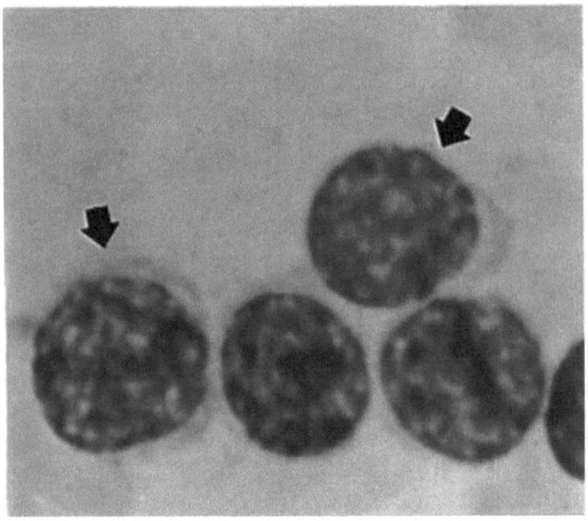

Fig. VIII/23

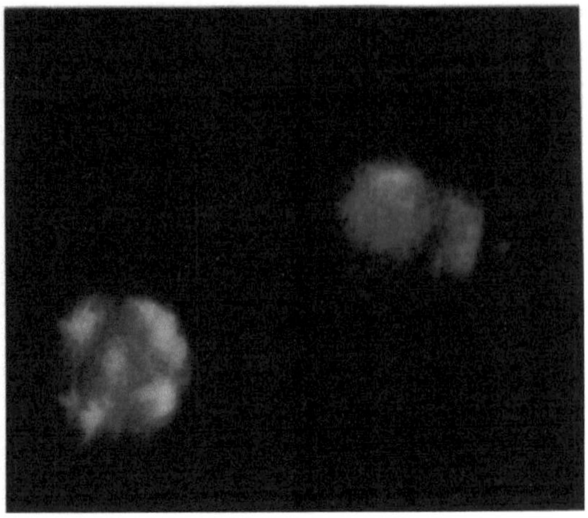

Fig. VIII/24

4. *TLP* (*T*otal *L*ight-chain *P*ool):

$$\text{TLP} = \frac{(\text{BJ urine g/day})}{24 \times \text{FPR/hour}}$$

5. *PROD* (Production)
   PROD = TLP × FMR;
   PROD/day = TLP × FMR × 24
6. Cell Mass:
   PROD/day: FMR/pg/cells/day.

0.6 or 0.7 × $10^{12}$ cells/$m^2$ is considered as low, while 1.2 × $10^{12}$ cells/$m^2$ as high cell mass. It is interesting to mention that no correlation could be observed during the cell mass analysis. This observation seems to support our general statement that different plasma clones have different secretion rates. Independently of these findings, all the studies showed that the myeloma growth rate can be described by the Gompertz equation (110, 111, 118, 124). An initial intensive growth rate later slows down until it reaches the clinically recognizable size 3 × $10^{11}$ cells (108, 118). The death is due to the progressive tumor growth when the plasmocyte count reaches 3-5 × $10^{12}$ (Fig. VIII/25).

The Figure shows that apart from a slow, 20-year-long development process (53, 54) and the evolution time calculated with the Gompertz equation (112, 114, 119, 124), another rapidly growing, so-called induced cell population is assumed. We consider this population as the most probable in the PCP -myeloma combination (133).

Only a mathematical model, which assumes two cell-populations (chemotherapy-sensitive and -resistant), can explain the variances of the tumor-cell count decrease and the kinetics of the relapses (56). These two populations might be similar to the proliferating plasmocytes and lymphocytes in myeloma suggested by us (Fig. VIII/25).

In clinical practice the evaluation of the above parameters is a useful tool for determination of the stage of the disease. Different mathematical approaches are documented in the literature since 1967. Retrospective analysis of the data shows that though neither approach is fully based on clinical experience, they can not be ranked, for only minor differences can be observed between the different methods. Thus, either of them can help in prognostic decisions by providing criteria for comparison of the patients examined in different parts of the world (4, 17, 18, 26, 36, 37, 44, 81, 84, 89, 95, 130, 138) (Table VIII/I).

This Table gives a chronological list of the methods documented in the literature.

Despite certain doubts, stage classification by Durie and Salmon is the most common (36).

Stage I:
1. Hb value > 10g%
2. serum calcium level is normal or less than 12 mg%.
3. no osteolytic signs, or solitary osteolysis.
4. low concentrations of monoclonal Ig:
   IgG < 5g%
   IgA < 3g%
   light-chain concentration in urine < 4g/24h

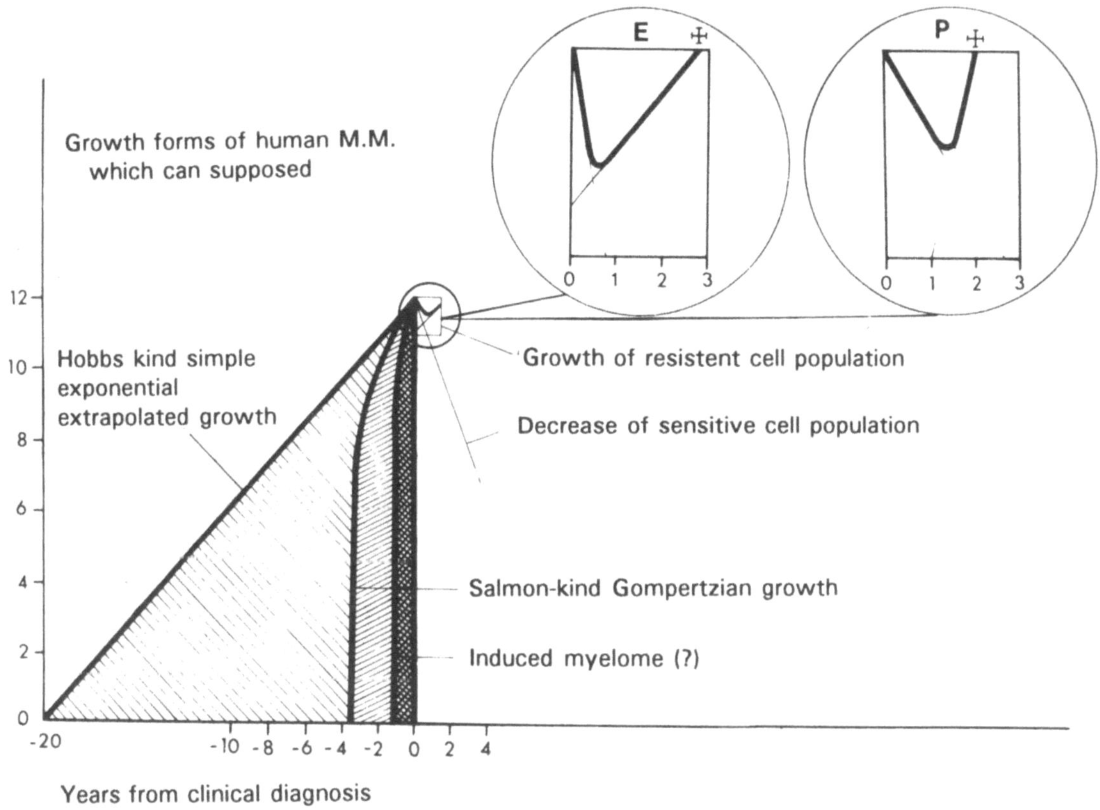


Fig. VIII/25

| | | |
|---|---|---|
| 1. | Carbone, P. P. Kellerhouse, L. E., Gehan E. A. | 1967 (26) |
| 2. | Durie, B. G. M., Salmon S. E. | 1975 (36) |
| 3. | Alexanian, R., Balcerzak, S., Bonnet, J. D., Gehan, E. A., Haut A., Hewelett, J. S., Monto, R. W. | 1975 (4) |
| 4. | Ösby, E., Carlmark, B., Reizenstein, P. | 1978 (89) |
| 5. | Woodruff, R. K., Wadsworth, J., Malpas, J. S., Tobias, J. S. | 1979 (138) |
| 6. | Durie, B. G. M., Salmon, S. E., Moon, T. E. | 1980 (37) |
| 7. | Merlini, G., Waldenström, J. G., Jayakar, S. D. | 1980 (84) |
| 8. | Medical Research Council's Working Party on leukemia in adults | 1980 (81) |
| 9. | Vercelli, D., Cozzolino, F., DiGuglielmo. R. | 1981 (130) |
| 10. | Bartl, R., Frisch, B., Burkhardt, R., Fateh-Moghadam, A., Mahl, G., Gierster, P., Sund, M., Ketner, G. | 1982 (17) |
| 11 | Pennec, Y., Mottier, D., Youinou, P., Asselain, B., Chavance, M., Le Goff, P., Le Prise, P. Y., Miossec, P. I., Le Menn, G. | 1983 (95) |
| 12. | Bataile, R., Grenier, J., Sany, J. | 1984 (18) |
| 13. | Gassmann, W., Pralle, H.,Haferlach, T., Pandurevic, S., Graubner, M., Scmitz, N., Löffler, H. | 1985 (44). |

Table VIII/I

Stage II: all the parameters are between the values typical of Stages I and III.

Stage III:
1. Hb < 10g%
2. Serum calcium level > 12mg%
3. Signs of multiplex osteolysis
4. High levels of monoclonal Ig:
   IgG > 7g%
   IgA > 5g%
   light-chain concentration in urine ≥ 12g/24h.

The state of the kidney classified herein is very important in all the three stages:
- A) relatively good renal function (serum creatinine concentration < 2 mg%)
- B) significant deterioration of the renal function (serum creatinine concentration < 2.0 mg%)

Therefore, the current state of a patient is assessed with two parameters, for example I A or III B. The roman numbers indicate the size of tumor, the letters - the functional state of the kidney.

In evaluation of tumor size, the bone alterations are assessed into 4 groups:
- 0 = no alterations
- 1 = osteoporosis or *a single* localized lesion!
- 2 = several destructive alterations
- 3 = spread multiple osteolytic areas and fracture(s)

Previously the survival rate was assessed by the pretreatment size of the tumor and the character of the remission. It is well known, however, that the proneness to therapy also depends on the characteristics of the cell kinetics, which need further investigation. The DNA concentration and the rate of its synthesis shows no correlations with the increased mass of the tumor. Experimental evidence indicates that at low DNA synthetic rate remission occurs more easily even if the tumor mass is large, whereas at high rate of DNA formation their incidence is low (55). The low values of the thymidin uptake index are independent of the initial size of the tumor, indicating favourable prognosis (73, 74). Such independence of the tumor size also suggests that the morphologically identified plasma cells do not participate in the tumor proliferation. Their increased count might be a product of an independently functioning clone, which is capable to prolong the life span of the plasmocytes (113).

To answer this question, the sternum puncture sample was analysed with a cellsorter and marked cell heterogeneity was identified (12). Fluorescent dye was used for determination of the DNA content of the cells. These findings were used for a quantitative evaluation of the chromosomes and the proliferative potential of the cells (10). The RNA content is an other valuable parameter: the DNA-RNA analysis was performed with acrydin-orange (128) and the aneuploid anomalies were revealed. Apart from the DNA-RNA and chromosomal studies, quantitative and qualitative analysis of the cytoplasmic immunoglobulins of the cells was also necessary for the correct conclusions. DNA, RNA, and the cytoplasmic immunoglobulin indices have been used for the definition of the tumor chromosomes, and the phenotype of the tumor cells respectively (11, 14, 15). The DNA-RNA and the DNA-CIg ratios indicate the extent of the tumor cell infiltration in the bone marrow. The RNA index seems to be of the highest prognostic value, while the DNA-CIg analysis is required for identification of the myeloma phenotype. The tumor character of the aneuploid cells revealed by the DNA analysis is unambiguous, however, the exact nature of the DNA-diploid (euploid) cells can be identified only if the RNA content and cytoplasmic immunoglobulins have been determined.

It is worth pointing out that the percent distribution of the cells obtained by aspiration or biopsy is not similar. In the aspirate the plasma cell/nuclear cell ratio is always higher than in the bone marrow biopsies. Plasma cells might have lower adhesion to the stroma or to the surface of the bone trabecules.

Cytogenetic studies of the tumor cell karyotype are usually not successful because of the low proliferative activity of the malignant cells (76).

The cellsorter studies are much more informative. The patients usually have 80% aneuploid, 5% hypodiploid, 25% hyperdiploid, and 5% mor than 2 aneuploid anomalies.

The hypodiploid and biclonal DNA lines are the most resistant to therapy (16).

The comparison of RNA to DNA content indicates the lability of the former and stability of the latter. Biochemical and ultrastructural studies explain this finding by the predominance of the ribosomal RNA involved in the protein synthesis. Thus the RNA content correlates with the protein synthetic potential, though the correlation is not linear (13). The non-linear character of this relationship might be due to the different rates of catabolism. The RNA content in the patients is a stable parameter, its changes indicating alteration of the tumor mass.

A so-called universal calculation, which can be applied to all the myeloma types, was used in our studies:

**Tumor cell count = (0.601 + 0.283 × indicator of bone lesions) + (0.031 × monoclonal Ig g/24$^h$ in urine) – (0.058 × Hgb + 0.051 × serum calcium level mg%) + (0.028 × monoclonal Ig level in serum g%) × $10^{12}/m^2$.**

The plasma cell count at different stages of the disease is: 0.6 × $10^{12}/m^2$ in the first stage, 0.7 - 1.2 × $10^{12}/m^2$ in the second stage, and more than 1.2 × $10^{12}/m^2$ in the third stage.

The reliability of this stage classification has been verified in the premyeloma (IMG) group, in which a non-tumor (?) plasma cell proliferation occurs. Our studies were based on the following:

1. osteoporosis generally occurs or can occur in the aged patients independently of myeloma - therefore it must be included into the equation;
2. extremely high monoclonal Ig levels occur in IMG, thus the existing view that the paraprotein concentration is low in IMG is erroneous;
3. paraproteinemia with long history - especially in the aged patients - will in some way (light chain, or the whole molecule excretion) show up in the urine.

The question arises, whether the pseudo-positive results of the patient examination would still support the stage classification of the disease? Besides the calculations, the results of sternum punctures were also evaluated. 337 inhabitants of the Budapest Methodological Retreatment Home (234 females, average age 77.2 years and 103 males, average age 71.8 years) have been studied (47). Immunochemical analysis revealed MG in 7 of 337 cases (Table VIII/II).

The average age of the elderly people (3 women and 4 men) with monoclonal gammopathy was 82.8 years. X-ray findings had proved osteoporosis in all cases. Their "tumor cell count" was lower than that expected at the first stage. High Hb level is, in our opinion, the cause of this finding. We failed to observe erroneous pseudopositivity, still the results yielding 1012 order of magnitude are debated, if only the bone marrow had been studied *without evaluation of the intestinal plasma cells.*

Finally, we would like to present a typical mistake: in the beginning of the Chapter it was emphasized that the plasma cells rather proliferate in the "progenitor" organs. There is a certain relationship between the *liver-spleen* and the bone marrow. The extramedullar type of hemopoiesis under normal and pathological conditions is

| | | Age | Hgb | MG | Stage | Bone marrow plasma cells |
|---|---|---|---|---|---|---|
| B.E | ♀ | 83 | 14.5 | IgA-kappa | I. 0.54 × | 10% |
| K.J.I | ♂ | 98 | 14.6 | $IgG_3$-lambda | I. 0.64 × | — |
| M.F. | ♂ | 80 | 15.4 | $IgG_1$-kappa | I. 0.47 × | 5% |
| L.L. | ♀ | 89 | 12.5 | $IgG_1$-kappa | I. 0.52 × | 9% |
| L.J | ♂ | 78 | 11.2 | IgA-kappa-$IgG_1$-lambda | I. 0.62 × | 11% |
| K.S. | ♂ | 70 | 15.4 | $IgG_2$-kappa | I. 0.48 × | 4% |
| L.J. | ♀ | 82 | 11.7 | IgA-kappa | I. 0.60 × | 7% |

**Table VIII/II**

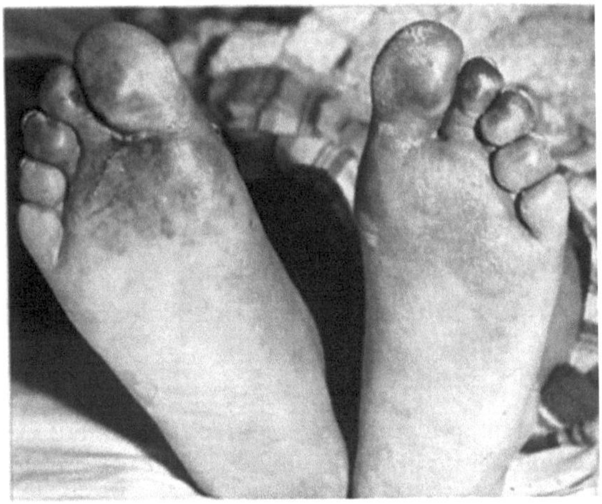

Fig. VIII/26

well known, however, only few data are available concerning the "migration" tendency of the liver cells.

M.V. male patient aged 69 had complaints of osteal pains since the summer of 1977 (cracked vertebrae). In November, 1977 malignant tumor had been suspected and the patient had been admitted to the local hospital. Loss of body weight, anemia and the bone lesions visualized by X-ray suggested MM.

Bone marrow cells obtained after sternum punction did not resemble those commonly known as proliferating plasma cells. Several days after the admission the patient became febrile and both feet showed signs of gangrene (Fig.VIII/26).

Microcirculatory impairment and hyperviscosity were assumed and the patient was transferred to our Clinic. A repeated bone marrow aspiration was performed (Fig.VIII/27, VIII/28).

Gel electrophoresis showed typical M component (Fig.VIII/29)

Other findings: sedimentation: 42 mm/h; Hb: 1.3 mmol/L; Ht: 0.26; WBC: 11 $\times 10^9$/L; Platelets: 240 $\times 10^9$/L; blood glucose: 3.7 mmol/L; CN: 36 mmol/L; Se creatinine: 60 mmol/L; Se bi: 31 mmol/L; AST: 81 U/L; ALT 11 U/L; ALP 28 U/L; Se Na: 137 mmol/L; Se K: 4.3 mmol/L; Se Ca: 2.15 mmol/L; Prothrombin: 65%; Blood viscosity (37 °C): 5.0 cP.

The patient had suddenly died before the termination of the studies. Diagnosis: multiple myeloma, monoclonal gammopathy, hyperviscosity syndrome.

Further evaluation failed to identify the M-component as Ig, therefore, further fractionation was necessary. Finally, the strip first taken as an M-component was identified as mucoprotein. Electronmicroscopic studies of the bone marrow aspirate showed that the cells first identified as plasmocytes proved to be liver cells (Fig.VIII/30).

At autopsy cirrhosis and primary hepatocellular carcinoma were found.

How can one evaluate the sequellae of this experimentum naturae? Are there any marrow-liver or liver-marrow relationships at the cellular level, or had it only been a "trick" of nature? (Fig. VIII/31).

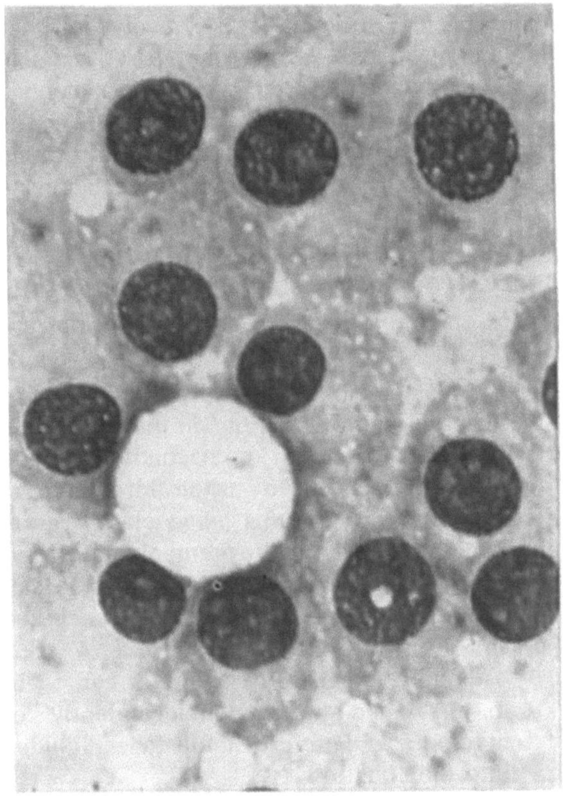

Fig. VIII/27

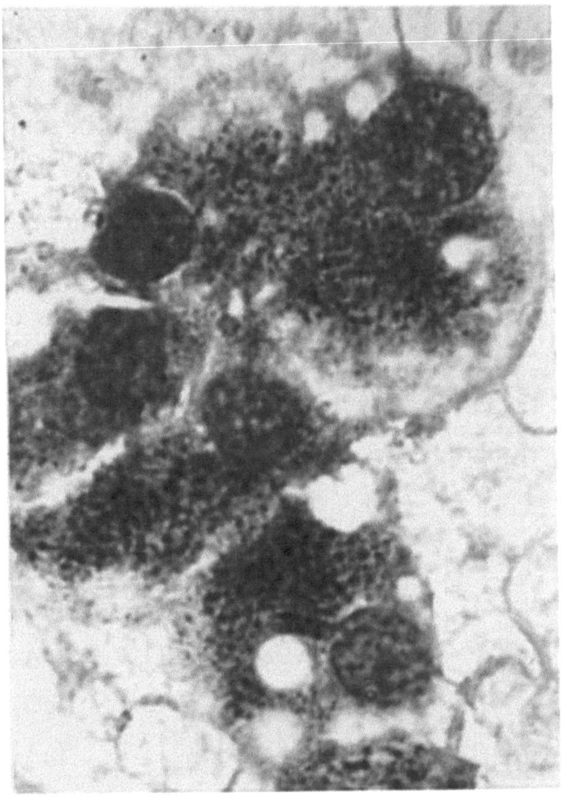

Fig. VIII/28

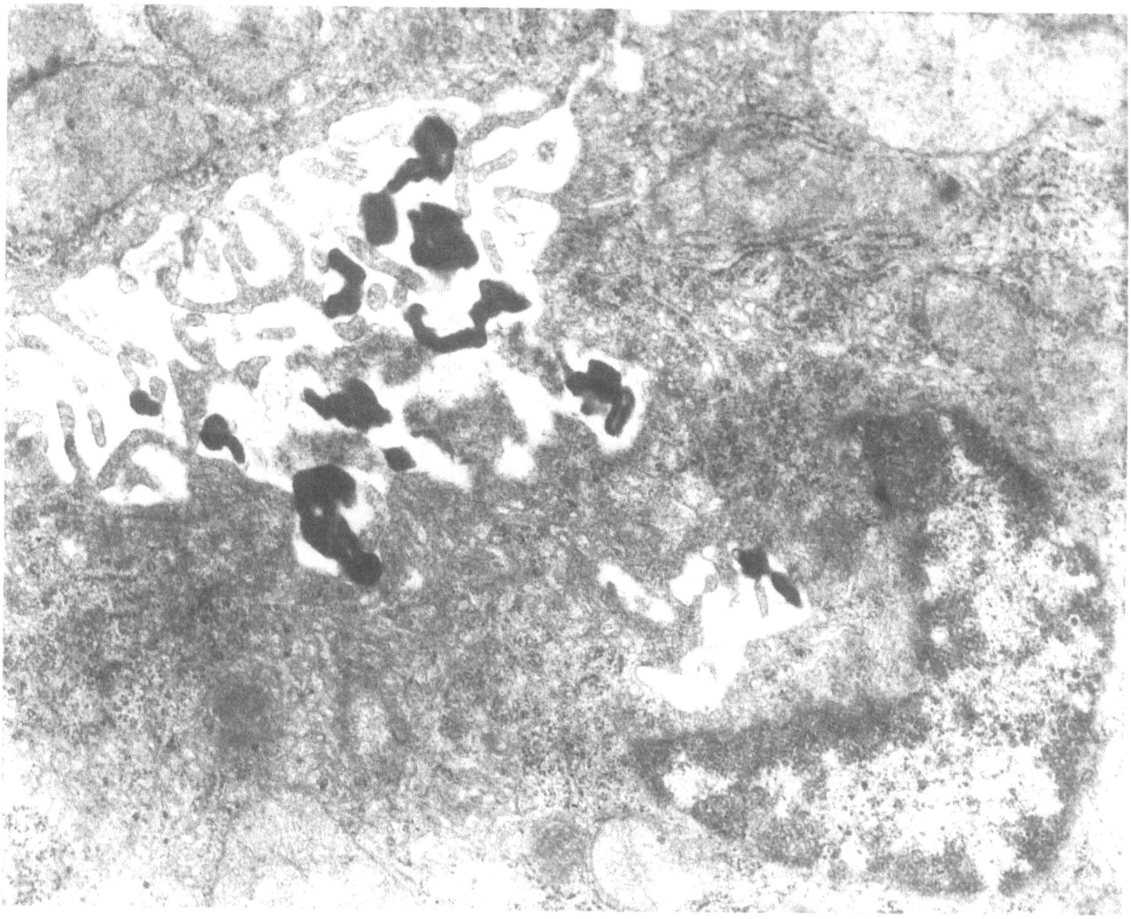

Fig. VIII/29

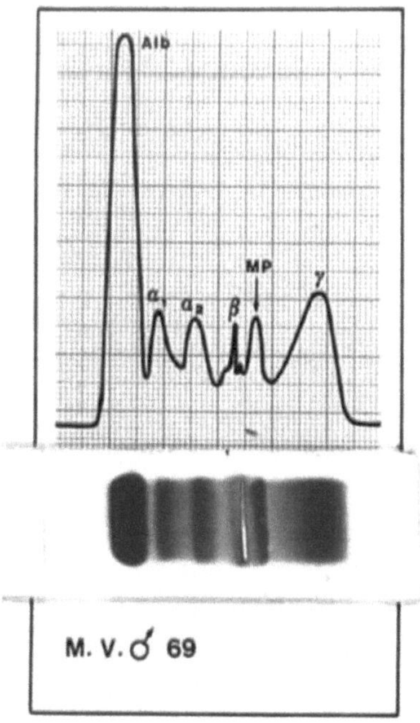


Fig. VIII/30

| PLASMOCYTE | HEPATOCYTE |
|---|---|
| Protein production (Ig)<br>PAS+<br>AP+<br>ER<br>Has precursor<br>Chalon<br>Storage - rare Chalon | Protein production (A.f.)<br>PAS+<br>AP+<br>ER<br>Might have precursor<br>(Cell modulation)<br>Chalon<br>Storage - frequently |
| Cell products are secreted in the surroundings | Cell products are secreted via definite routes, but without special structures (?!) |
| Extramedullar hemopoiesis in the liver. (no ICS is observed) | Site of metastases of the primary hepatocellular cc. Bone marrow - ostelytic signs |

Fig. VIII/31

In summary, this chapter leaves the reader with some unanswered questions:
- The existence of contacts (syncitial structures) between the cells is not proved. The following assumed sequence of events: plasmocytoma → multiple myeloma → plasmoblastoma - also needs further investigation.
- no straight answer is obtained concerning the assumed dual plasmocytoma, which starts as a precursor disease and finally develops from the proliferating mature plasma cells. These theoretical problems imply introduction of *routine* clinical studies, which can be used to answer these questions in practice.
- There is also doubt about the stage classifications presented in the literature. Though we try to comply with them, we feel that certain gaps do exist.
- The differences between the intramedullar and extramedullar forms might be explained by the different "homing" properties.
- The description of the bone marrow-liver, liver-bone marrow relationships might be of importance not only in terms of general hematology, but also to get a keener insight into the nature of the plasma cell tumors. The Chapters dealing with polyclonal immunoglobulin production and plasma exchange are also worth pointing out, though these problems also need further investigation.

We finish the Summary with the presentation of 7 cases in which therapy had triggered plasmoblastoma. In two cases plasmoblastoma had been due to cyclophosphamide and cyclophasphamide-steroid therapy. In other patients other medications had lead to this pathology. (Fig.VIII/32, VIII/33, VIII/34).

In conclusion, it can be generally assumed that cytostatic therapy might be responsible for the altered properties of the cells.

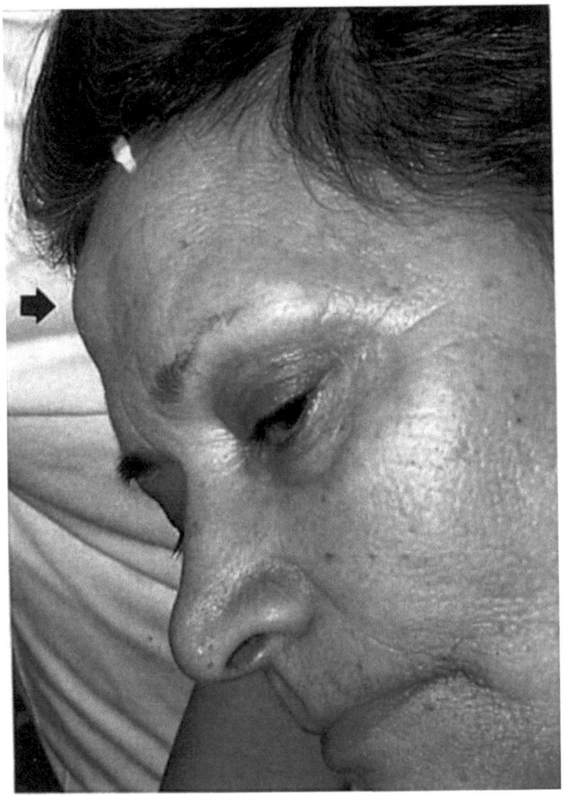

**Fig. VIII/32**

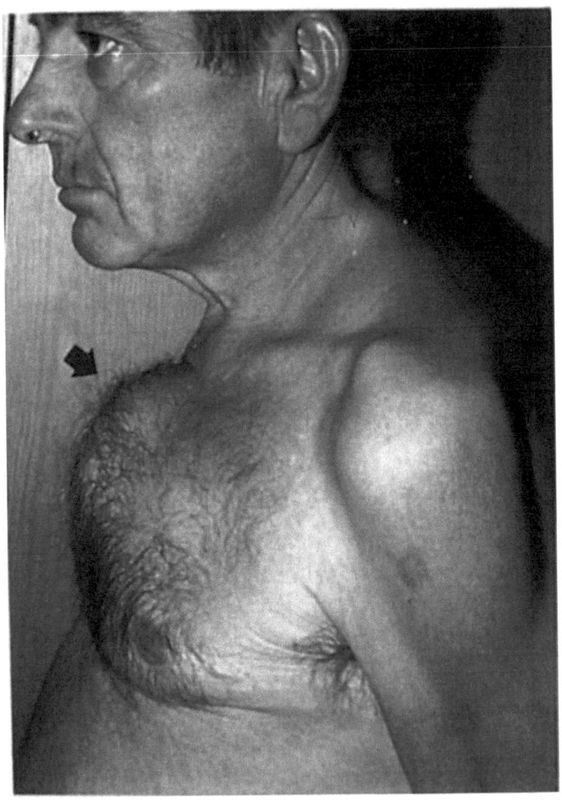

**Fig. VIII/33**

### References


1. Abdou, N. I., Abdou, N. L.
   The monoclonal nature of lymphocytes in multiplex myeloma.
   Ann. intern. Med. 83, 42. 1975.
2. Ackerman, L. V., Spjut, H. J.
   Tumors of bone and cartilage.
   In: Atlas of Tumor Pathology.
   Armed Forces Institute of Pathology. Washington, D. C. 152. 1962.

Fig. VIII/34


3. Alexanian, R.
Monoclonal gammopathy in lymphoma.
Arch. intern. Med. 135, 62. 1975.
4. Alexanian, R., Balcerzak, S., Bonnet, J. D., Gehan, E. A., Haut, A., Hewlett, J. S., Monto, R. W.
Prognostic factors in multiple myeloma.
Cancer. 36, 1192. 1975.
5. Alexanian, R., Gehan, E., Hant, A., Saiki, J., Weick, J.
Unmaintained remissions in multiple myeloma.
Blood. 51, 1005. 1978.
6. Andaloro, V. A. Jr., Baboll, D.
Testicular involvement in plasma-cell leukemia.
Urology. 3, 636. 1974.
7. Anderson, P. E.
Extramedullary plasmocytomas.
Acta Radiol. 32, 365. 1949.
8. Ash, J. E., Beck, M. R., Wilkes, J. D.
Tumors of the upper respiratory tract and ear.
In: Atlas of Tumor Pathology.
Armed Forces Institute of Pathology. Washington, D. C., 150. 1964.
9. Azar, H. A., Hill, W. T., Ossermann, E. F.
Malignant lymphoma and lymphocytic leukaemia associated with myeloma type serum proteins.
Amer. J. Med. 23, 239. 1957.
10. Barlogie, B., Gohde, W., Johnston, D. A., Smallwood, L., Schumann, J., Drewinko, B.
Determination of ploidy and proliferative characteristics of human solid tumors by pulse cytophotometry.
Cancer Research. 39, 3333. 1978.
11. Barlogie, B., Latreille, J., Freireich, E. J., Fu, C. T., Mellard, D., Meistrich, M., Andreeff, M.
Characterization of hematologic malignancies by flow cytometry.
Blood Cells. 6, 719, 1980.
12. Barlogie, B., Alexanian, R., Gehan, E. A., Smallwood, L., Smith, T., Drewinko, B.
Marrow cytometry and prognosis in myeloma.
J. Clin. Invest. 72, 853. 1983.
13. Barlogie, B.. Maddox, A. M., Johnston, D. A., Raber, M. N., Drewinko, B., Keating, M. J., Freireich, E. J.
Quantitative cytology in leukemia research.
Blood Cells. 9, 35. 1983.
14. Barlogie, B., Raber, M N., Schumann, J., Johnson, T. S., Drewinko, B., Swartzendruber, D. E., Gohde, W, Andreeff, M, Freireich, E. J.
Perspectives in cancer research, flow cytometry in clinical cancer research.
Cancer Research. 43, 3982. 1983.
15. Barlogie, B., Alexanian, R., Pershouse, M., Smallwood, L., Smith, L.
Cytoplasmic immunoglobulin content in multiple myeloma.
J. Clin Invest. 76, 765. 1984.
16. Barlogie, B., Alexanian, R., Smith, L., Dixon, D., Smallwood, L., Delasalle, K.
Prognostic implications of tumour cell DNA and RNA content in multiple myeloma.
Blood. 66, 335. 1985.
17 Bartl, R., Frisch, B., Burkhardt, R., Fateh-Moghadam, A., Mahl. G., Gierster, P., Sund, M., Kettner, G.
Bone marrow histology in myeloma: its importance in diagnosis, prognosis, classification and staging.
Brit, J. Haematol. 51, 361. 1982.
18. Bataille, R., Grenier, J., Sany, J.
Beta-2-microglobulin in myeloma: optimal use for staging, prognosis, and treatment – a prospective study of 160 patients.
Blood. 63, 468. 1984.
19. Batsakis, J. G., Faries, G. T., Goldman, R. T., Arlsberg, R. C.
Upper respiratory tract plasmacytoma.
Arch. Otolaryngol. 79, 613. 1964.
20. Bergsagel, D. E., Kyle, R., Bataille, R.
Plasma cell neoplasms.
International Congress ISH-ISBT.
Budapest. 1, 7. 1982.
21. Bernard, A., Boumsell, L., Dausset, J., Milstein, C., Schlossman, S. F.
Leucocyte Typing.
Springer-Verlag. Berlin. 1, 814. 1984.
22. Bichel, J., Kirketerp, P.
Notes on myeloma.
Acta Radiol. 19, 487. 1983.
23. Bodó, M., Pető, I., Sas, G., Juhász, J.
Reticulosarcomatous transformation of a primary recurrent gastric plasmocytoma.
Haematologia. 10, 213. 1976.
24. Bösken, W., Noltenius, H.
Grosszelliges Reticulumzellsarkom als Spätzustand einer endoxanbehandelten Makroglobulinaemie Waldenström.
Zbl. aellg. Path. 111, 407. 1968.
25. Caligaris-Cappio, F., Bergui, L., Tesio, L., Pizzolo, G., Malavasi, F., Chilosi, M., Campana, D., van Camp, B., Jánossy, G.
Identification of malignant plasma cell precursors in the bone marrow of multiple myeloma.
J. Clin. Invest. 76, 1243. 1985.
26. Carbone, P. P., Kellerhouse, L. E., Gehan, E. A.
Plasmacytic myeloma. A study of the relationship of survival to various clinical manifestations and anomalous protein type in 112 patients.
Amer. J. Med. 42, 937. 1967.
27. Carson, C. P., Ackerman, L. V., Maltby, J. D.
Plasma cell myeloma. A clinical pathologic and roentgenologic review of 90 cases.
Amer. J. clin. Path. 25, 849. 1955.

28. Christopherson, W. M., Miller, A. J.
A re-evaluation of solitary plasma-cell myeloma of bone.
Cancer. 3, 240. 1950.
29. Churg, J., Gordon, A. J.
Multiple myeloma. Lesions of the extra-osseous hematopoietic system.
Amer. J. Clin. Path. 20, 934. 1950.
30. Colls, B. M., Lorier, M. A.
Immunocytoma, cancer and other associations of monoclonal gammopathy: a review of 224 cases.
New Zealand Med. 7, 82. 1975.
31. Custer, R. P., Bernhard, W. G.
Interrelationship of Hodgkin's disease and other lymphatic tumors.
Amer. J. med. Sci. 216, 625. 1948.
32. Cutler, M., Buchke, F., Cantril, S. T.
Course of single myeloma of bone: Report of 20 cases.
Surg. Gynecol. Obstet. 62, 918. 1936.
33. Dao, C., Zittovn, R., Diebold, J., Cadiou, M., Reyner, M., Gandric, M., Tricot, G., Bousser, J.
Formes sarcomateuses et leukemiques de la macroglobulinémie de Waldenström.
Sem. Hop. Paris. 51, 2935. 1975.
34. Dolin, S., Dewar, J. P.
Extramedullary plasmocytoma.
Amer. J. Path. 32, 82. 1956.
35. Downey, H.
Handbook of Haematology.
Ed: Hoeber, P. B.
New York. 2, 1438. 1938.
36. Durie, B. G. M., Salmon, S. E.
A clinical staging system for multiple myeloma.
Correlation of measured myeloma cell mass with presenting clinical features, response to treatment, and survival.
Cancer. 36, 842. 1975.
37. Durie, B. G. M., Salmon, S. E., Moon, T. E.
Pretreatment tumor mass, cell kinetics, and prognosis in multiple myeloma.
Blood. 55, 364. 1980.
38. Durie, B. G. M., Grogan, T. M.
CALLA-positive myeloma: an agressive subtype with poor survival.
Blood. 66, 229. 1985.
39. Durrant, J. R., Barry, W. E., Learner, N.
The changing face of myeloma.
Lancet. i, 119. 1966.
40. Ewing, M. R., Fotte, F. W. Jr.
Plasma-cell tumours of mouth and upper air passages.
Cancer. 5, 499. 1952.
41. Farrer-Brown, G.
„A classification of the non-Hodgkin's lymphomas".
XI. Internat. Congr. Internat. Acad. Path. Washington, D. C. 80. 1976.
42. Frangione, B., Franklin, E. C.
Heavy chain diseases: clinical features and molecular Significance of the disordered immunoglobulin structure.
Semin. Hematol. 10, 53. 1973.
43. Frizzera, G., Moran, E. M., Rappaport, H.
Angio-immunoblastic lymphadenopathy with dysproteinaemia.
Lancet, ii, 1070. 1974.
44. Gassmann, W., Pralle, H., Haferlach, T., Pandurevic, S., Graubner, M., Schmitz, N., Löffler, H.
Staging systems for multiple myeloma: a comparison.
Brit. J. Haematol. 59, 703. 1985.
45. Givler, R.
Lymphocytic leukemia with coexistent localized reticulum cell sarcoma.
Cancer. 21, 1184. 1968.
46. Gonzalez, F., Trujillo, J. M., Alexanian, R.
Acute leukemia in multiple myeloma.
Ann. intern. Med. 86, 440. 1977.
47. Hallen, J.
Frequency of „abnormal" serum globulins (M-components) in the aged.
Acta, med. scand. 173, 737. 1963.
48. Hayes, D. W., Bennett, W. A., Heck, F. J.
Extra-medullary lesions in multiple myeloma: review of literature and pathologic studies.
Arch. Pathol. 53, 262. 1952.
49. Heller, P., Yakulis, V., Bhoopalam, N., Costea, N., Cabana, V., Nathon, R. D.
Surface immunoglobulins on circulating lymphocytes in mouse and human plasmocytoma.
Trans. Ass. Amer. Physns. 85, 192. 1972.
50. Helmus, C.
Extramedullary plasmacytoma of the head and neck.
Laringoscope. 74, 553. 1964.
51. Hijmans, W., Schuit, H. R. E., Hulsing-Hesselink, E.
An immunofluorescence study on intracellular immunoglobulins in human bone marrow cells.
Ann. N. Y. Acad. Sci. 177, 920. 1971.
52. Hijmans, W., Schuit, H. R. E.
Immunofluorescence studies on immunoglobulins on the lymphoid cells of human peripheral blood.
Clin. exp. Immunol. 11, 483, 1972.
53. Hobbs, J. R.
Growth rates and responses to treatment in human myelomatosis.
Brit, J. Haematol. 16, 607. 1969.
54. Hobbs, J. R.
Monitoring myelomatosis.
Arch. Intern. Med. 135, 125. 1975.
55. Hofmann, W., Salmon S., Durie, B. G. M.
Drug resistance in multiple myeloma associated with high in vitro incorporation of H-thymidine.
Blood, 58, 471. 1981.
56. Hokanson, J. A., Brown, B. W., Thompson, J. R., Drewinko, B., Alexanian, R.
Tumor growth patterns in multiple myeloma.
Cancer. 39, 1077. 1977.
57. Hollard, D., Müller, J. M., Leger, J.
Association myeloma, leucose myeloide et lymphosarcome: Reflexions nosologiques.
Lyon, Med. 213, 967. 1965.
58. Holm, G., Mellstedt, H., Petterson, D., Biberfeld, P.
Idiotypic immunoglobulin structures on blood lymphocytes in human plasma cell myeloma.
Immunol. Rev. 34, 139. 1977.
59. Holt, J. M., Robb-Smith, A. H. T.
Multiple myeloma: development of plasma cell sarcoma during apparently successful chemotherapy.
J. Clin. Pathol. 26, 649. 1973.
60. Jákó, J.
Die idiopathische monoclonale Gammopathie.
Kongress für Laboratoriumsmedizin.
Berlin. május 15–19. 1977.
61. Jákó, J., Virág, Sz., Riesz, T., Juhász, J.
Plasmoblastoma.
Magyar Pathológusok Társasága Nagygyűlése.
Szeged. május 25–28. 1977.
62. Keiser, G., Veklinger, E., Virieux, C.
2 cases of chronic lymphatic leukemia and Hodgkins's disease.
Acta Haemat. 26, 29. 1961.
63. Klein, E., Eskeland, T., Inove, M., Strom, R., Johannson, B.
Surface immunoglobulin – moieties on lymphoid cells.
Exp. Cell. Res. 62, 133. 1970.
64. Klemm, D., Merker, H., Westerhausen, M.
Akute Neuproliferation bei *paraproteinaemischen* Hämoblastosen unter langdauernder Zytostatikabehandlung.
Dtsch. med. Wschr. 43, 2063. 1968.
65. Knapp, W., Schuit, H. R. E., Bolhuis, R. L. K., Hijmans, W.

Surface immunoglobulins in chronic lymphatic leukaemia, macroglobulinaemia and myelomatosis.
Clin. exp. Immunol. 16, 541. 1974.

66. Krauss, S., Sokal, J. E.
Paraproteinemia in the lymphomas.
Amer. J. Med. 40, 400. 1966.
67. Kyle, R. A., Pierre, R. V., Bayrd, E. D.
Multiple myeloma and acute myelomonocytic leukemia.
N. Engl. J. Med. 283, 1121. 1970.
68. Kyle, R. A., Maldonado, J. E., Bayrd, E. D.
Plasma cell leukemia: report on 17 cases.
Arch. intern. Med. 134, 133. 1974.
69. Kyle, R. A., Pierre, R. V., Bayrd, E. D.
Multiple myeloma and acute leukemia associated with alkylating agents.
Arch. intern. Med. 135, 185. 1975.
70. Kyle, R. A., Bayrd, E. D.
The monoclonal gammopathies.
Charles C. Thomas. Publisher USA. 148. 1976.
71. Kyle, R. A., Greipp, P. R.
Smoldering multiple myeloma.
N. Engl. J. Med. 302, 1347. 1980.
72. Labedzki, L., Fuchs, H. J., Grips, H.
Multiple myeloma and terminal acute myeloid leukemia.
Dtsch. med. Wschr. 101, 108. 1976.
73. Latreille, J., Barlogie, B., Dosik, G., Johnston, D. A., Drewinko, B., Alexanian, R.
Cellular DNA content as a marker of human multiple myeloma.
Blood. 55, 403. 1980.
74. Latreille, J., Barlogie, B., Johnston, D. A., Drewinko, B., Alexanian, R.
Ploidy and proliferative characteristics in monoclonal gammopathies.
Blood. 59, 43. 1982.
75. Law, I. P.
Kaposi sarcoma and plasma cell dyscrasia.
J. Amer. Med. Ass. 229, 1329. 1974.
76. Liang, W., Hopper, J. E., Rowley, J. D.
Karyotypic abnormalities and clinical aspects of patients with multiple myeloma and related paraproteinemic disorders.
Cancer, 44, 630. 1979.
77. Lindström, F. B., Hardy, W. R., Eberle, B. J., Williams, R. C. Jr.
Multiple myeloma and benign monoclonal gammopathy: differentation by immunofluorescence of lymphocytes.
Ann. intern. Med., 78, 837. 1973.
78. Lortholary, P., Boiron, M., Ripault, P.
Leucémie lymphoide chronique secondairement associée à une reticulopathie maligne, syndrome de Richter.
Nouv. Rev. franc. Hémat. 4, 621. 1964.
79. Marchal, G., Duhamel, G., Weill-Fage, J. C., Blamoutier, B.
Maladie de Hodgkin avec tableau initial de leucémie lymphoide et association de tuberculose.
Bull. Soc. Med. hop. Paris. 69, 734. 1953.
80. Mathé, G., Belpomme, D., Dantchev, D., Pouillart, P.
Progress in the classification of lymphoid and/or monocytoid leukemias and of lympho- and reticulosarcomas.
Biomedicine. 22, 177. 1975.
81. Medical Research Council's Working Party on Leukaemia in Adults:
Prognostic features in the third MRC myelomatosis trial.
Brit. J. Cancer. 42, 831. 1980.
82. Mellstedt, H., Hammarström, S., Holm, G.
Monoclonal lymphocyte population in human plasma cell myeloma.
Clin. exp. Immunol. 17, 371. 1974.
83. Mellstedt, H., Pettersson, D., Holm, G.
Monoclonal B-lymphocytes in peripherial blood of patients with plasma cell myeloma.
Scand. J. Haematol. 16, 112. 1976.
84. Merlini, G., Waldenström, J. G., Jayakar, S. D.
A new improved clinical staging system for multiple myeloma based on analysis of 123 treated patients.
Blood. 55, 1011. 1980.
85. Moore, D. F., Migliore, P. S., Schullenberger, C. C., Alexanian, R.
Monoclonal macroglobulinemia in malignant lymphoma.
Ann. intern. Med. 72, 43. 1970.
86. Naidu, R. I., Rosner, F.
Combined multiple myeloma and chronic lymphatic leukemia.
N. Engl. J. Med. 284, 108, 1971.
87. Narasimhan, P., Jagathambal, K., Elizalde, A. M., Rosner, F.
Chronic lymphocytic leukemia and lymphosarcoma associated with multiple myeloma.
Arch. intern. Med. 135, 729. 1975.
88. Okano, H., Azar, H. A., Osserman, E. F.
Plasmocytic reticulum cell sarcoma. Case report with electron microscopic studies.
Amer, J. clin. Path. 46, 546. 1966.
89. Ösby, E., Carlmark, B., Reizenstein, P.
Staging of myeloma. A preliminary study of staging factors and treatment in different stages.
Recent Results in Cancer Research. Lymphoid Neoplasias.
In: II. Clinical and Therapeutic Aspects
(Ed. by Mathe, G., Seligman, M., Tubiana, M.,
Springer, Berlin. 21, 1978.
90. Österberg, G., Rausing, A.
Reticulum cell sarcoma in Waldenström's macroglobulinemia after Chlorambucil treatment.
Acta med. scand. 188, 497. 1970.
91. Palutke, M., McDonald, J. M.
Monoclonal gammopathies associated with malignant lymphomas.
Amer. J. Clin. Pathol. 60, 157. 1973.
92. Paul, L. W., Pohle, E. A.
Solitary myeloma of bone.
Radiol. 35, 651. 1940.
93. Pazmantier, M. W., Azar, H. A.
Extraskeletal spread in multiple myeloma.
Cancer. 23, 167. 1969.
94. Pedraza, M. A.
Plasma-cell leukemia with unusual immunoglobulin abnormalities.
Amer. J. clin. Pathol. 64, 410. 1975.
95. Pennec, Y., Mottier, D., Youinou, P., Asselain, B., Chavance, M., Le Goff, P., Le Prise, P. Y., Miossec, P., Le Menn, G.
Critical study of staging in multiple myeloma.
Scand. J. Haematol. 30, 183. 1983.
96. Pernis, B., Formi, L., Amante, L.
Immunoglobulin spots on the surface of rabbit lymphocytes.
J. exp. Med. 132, 1001. 1970.
97. Pierson, J., Darley, T., Stevenson, G. T., Virji, M.
Monoclonal immunoglobulin light chains in urine of patients with lymphoma.
Brit. J. Cancer. 41, 681. 1980.
98. Polliack, A., Naparsteck, Elizabeth, Eldor, A., Ben Zvi, A., Ben Bassat, H., Zlotnick, A.
Multiple myeloma terminating in lymphocytic leukemia with B-lymphocyte membrane markers.
Amer. J. Haematol. 3, 153. 1977.
99. Porez-Soler, R., Estebon, R., Allende, E., Tornos-Salomo, C., Julia, A., Quardia, J.
Live involvement, in multiple myeloma.
Amer. J. Haematol. 20, 25. 1985.
100. Potter, M.
Experimental plasma cell tumors and other immunoglobulin – producing lymphoreticular neoplasms in mice.
In: Multiple myeloma and related disorders.
Ed: Azar, H. A., Potter, M.
Hagerstown, Maryland, Harper and Row Publishers 1, 177. 1973.
101. Preud'homme, J. L., Seligmann, M.

Surface-bound immunoglobulins as cell markers in human lymphoproliferative diseases.
Blood. 40, 77. 1972.

102. Rabellino, E., Colous, S., Grey, H. M., Unanue, E. R.
Immunoglobulins on the surface of lymphocytes.
I. Distribution and quantitation,
J. Exp. Med., 133, 156. 1971.

103. Raff, M. C., Sternberg, M., Taylor, R. B.
Immunoglobulin determinants on the surface of mouse lymphoid cells.
Nature. 225, 553. 1970.

104. Ralph, P., Prichard, J., Cohn, M.
Reticulum cell sarcoma: an effector cell in antibodydependent cell mediated immunity.
J. Immunol. 114, 898. 1975.

105. Rappaport, H.
"Tumors of the hematopoietic system",
In: "Atlas of tumor pathology"
Washington, D. C., AFIP. 1966.

106. Rausing, A.:
Lymphatic leukemia and malignant lymphoma in the adult.
Acta med. scand. suppl. 595. 1976.

107. Renau-Piqueras, J., Wetter, O., Miragall, F., Miquel, A., Hertenstein, Ch., Cervera, J., Brandhorst, D.
Extramedullary multiple myeloma:
A qualitative and quantitative electronmicroscopic study.
Virchows Arch. (Cell Pathol) 40, 171. 1982.

108. Riesz, T., Jákó, J., Juhász, J.
Second malignant tumors accompanied with primary hepatocellular carcinoma.
Acta Hepato-Gastroenterol. 26, 729. 1979.

109. Sacks, P. V., Tavassoli, M., Eastlund, D. T.
Simultaneous accurence of myeloma and Hodgkin's disease.
Acta Haematol. 55, 118. 1976.

110. Salmon, S. E., Smith, B.
Immunoglobulin synthesis and total body tumor cell number in IgG multiple myeloma.
J. Clin. Invest. 49, 1114. 1970.

111. Salmon, S. E., McIntyre, O. R., Ogawa, M.
IgE myeloma: total body tumor cell number and synthesis of IgE and DNA.
Blood. 37, 696. 1971.

112. Salmon, S. E.
Immunoglobulin synthesis and tumor kinetics of multiple myeloma.
Semin. Hematol. 10, 135. 1973.

113. Salmon, S. E.
In vitro cloning and chemosensitivity of human myeloma stem cells.
Clinics in Haematology. 11, 47. 1982.

114. Schabel, F. M.
The use of tumor growth kinetics in planning curative chemotherapy of advanced solid tumors.
Cancer Res. 29, 2384. 1967.

115. Schechter, I., Wolf, O., Kantor, F., Schecher, B., Burstein, Y.
Immunoglobulin precursors: structure, function, gene-protein correlation and evolution.
Annals N. Y. Acad. Sci. 343, 218. 1980.

116. Schultz, D. R., Yunis, A. A.
Immunoblastic lymphadenopathy with mixed cryoglobulinemia.
N. Eng. J. Med. 292, 8. 1975.

117. Shuster, M., Causing, W. C., Young, A.
Chronic lymphatic leukemia and lymphosarcoma terminating in multiple myeloma.
J. Med. Soc. N. Y. 68, 365. 1971.

118. Skipper, H. E.
Thoughts on cancer chemotherapy and combination modality therapy.
J. Amer. Med. Ass. 230, 1033. 1974.

119. Simpson-Herran, I. Lloyd, H. H.
Kinetic parameters and growth curves for experimental tumor systems.
Cancer Chemother. Rep. 54, 143. 1970.

120. Snapper, I., Turner, L. B., Muscovitz, H. L.
Multiple myeloma.
Grune et Stratton, New York. 13. 49. 104. 109. 1953.

121. Spiegelber, H. L., Heath, V. C., Lang, J. E.
IgG half-molecules: clinical and immunologic features in a patient with plasma cell leukemia.
Blood. 45, 305. 1975.

122. Stobbe, H.
Syntropie von chronischer Lymphadenose und Plasmocytom.
Acta Haematol. 27, 180. 1962.

123. Sugai, S., Hirose, Y., Konaka, Y., Konda, S.
Leukemic lymphosarcoma (LLS) with monoclonal IgM: Idiotypic specificity on the cell surface and in the cytoplasm of lymphosarcoma cells.
Blood. 52, 922. 1978.

124. Sullivan, P. W., Salmon, S. E.
Kinetics of tumor growth and regression in IgG multiple myeloma.
J. Clin. Invest. 51, 1967. 1972.

125. Teodorescu, M., Mayer, E. P.
Surface immunoglobulin in immunproliferative diseases.
Ann. clin. Lab. Sci. 8, 353. 1978.

126. Togawa, A., Inove, N., Hiyamoto, K., Hyodo, H., Namba, M.
Establishment and characterization of a human myeloma cell line (KMM-1)
Int. J. Cancer. 29, 495, 1982.

127. Thomas, F. B., Clavsen, K. P., Greenberg, N. J.
Liver disease in multiple myeloma.
Arch. Intern. Med. 132, 195. 1973.

128. Traganos, F., Darzynkiewicz, Z., Sharpless, T., Melamed, M. R.
Simultaneous staining of ribonucleic and deoxyribonucleic acids in unfixed cells using acridine orange in a flow cytofluorometric system.
J. Histochem. Cytochem. 25, 46. 1977.

129. Vander, J. B., Johnson, H. A.
Chronic lymphocytic leukemia and multiple myeloma in the same patient.
Arch. intern. Med. 53, 1052. 1960.

130. Vercelli, D., Cozzolino, F., Di Guglielmo, R.
A comparison of two staging systems for myeloma.
Nouv. Rev. Franc. d'Hématol. 23, 107. 1981.

131. Warner, T. F. C. S., Krueger, R. G.
Circulating lymphocytes and the spread of myeloma.
Lancet. i. 1174. 1978.

132. Watanabe, S. H., Shimosato, Y., Shimoyama, M., Minato, K.
Studies with multiple markers on malignant lymphomas and lymphoid leukemias.
Cancer. 50, 2372, 1982.

133. Wegelius, O., Skrifvars, B., Andersson, L.
Rheumatoid arthritis terminating in plasmacytoma.
Acta med. scand. 187, 133. 1970.

134. Wildhack, R.
Über gleichzeitiges Vorkommen von Leukose und Lymphogranulomatose.
Folia haemat. Neue Folge. 7, 303. 1963.

135. Willis, R. A.
Solitary plasmocytoma of bone.
J. Path. Bact. 53, 77. 1941.

136. Willis, R. A.
Pathology of tumours.
Butterworth and Co., London. 106, 1948.

137. Wiltshaw, E.
The natural history of extramedullary plasmacytoma and its relation to solitary myeloma of bone and myelomatosis.
Medicine. 55, 217. 1976.

138. Woodruff, R. K. Wadsworth, J., Malpas, J. S., Tobias, J. S.
Clinical staging in multiple myeloma.
Brit. J. Haematol. 42, 199. 1979.

# IX. Prognosis and epidemiology of plasmocytomas

The introduction of the alkylating agents was a milestone in the therapy and prolongation of the post-diagnosis survival of the patients with plasmocytoma. The survival rate has been expanded from 6-12 months (32, 57) to 30-36 months (29). Numerous other mono- and polychemotherapy protocols failed to statistically significantly further prolong the survival rate (3, 9). The promising exceptionally beneficial results of some schemes could not, however, be adequately controlled. The analysis of the beneficial results and fiascos made the specialists *first* to look for a single (?) factor responsible for the prolongation of the survival rate and then to determine those clinical and laboratory tests, which can be applied for the evaluation of the tumour mass.

Empiric research of the plasma cells growth, identification of the abnormal changes responsible for the enhanced intracellular activity, as well as the studies of extracellular factors seem worthwhile. The other aspect of the current research is to study the derangements in different organs (primarily kidney deterioration) induced by complications of the main disease. The efficacy of the applied therapy is also extensively studied (for Stage Classification and critical remarks, see Chapter VIII).

## Proliferating tumour plasmocytes

As already stated above, the extent of *the bone marrow infiltration* is a factor of utmost importance (19, 50). Histometric and morphometric studies of the bone marrow biopsy smears are the most relevant methods of investigation (7, 36), for they provide not only the quantitative and distribution data, but also give the following information about the cell maturity:

- mature differentiated cells with low mitotic index;
- blasts with round or scarcely identified nucleolus;
- polymorphic, multinuclear plasmocytes with mitotic index.

Moreover, these methods also indicate *the extent of the marrow fibrosis,* which shows inverse relationship with the survival rate, but exhibits no significant correlation with the morphology of the plasma cells (though pronounced fibrosis frequently occurs in the blasts).

Histometric and morphometric analysis identifies three forms *of cell proliferation,* each form indicating worse prognosis:

- interstitial
- interstitial and nodular
- compact bone marrow.

The prognosis is worsened in the above presented order.

Apart from the above mentioned, aspiration cytology can also be used for prognostic purposes (33, 35).

The concept that the increasing tumor mass means an increasing load to the organism is not new, therefore, the efforts of the researchers were focused on quantitative evaluation of this parameter. The first successful result in this field is the proved linear correlation between the tumor weight and the paraprotein concentration in mouse plasmocytoma (55). However, human interpolations of this model did not hold true, due to the heterogeneity of the protein synthetic capacity of the clones and the differences in the rates of Ig catabolism (60). Based on approximate calculations, it can be accepted that by the time of clinical diagnosis the bone marrow comprises about $0.5x10^{12}$ (?) tumor cells (0.5 kg), their number increasing up to $3x10^{12}$ by the time of death (66). The efficacy of therapy is evaluated on the basis of the tumor mass decrease. The analysis of the residual tumor behaviour indicates that the tumor cells are heterogeneous, some of them being drug-resistant, others -

drug-sensitive, with the ratio of $1:10^6$ or $1:10^7$ (65). A 50% decrease of the IgG M-component means a 75% reduction of the tumor mass (4). Evidently, such therapy, which would markedly depress the tumor spreading, would result in better survival; still it is worth pointing out, that the therapy inducing fast regression, leads to short remission periods (1-2 months) with subsequent relapses (24). The susceptibility of tumors to therapy can be classified as follows:

| | |
|---|---|
| 1. Uneffective therapy that fails to affect the rate of the M-component synthesis; furthermore, the paraprotein concentration increases and the death occurs early | 5%. |
| 2. Fast remission followed by a fast relapse, short life span | 17%. |
| 3. Slow, but pronounced reduction of the tumor mass indicates good prognosis | 65%. |
| 4. Minor changes, unpredictable remission period, unexpected relapse (usually LCD), limited life span | 13% |
| TOTAL: | 100% (38). |

The proneness of plasmocytomas to the constantly developing therapy remains unchanged, indicating some different cell biological properties of the tumor cells. Tumor kinetic studies of the above mentioned findings also supported by others strongly indicate that the tumor cell clone consists of *two populations* with basically different properties (proliferating and non-proliferating), which behave as successors of a single clone (24). The need for accurate methods has lead to the introduction of the short-term cultures of proliferating plasma cells. It was assumed that the rate of paraprotein synthesis might serve as an indicator of their malignity (60). This concept is, however, erroneous, since the non-secreting or hyposecreting myelomas are usually more malignant than the hypersecreting ones.

The presented stage classification systems are widely applied in the literature. It is stressed that these results are in good *statistical* correlation with the survival rate (1, 70). However, *individual* data show marked deviations, which can not be explained by the size of the tumor mass (34). Therefore, there is no straight answer to this question, and it seems that the plasma cell mass can not be considered as a characteristic factor.

The kinetics of the tumor cell growth has been the next step in this field. $H^3$-thymidin uptake in tumor cell cultures indicates that the fast rate of DNA synthesis suggests poor prognosis, whereas the slow rate is typical of MGUS (13) or indolent myeloma (2, 45). These results were independent of the tumor mass (26). The cells with fast DNA metabolic rate are not homogeneous, but consist of a drug-sensitive and a drug-resistant population (39). During the following decade after that observation no attempts have been made to use these properties in therapy. It has also been found that the cultured plasma cells with fast thymidin uptake show either homogeneous distribution, or form colonies or "nests" (30). The latter property is considered as a sign of increased aggressiveness (10). Cellsorter was used to detect anomalies of the DNA metabolism: aneuploid cells were found in 75% of the myeloma cases; the presence of hyperdiploid cells unambiguously indicated poor prognosis (17). In contrast to the cellsorter studies, traditional chromosome analysis seldom revealed hypodiploid cells (71). This controversy is still unexplained.

It generally accepted, that the presence and heterogeneity of chromosomal aberrations indicates low survival rate (48), though data proving the opposite are also available.

Assessment of $\beta_2$-microglobulin is another method for evaluation of the proliferating cell mass and the outcome of the disease. $\beta_2$-microglobulin, as a small-sized structural element of the histocompatibility antigen in the cell membrane, is excreted by the kidney, therefore, its role must be considered if the renal function is impaired. determination of the $\beta_2$-microglobulin excreted in urine is a valuable method for the evaluation of tumor mass and the prognosis of the disease (8, 56, 63).

Monoclonal immunoglobulins as "markers" produced by the proliferating cells must be treated with special attention. The first research data considered high concentrations of plasma paraproteins as an indicator of malignancy. Further studies, as well as our data, failed to support this view. It is accepted that the paraprotein titers have no prognostic value (16).

No uniform view exists in the literature with respect to the role of the monoclonally produced immunoglobulins of different classes.

IgG gammopathy being a homogeneous (?) group shows the same proneness to therapy as the others (11, 50). Different subclasses, however, have different properties, the $IgG_3$ subtype being *the most nephrotoxic one* (64). In the course of the following 16 years this observation has neither been supported, nor opposed, probably because no further studies have been made in this field.

In the past the poor prognosis of the IgA-type myeloma stressed in the literature (25, 49) has been generally accepted, but recent investigations failed to confirm it (16, 46).

It is generally accepted that the life span of the patients with IgD monoclonal immunoglobulins is rather short - their survival rate is about 9 months (42).

Though the incidence of Kahler disease with IgE markers is extremely low, there are no data about its prognosis. Some researchers report short survival (31).

Plasmocytomas producing and excreting light chains only (LCD) are also of malignant character (6). The comparative studies of the kappa and lambda types unambiguously indicates poorer outcome for the latter (26, 43). It must be stressed that the light chain excretion does not necessarily mean LCD, for it can be a concomitant feature of any monoclonal gammopathy (11).

Renal failure is a main complication of the malignant proliferation of the plasma cells. The impaired renal function is clearly indicated by the abnormal values BUN. Renal damages are independ of the tumor mass, but play a major role in the outcome of the disease.

The following individual or complex mechanisms contribute to the development of anaemia: the mass of the plasma cells infiltrating the bone marrow with its direct and indirect effect; the extent of the kidney lesions; cytostatic therapy; stabilization of the tumor mass, the effect of α-interferon.

The extent of the bone lesions undoubtedly depends on the secretion of an osteoclast-stimulating factor by the proliferating malignant plasmocytes and shows no correlation with the tumor mass or the progress of the disease (27).

Data of the literature indicate that hypercalcemia occurs in 30% of the cases (44). Our analysis of twice as many patients could reveal hypercalcemia only 2.7% (41 cases studied). We could not explain this controversy. Our data support the general conclusion of other researchers, that the Ca level has a minor prognostic value (53, 54).

Changes of the Se albumin level are noteworthy. A reciprocal correlation exists between the albumin and paraprotein concentrations: elevation of the monoclonal gammaglobulin is accompanied by a decreased albumin level. In those forms of LCD, where no light chains are secreted into plasma (or their level is very low) the albumin concentration is significantly higher than in typical cases of paraproteinemia (50).

Similar to other parameters, there are multiple explanations of the increased Se uric acid level, for example the increased turnover of the tumor cells, deranged renal excretion capacity, cell-destructive effect of the cytostatics. Finally, the developing hyperuricemia damages the kidney (54).

The sensitivity to the applied therapy is one of the most important, though hardly evaluated, prognostic factors, for it can not be introduced into neither of the Stage Classifications. It is worth pointing out that the utmost resistance to therapy is typical of the final stages of the disease. Thus, positive results can be expected only in the cases of early diagnosis, and vice versa, beneficial effect of therapy precludes the final stage of the disease. Efficacy of therapy is frequently interpreted even by recognized researchers in terms of decreased paraprotein concentration (47). We used the same therapy protocol as presented in the publications, but failed to observe the cessation of monoclonal gammopathy or a titer decrease of more than 30%! The observed survival corresponded to the average European data.

It is well known that in cases with unpredictable course a fast regression of plasmocytoma implies early relapse and death (38). Though this finding is in controversy with the above data, we must consider this rare, but real situation.

The efficacy of the therapy as another prognostically important factor has inspired the researchers to test the drug-sensitivity of the cultured cells (37). 95% of the soft agar cultures are suitable for prognosis of drug resistance and 79% - for prognosis of a positive clinical course (61, 62).

Summary: the average survival rate of patients with plasmocytoma is 2-3 years. However these mean values imply survival ranging from several weeks to a decade. Despite the extreme importance of prognostic factors, the fate of a patient also depends on *the optimum selection and use* of the therapy and supportive treatment.

Very few data are available on the causes of the multiple myeloma, probably because the research in this field is insufficient. The intensification of the research is, however, expected, due to the increasing myeloma mortality statistics. It is assumed that this trend might be either due to the effect of a new unknown agent, or to the development of more accurate methods of study. Moreover, an increasing number of patients admitted to hospitals might also contribute to the accuracy of the statistical data (21, 58).

In 1980 in Europe plasmocytoma had comprised 1% of all the malignant pathology, and 10% of the malignant hematological diseases (12).

Between 1950 and 1970 in the UK plasmocytoma mortality showed an increasing tendency, so it was not surprising that in 1982 there had been 1613 victims of this disease, yielding 1.2% of all cases of malignancy and 20% (!!) of the malignant hematological diseases. Statistical analysis in different countries indicates that the mortality rate in Europe is 3.3/100,000 for females, and 2.5/100,000 for males (69).

The incidence of the plasma cell malignancy in Negroes, involving the younger age groups is higher than in the white population (59, 68). The opposite situation is observed for the orientals: their morbidity and mortality is markedly lower, compared to the Europeans. It is interesting to mention, that the emigrants (for example in the USA) from Africa and the Oriental countries maintain their typical (high or low) mortality rate, suggesting involvement of genetic mechanisms. Another proof in favor of this view is the familial occurrence of the disease (12, 14, 15).

Evidence exists that radiological load plays a leading role in the development of plasmocytomas (40). Among the population of Hirosima and Nagasaki, a risk factor of 4.7 was observed for those irradiated with more than 100 rad, and a factor of 1.5, if the load ranged between 1-100 rad. The disease has usually developed in 15 years (20). Statistical analysis of the women irradiated for cervix carcinoma gave similar results (with respect to doses); furthermore, intracavital irradiation seems to be more dangerous in terms of the later developing malignancy. Considering the above data, it must be strongly emphasized that the 15-20 -year results of the local and the Middle-European statistics of the Chernobil disaster must be thoroughly evaluated.

Occupational testing of workers dealing with chemical agents is also of importance. Asbestos, arsenic, cooling oil, products of petroleum chemistry, plastics are on the list of the suspected. Agricultural and timber workers, tanners also belong to the risk group (18, 22, 23, 51, 52).

Proliferating malignant plasma cells belong to the clone escaped from under the normal immune control. The exact mechanisms or disfunctions finally leading to the development of myeloma multiplex are still unknown. A constant antigen stimulus, an abnormal antigen presentation or both can first result in a polyclonal immunoglobulin production (IMG). The thorough long-lasting follow-up of these patients is the duty of a clinician.

In summary, similar to other malignancies, the etiology of the plasma cell tumors is unknown, though the presented possibilities seem plausible. The racial differences or the observed relationship between rheumatoid arthritis and MG (41) are interesting examples of experimentum naturae. The researchers should focus their attention on this pathology.

## References


1. Alexanian, R., Balberzak, S., Bonnet, J. D., Geham, E. A. Haut, A., Hewlett, J. S., Monto, R. W.
Prognostic factors in multiple myeloma.
Cancer. 36, 1192. 1975.
2. Alexanian, R.
Localised and indolent myeloma.
Blood. 56, 521. 1980.
3. Alexanian, R., Dreicer, R.
Chemotherapy for multiple myeloma.
Cancer. 55, 583. 1984.
4. Alexanian, R.
Treatment of multiple myeloma.
Acta Haematol. 63, 237. 1980.
5. Barlogie, B., Latrielle, J., Alexanian, R., Swartzengruber, D. E., Smallwood, L., Maddox, A. M., Raker, M. N., Drewinko, B.
Quantitative cytology in myeloma research.
Clin. Haematol. 11, 19. 1982.
6. Bartl, R., Frisch, B., Burkhardt, R.
Multiple myeloma.
Bone Marrow Biopsies Revisited.
Karger. Basel. 42. 1982.
7. Bartl, R., Frisch, B., Burkhardt, R., Fateh- Moghadam, A., Mahl, G., Gierster, P., Sund, M., Kettner, G.
Bone marrow histology in myeloma: its importance in diagnosis, prognosis, classification and staging.
Brit. J. Haematol. 51, 361. 1982.
8. Bataille, R., Durie, B. G. M., Grenier, J.
Serum béta-2 microglobulin and survival duration in multiple myeloma: a simple reliable marker for staging.
Brit. J. Haematol. 55, 439. 1983.
9. Bergsagel, D. E.
Progress in the treatment of plasma cell myeloma.
J. Clin. Oncol. 1, 510. 1983.
10. Bergsagel, D. E.
Prognostic factors in plasma cell myeloma.
Eur. J. Canc. Clin. Oncol. 19, 1493. 1983.
11. Bettini, R., Steidl, L., Rapazzani, P., Giardina, G.
Prognostic value of the staging system proposed by Merlini, Waldenström and Jayakar for multiple myeloma.
Acta Haematol. 70, 378. 1983.
12. Blattner, W. A.
Epidemiology of multiple myeloma and related plasma cell disorders: an analytic review.
In: Progress in myeloma.
Ed: by Potter, M.
Elsevier North-Holland. N. Y. 1, 1980.
13. Boccadoro, M., Gavarotti, P., Fossati, L. Pileri, A., Marmont, F., Nerotto, G., Gallamino, A., Volta, C., Tribalto, M., Testa, M. G., Amadori, S., Mandelli, F., Durie, B. G. M.
Low plasma cell 3(H) thymidine incorporation in monoclonal gammopathy of undetermined significance (MGUS), smouldering myeloma and remission phase myeloma: a relable indicator of patients not requiring therapy.
Brit. J. Haematol. 58, 689. 1984.
14. Boga, Marianna, Jákó, J., Domán, J., Magyar, Éva, Konyár, E.
Familial myeloma.
Folia Haematol. 3, 201. 1973.
15. Bourguet, C. Claire, Grufferman, S., Delzell, Elizabeth, DeLong, R. Elizabeth, Cohen, H. J.
Multiple myeloma and family history of cancer.
Cancer. 56, 2133. 1985.
16. Buckman, R., Guzick, J., Galton, D. A. G.
Long term survival in myelomatosis.
Brit. J. Haematol. 52, 589. 1982.
17. Bunn, P. A., Krasnow, S., Makuch, R. W., Selam, M. L., Schechter, G. P.
Flow cytometric analysis of DNA content of bone marrow cells in patients with plasma cell myeloma: clinical implications.
Blood. 59, 528. 1982.
18. Burmeister, L. F., Burmeister G. D., Van Lier, S. F., Isacson, P.
Selected cancer mortality and farm practices in Iowa.
Amer. J. Epidemiol. 118, 72. 1983.
19. Cavo, V., Baccarani, M., Gobbi, A. Lipizer, A., Tura, S.
Prognostic value of bone marrow plasma cell infiltration in stage I multiple myeloma.
Brit. J. Haematol. 55, 683. 1983.
20. Cuzick, J.
Radiation-induced myelomatosis.
N. Engl. J. Med. 304, 204. 1981.
21. Cuzick, J., Velez, R., Doll, R.
International variations and temporal trends in mortality from multiple myeloma.
Internat. J. Cancer. 32, 13. 1983.
22. Decoufle, P., Stanislawczyk, K., Houten, L.
A retrospective survey of cancer in relation to occupation.
DHEW (NIOSH) Publication. 77, 1977.
23. Dorken, H., Vollmer, I.
The epidemiology of multiple myeloma – investigation of 149 cases.
Arch. Geschwulstforsch. 31, 18. 1968.
24. Drewinko, B., Alexanian, R., Boyer, H.
The growth fraction of human myeloma cells.
Blood. 57, 333. 1981.
25. Durie, B. G. M., Salmon, S. E.
A clinical staging system for multiple myeloma.
Cancer. 36, 842. 1975.
26. Durie, B. G. M., Salmon, S. E., Moon, T. E.
Pretreatment tumour mass, cell kinetics and prognosis in multiple myeloma.
Blood. 55, 364. 1980.
27. Durie, B. G. M., Salmon, S. E., Murdy, G. R.
Relation of osteoclast growth: interrelationships between drug sensitivity, cell kinetics and activating factor production to extent of bone disease in multiple patient survival duration.
Blood. 61, 929. 1981.
28. Durie, B. G. M.
Staging and kinetics of multiple myeloma.
Clinics in Haematol. 11, 3. 1982.
29. Durie, B. G. M., Salmon, S. E.
The current status and future prospects of treatment for multiple myeloma.
Clinics in Haematol. 11, 181. 1982.
30. Durie, B. G. M., Young, L. A., Salmon, S. E.
Human myeloma in vitro colony myeloma.
Brit. J. Haematol. 47, 21. 1983.
31. Endo, T., Okumura, H., Kikuchi, K., Munakata, J., Otake, M., Nomura, T., Asakawa, H.
Immunoglobulin E (IgE) multiple myeloma. A case report and review of the literature.
Amer. J. Med. 70, 1127. 1981.
32. Feinleib, M., MacMahon, B.
Duration of survival in multiple myeloma.
J. Nat. Cancer Inst. 24, 1259. 1960.
33. Fritz, E., Ludwig, H., Kundi, M.
Prognostic relevance of cellular morphology in multiple myeloma.
Blood. 63, 1072. 1984.
34. Gassmann, W., Pralle, H., Haferlach, T., Pandurevič, S., Graubner, M., Schmitz, N., Loffler, H.
Staging system for multiple myeloma: a comparison.
Brit. J. Haematol. 59, 703. 1968.
35. Grepp, P. R., Raymond, N. M., Kyle, R. A., O'Fallon, W. M.
Multiple myeloma: significance of plasmablastic subtype in morphological classification.
Blood. 63, 305. 1985.
36. Gramont, A., Benitez, O., Smadja, N., Brissand, P., Sirenelli, A., Hubert, D., Krulik, M., Debray, J.
Bone marrow plasma cell infiltration in multiple myeloma.
Brit. J. Haematol. 57, 351. 1984.
37. Hamburger, A. W., Salmon, S. E.
Primary bioassay of human myeloma stem cells.
J. Clin. Invest. 60, 846. 1977.
38. Hobbs, J. R.
Growth rates and response to treatment in human myelomatosis.

Brit. J. Haematol. 16, 607. 1969.
39. Hoffman, V., Salmon, S. E., Durie, B. G. M.
Drug resistance in multiple myeloma associated with high in vitro incorporation of 3(H) thymidine.
Blood. 58, 471. 1981.
40. Ichimaru, M., Ishimaru, T., Mikami, M.
Multiple myeloma among atomic bomb survivors in Hiroshima and Nagasaki by dose 1950–1976.
Radiation Effects Research Report, Technical Report. 9. 1979.
41. Isomaki, H. A., Hakulinen, T.,·Joutsenlahti, U.
Excess risk of lymphoma, leukaemia and myeloma in patients with rheumatoid arthritis.
J. Chronic. Dis. 31, 691. 1978.
42. Jancelwiez, Z., Talatsuki, K., Sugai, S., Pruzanski, W.
IgD multiple myeloma.
Arch. Intern. Med. 135, 87. 1975.
43. Jákó, J., Virágh, Sz., Schopper, J., Sebők, J., Juhász, J.
Nierenerkrankungen bei monoclonalen Gammopathien – eine klinisch–pathologische Studie.
Immun. Infekt. 9, 60. 1981.
44. Kyle, R. A.
Multiple myeloma: a review of 869 cases.
Mayo Clinic Proc. 50, 29. 1975.
45. Kyle, R. A., Grepp, P. R.
Smouldering multiple myeloma.
N. Engl. J. Med. 302, 1374. 1980.
46. Kyle, R. A.
Long term survival in multiple myeloma.
N. Engl. J. Med. 308, 314. 1983.
47. Lee, B. J., Myers, J.
Long term survival in myeloma.
N. Engl. J. Med. 308, 243. 1983.
48. Liang, W., Hopper, J. E., Rowley, J. D.
Karyotypic abnormalities and clinical aspects of patients with multiple myeloma and related paraproteinaemic diseases.
Cancer. 44, 630. 1979.
49. Matzner, Y., Benbassat, J., Polliack, A.
Prognostic factors in multiple myeloma.
Acta Haematol. 60, 257. 1978.
50. Merlini, G., Waldenström, J. G., Jayakar, S. D.
A new improved clinical staging system for multiple myeloma based on analysis of 123 treated patients.
Blood. 55, 1011, 1980.
51. Milham, S.
Leukemia and multiple myeloma in farmers.
Amer. J. Epidemiol. 94, 307. 1971.
52. Milham, S.
Occupational mortality in Washington State, 1950–1971.
DHEW (NIOSH) Publication 76, 1976.
53. MRC Working Party on Leukaemia in Adults.
Report on the first myelomatosis trial.
part 1. Analysis of presenting features of prognostic importance.
Brit. J. Haematol. 24, 123. 1973.
54. MRC Working Party of Leukaemia in Adults.
Prognostic features in the third myelomatosis trial.
Brit. J. Cancer. 42, 831. 1980.
55. Nathans, D., Fahey, J. L., Potter, M.
The formation of myeloma protein by a mouse plasma cell tumour.
J. Exp. Med. 108, 121. 1958.
56. Norfolk, D., Child, J. A., Cooper, E. H., Kerruish, S., Milford Ward, A.
Serum beta-2 microglobulin in myelomatosis; potential value in stratification and monitoring.
Brit. J. Cancer. 42, 510. 1980.
57. Osgood, E. E.
The survival time of patients with plasmacytic myeloma.
Cancer Chemotherapy Reports. 9, 1. 1960.
58. Osmond, C., Gardner, M. J.
Age, period and cohort models applied to cancer mortality rates.
Statistical Medicine. 1, 245. 1982.
59. Otieno, L. S., Ogada, T.
Multiple myeloma at Kenyatta National Hospital, 1973–1976.
East African Med. J. 54, 574. 1977.
60. Salmon, S. E., Smith, B. A.
Immunoglobulin synthesis and total body tumour cell number in IgG multiple myeloma.
J. Clin. Invest. 49, 1114. 1970.
61. Salmon, S. E., Hamburger, A. W., Soehnlen, B. J., Durie, B. G. M., Alberts, D. S., Moon, T. E.
Quantitation of differential sensitivity of human tumour stem cells to anticancer drugs.
N. Engl. J. Med. 298, 1321. 1978.
62. Salmon, S. E.
Cloning and chemosensitivity of myeloma stem cells.
Clin. Haematol. 11, 47. 1982.
63. Scarffe, H., Anderson, H., Palmer, M. K., Crowther, D.
Prognostic significance of pretreatment serum beta-2 microglobulin levels in multiple myeloma.
Eur. J. Cancer Clin. Oncol. 19, 1361. 1983.
64. Schur, P. H., Kyle, R. A., Block, K. J., Hammack, W. J., Rivers, S. L., Sargent, A., Ritchie, R. F., McIntyre, O. R., Moloney, W. C., Wolfson, L.
IgG subclasses: relationship to clinical aspects of multiple myeloma and frequency distribution among M-components.
Scand. J. Haematol. 12, 60. 1974.
65. Skipper, H. E.
Thoughts on cancer chemotherapy and combination modality therapy.
J. Amer. Med. Ass. 230, 1033. 1974.
66. Sullivan, P. W., Salmon, S. E.
Kinetics of tumours growth and regression in IgG multiple myeloma.
J. Clin. Invest. 51, 1697. 1972
67. Velez, R., Beral, V., Cuzick, J.
Increasing trends of multiple myeloma mortality in England and Wales; 1950–79; are the changes real?
J. Nat. Cancer Inst. 69, 387. 1982.
68. Waldbaum, B., Gelfand, M.
Myelomatosis in the Rhodesian African.
Tropical Geographical Med. 26, 26. 1974.
69. Waterhouse, J. Muir, C., Shanmugaratnam, K., Powell, J.
Cancer incidence in five continents, vol IV.
IARC publication No 42. Lyon. 1982.
70. Woodruff, R. K., Wodsworth, J., Malpass, J. S., Tobias, J. S.
Clinical staging of multiple myeloma.
Brit. 7. Haematol. 42. 199. 1979.
71. Wurster-Hill, D. H., McIntyre, O. R., Cornwell, G. G.
Chromosome studies in myeloma.
Virch. Arch. 'B Cell Path. 29, 93. 1978.

# X. Forms and causes of coagulation disturbances in monoclonal gammopathies

The hemostasis disturbances occurring in different types of monoclonal gammopathies are well documented (4, 25, 32, 35); they can be assessed into three groups, however, the clinical character of the hemorrhagic symptoms is usually multicausal suggesting certain relationship between the groups:

a) interactions between the blood clotting factors;
b) mechanisms interfering with the platelet function;
c) pathological factors increasing capillary permeability.

ad a) Impaired conversion of Factor I (fibrinogen) into fibrin (14). "Competitive inhibition" is assumed in the background of this pathology: monoclonal antibodies abolish the end-to-end or side-to-side binding and aggregation of the fibrinogen and fibrin molecules (10, 24, 35), leading to formation of an unstable clot. The increased $Ca^{++}$ concentration compensates for the prolonged reptilase and thrombine time (10).

Inhibition of tissue thromboplastin (Factor III) activity is also reported (18). It is probably due to the direct or indirect (hapten) antibody effect of paraproteins. Furthermore, a certain absorptive binding is also assumed for this inhibitory effect. Inactivation of the antihemophilic globulin (Factor VIII, AHG) is due to an antigen-antibody reaction (7, 15).

The mentioned disturbances are typical of the IgM MG cases. Prothrombin (Factor II), accelerator globulin (Factor V), and proconvertin (Factor VII) form loose bonds with pathological immunoglobulins in IgA gammopathies causing coagulopathies (15, 37, 42).

Paraproteins also affect the physiological anticoagulant (antithrombin) system (13); in these cases clinical symptoms of chronic DIC occur (17, 41). Besides the clotting factors, other mechanisms can also be involved in the pathological process. In a case of treated multiple myeloma a cumarin necrosis-like cutaneous symptoms were observed. In this aged male patient with an $IgG_2$ λ-gammopathy the uniform immunoglobulins were proved to form complexes with protein-C. The extremely low titers of this important coagulation regulator protein had lead to the development of skin necrosis in the terminal phase of the disease (16) (Fig. X/1).

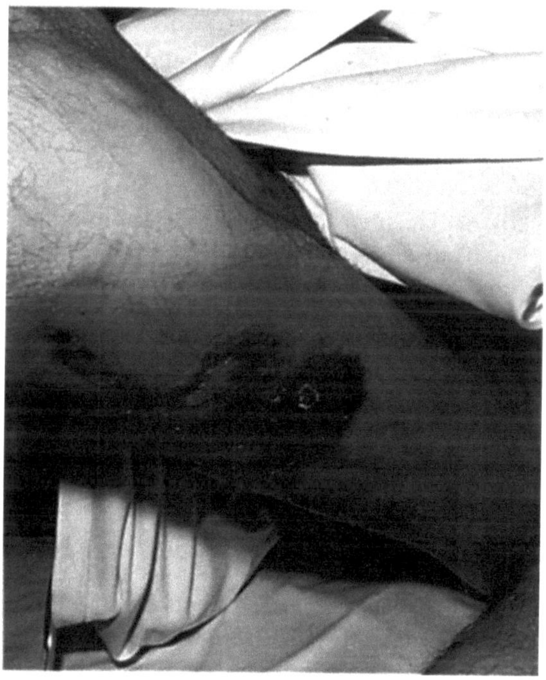

**Fig. X/1**

ad b) *Increased* or *decreased* platelet function both occur in disproteinemias.

1. Enhanced aggregation:
   - rare cases of direct anti-platelet antibodies usually occurring in IgM type monoclonal gammopathy (38).
   - antibodies showing no direct reactions with platelets, but affecting aggregation via the complex formation with platelet-fibrinogen (platelets contain substantial

amounts of plasma proteins, including fibrinogen) (14).
- platelets with high adhesion capability adhere on the high avidity surfaces of the large, cross-linked circulating immune complexes. This mechanism of coprecipitation (31) induces thrombocytopenia (30).

2. Decreased aggregation:
- An opposite reaction, when the IgM or IgG polymers bind to the platelets preventing their aggregation (12, 39). This process is frequently observed under pathological conditions. The prolonged bleeding time might also be due to a decreased activity of the platelet-3-factor. Furthermore, the extent of platelet retraction is also diminished. In contrast to the data of the literature, we failed to find any relationship between the concentration of the monoclonal immunoglobulins and the impairment of the platelet function (26).

ad. c) The mechanism of the capillary permeability changes is still unknown. The first step is the edematous loosening of the capillary walls. Since the nutrition of the endothelium occurs from the luminal site, trophic damages develop.
Consequently, the amount, composition and the direction of the fluid flux is altered (11).

Not only the passive functions of the vascular wall, but also the active - RES - properties are simultaneously impaired (23). With respect to the mechanism of this pathological process, endothelial cells cannot eliminate the growing "supply" of the immune complexes. The latter, stored in the endothelial cells triggers a release of vasoactive substances with the concomitant increase of the vascular permeability (9, 19, 20).

Basophils and mastocytes stimulated by the biogenic amines release the platelet activating factor (PAF), which would maintain the vicious circle by promoting the liberation of serotonin and histamine (5, 28).

Clinically the vascular permeability changes are usually accompanied by the expanded plasma volume, wich might *partly* contribute to hypervolemia observed in MG. In other cases plasma fluid flows into the interstitium, increasing the Htc value to 82% and eliciting shock. This state, well documented in the literature is termed the "sytem capillary leak syndrome" (2).

The above mentioned reactions are accepted and widely recognized coagulation damages occurring in gammopathies. However in other rare coagulapathies heparan-sulphate was responsible for the occurring hemorrhage (22, 33, 34). In these cases the prolonged prothrombin time correlated with the protamine sulphate levels (13, 33) (with toluidin blue) (22). The action mechanism of this proteoglycan named heperan-sulphate is similar to the commercially available heparin, both increasing the antithrombin-III activity. Paraproteins do not cause anticoagulation, for no clotting inhibitors can be extracted from these malignant plasma cells. This finding indicate that gammopathies elicit neither direct, not indirect effect on the blood coagulation system (8).

In minor permeability changes the process is usually localized in the skin and extravasation of the whole blood leading to typical purpura occurs (Fig. X/2).

It has already been mentioned in the Introduction, that besides the accumulation of monoclonal IgM, inhibition of platelet aggregation contributes to the development of the permeability changes (1), as illustrated by the following case:

B.I., male patient aged 64. Three years ago first noticed pin-head sized or larger confluent red spots on the hind limbs (21) (Fig. X/2).

These bright-red cutaneous symptoms usually persisted for 2-3 days and gradually disappeared leaving brownish spots behind. Confirming the patient's observation, we had also verified that the symptoms have always developed after consumption of certain types of canned food or beverages.

Acetyl-salicilic acid, sodium-benzoate, sodium-glutamate and potassium-sorbite all proved to be "effective".

Though the patient's serum contained kryoprecipitate, gel elctrophoresis of the serum failed to reveal any M component (Fig. X/3).

Kryoprecipitate was further fractionated into two fractions with a medium and slow mobility respectively (Fig. X/4).

The further immunochemical analysis of the precipitate revealed a mixed-type kryoglobulinemia. The IgM κ-type monoclonal product transformed the polyclonal sol-state IgG-s into gel (Fig. X/5).

The bone marrow analysis revealed malignant reticulosis confirming the data of immunochemical study (Fig. X/6).

The next logical step was to test the effect of different food preservatives on the aggregation-stimulation action of adrenalin, ADP and collagen.

The following materials have been used for the analysis:

1. 3 ml platelet-enriched plasma;
2. IgM-κ prepared from the serum with gel filtration (final concentration 0.25 mg/ml);
3. Food preservative in final concentration 250 gamma/ml.
4. Aggregation-inducing agent: either ADP in final concentration $2 \times 10^{-6}$ M, or collagen in the concentration inducing submaximal aggregation (0.3 ml) prepared by us from the Achilles tendon using the Hovig's method.

The reagents were mixed and the volume adjusted to 4 ml by 0.15 M physiological saline (Fig. X/7, X/8, X/9).

The three Figures unambiguously prove that IgM together with food preservative markedly decreases the platelet aggregation in vitro; these mechanisms might play a leading role in the development of the cutaneous symptoms in vivo. The similarity of the results obtained with the three different aggregating agents exclude the direct inhibitory ef-

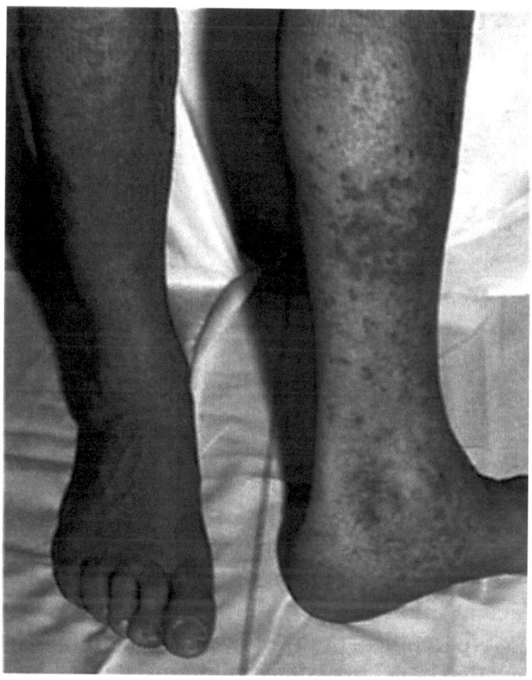

Fig. X/2

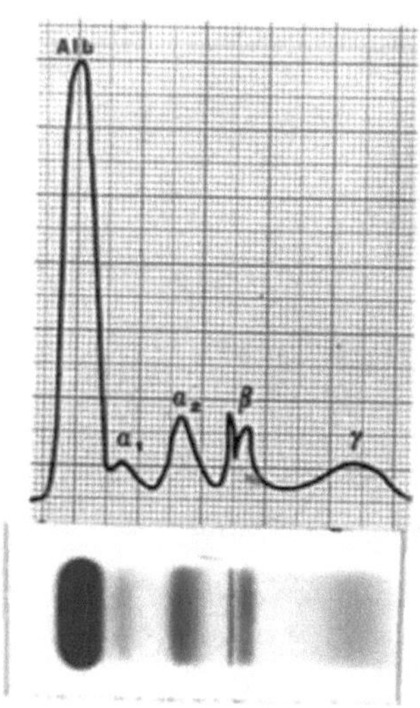


B. I. 64 ♂

Fig. X/3

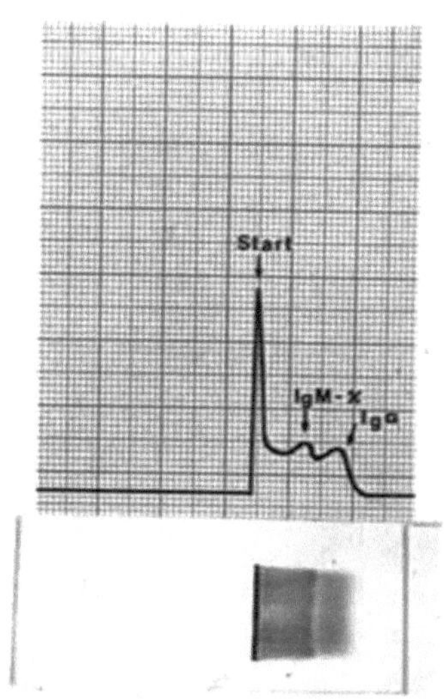


B. I. 64 ♂

KRYOPRAECIPITATUM

Fig. X/4

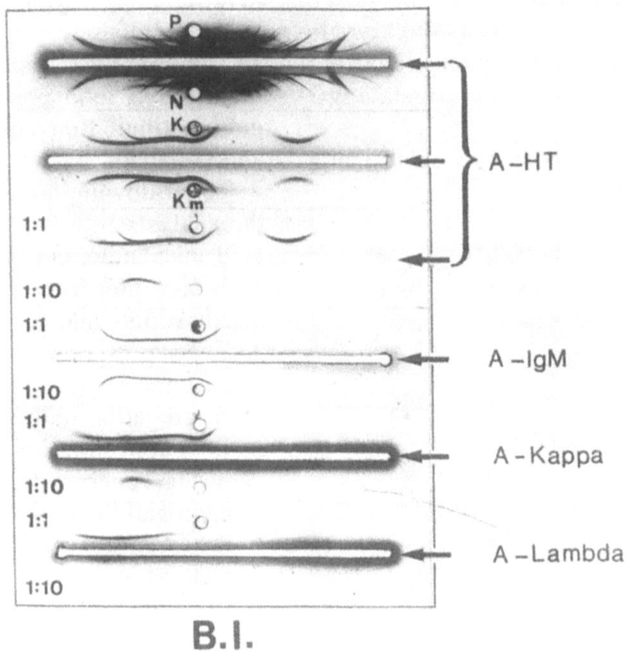


Fig. X/5

fect of IgM and food preservative. A direct interaction of the IgM-κ and the preservative with the platelet surface presumably inhibits the aggregation.

Passage disturbances and seizures occurred during the observation period. Studies of the organic background of this pathology indicated a stricture 17 cm from the anus caused by tissue proliferation of non-neoplastic character (Fig. X/10).

Histological study showed Crohn's disease, so the large intestine symptoms due to Waldenstrom's macro-globulinemia were precluded (3, 6, 36).

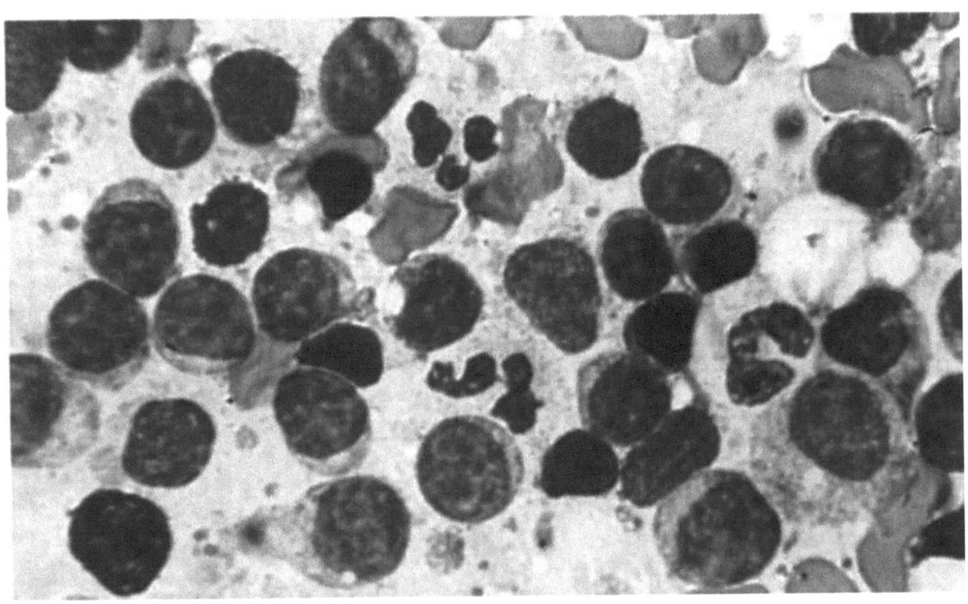

Fig. X/6

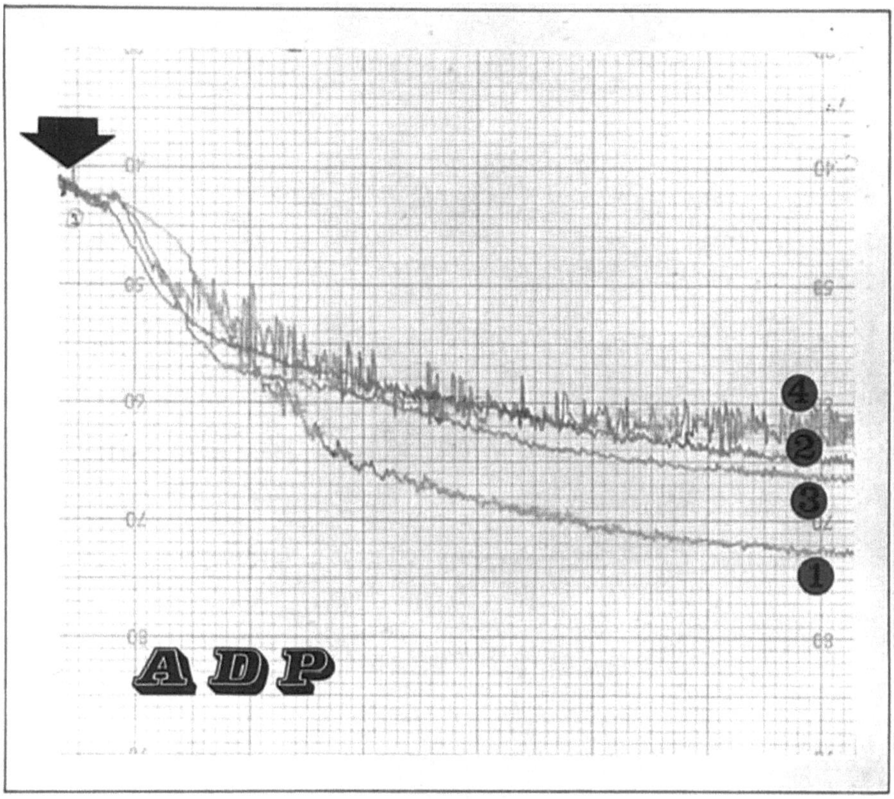


Fig. X/7

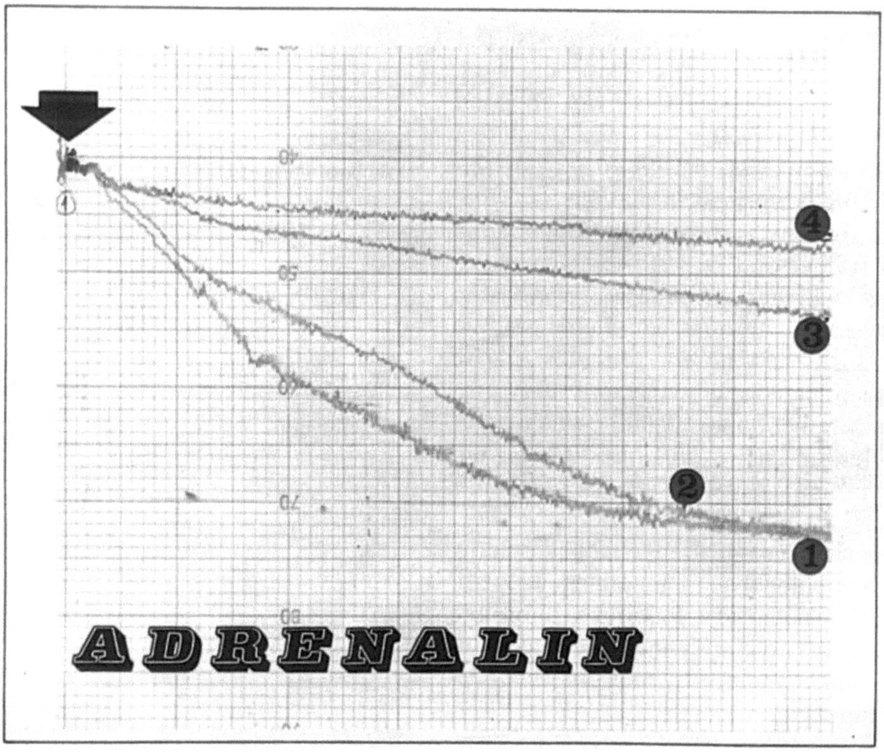


Fig. X/8

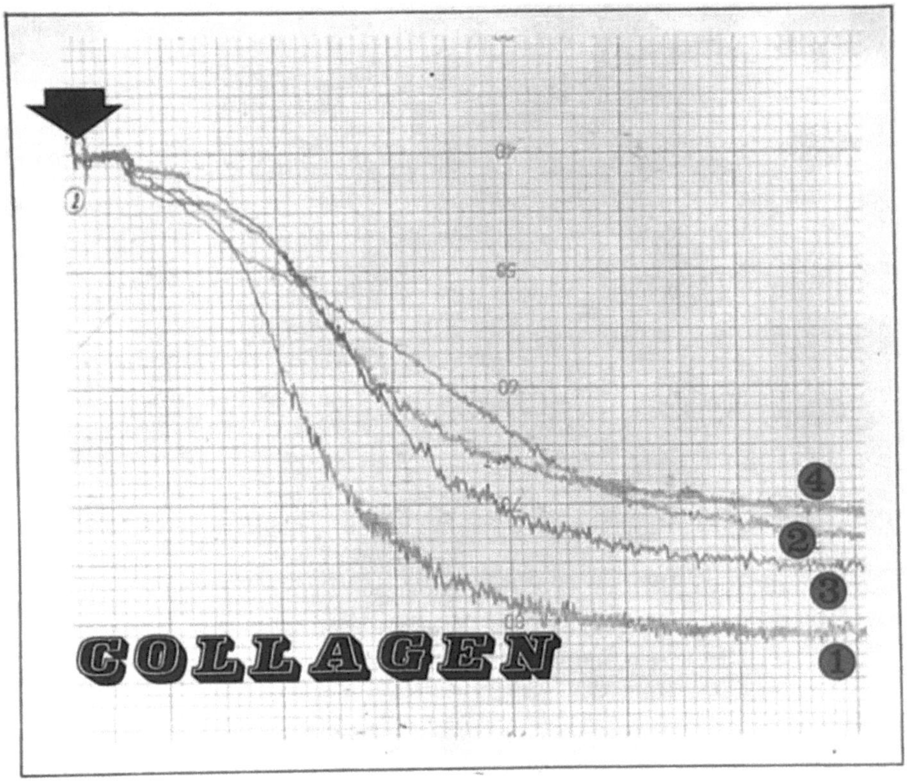


Fig. X/9

Therefore, our patient had kryomacroglobulinemia based on malignant reticulosis. The coagulation anomaly and fits of purpura were due to the *non-allergic sensitivity* to food preservatives. This case is of special interest because evidence exists that *in 40%* of the intestinal shunts performed in patients with pathological obesity, kryomacroglobulinemia and arthritis develop (29, 40). A logical question arises, whether any causal relationship exist between the large intestinal changes and the previously diagnosed reticulosis; can either of them be considered as a primary process, or is it rather an accidental coincidence?

Our failure to answer this question did not affect the primary choice of plasma exchange therapy (27).

In summary: coagulation changes seldom occur in monoclonal gammopathies. Identification of this pathology requires a thorough laboratory study to clarify the exact nature of the developed coagulopathy and to select adequate therapy.

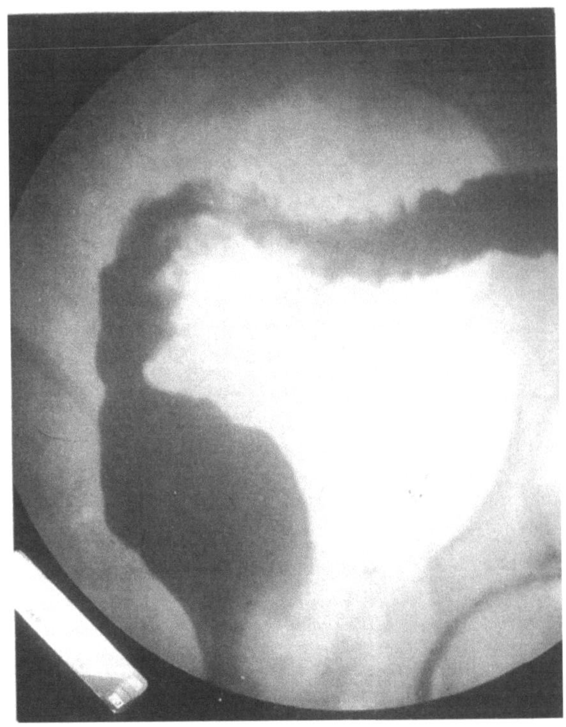

Fig. X/10

## References


1. Achenbach, W., Kanzow, V.
Zur hämorrhagischen Diathese bei der Makroglobulinämie Waldenström.
V. Kong. der. Europ. Ges. für Hämatol. Freiburg. 631, 1955.
2. Atkinson, J. P., Waldmann, Th. A., Stein, S. F., Gelfand, J. A., MacDonald, W. J., Heck, L. V., Cohen, E. L., Kaplan, A. P., Frank, M. M.
Systemic capillary lead syndrome and monoclonal IgG gammopathy.
Medicine. 56, 225. 1977.
3. Bedine, M. S., Yardley, J. H., Elliot, H. L., Banwell, J. G., Hendrix, T. R.
Intestinal involvement in Waldenström's macroglobulinemia.
Gastroenterology. 65, 308. 1973.
4. Benda, L., Deutsch, E., Mammen, E.
Über eine Gerinnungsstörung bei einem Patienten mit multiplem Myelom.
Wien. Klin. Wschr. 70, 559. 1958.
5. Benveniste, J., Henson, P. M., Cochrane, C. G.
Leukocyte-dependent histamine release from rabbit platelets. The role of IgE, basophils and a platelet activating factor.
J. Exp. Med. 136, 1356. 1973.
6. Cabreta, A., De la Pava, S., Pickeru, J. W.
Intestinal localisation of Waldenström's disease.
Arch. Intern. Med. 114, 399. 1964.
7. Castaldi, P. A., Penny, R.
A macroglobulin with inhibitory activity against coagulation Factor VIII.
Blood. 35, 370. 1970.
8. Chapman, G. S., George, Ch. B., Danley, D. L.
Heparin-like anticoagulant associated with plasma cell myeloma.
A. J. C. P. 34, 764. 1985.
9. Cochrane, C. G.
Mechanism involved in the deposition immune complexes in tissues.
J. Exp. Med. 134, 75. 1971.
10. Coleman, M., Vigliano, E. M., Weksler, M. E., Nachman, R. L.
Inhibition of fibrin monomer polymerization by lambda myeloma globulins.
Blood. 39, 210. 1972.
11. Deutsch, E., Lechner, K.
Gerinnungsströrungen bei paraproteinamischen Hämoblastosen.
Wien. Klin. Wschr. 84, 253. 1972.
12. Farbiszewski, R., Worowski, K.
Inhibition of blood platelet aggregation by serum IgM macroglobulin.
Haematologia. 4, 75. 1970.
13. Favre-Gilly, J., Creyssel, R., Thouverez, J. P., Revol, L., Croizat, P.
Antithrombine du type de l'héparine dans un gamma-myélome.
Considérations sur les anticoagulants spontanés dans la maladie de Kahler.
Hémostase. 3, 325. 1963.
14. Fick, P. G.
Inhibition of conversion of fibrinogen to fibrin by abnormal proteins in multiple myeloma.
Amer. J. Clin. Path. 25, 1263. 1955.
15. Glueck, H. I., Hong, R. A.
Circulating anticoagulant in IgA multiple myeloma: its modification by penicillin.
J. Clin. Invest. 44, 1866. 1965.
16. Gruber, A., Blaskó, Gy., Sas, G.
Functional deficiency of protein C and skin necrosis in multiple myeloma.
Thrombosis Research. 42, 579. 1986.
17. Harbaugh, M. E., Hill. E. M., Conn, R. B.
Antithrombin and antithromboplastin activity accompanying IgG myeloma: report of a case with a severe bleeding tendency.
Amer. J. Clin. Path. 63, 57. 1975.
18. Hensen, A., Loeliger, E. A.
Antithrombin and antithromboplastin activity accompanying IgG myeloma.
Amer. J. Clin. Path. 64, 820. 1975.
19. Henson, P. M., Cochrone, C. G.
Immune complex disease in rabbits. The role of complement and of a leukocyte-dependent release of vasoactive amines from platelets.
J. Exp. Med. 133, 554. 1971.
20. Humprey, J. H., Jagues, R.
Release of histamine and 5–HT from platelets by antigen-antibody reaction (in vitro).
J. Physiol. 128, 9. 1955.
21. Jákó, J., Nemesánszky, E., Bédi, J.
Kryoglobulinämie mit Purpuren, die sich auf die Auswirkung von Konservierungsmitteln entwickeln.
Allergologia et Immunopathologia. 5, 424. 1977.
22. Khoory, M. S., Nesheim, M. E., Bowie, E. J. W., Mann, K. G.
Circulating heparan sulfate proteoglycan anticoagulant from a patient with a plasma cell disorder.
J. Clin. Invest. 65, 666. 1980.
23. Kohler, P. F.
Manifestations in systemic lupus erythematosus and hepatitis B virus infection.
Medicine. 52, 419. 1973.
24. Krick, J. A., Ménaché, D.
Anomalié de la polymerisation de la fibrine dans un cas de gamma-A myélome.
Nouv. Rev. Franc. Hémat. 6, 744. 1966.
25. Lackner, H.
Hemostatic abnormalities associated with dysproteinemias.
Semin. Hematol. 10, 125. 1973.
26. László, E.
Trombocitafunkciók mielomában.
Magyar Belorvosi Archívum. 17. 5, 181. 1980.
27. Messmore, H. L., Fareed, J., Silberman, S., Gawlik, G. M., Bermes, E. W.
Macroglobulinemia of Waldenström diagnosis and management.
Ann. Clin. Lab. Sci. 8, 310. 1978.
28. Movat, H. Z., Mustard, J. F., Taichman, N. S., Urivhara, T.
Platelet aggregation and release of ADP, serotonin and histamine associated with phagocytosis of antigen-antibody complexes.
Proc. Soc. Exp. Biol. 120, 232. 1965.
29. Müller-Schoop, J.
Immunologie und Magen-Darm-Leiden. Fakten und Hypothesen.
Schweiz. Med. Wschr. 107, 1681. 1977.
30. Myllylä, G., Vaheri, A., Pentitten, K.
Detection and characterisation of immune complexes by platelet aggregation test. II. Circulating complexes.
Clin. Exp. Immunol. 8, 399. 1971.
31. Nelson, R. A. Jr., Nelson, D. S.
On the mechanism of immune adherence.
II. Analogy to mixed aggregation of sensitized antigens in the presence of complement: immune adherence with animal platelets.
Yale. J. Biol. Med. 31, 201. 1959.
32. Niléhn, J. E., Nilsson, S. M.
Coagulation studies in different types of myeloma.
Acta Med. Scand. 445, 194. 1966.
33. Palmer, R. N., Rick, M. E., Gralnick, H. R.
Characteristics of a heparin-like inhibitor.
Clin. Res. 29, 343. 1981.
34. Palmer, R. N.
Circulating heparan sulfate anticoagulant in a patient with a fatal bleeding.
N. Engl. J. Med. 310, 1696. 1984.

35. Perkins, H. A., MacKenzie, M. R., Fudenberg, H. H.
Haemostatic defects in dysproteinaemias.
Blood. 35, 695. 1970.
36. Roge, J., Marche, C., Camilleri, J. P., Dinet, P., Silvéréano-Rogé, F., Vernier, G.
Entéropathie exsudative et macroglobulinémie.
Gastroenterol. Clin. Biol. 2, 897. 1978.
37. Sanchez-Avalos, J., Soong, B. C., Miller, S. P.
Coagulation disorders in cancer. II. Multiple myeloma.
Cancer. 23, 1388. 1969.
38. Sandler, R. M., Evberson, C., Roberts, G. E., Voak, D., Darnborough, J., Heeley, A. F.
IgM platelet autoantibody due to sodium valroate.
Brit. Med. J. 2, 1583. 1978.
39. Wager, O., Penttinen, K., Myllylä, G.
Precipitation of IgG complexes and inhibition of complex-induced platelet aggregation by FR-active cryoglobulin M component.
Scand. J. Clin. Invest. 29, 122. 1972.
40. Wands, J. R., La Mont, J. T., Mann, E., Isselbacher, K. J.
Arthritis associated with intestinal – bypass procedure for morbid obesity.
N. Engl. J. Med. 294, 121. 1976.
41. Verstraete, M., Vermylen, C.
Recherches sur l'antithrombine gamma-dans la maladie de Kahler.
Acta Haematol. 22, 240. 1959.
42. Vigliano, E. M., Horowitz, H. I.
Bleeding syndrome in a patient with IgA myeloma: Interaction of protein and connective tissue.
Blood. 29, 823. 1967.

# XI. Organ destructive effect of the light chains. Myeloma kidney

Apart from infections, renal insufficiency is the most frequent cause of death in patients with monoclonal gammopathy (5, 8, 23, 87). Renal destruction develops very slowly, whereas the main disease still remains "symptom free". Explanation of both clinical findings and their relationship is discussed.

Before the detailed discussion of these mechanisms, relevant protein chemistry and laboratory findings are discussed.

## 1. The light chains

- At present only two light chain types - kappa and lambda - are known. Fig. XI/1 summarizes their structure and subtypes.

The fine structure of the $V_L$ has been analysed in the recent past. It has been shown that in the 6-15% of the cases the VL region contains not only amino acids, but also oligosaccharides (130).

The systematic analysis of the light chain structure, had been preceeded by more than a century-long period, when the characteristic behavior of these relatively low molecular weight proteins had been recognized as "bone diseases" (57, 73). After the first findings, fourty years later, studies of the cytologic background had associated the plasma cell proliferation with these pathologic proteins (58). The fine structure of immunoglobulins as the products of the plasma cells has been identified only in the 60ies (33, 127). Despite this long way, the name of the first researcher is still used to identify the light chains excreted with the urine as a Bence-Jones protein (BJ). If the light chains occur in plasma, the process is termed Bence-Jones proteinemia.

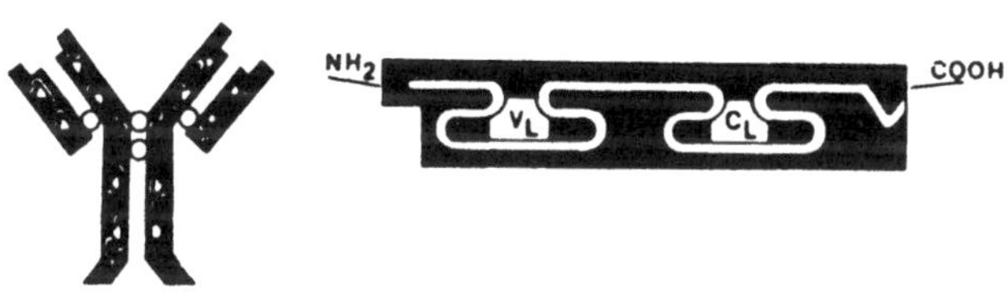


VL - variable domain;

CL - constant domain;

L - light chain;

LIGHT CHAIN STRUCTURE:

KAPPA = $V_\kappa$ + $C_\kappa$

$V_\kappa$: $V\kappa$ I., $V\kappa$ II., $V\kappa$ III., $V\kappa$ IV.

$C_\kappa$: Inv I., Inv II., Inv III.

LAMBDA = $V_\lambda$ + $C_\lambda$

$V_\lambda$: $V\lambda$ I., $V\lambda$ II., $V\lambda$ III., $V\lambda$ IV., $V\lambda$ V., $V\lambda$ VI.

$C_\lambda$: Oz, KERN, Mz, McG

**Fig. XI/1**

Besides the genetically determined primary damages of the amino acid sequence (see Fig. ), the differnces of the terciary structure seem to be more important. Due to the close relationship between the two mechanisms, both are responsible for the tissue and metabolic alterations presented herein.

The light chains occur as:

a) monomers
b) dissociating dimers
c) stable dimers
d) tetrameris.

For better understanding their tetrameric structure is first analyzed:

- *tetrameric* form is primarily identified in the serum as a protein with molecular weight of 4 × 22,000 Dalton, that can scarcely pass into the tubular lumen (44, 99). These cases are termed BJ proteinemia without proteinuria (7).
- *Stable dimeric* forms contain disulphide bonds in the C-terminal position in the kappa chains, and in penultimate position in the lambda chains (136).
- *The dissociating dimeric* forms are observed in the lambda light chains as products of partial reduction (136).
- *Secreted monomer* is the most frequent form: due to its 22,000 D molecular weight, it is easily filtered through the glomerular membrane. This form is mainly typical of the kappa light chains (14, 126).

The light chain molecules and their subtypes have different sensitivity to the pH, ionic strength and protein concentration. Physiologically (blood) these factors play only a minor role, however under non-physiological conditions (urine or tubular epithelium) their modulator role must be seriously considered.

Thus monomers can easily form dimers in the urine (79); these forms are less soluble and can form crystalline deposits in the tubular epithelium (34, 132).

The monomeric molecules in the urine can not only "grow" (form dimers), but also split into $V_L$ and $C_L$, due to the effect of the secreted endopeptidases (83).

Apart from splitting, which is not regarded as a "truely" pathological reaction, other non-physiological monomers are also described. For example, a so-called mutant BJ protein which differs from the original physiological one in a single amino acid is usually produced after chemotherapy (50). Another protein with BJ-like antigen properties and molecular weight lower than 22,000 D is known as a "degradation product" with (3, 125, 142). Prednisolon treatment modifies protein synthesis thus leading to a formation of BJ-like protein ($C_L^+$), that appears in the urine (128). This product has a longer amino acid chain compared to the $C_L^P$ obtained by peptic proteolysis (117). This finding proved that the mentioned molecule is not a degradation or proteolytic product, but is a result of a "false" synthesis (128).

## 2. Proteinuria

Proteinuria occurs in 50% of the MG (29, 30). Appearance of the protein molecules in the urine might be the first MM sign indicating renal damage. Proteinuria (usually, but not always prerenal) develops several weeks or even years prior to other symptoms of the Kahler's disease.

Recent results concerning the origin and biological effects of the proteins, as well as clinical and laboratory aspects of proteinuria should be considered in diagnostics. Fig. XI/2 summarizes the origine and distribution of the proteins in the urine:

The Tamm-Harsfall uroprotein is produced by the cells of the ascending part of the loop of Henle and the distal tubules (112). Its molecular weight is $7 \times 10^6$ Dalton. The molecule consists of several subunits, $10^5$ Daltons each (36) and is excreted at a concentration of 50 mg/L (10).

### *Physiological proteinuria*

Normal urine contains certain amounts of protein, however, it can not be detected without concentrating the sample. The daily excretion is under 100 mg/L (76). The A/G ratio 1.

The serum-derived proteins are:

albumin, 9% IgG, no transferrin! Both kappa and lambda light chains occur amongst the low molecular weight proteins (39, 133, 137). Their total amount normally never exceeds 50 mg/24h (27) About 50% of the excreted protein is of renal origin; 2% of the IgA are derived from the urogenital tract (85).

## PROTEINS OCCURRING IN THE URINE (SUMMARY):

**PLASMA-DERIVED PROTEINS:**

80–180,000 MW
ALBUMIN
TRANSFERRIN
IMMUNOGLOBULINS (IgG)

> 40,000 MW
LOW MOLECULAR WEIGHT
PROTEINS (LMW) ① ②
LIGHT CHAINS
LYSOZYMES ③
CARBOANHYDRASE ④
ALPHA-2-MICROGLOBULINS
BETA-2-MICROGLOBULINS:
c) IgG HEAVY CHAIN FROM THE 3-RD DOMAIN ⑤
b) FROM THE LYMPHOCYTE MEMBRANES ⑥
c) FROM THE TUMOR CELL MEMBRANES ⑦ ⑧

**PRODUCTS OF THE UROGENITAL TRACT**

TAMM-HORSFALL
GLYCOPROTEINS (IgA) ⑨
OTHER KIDNEY-DERIVED
PROTEINS ⑩

**References:**

1 = (78) 2 = (77) 3 = (48) 4 = (102) 5 = (101)
6 = (15) 7 = (91) 8 = (92) 9 = (82) 10 = (16)

**Fig. XI/2**

*Glomerular proteinuria occurs in two forms:*

a) selective, characterized by the A/G ratio1. Transferrin appears in the urine. The trans ferrin-gammaglobulin selectivity index IS At load the protein concentration in the urine increases.
b) non-selective: urine contains all the proteins at certain levels. The transferrin/gammaglobulin ratio exceeds 0.3.

*Tubular proteinuria*

Albumin comprises 25% of the proteins. The postgamma fraction, alpha-2-microglobulin, beta-2-microglobulin (60), lysozyme and many light chains are typical of this form.

*Prerenal proteimuria*

Many low- (rarely high) molecular weight proteins, which are accumulated in plasma can be excreted in the urine. The term prerenal proteinuria summarizes pathological alterations occurring in the MG-s (Fig. XI/3)

Renal complications in MG-s are well documented (9, 52, 26, 62, 64, 105, 147). Data of the literature and our own findings urged us to summarize the existing empiric data. In our opinion, close relationship exists between the type of the kidney-deteriorating molecules, site of the lesion and the clinical symptoms.

The following physico-chemical properties of the molecules are of importance:
1. quantity;
2. size;
3. tertiary and quaternal structure;
4. proneness to kryoprecipitation;
5. heavy chains type;
6. type and isoelectric point (pI) of the light chain;
7. antibody properties - proneness to form immune complexes;
8. interaction with non-immunoglobulin molecules and its sequellae

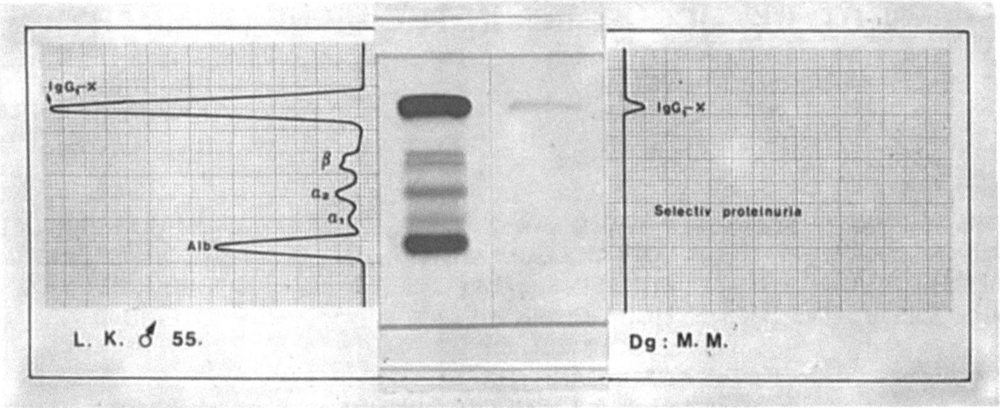


Fig. XI/3

9. effects of the non-immunoglobulin products of B cells.

1. High concentrations of paraprotein first induces a so-called prerenal proteinuriaa; later, in the course of a very slow process, it would precipitate in the mesangium (2, 141). Paraproteins are reported to be centers of the mesangial nodules in myelomal glomerulosclerosis (Fig. XI/4).
2. The largest immunoglobulin is the Igm, its molecular weight is about 1,000,000 Dalton. The renal damages occurring in the Waldenström's macroglobulinemia affect the glomeruli (148): their capillaries are filled with amorphous eosinophil thrombi (89), which contain IgM aggregates (Fig. XI/5). Tubular lesions are not typical; the rare cases are due to the nephrotoxic effect of the light chains, if Waldenström's macroglobulinemia is accompanied with BJ proteinuria (55). Sometimes lymphoid infiltration is observed in renal interstitium. Waldenström's macroglobulinemia can be induced by a special form of IgA monoclonal gammopathy (51).

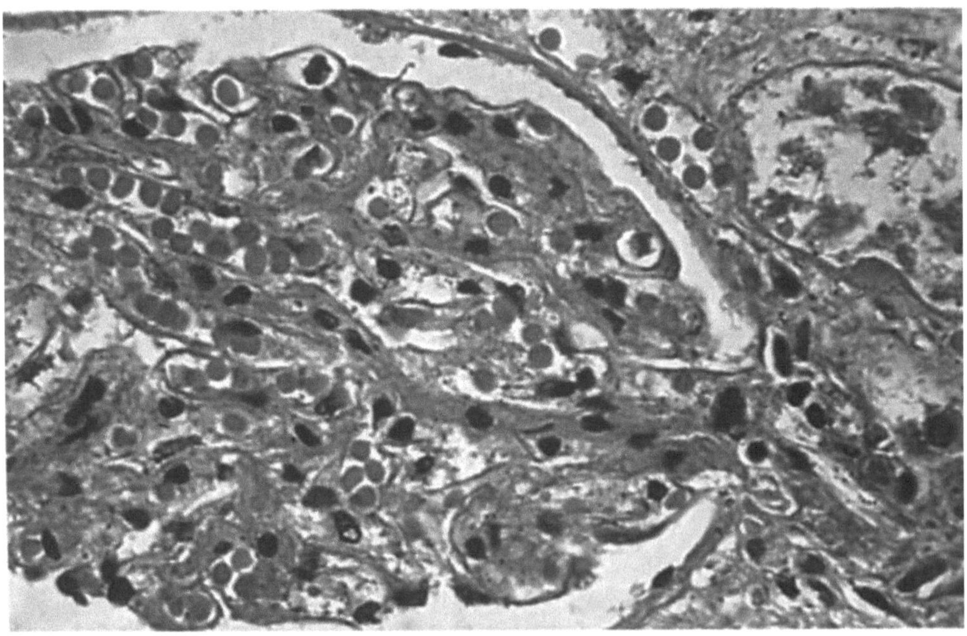

Mesangial proliferation

Fig. XI/4

The immunochemically proved heavy-chain disease, in which heavy chains comprise the M component, never damage the renal parenchyma (17, 37, 38, 40, 41, 96, 115). Heavy chains appear in the tubular lumen only if glomerular lesions already exists (135).

3. The tertiary and quaternal structure of the molecules determines their physico-chemical properties. As it has already been mentioned, the pathologically altered molecules increase the blood viscosity leading to deterioration of microcirculation. The kidney is extremely sensitive to circulatory impairment, hence, hyperviscosity has a renal component (138). The latter is, however, not specific for hyperviscosity.
4. Cold-aggregating kryoglobulins comprise a rare subgroup of monoclonal immunoglobulins (45, 146). These molecules primarily damage the glomeruli leading to glomerulosclerosis (18, 32, 42, 138). Interstitium is infiltrated with typical lymphoplasmatic, PAS+ cells, which are related to kryoglobulin macroglobulins (138). Histological findings do not always reveal glomerulosclerosis, but rather indicate membrane proliferative glomerulonephritis with its typical ultrastructural signs (88) (Fig. XI/6).
If a kryoprecipitating protein ($IgG_1$) passes through the glomerular membrane (46, 68) it might cause tubular damages (Fig. XI/7).
It is worth pointing out that the appearance of kryoproteins does not necessarily mean OMG diagnosis, for it also develops in IMG (75) and systemic diseases with enhanced synthesis of polyclonal immunoglobulins (LED, periarthritis nodosa, scleroderma) (4).
5. The type of the heavy chains is an other differential criterion for morphological and - to a certain extent - diagnostic prognosis of the clinically detected renal damages. Histological features of macroglobulins in cases of the IgM-induced renal damage are described in Item 2 (molecular size). In the literature these cases are known as macroglobulinemic nephropathies (1, 89). Renal insufficiency - a so-called IgA nephropathy - is a rare complication of IgA MG (81). Focal or generalized mesangial proliferation is a typical feature of this pathology (71). The former is diagnosed as focal segmental glomerulonephritis. Glomerular

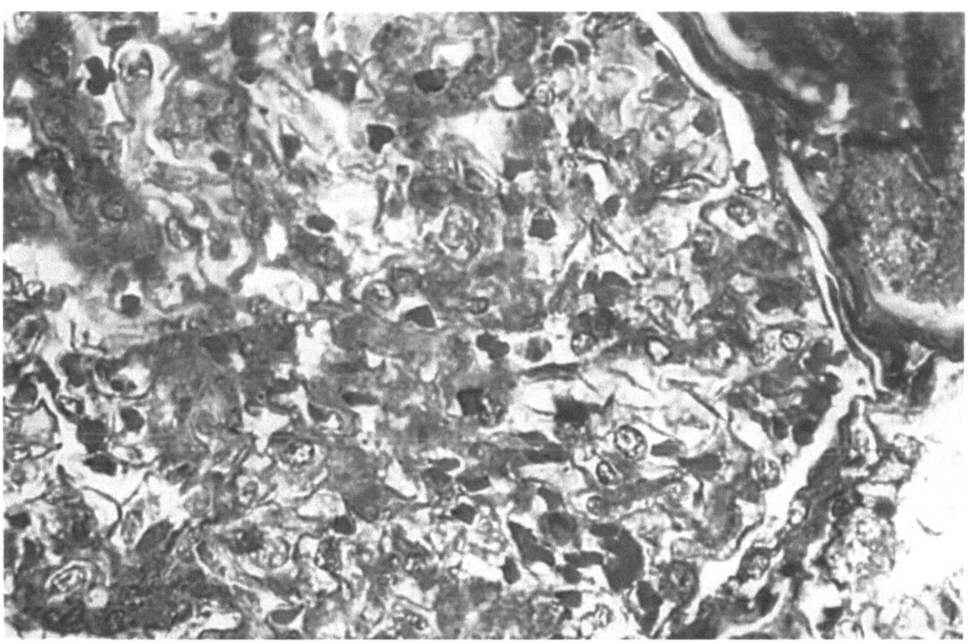

**Thrombi in the lumen of a glomerular capillary. Staining: Mallory trichrom; magnification: 300 ×**

**Fig. XI/5**

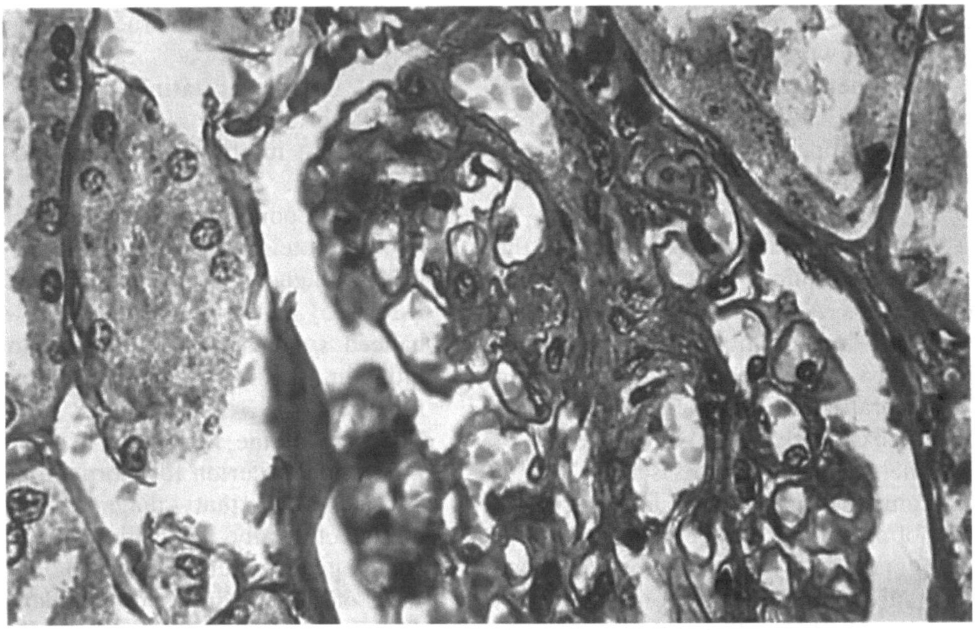

**Membrane-proliferative glomerulonephritis. Staining PAS. Magnification 300 ×**

**Fig. XI/6**

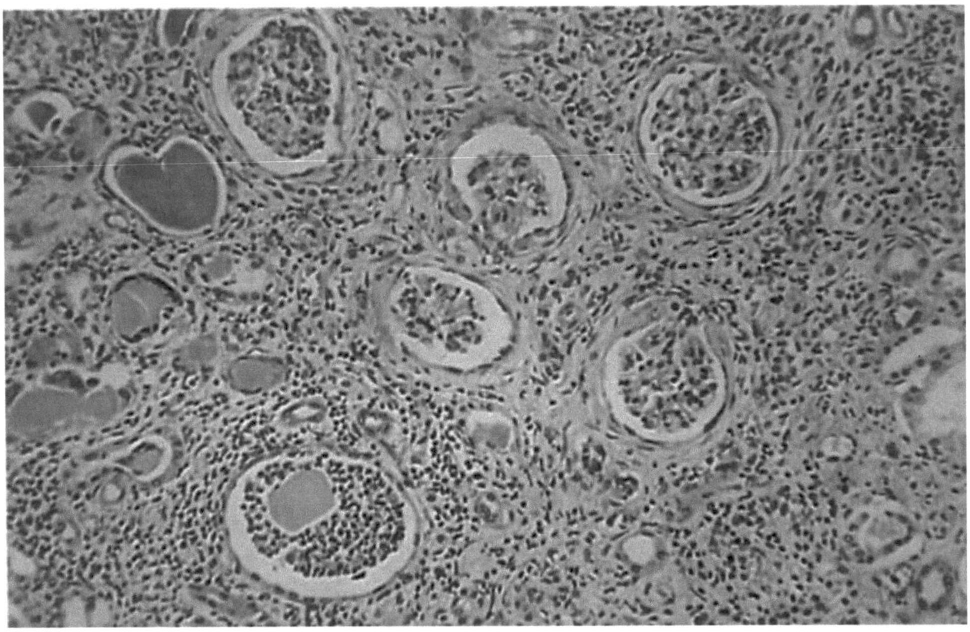

**Tubular changes. Amorphous eosinophilic substance in the lumen. Destructive cellular reaction. Staining PAS. Magnification 80 ×**

**Fig. XI/7**

alterations are accompanied by focal degeneration of the tubular epithelial cells. Mononuclear infiltration and moderate fibrosis are the concomitant signs of the basal membrane thickening (31) (Fig. XI/8).

Electronmicroscopic studies reveal electrondense particles and subendothelial deposites in the Bowman's capsule. The basal membrane thickens. Proliferation of the mesangial cells and expansion of

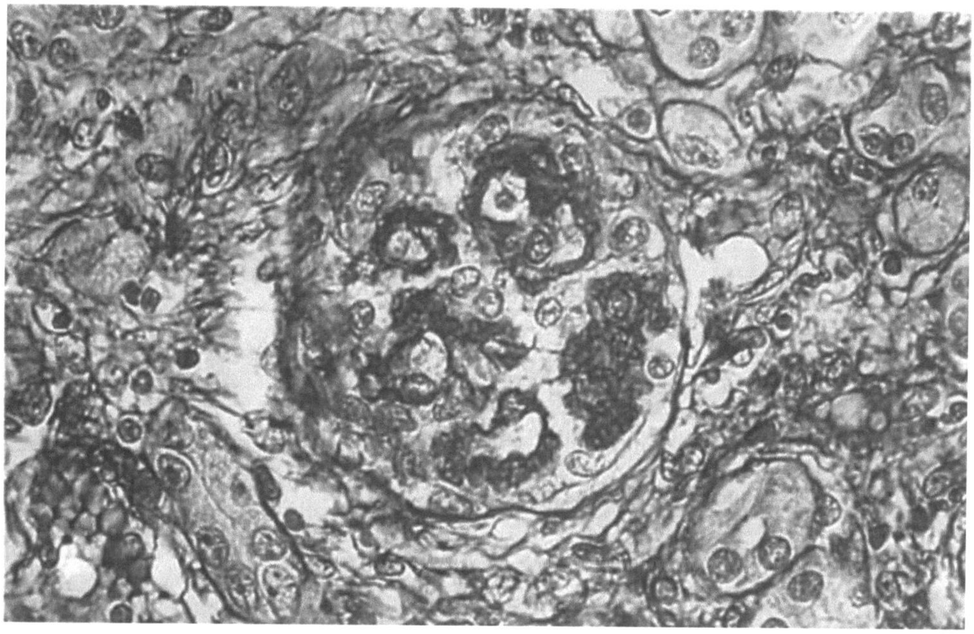

**Basal membrane thickening and signs of sustained growth of mesangial matrix. Staining: Mallory trichrom; magnification: 300 ×**

**Fig. XI/8**

the matrix indicate glomerular damage (31).

Myelomal kidney (a typical renal damage) is elicited by the IgG Mg (144). We are strongly convinced that the appearance of free light chains in the urine is obligatory and, therefore, is specially discussed later. Myelomal kidney rarely occurs in other diseases (86) not not related to monoclonal gammopathy. In these rare cases other excreted tumour products (for example mucoprotein) damage the renal parenchyma (49, 80).

6. Despite the substantial amount of laboratory and clinical data, the types and the role of the light chains in eliciting renal damage still remains unknown. Some researchers are strongly convinced that BJ proteinuria is a prerequisite (95) and the cause of the late renal damages (6, 120). An opposite statement claims that no relationship exists between BJ proteinuria and kidney lesions, since it can also occur in healthy individuals (63, 100). Moreover, light chain excretion sometimes occurs for decades without any clinical or laboratory signs of damaged renal function (129).

These rather empiric data concerning the light-chain excretion should be revised. Based on the differences of isoelectric points (pI), purified urine BJ proteins can be assessed into three groups: pI = 7.6 - 7.7 (kappa type); pI = 6.9 - 7.2 (kappa type); pI = 5.6 - 5.7 (lambda type). Proteins with pI = 7.6 - 7.7 damage the proximal-, while those with pI = 6.9 - 7.2 and 5.6-5.7 affect the distal tubules (26).

The light chain excretion showed no correlation with serum creatinin changes, suggesting that renal toxicity of the light chains does not depend on their tubular concentration (94); moreover, the significance of the duration of BJ proteinuria is also doubted (111). Therefore, recent data of the literature do not provide unambiguous conclusions about the role of the light chains.

The development of the myeloma kidney is a multifactorial process, and the mentioned light chain nephropathy (119) is induced not solely by BJ proteinuria. Due to the importance of the latter, the role of the light chains is discussed herein.

Light chains are produced by the plasma cells (84, 106, 122, 123, 124) and are catabolized by the renal tubular epithelium (141, 145). The low molecular weight

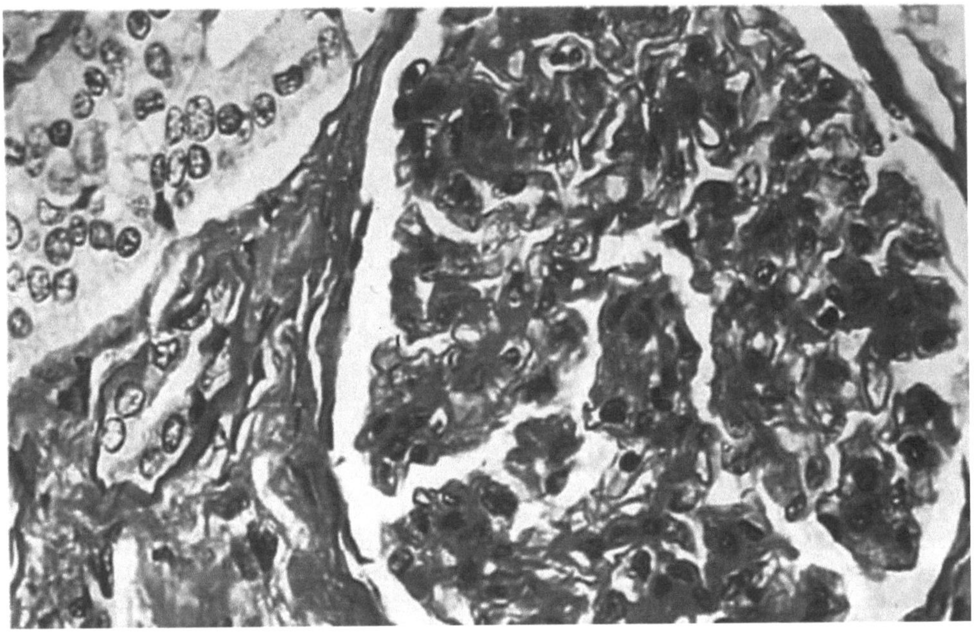

**Membrane proliferative glomerulonephritis. Staining PAS. Magnification 300 ×**

**Fig. XI/9**

monomers and dimers are filtered by the glomeruli. Proteins transported into the tubular lumen are reabsorbed by the energy-dependent carrier systems of the epithelial cells (72, 140).

The next step is the light chain catabolism in the cells

If excess proteins are reabsorbed from the lumen, the catabolic activity of the epithelial cells is diminished despite the in-

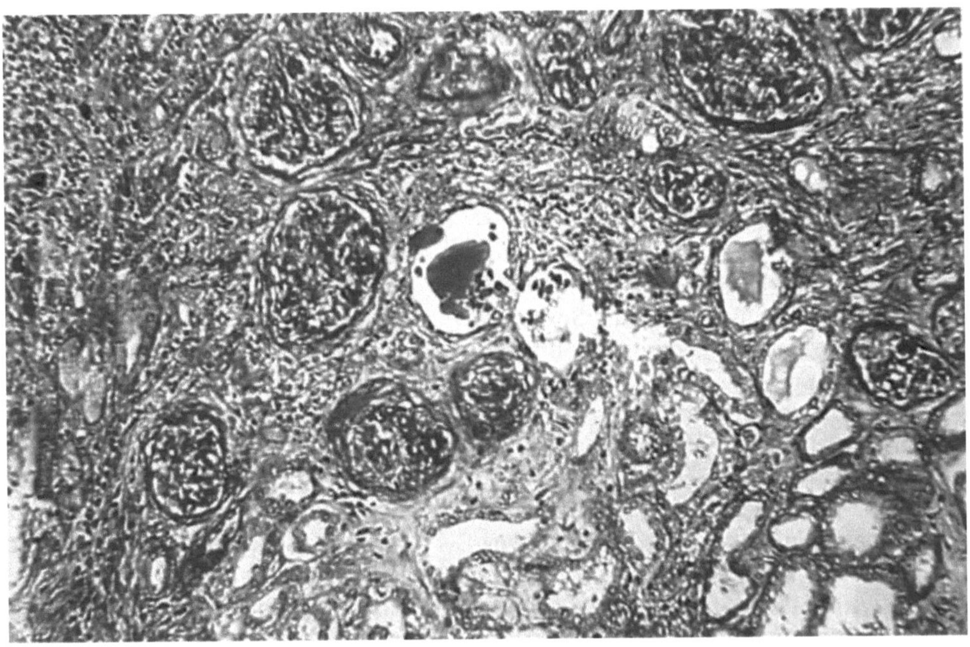

**Renal lesions with atrophic tubules with expanded lumen. Staining: Mallory trichrom. Magnification: 80 ×**

**Fig. XI/10**

creased "supply", thus less proteins are catabolized (103). Under experimental conditions this light chain effect proved to be specifically toxic for the epithelial cells (35, 70). The initial tubular cell damage is followed by the release of the lysosomal enzymes, which has a direct nephrotoxic effect (104). Besides the intraepithelial damage, the ligh chains also affect the interstitial sodium transport. Concomitantly light chains are excreted in the urine, indicating saturation of the reabsorbtive mechanisms on one hand, and impaired catabolism due to the progressive renal damage on the other (30).

Furthermore, the intraluminal light chains and certain predisposing factors trigger cast formation, which contribute to the aggregation of Tamm-Horseall mucoproteins and formation of special aggregates (119). The light chains are also very sensitive to the pH changes. Transformation of monomers into di- and trimers further promotes the clast formation. Rings of reactive epithelial macrophages would surround the aggregates finally leading to the occlusion of the tubular lumen (129).

Most researchers are inclined to explain the renal damages solely by the cast formation (114). It is, however, more probable that the latter indicates the final stage of the progressive renal lesion. The following overview correlates the morphological (but not pathophysiological) forms of renal damge with the types of produced and secreted light chains (97, 117).

- The mildest form is the increased glomerular permeability, which develops without any structural changes and is usually caused by kappa-monomers. This form clinically corresponds to the nephrotic syndrome (66, 108, 131).
- In special cases the light chain has kryo-properties and forms complexes with the circulating proteins. Histologically membrane proliferative glomerulonephritis occurs (19) (Fig. XI/9).
- Excess of the excreted kappa chains cause degeneration and functional damages of the proximal tubular epithelium. A typical "swan-neck" morphological picture is observed due to the shortening and narrowing of the proximal tubules; these alterations are accompanied by non-specific degeneration. Needle-like enclusions appear in the tubular cells (24). Laboratory and clinical picture corresponds to the acquired Fanconi syndrome of adults (decreased $PO_4$ reabsorption in the proximal tubules, glucosuria, aminoaciduria, albuminuria, and vitamine-D resistant rickets (25, 47, 59, 69, 74, 139). Experimental evidence exists that 80% of the kappa light chains are catabolized in the proximal tubules (110).
- Primary damage of the distal tubules. With the progression of the diseases the leading alterations - tubular atrophy and degeneration - are usually accompanied by interstitial sclerosis (70) (Fig. XI/10).
  The lumen is filled with casts. The lamda-chain excretion is responsible for these morphological and functional changes (53). It is not yet known, whether the obstructive renal lesions are rather due to the direct effect of the light chains infiltrating tubular epithelium, or to the indirect effect of the luminal casts.
- Combined lesions of the proximal and distal tubules cause the so-called combined light chain nephropathy. In this renal pathology the kappa chain excretion is accompanied by glucosuria, phosphaturia, uricosuria, aminoaciduria, renal acidosis in the proximal tubules and nephrogenic diabetes insipidus, due to the impairement of the distal tubules (90, 119) (Fig. XI/11).
- Concomitant glomerular and vascular changes develop if the stored lambda light chains induce amyloidosis (107).

7. Monoclonal antibodies form complexes with the circulating proteins. These immune complexes alter glomerular permeability while passing through the kidney (56, 121). Such sites of low resistance contribute to further pathological reactions. Apart from this mechanism, the complexes can easily cause a "pure" immune complex nephritis (3, 134, 143).
8. Interaction of protein with proteins or

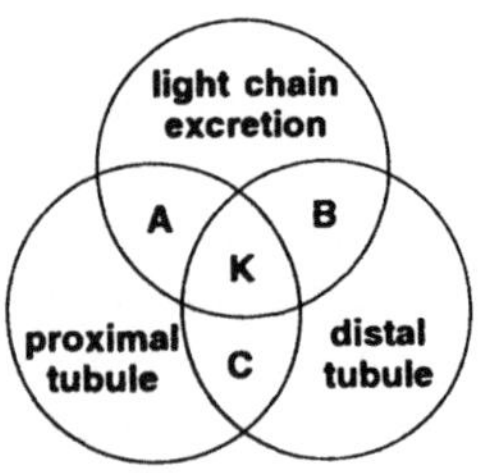


A – Fanconi syndrome
B – decreased capability to concentrate urine (1)
K – combined light chain nephropathy
C – nephropathy without light chain excretion (myelomal kidney) (2)

**(1) - corresponds to reference 28**
**(2) - to reference 59**

**Fig. XI/11**

other chemical substances does not necessarily mean antigen-antibody binding. Momentary formation of casts (67), for example, between immunoglobulins (or light chains) and the Tamm-Horsfall uroproteins or pyelographic contrast substances, leads to the development of the mostly feared tubular insufficiency.

9. Non-immunoglobulin products of B cells can be also highly nephrotoxic (116). In renal lesions occurring in malignant lymphomas immunoglobulins form no complexes with proteins and, therefore, can not attribute to the clinical and histological changes observed (20).
It is documented that B cells differentiating into plasmcytes release a calcium-mobilizing substance. Thereby, hypercalcemia and later hypercalciuria are followed by renal insufficiency (90).
Despite the wide variety of forms and causes, the clinical course of renal diseases can be acute or chronic. Renal lesions in MG usually develop very slowly, for years or decades (147); acute forms are extremely rare. Available literature presents only 33 acute cases (12, 29). The post-diagnosis survival period of the patients excreting kappa or lambda chains is 30 and 10 months respectively. 42.4% of the lambda-excreting patients would die within the 6 months after diagnosis. The main laboratory symptoms are as followes: azotemia, proteinuria and anemia. Amyloidosis occurs in 30% of these patients (28, 118).
We were among the first to emphasize the nephrotoxic effect of the lambda light chains (53, 54). A detailed analysis of our clinical material shows that of 1538 MG cases 33 patients have been treated due to renal lesions induced by the light chain excretion. The lethal outcomes were caused by renal insufficiency. Histological studies showed combined renal lesions involving glomeruli, tubules and interstitium. In our opinion such generalization is typical of the terminal stage (Fig. XI/12-14 XI/15 a-b XI/16).

Most of the patients excreted lambda light chains. Both theoretical and clinical data mentioned in the Introduction suggested plasma exchange as a part of complex therapy (13). Plasma exchange elicit a series of important beneficial changes: lowers the level of monoclonal immunoglobulins (or light chains), which play a leading role in the pathogenesis of the disease and lowers the accumulated serum creatinin, UN etc. Based on the above mentioned principles, the data of 4 patients excreting lambda light chain are presented; clinical and laboratory signs of renal lesion necessitated plasma exchange in these cases (Fig. XI/17).

The patients were cachetic, in the stage of compensatory polyuria. Therefore, plasma exchange could be performed very slowly, using the small centrifuge rotor.

Since the group-matched plasma exchange usually caused complications (shivering, febrility), it was replaced with a 5% albumin solution.

Considering the plasma flow rate, a relatively simple mathematical equation yields the concentrations of the post-exchange plasma components as a function of the exchanged plasma volume:

$$C = (C_o - C_s^i) \times e - V_{ou} / V_o + Cis$$

$C$ = concentration of plasma component after plasma exchange;
$C_o$ = concentration of the same component before plasma exchange;
$C_s^i$ = concentration of the same component in the fluid used for the exchange;
$V_{ou}$ = the withdrawn plasma volume;
$V_o$ = calculated plasma volume of the patient.

Concentrations calculated on the basis of the plotted curves were lower than the measured values, which - in our opinion - might be due

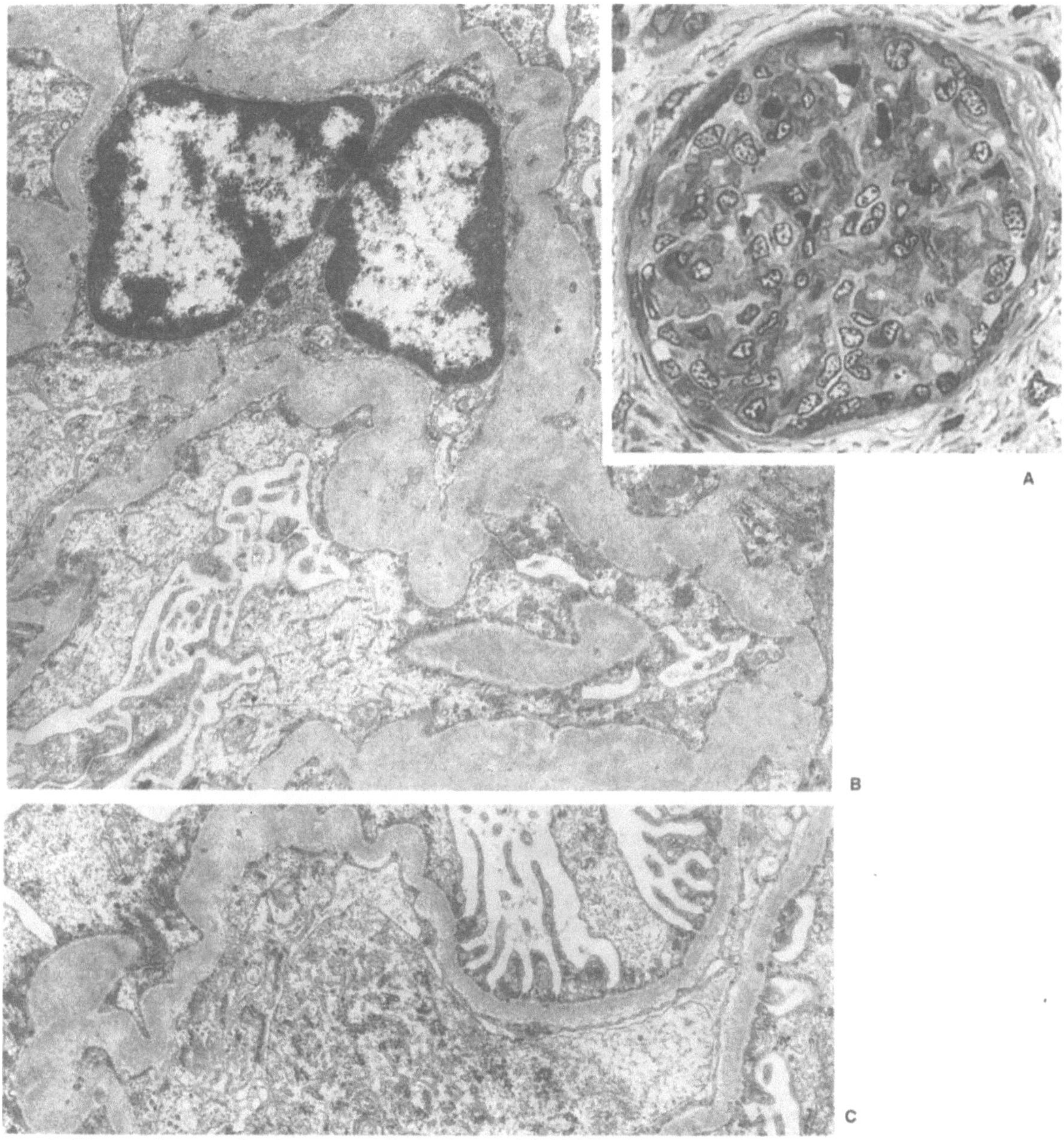


A. Glomerular vascular loops are separated from each other and accumulation of mesangial cells is observed. Substance corresponding to the basal membrane is seen in the homogeneous areas between the proliferating mesangial cells. The cells of the Bowman's capsule thicken and exhibit signs of discontinuity. Magnification 560 ×. B. EM study: enlargement of basal membrane, confluent processes of the podocytes. C. Network of well developed rough ER in the cytoplasm of a mesangial cell. Magnification 13.000 ×

**Fig. XI/12**

to the effect of some known and unknown factors.

The Figure shows the changes of serum potassium, creatinin and urea nitrogen (UN), which indicate the efficacy of the applied therapy (Fig. XI/18).

The following figure presents the platelet count, hemoglobin values and the total protein concentration. The curves show no significant changes in these parameters (Fig. XI/19).

Concentration changes of the lambda light chain immunoglobulins and the free light

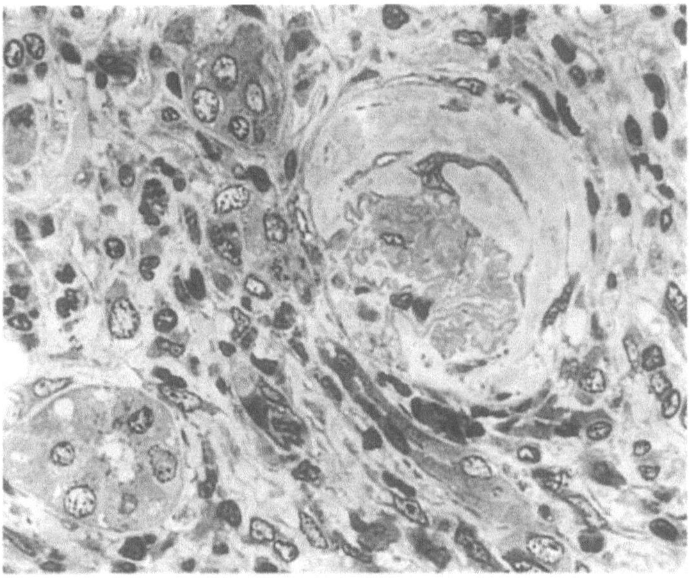

Glomerular and interstitial fibrosis with mononuclear infiltration. Only a small part of the glomerular vascular loops is seen in the enlarged Bowman's capsule. Magnification 560 ×.

Fig. XI/13

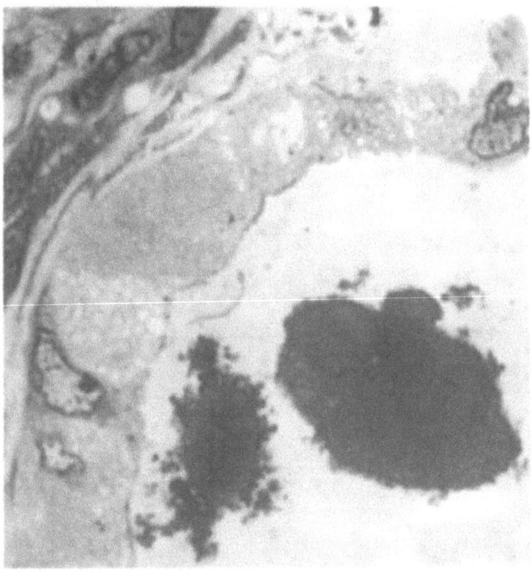

B

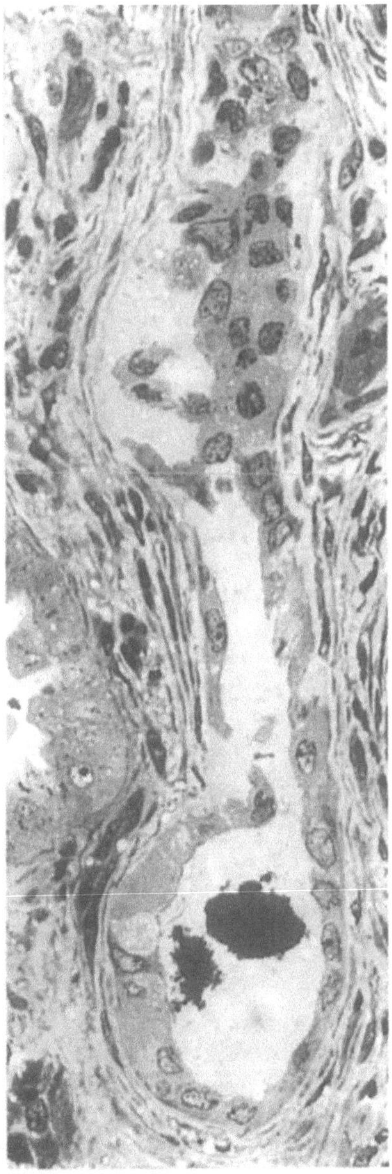

A

C

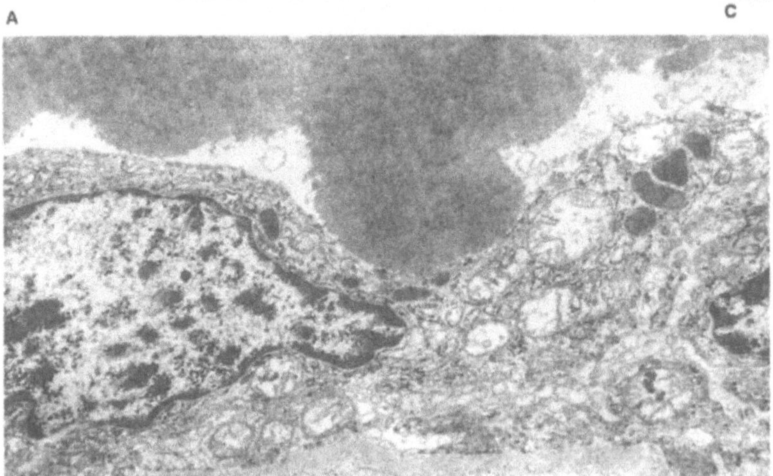

A-B: Vacuolization of cytoplasm and discontinuity of the epithelial layer indicate tubular lesion. Casts and desquamated epithelial cells of various density appear in the lumen (1400 ×).

C: EM shows degeneration of tubular epithelium and casts in the tubular lumen. Magnification 13.000 ×.

Fig. XI/14

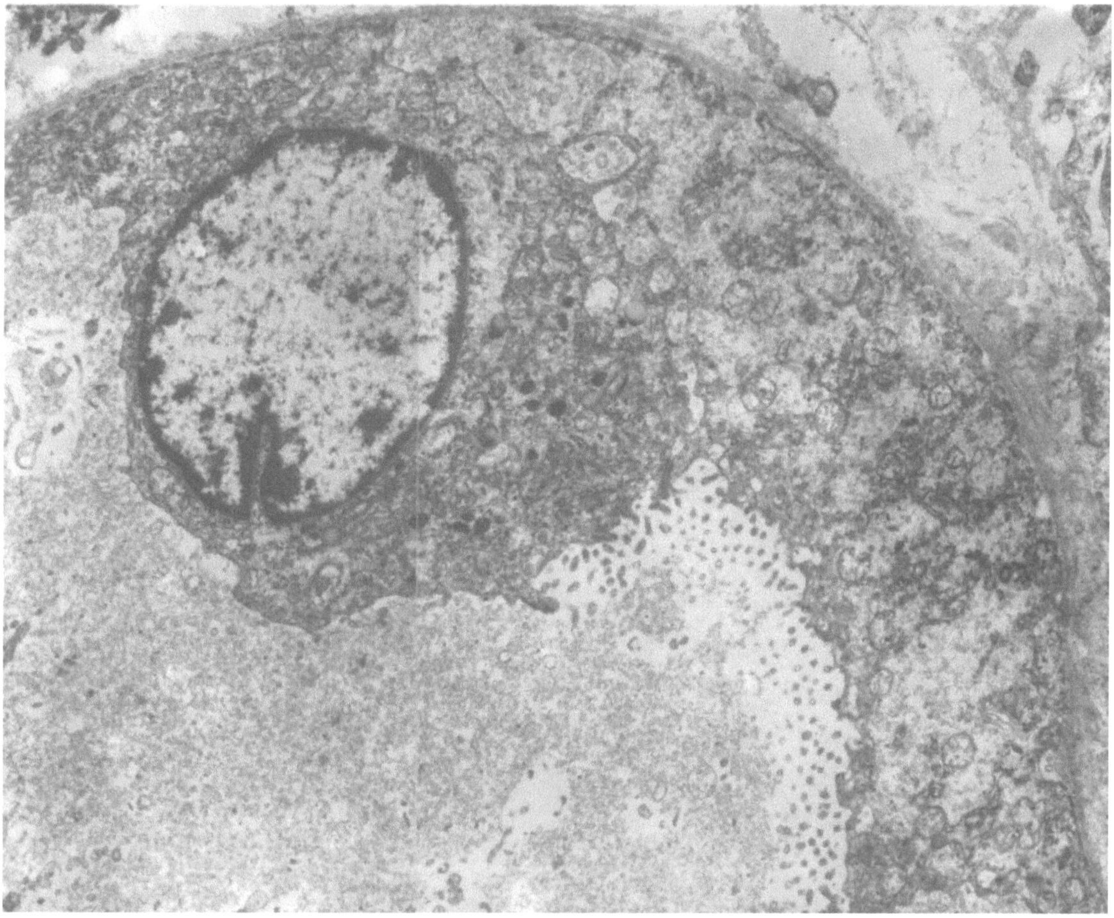

a

b

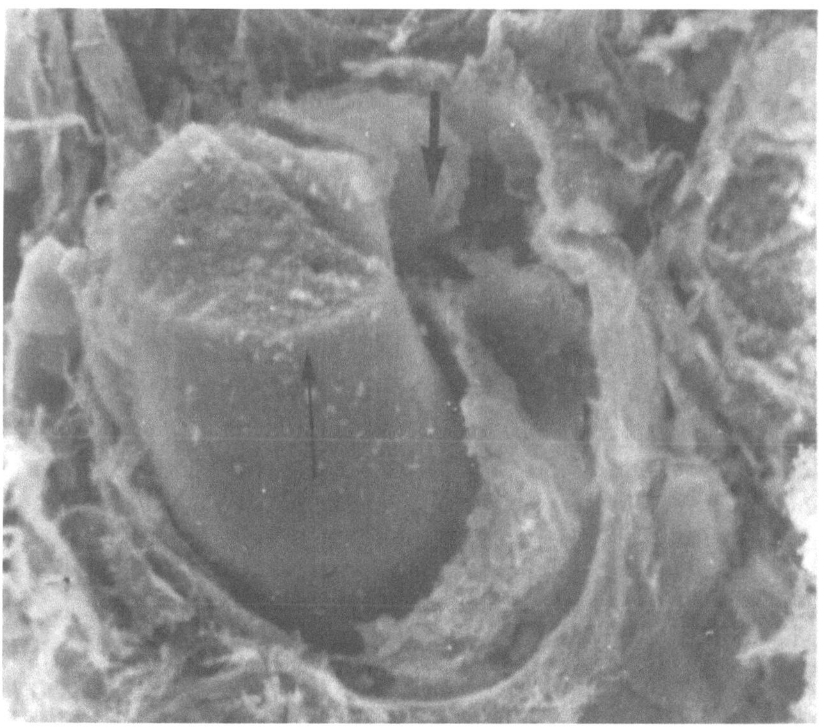

a. substance of similar structure in the tubular epithelial cells and lumen. Enlarged intercellular gaps between the epithelial cells. Discontinuity of the epithelial layer (top of the Fig. ). The tubular contents can get into the interstitium through the basal membrane. Basal membrane is made up of several layers. Magnification 7000 ×.

b. Scanning EM: a rigid cast protruding from the tubular lumen.

Fig. XI/15

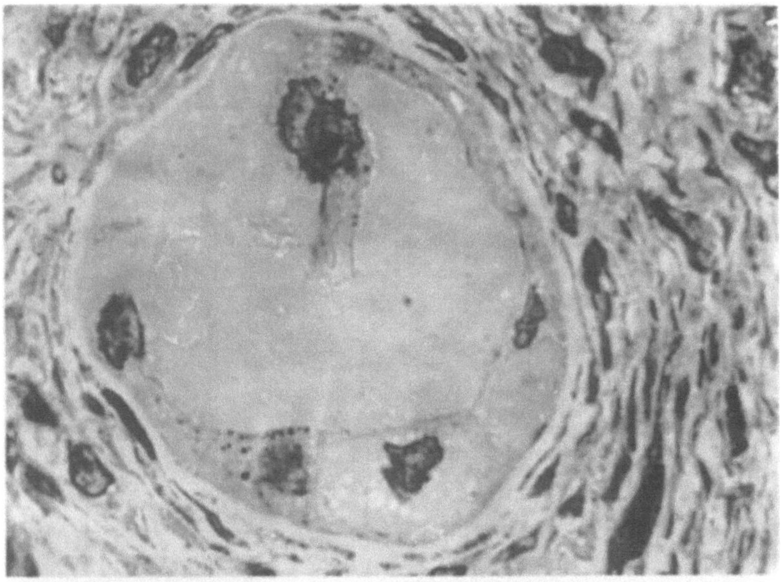

**A: a tubule in fibrotic interstitium. Epithelial cytoplasm is partly homogeneous (soaked), partly vacuolized containing dark lysosome-like structures. Detachment of the epithelial layer at about 1/4th of the tubular diameter. The tubular contents get into the interstitium through the basal membrane. Magnification 1400 ×.**

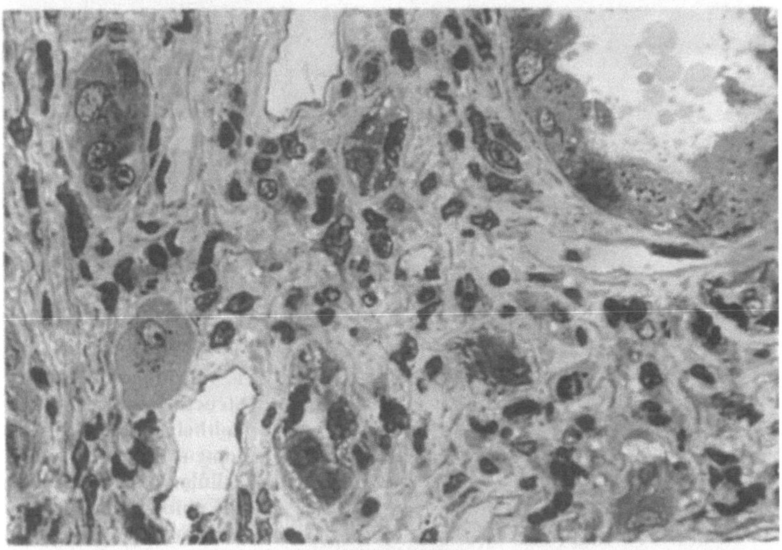

**B. Degenerated tubular epithelium. Epithelial cells contain dark lysosome-like structures. Nuclei of the dead cells and cytoplasm inclusions in the tubular lumen. Mononuclear infiltration in the inflammated interstitium. A vacuolized phagocyte is clearly seen. Magnification 560 ×.**

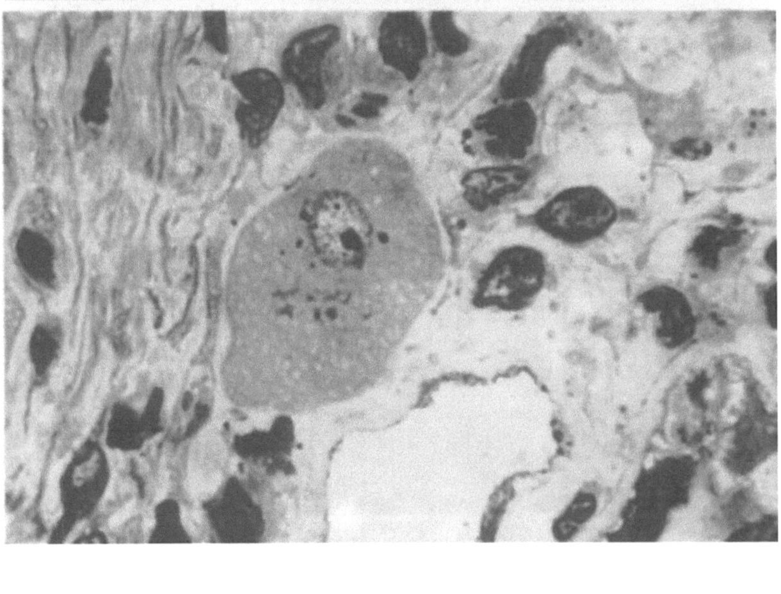

**C. A magnified vacuolized phagocyte. Cytoplasm contains dark lysosome-like structures. Magnification 1500 ×.**

**Fig. XI/16**

| Name | Sex | Age | Diagnosis | Light chain in the urine | Plasma volume (ml) | Volume of the exchanged plasma | Number of exchanges |
|---|---|---|---|---|---|---|---|
| I. K. | ♀ | 49 | IgG1-λ | λ | 2832 | 3400 | 4 × |
| J. A. | ○ | 52 | IgA-λ | λ | 2631 | 2700 | 3 ö |
| J. K. | ♀ | 64 | IgG1-λ | l | 2335 | 3700 | 5 × |
| P. T. | ♀ | 59 | IgG1-λ | l | 2718 | 3500 | 4 × |

Fig. XI/17

chains in the serum and urine are summarized in Fig. XI/20.

Clinical and laboratory findings indicate marked improvement in the general state of the patients, which lasted for about 4-10 weeks, thereby, we conclude that plasma exchanges should be monthly repeated.

In summary, the clinico-pathologic concept

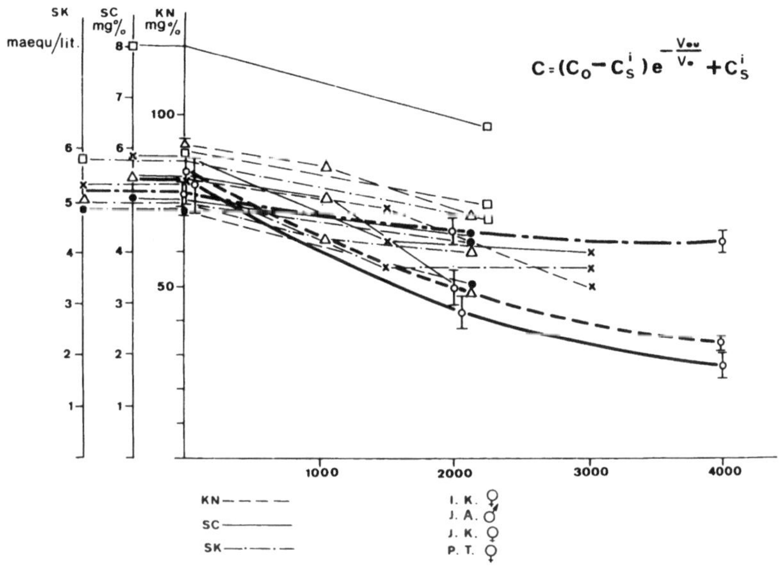


Fig. XI/18

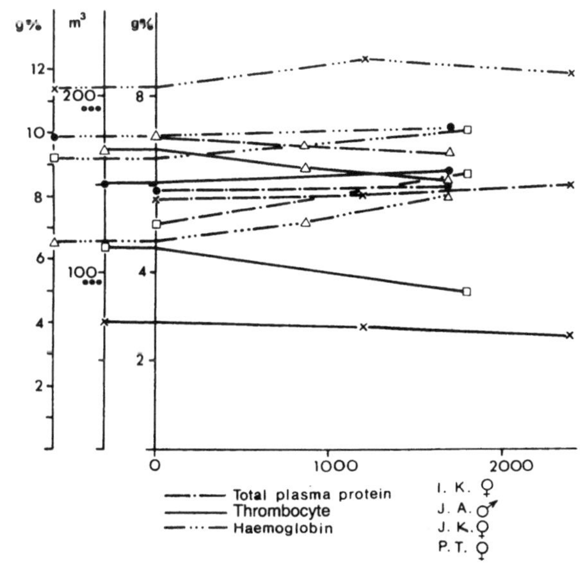


Fig. XI/19

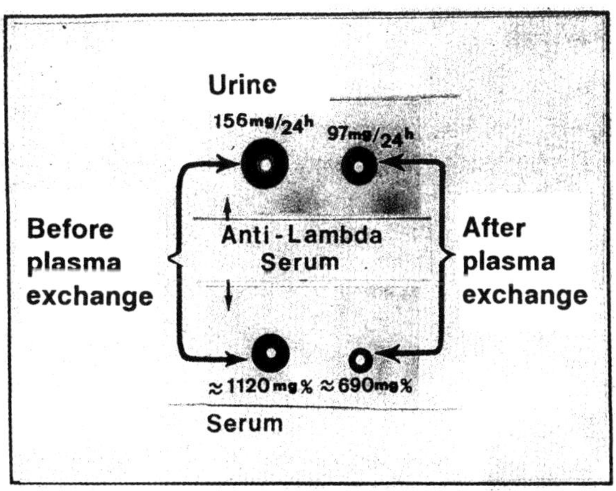


Fig. XI/20

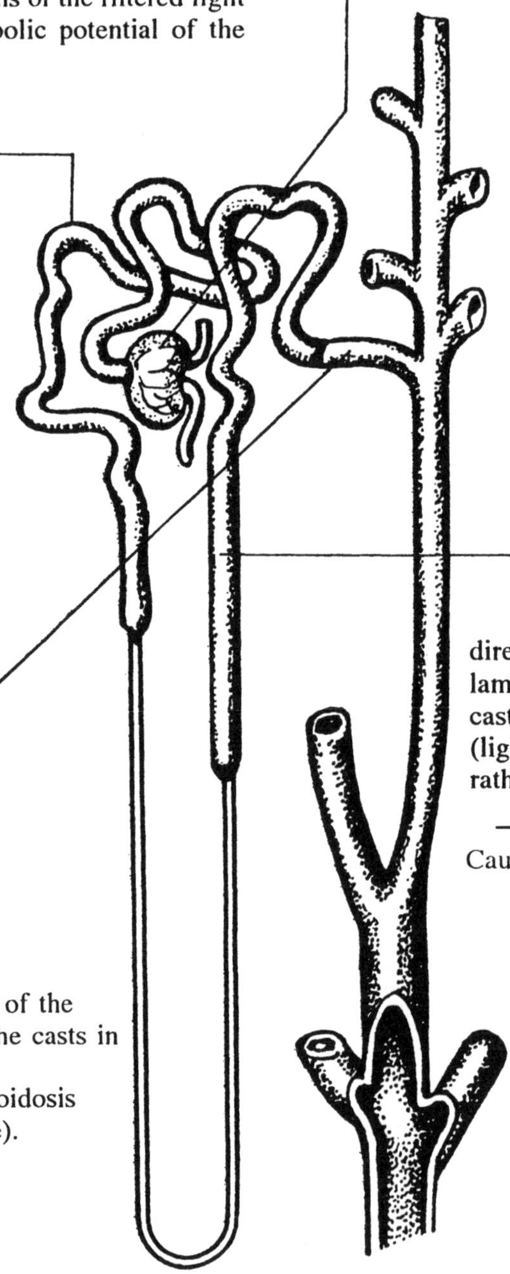


**Fig. XI/21**

assesses renal lesions in gammopathies into two large groups: cylinder nephropathy and damages caused by the light chain deposits (LCDD).

Cylinder nephropathy is a more important form of the disease - the results of renal biopsies indicate that its mild or severe form occurs in every third patient with myeloma (52).

By the time of diagnosis, unfortunately, azotemia is already present, thereby this form can be considered as the most frequent cause of renal insufficiency. Apart from its chronic form, obstructive pathomechanism can elicit acute renal insufficiency. In these cases hypercalcemia, dehydration, appearance of Tamm-Horsfall uroproteins and the effect of some tubulotoxic drugs can contribute to the tubular lesions leaving the glomeruli intact. Cast formation in both acute and chronic forms undoubtedly indicates renal lesion. The capability of the casts to pass across the tubular wall promotes granulomatous inflammation of the interstitium (11, 65).

Light chain nephropathy and the development of myeloid kidney is a special rare (?) form of the disease. The details are presented in the Chapter entitled "Amyloidosis". The following section deals only with the renal pathology induced by the light chain deposits.

The term "myeloma kidney" is generally used to define a recently accepted special subgroup of renal pathology, in which glomeruli are primarily damaged. Clinically it corresponds to glomerulonephritis. The light chain deposition (LCDD) occurs not only in kidneys, but also in other tissues. Data of autopsy indicate that the renal lesions developing in the terminal stage of the disease are never localized solely in the glomeruli or the tubules. There is evidence that in LCCD the deposits occur in the glomeruli, basal mebrane of the tubules, mesangium, and even in the smooth muscles of the renal vessels. Despite the extreme heterogeneity of the glomerular changes, nodular glomerulosclerosis with the thickening of the glomerular stem, resembling the Kimmelstil-Wilson disease, is the most frequently observed form (98). Mesangium is also involved in the pathological process, since protein deposits are usually found; however, cell proliferation is not, or only rarely observed (61, 93). The thickening of the Bowman's capsule is a characteristic finding observed together with tubulo-interstitial alterations, such as tubular atrophy, degeneration, focal dilation and destruction of the basal membrane. Clinically nephrotic syndrome with all the dangerous sequellae (thrombosis, embolism due to low AT III levels) is predominant.

Finally, it is worth pointing out that pathomorphology of the kidney does not affect the therapy: in our opinion, renal biopsy in myelomal patients is not etical, therefore, we don't have sufficient histological data for accurate conclusions. Immunochemical and laboratory studies, however, provide ample information for the localization and evaluation of the lesion, its character and extent. Therefore, adequate prognosis can be made and therapy indicated (Fig. XI/21).

## References


1. Argani, I., Kipkie, G. F.
Macroglobulinemic nephropathia. Acute renal failure in macroglobulinemia of Waldenström.
Amer. J. Med. 36, 151. 1964.
2. Avasthi, P. S., Erickson, D. G., Williams, R. C., Tung, K. S. K.
Benign monoclonal gammaglobulinemia and glomerulonephritis.
Amer. J. Med. 62, 324. 1977.
3. Baglioni, C., Cioli, D., Gorini, G.
Studies on fragments of light chains of human immunoglobulins: genetic and biochemical implications.
Cold Spring. Harbor Symp. Quant. Biol. 32, 147. 1967.
4. Baldwin, D. S., McCluskey, R. T.
Renal involvement in systemic lupus erythematosus, periarteritis nodosa, scleroderma, and cryoglobulinemia.
In: Structural basis of Renal Disease.
Ed: by Bekre, A. L.
New York, Harper and Row. 428. 1968.
5. Balogh, E., Krizsa, F., Varga, Gy., Cserháti, J., Husz, S.
Myeloma multiplexben szerzett tapasztalataink 25 beteg észlelése során.
Orv. Hetilap. 121, 2505. 1980.
6. Bargenter, A. R., Murphy, W. M.
Kappa light chain nepropathy.
Wirchow Arch. A. 410, 531. 1987.
7. Bauer, H. J., Lamerz, R., Gokel, J. M., Remberger, K.
Bence Jones Proteinämie und Amyloidosis.
Münch. med. Wschr. 118, 361. 1976.
8. Bayrd, E. D., Heck, F. J.
Multiple myeloma: Review of eighty-three proved cases.
J. Amer. Med. Ass. 133, 147. 1947.
9. Beaufils, M., Morel-Maroger, L.
Pathogenesis of renal disease in monoclonal gammopathies: current concepts.
Nephron. 20, 125. 1975.
10. Berdon, W. E., Schwartz, R. H., Becker, J.
Tamm–Horsfall proteinuria.
Radiology. 92, 714. 1969.
11. Bergesio, F., Salvatori, M., Lombardi, M., Mishelassi, S., Salerno, A., Imreh, F., Guidi, S., Mousoui, G., Florian, A.
Renal involvement in IgD myeloma.
Scand. J. Urol. Nephrol. 22, 309. 1988.
12. Berkel, J., Granillo-Bodansky, C., Borne, A. E. G.
Acute renal failure associated with a malignant lymphoproliferative disorder with monoclonal light chain immunoglobulin production.
Scand, J. Haematol. 20, 377. 1978.
13. Berkessy, S., Balyi, G., Molnár I., Nagy, Zs., Pauker, Zs., Radványi, G., Végh, I.
Tapasztalataink myeloma multiplexben.
Orv. Hetilap. 128, 1975. 1987.
14. Bernier, G. M., Putnam, F. W.

Monomer-dimer forms of Bence Jones proteins.
Nature. 200, 223. 1963.
15. Bernier, G. M., Fanger, M. W.
Synthesis of beta-2-microglobulin by stimulated lymphocytes.
J. Immunol. 109, 407. 1972.
16. Bienenstock, J., Poortmans, J.
Synthesis and assembly of IgA and other urinary proteins.
In: Protides of the Biological Fluids 21st Colloquium.
Ed: by Peeters, H.
Pergamon Press. 104. 1973.
17. Bloch, K. J.
Gamma heavy chain disease:
An expanding clinical and laboratory spectrum.
Amer. J. Med. 55, 61. 1973.
18. Boros, Gy., Gofman, L., Romhányi, M., Ambrus, M., Ottó, Sz., Deák, Gy.
Kryoglobulinaemia, illetve kryofibrinogenaemia és vese.
Orv. Hetilap. 115, 1048. 1974.
19. Brovet, J. C., Clauvel, J. P., Danon, F., Klein, M., Seligman, M.
Biologic and clinical significance of cryoglobulins.
Amer. J. Med. 57, 775. 1974.
20. Burke, F. J., Flis, R., Lasker, N., Simenhoff, M.
Malignant lymphoma with „myeloma kidney" acute renal failure.
Amer. J. Med. 60, 1055. 1976.
21. Butler, W. T., Rossen, R. D.
Effects of corticostreroids on immunity in man.
I. Decreased serum IgG concentration caused by 3 or 5 days of high does of methylprednisolone.
J. Clin. Invest. 52, 2629. 1973.
22. Butler, W. T., Couch, R. B., Rossen, R. D.
Methlyprednisolone fails to inhibit primary and secondary antibody responses but causes marked suppression of ongoing antibody formation in man.
J. Clin. Invest. 53, 14. 1974.
23. Carson, C. P., Ackerman, L. V., Maltby, J. D.
Plasma cell myeloma: Clinical, pathologic, and roentgenologic review of 90 cases.
Amer. J. Clin. Path. 25, 849. 1955.
24. Clyne, D. H., Brendstrup, B., First, M. R., Pesce, A. J., Finkel, P. N., Pollah, V. E., Pirani, C. R.
Renal effects of intraperitoneal kappa chain infection. Induction of crystals in renal tubular cells.
Lab. Invest. 31, 131. 1974.
25. Constanza, D. J., Smoller, M.
Multiple myeloma with Fanconi syndrome.
Study of a case, with electron microscopy of the kidney.
Amer. J. Med. 34, 125. 1963.
26. Coward, R. A., Delamore, I. W., Mallick, N. P., Robinson, E. L.
The Importance of urinary immunoglobulin light chain isoelectric point (pJ) in nephrotoxicity in multiple myeloma.
Clin. Sci. 66, 229. 1984.
27. Dammacco, F., Waldenström, J.
Bence Jones proteinuria in benign monoclonal gammopathies: incidence and characteristics.
Acta. Med. Scand. 184, 403. 1968.
28. DeFronzo, R. A., Cooke, C. R., Wright, J. R.
Bence Jones proteinuria and renal failure in multiple myeloma.
Clin. Res. 22, 486. 1974.
29. DeFronzo, R. A., Humprey, R. L., Wright, J. R., Cooke, C. R.
Acute renal failure in multiple myeloma.
Medicine. 54, 208. 1975.
30. DeFronzo, R. A., Cooke, C. R., Wright, J. R., Humphrey, R. L.
Renal function in patients with multiple myeloma.
Medicine. 57, 151. 1978.
31. Doyle, G. D., O'Dwyer, W. F., Carmody, M., Cambell, Eileen, Browne, O.
IgA nephropathy – an immunpathological study.
Irish J. Med. Sci. 145, 292. 1976.
32. Druet, P., Letonturier, P., Contet, P., Mandet, C.
Cryoglobulinaemia in human renal diseases.
A study of seventy-six cases.
Clin. Exp. Immunol. 15, 483. 1973.
33. Edelman, G. M., Gally, J. A.
The nature of Bence Jones proteins: chemical similarities to polypeptide chains of myeloma globulins and normal gammaglobulins.
J. Exp. Med. 116, 207. 1962.
34. Engle, R. L. Jr., Wallis, L. A.
Multiple myeloma and adult Fanconi syndrome: I. Report of a case with crystal-like deposit in the tumor cells and in the epithelial cells of the kidney.
Amer. J. Med. 22, 5. 1957.
35. Fine, J. D., Rees, E. D.
Renal cytotoxic effects of Bence Jones proteins.
Lancet i. 150. 1977.
36. Fletcher, A. P., McLanghlin, J. E., Ratcliffe, W. A., Woods, D. A.
The chemical composition and electron microscopic appearance of a protein derived from urinary casts.
Biochim, biophys. Acta. 214, 299. 1970.
37. Forte, F. A.
Heavy chain disease of the $\mu$/gamma M/type: Report of the first case.
Blood. 36, 137. 1970.
38. Frangione, B., Franklin, E. C.
Heavy chain disease:
Clinical features and molecular significante of the disordered immunoglobulin structure.
Semin. Hematol. 1, 53. 1973.
39. Franklin, E. C.
Physicochemical an immunologic studies of gammaglobulins of normal human urine.
J. Clin-Invest. 38, 2159. 1959.
40. Franklin, E. C.
Heavy chain disease: A new disorder of serum gammaglobulins.
Report of a first case.
Amer. J. Med. 37, 332. 1964.
41. Franklin, E. C.
Mü-chain disease.
Arch. Intern. Med. 135, 71. 1975.
42. Golde, D., Epstein, W.
Mixed cryoglobulins and glomerulonephritis.
Ann. Intern. Med. 69, 1221. 1968.
43. Green, R. W.
Conformation and association of the light chain from a homogeneous human immunoglobulin.
Biochemistry. 12, 3225. 1973.
44. Grey, H. M., Kohler, P. F.
A case of tetramer Bence Jones proteinaemia.
Clin. Exp. Immunol. 3, 277. 1968.
45. Grey, H. M., Kohler, P. F.
Cryoimmunoglobulins.
Semin. Hematol. 10, 87. 1973.
46. Hardwicke, J.
Glomerular filtration of macromolecules.
Adv. Nephrol. 2, 61. 1972.
47. Harrison, J. F., Blainey, J. D.
Adult Fanconi syndrome with monoclonal abnormality of immunoglobulin light chain.
J. Clin. Path. 20, 42. 1967.
48. Harrison, J. F., Lunt, G. S., Scott, P.
Urinary lysozyme, ribonuclease and low molecular weight protein in renal disease.
Lancet, i. 371. 1968.
49. Hobbs, J. R., Evans, D. J., Wrong, O. M.
Renal tubular obstruction by mucoproteins from adenocarcinoma of pancreas.
Brit. Med. J. 2, 87. 1974.

50. Hobbs, J. R.
Monitoring myelomatosis.
Arch. Intern. Med. 135, 125. 1975.
51. Hymans, W.
Waldenström's disease with an IgA paraprotein.
Acta Med. Scand. 198, 519. 1975.
52. Iványi, B.
A myeloma multiplex veseszövődményei.
Orv. Hetilap. 38, 2041. 1989.
53. Jákó, J., Virágh, Sz.
Nierenschädigung durch Lambda-Leichtketten.
J. Donausymposium für Nephrologie.
Linz. 252. 1976.
54. Jákó, J., Virágh, Sz.
Nierenerkrankungen bei monoklonalen Gammopathien.
Eine klinisch-pathologische Studie.
Immunität und Infektion. 9, 60. 1981.
55. Jensen, K.
Two Bence Jones proteins in a macroglobulinaemic patient.
Scand. J. Clin. Lab. Invest. 26, 23. 1970.
56. Jensen, H., Wiik, A.
Monoclonal immunoglobulinemia associated with glomerulopathy.
Acta Med. Scand. 197, 265. 1975.
57. Jones, H. B.
On a new substance occurring in the urine of a patient with mollities ossium.
Philos. Trans. R. Soc. London. 55, 62. 1848.
58. Kahler, O.
Zur Symptomatologie des multiplen Myelomas: Beobachtung von Albumosurie.
Prag. med. Wochenschr. 14, 33. 1889.
59. Kamm, D. E., Fisher, M. S.
Proximal renal tubular acidosis and the Fanconi syndrome in a patient with hypergammaglobulinemia.
Nephron. 9, 208. 1972.
60. Kin, K., Sakurabayashi, I., Kawai, T.
Beta-2 mikroglobulin levels of serum and ascites in malignant diseases.
Gann. 68, 427. 1977.
61. Kirkpatrick, C. J., Curry, A., Galle, J., Melzner, J.
Systhemic kappa light chain deposition and amyloidosis in multiple myeloma: novel morphological observation.
Histopathology. 10, 1065. 1986.
62. Koss, M. N., Pirani, C., Osserman, E. F.
Experimental Bence Jones cast nephropathy.
Lab. Invest. 34, 579. 1976.
63. Kyle, R. A., Maldonado, J. E., Bayrd, E. D.
Idiopathic Bence Jones proteinuria: a distinct entity.
Amer. J. Med. 55, 222. 1973.
64. Kyle, R. A.
Multiple myeloma: review of 869 cases.
Mayo Clin. Proc. 50, 29. 1975.
65. Lapis K., Beregi E.
A vesebetegségek klinikopathológiája.
Akadémia Kiadó. Budapest. 149. 1988.
66. Larcan, A.
Syndrome néphrotique et maladie de Kahler.
A propos de 5 observations.
Sem. Hop. Paris. 40, 69. 1964.
67. Lawler, W., Tarpex, P., Williams, G., Acheson, E. J., Mallick, N. P.
Diseases and histological normality of the renal glomerulus: a clinico-pathological study.
J. Clin. Path. 29, 380. 1976.
68. Leathem, A., Wiggins, R.
Cryoglobulins in urine.
Clin. Chim. Acta. 71, 353. 1976.
69. Lee, D. B. N., Drinkard, J. P., Rosen, V. J., Gonick, K. C.
The adult Fanconi syndrome.
Medicine. 51, 107. 1972.
70. Levi, D. F., Williams, R. C., Linstrom, F. D.
Immunofluerescent studies of the myeloma kidney with special reference to light chain disease.
Amer. J. Med. 44, 922. 1968.
71. Lindeman, R. D.
Drug treatment of the chronic idiopathic glomerulopathies.
Postgrad. Med. 62, 135. 1977.
72. Maack, T.
Renal handling of low molecular weight proteins.
Amer. J. Med. 58, 57. 1975.
73. MacIntyre, W.
Case of mollities and fragilitas ossium accompanied with urine strongly charged with animal matter.
Med. Chir. Soc. Tr. 33, 211. 1850.
74. Maldonado, J. E., Velosa, J. A., Kyle, R. A., Wagoner, R. D., Holley, K. E., Salassa, R. M.
Fanconi syndrome in adults.
A manifestation of a latens form of myeloma.
Amer. J. Med. 58, 354. 1975.
75. Mallick, N. P., Dosa, S., Acheson, E. J., Delamore, I. W., McFarlane, H., Seneviratne, C. J., Williams, G.
Detection significance and treatment of paraprotein in patients with „idiopathic" proteinuria without myeloma.
Quat. J. Med. 47, 186. 1977.
76. Manuel, Y., Greenland, T. B.
Immunoelectrophoresis: a convenient method of studying „minimal changes" in proteinuria.
In: Protides of the Biological Fluids 21st Colloquium.
Ed: by Peeters, H.
Pergamon Press. 393. 1973.
77. Manuel, Y., Leclercq, M., Colle, A., Tonnelle, C.
Low molecular weight proteins in tubular proteinuria.
In: Protides of the Biological Fluids 21st Colloquium.
Ed.: by Peeters, H.
Pergamon Press. 505. 1973.
78. Manuel, Y., Colle, A., Leclercq, M., Tonnelle, C.
Low molecular weight proteinuria.
Contributions to Nephrology. 1, 156. 1975.
79. Martinez-Maldonado, M., Yium, J., Sulo, W. N., Eknoyan, G.
Renal complications in multiple myeloma: pathophysiology and some aspects of clinical management.
J. Chron. Dis. 24, 221. 1971.
80. Mauer, S. M., Fich, A. J., Day, N. K., Michael, A. F.
The glomerular mesangium. II. Studies of macromolecular uptake in nephrotoxic nepritis in rats.
J. Clin. Invest. 53, 4321. 1974.
81. McCoy, R. C., Abromowsky, C. R., Tisher, C. C.
IgA-nephropathy.
Amer. J. Path. 76, 123. 1974.
82. McGiven, A. R., Hunt, J. S., Day, W. A., Bailey, R. R.
Tamm–Horsfall protein in the glomerular capsular space.
J. Clin. Path. 31, 620. 1978.
83. McLaughlin, C. L., Solomon, A.
A hidden antigenic site localized to the constant region of light chains of immunoglobulins.
Science. 179, 580. 1973.
84. Meyer, F.
In vitro incorporation of $C^{14}$-lysine into Bence Jones protein.
Proc. Soc. Exp. Biol. 110, 106. 1962.
85. Miano, L.
Le proteinurie infantili.
Inquandramento clinico-patogenetico.
Progr. Med. Roma. 30, 357. 1974.
86. Min, K. W., Cain, D. G., Györkey, Ph., Györkey, F.
Myeloma-like lesions of the kidney.
Amer. Med. Assoc. 136, 1299. 1976.
87. Miszlai, Zs., Olasz, É., Nagy, I., Losonczy, H., Jávor, T.
Myeloma multiplexben szenvedő betegeink adatelemzése 1977–85 között.
Orv. Hetilap. 128, 297. 1987.
88. Monga, G., Mazzucco, G., Coppo, R., Piccoli, G., Coda, R.
Glomerular finding in mixed IgG-IgM cryoglobulinemia.
Virschows Arch. B. Cell. Path. 20, 185. 1976.
89. Morel-Maroger, L., Basch, A., Danon, F., Verroust, P., Richet, G.
Pathology of the kidney in Waldenström's.

N. Engl. J. Med. 283, 123. 1970.
90. Morris, R. C., Sebastian, A., McSherry, E.
Renal acidosis.
Kidney Int. 1, 332. 1972.
91. Nakamuro, K., Tanigaki, N., Pressman, D.
Multiple common properties of human beta-2-microglobulin and the common portion fragment derived from HL-A antigen molecules.
Proc. Nat. Acad. Sci. 70, 2863. 1973.
92. Neauport-Sautes, C., Bismuth, A., Kourilsky, F., Manuel, Y.
Relationship between HL-A antigens and beta-2-microglobulin as studied by immunofluorescence on the lymphocyte membrane.
J. Exp. Med. 139, 957. 1974.
93. Noel, L. H., Droz, D., Gavenal, D., Grünfeld, J. P.
Renal granular monoclonal light chain deposits.
Clin. Nephrology. 21, 263. 1984.
94. Norden, A. G. W., Flynn, F. V., Fulcher, L. M., Richards, J. D. M.
Renal impairment in myeloma: negativ accosiation with isoelectric point of excreted Bence Jones protein.
J. Clin. Pathol. 42, 59. 1989.
95. Osserman, E. F.
Plasma cell myeloma: II. Clinical aspects.
N. Engl. J. Med. 261, 1006. 1959.
96. Osserman, E. F., Takatsuki, K.
Clinical and immunochemical studies of four cases of heavy (H gamma-2) chain disease.
Amer. J. Med. 37, 351. 1964.
97. Osterland, C. K., Espinoza, L. R.
Biological properties of myeloma proteins.
Arch. Intern. Med. 135, 32. 1975.
98. Palant, C. E., Bonitati, J., Bartholomew, W. R., Bretjens, J. R.
Nodular glomerulosclerosis associated with multiplex myeloma.
Amer. J. Med. 80, 98. 1986.
99. Parr, D. M., Pruzanski, W., Scott, J. G., Mills, D. M.
Primary amyloidosis with plasmacytic dyscrasia and a tetramer of Bence Jones type lambda globulin in the serum and urine.
Blood. 37, 473. 1971.
100. Perry, M. C., Kyle, R. A.
The clinical significance of Bence Jones proteinuria.
Mayo Clin. Proc. 50, 234. 1975.
101. Peterson, P. A., Cunningham, B. A., Berggard, I., Edelman, G. M.
Beta-2 microglobulin. A free immunoglobulin domain.
Proc. Nat. Acad. Sci. 69, 1967. 1972.
102. Poortmans, J. R., Jeanloz, R. W., Schmid, K.
3 S gamma-1 globulin levels of normal human serum and urine.
Biochim. Biophys. Acta. 133, 363. 1967.
103. Preuss, H. G., Hammack, W. J., Murdaugh, H. V.
The effect of Bence Jones protein on the in vitro function of rabbit renal cortex.
Nephron. 5, 210. 1967.
104. Preuss, H. G., Weiss, F. R., Tammarino, R. M., Hammack, W. J., Murdaugh, H. V.
Effects of rat kidney slice function in vitro of proteins from the urines of patients with myelomatosis and nephrosis.
Clin. Sci. Mol. Med. 46, 283. 1974.
105. Pruzanski, W.
Clinical manifestations of multiple myeloma: relation to class and type of M component.
Can. Med. Ass. J. 114, 896. 1976.
106. Putnam, R. W., Hardy, S.
Proteins in multiple myeloma:
III. Origin of Bence Jones protein.
J. Biol. Chem. 212, 361. 1955.
107. Randall, R. E., Williamson, C. Jr., Mullinax, F., Tung, M. Y., Still, W. J.
Manifestations of systemic light chain deposition.
Amer. J. Med. 60, 293. 1976.
108. Renner, E.
Glomeruläre Filtration und nephrotisches Syndrom.
In Klinische Pathophysiologie.
Ed. by K. Siegenthaler.
Thieme. Stuttgart 3, 865. 1976.
109. Rota, S., Mougenot, B., Baudouin, B., de Meyer-Brausseur, M., Lemaitre, V., Michel, C.
Multiple myeloma and severe Renal Failure:
A clinicopathologic study of outcome and prognosis in 34 patients.
Medicine. 66, 126. 1987.
110. Sandres, P. W., Herrera, G. A., Chen, Ann, Booker, B. B., Galla, J. H.
Differential nephrotoxicity of low molecular weight proteins including Bence Jones proteins in the perfused rat nephron in vivo.
J. Clin. Invest. 82, 2086. 1988.
111. Scarlett, J. D., Wiley, J. S., Miach, P. J.
Long-term survival in light-chain myeloma with renal failure.
Aust. N. Z. J. Med. 15, 361. 1985.
112. Schenk, E. A., Schwarz, R. H., Lewis, R. A.
Tamm–Horsfall mucoprotein. 1. Localization in the kidney.
Lab. Invest. 25, 92. 1971.
113. Schubert, S. E., Adam, A.
Glomerular nodules and long-spacing collagen in kidneys of patients with multiple myeloma.
J. Clin. Path. 27, 800. 1974.
114. Schubert, G. E., Veigel, J., Lennert, K.
Structure and function of the kidney in multiple myeloma.
Virchows Archs. Path. Anat. 335, 135. 1972.
115. Seligmann, M.
Alpha chain disease: A new immunoglobulin abnormality.
Science. 162, 1396. 1968.
116. Seligmann, M., Clauvel, J. P.
Données actuelles sur les proliférations de cellules B et les dysglobulinémies.
Actualités Néphrologiques de l'Hospital, Necker.
Medicales Flammarion. Paris. 289. 1977.
117. Seon, B-K., Grossberg, A. L., Roholt, O. A.
Position of the selective peptic cleavage of a human kappa-chain between the V-kappa and C-kappa domains.
J. Immunol. 111, 269. 1973.
118. Shustik, C., Bergsagel, D. E., Pruzanski, W.
Kappa and lambda light chain disease:
survival rate and clinical manifestations.
Blood. 48, 41. 1976.
119. Smithline, N., Kassirer, J. P., Cohen, J. J.
Light chain nepropathy: tubular dysfunction.
N. Engl. J. Med. 294, 71. 1976.
120. Snapper, I., Kahn, A.
Myelomatosis: Fundamentals and clinical features.
Baltimore, University Park Press. 189. 1971.
121. Sobel, A. T., Antonucci, M., Intrator, L., Bernard, D., Beaumont, J. L., Lagrue, G.
Association d'une gammapathie monoclonale, d'une glomérulopathie chronique et d'une hyperlipidémie.
Nouv. Presse Méd. 5, 2375. 1976.
122. Solomon, A., Fahey, J. L., Malmgren, R. A.
Immunohistologic localization of gamma-1 macroglobulins, beta-2A myeloma proteins, 6,6 S gamma myeloma proteins and Bence Jones proteins.
Blood. 21, 403. 1963.
123. Solomon, A., Fahey, J. L.,
Metabolism of Bence Jones proteins.
Amer. J. Med. 37, 206. 1964.
124. Solomon, A., Waldman, T. A., Fahey, J. L., McFarlane, A. S.
Metabolism of Bence Jones protein.
J. Clin. Invest. 43, 103. 1964.
125. Solomon, A., Killander, J., Gray, H. M.
Low-molecular-weight proteins related to Bence Jones proteins in multiple myeloma.
Science. 151, 1237. 1966.
126. Solomon, A., McLaughlin, C. L.
Bence Jones proteins and light chains of immunoglobulins.

I. Formation and characterization of amino-terminal (variant) and carboxyl-terminal (constant) halves.
J. Biol. Chem. 244, 3393. 1969.
127. Solomon, A.
Bence Jones protein: a molecular entity for studies on the structural and genetic control of antibody specificity.
Mt. Sinai. J. Med. N. Y. 37, 466. 1970.
128. Solomon, A., McLaughlin, C. L., Capra, J. D.
Bence Jones proteins and light chains of immunoglobulins. XI. A transient Bence Jones-related protein associated with corticosteroid therapy.
J. Clin. Invest. 55, 579. 1975.
129. Solomon, A.
Bence Jones proteins and light chains of immunoglobulins.
N. Engl. J. Med. 294, 17. 1976.
130. Sox, H. C. Jr., Hood, L.
Attachment of carbohydrate to the variable region of myeloma immunoglobulin light chain.
Proc. Nat. Acad. Sci. 66, 975. 1970.
131. Squire, J. R., Blainey, J. D., Hardwicke, J.
The nephrotic syndrome.
Brit. Med. Bull. 13, 43. 1957.
132. Stekhoven, Sch., Haelst, J. H., Van. U. JG. M.
Unusual findings in the human renal glomerulus in multiple myeloma. A light- and electron-microscopyc study.
Cell. Path. 9, 311. 1971.
133. Stevenson, G. T.
Detection in normal urine or protein resembling Bence Jones protein.
J. Clin. Invest. 39, 1192. 1960.
134. Strife, C. F., McDonald, B. M., Ruley, E. J., McAdams, A. J.,. West, C. D.
Shunt nephritis – nature of serum cryoglobulins and their relation to complement profile.
J. Pediat. 88, 403. 1976.
135. Strober, W., Waldman, T. A.
The role of the kidney in the metabolism of plasma protein.
Nephron. 13, 35. 1974.
136. Tanaka, Y., Azuma, T., Hamgucki, K.
Ionization and reactivities of the thiol groups which participate in the formation of interchain disulfide bonds of Bence Jones proteins and an Fab(t)fragment.
J. Biochem. 83, 1249. 1978.
137. Vaughan, J. H., Jacox, R. F., Gray, B. A.
Light and heavy chain components of gamma globulins in urines of normal persons and patients with agammaglobulinemia.
J. Clin. Invest. 46, 266. 1967.
138. Verroust, P.
Les Lésions glomérulaires des gammopathies monoclonales et des cryoglobulinémies idiopathiques IgG-IgM.
Actualités Nephrologiques de l'Hopital, Necker.
Médicales Flammarion. Paris. 167, 1971.
139. von Schéele, Ch.
Light chain myeloma with features of the adult Fanconi syndrome: six years remission following one course of melphalan.
Acta. Med. Scand. 199, 533. 1976.
140. Waldmann, T. A., Strober, W., Mogielnicki, R. P.
The renal handling of low molecular weight proteins. II. Disorders of serum protein catabolism in patients with tubular proteinuria, the nephrotic syndrome or uremia.
J. Clin. Invest. 51, 2162. 1972.
141. Waldmann, T. A., Strober, W.
Renal regulation of serum protein metabolism.
In: Protides of the Biological Fluids 21st Colloquium.
Ed: by Peeters, H.
Pergamon Press. 419. 1973.
142. Williams, R. C. Jr., Pinnell, S. R., Bratt, G. T.
Low molecular weight L-chain components related to Bence Jones proteins.
J. Lab. Clin. Med. 68, 81. 1966.
143. Winer, R. L., Cohen, A. H., Sawhney, A. S., Gorman, J. T.
Sjögren syndrome with immune-complex tubulointerstitial renal disease.
Clin. Immunol. Immunpath. 8, 494. 1977.
144. Witts, L. J., Bladkurn, E. K., Callender, S. T.
Report on the first myelomatosis trial: Part I. Analysis of presenting features of prognostic importance.
Brit. J. Haemat. 24, 123. 1973.
145. Wochner, R. D., Strober, W., Waldmann, T. A.
The role of the kidney in the catabolism of Bence Jones proteins and immunoglobulin fragments.
J. Exp. Med. 126, 207. 1967.
146. Zinneman, H. H., Levi, D., Seal, U. S.
On the nature of cryoglobulins.
J. Immunol. 100, 594. 1968.
147. Zlotnick, A., Rosenmann, E.
Renal pathologic findings associated with monoclonal gammopathies.
Arch. Intern. Med. 135, 40. 1975.
148. Zollinger, H. V.
Die pathologische Anatomie der Macroglobulinaemie Waldenström.
Helv. Med. Acta. 25, 153. 1958.

# XII. Amyloidosis

Amyloidosis was probably first described by Nicolaus Fontanus, however, he certainly did not use this term.In his book edited in 1639 and entitled "Responsinum et curationum medicinalium liber unus" he presented an autopsy record of a young man who had jaundice, nasal bleeding and protrusion of the belly due to ascites. Macroscopic study of the organs revealed hepatomegaly - with liver abscess and splenomegaly - with "white stones" in the spleen. This phenomenon termed "sago spleen" is still used.*

In 1657 Thomas Bartolin - the discoverer of the lymphatic system - described the autopsy results of a woman and reported indurations in the spleen and muscles. Cutting spleen was compared by the author to "the cutting of spongy timbers". Theophile Bonet in his book entitled "Sepulchretum sive anatomia practica" - edited in 1679 - collected about 3000 data of autopsies from "autopsy reports " of that period. Schwartz (1970) consideres these two reports as the first publications on amyloidosis. (322) It should be, however, emphasized that without histological studies these results can not be taken for guaranteed. Rokitansky in 1842 reported that in some - cases of syphilis, rickets, tuberculosis or osteomyelitis - autopsy revealed waxy liver with fatty degeneration (301). His report seems to be more authentic and it is also in agreement with the later experiences.

The term "amyloid" has been first introduced by the botanist Schleiden for the designation of plant components containing starch. The Greek word amyloid means "starch-like" (318) Virchow in 1854 used the definition "amyloid-(like) alteration" because of the similarity between the reactions of amyloid and cellulose with iodine and sulphuric acid (377). Meckel furnished further proofs of the involvement of various organs by detecting "fatty" lesions in the aorta, arteries and intestinal wall in addition to the previously discovered *degeneration* of the liver, spleen and kidneys (248). Budd pointed out that the waxy consistence of the organs is caused not by fat, but rather by a protein-like substance (36). According to Fridreich and Kekule (1859) amyloid is rather albuminoid (that is a protein) than fat but nevertheless, the term amyloid has been kept (110).

It was perhaps Wilks in 1856 who was the first to describe primary amyloidosis. He observed the typical "fatty degeneration" without association with any other disease known by that time (398). Some years later - in 1872 -, as a logical consequence of the previously mentioned observations, Adams found amyloidosis in an autopsy of multiple myeloma (2). In spite of the above mentioned reports the first description of primary amyloidosis in the medical literaturc had been attributed to Wild (1866) (397).

In 1895 experimental Staphylococcus aureus infection was the first animal model of the diseases (206). Much later - in 1923 - injection of casein to nice lead to the development of amyloidosis (207); it should be noted that the same method has been regarded until recently as the most generally accepted procedure for experimental induction of this strange metabolic disease.

The metachromasia of amyloid was first reported in 1875 (6). It was observed that the colour of methyl violet and crystal violet becomes altered when staining amyloid. Staining with Congo red is considered to be most specific method for detecting this degenerative process. Staining with Congo red was first used by Bennhold in 1922 as the first method employing polarized light (16).

* Sago is a Malayan-Portuguese word referring to a mealy matter extracted from the starchy pith of the so-called sago palm-trees. When cooked, the grains of this matter are closely sticked together resembling amyloid deposits.

Amyloid is apparently homogeneous and amorphous when studied with light microscope. Amyloid is stained pink with haematoxyline-eosin and shows metachromasia with the above mentioned dyes. In polarized light the colour of the Congo red dye looks apple-green and can be characterized by double refraction of light (302, 303). It is however, worth pointing out that sometimes false-positive results might occur,therefore it is essential to try to obtain unambiguous results by using, parallel control methods. Electron microscopic examination (1959) can be regarded as the most specific method in this respect: the linear, non-ramified 7.5 × 10 nm wide and illimitably long, aggregated fibrillar structure of amyloid is visualized as the hollow nuclei (49, 350) (see Fig. XII/1 and XII/2). The final break-through happened in 1968 when it became possible to isolate fibrilles from the tissues and purify them for the chemical analysis (285).

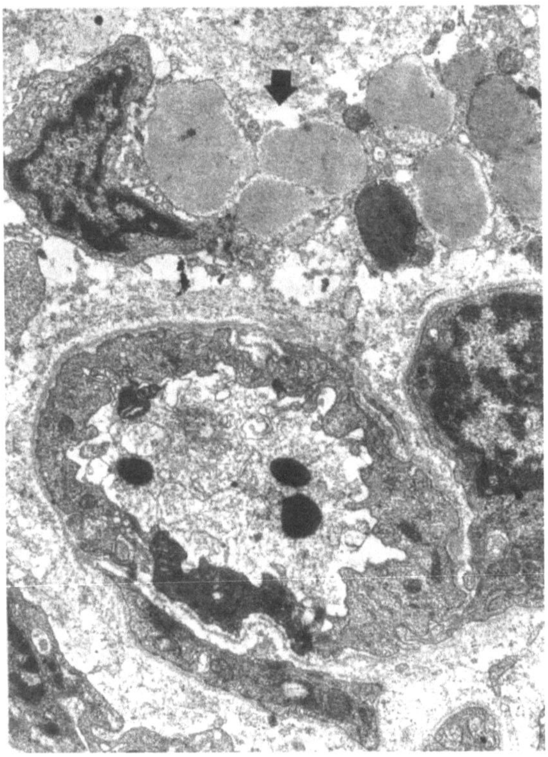

**Electromicrograph of macrophages storing amyloid (magnification 8000 ×)**

**Fig. XII/1**

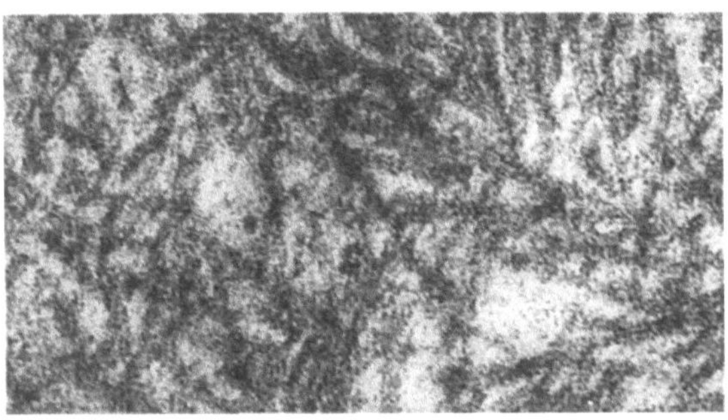

**Characteristic filamentous structure of amyloid shown on a high power electromicrograph. Note periodic structure of filaments (magnification 13000 ×)**

**Fig. XII/2**

The possible connection between the Bence Jones proteinuria, abnormal plasma proteins and the plasma cell dyscrasia - as a possible clinicopathological background of the disease - has been first proposed in 1931 (240), thereafter - for the second time - in 1964 (264), but it was proved only in 1971 (127). The amino acid sequence analysis of the AA protein inducing secondary amyloidosis was performed in 1972 (225). Later new results have expanded our knowledge regarding the amyloid precursors (52, 66, 130).

The results internationally recognized amyloid research in Hungary are mainly attributed to Romhányi. He found that amyloid stained with Congo red shows a green-coloured double refraction in polarized light, therefore, it has an oriented structure/it is of oriented structure. It was also Romhányi who had introduced the concept of permanganate sensibility and resistance, often mentioned by handbooks without, however, mentioning the name of the author (302, 303).

Substancial amount of the literature on amyloides associated or accidentally connected with various diseases is available, but can not be unambiguously analyse (55, 175, 244). A chronological list should, however, be disclosed to provide a survey for the reader. The recently published more accurate studies are, on the hand based on the development of laboratory tests; on the other hand, these studies try to differentiate between accidental coincidences and the causal relationship.

| | | |
|---|---|---|
| Auerbach | 1944 (7) | of 468 tuberculotic patients treated in a T.B. hospital only 5 had amyloidosis; |
| Kimball | 1961 (198) | found only 16 cases (0.4%) of amyloidosis in 4033 auto- |

| | | |
|---|---|---|
| | | psies of patients with malignant tumours; |
| Williams | 1965 (400) | in primary leprosy amyloidosis was found in 6% of the patients; |
| Kuhlbäck | 1966 (208) | observed amyloidosis in patients with bronchiectasis; |
| Case Records of MGH | 1968 (41) | description of Niemann-Pick disease and amyloidosis; |
| Sharma | 1970 (329) | haemophilia (maybe due to many blood transfusions?) and amyloidosis; |
| Prentice | 1971 (292) | Christmas-diseases associated with amyloidosis; |
| Penman | 1972 (276) | of 59 patients with hypernephroma only one had generalized and two liver localized amyloidosis; |
| Elcock | 1972 (88) | Waldenstrm's macroglobulinaemia and amyloidosis; |
| Scott | 1973 (325) | Waldenströms macroglobulinaemia complicated with peripheral neuropathy and amyloidosis; |
| Falkson | 1973 (94) | 53 patients with Hodgkin's disease and amyloidosis - kidney was affected in most of the cases; |
| Husby | 1975 (167) | 10-15% of patients with rheumatoid arthritis had amyloidosis; |
| Rosenthal | 1975 (307) | concomitant occurrence of Behcet's syndrome and amyloidosis; |
| Tschang | 1976 (366) | amyloidosis connected to non-Hodgkin's lymphoma; |
| Stevens | 1976 (353) | Waldenström's macroglobulinaemia and amyloidosis; |
| Hackler | 1977 (146) | development of renal amyloidosis in paraplegic patients was proportional to the development of decubitus ulcers, suggesting that the latter were involved in the development of amyloidosis; |
| Dikman | 1977 (75) | a 6-year history of chronic purulent skin lesions following burns lead to renal amyloidosis; |
| Shorvon | 1977 (337) | amyloidosis was detected in two cases of ulcerative colitis; |
| Clark | 1977 (47) | X-chromosome-linked anhydrotic ectodermal dysplasia complicated with amyloidosis; |
| Gaffney | 1978 (116) | primary hypogammaglobulinaemia and X-chromosome-linked hypogammaglobulinaemia associated with primary amyloidosis - these cases require special attention; |
| Joachim | 1978 (171) | cryoglobulinaemia and amyloidosis; |
| Pruzanski | 1978 (295) | μ-heavy chain disease and amyloidosis; |
| Knowles | 1978(203) | reticulum cell sarcoma and lymphocytic leukaemia associated with amyloidosis as a cause of death; |
| Dikman | 1978 (76) | Gaucher's disease and amyloidosis; |
| Hanash | 1978 (149) | Gaucher's disease and amyloidosis; |
| Strausbaugh | 1978 (356) | in four children with Schistosoma mansoni, in one adult with Schistosoma haematobium and two other adults with Schistosoma mansoni and Salmonella paratyphi A the infection was associated with amyloidosis; |
| Hunter | 1979 (166) | protein-losing enteropathy and amyloidosis; |
| Champion | 1979 (44) | four patients with Hodgkin's disease developed nephrotic syndrome due to amyloidosis; |
| Miller | 1979 (251) | in Bechterew's disease, in juvenile rheumatoid arthritis, in psoriatic arthritis, in Sjögrens syndrome and in Reiter's syndrome amyloidosis was detected (in one case in each disease); |
| Westermark | 1979 (390) | in 4 of the 20 examined patients with renal diseases amyloid was found in the blood-vessels, only slightly affecting glomeruli; considerable proteinuria was, however, detected only in one case; |
| Verbanck | 1979 (374) | regional enteritis and amyloidosis; |
| Scholes | 1979 (320) | drug abuse, several loci of purulent inflammation of the skin corresponding to drug administration sites + amyloidosis; |
| Shiomura | 1979 (332) | cyclic neutropenia and amyloidosis; |
| Rey | 1979 (299) | retroperitoneal paraganglioma associated with amyloidosis; |
| Case Records of MGH | 1980 (40) | hypernephroma and amyloidosis; |
| Moyner | 1980 (254) | Waldenströms macroglobulinaemia + AA amyloid!! |
| Plavnick | 1980 (274) | immunoblastic lymphadenopathia + amyloidosis; |
| Huston | 1981 (170) | lupus erythematosus and amyloidosis; |
| Kyle | 1981 (213) | gamma-heavy chain disease and amyloidosis; |

| | | |
|---|---|---|
| Troussard | 1989 (365) | systemic amyloidosis and light-chain deposition; |
| Sargent | 1989 (315) | amyloidosis related to bone cyst and haemodialysis; |
| Perez-Villa | 1989 (280) | renal amyloidosis and acne conglobata; |
| Friedman | 1989 (109) | amyloidosis associated with astrocytoma.The structure of the amyloid |

## Structure of amyloid

Amyloidosis is a "metabolic anomaly generally characterized by extracellular protein deposition and by specific staining and ultrastructure of the abnormal protein. The deposits may be found in systemic or localized arrangement. If vital organs are involved, a progressive severe disease will develop. The deposited substance - according to our present knowledge - has been found insoluble. Since no effective therapeutic measures are available, systhemic amyloidosis usually has fatal outcome. The local form of the disease, usually develops with ageing and is asymptomatic.

The amyloid deposits contain protein fibrils consisting of non-glycosilated peptides of various - but a for a certain deposit, however, of similar - origin. The two major clinical forms of the disease - primary and secondary amyloidosis - can be classified on the basis of the type(s) of precursor(s). In addition to the main factor - the amyloid ogenous peptide - in both forms of the disease a non-fibrillar glycoprotein, the so-called component P (AP) (277), contributes in 4-14% to the formation of the minor structure. Apart from the above mentioned typical form of amyloidosis, another form with a typical intracellular deposition of the amyloid (in the brain cells) lacking the P component also occurs (394).

The AP protein belongs to the pentraxine-group of plasma proteins (278), in the English bibliography it is mentioned as a doughnut-shaped protein, and is structurally closely related to C-reactive protein. Some researchers consider this protein as a "basement for fibril deposits (269). This, however, is only a hypothesis and, at present, its real involvement in amyloid formation and maintenance has not yet been elucidated. Recently a "specific ligand" has been synthetized (155), which is capable of vitro dissolving the SAP, - or rather the AP - incorporated in vivo. The SAP is a plasma component requiring calcium ions ($Ca^{++}$) for its attachment to the amyloid structure. The molecule is a glycoprotein and migrates with the $\alpha_1$-globulin fraction. The N-terminal amino acid sequence of the so-called $\alpha_1$-glycoprotein of the plasma is identical with the structure of component P extracted from amyloid (23). It should be emphasized that this protein can be found not only in the serum of patients with amyloidosis, but also as a constant component in the blood of healthy individuals. Moreover it is detected in the normal glomerular basal membrane (85). There has been debate about the abnormal nature of this substance. Its possible pathogenetic role has not yet been elicudated.

In all forms of the disease the amyloid fibrils are rigid and non-fragile, with 10-15 nm in diameter their length can not be defined (103). The diameter of a filament is about 70 Å, that of its subunit is approximately 30-30 Å (51). Each fibril comprises an aggregate of 2-5 filaments with a configuration of a twisted band, in other words, of a double β-helix (60). The longitudinal axis of the polypeptide chains of the filaments is perpendicular to that of the fibrils, exhibiting a so-called antiparallel or β-folded structure. The recent observations are in some respect inconsistent with this generally accepted model. It is debated the fibrils have β-structures formed from planes with systematic cross-folding (367)

Several aspects must be taken into consideration in classification of the types of amyloidosis, therefore, such classification can not be regarded as uniform or definitive. It will, probably be suitable or at least more or less definite when our knowledge concerning this "metabolic" disease will be more profound. The present classification is based on the biochemically identified nature of molecules forming the amyloid (52, 130). For example on the analyses proving the presence of the precursor in the serum and its identity with the constituent of the amyloid. There are, however, certain cases in which the precursor-product relationship cannot be proved and in some cases even nature of the precursor can not be identified. (AUO = *a*myloid of *u*n-known origin.), (53, 105, 225, 279). The main pitfall of the first classification attempts is that they consider "only" the clinical and pathological aspects (50). However, due to the lack of biochemical results a better classification could not have been introduced.

In primary amyloidosis the heart, tongue, gastrointestinal tract, nerves and the skin are infiltrated, in the secondary type, however, the deposits are mainly found in the tissues of the liver, kidneys and spleen.

A close analysis of between the involvement of various organs or by histochemical methods - including electronmicroscopy - indicates that the above two forms of the disease could not be separated by these criteria (51).

The accurate diagnosis has to be based on immunochemical methods; the results of this analysis helps to identify the precursor and, hence, to establish a real classification.

The precursor of primary amyloidosis is the light chain of immunoglobulins; the AL fibrill-protein, fragments of the monoclonally produced light chains or their mixtures comprise the charasteristic structure of the deposits. The fragment belongs to the N-terminal region of the molecule and usually contains the $V_2$ domain. This is the why the molecular weight of AL proteins varies between 8000 and 30,000 dalton. Amyloids containing intact, not fragmented light chains are rarely detected. Therefore, studies of the light chains of the circulating or urine monoclonal proved that they are identical with the protein extracted from amyloid or with its fragments, suggesting that the isolated protein plays a precursor role (399). The above statement could also be proved by: in vitro experiments, since normal amyloid fibrils could be produced by pepsine-digestion of the light chains (128). The AL are of lambda light chain origin (22) and are only rarely derived from the kappa light chain (215, 256, 281). It should be definitely emphasized that mostly the lambda light chains and their subtype VI can be detected in the amyloid extracts. On the other hand, however, in monoclonal gammopathies and even also among the normally occuring immunoglobulins the κ light chain can be found in a larger proportion than the λ type and, furthermore, the subtype VI of the λ light chains occurs extremely rarely. Therefore, a concept of amyloidogeneous molecule seems to be reasonable (348). It is also discussed whether the amyloid *building material* is produced intracellulary (79, 261, 409). The observation that anti-amyloid monoclonal antibodies bind to the plasma cell membrane as well as the recent concept that the light chain fragments are glycosilated are insufficient to solve this problem (97, 293). Despite the above mentioned observations we have to agree with the opinion of certain specialists emphasizing that the amyloid can not be formed from the precursor alone, but also one or several unknown factors - or some unknown factors - is/are required (129, 291, 349, 354).

In the future it has to be found out whether an idiotype-identity is involved; that is the specific immune response to a certain unknown antigen would also indicate of the λ type immunoglobulins with subsequent formation of amyloidogeneous molecules via an idiotype determined by the initial antigen molecule. The investigation of primary amyloidosis is mainly hindered by the lack of a suitable animal model.

The connection between the accumulation of light chains, their presence in the bloodstream and the development of amyloidosis has remained questionable despite some conclusive publications (128, 161). Still, it has been ascertained from the aspect of experimentum naturae that:

1. The light chain disease may occur *without* multiple myeloma (297).
2. Light chain deposits may be found on the tubular and glomerular basal membranes of the kidney *without* the presence of amyloid structure (117, 373).
3. At the same time in the blood-vessels *amyloid* can be detected *precluding* the precursor role of light chains and proving the presence of AA.
4. The variable part of the heavy chains of immunoglobulins may actively affect the cells. For example, the free heavy chains may possibly play an inhibiting role in the development of certain B-cell malignancies. This observation may perhaps be related to the well-known fact that if the malignant B-cells produce free heavy chains, they are deficient in the variable part

(104). The amyloidogeneous light chain fragments, however, always contain variable regions or parts of them, insofar as the lack of the OZ-marker can be proved (233). It is also well-known fact that in heavy chain disease, which otherwise occurs very rarely amyloidosis can only sporadically be detected (213).

There is also evidence that in some patients with monoclonal gammopathy SAA was the amyloid precursor (61, 286).

Among secondary amyloidosis the deposits containing AA fibrill-proteins might play the most important pathogenetic role. The AA protein of amyloid is a peptide comprising 75 amino acids with a molecular weight of approximately ~ 8500 dalton (225). The results of the recent studies suggest the existence of smaller and larger peptide as well (168, 173). The amino acid chain of the AA protein prepared from amyloid is identical with one of the serum acute-phase globulins therefore, the precursor seems to be a normal plasma component (SAA) detectable also under physiological conditions. Its Se concentration is 200 μg/ml and its function has not yet been elucidated, although some evidences exists that it might modulate the collagenase synthesis (33). Presumably, the chronic inflammations in which amyloid is found in the AA subunits not only increases the production of this precursor protein, but also simultaneously inhibits the breakdown and/or elimination of the old molecules. Accordingly, the formation of amyloid deposits goes on in two different ways with the same final outcome (165). High level of the acute-phase globulins can be observed in several diseases, while the prevalence of amyloidosis is rather low. Furthermore its incidence is even lower in lupus erythematosus or ulcerative colitis, for in these two diseases the otherwise increased concentration of SAA is lower than the its levels measured in other diseases (72). The previously described two observations suggest that the permanently high SAA concentrations are related with the formation of amyloid and be responsible for a fast progression in cases of already existing amyloidosis (91). The SAA is probably not a homogeneous substance, but similarly to certain experimental results observed in mice has isotypes, one of which is amyloidogeneous (160). There are certain data available referring to the existance of 6 polymorphous formations humans (13). In addition to the AA substance (AA protein) a so-called ASF (acid soluble fraction) can also be detected, however, not too much is known about it, apart from the characteristics implied in its name (225). It can, however, be definitely stated that this substance is not identical with either of the presently known peptides. The SAA strongly binds to the high density lipoprotein (HDL), (15, 90). It seems to be a strange contradiction that the SAA concentration is generally high in both primary and secondary amyloidosis, but has been found to be normal in the heredo-familial cases (18). The SAA levels are usually normal in ALL (acute lymphobastic leukaemia) and in CLL (chronic lymphocytic leukaemia), but have been found higher in case of metastatic tumours, in AML (acute myeloblastic leukaemia), in CML (chronic myeloid leukaemia) and in various types of lymphoma (306). It should, however, be taken into consideration that in elderly people the SAA concentration is usually higher than in the young population (19), although there are other publications representing the opposite opinion (153). The SAA may originate from the macrophage cells, the polymorphonuclear leukocytes (305), and mainly from the liver (199), The decomposition of amyloid fibrils – inhibition of their catabolism – may contribute to the development of the disease. There are certain publications reporting the lack of SAA decomposition in patients with amyloidosis (222). Therefore the following principle is assumed: the substance which can not be decomposed by the organism is stored. The underlying factor is probably the monocyte dysfunction, as the elastase of the monocytes is usually impaired or originally abnormal (hereditary weakness or excessive utilization may be the reasons), (223). At the same time it has also been observed that in mononuclear leukocyta cultures the SAA substance has degraded to AA protein, that is, in other words, the enzymatic effect of the leukocytes instead of protecting the organism from amyloidosis may even precipitate it (344).

AF, $ASc_1$ – Prealbumin is the precursor protein of familial amyloidosis (AF), however, apart from the deposits containing intact molecules, various fragments of prealbumin could also be detected as constituents (66). It is a negative acute-phase reactant, that is its concentration does not increase, but rather decreases in the inflammatory processes (395). Its normal concentration is about 20-40 µg/ml and it is synthetized by the liver. Prealbumin - as precursor of familial amyloidosis - has been described in 1978, (66). Somewhat later it was found that the peptide detectable in senile amyloidosis ($ASc_1$) also originates from prealbumin (64, 282, 346). Although the familial amyloidose polyneuropathy (FAP) and the senile systhemic amyloidosis ($ASc_1$) are similar in certain respect (fibrillar structure and, especially, the precursors), in view of the clinical symptoms, however, these diseases are different and, furthermore, while FAP mainly occurs in young and middle-aged individuals, $ASc_1$ rather develops in elderly people. The production of prealbumin is connected with the chromosome 18 (is chromosome 18-linked), (381). The molecular weight of prealbumin is 55 kD, it is a tetrameric protein with non-covalent bonds, consisting 4 chains of 127 amino acids each, showing a tetraedric symmetric arrangement (187), (see Fig. XII/3). Thus, a narrow channel is formed in the middle of the molecule with two binding sites for thyroxine (27); this carrier protein, however, binds not only thyroxine, but also vitamin A (186). Each dimer comprising 8 bands (26) forms two beta folds such structure is unusually stable and resistant to forces causing dissociation. The shape and characteristics of the molecule render an ideal precursor for amyloid. It contains a P component and reacts with anti-prealbumin or with anti-$ASc_1$ monovalent immunoserum; the specificity of binding can be proved by the fact that both reactions can be blocked with purified human prealbumin or extracted $Sc_1$. Of course, these immunosera will not react with AL and AA proteins. The enzyme-digestion of Sc1 will result in the production of peptides homologous with the amino acid chain sequences 70-90, 96-107, 109-115, 121-127 of the prealbumin (346). The $ASc_1$ is insoluble in water, while the prealbumin, can be easily solved. The electrophoretic analysis of $ASc_1$ has shown 3 components of different mobility (98).

I. The amyloid precursor nature of prealbumin can be confirmed by the following observations:
   - Identical molecular weights, as shown by SDS-PAGE method;
   - reacts with anti-human prealbumin immunoserum;
   - contains only one cysteine residue;
   - non-covalent bonds hold together the subunits forming the molecules both in prealbumin and amyloid.

II. The subunit smaller than prealbumin (of 13 kD):
   - strongly reacts with anti-$ASc_1$ and very slightly with anti- prealbumin immunoserum (335);
   - does not contain cysteine;
   - is the major component of $ASc_1$.

III. The characteristics of the smallest subunit are as follows:
   - contains cystein;
   - strongly reacts with anti-$Sc_1$ immunoserum, but scarcely with anti-prealbumin serum (346).

On the basis of the characteristics described the peptides mentioned in items II and III can be regarded as the fragmentation products of prealbumin (63).

$AE_t$ - Among the hormones produced by the endocrine glands some might be the precursors of the fibrils (389.) Amyloidosis has most frequently been associated with the medullary carcinoma of the thyroid gland, where precalcitonin or calcitonin synthetized by the proliferating tumour cells would become a precursor (71, 345). Namely, it has been proved that the amino acid sequence of the amyloidogeneous peptide is identical with that of the above mentioned substances (345). The hormones of the pancreas and those of the pituitary gland have also been found to exert a similar effect (128, 215), the islet-cell adenoma produces the so-called islet amyloid polypeptide (IAPP) - a molecule which differs from both proinsulin and insulin (180). Irrespectively of the causes of accumulation of the endocrine amyloid, the undesirable, extracellular protein serves as a tumour matrix.

One of the forms of the senile cardiac amyloidosis affects the atria of the heart, therefore, it is called isolated atrial amyloidosis (IAA). It should be mentioned in this context because its precursor is the a-atrial natriuretic factor (α-ANF), (179). The prohormon is, on the other hand, a larger molecule. The active diuretic end-product containing 126 amino acids splits of the C-terminal ending. Still the amyloidogeneous peptide of the highest molecular weight could not be identified in these cases (56). Examinations of tumours of other endocrine organs also revealed a connection between the subunits and the specific hormonal product of the respective organ, the final biochemical and immunological identification, however, has not yet been performed. The following experimental result should absolutely be taken into consideration: some polypeptide hormones (HGH - human growth hormone -, gastrin and parathyroid hormone) acquire in vitro a β-helix modification due to acidic or enzymatic effects (193). According to the above observation local amyloid production in certain organs may perhaps be initiated by a local undesirable proteolytic effect, the latter, however, can not be proved.

AH - In patients dialysed for longer periods the prevalence of amyloid deposition has been found to be markedly higher than in those treated otherwise (e.g. plasma substitution), or compared to the normal population (183, 209). The clinical symptoms, such as carpal tunnel syndrome, cystic bone lesions and other rheumatological diseases (99, 385) occur. These symptoms differ from those usually caused by the main disease requiring the dialysis.

Data of HPLC of the amyloid and the N-terminal amino acid sequence analysis of the 11,800 dalton molecular weight basic protein suggest that a $\beta_2$-microglobulin (336) can be a possible precursor. Under physiological conditions this substance is a component of the cell membrane, connected to the first-class HLA antigens. Its structure is homologous with a part of the constant region of the heavy and light chains (68). It often binds to collagen. It is assumed that this characteristic ensures the amyloidogeneous nature of the sub-

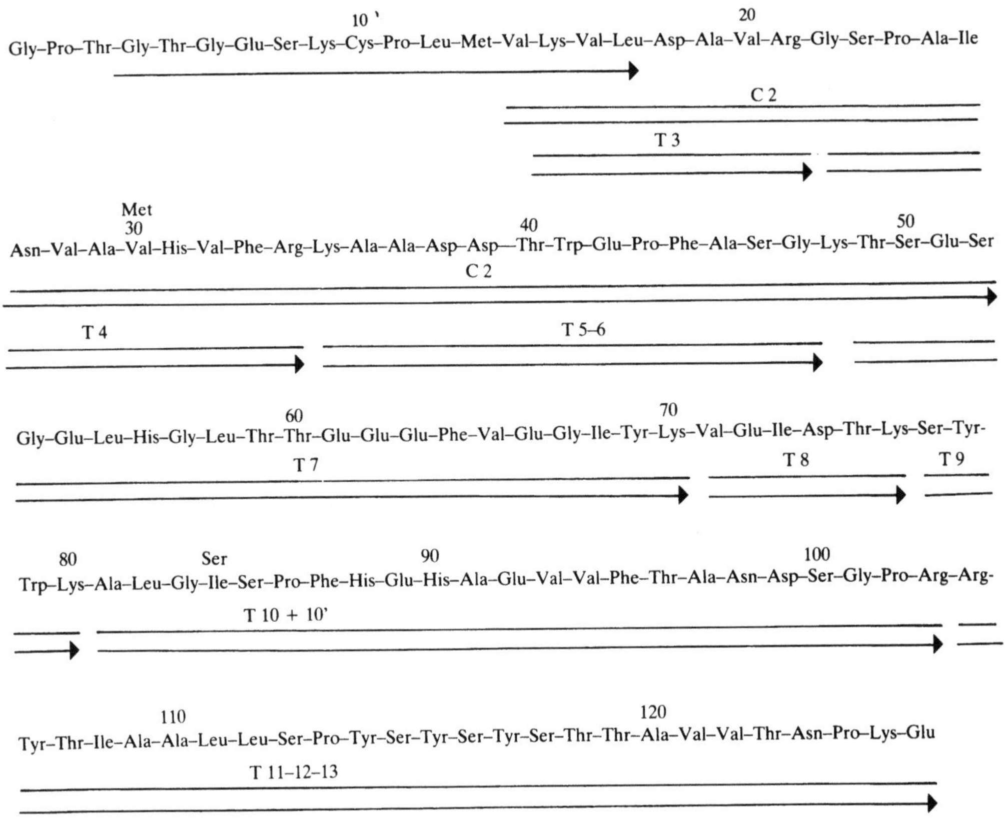


**Amino acid sequence of prealbumin**

**Fig. XII/3**

stance (163) and similarly to the SAA it may also participate in the regulation of the decomposition of collagen (33). Its is filtrated through the glomeruli and is reabsorbed or catabolized in the proximal tubuli. (201, 368)

The critically high serum concentration of the $\beta_2$-microglobulin might be the real reason as the damaged kidney is unable to decompose the above molecules, or the molecule itself can not permeate through the pores of the dialyzing membrane and, thereby remains in the circulation contributing to the development of the final stage of systhemic amyloidosis. The above findings referring to the nature of amyloidogenesis are merely theoretical, as other publications fail to support the above concentration theory (201, 368).

*Gamma trace microprotein* - Amyloidosis affecting the blood-vessels of the central nervous system and creating a "locus minoris resistentiae" might contribute to the hereditary cerebral hemorrhage (HCHWA). The so-called γ-trace protein is the precursor of amyloid and it has been proved to be identical with the C-terminal 36 amino acids of cystatin C, cystein protease inhibitor, a protein with a molecular weight of 12,000 dalton containing 120 amino acids.

This protein is present in the serum and the cerebrospinal fluid (54, 126).

*β-protein* - In certain cases of the Alzheimer's disease the identification of the amyloid fibrils which play an essential role in the pathogenesis of the disease was not successful, therefore, the molecule was specially termed β-protein (133). It consists of 28-34 amino acids and its molecular weight is 4200 dalton (133, 135, 376). The formation of β-protein is linked to the chromosome 21 and since the Down's syndrome is a chromosome 21 trisomy *(trisomy 21)* possible genetical relationship might be established. The involvement of 3 different genes: the gene of the β-protein, that of the Alzheimer's disease and the gene of the Down's- syndrome (135, 359, 360) are suggested. The precursor of β-protein - the beta-APP - comprises 695 amino acids, its molecular weight is 79.000 dalton. The amino acid sequence of the β-protein is similar to the C-terminal of the precursor molecule. Data of biochemical studies suggest that the precursor is a glycosilated cell receptor (188) and is similar to the nuclear protein of the heparane sulphate proteoglycane (321). Its origin has not yet been elucidated and there are no data whether it exists in plasma or only in the cerebrospinal fluid. Another possibility is that it originates from some unknown infection and perhaps it is a viral component (56, 294). The cellular origin and function of β-APP, however, has to be determined (327). It is important to emphasize that no data are available whether a pathogenetic form (135) of β-APP as the amyloid precursor exists. For this pathogenetic factor a catabolic impairement is assumed. Although exidence exists that a serine protease inhibitor, the so-called $\alpha_1$-chymotrypsin (1) is one of the accessory factors of the amyloid formed by β-protein, two possibilities acise, namely, 1) if the concentration of the inhibitor is low, its decreased activity enhances the function of the proteolytic enzymes and thereby promotes the formation of amyloidogeneous molecules from the precursor. 2) at high concentrations the activity may increase, stimulating the deposition by hindering the total decomposition (Fig. XII/4).

*Keratin* - In case of isolated amyloidosis of the skin specific degeneration of the epithelium is elicited by the amyloidotic "transformation" of the tonofilaments (151). Immunochemical studies indicate that the keratin (204, 263) is a possible precursor. It is, however, obvious that in this special case only more information and a larger scale of experimental results may promote our knowledge.

## The pathogenesis of amyloidosis

The above mentioned data clearely show that the amyloid has to be regarded as a protein-like, special substance originating from either abnormal precursors or proteolytic splitting of the normal plasma components. This amyloidogeneous molecule will be polymerized to fibrils and deposited forming an special extracellular (rarely intracellular) structure.

It is generally accepted that the light chains of immunoglobulins serve as precursors of amyloid, at the same time, however, the relationship between the SAA (a normal plasma component) and AA has not yet been established.

Under experimental conditions amyloidogeneous fragments similar to $V_L$ can be obtained from the Bence Jones protein with the lysosomal enzymes pepsin(e), trypsin and the leukocyte derived ones (128, 232), furthermore, in myeloma cell cultures amyloid fibrils, may be produced in the presence of macrophages (79, 410). The above mentioned observations suggest a probability of a similar in vivo processes.

The amyloid analogues (SAA) are rather sensitive to the enzyme effects and can easily be split by proteinase and serine protease enzymes originating from segmentated leukocytes (222, 338). Though the role of monocytes and other cells of the immune system in the formation of amyloid has been investigated by several authors, the results, are still controversial, so definite conclusions can not yet be drawn in this matter.

According to our present knowledge the suppression of T cells and the proliferation of B cells are of crucial importance, for the tissue macrophages exert a direct stimulatory effect on the B cells and initiate a monoclonal proliferation, which may be a precondition of the primary amyloidosis (317). It is an unquestionable fact that the normalization or activation of the T cell function with thymus hormone will prevent the development of amyloidosis. In animal experiments the immunosuppressive agents promote amyloid deposition (106). In discussion of the role, of the immune system in the formation of amyloid interleukin 1 and interleukin 6 should be mentioned. They control the function of the immune system by regulating the production of immunoglobulins and, by increasing the concentration of the acute-phase globulins. The production of these precursors a primary condition of the formation of deposits (157).

Apart from a certain relationship between precursors and the amyloidogeneous substance, predisposing factors, should also exist. The results of the literature report, that amyloidosis can be detected only in 15% of patients with multiple myeloma. These results are contradictory to our observations (only 1% of our patients with monoclonal gammopathy that is not only with myeloma has developed the disease) and to those of other researchers: in families of Armenian and Askenasi Jewish origin suffering from familial Mediterranean fever (FMF) no amyloidosis can be detected at all, while in families of other origin suffering from the same disease, the incidence of amyloidosis is very high. The above mentioned experiences suggest that genetic factors must certainly play on important role in this pathology (288), there is assumed the executor molecule itself an enzyme variant, performs, the special amyloidogenic splitting of the precursors.

Component P has been mentioned previously. It turned to be a normal component of the glomerular basal membrane, endothelium of the blood-vessels and of the elastic fibres found everywhere in the body. Conse-

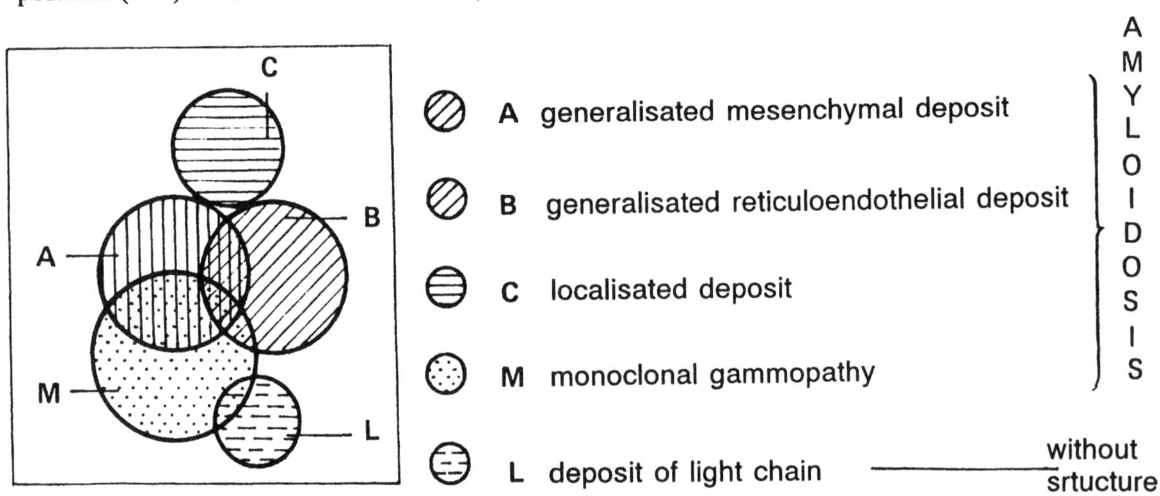


Fig. XII/4

quently, component P might catalyze the amyloid deposition, since it is simply present (32, 86) but will never become a part of the "unsoluble" deposits.

The existence and function of two further factors in the process may further complicate the problem of amyloidosis. The amyloid enhancing factor (AEF) and the amyloid degrading factor (ADF) are theoretically rather important "constituents", although besides the function mentioned in their names (200, 388). AEF is - in all likelihood - a glycoprotein of medium molecular weight. Under experimental conditions ADF has been found to dissolve AA fibrils suspended in agar (191). The studies of the decomposing effect show that it is similar to EDTA (342). This property can be attributed to binding of the calcium ions ($Ca^{++}$) necessary for the attachment of component P, in other words the elimination of calcium from the system occurs.After these preliminary remarks we have to return to the clinical and histopathological experiences strongly indicating that: amyloid is an insoluble substance, and therefore, is a fatal metabolic disease. In a few cases of AA amyloidosis, however, regression of the diseases - "recovery" - could be attained by applying aggressive therapy. Serial histological examinations were performed on biopsy samples of the liver and the rectal mucous membrane to verify the above mentioned "recovery" (238). As and supporting experimental results no similar cases have been reported in the literature, the truthfulness of the above observations is doubted (334). The above empiric results and, the existence of ADF suggest that the formation of amyloidogeneous molecules is a permanent process occurring in the course of degradation of certain proteins also under physiological conditions; these products are decomposed "in statu nascendi" and will never become deposits (194, 195). It has been shown that in secondary amyloidosis the ADF concentration is rather low and, its degrading activity changes in parallel with the albumin concentration (246). Therefore, at present the attention is focused on albumin. Some studies have proved that the dissolving (clearing) of the opalescent AA suspensions instantly occurs not following the kinetics of enzyme reactions. This is actually an optical effect of the albumin solution (43). It is interesting to mention that a high activity elastase-like substance can bind to the amyloid fibrils. It may perhaps reflect a "trial" of segmented leukocytes to eliminate amyloid (195, 343). Another theory of a dual role of the elastase of the neutrophil leukocytes elain that the above enzyme may also take part in the formation of the amyloidogeneous (222).

In addition to the above mentioned in the pathogenesis of secondary amyloidosis the role of the free radicals has also been proposed. Evidence exist that in the inflammatory free radicals are released by the vascular. Oxidative degeneration of these elements by the connective tissue leads to the AA formation (96).

*Summarizing* the above results it can be stated that at present the pathogenesis of amyloidosis is far from being understood and many unknown factors need further investigation. A certain attitude, however, is required for adequate medical treatment. It is worth pointing out that the high precursor concentration seems to be the first prerequisite of the amyloid deposition, however, this factor alone is in sufficient, since amyloid rarely occurs in patients risk. Therefore, the following can be assumed:

- amyloidogeneous molecules can easily be transformed
- deviation of the precursor molecules from the normal configuration can be a preliminary condition
- a genetic disorder of the fibrils decomposition
- specific enzyme pattern of the monocytes which destroys the molecules to be eliminated in a way promoting the formation of amyloidogeneous structures. According to experimental results monocytes are able, for example, to (1) completely decompose SAA or (2) degrade it to peptides of low molecular weight or (3) produce typical amyloid precursors. On the basis of the above findings it can be hypothetized that the decomposition of SAA by macrophages is a multi-stage process beginning not with the pinocytosis, phagocytosis or secretion of the enzymes into the medium, but a with the binding of the membrane-bound enzymes to the protein molecules to be decomposed. The substance itself will be degraded on the cell membranes. The

enzymes belong to the esterase group and show elastase properties. Since predisposition affects proteolysis, further genetic research is worth while (Fig. XII/5).

## The diagnosis of amyloidosis

A definite diagnosis of amyloidosis is based on the histological data. Therefore, in clinical practice tissue samples are usually taken in case of a well-founded suspicion. Tissue deposits stained with Congo red have apple-green colour in polarized light due to their double refraction (302). The immunohistochemical methods are not widely introduced and at present staining with Congo red seems to be less expensive and more specific (296), (Fig. XII/6).

If "more than a suspicion arises it would be very convenient to take a biopsy from the affected organ. Though the positivity of this method is 90-100%, this can not be performed in most of the cases because of the high risk of haemorrhages and other factors. Hence, in systemic forms rectal biopsy has become the most important and reliable procedure accepted all over the world. Regrettably, practice this procedure is not completely reliable, because the doctor taking biopsy from the rectum usually does not dare to to take a deep sample from the submucous tissue (absolutely necessary for a valuable examination).Consequently, biopsy taken from the gingiva have proved to be more advantageous in practice; in our opinion the results obtained by this method do not differ from those of a successful rectal biopsy (Fig. XII/7).

Needle biopsy from the adipose tissue of the abdominal wall is a further possibility; it is reported to give positive results in 88% of patients with amyloidosis (80, 81, 228).

Biopsies taken from the bone marrow, small intestine, prostate, endocardium as well as samples obtained by aspiration from the spleen and tissues removed in the course of decompression surgery in carpal tunnel syndrome have been found to be more or less informative, reflecting the essential pathological processes of the organism.

In polyneuropathy the sural nerve biopsy (nervus suralis) is strongly advised.

After histological diagnosis, the next step will be the examination of the biochemical characteristics of fibril-protein and of the etiology. Separation of the AA protein from the others comprising the primary and secondary amyloid is performed with the potassium-permanganate method. Due to the effect of the above agent the AA protein will lose its capability to take up Congo red. Thus, in contrast to deposits formed from other precursors, the AA is. The rest will remain Congo red positive following the pretreatment (371). Some publications, however, report cases of AA protein-insensitivity. These data suggest that the AA can not be considered as a homogeneous type, existing at least in 2 forms (221).

Immunohistochemical methods could identify the fibrils containing various precursors in the most specific way, corresponding to the modern requirements. Unfortunately, it proved to be a rather difficult task to produce a suitable immunoserum binding to AA protein (235, 236.) The use of anti-prealbumin as a reagent though evident, but it can only be

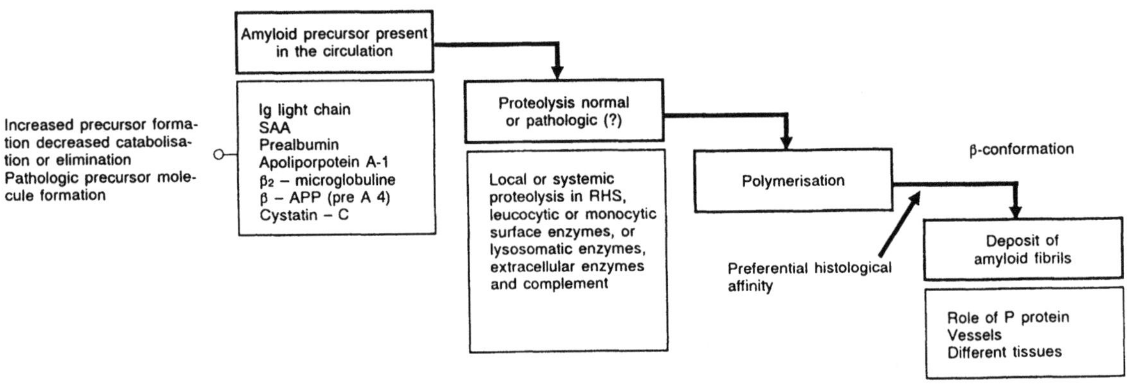


Fig. XII/5

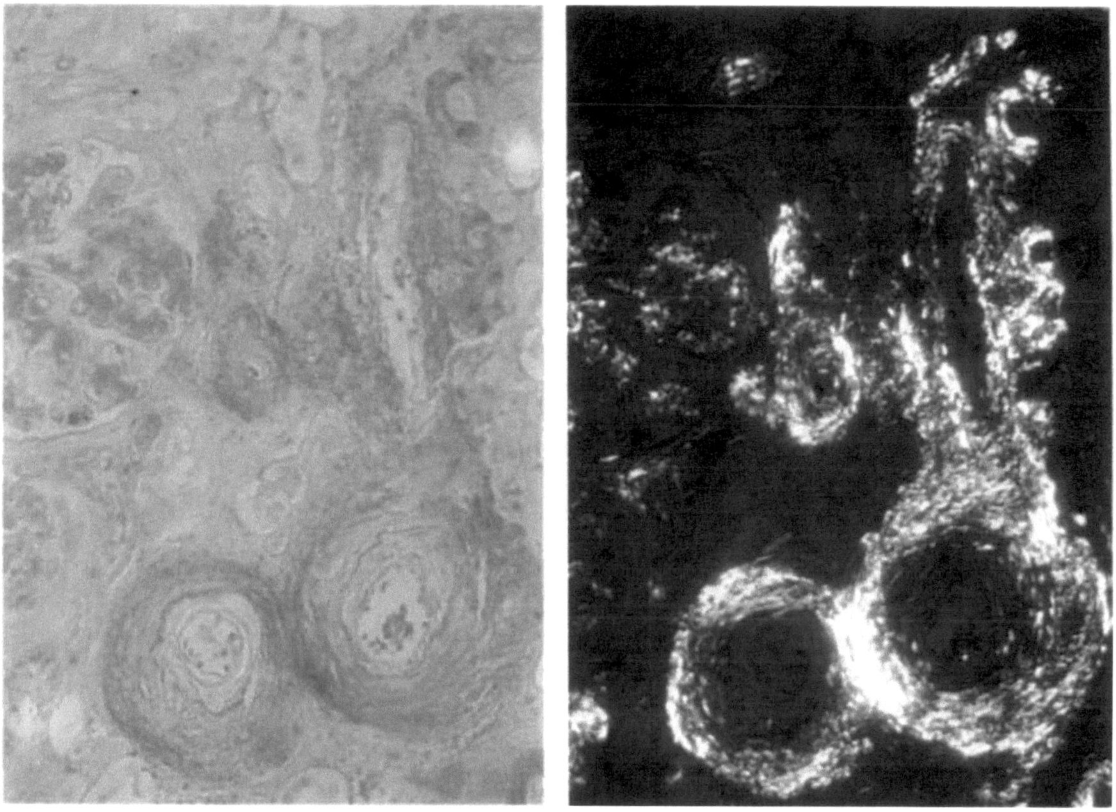

Detection of tissue amyloid using congo red stain (A) - same area in polarized light. Amyloid - containing areas show yellowish green double fraction (B) (magnification 125 ×)
Fig. XII/6

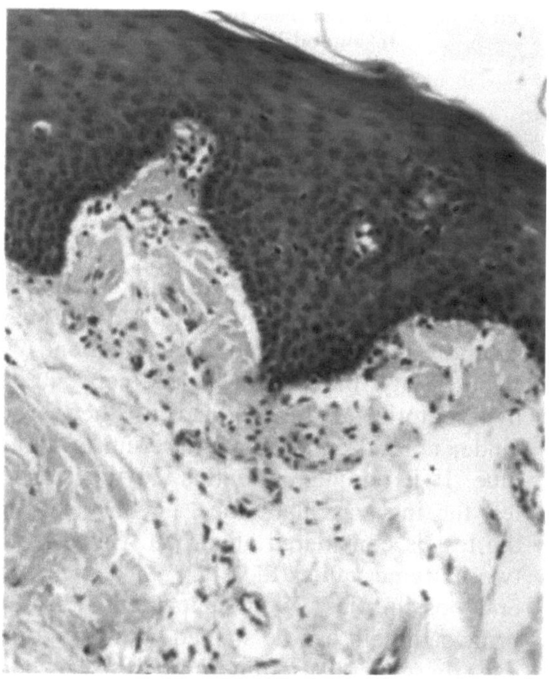

Gingiva H. E. (80 ×)
Fig. XII/7

rarely applied, for it reacts only with the amyloid of familial polyneuropathy. The identification of AL seems to be even more difficult, as the λ- or κ-chain precursors when becoming "building-stones" lose the antigen determinant which have been their isotopic epitope. Therefore, the specific anti-light chain sera otherwise used only for the separation of the subgroups have proved to be more suitable (112). With the exception of presenile and senile amyloidosis of the central nervous system, P-protein can be found in all types of amyloid (although not as its constituent). It provides a possibility for the production of a special "universal" immunoserum suitable for the detection of amyloid (154).

## Non-invasive methods

If amyloidosis is suspected or proved by histological methods, analysis of all protein generally applied in of gammopathies have to be performed in the serum and the urine. In 82-88% of patients with AL amyloidosis some of monoclonally produced proteins be detected (215). Hence, in about 10-20% of the patients no proteins with be found despite the already existing plasma cell dyscrasia. The proteins produced by these cells (the precursor) has not yet reached a concentration detectable in serum, or urine. This finding based on the statistical data and the everyday observations also implies that the precipitation of amyloidosis is not provoked by the permanently high precursor concentration, *but rather by a special form of protein degradation in the formation of amyloidogeneous molecules.* Bone marrow biopsy and X-ray examination of the bones also has to be performed.

In addition to P-protein the amyloid deposits also contain calcium, therefore, isotopes binding to calcium, for example $^{99m}$Tc-pyrophosphate or diphosphonate can also be used for diagnostic purposes (93, 412) (Fig. XII/8). In case of myocardial amyloidosis M-echocardiography is a suitable tool, because by this method can differentiate between the restrictive amyloidous cardiopathy and the constrictive pericarditis often exhibiting similar clinical symptoms. Furthermore, the two dimension echocardiography can detect the hyperreflexive shining of the myocardium (340) when amyloidosis deposits are present (Fig. XII/9 and XII/10).

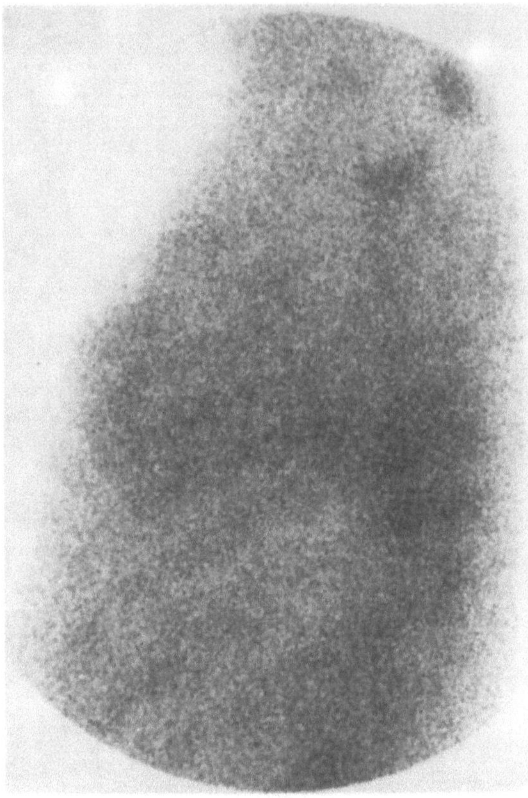

**Isotope scan of myocardium; Following the i.v. administration of 600 MBq 99 mTc-pyrophosphate diffuse enrichment of radioactivity can be observed at the projection site of myocardium as shown on an X-ray film taken from the LL direction, Dr. M. Józan and Dr. E. Pasztarak**

**Fig. XII/8**

The direct and indirect diagnostic findings should be considered if amyloidosis is suspected. Its most important clinical signs are as follows:

- light chains excreted with the urine
- monoclonal gammopathy
- neuromuscular disease of unknown origin;
- hepatomegaly, liver failure - without laboratory signs
- hepatomegaly associated with chronic inflammation
- malabsorption
- nephrotic syndrome
- carpal tunnel syndrome
- congestive heart failure.

The above list though incomplete, but is good for practical purposes. If amyloidosis is suspected, two further laboratory tests have to be performed. Their positivity or negativity, never indicates a certain presence or absence of the deposits, still our suspicion may be supported or disproved. The Congo-red test has to be mentioned first. (According to Paunz 0,2 ml/kilogramm body weight of 0,6% Congo-solution is administered to the patient.) 60 minutes later blood is taken and concentrated hydrochloric acid is added to the serum. Under normal conditions the serum is stained blue. If it remains colourless, - the dye has left the intravascular space and the result of the test is considered abnormal. Between the two extreme values there are, of course, several possibilities, given in percent. Also the urine sediment can be stained with Congo-red dye, the general usefulness of this test, however, could not be verified by further investigations.

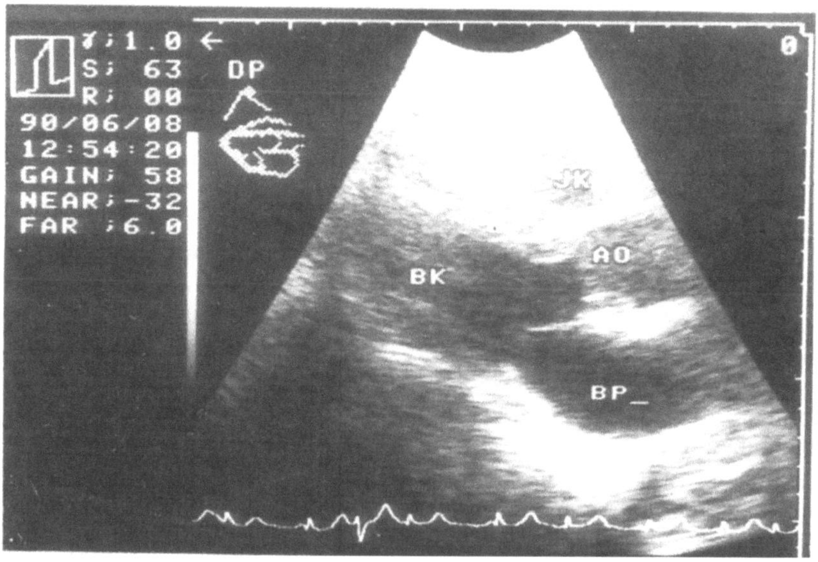


Fig. XII/9

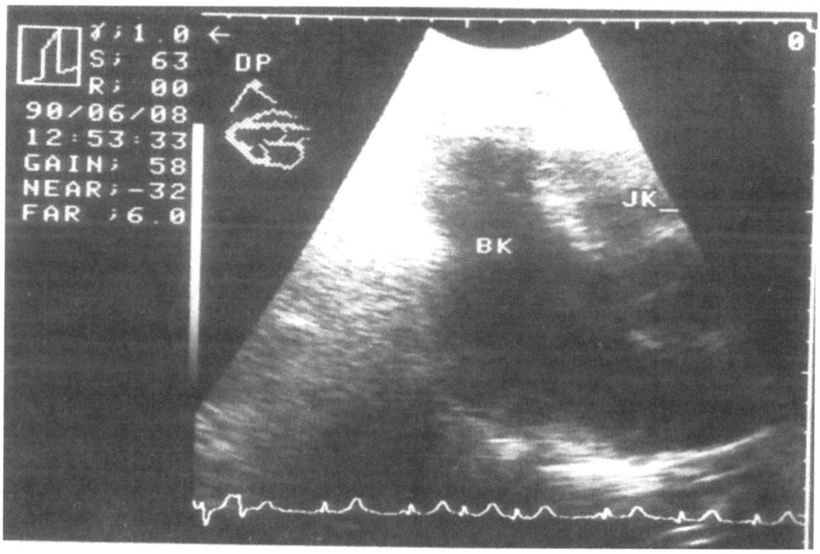


**On the apical, longitudinal , and septal sections and to a lesser extent diffusely in the entire myocardium many small sparkling can be observed (Dr. L. Lascsik)**

**Fig. XII/10**

## Clinical amyloidosis syndromes

### *Systhemic amyloidosis*

This type of amyloidosis involves all the organs of the whole organism (Abercombie's syndrome) and can develop as a result of acquired or inherited metabolic disorders. The acquired from induced by chronic infections, rheumatic diseases, occur more frequently all over the world than the inherited form. Still in certain geographic areas (Portugal, Japan, etc.) the latter type is predominant as a result of the prevalence of abnormal genes (129).

## PRIMARY AMYLOIDOSIS

### Amyloidosis associated with plasma cell dyscrasias (AL)

The role of precursor should be first emphasized in the so-called primary amyloidosis (AL). Monoclonally produced immunoglobulins with a concomitant metabolic disorder of unknown origin may induce amyloidosis. The urinary excretion of free light chains, including λ light chains, independently of whether the above phenomenon was associated with multiple myeloma or idiopathic monoclonal gammopathy is regarded as the most important pathogenic factor. It occurs mainly in males in their fifties and the deposits are predominantly localized in the mesenchyma and the visceral organs the kidneys, liver, spleen and the heart. In patients with multiple myeloma the prevalence of amyloidosis is reported to be about 6-15 % (215). It is important to emphasize that the number of the bone marrow plasma cells and their morphological characteristics are not correlated with the way of the development of amyloidosis.

## SECONDARY AMYLOIDOSIS

### Reactive systhemic amyloidosis (AA)

In this chapter not all the types of the secondary amyloidosis will be discussed. Amyloidosis can be detected in 5-20% of the acute recurrent diseases, the chronic degenerative inflammations and the tumours, aggravating them (387). AA amyloid deposits predominantly localize in the parenchymal organs, therefore, this possibility should be considered in cases of proteinuria, nephrotic syndrome or hepatosplenomegaly of unknown origin.

### Heredofamilial systhemic amyloidosis

The protein precursor of the disease can not be verified in most of the cases. This form of amyloidosis can be rarely detected, compared to the other types and can be characterized by a relatively early appearance (in the thirties and forties).

The disease shows autosomal dominant inheritance. The leading symptoms exist in four forms: neuropathic, nephrotic, cardial and mixed (158).

The neuropathic (FAP), so-called Portuguese amyloidosis can be characterized by the following symptoms: at the beginning analgesia and loss of heat and cold sensation, paraesthesia in the lower limbs, and the loss of vibratory sense; later followed by myasthenia, bladder and gastointestinal disorders and impotence. The skin of the extremities becomes ulcerated. These clinical signs appear in the thirties and the fatal outcome usually occurs within 10 years following the appearance. The amyloid fibrils are formed of prealbumin $AF_p$ (66). A similar syndrome has also been described in Japan. The following types can be considered as mitigated forms of the neuropathic type:

- Indiana-type (311) - It was observed in a family of Swiss origin. The disease was benign and did not shorten the life of the patients;
- Maryland-type (242)
- Iowa-type (369) - The cause of death in this case was renal failure;
- Ohio-type (138)
- Swedish type (4, 19)
- Finnish type (249) - Actually all the above types are various forms of secondary amyloidosis they may appear earlier or later and may or may not affect the kidney. In summary: Apart from the familial occurrence and affecting the nervous system no other similarities exist.

- Nephrotic type - Recurrent fever and abdominal pain are characteristic of this form (347). The disease shows autosomal recessive inheritance, the main symptom is renal failure. Its prevalence is high among in the Turkish, Sephardic Jewish (in lower percentage in Ashkenazi; Jews), Lebanese and Armenian population (250).
- Cardiac type - This form can also be termed "the disease of twins". It usually begins in the forties and fifties and may cause death within 2-6 years (3, 107).
- Mixed type - It is usually associated with

the medullary carcinoma of the thyroid gland, pheochromocytoma or parathyroid hyperplasia. In the latter form the basic structure of amyloid fibrils comprises thyreocalcitonin (345).

- Familial occurence of cutaneous amyloidosis has also been described (372).

## Amyloidosis associated with familial Mediterranean fever

Familial Mediterranean fever - or recurrent polyserositis - is a disease characterized by autosomal recessive inheritance of unknown pathogenesis. The clinical symptoms are as follows: fever, vomiting, absence of bowel sounds and fits of abdominal colic usually lasting for 48 hours. This disease appears sometimes already at the age of 10, but most frequently (in 80-90% of the cases) at the age of 20. Polyserositis, arthritis and cutaneous eruptions are also typical. The precursor of this type of amyloidosis is the AA substance belonging to the group of acute-phase globulins (89).

## Localized amyloidosis

This type of amyloidosis is localized in certain organs and tissues and can be characterized by focal or homogeneous distribution of amyloid. The disease exists in various forms. This type of amyloidosis comprises about 9% of the disease (215). In the half of these cases amyloid deposits can be found in the bladder, lungs, skin and larynx. The primary criterium is to verify the absence of generalisation and monoclonal immunoglobulins.

## Cutaneous amyloidosis

Among the forms of cutaneous amyloidosis, local forms occur more frequently than those affecting the whole epidermis (392). The localized forms can be of papular or macular type. The *papular* type is the lichen amyloidosis mainly affecting middle aged people. It develops always as a consequence of permanent pruritus. Small cone-shaped papules of 3-4 millimeters can be found of the extensor surfaces of the legs and tighs. Histologically it is always confined to the dermis and never involves vascular walls.

The *macular* type is more rare than the papular one and occurs on the chest and the trunk. The clusters of macules of 2-3 millimeters in diameters contain pigment and are rather characteristic. It occurs very rarely in the Europeans. The nodular and tumefacient cutaneous amyloidosis can be scarcely differentiated from the cutaneous lesions observed in generalized amyloidosis (45).

## Cerebral amyloidosis

Though the brain is frequently affected by amyloid deposition, it is diagnosed only at the autopsy. Two forms of the above type of amyloidosis exist:senile or neurotic, and a presenile type or Alzheimer's disease. In the first case amyloid can be detected as cortical plaques, destruction of the fibrillar plexus is typical of the second form (132). In this type amyloidotic lesions of the medium and small arteries of the leptomeninx and the cortex can be found. The damages in the temporal and the occipital lobes are especially typical. In Creutzfeldt-Jacob disease (subacute spongiform encephalopathy: Kuru-disease) caused by a slow virus in fection the histological examination of the brain reveals the presence of amyloid deposits (245).

## Senile cardiac amyloidosis

The usual cardiological diseases of elderly people often distract the attention of the physician from this rare metabolic disorder (64). 25% of the patients suffering from this disease are older than 60, but in 75% they are only in their fifties. The disease is usually observed as an incidental finding during autopsy, but can also be assumed if the patient is insensitive to digitalis. Two types of the disease are known:

1. isolated atrial amyloidosis (IAA) and
2. senile cardial amyloidosis (SCA).

The first type is the prevalent one (391). In case of SCA amyloid deposits contain a characteristically high amount of tryptophan, referring also to the prealbumin as a precursor substance. The significance of cardial amyloidosis is not yet known, it has, however, to be mentioned that the occurence of senile atrial fibrillations and congestive cardiomyopathies are correlating with each other (64, 159, 230, 252).

## Endocrine amyloidosis

The Langerhans' islets of elderly people and those suffering fro diabetes mellitus independently of their age often contain amyloid (65). In other endocrine organs especially in case of tumours localized amyloidosis can easily be recognized, for example in case of apudoma or the medullary carcinoma of the thyroid gland (389) (Fig. XII/11).

In addition to the above mentioned important types, other forms, of amyloidosis with sporadic symptoms, which are revealed only at the autopsy exist. For example in the type affecting the joints (136) the locally occuring amyloid is mainly of the AL type and can be found in the urogenital (113) and respiratory tract or sometimes in the orbit of eye. Many types from the microscopic size to the extensive, massive infiltration of the tissues. Extramedullary plasmocytoma (131) is a mutual feature of these forms. The main symptom of bladder amyloidosis is haematuria, which occurs in 85% of the cases. The tissue infiltration, can be partial in the bladder, or can affect the ureter and the renal pelvis (38). In the respiratory system laryngeal and pulmonary forms can be identified depending on the localization. The latter form may be tracheobronchial, multiple nodular or diffuse pulmonary (308). Inspiratory or expiratory dyspnoe depending on the localization is an important sympton of the disease lough and later haemoptysis are also of diagnostic importance. X-ray examination of the lungs reveals typical atelectatic areas and multiple, noncalcificated foci. If the whole area of the lungs is involved, dyspnoe becomes progressive and the fatal outcome is inevitable. Periorbital form reveals the involvement of the trigeminal nerve, Gasser's ganglion, infraorbital nerve, vitreous body and the conjunctiva. Isolated amyloidosis rarely affects the following organs: breast, mesenterial lymphatic glands, tongue, tonsils, palate, hip-joint, abdominal wall, inguinal region, sacrum, pancreas, aorta (that is the isolated amyloidosis of the above organs) (73, 220, 237, 257). The significance of the above described rare forms is not yet known, and therefore can not be interpreted, still it might play some.

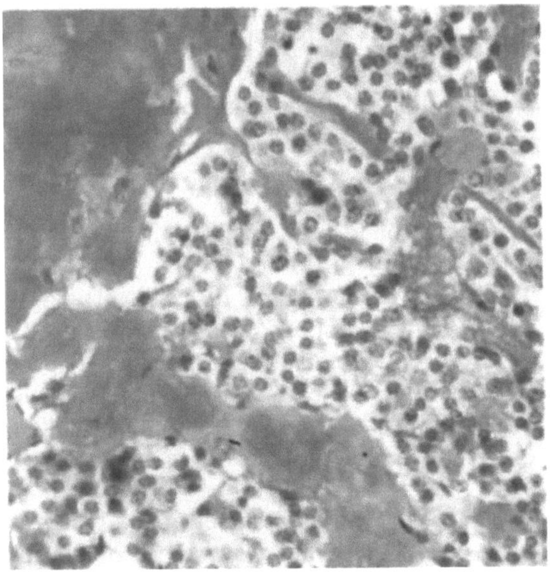

**Endocrine amyloid in medullary carcinoma of the thyroid gland. Amyloid deposits are homogeneous eosinophilic areas. Hematoxilin and eosin staining. (Magnification 125 ×)**

**Fig. XII/11**

## Classification of amyloidosis

No classification meeting all the special requirements and based on the principle "we have to speak the same lanquage" exists for amyloidosis, according to the present conception the biochemical classification has to be accepted (Table XII/I).

## Clinical aspects of systhemic amyloidosis

The symptoms of systhemic amyloidosis homogeneously affecting the whole body, and all the tissues of the organism is not characteristic of the amyloidosis itself, with the exception of petechiae of the eyelids. Functional disorders of the most important organs can be ob-

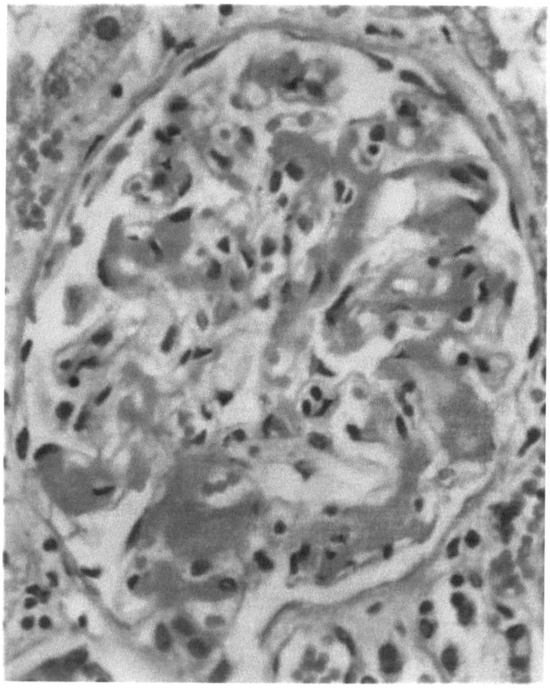

Glomerular deposition of amyloid. Hematoxilin and eosin staining. (Magnification 300 ×)

Fig. XII/12

served also without the existence of this metabolic disorder. The following part of this chapter presents a detailed description of the organic lesions, emphasizing their significance and their possible role as a cause of death.

## Kidney

This organ is involved in 75–90% of the cases with amyloidosis. Four stages can be differentiated:

I. Histologically verified renal amyloidosis without clinical symptoms
II. Slight proteinuria with amyloidosis
III. Nephrotic syndrome with underlying amyloidosis
IV. Uraemia as a consequence of amyloidosis (147)

Nephrotic syndrome have been found to be associated with Fredrickson's II type hyperlipoproteinaemia.

64-72% of the early symptoms are due to the functional disorder of the kidney (181, 215). Amyloid is at first deposited in the glomerular mesangium later, however, around the basal membrane. The laboratory and morphological symptoms of renal failure detectable in primary and secondary amyloidosis do not differ, Fanconi's syndrome can sometimes be observed in adult patients (109) (acquired Fanconi's syndrome). Because of the presence of deposits in the glomeruli generally *proteinuria* can be observed. The amyloid penetrates through the basal membrane and strips it by destroying the epithelium (378) (Fig. XII/13). Evidence exists that the proteinuria is not selective, therefore tubuli are also affected. Nephrotic syndrome can frequently be observed, but it has to be emphasized that the extent of proteinuria does not correlate with the degree of amyloid deposition (74, 379). *Haematuria* is also often detected, but its macroscopic type is not typical. The latter can be evoked by the thrombosis of the renal vein caused by massive proteinuria. *Hypertension* develops in about half of the cases. Though diagnostic value of the fibrillar structures appearing in the urine is of certain importance, this finding is positive only in 40% of the cases and, furthermore, it could be sometimes in healthy controls (330). On rare occasions tubular damage may results in nephrogenic diabetes or renal acidosis. X-ray study reveals enlarged, normal or atrophic kidneys. Calculation of the kidneys is a rarely occuring sign (87).

It has to be emphasized that the increasing amyloid deposition in the kidneys causes a progressive decrease of glomerular filtration and, as a consequence an illusory "improvement" of the nephrotic syndrome can be observed before the total renal failure develops (181). In 43-65% of systemic AA amyloidosis (secondary type) and in 16–25% of the primary form (AL amyloidosis) renal failure has been the cause of death.

Sometimes acute renal failure is revealed, which always means the thrombosis of the renal vein. It is important that is state can be provoked by dehydration. The concomitant symptom, priapism is a serious problem.

## The heart

Amyloidosis occuring in monoclonal gammopathy (AL amyloidosis) causes heart damage in 90% of the cases, while the amyloid anal-

## PRIMARY AMYLOIDOSIS

*Disease associated with amyloidosis*

- monoclonal gammopathy
  - idiopathic – (118, 212)
  - facultative
    - associated with tumours
    - associated with store disturbances (149)
  - obligatoric
    - macroglobulinaemia Waldenström (63, 97, 209, 222, 278, 305)
    - Diffuse plasmocytoma (172, 272)
    - Multiple myeloma (62, 210, 241)

- localisated
- systemic – mesenchymal

## SECUNDARY AMYLOIDOSIS

- Reactive states associated with inflammation
  - chronic inflammations
    - TBC (7), leprosy (400), osteomyelitis, fibrosis cystica, acne conglobata (280), hypogammaglobulinemy (116)
    - paraplegy, decubitus, chronic infected burn (146)
    - RA (232), juvenile RA (200), morbus-Bechterew (251), SLE (78), dermatomyositis (408), Sjögren-syndrome (85, 239), Wegener granulomatosis, Goodpasture syndrome, Behĉet syndrome (307), Crohn disease (374)
    - Whipple disease (120), Colitis ulcerosa (337), Psoriatic arthropathy, Reiter syndrome (251), (170, 268, 319, 365, 403)
  - neoplasy
    - Morbus Hodgkin (189), non-Hodgkin lymphomas (355), HCL (231), melanoma malignum,
    - Hypernephroma (276), neoplasies of intestines, lung, and urogenital tract (198, 128, 168, 289, 348)
  - haematologic disorders
    - cyclic neutropeny (72, 332)
    - hypospleny (19, 29, 102, 120, 237, 306, 355)
    - absence of X factor (24, 39, 64, 114, 247, 248, 275, 282, 304, 344, 346, 381)
  - store disturbances
    - Nieman-Pick disease (41)
    - Gaucher disease (149, 165)
- Localisated
- Systemic – *reticuloendothelial*
- Heredofamiliar (242, 382) — Muckle-Weiss syndrome (urticaria, deafness, neutropathy) (234, 288)
  - perireticular recessive autosomal (28, 280) — FMF – sephard jewish (336), turkish, armenian (102, 338, 410)
  - dominant ausomal — FAP (382)
    - I. typ. – Japanese (5, 6, 16, 174, 185, 361, 362, 377), jewish (255, 285, 287, 290), English (229), German, Greek, Portuguese (102, 302, 314, 371) Swedish (83, 207, 341, 398), French (182, 397)
    - II. typ – Swiss (96), German, Indiana/Swiss (36, 303, 21, 84, 285, 320),
    - III. typ – Finnish (194), Danish (169, 370, 310, 343), Dutsch (87), Scottish-Irish (334, 351)
  - pericollagenal — FAC (375, 386); HCHWA Icelandic (144, 387)
- Senilic systhemic amyloidosis (SSA) (379)
  Senilic cardiac amyloidosis
- Amyloid depositum producing in CNS
  - Alzheimer disease, I–II–III (40, 89, 108, 131, 132, 135, 136, 188, 253, 270, 326, 328, 376, 392)
  - Ewan syndrome (65, 125, 312)
  - Pick disease
  - PSP
  - ALP/PD
  - Result of virus induced disease
    - Scrapie
    - Kuru
    - Creutzfeld-Jakob (141, 245, 387)
    - Gerstmann-Straussler (245)
- Endocrine correlated amyloidosis (70)
- Amyloidosis of haemodialised patients (10, 42, 85, 103, 112, 119, 139, 140, 154, 201, 221, 235, 340, 371, 412)
- Cutan-localised amyloidosis
  - (46, 263, 411)
  - Subcutan applicated heroin (91, 258)

**P protein** (30) ADF = EDTA

Table XII/I

| *Amyloidogen molecule* | *Precursor molecule* | | |
|---|---|---|---|
| AL (lambda or kappa) mostly $\kappa$ I. and $\lambda$ VI. but yet not identified only $\kappa$ II. and $\lambda$ V. (410) | Immunoglobulines first of all ligth chains | | |
| AA | SAA (acut phasis protein) | | |
| AA | SAA 1, 2, 3, 4, 5, 6 – HDL apoprotein (260, 271) | | |
| AA | SAA 1, 2, 3, 4, 5, 6 – HDL apoprotein (1, 14, 180) | | |
| | | P, J, S | 30 (MET-VAL) (24, 248) |
| | | Je | 33 (ILE-PHE) (17, 206) |
| AF$_p$ (342) | Praealbumin or its variant | | 49 (GLY-TRE) |
| | TTR-transthyretin (20) | G–Sw | 84 (SER-ILE) (82, 350) |
| | | Ir | 60 (ALA-TRE) (14) |
| | | | 77 (TYR- *) (7, 383) |
| | | | 111 (TYR- ) (186) |
| AF$_c$ | Prealbumin like protein | | 122 (TYR- ) (191, 246) |
| AF$_b$ | | | |
| AS$_c$ | Gamma-trace microprotein (156, 331) cystein protease inhibitor (Cystatin C) (130, 176) Prealbumin like protein ANF-isolated atrial amyloid (IAA) (211) | | |
| $\beta$ – protein A4 (321, 359, 360) | ? $\beta$ – APP (74, 107, 131, 181, 245, 330, 378) | | |
| ? | ? | | |
| AE – AE$_t$ (71, 281, 363) | thyreocalcitonin (procalcitonin) (120, 152, 184, 215) | | |
| AE – AE$_i$ (53, 105) | insulin-glukagon Islet amyloid polypeptid (IAA) (179, 211) | | |
| AE – AE$_{other}$ | other hormones (IAPP) (48, 180) | | |
| AH | B$_2$ – microglobulin | | |
| AD | Keratin | | |

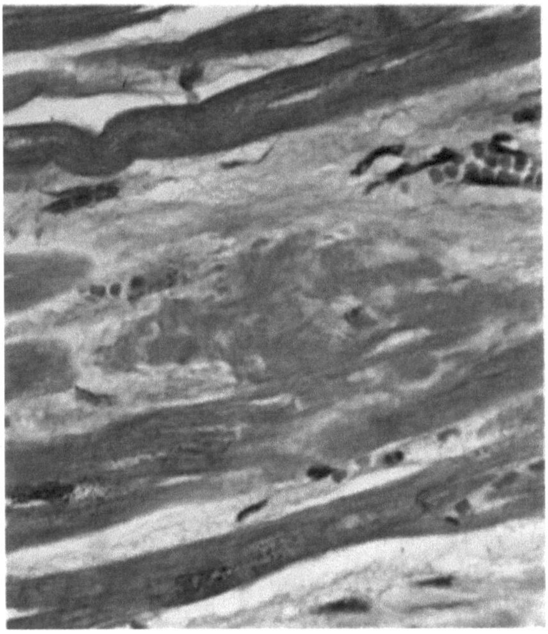

Amyloid deposits in myocardium. Hematoxilin and eosin staining. (Magnification: 300 ×)

Fig. XII/13

ogues, affect the cardiac muscle in 50-60% the cases (215) (Fig. XII/13). In 33-50% of the cases of the latter group this was the cause of death. The damage itself is the biventricular restrictive cardiomyopathy or symptoms resembling those of the constrictive pericarditis. Amyloidosis is one of the cardiac diseases not causing cardiomegaly detectable by the X-ray study. Despite the delusive symptoms constrictive pericarditis can be excluded, with the ultrasound study (67). In the thickened myocardium the physiological thickening does not occur in the systole (156).

The shining, granular appearance is characteristic of the myocardium, this phenomenon can not be detected in other diseases. The electrocardiogram shows low QRS complexes in the standard leads, and the left deviation of the "R axis. In the chest leads the electrocardiogram is similar to that of the anterior in-

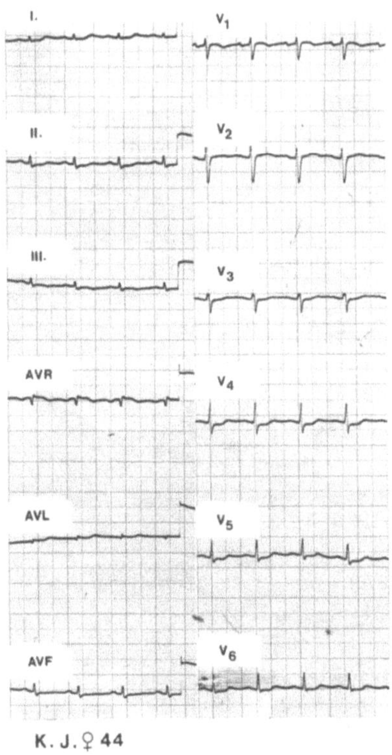


Sinus rhythm and normal atrioventricular conduction. Low voltage in all leads. Biphasic T-waves and ST-depression of 0.5 mm in leads I-II, aVL, and $V_3$ - $V_6$

Fig. XII/14

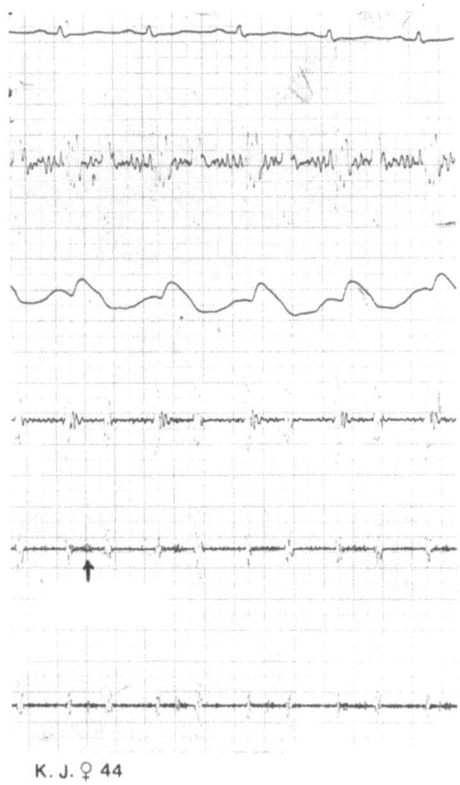


PKG. Small amplitude atrial sound (↑). Mid-systolic click both in the high and low frequency intervals. Small amplitude crescendo-decrescendo murmur in the high frequency interval.

Fig. XII/15

farction (Fig. XII/14, XII/15). Conduction disorders for example the first degree antrioventricular block or arrhytmias usually occur. In the node is infiltrated by the amyloid, sick syndrome will develop (176). The thickening of the left ventricular wall (left ventricular hypertrophy) can not be proved by electrocardiogram, but can be seen at the ultrasound examination. The valves of the heart may be damaged, but no substantial haemodynamic consequences occur (59, 331). Amyloid may sometimes narrow the diameter of the coronary blood vessels with all the clinical consequences of myocardial ischaemia. The damage of the pericardiac vessels may result in pericardiac tamponade. Some publications suggest that cardiac amyloidosis is not refractory to digitalis therapy on the contrary the myocardium sensitivity increases leading to fatal arrhythmias. In vitro studies show that digitalis may be bound to the amyloid fibrils and this "storage" may result in arrhythmias (310). Diseases of the peripheral blood vessels, claudication is a well-known characteristic symptom. In 15% of patients with amyloidosis orthostatic hypotension may be observed as a leading symptom (211). Diarrhea, lack of perspiration and impotence are also characteristic signs referring to the involvement of the respective organs.

## Lungs

Amyloid deposits may be formed in all regions of the respiratory tract (215), therefore verified pulmonary amyloidosis may be a part of the systemic type or "only" a local disease with deposits in the lungs (363). The latter type may often be asymptomatic. Two different forms are known: damaging the lower and the upper respiratory tract, respectively (46). Haemorrhage of the nasal airways and their perforation are due to submucous accumulation. The most feared type of this process is the lesion of the vocal chords. Dyspnoe is a typical symptom of the pulmonar damage (46). Amyloid may infiltrate the alveolar septa (283), and in extreme cases diffuse interstitial pulmonary amyloidosis may concomitantly occur with a serious ventilation disorder (150, 313). Tracheobronchial deposits may be found in the lobes in diffuse or nodular forms. Submucous plaques of the mucous membrane may cause coughing, dyspnea, stridor or haemoptysis (184, 224).

## Gastrointestinal tract

The gastrointestinal tract is very frequently involved in both forms of systemic amyloidosis. The Congo-red positive substance may be found everywhere from the tongue to the rectum. Sometimes no symptoms can be detected, but in other cases a various dysfunctions may be evoked (215). Macroglossia can generally be (but not exclusively) seen in primary amyloidosis, in 20% of the cases. In extreme situations the patient is unable to close the mouth and the teeth may be sunk in the tongue often causing decubitus (Fig. XII/16). It is reported that a patient had suffocated asleep because his tongue falled backwards and caused airway obstruction. Very rarely small nodules may develop on the tongue. Oesophageal lesions result in dysphagia (Fig. XII/17). Examination of the stomach often reveales tumour-like findings, sometimes with prepyloric obstruction with all the clinical consequences (227).

The typical motility disorders (constipation, diarrhea) are due to the lesions of the nerves and nerve-endings (12). Radiological signs are similar to those caused by tumours and complications, (ulcerations, haemorrhages (226), malabsorption, intestinal necrosis) and may further aggravate the situation. The above list does not rank the symptoms according to their importance or incidence.

If the colon is also involved, the clinical signs may be similar to those observed in ulcerative colitis (diarrhea and rectal bleeding). In some cases the symptoms suggested the possibility of Crohn's disease or rectal carcinoma (401).

The liver has been found infiltrated in 90% of the cases, but the above histological finding is accompanied by clinically detectable hepatomegaly only in 50% of the cases (124). Portal hypertension, hepatic failure can seldom be observed, the liver function tests usually nor-

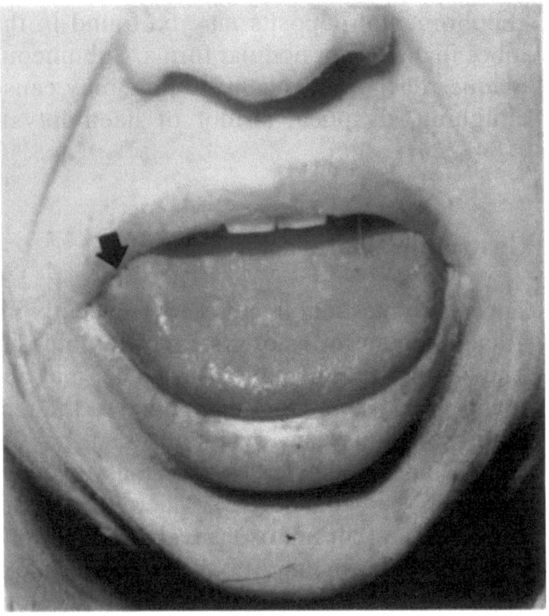

The arrow points at a decubitus on the lateral surface of the enlarged tonque caused by a tooth.
Fig. XII/16

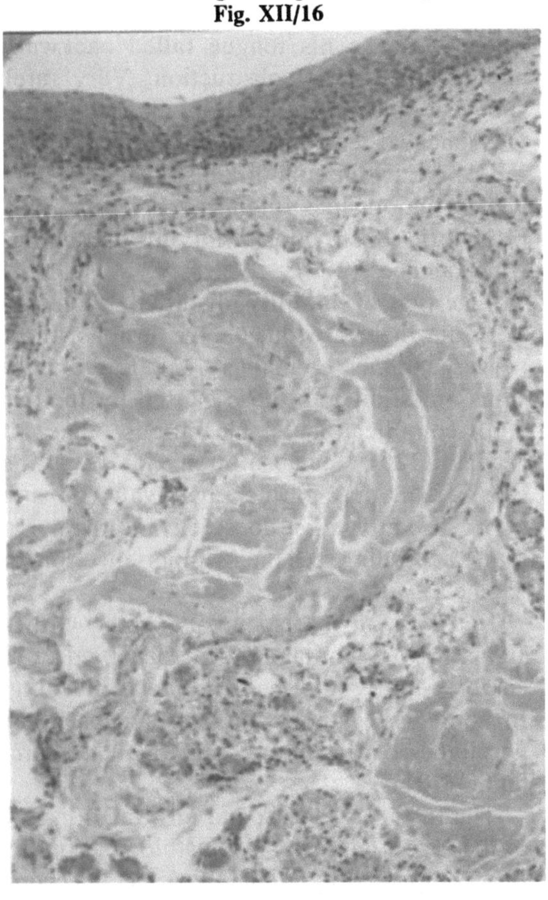

Histological section of the tongue. Spheroidshaped amyloid deposit can be observed. Hematoxilin and eosin staining. (Magnification 80 ×)
Fig. XII/17

mal, although the alkaline phosphatase may be slightly increased indicating intrahepatic obstruction (Fig. XII/18). Cholestasis or ascites, however, can be rarely found (309).

Infiltration of the spleen may result in moderate splenomegaly, associated with hepatomegaly in 10-20% of the cases. Comparing the previously mentioned findings with the latter, it can be stated that hepatomegaly as a symptom of amyloidosis can rarely be observed. Spontaneous rupture of the spleen has been reported on only once (284). The presence of Howell-Jolly's bodies in the peripheral blood refers to functional hypersplenism suggesting the possibility of amyloidosis (29). In the spleen the amyloid deposited in the Malpighi's follicles continually enlarges compressing the lymphatic tissue inducing its "atrophy". Vitreous, semi-transparent globules can be macroscopically seen at the sits of the follicles forming the so-called sago-spleen. If the amyloid deposits in the connective tissue of the sinus walls or the reticular fibers of the pulp, the sinus walls will be thickened because of the amyloid "impregnation". A similar thickening can be also observed in the vascular vessel walls. Due to the amyloid stored in the pulp, the spleen resembles the raw smoked and it is often referred to as "ham-like" spleen.

## Skin

Apart from the other symptoms diseases of the skin can be found in 40% of the cases. It general occurs in the primary form, than in the secondary one. The most common histological alteration can be observed in the vascular wall resulting in the development of purpurae, petechiae and echymosis spontaneously or after weak provocations (Fig. XII/19). Periorbital purpurae of the eyelids refer to the undoubted diagnosis of amyloidosis, as this symptom can not be observed in any other disease. The hypothesis that the above symptom develops after rectoscopy, vomiting or positive Valsalva-test does not hold true (Fig. XII/20). Papules, plaques and nodules appear mainly on the face and the skin of the chest. The accumulated amyloid in the skin is similar to scleroderma or myxoedema.

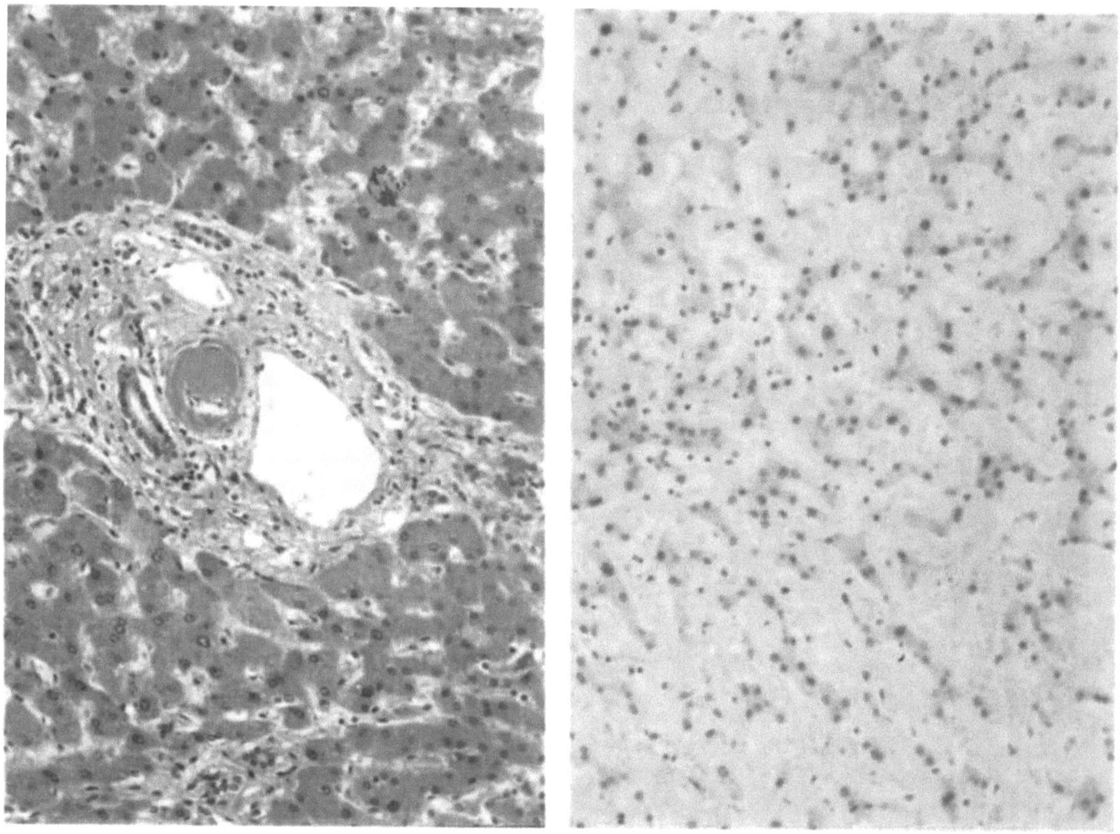

**Deposition of amyloid in the wall of an arteriole in the liver (A) and in the walls of sinusoids; hepatocytes are compressed. Hematoxilin and eosin staining. (Magnification 125 ×)**

**Fig. XII/18**

Alopecia and nail disorders are rare symptoms, but can not be considered negligible (393), (Fig. XII/21). In case of pigmentous urticaria the underlying disease is often amyloidosis (31).

## Peripheral nerves

Peripheral neuropathy can be diagnosed in 17% of the cases. It is characteristically symmetric, progressive and more pronounced at the distal parts. In the beginning the dysfunction of the sensory nerves numbness and the paraesthesia are the dominant symptoms. Temperature and pain sensation is earlier than the epicritical sensibilitiy (196).

Peripheral neuropathy can be found mainly in primary amyloidosis. It is a progressive, predominantly sensory polyneuropathy characterized by pain and autonomic dysfunction (impotence, orthostatic hypotension, gastrointestinal disturbances, *anhydrosis*). In the later stage of the disease the impaired motor function of the lower limbs develops. During this period thick nerves can be palpated . Sural nerve biopsy has a diagnostic value in most of the cases.

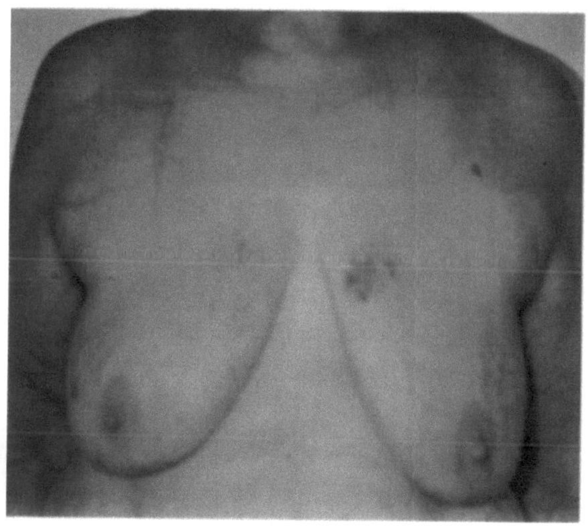

**Fig. XII/19**

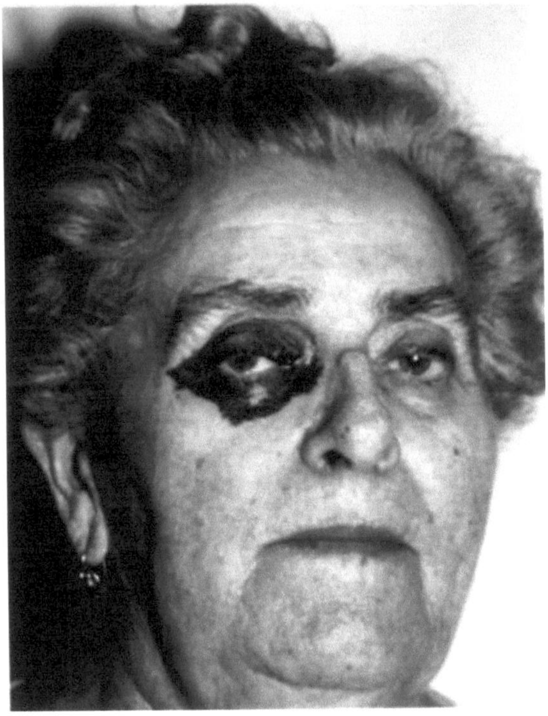

Fig. XII/20

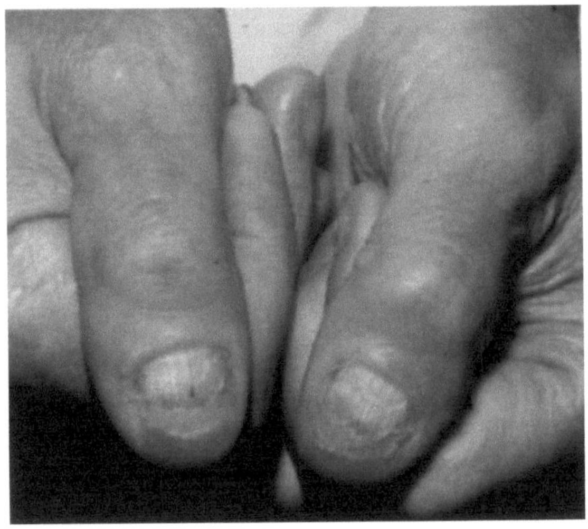

Fig. XII/21

Carpal-tunnel syndrome is observed in 20% of patients with primary amyloidosis (Fig. XII/22). The deposits in the muscles cause weakness and pseudohypertrophy (300). In some cases the opposite process occurs, in which the blood-vessel damages result in angiopathy and secondary atrophy (35). The eyes are seldom affected (247) (Fig. XII/23). In this case not only the structures of the eye playing part in the vision, but also the oculomotor muscles may be damaged contributing visual disorders (202).

## Joints

In about 5% of the cases arthropathy is caused by protein products formed from light chain precursors (324, 396). When it is associated with rheumatoid arthritis, asymmetric inflammation of the larger joints (e.g. hip, shoulder) can be observed (Fig. XII/24). The differential diagnosis or the assumption of amyloidosis are quite difficult, since the subcutaneous nodules found in 50% of the cases are very simi-

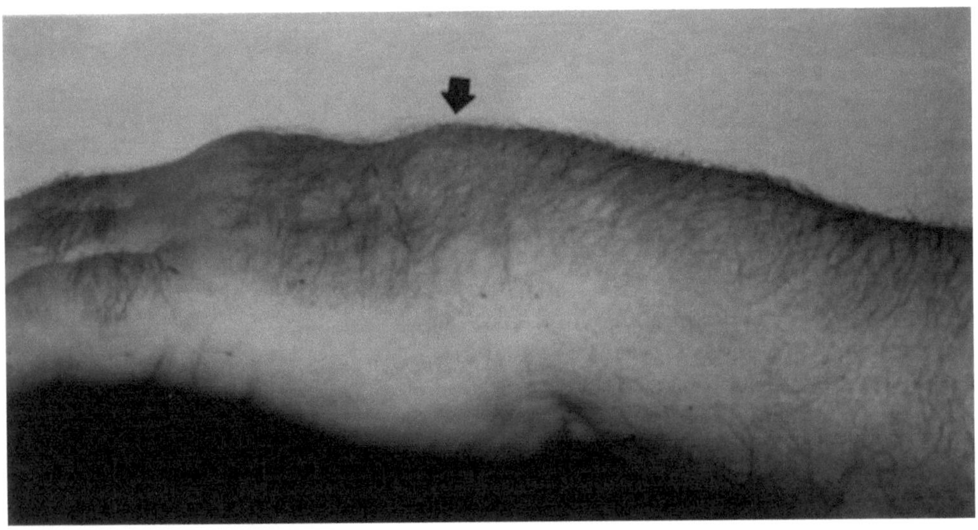

Fig. XII/22

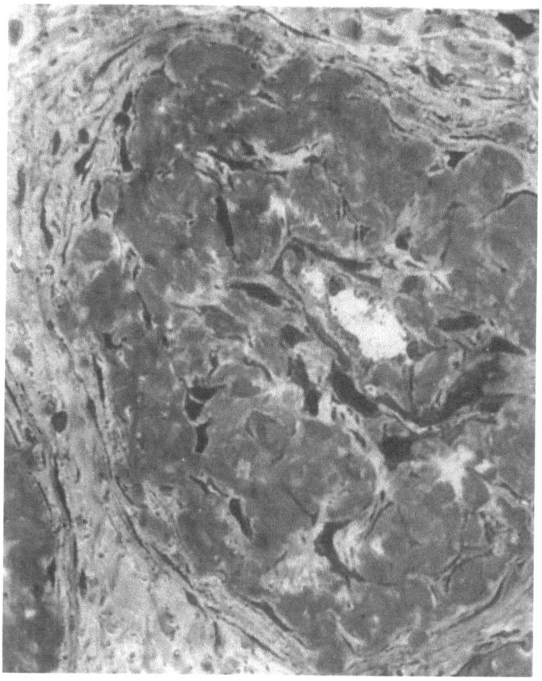

Deposition of extreme amount of amyloid in the wall of a blood vesel. The lumen is narrowed and the stucture of the vessel wall has deteriorated. (Magnification 560 ×)

Fig. XII/23

lar to those of rheumatic origin (Fig. XII/25). The infiltration of the shoulder-joint (infiltratio glenohumeralis) produces the typical shoulder-pad symptom, such patients look like rugby-players (190). The sediment of the synovial fluid can be stained by Congo-red dye proving the presence of amyloid fragments in the joints (137). Moreover, the amyloid may also cause osteolytic lesions leading to pathologic fractures (197).

## Coagulation disorders

Haemorrhagic diathesis can be considered the most threatening symptom of primary amyloidosis (143, 413). It may be caused by several factors: decrease in the concentration of vitamin K-dependant factors (IX, VII, II), increased antithrombin activity, increased fibrinolysis (DIC). These may be the consequences of interactions between the coagulation factors and the amyloid fibrils; this hypothesis also holds true for the factor X (115, 143). Petechiae, echymosis,

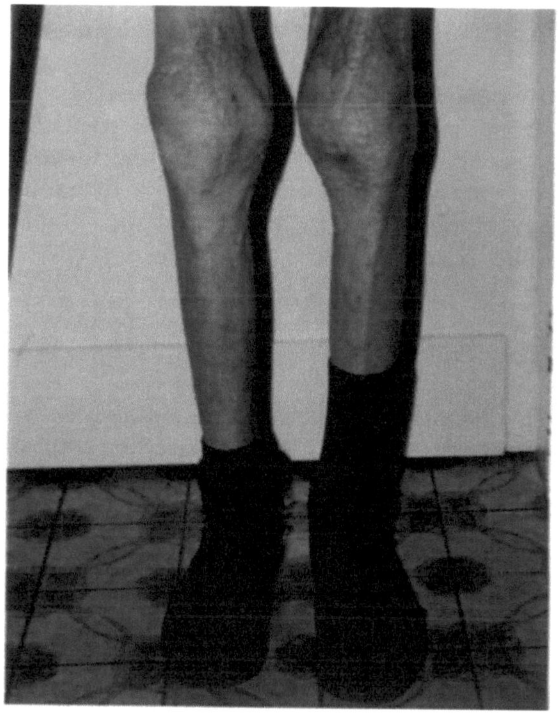

Fig. XII/24

suffusions appear on the skin; the increased permeability of the damaged capillary walls (413) causes in some cases pericardial tamponade (43).

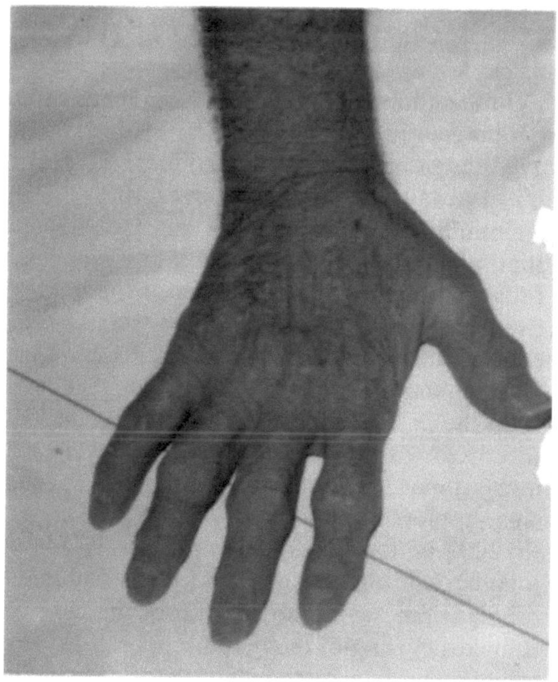

Fig. XII/25

### Abnormalities of the lymphatic glands

Lymphadenopathy and the related Sicca syndrome is not a rarity (339) if the deposits are present in the lymph nodes. In one form of amyloidosis the disease is mainly or exclusively localized in the lymphatic glands (205, 259).

### Endocrine organs

Involvement of the thyroid gland and the pituitary gland is well-known together with all the typical consequences of it.

### Symptomatology of amyloidosis

At first sight it may seem unnecessary for the reader to repeat certain aspects disease in this chapter, but since amyloidosis is a rarely observed disease often diagnosed only after autopsy it should be important to summarize the most significant aspects concerning the disease.

Amyloidosis is only occasionally detected in children: until 1989, 92 cases have been described: 43% of them associated with rheumatoid arthritis, 39% - with osteomyelitis and 18% - with Mediterranean fever (11, 147, 352).

In the adulthood the disease can more often be diagnosed in men than in women. The average age of patients is about 60-61 years, but it always affects people over 40.

Similarly to several other diseases the initial symptoms are not specific: weakness, fatigue, loss of weight, oedema of the ankles, dyspnoe, paraesthesia, syncope, hoarseness, or low-voice, excessively low or high tone may be warning signs.

In the *in floribus* (fully developed) stage of the disease the oedema of the ankles is the most important symptom, but other special signs typical of amyloidosis can be observed. In 20% of the cases macroglossia is found together with the enlargement of the submandibular tissues. - In about 10% of the cases - it makes closing of the mouth impossible. The tongue may become dentate. Petechiae of the eyelids - the so-called "raccoon-eye" (PPPP) have also an absolute diagnostic value. Haemorrhages and hepatomegaly are often seen. Rarely (in about 10% of the cases) splenomegaly is also present. The skin is often involved blisters, develop on the epidermis mainly at the sites under increased pressure, e.g. on the palms, below the breasts, etc. (Fig. XII/26 and XII/27).

Cardiomyopathy, nephrotic syndrome, malabsorption caused by amyloidotic infiltration can be detected in about 32% of the cases. In about 17% of the patients peripheral neuropathy, carpal tunnel syndrome and orthostatic hypotension may cause serious therapeutic problems. The amyloidotic infiltration of the organs may result in stiffness, stenosis, rigidity, obstruction, distortion of the size, obstacled, restricted locomotion, etc. - therefore, the functional disorders of the vital organs are mainly due to mechanical,and not biochemical alterations.

*The laboratory test* results are as follows: anaemia can be observed in about 50% of the cases, leukocyte count is usually withing the normal range and thrombocytosis can only very rarely be observed (402). Functional disorders of the kidneys are detected in about 50% of the cases, but serious renal failure can only be observed in the late phase of the disease. Proteinuria is almost always a characteristic sign, but hyperuricaemia only develop in multiple myeloma. Alkaline phosphatase and SGOT values are usually increased, and serum bilirubin is abnormal in 10-15% of the cases.

In 25% of the cases "M" components can be detected by serum electrophoresis. An increased level of polyclonal immunoglobulins can be verified in 25% of the cases. A half of the examinations yield insignificant or negative results.

Serum immunoelectrophoresis has a diagnostic value 50% of patients with primary amyloidosis, while is the amyloidosis and myeloma simmultaneously occur, the pathognomonic value of the above method reaches 74% (Fig. XII/28).

Immunoelectrophoresis can also be considered as the most suitable method for testing urine samples. It can be much better used for the detection of free light chains or immunoglobulin fragments than gel-electrophoresis or even SDS-PEG. Immunoelectrophoretic exa-

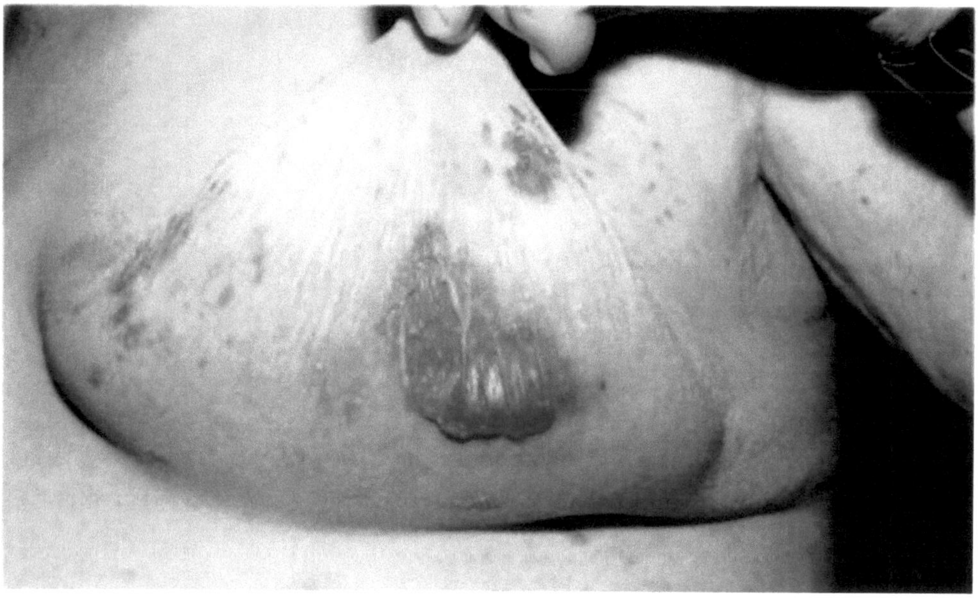

Fig. XII/26

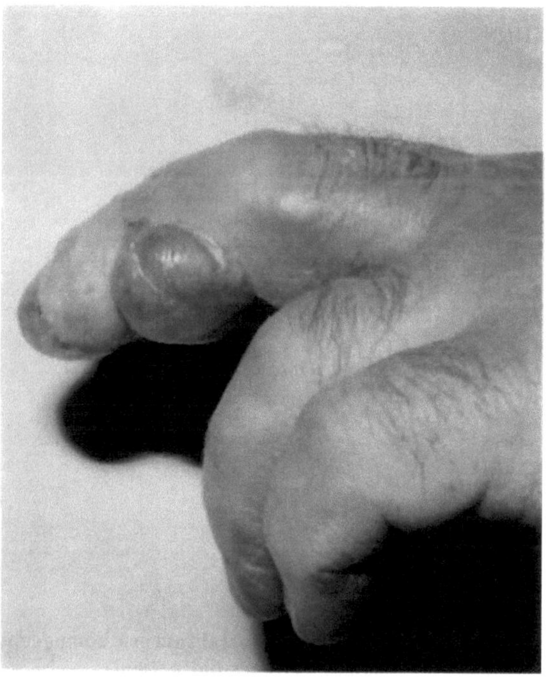

Fig. XII/27

mination of the serum or urine samples in primary amyloidosis can verify monoclonal gammopathy in 88% of the cases, this results is, however, 100% when both amyloidosis and Kahler's disease are present.

The lambda light chains can be considered pathogenetic factors in about 2/3 of the cases, the kappa light chains, however, in 1/3 of them. This remarkable because the physiological $\kappa/\lambda$ proportion is 3:2 and a similar light chain distribution is characteristic of those suffering from multiple myeloma (Fig. XII/29). The bone marrow plasma cell count is higher, though not significantly,in primary than in secondary amyloidosis.

Osteolytic lesions are very rarely observed without multiple myeloma, while osteoporosis, is present in each case.

## Prognosis

Systhemic amyloidosis still remains an incurable disease and only symptomatic treatment can support the patient before the fatal outcome due to renal or heart failure. The average survival period of primary amyloidosis not associated with myeloma is about 12-15 months, that occurring in multiple myeloma is between 4 and 9 months. The survival in amyloidosis associated with Waldeström's macroglobulinaemia ranges between these two values. These data are much worse than the survival period of patients with multiple myeloma without amyloidosis. Evidence exists that the prognosis of the disease is very poor, if cardiomyopathy or hypotension develop.

Patients suffering from secondary amyloi-

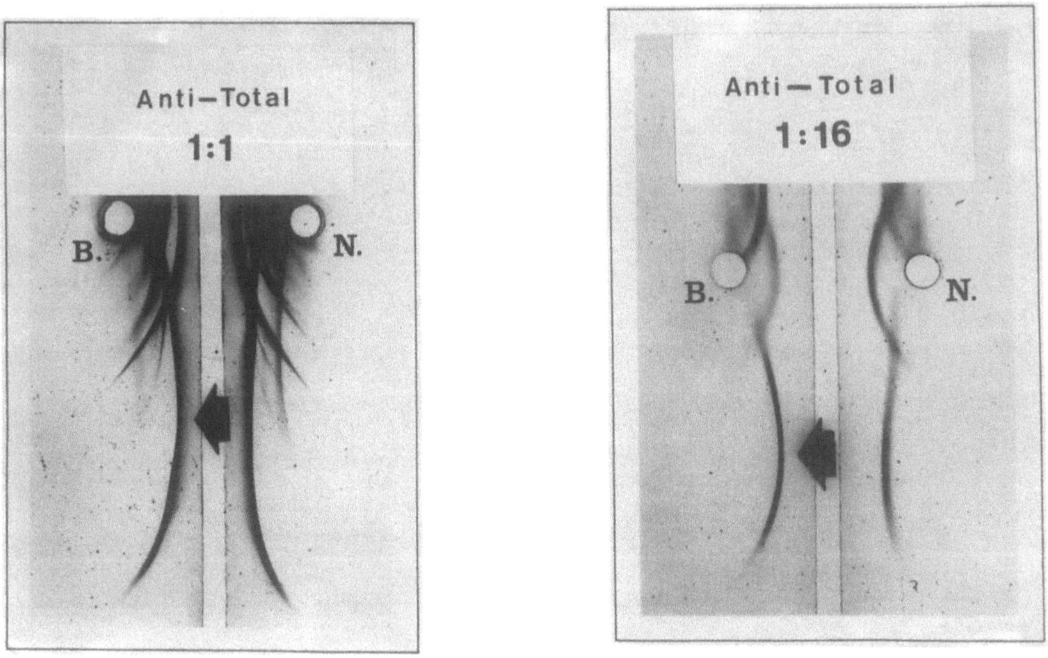


IMG, $IgG_3$-lambda (J.K., 44 years old patient)

Fig. XII/28

dosis comprise a more heterogeneous group, as the prognosis may fluctuate within a larger range according to the type of abnormality or the origin of the precursor. The survival period in these cases may reach 1-5 years, but some researchers report a period of 7-8 years. Furthermore, cure has been observed in some cases, following adequate treatment of the primary disease, (nephrectomy because of hypernephroma, lobectomy in lung with bronchiectasis or bowel resection in Crohn's disease) (243); following these interventions proteinuria ceased, hepatomegaly improved and amyloid deposits became of smaller extent. Histologically proved regression and even disappearance of amyloidosis verified by serial rectal or liver biopsies (238) is reported. Unfortunately, however, such of cases are rarely observed. Such observations could be used for a development of a rational basis for aggressive therapy of the primary disease. Nevertheless, it has to be concluded that the primary disease underlying amyloidosis must be treated by all possible means.

In amyloidosis associated with FMF nephrosis is the main symptom of the disease.

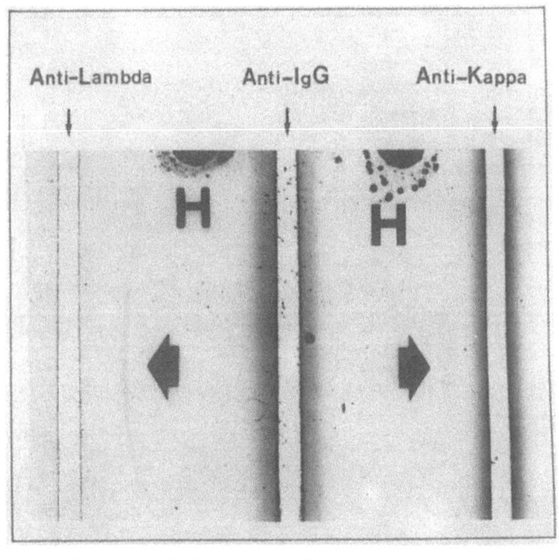


Lambda light chain in rectal mucosa homogenate (J.K., 44 years old patient)

Fig. XII/29

The period between the detection of proteinuria and the stage of uraemia is usually about 7 years. Naturally, the survival period may also depend on dialysis and the success of a possible kidney transplantation (89, 403).

## Therapy

Despite the inertness of the amyloid fibrils and their inaccessibility in the extracellular compartment, the generally irreversible storage of amyloid deposits can sometimes be reserved (75, 101, 364, 380). Nevertheless, no effective therapy of amyloidosis has been known till now. The observations mentioned in the first part of the present review, (surgical interventions performed in local amyloidosis "forced" kidney transplantation in renal failure) can be regarded as actual and definite solutions. In cases of developed amyloidosis there is no use to eliminate precursors detected in the serum or to try to decrease their production, because it is a metabolic disease, in which prevention of the amyloidogeneous molecules should be the goal. On the other hand, the amyloidogeneous molecules are possibly the results of genetically determined enzyme functions unknown at present. Nevertheless, the main task of the physicians is the decrease and possible elimination of the precursor production to try to delay the fatal outcome.

In primary amyloidosis the cytostatic therapy independently of the combinations applied does not significant by affect the survival period of the treated and untreated groups in the double-blind trials designed to control the efficacy of a therapeutic intervention. The situation is quite different in the secondary forms, where the adequate therapy may prolong the life of the patient (8). Unfortunately, cytostatic drugs have no effects on the previously deposited amyloid (100, 123, 214, 217, 218, 316, 323). The patients rarely survive a period of 12 months after the diagnosis of amyloidosis, although in some cases even longer survival periods have been reported (111). The symptoms referring to a poor prognosis in the first year are as follows: congestive cardiomyopathy, light chain elimination with the urine, hepatomegaly and definite symptoms of myeloma, regularly followed by an increased serum creatinin level and orthostatic hypotension as symptoms of the fatal terminal stage (159).

Chemotherapy mainly supports the renal function (sometimes even improving it to a certain extent) it may, furthermore, decrease hepatomegaly, but has no influence on the heart (217).

Trials using a combination of cytostatic drugs and colchicin, show that this treatment is not successful (216). Administration of dimethyl-sulphoxide a resulted in a similar failure (266).

Colchicin may prevent the development of AA amyloidosis in mice treated with casein, possibly because colchicin administration inhibits the increase of the acute-phase globulins level as precursor proteins (192, 333). On the basis of the above observation in FMF the prophylactic administration of a daily dose of 1-1.5 mg colchicin decreased the occurence of febrile attacks (77, 404, 405) and according to our present knowledge it may prevent the development of the primary and secondary amyloidosis (57), if the treatment had been commenced early enough (298, 407). Therefore, the FMF requires prophylactic treatment (405). Up till now no other data are available for other diseases inducing amyloidosis. It is an empirical observation that colchicin has in efficient in cases of verified amyloidosis (391) It is interesting that in FMF colchicin can prevent the formation of amyloid deposits in the transplanted kidney (178).

Dimethyl-sulphoxide (DMSO) - In animal experiments large doses may increase - the solubility of amyloid fibrils or facilitate the decomposition of the previously deposited amyloid. An actual regression of the amyloid content can be detected in the liver and spleen (141, 148, 265, 267). This drug has also been applied with a transient success in clinical practice (92). It has been rejected because of the transient character of the improvement, and due to the unbearable odour of the patients taking the drug. In summary, at present DMSO is no more used in therapy.

Vitamin C - It promotes the so-called amyloid degradation activity in "in vitro" experiments (195), therefore, attempts were made to use it as a therapeutic measure in amyloidosis. No results could, unfortunately, be obtained in either stage of the disease. Cytostatic treatment, melphalan, cyclophosphamide and prednisolon therapy in some cases proved to be effective due to their survival-prolonging effect.

## Supportive treatment

For the supportive treatment of the complaints caused by amyloidosis digitalis is frequently, however, its role is rather doubtful. On the one hand the unresponsiveness to digitalis suggests the possibility of amyloidosis, on the other hand, however, these patients may be rather sensitive to digitalis. Presumably, the degree of amyloidosis explains the difference (310). It is rather important to emphasize that administration of the calcium entry blockers may aggravate the congestive cardiomyopathy in amyloidosis (121). This effect might be due to the fact the binding of component P (SAP) requires the presence of calcium ions (Ca++), therefore, both calcium ions and the blockers can be linked to the deposits. The binding itself is an irreversible process and does not depend on the calcium ions (122). In desperate cases heart transplantation was also attempted (164). Postural hypotension could be improved by the administration of mineralocorticoids. Orthostatic hypotension could be prevented by the precursor of norepinephrine, L-threo-3.4-dihydroxy-phenyl-serine (357, 358). The symptoms of peripheral neuropathy and of painful paraesthesia can be successfully treated symptomatically by the administration of Carbamazepin in a dose of 200 mg/day (9). Dysphagia due to the dryness of the mouth can be alleviated by the use of coating materials. An extreme degree of macroglossia may result in a serious obstruction of food intake, therefore, in certain cases nasogastric tube-feeding and also surgical interventions (gastrostomy) become necessary. Diarrhoea indicating the involvement of the gastrointestinal tract can be ceased by the administration of somatostatin analogues (262). In case of airway-obstruction tracheostomy may be necessary to prevent suffocation.

Coagulation disorders and the subsequent pulmonary haemorrhage resulting in extreme blood loss have to be considered serious complications. Though other sources of haemorrhages are also possible, pulmonary lesions is the most frequently observed complication. According to the data of the literature coagulopathy could be ceased by a surgical removal of the enlarged, amyloidotic spleen (142).

Renal functions is maintained similar to the cases of renal diseases of other origin.

Haemodialysis has been applied several patients with secondary amyloidosis (145). Statistical data indicate that the above treatment prolonged the life span of the patients. Evidence exists that the carpal tunnel syndrome and cystic bone lesions can occur as complications of the above mentioned intervention (99, 219). Although haemodialysis may have a dangerous consequence increasing the concentration of $\beta_2$-microglobulin it can be overcome with filters made of polysulphone or by an extracorporal absorbent (201, 361).

The continuous ambulatory peritoneal dialysis (CAPD) seems to be an ideal method, but its results and possible complications are not yet finally evaluated.

In case of kidney transplants due to amyloidosis the results of the six monthès survival are worse than in patients with renal transplants with other indications. In the bloodvessels of the transplanted kidneys amyloid deposits will be stored within a short period (24, 177, 181, 273).

In case of high serum precursor level (69) or renal functional disorders due to light chain excretion, plasma exchange seems to be a suitable procedure (95, 349). Recent efforts are aimed at the selective elimination of the precursors with immunoabsorbent materials (134), similarly to the columns used for $\beta_2$-microglobulin bindings.

Preventive identification of persons at risk seems is performed with DNA testing which reveals amyloidosis caused by familial prealbumin as precursor (25, 162, 383, 384).

*Summarizing* the above mentioned results it can be stated that therapeutic trials applied in the treatment of amyloidosis can be assessed into following groups:

1. inhibition of the production of the amyloid precursor molecule,
2. prevention of the deposition of the previously formed pathogenic peptide,
3. dissolving of the previously formed deposits.

Evaluating of the results obtained in the 3 groups, it has to be declared that all the attempts resulted in failure.

## References


1. Abraham, C. R., Selkoe, D. J., Potter, H. Immunochemical identification of the serine protease inhibitor alpha-1-antichymotrypsin in the brain amyloid deposits of Alzheimer's disease. Cell. 52, 487. 1988.
2. Adams, W. Mollities ossium. Transactions of the Pathological Society of London. 23, 186. 1872.
3. Allensworth, D. C., Rice, G. J., Lowe, G. W. Persistent atrial standstill in a family with myocardial disease. Amer. J. Med. 47, 775. 1969.
4. Andersson, R. Familial amyloidosis with polyneuropathy. A clinical study based on patients living in Northern Sweden. Acta Med. Scand. (Suppl.). 590, 1. 1976.
5. Araki, S., Mawatari, S., Ohta, M., Nakjima, A., Kuroiwa, Y. Polyneurotic amyloidosis in a Japanese family. Arch. Neurol. 18, 593. 1968.
6. Aterman, K. A pretty a vista reaction for tissues with amyloid degeneration. 1875: An important year for pathology. Journal of the History of Medicine and Allied Sci. 31, 431. 1976.
7. Averbach, O., Stemmerman, M. G. Renal amyloidosis. Arch. Int. Med. 74, 244. 1944.
8. Bacon, P. A., Cadge, B., Harrison, P., Tribe, C. R. Treatment regimes for AA amyloidosis. In: Amyloidosis. Ed.: by Tribe, C. R., Bacon, P. A. E. A. R. S. Wright, Bristol. 131. 1983.
9. Bada, J. L., Cervera, C., Padro, L. Carbamazepine for amyloid neuropathy. N. Engl. J. Med. 296, 396. 1977.
10. Bardin, T., Zingraff, J., Shirahama, T., Noel, L. H., Droz, D., Voisin, M. C., Drueke, T., Dryll, A., Skinner, M., Cohen, A. S., Kuntz, D. Hemodialysis-associated amyloidosis and beta-2-microglobulin. Clinical and immunohistochemical study. Am J Med. 83, 419. 1987.
11. Barsaoui, S., Aissaoui, R., Bousnina, S., Tabbane, C., Ben-Moussa, F., Ben-Maiz, H. Renal amyloidosis in Tunisian children. Apropos of 14 cases. Tunis. Med. 67(3), 155. 1989.
12. Battle, W. M., Rubin, M. R., Cohen, R., Snape, W. J. Gastrointestinal motility dysfunction in amyloidosis. N. Engl. J. Med. 301, 24. 1979.
13. Bausserman, Linda, L., Herbert, P. N., McAdam, K. P. W. J. Heterogeneity of human serum amyloid A proteins. J. Exp. Med. 152, 641. 1980.
14. Black, M. M. Primary localised amyloidosis of the skin: clinical variants histochemistry and ultrastructure. In: Amyloidosis. Ed. by Wegelius, O., Pasternak, A. Academic Press. London. 479. 1976.
15. Benditt, E. P., Eriksen, N., Hanson, R. H. Amyloid protein SAA is an apoprotein of mouse plasma high density lipoprotein. Proc. Natl. Acad. Sci. USA. 76, 4092. 1979.
16. Bennhold, H. Eine specifische Amyloid Färbung mit Kongorot. Münch. med. Wschr. 69, 1537. 1922.
17. Benson, M. D. Partial amino acid sequence homology between an heredofamilial amyloid protein and human plasma prealbumin. J. Clin. Invest. 67, 1035. 1981.
18. Benson, M. D., Cohen, A. S. Serum amyloid A protein in amyloidosis, rheumatic and neoplastic diseases. Arth. Rheum. 22, 36. 1979.
19. Benson, M. D., Cohen, A. S. Generalized amyloid in a family of Swedish origin: A study of 426 family members in seven generations of a newkinship with neuropathy, nephropathy and central nervous systhem involvemtent. Ann. Intern. Med. 86, 119. 1977.
20. Benson, M. D., Dwulet, F. E. Identifications of carriers of a variant prealbumin (transthyretin) associated with familial amyloidotic polyneuropathy type I. J. Clin. Invest. 75, 71. 1985.
21. Benson, M. D., Dwulet, F. E. Prealbumin and retinol binding protein serum concentrations in the Indiana type hereditary amyloidosis. Arthritis Rheum. 26, 1493. 1983.
22. Benson, M. D., Dwulet, F. E., Madura, D., Wheeler, G. Structural analysis of a lambda IV primary amyloid protein. In: Amyloid and Amyloidosis. Ed.: by Isobe, T., Araki, S., Uchino, F., Kito, S., Tsubura, F. New York, NY. Plenum. 163. 1988.
23. Benson, M. D., Skinner, M. Amino terminal sequence of amyloid P-component isolated from serum. J. Lab. Clin. Med. 92, 78. 1978.
24. Benson, M. D., Skinner, M., Cohen, A. S. Amyloid deposition in a renal transplant in familial Mediterranean fever. Ann. Intern. Med. 87, 31. 1977.
25. Benson, M. D., Wallace, M. R. Amyloidosis. In: The Metabolic Basis of Inherited Disease. Ed.: by Scriver, C. R., Beaudet, A. L., Sly, W. S., Valle, D. New York, NY. McGraw-Hill. 2, 2439. 1989.
26. Blake, C. C. F., Geison, M. J., Oatley, S. J., Rérat, B., Rérat, C. Structure of prealbumin: secondary, tertiary and quarternary interactions determined by Fourier refinement at 1.8 A. J. Mol. Biol. 121, 339. 1978.
27. Blake, C. C. F., Geison, M. J., Swan, I. D. A., Rérat, C., Rérat, B. Structure of human plasma prealbumin at 2.5 A resolution. A preliminary report on the polypeptide chain confirmation, quarternary structure and thyroxin binding. J. Mol. Biol. 88, 1. 1974.
28. Blum, A., Gafni, J., Sohar, E., Shibolet, S., Heller, H. Amyloidosis as the sole manifestation of familial Mediterranean fever (FMF). Further evidence of its genetic nature. Ann. Intern. Med. 57, 795. 1962.
29. Boyko, W. J., Pratt, R., Wass, H. Functional hyposplenism, a diagnostic clue in amyloidosis. Amer. J. Clin. Path. 77, 745. 1981.
30. Breathnach, S. M., Bhogal, B., Dyck, R. F., de Beer, F. C., Black, M. M., Pepys, M. B. Immunohistochemical demonstration of amyloid P component in skin of normal subjects and patients with cutaneous amyloidosis. Brit. J. Dermatol. 105, 115. 1981.
31. Breathnach, S. M., Black, M. M. Systemic amyloidosis and the skin: A review with special emphasis on clinical features and therapy. Clin. Exp. Dermatol. 4, 517. 1979.
32. Breathnach, S. M., Melrose, S. M., Bhogal, B., de Beer, F. C., Dyck, R. F., Tennent, G., Black, M. M., Pepys, M. B. Amyloid P component is located on elastic fibre microfibrils of normal human tissues. Nature. 293, 652. 1981.

33. Brinckerhoff, C. E., Mitchell, T. I., Karmilowicz, M. J., Kluve-Beckerman, B., Benson, M. D.
Autocrine induction of collagenase by serum amyloid A-like and béta$_2$-microglobulin-like proteins.
Science. 243, 655. 1989.
34. Brodarick, S., Paine, R., Higa, E., Carmichael, K. A.
Pericardial tamponade, a new complication of amyloid heart disease.
Amer. J. Med. 73, 133. 1982.
35. Bruni, J., Bilbao, J. M., Pritzker, K. P. H.
Myopathy associated with amyloid angiopathy.
Canad. J. Neurol. Sci. 4, 77. 1977.
36. Budd, G.
On diseases of the liver.
John Churchill London. 243. 1845.
37. Cabili, S., Pras, M.
The course of amyloid nephropathy in familial Mediterranean fever (FMF).
In: Amyloidosis.
Ed.: by Tribe, C. R., Bacon, P. A.
E. A. R. S. Wright Bristol. 105. 1983.
38. Caldamone, A. A., Elbadawi, A., Moshtagi, A., Frank, I. N.
Primary localized amyloidosis of urinary bladder.
Urology. 15, 174. 1980.
39. Camoirano, J. K., Greipp, P. R., Bayer, G. K., Bowie, E. J. W.
Resolution of acquired factor X deficiency and amyloidosis with melphalan and prednisone therapy.
N. Engl. J. Med. 316, 1133. 1987.
40. Case Records of the Massachusetts General Hospital.
N. Engl. J. Med. 303, 985. 1980.
41. Case Records of the Massachusetts General Hospital.
N. Engl. J. Med. 278, 1276. 1968.
42. Casey, T. T., Stone, W. J., DiRaimondo, C. R., Brantley, B. D., DiRaimondo, C. V., Gorevic, P. H., Page, D. L.
Tumoral amyloidosis of bone of beta-2-microglobulin origin in association with long-term hemodialysis. A new type of amyloid disease.
Hum. Pathol. 17, 731. 1986.
43. Caspi, D., Baltz, M. L., Munn, E. A., Feinstein, A., Pepys, M. B.
'Amyloid degrading activity' of human serum, an in vitro clearing effect which does not involve degradation of amyloid fibrils.
Clin. Exp. Immunol. 57, 647. 1984.
44. Champion, M., Richards, R. L.
Amyloidosis in Hodgkin's disease. A Scottish survey.
Scot. Med. J. 24, 9. 1979.
45. Chapel, T. A., Birmingham, D. J., Malinowski, Y. E.
Nodular primary localized cutaneous amyloidosis.
Arch. Dermatol. 113, 1248. 1977.
46. Celli, B. R., Rubinow, A., Cohen, A. S., Brody, J. S.
Patterns of pulmonary involvement in systemic amyloidosis.
Chest. 74, 543. 1978.
47. Clark, R. A., Omenn, G. S.
Anhidrotic ectodermal dysplasia with frequent infections and amyloidosis.
West. J. Med. 126, 225. 1977.
48. Clark, A., Cooper, G. J. S., Lewis, C. E., Morris, J. F., Willis, A. C., Reid, K. M. B., Turner, R. C.
Islet amyloid formed from diabetes-associated peptide may be pathogenic in type 2 diabetes.
Lancet. ii, 231. 1987.
49. Cohen, A. S., Calkins, E.
Electron microscopic observations on a fibrous component in amyloid of diverse origins.
Nature. 183, 1202. 1959.
50. Cohen, A. S.
Amyloidosis.
N. Engl. Med. 277, 522. 574. 628. 1967.
51. Cohen, A. S., Shirahama, T.
Electron microscopic analysis of isolated amyloid fibrils from patients with primary, secondary and myeloma-associated disease. A study utilizing shadowing and negative staining techniques.
Israel J. Med. Sci. 9, 849. 1973.
52. Cohen, A. S., Wegelius, O.
Classification of amyloid 1979–1980.
Arth. Rheum. 23, 644. 1980.
53. Cohen, H. J., Crawford, J.
Disorders of amyloid deposition.
In: Pathology of immunoglobulins: diagnostic and clinical aspects.
Liss. New York. 293. 1982.
54. Cohen, D. H., Feiner, H., Jensson, O., Frangione, B.
Amyloid fibril in hereditary cerebral haemorrhage with amyloidosis (HCHWA) is related to the gastroentero-pancreatic neuroendocrine protein, gamma trace.
J. Exp. Med. 158, 623. 1983.
55. Cohen, A. S.
General introduction and a brief history of amyloidosis.
In: Amyloidosis.
Ed.: by Marrink, J., van Rijswijk, M. H.
Dordrecht, Martinus Nijhoff. 3. 1986.
56. Cohen, A. S., Connors, L. H.
The pathogenesis and biochemistry of amyloidosis.
J. Pathol. 151, 1. 1987.
57. Cohen, A. S., Rubinow, A., Anderson, J. J., Skinner, M., Mason, J. H., Libbey, C., Kayne, H.
Survival of patients with primary (AL) amyloidosis.
Colchicine-treated cases from 1976 to 1983 compared with cases seen in previous years (1961 to 1973).
Am. J. Med. 82, 1182. 1987.
58. Cohen, S. L., Boner, G., Shmueli, D., Yusim, A., Rosenfeld, J., Shapira, Z.
Cyclosporin poorly tolerated in familial Mediterranean fever.
Nephrol. Dial. Transplant. 4, 201. 1989.
59. Cooper, J. H.
Localized dystrophic amyloidosis of heart valves.
Hum. Pathol. 14, 649. 1983.
60. Cooper, J. H.
A histochemical construct of the amyloid fibril.
In: Amyloidosis.
Ed.: by Tribe, C. R., Bacon, P. A.
E. A. R. S. Wright, Bristol. 31. 1983.
61. Cornwell, G. G., Thomas, B., Kyle, R. A., Sletten, K., Westermark, P.
Immunohistochemical typing of tissues from 100 patients with systemic amyloidosis.
In: Amyloid and Amyloidosis.
Ed.: by Isobe, T., Araki, S., Uchino, F., Kito, S., Tsubura, E.
New York, NY. Plenum. 179. 1988.
62. Cornwell, G. G.
Progress in multiple myeloma, amyloidosis, and related monoclonal dysproteinemias.
Curr. Hematol. Oncol. 5, 121. 1987.
63. Cornwell, G. G. III., Sletten, K., Olofsson, B. O., Johansson, B., Westermark, P.
Prealbumin: its association with amyloid.
J. Clin. Pathol. 40, 226. 1987.
64. Cornweel, G. G., Murdoch, W. L., Kyle, R. A., Westermark, P., Pitkanen, P.
Frequency and distribution of senile cardiovascular amyloid. A clinicopathological correlation.
Am. J. Med. 75, 618. 1983.
65. Cornwell, G. G., Westermark, P.
Senile amyloidosis: a protein manifestation of the aging process.
J. Clin. Pathol. 33, 1146. 1980.
66. Costa, P. P., Figueira, A. S., Bravo, F. R.
Amyloid fibril protein related to prealbumin in familial amyloidotic polyneuropathy.
Proc. Natl. Acad. Sci. USA. 75, 4499. 1978.
67. Cueto-Garcia, L., Tajik, A. J., Kyle, R. A., Edwards, W. D., Greipp, P. R., Callahan, J. A., Shub, C., Seward, J. B.
Serial echocardiographic observations in patients with primary systemic amyloidosis. An introduction to the con-

cept of early (asymptomatic) amyloid infiltration of the heart.
Mayo Clin. Proc. 59, 589. 1984.
68. Cunningham, B. A., Wang, S. L., Berggard, I., Peterson, P. A.
The complete amino acid sequence of beta-2-microglobulin.
Biochemistry. 12, 4811. 1973.
69. Coutinho, P., Macedo, E., Estacio, A., Saraiva, A., Costa, P. P., Saraiva, J. M.
Periodic plasma exchanges in the treatment of familial amyloid polyneuropathy. Preliminary results.
In: Amyloid and Amyloidosis.
Ed.: by Isobe, T., Araki, S., Uchino, F., Kito, S., Tsubura, E.
New York, NY. Plenum. 845. 1988.
70. Daoud, F. S., Nieman, R. E., Vilter, R. W.
Amyloid goiter in a case of generalized primary amyloidosis.
Am. J. Med. 43, 604. 1967.
71. Dayal, Y., Tahjian, A. H., Wolfe, H. J.
Immunocytochemical localization of calcitonin-producing cells in a stromal carcinoid with amyloid stroma.
Cancer. 43, 1331. 1979.
72. de Beer, F. C., Mallya, R. K., Fagan, E. A., Lanham, J. G., Hughes, G. R. V., Pepys, M. B.
Serum amyloid A protein concentration in inflammatory diseases and its relationship to the incidence of reactive systemic amyloidosis.
Lancet. i, 231. 1982.
73. DeCastro, S., Sparks, J. T., Lapey, J. D., Freidberg, S. R.
Amyloidoma of the gasserian ganglion.
Surg. Neurol. 6, 357. 1976.
74. Dikman, S. H., Churg, J., Kahn, T.
Morphologic and clinical correlates in renal amyloidosis.
Hum. Pathol. 12, 160. 1981.
75. Dikman, S. H., Kahn, T., Gribetz, D., Churg, J.
Resolution of renal amyloidosis.
Am. J. Med. 63, 430. 1977.
76. Dikman, S. H., Goldstein, M., Kahn, T., Leo, M. A., Weinreb, N.
Amyloidosis. An unusual complication of Gaucher's disease.
Arch. Pathol. Lab. Med. 102, 460. 1978.
77. Dinarello, C. A., Wolff, S. M., Goldfinger, S. E., Dale, D. C., Alling, D. W.
Colchicine therapy for familial Mediterranean fever. A doubleblind trial.
N. Engl. J. Med. 291. 934. 1974.
78. Ducret, F., Pointet, Ph., Martin, C., Darsy, P., Deplante, J. P., Roda, L.
Lupus érythémateux disséminé, maladie de Basedow, gammapathie monoclonale et amylose associée chez une même malade.
Prese. Méd. 19, 1100. 1985.
79. Durie, B. G. M., Persky, B., Soehnlen, B. J., Grogan, T. M., Salmon, S. E.
Amyloid production in human myeloma stem-cell culture, with morphologic evidence of amyloid secretion by associated macrophages.
N. Engl. J. Med. 307, 1689. 1982.
80. Duston, M. A., Skinner, M., Shirahama, T., Cohen, A. S.
Diagnosis of amyloidosis by abdominal fat aspiration.
Analysis of four years' experience.
Am. J. Med. 82, 412. 1987.
81. Duston, M. A., Skinner, M., Meenan, R. F., Cohen, A. S.
Sensitivity, specificity, and predictive value of abdominal fat aspiration for the diagnosis of amyloidosis.
Arth. Rheum. 32, 82. 1989.
82. Dwulet, F. E., Benson, M. D.
Polymorphism of human plasma thyroxine binding prealbumin.
Biochem. Biophys. Res. Commun. 114, 657. 1983.
83. Dwulet, F. E., Benson, M. D.
Primary structure of an amyloid prealbumin and its plasma precursor in a heredofamilial polyneurophaty of Swedish origin.
Proc. Natl. Acad. Sci. USA. 81, 694. 1984.
84. Dwulet, F. E., Benson, M. D.
Characterization of a transthyretin (prealbumin) variant associated with familial amyloidotic polyneuropathy type II (Indiana/Swiss).
J. Clin. Invest. 78, 880. 1986.
85. Dyck, R. F., Evans, D. J., Lockwood, C. M., Rees, A. J., Turner, D., Pepys, M. B.
Amyloid P-component is a constituent of normal human membrane.
Abnormal patterns of immunofluorescent staining in glomerular disease.
Lancet. ii, 606. 1980.
86. Dyck, R. F., Lockwood, M., Kershaw, M., McHugh, N., Duance, V., Baltz, M. L., Pepys, M. B.
Amyloid P–component is a constituent of normal human glomerular basement membrane.
J. Exp. Med. 152, 1162. 1980.
87. Ekelund, L.
Radiologic features in renal amyloidosis.
Amer. J. Roentgenol. 129, 851. 1977.
88. Elcock, H. W., Grimadi, P. M. G. B.
Post-nasal obstruction by amyloid secondary to Waldenström's macroglobulinaemia.
J. Laryngol. Otol. 86, 1075. 1972.
89. Eliakim, M., Levy, M., Ehrenfeld, M.
Recurrent polyserositis. Familial Mediteranean fever, periodic disease.
Elsevier/North Holland, Amsterdam, 1981.
90. Eriksen, N., Benditt, E. P.
Isolation and characterization of the amyloid-related apoprotein (SAA) from human high density lipoprotein.
Proc. Natl. Acad. Sci. USA. 77, 6860. 1980.
91. Falck, H. M., Maury, C. P. J., Teppo, A. M., Wegelius, O.
Correlation of persistently high serum amyloid A protein and C-reactive protein concentrations with rapid progression of secondary amyloidosis.
Brit. Med. J. 286. 1391. 1983.
92. Falck, H. M., Wegelius, O.
Treatment of secondary renal amyloidosis with dimethyl sulphoxide.
In: Amyloidosis.
Ed.: by Tribe, C. R., Bacon, P. A.
E. A. R. S. Wright, Bristol. 128. 1983.
93. Falk, R. H., Lee, V. W., Rubinow, A., Hood, W. B., Cohen, A. S.
Sensitivity of technetium-99m-prophosphate scintigraphy in diagnosing cardiac amyloidosis.
Amer. J. Cardiol. 51, 826. 1983.
94. Falkson, G., Falkson, H. C.
Amyloidosis in Hodgkin's disease.
South African Med. J. 47, 62. 1973.
95. Fang, L. S. T.
Light-chain nephropathy.
Kidney Int. 27, 582. 1985.
96. Fehér, J., Csomós, G., Vereckei, A.
Free radical reactions in medicine.
Springer-Verlag Berlin. 125. 1988.
97. Feiner, H. D.
Pathology of dysproteinemia. Light chain amyloidosis, nonamyloid immunoglobulin deposition disease, cryoglobulinemia syndromes, and macroglobulinemia of Waldenström.
Hum. Pathol. 19, 1255. 1988.
98. Felding, P., Fex, G., Westermark, P., Olofsson, B. O., Pitkänen, P., Benson, L.
Prealbumin in Swedish patients with senile systemic amyloidosis and familial amyloidotic polyneuropathy.
Scand. J. Immunol. 21, 133. 1985.
99. Fenves, A. Z., Emmett, M., White, M. G., Greeway, G.,

Michaels, D. B.
Carpal tunnel syndrome with cystic bone lesions secondary to amyloidosis in chronic hemodialysis patients.
Am. J. Kidney Dis. 7, 130. 1986.

100. Fielder, K., Durie, B. G. M.
Primary amyloidosis associated with multiple myeloma. Predictors of successful therapy.
Am. J. Med. 80. 413. 1986.

101. Fields, M., Polliack, A., Laufer, A.
Resorption of amyloid and enzymatic studies in amyloidosis.
Israel J. Med. Sci. 9, 875. 1973.

102. Frank, J. M., Palomino, N. J.
Primary amyloidosis with diffuse splenic infiltration presenting as fulminant pneumococcal sepsis.
Am. J. Clin. Pathol. 87, 405. 1987.

103. Franklin, E. C., Zucker-Franklin, D.
Current concepts of amyloid.
Advances in Immunology. 15, 249. 1972.

104. Franklin, E. C., Buxbaum, J.
Immunoglobulins in the normal state and in neoplasias of B cells.
Compreh. Immunolog. 4, 361. 1978.

105. Franklin, E. C.
The amyloid diseases.
In: Clinical aspects of immunology, 4th edn.
Ed. by Lachmann, P. J., Peters, D. K.
Blackwell, Oxford. 1231. 1982.

106. Franklin, E. C.
The Immunopathology of amyloid diseases.
In: The biology of immunologic disease.
Ed.: by Dixon, F. J., Fisher, D. W.
A hospital practice book.
Sinaer Associates, Inc. 247. 1983.

107. Frederiksen, L., Götzsche, M., Harboe, N., Kmet, W. N., Mellemgoard, K.
Familial primary amyloidosis with severe amyloid heart disease.
Amer. J. Med. 33, 328. 1962.

108. Friedland, R. P., Koss, E., Haxby, J. V., Grady, C. L., Luxenberg, J., Schapiro, M. B., Kaye, J.
Alzheimer disease. Clinical and biological heterogeneity.
Ann. Intern. Med. 109, 298. 1988.

109. Friedman, E., Farfel, Z., Goldfarb, D., Gafni, J.
Renal tubular acidosis after prolonged remission of nephrotic syndrome in amyloidosis associated with astrocytoma.
South. Med. J. 82(6), 765. 1989.

110. Friedrich, N., Kékulé, A.
Zur Amyloidfrage.
Archiv für Pathologische Anatomie und Physiologie und für Klinische Medizin. 16, 50. 1859.

111. Fritz, D. A., Luggen, M. E., Hess, E. V.
Unusual longevity in primary systemic amyloidosis. A 19-year survivor.
Am. J. Med. 86, 245. 1989.

112. Fujihara, S., Balow, J. E., Costa, J. C., Glenner, G. G.
Identification and classification of amyloid in formalin-fixed paraffin-embedded tissue sections by the unlabelled immunoperoxidase method.
Lab. Invest. 43, 358. 1980.

113. Fujihara, S., Glenner, G. G.
Primary localised amyloidosis of the genitourinary tract. Immunohistochemical study on eleven cases.
Lab. Invest. 44, 55. 1981.

114. Furie, B., Greene, E., Furie, B. C.
Syndrome of acquired factor X deficiency and systemic amyloidosis.
In vivo studies of the metabolic fate of factor X.
N. Engl. J. Med. 297, 81. 1977.

115. Furie, B., Voo, L., McAdam, K. P. W. J., Furie, B. C.
Mechanism of factor X deficiency in systemic amyloidosis.
N. Engl. J. Med. 304, 827. 1981.

116. Gaffney, E. F., Lee, J. C. K.
Systemic amyloidosis and hypogammaglobulinaemia.
Arch. Path. Lab. Med. 102, 558. 1978.

117. Gallo, G., Picken, M., Buxbaum, J., Frangione, B.
The spectrum of monoclonal immunoglobulin deposition disease associated with immunocytic dyscrasias.
Semin. Hematol. 26, 234. 1989.

118. Ganeval, D., Noel, L. H., Preud'homme, J. L., Droz, D., Grunfeld, J. P.
Light-chain deposition disease. Its relation with AL-type amyloidosis.
Kidney Int. 26, 1. 1984.

119. Gejyo, F., Yamada, T., Odani, S., Nakagawa, Y., Arakawa, M., Kunitomo, T., Kataoka, H., Suzuki, M., Hirasawa, Y., Shirahama, T., Cohen, A. S., Schmid, K.
A new form of amyloid protein associated with chronic hemodialysis was identified as beta$_2$-microglobulin.
Biochem. Biohyps. Res. Commun. 129, 701. 1985.

120. Gertz, M. A., Kyle, R. A., Greipp, P. R.
Hyposplenism in primary systemic amyloidosis.
Ann. Intern. Med. 98, 475. 1983.

121. Gertz M. A., Falk, R. H., Skinner, M., Cohen, A. S., Kyle, R. A.
Worsening of congestive heart failure in amyloid heart disease treated by calcium channel-blocking agents.
Am. J. Cardiol. 55, 1645. 1985.

122. Gertz, M. A., Skinner, M., Connors, L. H., Falk, R. H., Cohen, A. S., Kyle, R. A.
Selective binding of nifedipine to amyloid fibrils.
Am. J. Cardiol. 55, 1646. 1985.

123. Gertz, M. A., Kyle, R. A.
Response of primary hepatic amyloidosis to melphalan and prednisone. A case report and review of the literature.
Mayo Clin. Proc. 61, 218. 1986.

124. Gertz, M. A., Kyle, R. A.
Hepatic amyloidosis (primary (AL), immunoglobulin light chain).
The natural history in 80 patients.
Am. J. Med. 85, 73. 1988.

125. Giaccone, G., Tagliavini, F., Linoli, G., Bouras, C., Frigerio, L., Frangione, B., Bugiani, O.
Down patients. Extracellular preamyloid deposits precede neuritic degeneration and senile plaques.
Neurosci. Lett. 97, 232. 1989.

126. Chiso, J., Jensson, O., Frangione, B.
Amyloid fibrils in hereditary cerebral hemorrhage with amyloidosis of Icelandic type is a variant of gamma-trace basis protein (cystatin C).
Proc. Natl. Acad. Sci. USA. 83, 2974. 1986.

127. Glenner, G. G., Terry, W., Harada, M., Isersky, C., Page, D.
Amyloid fibril proteins: proof of homology with immunoglobulin light chains by sequence analyses.
Science. 172, 1150. 1971.

128. Glenner, G. G., Ein, D., Eanes, E. D., Bladen, H. A., Terry, W., Page, D.
Creation of 'amyloid' fibrils from Bence Jones protein in vitro.
Science. 174, 712. 1971.

129. Glenner, G. G.
Amyloid deposits and amyloidosis. The beta-fibrilloses.
N. Engl. J. Med. 302, 1283. 1980.

130. Glenner, G. G., Pinho e Costa, P., Falcao de Freitas, A.
Amyloid and Amyloidosis.
Excerpta Medica. Amsterdam. 1980.

131. Glenner, G. G.
The bases of the staining of amyloid fibers: their physicochemical nature and the mechanisms of their dye-substrate interaction.
Progr. Histochem. Cytochem. 13, 1. 1981.

132. Glenner, G. G.
Alzheimer's disease: the commonest form of amyloidosis.
Arch. Path. Lab. Med. 107, 281. 1983.

133. Glenner, G. G., Wong, C. W.
Alzheimer's disease. Initial report of the purification and characterization of a novel cerebrovascular amyloid protein.

Biochem. Biophys. Res. Commun. 120, 885. 1984.
134. Glenner, G. G.
Future directions in amyloid research.
In: Amyloidosis.
Ed.: by Marrink, J., van Rijswijk, M. H.
Dordrecht. Martinus Nijhoff. 357. 1986.
135. Glenner, G. G.
The pathobiology of Alzheimer's disease.
Ann. Rev. Med. 40, 45. 1989.
136. Goffin, Y. A., Thona, Y., Potvliege, P. R.
Microdeposition of amyloid in the joints.
Ann. Rheum. Dis. 40, 27. 1981.
137. Gordon, D. A., Pruzanski, W., Ogryzlo, M. A.
Synovial fluid examination for the diagnosis of amyloidosis.
Ann. Rheum. Dis. 32, 428. 1973.
138. Goren, H., Steinberg, M. C., Farboody, G. H.
Familial oculoleptomeningeal amyloidosis.
Brain. 103, 473. 1980.
139. Gorevic, P. D., Casey, T. T., Stone, W. J., DiRaimondo, C. R., Prelli, F. C., Frangione, B.
Beta–2–microglobulin is an amyloidogenic protein in man.
J. Clin. Invest. 76, 2425. 1985.
140. Gorevic, P. D., Munoz, P. C., Casey, T. T., DiRaimondo, C. R., Stone, W. J., Prelli, F. C., Rodrigues, M. M., Poulik, M. D., Frangione, B.
Polymerization of intact beta-2 microglobulin in tissue causes amyloidosis in patients on chronic hemodialysis..
Proc. Natl. Acad. Sci. USA. 83, 7908. 1986.
141. Greenwald, M., Kedar, I., Sohar, E.
The effect of colchicine and dimethylsulfoxide (DMSO) on the genetic amyloidosis of the white Peking duck.
In: Amyloid and Amyloidosis.
Ed.: by Glenner, G. G., Pinho e Costa, P., Falcao de Freitas, A.
Excerpta Medica. Amsterdam. 584. 1980.
142. Greipp, P. R., Kyle, R. A., Bowie, E. J. W.
Factor X deficiency in primary amyloidosis: resolution after splenectomy.
N. Engl. J. Med. 301, 1050. 1979.
143. Greipp, P. R., Kyle, R. A., Bowie, E. J. W.
Factor X deficiency in amyloidosis. A critical review.
Am. J. Hematol. 11, 443. 1981.
144. Gudmundsson, G., Hallgrimsson, J., Jonasson, T. A., Bjarnason, O.
Hereditary cerebral hemorrhage with amyloidosis.
Brain. 95, 387. 1972.
145. Gurland, H. J.
Combined report on regular dialysis and transplatation in Europe, VI, 1975.
Proceedings of the European Dialysis and Transplant Association. 13, 3. 1975.
146. Hackler, R. H.
A 25-year prospective mortality study in the spinal cord injured patient: comparison with the long-term living paraplegic.
J. Urol. 117, 486. 1977.
147. Hake, H., Goll, U., Thoenes, W.
Primary perireticulin amyloidosis in a 14-years-old girl.
Europ. J. Pediat. 124, 43. 1976.
148. Hanai, N., Ishihara, T., Uchino, F., Imada, N., Fujihara, S., Ikegami, J.
Effects of dimethyl sulfoxide and colchicine on the resorption of experimental amyloid.
Virch. Arc. Abt. A, Pathol. Anat. 384, 45. 1979.
149. Hanash, S. M., Rucknagel, D. L., Heidelberger, K. P., Radin, N. S.
Primary amyloidosis associated with Gaucher's disease.
Ann. Intern. Med. 89, 639. 1978.
150. Hardy, T. J., Myerowitz, R. L., Bender, B. L.
Diffuse parenchymal amyloidosis of lungs and breast. Its association with diffuse plasmacytosis and – chain gammopathy. Arch. Path. Lab. Med. 103, 583. 1979.
151. Hashimoto, K.
Skin function and amyloidosis: Ultrastructural studies.
In: Amyloidosis.
Ed.: by Wegelius, O., Pasternack, A.
Academic Press. 512, 1976.
152. Hill, C. S., Ibanez, M. I., Samaan, N. A., Ahearn, M. J., Clark, R. L.
Medullary (solid) carcinoma of the thyroid gland: An analysis of the MD Anderson Hospital experience with patients with the tumor, its special features, and its histogenesis.
Medicine. 52, 141. 1973.
153. Hijmans, W., Sipe, J. D.
Levels of the serum amyloid A protein (SAA) in normal persons of different age groups.
Clin. Exp. Immunol. 35, 96. 1979.
154. Hind, C. R. K., Tennent, G. A., Evans, D. J., Pepys, M. B.
Demostration of amyloid A (AA) protein and amyloid P component (AP) in deposits of systemic amyloidosis associated with renal adenocarcinoma.
J. Pathol. 139, 159. 1983.
155. Hind, C. R. K., Collins, P. M., Renn, D., Cook, R. B., Caspi, D., Baltz, M. L., Pepys, M. B.
Binding specificity of serum amyloid P component for the pyruvate acetat. of galactose.
J. Exp. Med. 159, 1059. 1984.
156. Hind, C. R. K., Gibson, D. G., Lavender, J. P., Pepys, M. B.
Non-invasive demonstration of cardiac involvement in acquired forms of systemic amyloidosis.
Lancet. i, 1417. 1984.
157. Hirano, T., Kishimoto, T.
Interleukin 6 and plasma cell neoplasias.
Progress in growth factor reasearch. 1, 133. 1989.
158. Hobbs, J. R.
Genetic disease and amyloid.
J. Clin. Pathol. (Suppl.) 12, 128. 1978.
159. Hodkinson, H. M., Pomerance, A.
The clinical significance of senile cardiac amyloidosis:
A prospective clinico-pathological study.
Quarterly J. Med. 46, 381. 1977.
160. Hoffman, J. S. Ericsson, L. H., Eriksen, N., Walsh, K. A., Benditt, E. P.
Murine Tissue amyloid protein AA: $NH_2$-terminal sequence identity with only one of two serum amyloid protein (Apo-SAA) gene products.
J. Exp. Med. 159, 641. 1984.
161 Hofmann-Guilaine, C., Nochy, D., Jacquot, C., Tricottet, V., Bariety, J., Camilleri, J. P.
Association light chain deposition disease (LCDD) and amyloidosis. One case.
Path. Res. Pract. 180, 214. 1985.
162. Holt, I. J., Harding, A. E., Middleton, L., Chrysostomou, G., Said, G., King, R. H. M., Thomas, P. K.
Molecular genetics of amyloid neuropathy in Europe.
Lancet. i, 525. 1989.
163. Homman, N., Gejyo, F., Isemura, M., Arakawa, M.
Beta-2-microglobulin binding to collagen. An amyloidogenic factor in chronic hemodialysis patients.
In: Amyloid and Amyloidosis.
Ed.: by Isobe, T., Araki, S., Uchino, F., Kito, S., Tsubura, E.
New York, NY. Plenum. 623. 1988.
164. Hosenpud, J. D., Uretsky, B. F., O'Connell, J. B., Olivari, M.
Cardiac transplantation for amyloidosis: Results of a multicenter survey.
J. Heart. Transplant. 8, 99. 1989.
165. Howard, A., Gwyther, M., Ansell, B. M.
The acute phase proteins in amyloidosis of juvenile chronic arthritis.
In: Amyloidosis.
Ed.: by Tribe, C. R., Bacon, P. A.
E. A. R. S. Wright Bristol. 120. 1983.
166. Hunter, A. M., Borsey, D. Q., Cambell, I. W., Macaulay, R. A. A.
Protein-losing enteropathy due to gastrointestinal amyloidosis.

Postgrad. Med. J. 55, 822. 1979.
167. Husby, G.
Amyloidosis in rheumatoid arthritis.
Ann. Clin. Research. 7, 154. 1975.
168. Husby, G., Sletten, K.
The structure of amyloid proteins.
In: Amyloid and Amyloidosis.
Ed.: by Glenner, G. G., Pinho e Costa, P., Falcao de Freitas, A.
Excerpta Medica. Amsterdam. 266. 1980.
169. Husby, G., Ranlov, R. J., Sletten, K., Marhaug, G.
The amyloid in familial amyloidotic cardiomyopathy of Danish origin is related to pre-albumin.
Clin. Exp. Immunol. 60, 207. 1985.
170. Huston, D. P., McAdam, K. P. W. J., Balow, J. B., Bass, R., Debellis, R. A.
Amyloidosis in systemic lupus erythematosus.
Am. J. Med. 70, 320. 1981.
171. Ioachim, N. J., McKenna, P. J., Delaney, W. E., Tóth, J., Chung, F.
Cryoglobulinemia and amyloidosis associated with intestinal lymphoma.
Arch. Inter. Med. 138, 1158. 1978.
172. Isobe, T., Osserman, E. F.
Patterns of amyloidosis and their association with plasma-cell dyscrasia, monoclonal immunoglobulins and Bence Jones proteins.
N. Engl. J. Med. 290, 473. 1974.
173. Isobe, T., Husby, G., Sletten, K.
Characterisation of an amyloid protein AA similar to SAA.
In: Amyloid and Amyloidosis.
Ed.: by Glenner, G. G., Pinho e Costa, P., Falcao de Freitas, A.
Excerpta Medica. Amsterdam. 331. 1980.
174. Isobe, T.
Lambda predominance of Bence Jones protein (BJP) in AL amyloidosis among Japanese patients.
In: Amyloid and Amyloidosis.
Ed.: by Isobe, T., Araki, S., Uchino, F., Kito, S., Tsubura, E.
New York, NY. Plenum. 175. 1988.
175. Isobe, T., Araki, S., Uchino, F., Kito, S., Tsubura, E.
Amyloid and Amyloidosis.
New York. NY. Plenum. 1988.
176. Isokane, N., Fukushima, N., Miyazaki, T., Dohi, I.
A case of sick sinus syndrome in primary systemic amyloidosis.
J. Electrocardiol. 11, 191. 1978.
177. Isoniemi, H., Eklund, B., Hockerstedt, K., Salmela, K., Ahonen, J.
Renal transplantation in amyloidosis.
Transplant. Proc. 21, 2039. 1989.
178. Jacob, E., Shapiro, Z., Zemer, D., Gafni, J.
Renal transplantation in FMF amyloidosis.
In: Amyloid and Amyloidosis.
Ed.: by Glenner, G. G., Pinho e Costa, P., Falcao de Freitas, A.
Excerpta Medica. Amsterdam. 587. 1980.
179. Johansson, B., Wernstedt, C., Westermark, P.
Isolated atrial amyloid (IAA): A cardiac amyloid of endocrine origin.
In: Amyloid and Amyloidosis.
Ed.: by Isobe, T., Araki, S., Uchino, F., Kito, S., Tsubura, E.
New York. NY. Plenum. 661. 1988.
180. Johnson, K. H., O'Brien, T. D., Betsholtz, C., Westermark, P.
Islet amyloid, islet-amyloid polypeptide, and diabetes mellitus.
N. Engl. J. Med. 321, 513. 1989.
181. Jones, N. F.
Renal amyloidosis.
J. Clin. Pathol. 34, 1228. 1981.
182. Julien, J., Vital, C., Vallat, J. M., Lagueny, A., Ferrer, X.
Neuropathies amyloides familiales dans trois familles d'origine Française.
Rev. Neurol. (Paris). 139, 259. 1983.
183. Kachel, H. G., Altmeyer, P., Baldamus, C. A., Koch, K. M.
Deposition of an amyloid-like substance as a possible complication of regular dialysis treatment.
Contr. Nephrol. 36, 127. 1983.
184. Kanada, D. J., Sharma, O. P.
Long-term survival with diffuse interstitial pulmonary amyloidosis.
Amer. J. Med. 67, 879. 1979.
185. Kametani, F., Tonoike, H., Hoshi, A., Shinoda, T., Kito, S.
A variant prealbumin-related low molecular weight amyloid fibril protein in familial amyloid polyneuropathy of Japanese origin.
Biochem. Biophys. Res. Commun. 125, 622. 1984.
186. Kanai, M., Raz, A., Goodman, D. S.
Retinol-binding protein: the transport protein for vitamin A in human plasma.
J. Clin. Invest. 47, 2025. 1968.
187. Kanda, Y., Goodman, D. W. S., Canfield, R. E., Morgan, F. J.
The shoulderpad sign – a diagnostic feature of amyloid arthropathy.
188. Kang, J., Lemaire, H. G., Unterbeck, A., Salbaum, J. M., Masters, C. L., Grzeschik, K. H., Multhaup, G., Beyreuther, K., Muller-Hill, B.
The precursor of Alzheimer's disease amyloid A4 protein resembles a cell-surface receptor.
Nature. 325, 733. 1987.
189. Kaplan, H. S.
Hodgkin's Disease.
2nd ed. Cambridge, Harvard University Press. 220. 1980.
190. Katz, G. A., Peter, J. B., Pearson, C. M., Adams, W. S.
The shoulderpad sign – a diagnostic feature of amyloid arthropathy.
N. Engl. J. Med. 288, 354. 1973.
191. Kedar, I., Sohar, E., Gafni, J.
Demonstration of amyloid-degrading activity in normal human serum.
Proc. Soc. Exp. Biol. Med. 145, 343. 1974.
192. Kedar, I., Ravid, M., Sohar, E., Gafni, J.
Colchicine inhibition of casein induced amyloidosis in mice.
Israel J. Med. Sci. 10, 787. 1974.
193. Kedar, I., Ravid, M., Sohar, E.
In vitro synthesis of 'amyloid' fibrils from insulin, calcitonin and parathormone.
Israel J. Med. Sci. 12, 1137. 1976.
194. Kedar, I., Ravid, M., Sohar, E.
Amyloid degrading activity of human serum.
In: Amyloid and Amyloidosis.
Ed.: by Glenner, G. G., Pinho e Costa, P., Falcao de Freitas, A.
Excerpta Medica. Amsterdam. 60. 1980.
195. Kedar, I., Sohar, E., Ravid, M.
Degradation of amyloid by a serum component and inhibition of degradation. A dynamic concept of amyloid deposition.
J. Lab. Clin. Med. 99, 693. 1982.
196. Kelly, J. J., Jr., Kyle, R. A., O'Brien, P. C., Dyck, P. J.
The natural history of peripheral neuropathy in primary systemic amyloidosis.
Ann. Neurol. 6, 1. 1979.
197. Khojasteh, A., Arnold, L. K., Farhangi, M.
Bone lesions in primary amyloidosis.
Amer. J. Haematol. 7, 77. 1979.
198. Kimball, K. G.
Amyloidosis in association with neoplastic disease:
Report of an unusual case and clinico-pathological experience at Memorial Center for Cancer and Allied Diseases during eleven years (1948–1958).
Ann. Inter. Med. 55, 958. 1961.
199. Kisilevsky, R., Benson, M. D., Axelrad, M. A., Boudreau, L.
The effect of a liver protein synthesis inhibitor on plasma SAA levels in a model of accelerated amyloid deposition.
Lab. Invest. 41, 206. 1979.
200. Kisilevsky, R., Boudreau, L.
Kinetics of amyloid deposition. 1. The effects of amyloid-enhancing factor and splenectomy.

Lab. Invest. 48, 53. 1983.
201. Kleinman, K. S., Coburn, J. W.
Amyloid syndromes associated with hemodialysis.
Kidney Int. 35, 567. 1989.
202. Knowles, D. M. II., Jakobiec, F. A., Rosen, M., Howard, G.
Amyloidosis of the orbit and adnexae.
Surv. Ophthalmol. 19, 367. 1975.
203. Knowles, D. M. II., Shevehuk, M.
Pleomorphic reticulum cell sarcoma, monoclonal gammopathy and amyloidosis: an immunoperoxidase study.
Cancer. 41, 1883. 1978.
204. Kobayaski, H., Hashimoto, K.
Antigenic identity of amyloid in localized cutaneous amyloidosis with keratin.
J. Invest. Dermatol. 76, 320. 1981.
205. Kosaka, M., Iishi, Y., Okakawa, K., Saito, S., Sugihara, J., Muto, Y.
Tetramer Bence Jones protein in the immunoproliferative diseases.
Angioimmunoblastic lymphadenopathy, primary amyloidosis, and multiple myeloma.
Am. J. Clin. Pathol. 91, 639. 1989.
206. Krakow, N. P.
Ueber bei Thieren experimentell hervorgerufenes Amyloid.
Zentralblatt für Allgemeine Pathologie und Pathologische Anatomie. 6, 337. 1895.
207. Kuczynski, M. H.
Neue Beiträge zur Lehre vom Amyloid.
Klin. Wschr. 2, 727. 1923.
208. Kuhlbäck, B., Wegelius, O.
Secondary amyloidosis. A study of clinical and pathological findings.
Acta Med. Scand. 180, 737. 1966.
209. Kuntz, B., Bardin, T., Voisin, M. C., Zingraff, J.
Amyloidosis in patients on long-term hemodialysis.
Arthritis Rheum. 27, 57. 1984.
210. Kyle, R. A., Bayrd, E. D.
„Primary" systemic amyloidosis and myeloma. Discussion of relationship and review of 81 cases.
Arch. Intern. Med. 107, 344. 1961.
211. Kyle, R. A., Kottke, B. A., Schirger, A.
Orthostatic hypotension as a clue to primary systemic amyloidosis.
Circulation. 34, 883. 1966.
212. Kyle, R. A., Greipp, P. R.
Primary systemic amyloidosis: Comparison of melphalan and prednisone versus placebo.
Blood. 52, 818. 1978.
213. Kyle, R. A., Greipp, P. R., Banks, P. M.
The diverse picture of gamma heavy-chain disease. Report of seven cases and review of literature.
Mayo Clin. Proc. 56, 439. 1981.
214. Kyle, R. A., Wagoner, R. D., Holley, K. E.
Primary systemic amyloidosis. Resolution of the nephrotic syndrome with melphalan and prednisone.
Arch. Intern. Med. 142, 1445. 1982.
215. Kyle, R. A., Greipp. P. R.
Amyloidosis (AL). Clinical and laboratory features in 229 cases.
Mayo Clin. Proc. 58, 665. 1983.
216. Kyle R. A., Greipp, P. R., Garton, J. P., Gertz, M. A.
Primary systemic amyloidosis. Comparison of melphalan/prednisone versus colchicine.
Am. J. Med. 79, 708. 1985.
217. Kyle, R. A.
Diagnosis and management of multiple myeloma and related disorders.
Prog. Hematol. 14, 257. 1986.
218. Kyle, R. A., Greipp, P. R., O'Fallon, W. M.
Primary systemic amyloidosis. Multivariate analysis for prognostic factors in 168 cases.
Blood. 68, 220. 1986.
219. Kyle, R. A., Eilers, S. G., Linscheid, R. L., Gaffey, T. A.
Amyloid localized to tenosynovium at carpal tunnel release.
Am. J. Clin. Pathol. 91, 393. 1989.
220. Ladefoged, C., Christensen, H. E.
Congophilie substance with green dichroism in hip joints in autopsy material.
Acta Path. Microbiol. Scand. 88, 55. 1980.
221. Lanham, J. G., Meltzer, M. L., de Beer, F. C., Hughes, G. R. V., Pepys, M. B.
Familial amyloidosis of Ostertag.
Quarterly J. Med. 51, 25. 1982.
222. Lavie, G., Zucker-Franklin, D., Franklin, E. C.
Degradation of serum amyloid A protein by surface-associated enzymes of human blood monocytes.
J. Exp. Med. 148, 1020. 1978.
223. Lavie, G., Zucker-Franklin, D., Franklin, E. C.
Elastase-type proteases on the surface of human blood monocytes.
Possible role in amyloid formation.
J. Immunol. 125, 175. 1980.
224. Lesser, B. A., Leeper, K. V., Conway, W.
Obstructive sleep apnea in amyloidosis treated with nasal continuous positive airway pressure.
Arch. Intern. Med. 148, 2285. 1988.
225. Levin, M., Franklin, E. C., Frangione, B., Pras, M.
The amino acid sequence of a major nonimmunoglobulin component of some amyloid fibrils.
J. Clin. Invest. 51, 2773. 1972.
226. Levy, D. J., Franklin, G. O., Rosenthal, W. S.
Gastrointestinal bleeding and amyloidosis.
Amer. J. Gastroenterol. 77, 422. 1982.
227. Lewis, M.
Gastrocamera findings in a patient with gastric amyloidosis and multiple myeloma.
Gastrointes. Endoscopy. 14, 190. 1968.
228. Libbey, C. A., Skinner, M., Cohen, A. S.
Use of abdominal fat tissue aspirate in the diagnosis of systemic amyloidosis.
Arch. Inter. Med. 143. 1549. 1983.
229. Libbey, C. A., Rubinow, A., Shirahama, T., Deal, C., Cohen, A. S.
Familial amyloid polyneuropathy. Demonstration of prealbumin in a kinship of German/English ancestry with onset in the seventh decade.
Am. J. Med. 76, 18. 1984.
230. Lie, J. T., Hammond, P. I.
Pathology of the senescert heart: Anatomic observations on 237 autopsy studies of patients 90 to 105 years old.
Mayo Clin. Proc. 63, 552. 1988.
231. Linder, J., Silberman, H. R., Croker, B. P.
Amyloidosis complicating hairy cell leukemia.
Am. J. Clin. Pathol. 78, 864. 1982.
232. Linke, R. P., Zucker-Franklin, D., Franklin, E. C.
Morphologic-chemical and immunologic studies of amyloid-like fibrils formed from Bence Jones proteins by proteolysis.
J. Immunol. 111, 10. 1973.
233. Linke, R. P., Tischendorf, F. W., Zucker-Franklin, D., Franklin, E. C.
The formation of amyloid-like fibrils in vitro from Bence Jones proteins of the VI. subclass.
J. Immunol. 111, 24. 1973.
234. Linke, R. P., Heilman, K. L., Nathrath, W. B. J., Eulitz, M.
Identification of amyloid A protein in a sporadic Muckle-Wells syndrome.
Lab. Invest. 48, 698. 1983.
235. Linke, R. P.
Monoclonal antibodies against amyloid fibril protein AA.
J. Histochem. Cytochem. 32, 322. 1984.
236. Linke, R. P., Huhn, D., Nathrath, W. B. J.
Diagnosis of AA-amyloidosis using monoclonal antibodies: Application to paraffin and ultrathin sections as well as to urinary sediments.
In: Amyloid and Amyloidosis.
Ed.: by Isobe, T., Araki, S., Uchino, F., Kito, S., Tsubura, E.

New York. NY. Plenum. 247. 1988.
237. Lipper, S., Kahn, L. B.
Amyloid tumor: a clinicopathologic study of four cases.
Amer. Surg. Path. 2, 141. 1978.
238. Lowenstein, J. Gallo, G.
Remission of the nephrotic syndrome in renal amyloidosis.
N. Engl. J. Med. 282. 128. 1970.
239. Ludmerer, K. M., Kissane, J. M.
Subcutaneous masses and adenopathy in a 77 year-old man with Sjögren's syndrome and amyloidosis.
Am. J. Med. 86, 585. 1989.
240. Magnus-Levy, A.
Bence Jones Eiweiss und Amyloid.
Z. Klin. Med. 116, 510. 1931.
241. Magnus-Levy, A.
Amyloidosis in multiple myeloma. Progress noted in 50 years of personal observation.
J. Mt. Sinai Hosp. 19, 8. 1952.
242. Mahloudji, M., Teasdall, R. D., Adamkiewicz, J. J., Hartmann, W. H., Lambird, P. A., McKusick, V. A.
The genetic amyloidosis with particular reference to hereditary neuropathic amyloidosis, Type II (Indiana or Rukavina Type). Medicine. 48, 1. 1969.
243. Mandelstam, P., Simmons, D. E., Mitchell, B.
Regression of amyloid in Crohn's disease after bowel resection.
A 19 year follow-up.
J. Clin. Gastroenterol. 11, 324. 1989.
244. Marrink, J., van Rijswijk, M. H.
Amyloidosis.
Dordrecht, Martinus Nijhoff. 1986.
245. Masters, C. L., Gajdusek, D. C., Gibbs, C. J. Jr.
Creutzfeldt-Jacob disease virus isolations for the Gerstmann-Straussler syndrome with in analysis of the various forms of amyloid plaque deposition in the virus-induced spongiform encephalopathies.
Brain. 104, 559. 1981.
246. Maury, C. P. J., Teppo, A. M.
Mechanism of reduced amyloid A-degrading activity in serum of patients with secondary amyloidosis.
Lancet. ii, 134. 1982.
247. Maury, C. P. J., Teppo, A. M., Karinemi, A. L., Koeppen, A. H.
Amyloid fibril protein in familial amyloidosis with cranial neuropathy and corneal lattice dystrophy (FAP type IV) is related to transthyretin.
Am. J. Clin. Pathol. 89, 359. 1988.
248. Meckel, H.
Die Speck-oder Cholesterin-krankheit.
Ann. Charite-Krankenhauses, Berlin. 4, 264. 1853.
249. Meretoja, J.
Familial systemic paramyloidosis with lattice dystrophy of the cornea, progressive cranial neuropathy, skin changes and various internal symptoms: A previously unrecognized heritable syndrome.
Ann. Clin. Res. 1, 314. 1969.
250. Meyerhoff, J.
Familial Mediterranean fever: Report of a large family, review of the literature, and discussion of the frequency of amyloidosis.
Medicine. 59, 66. 1980.
251. Miller, L. D., Brown, E. C. Jr., Arnett, F. C.
Amyloidosis in Reiter's syndrome.
J. Rheumatol. 6, 225. 1979.
252. Morley, J. E., Reese, S. S.
Clinical implications of the aging heart.
Am. J. Med. 86, 77. 1989.
253. Mozar, H. N., Bal, D. G., Howard, J. T.
Perspectives on the etiology of Alzheimer's disease.
JAMA. 257, 1503. 1987.
254. Moyner, K., Sletten, K., Husby, G., Natvig, J. B.
An unusually large (83 amino acid recidues) amyloid fibril protein AA from a patient with Waldenström's macroglobulinemia and amyloidosis.
Scand. J. Immunol. 11, 549. 1980.
255. Nakazato, M., Kangawa, K., Minamino, N., Tawara, S., Matsuo, H., Araki, S.
Revised analysis of amino acid replacement in a prealbumin variant (SKO-III) associated with familial amyloidotic polyneuropathy of Jewish origin.
Biochem. Biophys. Res. Commun. 123, 921. 1984.
256. Natvig, J. B., Westermark, P., Sletten, K., Husby, G., Michaelsen, T.
Further structural and antigenic studies of light chain amylioid proteins.
Scand. J. Immunol. 14, 89. 1981.
257. Nehen, J. H.
Primary localized orbital amyloidosis.
Acta Ophthalmol. 57, 287. 1979.
258. Neugarten, J., Gallo, G. R., Buxbaum, J., Katz, L. A., Rubenstein, J., Baldwin, D. S.
Amyloidosis in subcutaneous heroin abusers („skin poppers' amyloidosis").
Am. J. Med. 81, 635. 1986.
259. Newland, J. R., Linke, R. P., Lennert, K.
Amyloid deposits in lymph nodes: A morphologic and immunohistochemical study.
Hum. Pathol. 17, 1245. 1986.
260. Nichols, W. C., Dwulet, F. E., Liepnieks, J., Benson, M. D.
Variant apoliprotein AI as a major constituent of a human hereditary amyloid.
Biochem. Biophys. Res. Commun. 156, 762. 1988.
261. Nomura, S., Kanoh, T., Uchino, H.
Intracellular formation of amyloid fibrils in myeloma.
Cytochemical, immunochemical, and electron microscopic observations.
Cancer, 54, 303. 1984.
262. O'Conner, C. R., O'Dorisio, T. M.
Amyloidosis, diarrhoea and a somatostatin analogue.
Ann. Intern. Med. 110, 665. 1989.
263. Okuzono, Y., Gondoh, T., Kawano, H., Nagasawa, T., Uchino, F.
Histopathology of cutaneous amyloid: A comparative study on 144 cases of localized cutaneous amyloidosis and 20 cases of systemic amyloidosis.
In: Amyloidosis.
Ed.: by Glenner, G. G., Osserman, E. F., Benditt, E. P., Calkins, E., Cohen, A. S., Zucker-Franklin, D.
New York. NY. Plenum. 821. 1986.
264. Osserman, E. F., Takatsuki, K., Talal, N.
The pathogenesis of 'amyloidosis'. Studies on the role of abnormal gammaglobulins and gammaglobulin fragments of the Bence Jones (L-polypeptide) types in the pathogenesis of 'primary' and 'secondary' amyloidosis associated with plasma cell myeloma.
Sem. Haematol. 1, 3. 1964.
265. Osserman, E. F., Isobe, T., Farhangi, M.
Effect of dimethyl-sulfoxide (DMSO) in the treatment of amyloidosis.
In: Amyloidosis.
Ed.: by Wegelius, O., Pasternack, A.
Academic Press. London. 553, 1976.
266. Osserman, E. F., Takatsuki, K., Talal, N.
The pathogenesis of „amyloidosis": studies on the role of abnormal gammaglobulins and gammaglobulin fragments of the Bence Jones (L-polypeptide) type in the pathogenesis of „primary" and „secundary" amyloidosis, and the amyloidosis associated with plasma cell myeloma.
Semin. Hematol. 1, 3. 1964.
267. Osserman, E. F., Sherman, W. H., Kyle, R. A.
Further studies of therapy of amyloidosis with dimethyl-sulfoxide (DMSO).
In: Amyloid and amyloidosis.
Ed.: by Glenner, G. G., Pinho e Costa, P., Falcao de Freitas, A.
Excerpta Medica. Amsterdam. 563. 1980.
268. Ozdemir, A. I., Wright, J. R., Calkins, E.
Influence of rheumatoid arthritis on amyloidosis of aging.

Comparison of 47 rheumatoid patients and 47 controls matched for age and sex.
N. Engl. J. Med. 285, 534. 1971.
269. Pinteric, L., Painter, R. H.
Electron microscopy of serum amyloid protein in the presence of calcium; alternative forms of assembly of pentagonal molecular in two-dimensional lattices.
Can. J. Biochem. 57, 727. 1979.
270. Palmert, M. R., Golde, T. E., Cohen, M. L., Kovacs, D. M., Tanzi, R. E., Gusella, J. F., Usiak, M. F., Younkin, L. H., Younkin, S. G.
Amyloid protein precursor messenger RNAs: Differential expression in Alzheimer's disease.
Science. 241, 1080. 1988.
271. Parmelee, D. C., Titani, K., Ericsson, L. H., Eriksen, N., Benditt, E. P., Walsh, K. A.
Amino acid sequence of amyloid-related apoprotein (apo-SAA[1]) from human high density lipoprotein.
Biochemistry. 21, 3298. 1982.
272. Parr, D. M. Pruzanski, W., Scott, J. G., Mills, D. M.
Primary amyloidosis with plasmacytic dyscrasia and a tetramer of Bence Jones lambda globulin in the serum and urine.
Blood. 37, 473. 1971.
273. Pasternack, A., Ahonen, J., Kuhlback, B.
Renal transplantation in 45 patients with amyloidosis.
Transplantation. 42, 598. 1986.
274. Plavnick, L., Weiner, P., Pevzner, S., Lichtig, C.
Amyloidosis in immunoblastic lymphadenopathy.
Israel J. Med. Sci. 16, 599. 1980.
275. Pechet, L., Kastrul, J. J.
Amyloidosis associated with factor X (Stuart) deficiency.
Ann. Intern. Med. 61, 315. 1964.
276. Penman, H. G., Thomson, K. J.
Amyloidosis and renal adenocarcinoma: A post-mortem study.
J. Pathol. 107, 45. 1972.
277. Pepys, M. B., Baltz, M., de Beer, F. C., Dyck, R. F., Holford, S., Breathnach, S. M., Black, M. M., Tribe, C. R., Evans, D. J., Feinstein, A.
Biology of serum amyloid P component.
Ann. N. Y. Acad. Sci. 389, 286. 1982.
278. Pepys, M. B., Baltz, M. L.
Acute phase proteins with special reference to C-reactive protein and related proteins (pentaxins) and serum amyloid A protein.
Adv. Immunol. 34, 141. 1983.
279. Pepys, M. B., Baltz, M. L.
Amyloidosis.
In: Coperman's Textbook of the rheumatic diseases, 6th edn.
Ed.: by Scott, J. T.
Churchill Livingstone, Edinburgh. 1984.
280. Perez-Villa, F., Campistol, J. M., Ferrando, J., Botey, A.
Renal amyloidosis secondary to acne conglobata.
Int. J. Dermatol. 28, 132. 1989.
281. Pick, A. I., Frohlichmann, R., Lavie, G., Duczyminer, M., Skvaril, F.
Clinical and haematological studies of 20 patients with amyloidosis and plasma cell dyscrasia.
Acta Haematol. 66, 154. 1981.
282. Pitkänen, P., Westermark, P., Cornwell, G. G. III.
Senile systemic amyloidosis.
Am. J. Pathol. 117, 391. 1984.
283. Poh, S. C.
Primary diffuse alveolar septal amyloidosis.
Thorax. 30, 186. 1975.
284. Polliack, A., Hershko, C.
Spontaneous rupture of the spleen in amyloidosis.
Israel J. Med. Sci. 8, 57. 1972.
285. Pras, M., Schubert, M., Zucker-Franklin, D., Rimon, A., Franklin, E. C.
The characterisation of soluble amyloid prepared in water.
J. Clin. Invest. 47, 924. 1968.
286. Pras, M., Zaretzky, J., Frangione, B., Franklin, E. C.
AA protein in a case of „primary" or „idiopathic" amyloidosis.
Am. J. Med. 68, 291. 1980.
287. Pras, M., Franklin, E. C., Prelli, F., Frangione, B.
A variant of prealbumin from amyloid fibrils in familial polyneuropathy of Jewish origin.
J. Exp. Med. 154, 989. 1981.
288. Pras, M., Bronshpigel, N., Zemer, D., Gafni, J.
Variable incidence of amyloidosis in familial Mediterranean fever among different ethnic groups.
Johns Hopkins Medical Journal. 150, 22. 1982.
289. Pras, M., Franklin, E. C., Shibolet, S., Frangione, B.
Amyloidosis associated with renal carcinoma of the AA type.
Am. J. Med. 73, 426. 1982.
290. Pras, M., Prelli, F., Franklin, E. C., Frangione, B.
Primary structure of an amyloid prealbumin variant in familial polyneuropathy of Jewish origin.
Proc. Natl. Acad. Sci, 80, 539. 1983.
291. Pras, M.
Systemic amyloidosis: Tissue specific variability in the primary structure of amyloid proteins.
In: Amyloid and Amyloidosis.
Ed.: by Isobe, T., Araki, S., Uchino, F., Kito, S., Tsubura, E.
New York, NY. Plenum. 41. 1988.
292. Prentice, C. R. M., Izatt, M. M., Adams, J. F., McNicol, G. P., Douglas, A. S.
Amyloidosis associated with the nephrotic syndrome and transfusion reactions in a haemophiliae.
Brit. J. Haematol. 21, 305. 1971.
293. Preud'homme, J. L., Morel-Maroger, L., Bronet, J. C., Cerf, M., Mignon, F., Guglielmo, P., Seligmann, M.
Synthesis of abnormal immunoglobulins in lymphoplasmacytic disorder with visceral light chain deposition.
Amer. J. Med. 69, 703. 1980.
294. Prusiner, S. B., McKinley, M. P., Bowman, K. A.
Scrapie prions aggregate to form amyloid-like birefringent rods.
Cell. 35, 349. 1983.
295. Pruzanski, W., Hasselback, R., Katz, A., Parr, D. M.
Multiple myeloma (light chain disease) with rheumatoid-like amyloid arthropathy and $\mu$-heavy chain fragment in the serum.
Amer. J. Med. 65, 334. 1978.
296. Puchtler, H., Sweat, F., Levine, M.
On the binding of Congo red by amyloid.
J. Histochem. Cytochem. 10, 355. 1962.
297. Randall, R. E., Williamson, W. C., Mullinay, F., Tung, M. Y., Still, W. J. S.
Manifestation of systemic light chain deposition.
Amer. J. Med. 60, 293. 1976.
298. Ravid, M., Robson, M., Kedar, I.
Prolonged colchicine treatment in 4 patients with amyloidosis.
Ann. Intern. Med. 87, 568. 1977.
299. Rey, C., Escribano, J. C., Vidal, M. T.
Retroperitoneal paraganglioma and systhemic amyloidosis: A case report.
Cancer. 43, 702. 1979.
300. Ringel, S. P., Claman, H. N.
Amyloid-associated muscle pseudohypertrophy.
Arch. Neurol. 39, 413. 1982.
301. Rokitansky, K. F.
Handbuch der Pathologischen Anatomie. Vol. III.
Baumuller and Seidel, Berlin. 1842.
302. Romhányi, G.
Selective differentiation between amyloid and connective tissue structures based on the collagen specific topo optical staining reaction with congo red.
Wirchows. Arch. Pathol. Anat. 354, 209. 1971.
303. Romhányi, G.
Differences in ultrastructural organization of amyloid as relevated by sensitivity of resistance to induced proteolysis.

Virchows. Arch. Pathol. Anat. 357, 29. 1972.
304. Rosenstein, E. D., Itzkowitz, S. H., Penziner, A. S., Cohen, J. I., Mornaghi, R. A.
Resolution of factor X deficiency in primary amyloidosis following splenectomy.
Arch. Intern. Med. 143, 597. 1983.
305. Rosenthal, C. J., Sullivan, L. M.
Serum amyloid A: Evidence for its origin in polymorphonuclear leukocytes.
J. Clin. Invest. 62, 1181. 1978.
306. Rosenthal, C. J., Sullivan, L. M.
Serum amyloid A to monitor cancer dissemination.
Ann. Inter. Med. 91, 383. 1979.
307. Rosenthal, T., Bank, H., Aladjem, M., David, R., Gafni, J.
Systemic amyloidosis in Behcet's disease.
Ann. Inter. Med. 83, 220. 1975.
308. Rubinow, A., Celli, B. R., Cohen, A. S., Ridgen, B. G., Brody, J. S.
Localized amyloidosis of the lower respiratory tract.
American Review of Respiratory Disease. 118, 603. 1978.
309. Rubinow, A., Koff, R. S., Cohen, A. S.
Severe intrahepatic cholestasis in primary amyloidosis.
Amer. J. Med. 64, 937. 1978.
310. Rubinow, A., Skinner, M., Cohen, A. S.
Digoxin sensitivity in amyloid cardiomyopathy.
Circulation. 63, 1285. 1981.
311. Rukavina, J. G., Block, W. D., Curtis, A. C.
Familial primary systemic amyloidosis: An experimental, genetic and clinical study.
J. Invest. Dermatol. 27, 111. 1956.
312. Rumble, B., Retallack, R., Hilbich, C., Simms, G., Multhaup, G., Martin, R., Hockey, A., Montgomery, P., Beyreuther, K., Masters, C. L.
Amyloid A4 protein and its precursor in Down's syndrome and Alzheimer's disease.
N. Engl. J. Med. 320, 1446. 1989.
313. Santiago, R. M., Scharnhorst, D., Ratkin, G., Crouch, E. C.
Respiratory muscle weakness and ventilatory failure in AL amyloidosis with muscular pseudohypertrophy.
Amer. J. Med. 83, 175. 1987.
314. Saraiva, M. J. M., Birken, S., Cost, P. P., Goodman, D. S.
Amyloid fibril protein in familial amyloidotic polyneuropathy, Portuguese type. Definition of molecular abnormality in transthyretin (prealbumin).
J. Clin. Invest. 74, 104. 1984.
315. Sargent, M. A., Fleming, S. J., Chattopadhyay, C., Ackrill, P., Sambrook, P.
Bone cysts and haemodialysis-related amyloidosis.
Clin. Radiol. 40, 277. 1989.
316. Schattner, A., Varon, D., Green, L., Hurwitz, N., Bentwich, Z.
Primary amyloidosis with unusual bone involvement: Reversibility with melphalan, prednisone, and colchicine.
Amer. J. Med. 86, 347. 1989.
317. Scheinberg, M. A., Catheart, E. S.
New concepts in the pathogenesis of primary and secondary amyloid disease.
Clin. Exp. Immunol. 33, 185. 1978.
318. Schleiden, M. J.
Ann. Physik. 43, 391. 1838.
319. Schnitzer, T. J., Ansell, B. M.
Amyloidosis in juvenile chronic polyarthritis.
Arthritis Rheum. 20, 245. 1977.
320. Scholes, J., Derosena, R., Appel, G. B., Jao, W., Boyd, M. T., Pirani, C. L.
Amyloidosis in chronic heroin addicts with the nephrotic syndrome.
Ann. Inter. Med. 91, 26. 1979.
321. Schubert, D., Schroeder, R., LaCorbiere, Saitoh, T., Cole, G.
Amyloid beta-protein precursor is possibly a heparan sulfate proteoglycan core protein.
Science. 241, 223. 1988.
322. Schwartz, P.
Amyloidosis: cause and manifestation of senile deterioration.
Thomas, Illinois. 5. 1970.
323. Schwartz, R. S., Cohen, J. R., Schrier, S. L.
Therapy of primary amyloidosis with melphalan and prednisone.
Arch. Inter. Med. 139, 1144. 1979.
324. Schwartz, Y., Tamse, A., Kissin, E., Shani, M.
An unusual case of temporomandibular joint arthropathy in systemic primary amyloidosis.
Journal of Oral Medicine. 34, 40. 1979.
325. Scott, R. B., Elmore, S. McD., Brackett, N. C. Jr., Harris, W. O. Jr., Still, W. J. S.
Neuropathic joint aisease (Charcot joints) in Waldenström's macroglobulinemia with amyloidosis.
Amer. J. Med. 54, 535. 1973.
326. Selkoe, D. J.
Altered structural proteins in plaques and tangles: What do they tell us about the biology of Alzheimer's disease?
Neurobiol. Aging. 7, 425. 1986.
327. Selkoe, D. J., Podlisny, M. B., Joachim, C. L., Vickers, E. A., Lee, G., Fritz, L. C., Oltersdorf, T.
Beta-Amyloid precursor protein of Alzheimer disease occurs as 110- to 135-kilodalton membrane-associated proteins in neural and nonneural tissues.
Proc. Natl. Acad. Sci. USA. 85, 7341. 1988.
328. Selkoe, D. J.
Aging, amyloid, and Alzheimer's disease.
N. Engl. J. Med. 320, 1484. 1989.
329. Sharma, H. M., Geer, J. C.
Multiple transfusions with sensitization associated with amyloidosis: report of a case.
Arch. Pathol. 89, 473. 1970.
330. Shemer, J., Messer, G. Y., Pras, M., Gafni, J.
Amyloid in urinary sediments as a diagnostic technique.
Ann. Inter. Med. 90, 61. 1979.
331. Shenoy, U. A., Peric-Golie, L.
Primary cardiovascular amyloidosis manifesting as mitral stenosis.
Hum. Pathol. 13, 768. 1982.
332. Shiomura, T., Ishida, Y., Matsumoto, N., Sasaki, K., Ishihara, T., Miwa, S.
A case of generalized amyloidosis associated with cyclic neutropenia.
Blood. 54, 628. 1979.
333. Shirahama, T., Cohen, A. S.
Blockage of amyloid induction of colchicine in an animal model.
J. Exp. Med. 140, 1102. 1974.
334. Shirahama, T., Cohen, A. S.
Redistribution of amyloid deposits.
Amer. J. Pathol. 99, 539. 1980.
335. Shirahama, T., Skinner, M., Westermark, P., Rubinow, A., Cohen, A. S., Brun, A., Kemper, T. L.
Senile cerebral amyloid. Prealbumin as a common constituent in the neuritic plaque, in the neurofibrillary tangle, and in the microangiopathic lesion.
Amer. J. Pathol. 107, 41. 1982.
336. Shirahama, T., Skinner, M., Cohen, A. S.
Histochemical and immunohistochemical characterization of amyloid associated with chronic haemodialysis as $beta_2$-microglobulin.
Lab. Invest. 53, 705. 1985.
337. Shorvon, P. J.
Amyloidosis and inflammatory bowel disease.
Amer. J. Digest. Dis. 22, 209. 1977.
338. Silverman, S. L., Cathcart, E. C., Skinner, M., Cohen, A. S.
The degradation of serum amyloid A protein by activated polymorphonuclear leucoytes: participation of granulocyte elastase.
Immunology. 46, 737. 1982.
339. Simon, B. G., Moutsopoulos, H. M.
Primary amyloidosis resembling sicca syndrome.

Arthritis Rheum. 2, 932. 1978.
340. Siqueiro-Filho, A. G., Cunha, C. L. P., Tajik, A. J., Seward, J. B., Schattenberg, T. T., Giuliani, E. R.
M-mode and two-dimensional echocardiographic features in cardiac amyloidosis.
Circulation. 63, 189. 1981.
341. Skinner, M., Cohen, A. S.
The prealbumin nature of the amyloid protein in familial amyloid polyneuropathy (FAP) – Swedish variety.
Biochem. Biophys. Res. Comm. 99, 1326. 1981.
342. Skinner, M., Schneller, S. I., Connors, L. H.
Observations on the amyloid-degrading activity of serum and its relationship to human neutrophil elastase.
Trans. Assoc. Am. Physicians. 96, 437. 1983.
343. Skinner, M., Stone, P., Shirahama, T., Connors, L. H., Calore, J., Cohen, A. S.
The association of an elastase with amyloid fibrils.
Proc. Soc. Exp. Biol. Med. 181, 211. 1986.
344. Skogen, B., Thorsteinsson, L., Natvig, J. B.
Degradation of protein SAA to an AA-like fragment by enzymes of monocytic origin.
Scand. J. Immunol. 11, 533. 1980.
345. Sletten, K., Westermark, P., Natvig, J. B.
Characterization of amyloid fibril proteins from medullary carcinoma of the thyroid.
J. Exp. Med. 143, 993. 1976.
346. Sletten, K., Westermark, P., Natvig, J. B.
Senile cardiac amyloid is related to prealbumin.
Scand. J. Immunol. 12, 503. 1980.
347. Sohar, E., Gafni, J., Pras, M., Heller, H.
Familial Mediterranean fever: A survey of 470 cases and review of literature.
Amer. J. Med. 43, 227. 1967.
348. Solomon, A., Frangione, B., Franklin, E. C.
Bence Jones proteins and light chains of immunoglobulins. Preferential association of $V_{VI}$. subgroup of human light chains with amyloidosis AL.
J. Clin. Invest. 70, 453. 1982.
349. Solomon, A.
Clinical implications of monoclonal light chains.
Semin. Oncol. 13, 341. 1986.
350. Spiro, D.
The structural basis of proteinuria in man. Electron microscopic observations of renal biopsy specimens from patients with lipid nephrosis, amyloidosis, and subacute and chronic glomerulonephritis.
Amer. J. Pathol. 35, 47. 1959.
351. Staunton, H., Dervan, P., Kale, R., Linke, R. P., Kelly, P.
Hereditary amyloid polyneuropathy in northwest Ireland.
Brain. 110, 1231. 1987.
352. Steinitz, L, Schreiter, G.
Zur Amyloidose im Kindesalter.
Pädiatr. Grenzgeb. 10, 261. 1971.
353. Stevens, D. B., Whitehouse, G. H.
Waldenström's macroglobulinaemia with amyloidosis-lymphographic findings.
Lymphology. 9, 142. 1976.
354. Stone, M. J., Frenkel, E. P.
The clinical spectrum of light chain myeloma. A study of 35 patients with special reference to the occurrence of amyloidosis.
Amer. J. Med. 58, 601. 1975.
355. Stone, M. J., Hirsch, V. J.
Splenic function in amyloidosis.
In: Amyloidosis.
Ed.: by Glenner, G. G., Osserman, E. F., Benditt, E. P., Calkins, E., Cohen, A. S., Zucker-Franklin, D.
New York. NY. Plenum. 583. 1986.
356. Strausbaugh, L. J., Farid, Z., Higashi, G. J., Bassily, S.
Renal amyloidosis with nephrotic syndrome in two patients with Schistosomiasis mansoni and chronic salmonellosis.
Amer. J. Tropical Med. and Hyg. 27, 919. 1978.
357. Suzuki, T., Higa, S., Sakoda, S., Hagashi, A., Yamamura, Y., Takaba, Y., Nakajima, A.
Orthostatic hypotension in familial amyloid polyneuropathy: treatment with DL-threo-3,4-dihydroxyphenylserine.
Neurology. 31, 1323. 1981.
358. Suzuki, T., Azuma, T., Araki, S., Ikegawa, S., Kito, S., Matsubayashi, H., Yanagisawa, N., Ikeda, S.
Treatment of autonomic dysfunction with L-threo-3,4-dihydroxyphenyl-serine in patients with familial amyloidotic polyneuropathy. A multicenter study.
In: Amyloid and Amyloidosis.
Ed.: by Isobe, T., Araki, S., Uchino, F., Kito, S., Tsubura, E.
New York. NY. Plenum. 851. 1988.
359. Tanzi, R. E., Bird, E. D., Latt, S. A., Neve, R. L.
The amyloid beta-protein gene in not duplicated in brains from patients with Alzheimer's disease.
Science. 238, 666. 1987.
360. Tanzi, R. E., Gusella, J. F., Watkins, P. C., Bruns, G. A. P., St George-Hyslop, P., Van Keuren, M. L., Patterson, D., Pagan, S., Kumit, D. M., Neve, R. L.
Amyloid beta-protein gene: cDNA, mRNA distribution and genetic linkage near the Alzheimer locus.
Science. 235, 880. 1987.
361. Tawara, S., Araki, S., Toshimori, K., Nakagawa, H., Ohtaki, S.
Amyloid fibril protein in type I familial amyloidotic polyneuropathy in Japanese.
J. Lab. Clin. Med. 98, 811. 1981.
362. Tawara, S., Nakazato, M., Kangawa, K., Matsuo, H., Araki, S.
Identification of amyloid prealbumin variant in familial amyloidotic polyneuropathy (Japanese Type).
Biochem. Biophys. Res. Commun. 116, 880. 1983.
363. Thompson, P. J., Citron, K. M.
Amyloid and lower respiratory tract.
Thorax. 38, 84. 1983.
364. Triger, D. R., Joekes, A. M.
Renal amyloidosis – A fourteen year follow-up.
Q. J. Med. 42, 15. 1973.
365. Troussard, X., Hurault-de-Ligny, B., Gallet, B., Ganeval, D., Mandard, J. C., Ryckelynck, J. P., Leporrier, M.
Massive systemic amyloidosis associated with light-chain deposition disease.
Nephron. 52, 139. 1989.
366. Tschang, T. P.
Nodular malignant lymphoma and amyloidosis: A case report.
Cancer. 38, 2192. 1976.
367. Turnell, W. G., Sarra, R., Glover, L.
In: Amyloidosis.
Ed.: by Glenner, G. G., Osserman, E. F., Benditt, E. P., Colkins, E., Cohen, A. S., Zucker-Franklin, D.
New York. NY. Plenum 49. 1986.
368. Ullian, M. E., Hammond, W. S., Alfrey, A. C., Schultz, A., Molitoris, B. A.
Béta-2-microglobulin-associated amyloidosis in chronic hemodialysis patients with carpal tunnel syndrome.
Medicine. 68, 107. 1989.
369. Van Allen, M. W., Frohheh, J. A., Davis, J. R.
Inherited predisposition to generalized amyloidosis. Clinical and pathological study of a family with neuropathy, nephrophathy, and peptic ulcer.
Neurology. 19, 10. 1969.
370. Van Duinen, S. G., Gastano, E. M., Prelli, F., Bots, G. T. A. B., Luyendijk, W., Frangione, B.
Hereditary cerebral hemorrhage with amyloidosis in patients of Dutch origin is related to Alzheimer's disease.
Proc. Natl. Acad. Sci. USA. 84, 5991. 1987.
371. van Rijswijk, M. H., van Heusden, C. W. G. J.
The potassium permanganate method.
Amer. J. Pathol. 97, 43. 1979.
372. Vasily, D. B., Bhatia, S. G., Uhlin, S. R.
Familial primary cutaneous amyloidosis: Clinical, genetic, and immunofluorescent studies.
Arch. Dermatol. 114, 1173. 1978.
373. Venkataseshan, V. S., Faraggiana, T., Hugshon, M. D.,

Buchwald, D., Olesnicky, L., Goldstein, M. H.
Morphologic variant of light-chain deposition disease in the kidney.
Am. J. Nephrol. 8, 272. 1988.

374. Verbauck, J., Lameire, N., Praet, M., Ringoir, S., Elawant, A., Barbier, F.
Renal amyloidosis as complication of Crohn's disease.
Acta Clinica Belgica 34, 6. 1979.

375. Vinters, H. V.
Cerebral amyloid angiopathy. A critical review.
Stroke. 18, 311. 1987.

376. Vinters, H. V., Miller, B. L., Pardridge, W. M.
Brain amyloid and Alzheimer's disease.
Ann. Intern. Med. 109, 41. 1988.

377. Virchow, R.
Über eine im Gehirn und Rückenmark des Menschen aufgefundene Substanz mit der chemischen Reaction der Cellulose.
Virchows Archiv für Pathologische Anatomie. 6, 135. 1854.

378. Von Gise, H., Helmehen, U., Mikeler, E., Brüning, L., Walther, C., Christ, H., Mackensen, S., Bohle, A.
Correlations between the morphological and clinical findings in a patient recovering from secondary generalised amyloidosis with renal involvement: light- and electron microscopic investigations on serial biopsies.
Virchows Archiv. A. Pathological Anatomy and Histology. 379, 119. 1978.

379. Von Gise, H., Von Gise, V., Stark, B., Bohle, A.
Nephrotic syndrome and renal insufficiency in association with amyloidosis: a correlation between structure and function.
Klin. Wschr. 59, 75. 1981.

380. Waldenström, H.
On the formation and disappearance of amyloid in man.
Acta Chir. Scand. 63, 479. 1928.

381. Wallace, M. R., Naylor, S. L., Kluve-Beckerman, B., Long, G. L., McDonald, L., Shows, T. B., Benson, M. D.
Localization of the human prealbumin gene to chromosome 18.
Biochem. Biophys. Res. Commun. 129, 753. 1985.

382. Wallace, M. R., Dwulet, F. E., Conneally, P. M., Benson, M. D.
Biochemical and molecular genetic characteristic of a new variant prealbumin associated with hereditary amyloidosis.
J. Clin. Invest. 78, 6. 1986.

383. Wallace, M. R., Dwulet, F. E., Williams, E. C., Conneally, P. M., Benson, M. D.
Identification of a new hereditary amyloidosis prealbumin variant, Tyr-77, and detection of the gene by DNA analysis.
J. Clin. Invest. 81, 189. 1988.

384. Wallace, M. R., Benson, M. D.
DNA tests for four prealbumin mutations in hereditary amyloidosis.
In: Amyloid and Amyloidosis.
Ed.: by Isobe, T., Araki, S., Uchino, F., Kito, S., Tsubura, E.
New York, NY. Plenum. 355. 1988.

385. Walts, A. E., Goodman, M. D., Matorin, P. A.
Amyloid, Carpal Tunnel Syndrome, and chronic hemodialysis.
Am. J. Nephrol. 5, 225. 1985.

386. Wattendorf, A. F., Botts, G. T. A. M., Went, L. N., Endtz, L. J.
Familial cerebral amyloid angiopathy presenting as recurrent cerebral hemorrhage.
J. Neurol. Sci. 55, 121. 1982.

387. Wegelius, O., Wafin, F., Falck, H. M., Tornroth, T.
Follow-up study of amyloidosis secondary to rheumatic disease.
In: Amyloid and Amyloidosis.
Ed.: by Glenner, G. G., Pinho e Costa, P., Falcao de Freitas, A.
Excerpta Medica. Amsterdam. 183. 1980.

388. Werdelin, O., Ranlow, P.
Amyloidosis in mice produced by transplantation of spleen cells from casein treated mice.
Acta Pathol. Microbiol. Scand. 68, 1. 1966.

389. Westermark, P., Grimelius, L., Polak, J. M., Larsson, L. I., van Noorden, S., Wilander, E., Pearse, A. G. E.
Amyloid in polypeptide hormone-producing tumours.
Lab. Invest. 37, 212. 1977.

390. Westermark, P., Sletten, K., Eriksson, M.
Morphologic and chemical variation of the kidney lesions in amyloidosis secondary rheumatoid arthritis.
Lab. Invest. 41, 427. 1979.

391. Westermark, P., Johansson, B., Natvig, J. B.
Senile cardiac amyloidosis: Evidence of two different amyloid substances in the ageing heart.
Scand. Immunol. 10, 303. 1979.

392. Westermark, P.
Amyloidosis of the skin: a comparison between localised and systemic amyloidosis.
Acta Dermato-Venereologica. 59, 341. 1979.

393. Westermark, P., Ohman, S., Domar, M., Sletten, K.
Bullous amyloidosis.
Arch. Dermatol. 117, 782. 1981.

394. Westermark, P., Shirahama, T., Skinner, M., Brun, A., Cameron, R., Cohen, A. S.
Immunohistochemical evidence for the lack of amyloid P component in some intracerebral amyloids.
Lab. Invest. 46, 457. 1982.

395. Westermark, P., Pitkanen, P., Benson, L., Vahlquist, A., Olofsson, B. O., Cornweel, G. G.
Serum prealbumin and retinol-binding protein in the prealbumin-related senile and familial forms of systemic amyloidosis.
Lab. Invest. 52, 314. 1985.

396. Wiernik, P. H.
Amyloid joint disease.
Medicine. 51, 465. 1972.

397. Wild, C.
Beitr. Pathol. Anat. 1, 177. 1886.

398. Wilks, S.
Cases of lardaceous disease and some allied affections: with remarks.
Guy's Hospital Reports Series 3. 2, 103. 1856.

399. Wille, L. E., Olsen, E., Forre, O., Sletten, K., Jentoft, H.
Demonstration and partial characterisation of an atypical protein in the urine of a patient with primary amyloidosis.
Acta Medica Scandinavica. 208, 177. 1980.

400. Williams, R. C. Jr., Cathcart, E. S., Calkins, E., Fite, G. L., Rutio, Barbara, J., Cohen, A. S.
Secondary amyloidosis in lepromatous leprosy: Possible relationships of diet and environment.
Ann. Inter. Med. 62, 1000. 1965.

401. Williamson, R. C. N.
Primary amyloidosis of the rectum.
Proc. Roy. Soc. Med. 65, 74. 1972.

402. Wolf, B. C., Kumar, A., Vera, J. C., Nieman, R. S.
Bone marrow morphology and immunology in systemic amyloidosis.
Am. J. Clin. Pathol. 86, 84. 1986.

403. Woo, P., O'Brien, J., Robson, M., Ansell, B. M.
A genetic marker for systemic amyloidosis in juvenile arthritis.
Lancet, ii, 767. 1987.

404. Wright, D. G., Wolff, S. M., Fauci, A. S., Alling, D. W.
Efficacy of intermittent colchicine therapy in familial Mediterranean fever.
Ann. Intern. Med. 86, 162. 1977.

405. Zemer, D., Revach, M., Pras, M., Modan, B., Schor, S., Sohar, E., Gafni, J.
A controlled trial of colchicine in preventing attacks of familial Mediterranean fever.
N. Engl. J. Med. 291, 932. 1974.

406. Zemer, D., Pras, M., Shenner, Y., Sohar, E., Gafni, J.
Daily prophylactic colchicine in familial Mediterranean fever.
In: Amyloid and amyloidosis.

Ed.: by Glenner, G. G., Pinho e Costa, P., Falcao de Freitas, A.
Excerpta Medica. Amsterdam. 580. 1980.
407. Zemer, D., Pras, M., Sohar, E., Modan, M., Cabili, S., Gafni, J.
Colchicine in the prevention and treatment of the amyloidosis of familial Mediterranean fever.
N. Engl. J. Med. 314, 1001. 1986.
408. Zilko, P. J., Dawkins, R. L.
Amyloidosis associated with dermatomyositis and features of multiple myeloma. The progression of amyloidosis associated with corticosteroid and cytotoxic drug therapy.
Am. J. Med. 59, 448. 1975.
409. Zucker-Franklin, D., Franklin, E. C.
Intracellular localization of human amyloid by fluorescence and electron microscopy.
Am. J. Pathol. 59, 23. 1970.
410. Zucker-Franklin, D., Frangione, B.
Amyloid any myeloma stem-cell culture.
N. Engl. J. Med. 308, 1164. 1983.
411. Yoneda, K., Watanabe, H., Yanagihara, M., Mori, S.
Immunohistochemical staining properties of cutaneous amyloidosis using anti-keratin monoclonal antibodies.
In: Amyloid and Amyloidosis.
Ed.: by Isobe, T., Araki, S., Uchino, F., Kito, S., Tsubura, E.
New York. NY. Plenum 709. 1988.
412. Yood, R. A., Skinner, M., Cohen, A. S., Lee, V. W.
Soft tissue uptake of bone seeking radionuclide in amyloidosis.
J. Rheum. 8, 760. 1981.
413. Yood, R. A., Skinner, M., Rubinow, A., Talarico, L., Cohen, A. S.
Bleeding manifestations in 100 patients with amyloidosis.
J. Amer. Med. Ass. 249, 1322. 1983.

# XIII. Neurologic symptoms and complications of monoclonal gammophaties

## The hyperviscosity syndrome

Neurologic complications of monoclonal gammopaties are very divergent, since the alterations in the central and peripheral nervous system, as well as hemorrhagic changes in amyloidosis belong to this group, similar to the symptoms of hyperviscosity or palsy caused by vertebral fractures. Nevertheless, it is advisable to distinguish a well circumscribed field, which is either the consequence of the local lesions caused by tumorous plasma cells or is due to a direct effect of paraproteins. However, a more thorough analysis of these alterations often fails to distinguish between the above mentioned alterations.

The occurrence of neurologic complications in plasmacytoma and macroglobulinaemia varies between 33-40% (86) and 25% (46), respectively. However, published data do not specify, the criteria used for assessment of the patients to the affected or complication-free group. Accepting the conditions discussed in the first paragraph, neurophatic complications were found in 29 patients of 1538 patients (1.88%) observed by us. Such difference clearly shows that our criteria are far from those described in the literature. This assumption is supported by a publication, in which pathological symptoms were reported in 166 patients of 277 cases of multiple myeloma (78). Using our criteria these results seem quite absurd.

I. Irreversible and rarely reversible damages of the central nervous system are either due to the local compression of certain parts of the brain, "infiltration" caused by the cells (87), or to the effect of paraproteins (10). Local symptoms, impairment of hearing, vision, or smell belong to the first, while the decreased voluntary muscle performance - to the second type. Epilepsy or epileptiform fits can occur in both cases. Unfortunately all these symptoms accompany not only monoclonal gammopathies but also any other neoplastic disease, hence adequate therapy requires precise diagnosis.

Dysfunction of the central nervous system, as the first symptom of plasmocytoma is well known. Lack of immunochemical findings suggests a local process, for example, solitary plasmacytoma starting from the petrosum bone or dura (81). If all the symptoms of myeloma develop, diffuse cellular meningeal infiltration might be in the background. Three main forms can be distinguished according to the most frequently affected regions:

- intracranial processes
- orbital involvement
- spinal compression.

Solitary medullar or extramedullar plasmocytoma (20), multiple myeloma or plasmoblastoma can affect all the three regions of the nervous system (9, 54).

- Intracranial involvement: it must be emphasized that osteolytic lesions of the calvaria in Kahler's disease is almost a triviality. However, in agreement with the literature, we failed to observe increased intracranial pressure or indirect injury of the brain due to the meningeal infiltration (88). At the same time, the loci in the base of the skull (sphenoid or petrosum bones) usually injure the cranial nerves VIII, VI and V (86) (Fig. XIII/1). The sella or other areas can also be involved (90).

Plasmocytoma infiltrating the meninx is extremely rare. Sudden deterioration of general condition, headaches, sleepiness, loss of consciousness - the later can be periodical - indicates leptomeningeal involvement (89). The diagnosis is, however, not so easy, since similar symptoms are also caused by HVS, bacterial or mycotic infections with the same incidence as that of the cellular infiltration. Only two papers published so far, deal with the isolated cerebral plasmacell tumors. The first was diagnosed in the corpus callosum and with respect to aggresivity of its growth

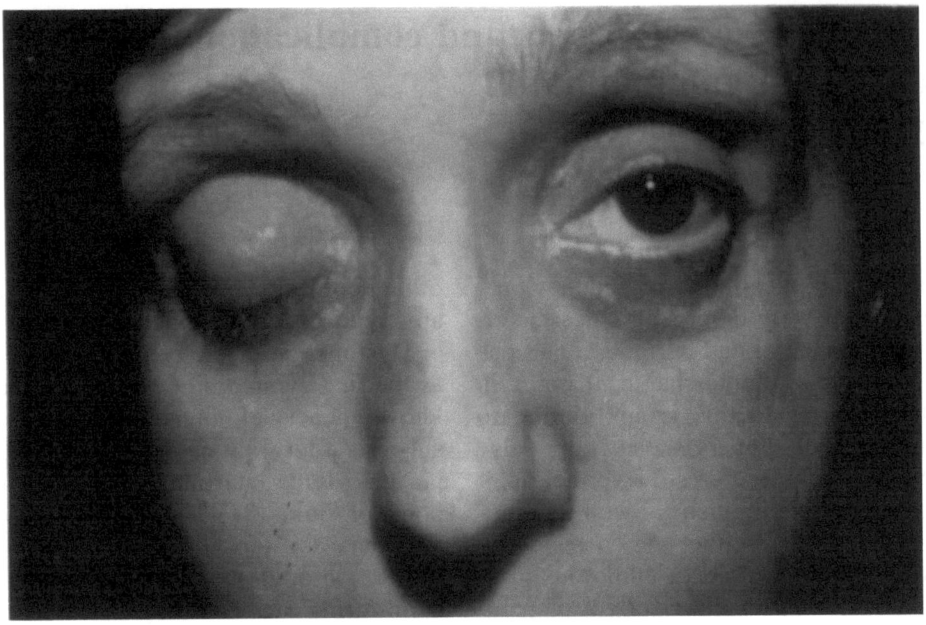

**The patient was unable to open his right eye. After treatment he could lift his eyelid.**

**Fig. XIII/1**

could be compared to astrocytoma. Plasmocytoma diagnosis has been verified by immunohistochemical methods (2). The second publication presented a patient with plasma cell leukaemia developing a metastasis in the pineal corpus (34). Other papers on intracranial processes do not convince the reader that there is no relationship between the site of the tumor and the bones of the cranial base (6, 19, 41, 54, 59, 85, 86, 88).

There is one publication on a special form of multifocal leucoencephalopathy complicated by plasmocytoma (25). This disease is a rare form of a cerebral viral infection, in which subcortical multifocal loci are formed. Antibody deficiency syndrome may naturally be the background, so this from of lencoencephalogpathy is rather the result of a coincidence, that a special disease occuring in myeloma.

It is debated whether paraprotein is present in the cerebrospiral fluid (32) in monoclonal gammopathies. Some authors believe that it does. It can either appear due to the damages of the blood-brain barrier, which makes the appearance of pathological plasma immunoglobulins in the cerebrospiral fluid possible, or due to some kind of in situ formation (38, 75, 93). Our experiences show that in monoclonal gammopathies the discussed protein can be found in the cerebraspiral fluid, but its detection is not always simple, as its concentration can be extremely low. If paraprotein occurs only in the cerebrospiral fluid, but not in the plasma and urine, this might suggest an intracerebral process accompanied by in situ formation of paraprotein.

- Orbital involvement is rare. Ophtalmoplegia, protrusion and the compression of the optic nerve are the most frequent syndromes (77) caused by the plasmocytomal damage of the orbital bones. Depending on the localization of the process, (primary affection of the sphenoid bone, the base or the lateral bone of the skull) one of the above mentioned clinical symptoms will develop. They react surprisingly well to cytostatic therapy and even in case of relapses more than one remission can be induced by repeated treatment (Fig. XIII/2, XIII/3).
- Spinal compression is the most ordinary neurologic complication (16, 21). It can occur anywhere from the neck to the sacral area. The symptoms depend on the affected segment. General early symptoms involve pain followed by paresis, impairment of perception and sphincter function, paraplegia and finally symptoms of the transversal spinal lesion. The symptoms can either de-

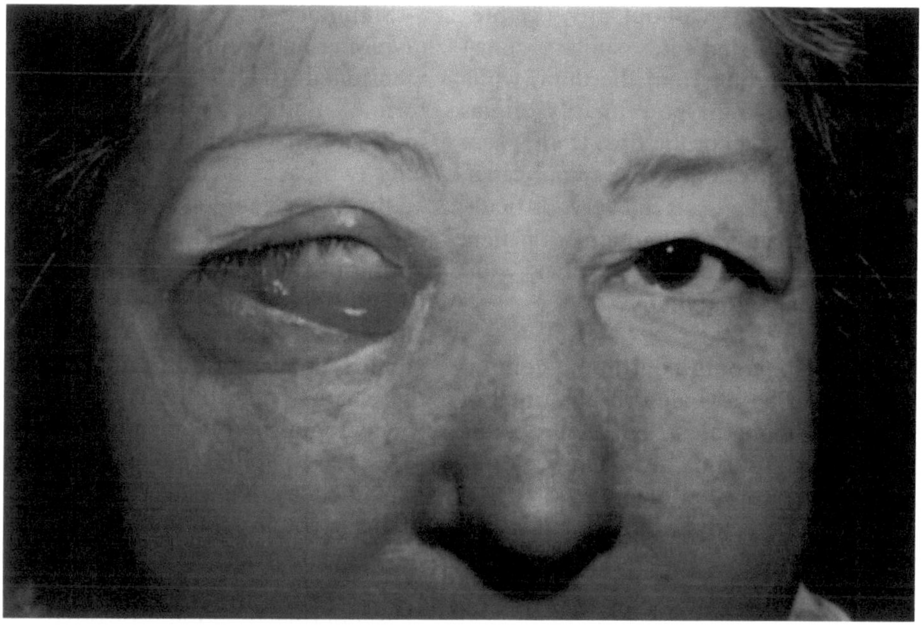

**Plasmoblastous infiltration starting from the right orbita causing protrusion.**

**Fig. XIII/2**

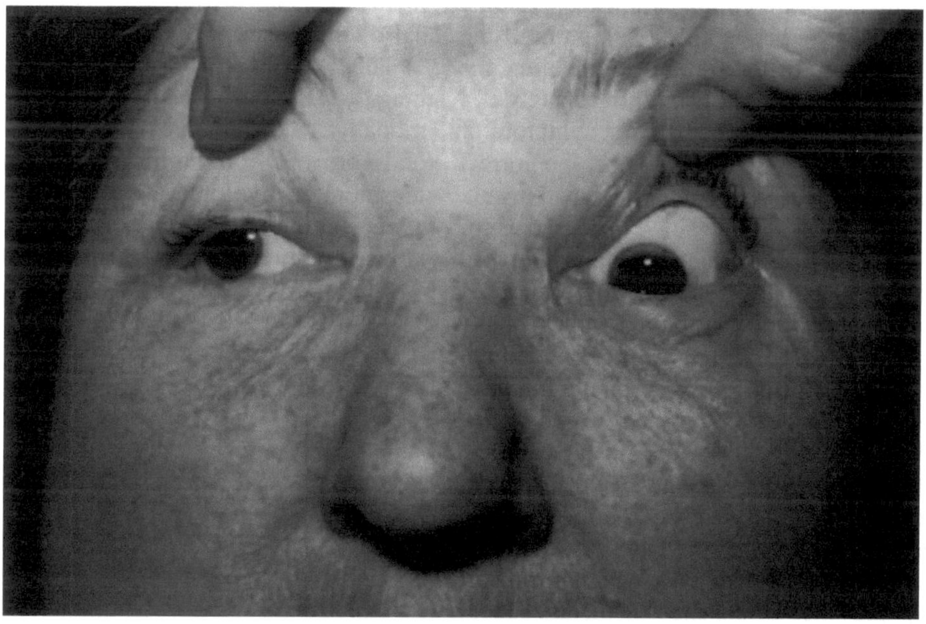

**Normalization following treatment.**

**Fig. XIII/3**

velop very slowly or occur within a few hours. Complications caused by the cervical spinal compression can scarcely be treated, compared to those of the lower segments (21). Symptoms indicating spinal cord injury might be the first manifestations of plasmacytoma, or - in case of existing diagnosis - indicate the final stage and widespread dissemination. Suddenly developing compression is more responsive to therapy; in most cases an immediate operation prolongs life. Development of a sudden compression without any (?) previous sign usually indicates solitary spinal plasmocytoma starting from the dura, while the slowly developing type primarily originates from the vertebra or the ribs, from where it grows in the direction of the spinal canal. The differential diagnosis is very difficult, since the examination of a motionless patient is not an easy task and cannot be solved without modern imaging methods.

II. Neuropathy is the only neurologic complication, which can be revealed in all forms of monoclonal gammopathies. The neuropathic symptoms slightly differ in plasmocytoma, Waldenström macroglobulinaemia or IMG and, therefore, it is advisable to study the three groups separately.

**Polyneuropathy in plasmocytoma**

Neuropathy is a rare complication of myeloma that can appear either as a first symptom of the disease, or as a factor further aggravating the already diagnosed state (42). Unfortunately if a patient with neuropathy as the first symptom of plasmocytoma addresses the doctor, the correct diagnosis of one case of ten is a good result. The neuropathy syndrome slowly progresses for years concealing the basic disease which had started much earlier. Usually progressive ascendent sensomotoric symptoms might finally lead to total loss of voluntary motion. Similar to the Guillain-Barré syndrome (acute infective polyneuritis) these symptoms are accompanied with the paralysis of the trunk, respiratory and cranial muscles (83). Weakness and fasciculation of the tongue has also been noticed (23).

Examination of the patients revealed peripheral neuropathies, distal weakness, muscle atrophy, presence or absence of distal sensory dysfunctions and the lack of reflexes. It is worth pointing out that the presence of these symptoms, without lack of reflexes suggests amyloidosis (42). Impotence, postural hypotension indicate endocrine involvement (60). Postural anomalies are symmetric and sometimes amyotrophia and isolated atrophia also occur. When the symptoms indicate compression or infiltration of a radix or pathway probably no neuropathy is present.

Simultaneous occurrence of osteosclerotic myeloma and neuropathy has to be specially mentioned (67, 91, 92). The osteosclerotic form of multiple myeloma is claimed to occur in 3% of the plasmocytomas (68). We found only two cases of this type suggesting a markedly lower incidence. Despite the rarity of osteosclerotic myeloma, polyneuropathy occuring in such cases is very frequent (37), especially among the male patients with worse chances of survival (27). The monoclonal gamma concentration and the percentage of the plasma cells in the bone marrow are both low. Cranial involvement is also rare (43). In few cases papillar edema that could not be explained by local factors was observed.The higher protein content of the liquor might have contributed to it. Knuckled fingers are also typical (Fig. XIII/4).

Endocrine functional disorders are not rare (70). Cytostatic treatment was very effective and brought a surprisingly rapid improvement. Resistance to therapy is otherwise typical of this form.

The POEMS (Polyneuropathy, Organomegaly, Endocrinopathy, Monoclonal gammopathy, Skin changes ) syndrome is a special form of concomitant occurrence of osteosclerotic myeloma and neuropathy (8).

Its symptoms comprise sclerotic bone lesions, symmetric distal sensomotoric neuro-

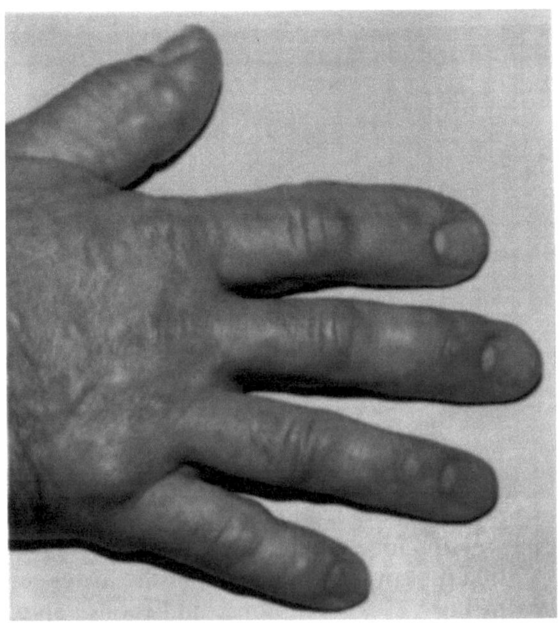

**Fig. XIII/4**

pathy, low monoclonal immunoglobulin level, moderately raised cell count in the bone marrow (36). Other manifestations are the hyperpigmentation of the skin, hirsutism, pleural effusion, ascites, articular pain, hepatosplenomegaly (it should be emphasized that splenomegaly is a primary symptom of myeloma), lymphadenopathy and increased perspiration. Gynaecomastia, testicular atrophy, impotence, amenorrhaea and sensitivity of breasts in women indicate endocrine disfunction. Other manifestations are: diabetes mellitus, hypothyreodism, subfebrility and slight fever (8, 36, 37). It was, however, found that the simultaneous presence of all the manifestations of the syndrome is almost impossible, thus diagnosis meets certain difficulties.

## Polyneuropathy in Waldenström's macroglobulinaemia

Except the forms caused by HVS, neuropathy is rarer in WM than in plasmocytoma, but in contrast the latter, these symptoms make the patient to consult a doctor. Most often the complaints are of a sensomotoric nature, but purely sensory neuropathies also occur (39). A typical asymmetric onset later generalizes together with other complications of the underlying therapy disease (5). This form is treatment-resistant, so it can be regarded as uncurable. Nevertheless some cases reacted well to the usual cytostatic therapy and remained symptom-free for a long time. It has been proved that in amyloidosis neuropathy is refractory to therapy, whereas patients reacting to treatment have no amyloidosis.

## Neuropathy in idiopathic monoclonal gammopathy

Polyneuropathy was reported in about 25% of the cases (40). In our opinion and according to our criteria this is not completely tone. It seems that the contribution of various immunoglobulin molecules to this pathology cannot be judged by the same standard, since neural damages develop more often in IgM-MG, compared to IgG- or IgA-forms (80). In most cases neuropathy is symmetric. In cases of IgM the Raynaud syndrome accompained by tremor and ataxia often develops. Electrophysiologic tests indicate demyelinisation type neuropathy in all IMGs (66). Therapy is based on steroid administration though sometimes plasma exchange and cytostatic treatment were also applied though the results were disappointing (1).

Gammopathy is a common feature of the three groups mentioned above, therefore, similar pathological and histological changes could be expected in the background of neuropathy.

In multiple myeloma in addition to the cellular infiltration of the peripheral nerves (55), segmental demyelinisation and axonal degeneration occur (33). As a matter of fact segmental demyelinisation is a consequence of axonal degeneration (56), thus the two events are not parallel but follow each other.

In Waldenström's macroglobulinaemia the pathogenetic role of IgM seemed to be proved by the binding of IgM to the myelin sheath (63). Later it was discovered that this is a true antigen-antibody complex in which *m*yelin-*a*ssociated *g*lycoprotein (MAG) acts as an antigen, which reacts with the monoclonally produced IgM antibodies (57). A more thorough analysis suggests that this might be a sulphated sphyngoglycolipid (35).

Anti-MAG antibodies found in idiopathic monoclonal gammopathies were not only of the IgM-type (61, 66). In later experiments neuropathy was transferred to mice with IgG paraproteins; this finding indicates that the antibodies react with the uncovered structural elements of the nervous fibres and the reaction is not species specific (11, 24).

In summary: Patients with diagnosed monoclonal gammopathy neuropathy can be, for practical reasons, assessed into three groups:

I. Patients without signs of neuropathy. The latter is suggested by the EMG-ENG findings.
II. Patients mentioning complaints suggesting polyneuropathy in the anamnesis but complaint-free at admittance
III. Patients with polyneuropathy.

It has already been mentioned that only few patients belong to the third group, nevertheless, practitioners should be aware of this possibility. Another group requiring special attention comprises patients with pain or palsy due to the bone lesions. If it is technically possible, these cases undergo surgery. Surgery of extremities implies nailing or arthrodesis. After the operation the wounds heal almost normally, so that the surgeon is not facing a higher risk in the postoperative period. Vertebral fractures have several consequences. A rare case is shown as an example: despite the fact that the X-ray study revealed no intact vertebrae, the patient had no pains and even bending down caused no complaints (Figs XIII/5, XIII/6).

Fractures of one or two, usually neighbouring vertebrae with lasting pain that can be hardly influenced are typical. According to the data of the literature, this form is not dangerous, provided the underlying disease can be stabilized, otherwise, or if the danger of transversal lesion due to vertebral destruction is suspected decompressive laminectomy or corpodesis followed by two three months in a plaster bed is the last possibility.

## Diagnosis and treatment of the hyperviscosity syndrome (HVS)

Clinical and laboratory findings in the MG-s raise the question whether the nature of the of tumor cell products can help in the prognosis of the disease or in recognition of its special symptoms. HVS gives a positive answer to this question, despite the fact that its clinical manifestations have been scarcely investigated.

The first relevant publication presenting this concept was published in 1929 (7). The description of the syndrome accepted at present appeared only in 1965 (28). It typically occurs in WM (84). It has been observed that though high viscosity occurs in 55% of the patients, its clinical manifestation appears only in 18-20% of the cases (51). Data of the literature show that in MM the incidence of HVS-induced symptoms is only 10% (84). In our material 9 cases (less than 1%) with serious complications that required plasma ex-

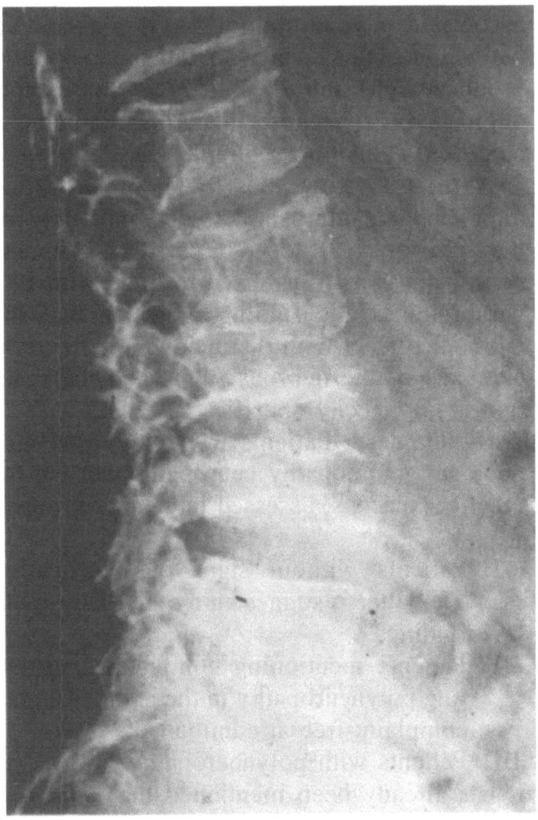

Lateral image of lumbar vertebra

Fig. XIII/5

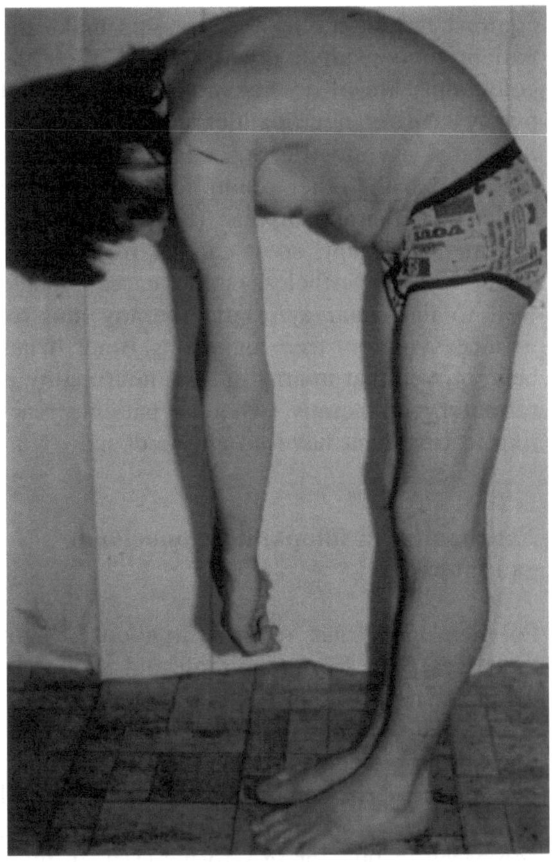

Fig. XIII/6

change were observed. The clinical picture can be attributed to a capillary circulatory disorder, consequently the character of the symptoms indicates the affected organ.

*Lesions of the skin are the most pronounced*

In chronic cases ulcers similar to the cutis marmorata develop, but in contrast with the latter persisting skin patterns appear (Fig. XIII/7). Other typical symptom is the appearance of the trophic ulcers on the leg (Fig. XIII/8). The latter is further aggravated by the orthostatic load and cooling. If cryoglobulinaemia is also present the chance of healing is low (Fig. XIII/9).

Cryoglobulins cause a typical Raynaud syndrome on the hand without any lasting cutaneous changes (Fig. XIII/10).

*Mucosal haemorrhage*

Nasal bleeding is the most characteristic form; gastrointestinal bleeding is more rare, but is difficult to control and almost impossible to localize. The choice of conservative or surgical treatment is extremely difficult. Intracranial bleeding is rare and its true cause, that is HVS almost always remains undiscovered if the disease is not diagnosed.

*Coronary disorders*

The ECG indicates irregularities of repolarisation. Unfortunately, the changes are not characteristic and as the patients belong to the elderly group, in which similar alterations are not rare even without HVS, its evaluation is very difficult. The acute form of the hyperviscosity syndrome can mimic that of congestive cardiomyopathy (Fig. XIII/11 and XIII/12).

Functional disturbances of the sensory organs are very characteristic, for example retinopathy is a general symptom, always accompanying HVS. At the same time it is non-specific as it occurs in several other diseases (48). Sequential retinopathy is, however, typical of HVS (28, 48, 82). Plasmapheresis might greatly improve vision in not more than one night (76, 79). The fundus is paraproteinaemic, retinal veins are dilated as a consequence of arteriovenous shunts. High viscosity causes spotty, patch- and flame-shaped haemorrhages, which finally lead to papilla-edema or even retinal venous thrombosis (Fig. XIII/13).

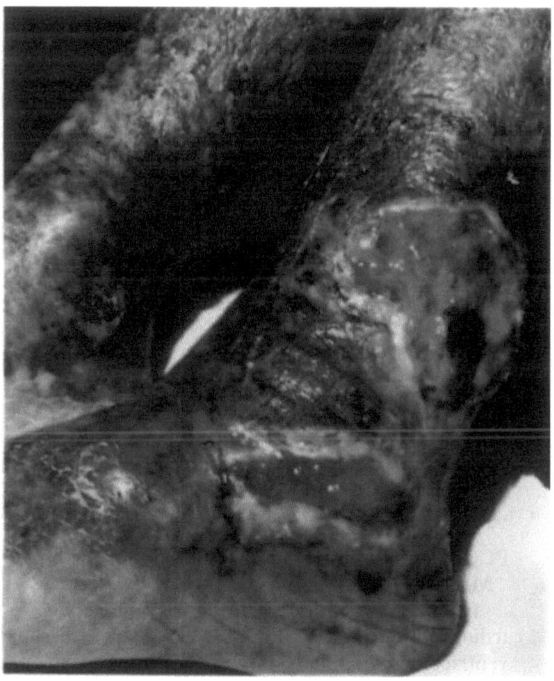

**Fig. XIII/7**

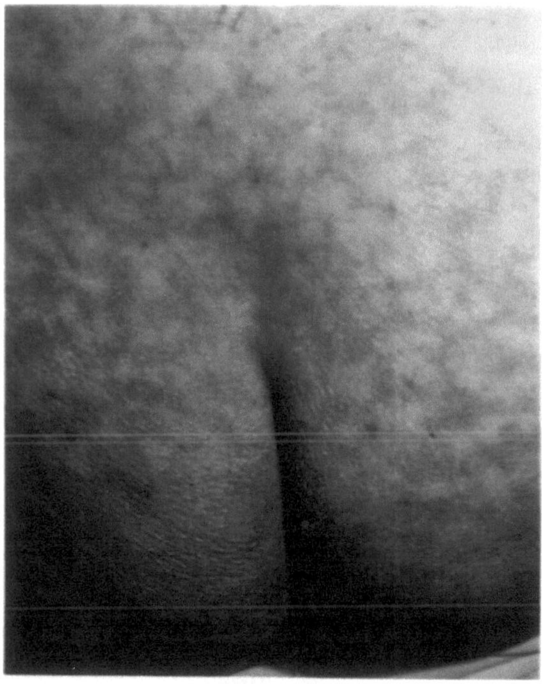

**Fig. XIII/8**

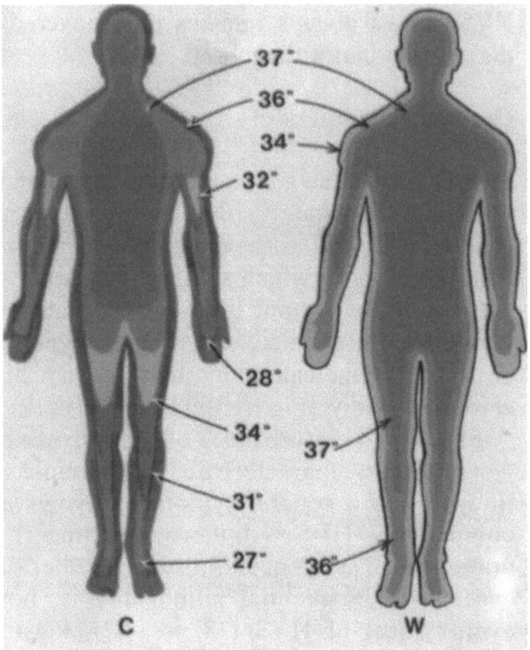


C = cold, W = warm

Fig. XIII/9

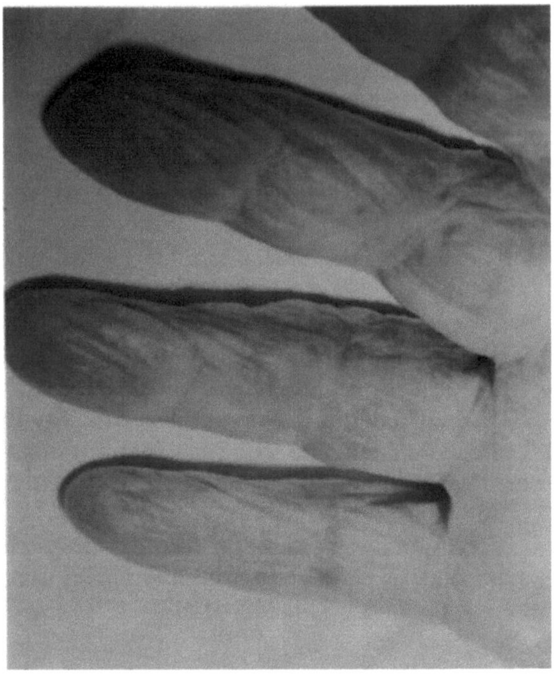

Fig. XIII/10

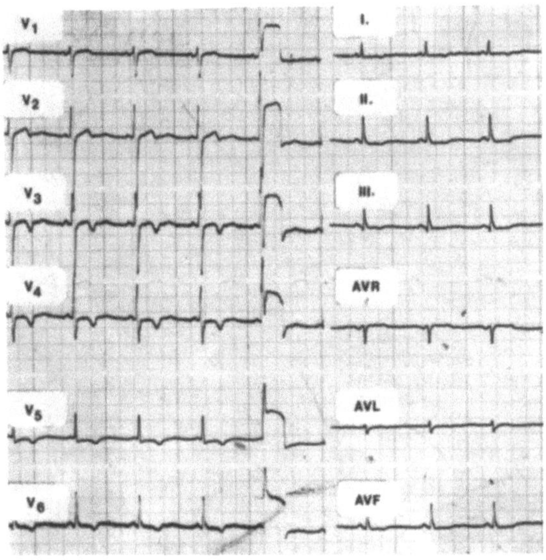


**ECG: sinus rhythm, Short atrioventricular conduction (P-Q 0.12 sec. Normal electric axis. Negative T-waves in the I-II-AVL-AFV-$V_{3-6}$ leads, 0.5 mm ST depression in $V_{4-5}$ leads).**

**Fig. XIII/11**

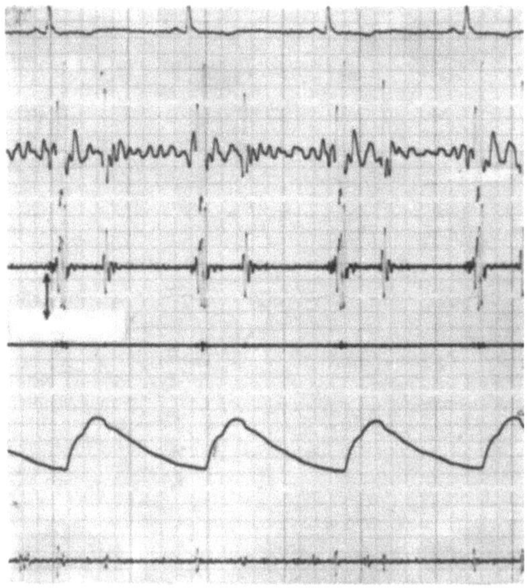

**PCG. Low-amplitude atrial sound (↑) on the apex. Mechanography: anacrotic carotid pulse curve, normal ejection time. On the left ventricular cardiogram normal IVCP and IVRP, increased rapid ventricular filling, late systolic paradox movement.**

**Fig. XIII/12**

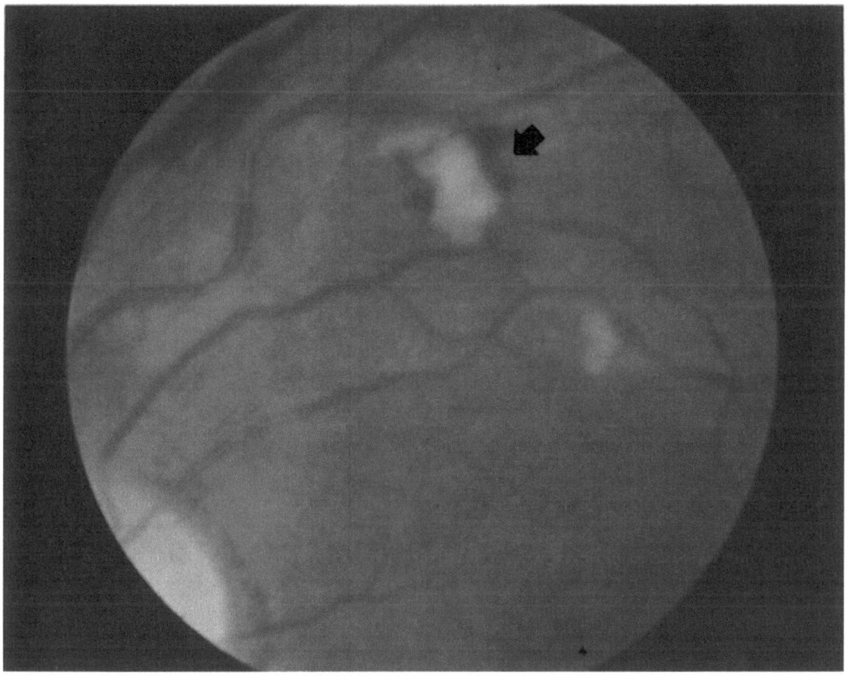

**Paraproteinaemic fundus. Hemorrhage from the dilated vein; ischaemic infarction surrounded by bleeding at the bifurcation of a tortuous arteriole.**

**Fig. XIII/13**

Impaired hearing might involve complete slow or sudden deafness (71).

Nervous system symptoms are manifested in headache, nystagmus, ataxia, disbalance, deteriorated mental activity, postural hypotension, paresthesia, progressive peripheral neuropathy, Bing-Neel (Bichel, Bing, Harboe) syndrome (12, 13, 14), diplopia, which can occur together, or alone as a leading symptom. Severe damages of the capillary circulation are responsible for somnolence, stupor and coma which can be easily taken for a cerebrovascular insult (rhexis). It is always accompanied by moderate or serious EEG anomalies (4, 22, 29).

Identification of the above mentioned symptoms does not require specially equiped internal medical wards, nevertheless, the blood viscosity is rarely measured.

Hyperviscosity syndrome occurs not only in monoclonal gammopathies (64), but can be also caused by the increased blood cell count or by the pathologic composition of the plasma. The latter possibility is discussed herein. Pathogenesis of the syndrome is due to the aggregation of immunoglobulins that form giant complexes of two, three, or sometimes even 10 molecules (10, 45). It has been first assumed that $IgG_3$s are prone to form aggregates (18) but later it was proved that it is rather the $IgG_1$ molecules play a more important role in increased viscosity, since in most myelomas with HVS complications the $IgG_1$ type is present (64, 72). Although the IgA molecule has an increased tendency for polymerization that can theoretically cause HVS, hyperviscosity was only rarely found in IgA-MG (26, 30, 94). As already mentioned, HVS develops in patients with macroglobulinaemia (52, 53). Paraprotein concentration and the total protein level might indicate the possibility of HVS, however, no linear correlation could be found: at lower total protein levels the viscosity of the blood might still be high (10).

The highest value of normal plasma viscosity is 1.8–2 centipoise (cP). Above 4 cP the pathological symptoms might, but not necessarily, develop, hence, some still unknown factor which promotes the appearance of the symptoms should exist. For example its presence or absence can contribute to the symptom-free state even at blood viscosity values over 15 cP. This finding indicates that blood viscosity does not correlate with the clinical symptoms. The known factors are: the state of the capillaries (69), flow rate, haematocrit value, the inotropic state of the heart, local tissue pH and ionic strength (15). It is well known that concomitantly with the development of hyperviscosity a compensatory plasma dilution takes place (3, 49, 73, 84). The latter is probably due to the aldosteron-independent retention of $Na^+$ (49). The extent of dilution might contribute or prevent the symptoms.The intravasal volume correlates with the increase of the relative serum viscosity (50).

The review of the published data makes it obvious that hyperviscosity needs further study. Our measurements were performed with the Ostwald viscosimeter which ensures laminar flow and the Couette-type rotation viscosimeter (Fig. XIII/14 and XIII/15).

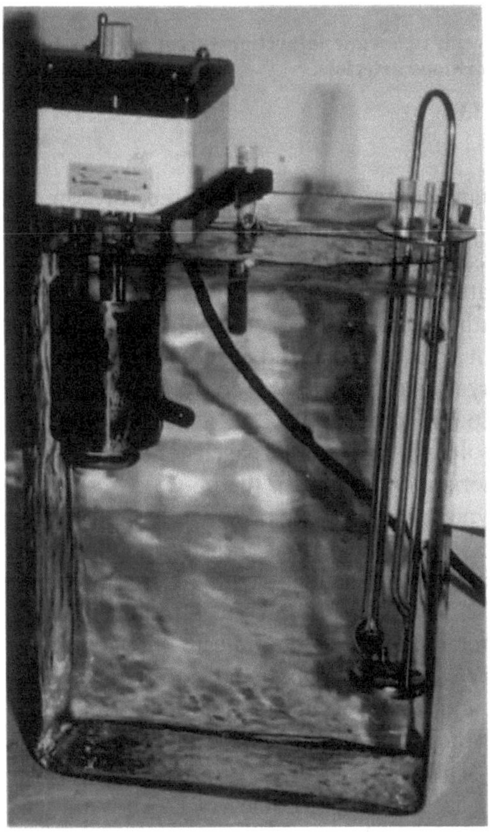

Temperature regulating device is inserted into a flat-high glass chamber filled with distilled water, which keeps the temperature of the water within 0.1 °C of the set value. The viscosimeter ensures the optimal possibility for measuring (k = 0.09421).

Fig. XIII/14

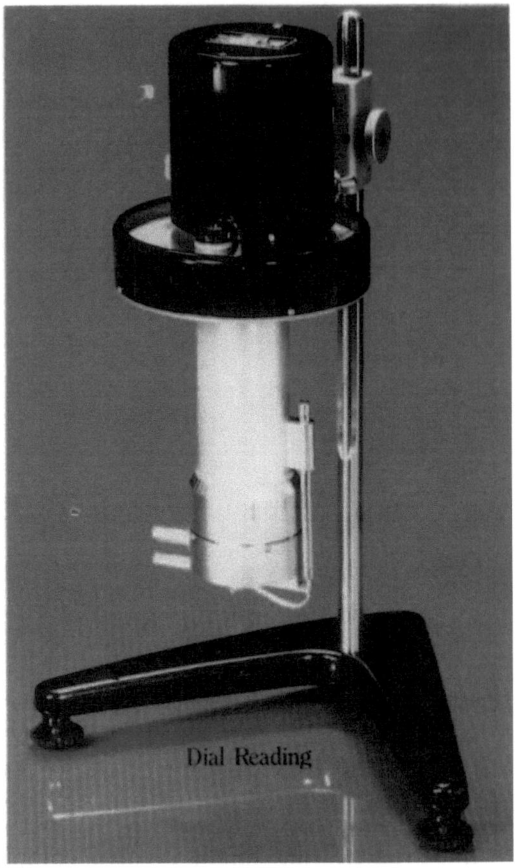


Fig. XIII/15

Relative viscosity η (in cP) was determined as shown in the equation XIII/16.

$$\eta = \frac{\eta_1(\varsigma \times t)}{\varsigma_1 \times t_1}$$

$\eta_1$ = viscosity of distilled water
$t_1$ = flow time of distilled water
$t$ = flow time of blood
$\varsigma_1$ = density of distilled water
$\varsigma$ = density of blood

Fig. XIII/16

Before going into the details of therapy, some general remarks seem worth pointing out. The capillary circulation is affected by the following factors:

a) composition of the plasma,
b) the percentage of the red blood cells and their aggregation properties under given conditions,
c) flow rate in the capillaries.

the cases the concentration of protein exceeded 3 g%) (50). Other properties of the suspending medium (point 2, 3 and 4) become obvious mainly in the presence of the red blood cells.

a) composition of the plasma as a **suspending** medium is important because of the following:
   1. the percentage of protein or substituting molecules in the liquid phase;
   2. presence, charge and shape of the molecules with MW exceeding 300,000 D, their
   3. irregular physiochemical properties of some protein particles, for example, cryoprecipitation tendency;
   4. binding (antibody) properties of the immunoglobulin molecules.

The results shown in Fig. XIII/17 prove that a 10% solution of Dextran RMI (Rheomacrodex) with an average molecular weight of 40,000 has high viscosity, while the viscosity of 5% Dextran RMI or 5.5% Oxypolygelatine (Gelifundol) solutions is similar to that of the plasma. Thus, the concentration of the dissolved particles is an important factor which increases the viscosity of a solution. Plasma and serum with M-component, have almost the same viscosity as the normal plasma (in none of

b) the % of the red blood cells, or the haematocrit value is a very important factor (Fig. XIII/18). Both their count and the proneness to aggregate is of importance. Aggregation is a reversible physiological process. It has been proved that the presence of RBC and serum proteins is necessary for the aggregation. In other words: aggregation is a result of RBC - plasma protein interaction, and is definitely temperature dependent (Fig. XIII/19). Our studies, however, failed to prove this statement, since in Gelifundol solutions with the same haematocrit the changes of viscosity with temperature are similar to plasma. This finding suggests that the interaction between the molecules of a certain size and charge with the RBC is essential (Fig. XIII/20).

The high viscosity of RBC suspended in Rheomacrodex is interesting but not surprising if Fig. XIII/20 is considered. This situation cannot occur in vivo. We, therefore, studied the viscosity of physiological Dextran RMI concentrations. The results altered from those of native plasma.

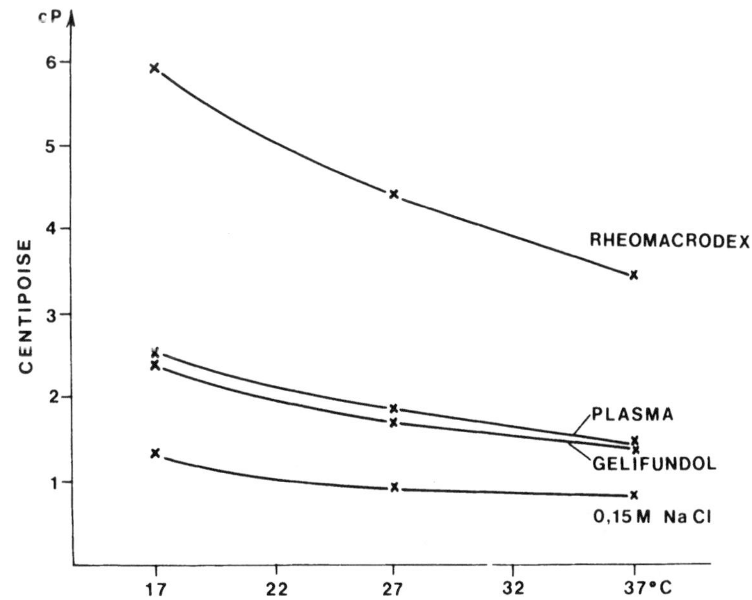


Fig. XIII/17

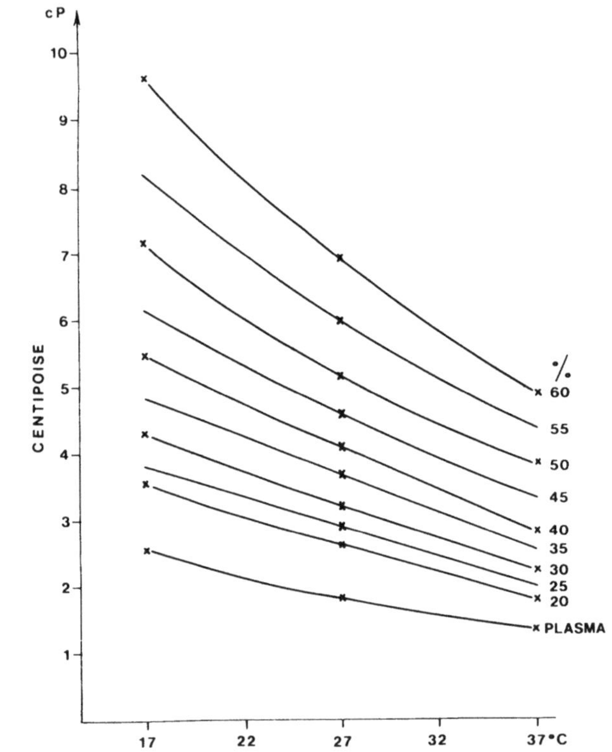


Viscosities measured in the blood of healthy individuals with artificially altered haematocrit values as a function of temperature.

**Fig. XIII/18**

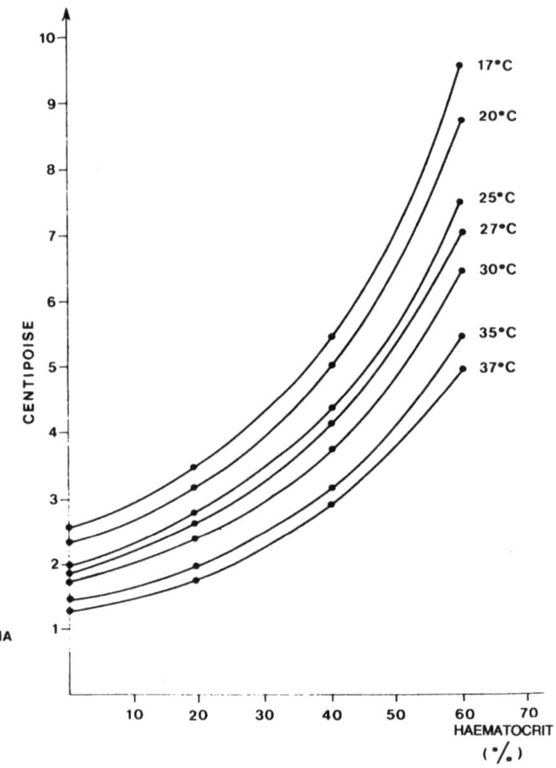


Viscosity of normal blood as a function of haematocrit and temperature.

**Fig. XIII/19**

Suspensions are usually dispersed systems and the shape of the particles greatly affects viscosity. Such systems are non-newtonian fluids, that is, instead of a linear relationship the points obtained by measurements fit to a monotonously increasingly steep curve. Blood is a *non-newtonian* fluid and aggregation further prevents solvation. Temperature changes might cause aggregation between molecules, which - though extremely labile - can cause aggregation of the RBC.

An enhanced form of this phenomenon occurs in the presence of cryo-type proteins (for example, cryoglobulinaemia). In this case precipitation is far more stable, therefore, its effect promoting RBC aggregation is far more pronounced and lasts for a much longer time (Fig. XIII/21). In another form of aggregation the antigen - antibody reaction, in other words immunocomplex formation is in the background. In this special case the complex is formed by immunoglobulin molecules. Immunoglobulins with antibody activity are of the IgM (68) or IgG (62) type. Furthermore, similar to the above mentioned possibility, the monoclonally formed IgM molecules with cryoglobulin properties usually posses RF activity (58).

Aggregation in which the erythrocytes are no longer innocent bystanders is far more serious and plays no role in the development of the discussed clinical picture. If it happens acute HVS will develop.

c) The capillary flow conditions are extremely important. Aggregates are formed at low flow velocity that greatly increases the viscosity. In the peripheral capillary system blood slows down and the aggregates together with the other known factors aggravate the microcirculatory disorder. At higher flow rates viscosity is obviously

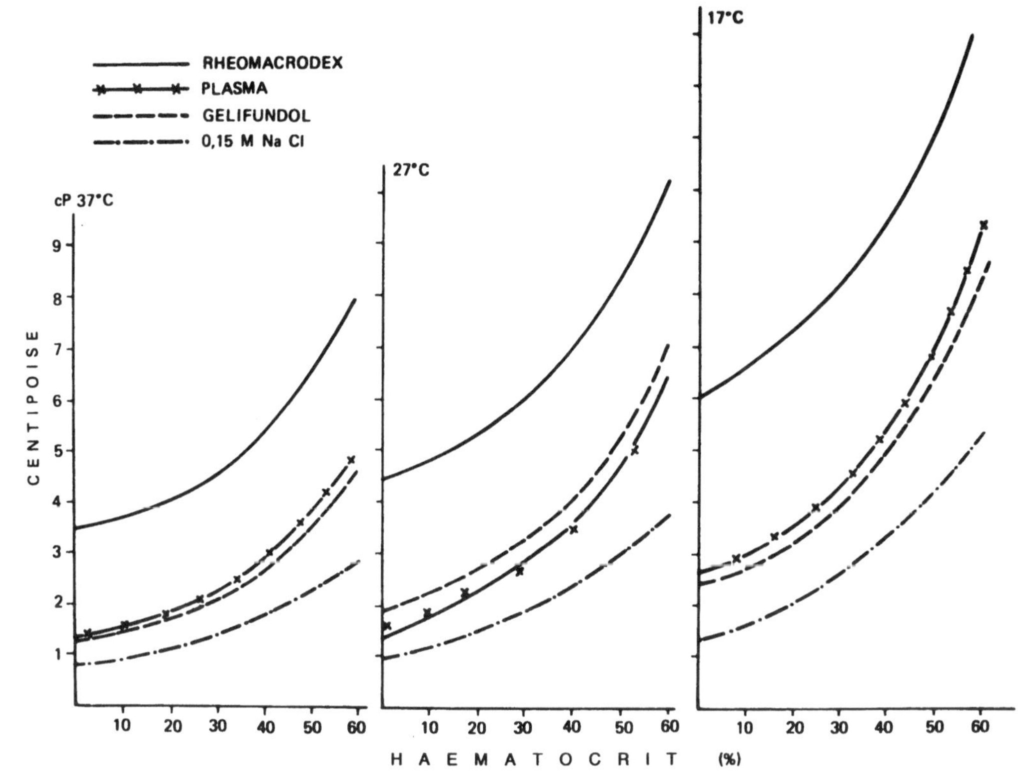


**Viscosity of RBC suspended in plasma or plasma-replacing fluids at 17, 27 and 37 °C as a function of haematocrit.**

**Fig. XIII/20**

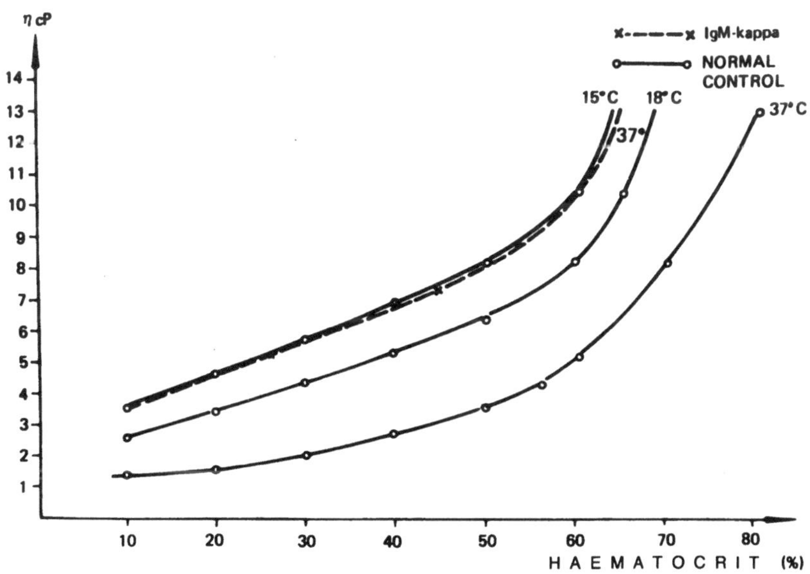


**Viscosity of the blood of a cryoglobulinaemic patient at 37 °C was similar to that of normal blood at 15 °C.**

**Fig. XIII/21**

lower (74), since high shear rates decrease formation of the aggregates (17, 74).

These quite simple tests emphasize the importance of plasma exchange in the treatment of patients with high blood viscosity.

*In summary:* although cases of MGs complicated with hyperviscosity seldom occur, they should nevertheless be takem into consideration, partly because the symptoms cannot be controlled by cytostatic therapy in contrast to the usual paraproteinaemic cases and partly because the absolutely necessary plasma exchange (and not plasmapheresis) (47) has an almost immediate effect. This observation proves that the pathological changes of the capillary circulation, which can be explained with physical laws, can be normalized - though only temporarily - with the technical tools. Sometimes HVS develops without MG. In such cases a special tertier and quaternary molecular structure is assumed which alone is sufficient to cause hyperviscosity (31, 50).

### References


1. Abrams, G. M., Latov, N., Hays, A. P., Sherman, W., Zimmerman, E. A.
Immunocytochemical studies of human peripheral nerve with serum from patients with polyneuropathy and paraproteinaemia.
Neurology. 325, 821. 1982.
2. Adams, R. D.
Case records of Massachusetts General Hospital.
N. Engl. J. Med. 228, 150. 1973.
3. Alexanian, R.
Blood volume in monoclonal gammopathy.
Blood. 49, 301. 1977.
4. Azzena, D., Costa, M., Ghigliotti, G., Astegno, F.
Serum hyperviscosity in multiple myeloma.
Confin. Neur. 30, 65. 1968.
5. Bajada, S., Mastaglia, F. C., Fisher, A.
Amyloid neuropathy and tremor in Waldenström's macroglobulinaemia.
Arch. Neurol. 37, 240. 1980.
6. Bannerjee, A. K., Blackwood, W.
A subfrontal tumour with the features of plasmacytoma and meningioma.
Acta Neuropath. 18, 84. 1971.
7. Bannick, E. G., Greene, CH.
Renal insufficiency associated with Bence Jones proteinuria: report of thirteen cases with a note on the changes in the serum proteins.
Arch. Intern. Med. 44, 486. 1929.
8. Bardwick, P. A., Zvaitler, N. J., Gill, G. N., Newman, D.
Plasma cell dyscrasia with polyneuropathy organomegaly, endorcrinopathy, M. Protein and skin changes. The POEMS syndrome.
Medicine. 59, 311. 1980.
9. Bataille, R., Sany, J.
Solitary myeloma. Clinical and prognostic features of a review of 114 cases.
Cancer. 48, 845. 1981.
10. Benninger, G. W., Kerps, S. I.
Aggregation phenomenon in and IgG multiple myeloma resulting in the hyperviscosity syndrome.
Amer. J. Med. 51, 287. 1971.
11. Besinger, U. A., Anzil, A., Toyka, K. V., Fateh-Moghadam, A., Rauscher, R., Struppler, A.
Neuropathie bei monoklonaler Gammopathie: Passiver Transfer von Mensch zu Maus.
Jahrestagung der Deutschen Gesellschaft für Neurologie 1980 in Verh. dtsch. Gesell. Neurologie 1980.
12. Bichel, J., Bing, J., Harboe, N.
Another case of hyperglobulinaemia and affection of central nervous system.
Acta Med. Scand. 138, 1. 1950.
13. Bing, J., Neel, A. V.
Two cases of hyperglobulinaemia with affection of the central nervous system on a toxi-infectious basis.
Acta Med. Scand. 88, 492. 1936.
14. Bing, J., Fog, M., Neel, A. V.
Report of a third case of hyperglobulinaemia with affection of the central nervous system on a toxi-infectious basis.
Acta Med. Scand. 91, 409. 1937.
15. Bloch, K. J., Maki, D. G.
Hyperviscosity syndrome associated with immunoglobulin abnormalities.
Sem. Hematol. 10, 113. 1973.
16. Brenner, B., Carter, A., Tatarsky, I., Gruszkiewicz, J., Peyser, E.
Incidence, prognostic significance and therapeutic modalities of CNS involvement in multiple myeloma.
Acta Haematol. 68, 77. 1982.
17. Brundage, J. T.
Blood and plasma viscosity determined by the method of concentric cylinders.
Amer. J. Physiol. 110, 659. 1935.
18. Capra, J. D., Kunkel, H. G.
Aggregation of gamma $G_3$ proteins: relevance to the hyperviscosity syndrome.
J. Clin. Invest. 49, 610. 1970.
19. Clarke, E.
Cranial and intracranial myeloma.
Brain. 77, 61. 1954.
20. Corwin, J., Lindberg, R. D.
Solitari plasmocytoma and their relationship to multiple myeloma.
Cancer. 43, 1007. 1979.
21. Dahlstrom, V., Lindstrom, F. D.
Paraplegia in myelomatosis. A study of 20 cases.
Acta Med. Scand. 205, 173. 1979.
22. Davis, L. E., Drachman, D. B.
Myeloma neuropathy, successful treatment of 2 cases and review of literature.
Arch. Neurol. 27, 567. 1972.
23. Davison, C., Balser, B. H.
Myeloma and its neural complication.
Archives of Surgery. 35, 913. 1937.
24. Dayan, A. D., Stokes, M. I.
Peripheral neuropathy and experimental myeloma in the mouse.
Nature. 236, 117. 1972.
25. Delduca, V., Morningstar, W. A.
Multiple myeloma associated with multifocal leuco-encephalopathy.
J. Amer. Med. Ass. 199, 671. 1967.
26. Dine, M. E., Guay, A. T., Snyder, M.
Hyperviscosity syndrome with IgA myeloma.
Amer. J. Med. Sci. 264, 111. 1972.
27. Dreidger, H., Prozanski, W. A.
Plasma cell neoplasia with peripheral neuropathy. A study of 5 cases and review of literature.
Medicine. 59, 301. 1980.
28. Fahey, J. L., Barth, W. F., Solomon, A.
Serum hyperviscosity syndrome.

J. Amer. Med. Ass. 192, 464. 1965.
29. Fraser, D. M., Parker, A. C., Amer, S., Cambell, I. W. Mononeuritis multiplex in a patient with macroglobulinaemia. J. N. N. s. P. 39, 711. 1976.
30. Freel, R. J., Maldonado, J. E., Gleich, G. J. Hyperviscosity syndrome associated with immunoglobulin a myeloma. Amer. J. Med. Sci. 264, 117. 1972.
31. Hadler, M. N., Gabriel, D., Chung, K. S., Teague, Patricia, Nepier, A. Mary. Polyclonal hyperviscosity syndrome. Arthr. Rheum. 20, 1388. 1977.
32. Hansotia, P., Gani, K., Friedenberg, W. Cerebrospinal fluid monoclonal gammopathy in multiple myeloma and Waldenström's macroglobulinaemia. Neurology. 33, 1411. 1983.
33. Hesselvik, M. Neuropathological studies in myelomatosis. Acta Neurol. Scand. 45, 95. 1969.
34. Holness, R. O., Sangalang, V. E. Myelomatous metastasis to the pineal body. Surgical Neurology. 5, 97. 1976.
35. Ilyas, A. A., Quarles, R. H., Dalakas, M. C., Brady, R. O. Polyneuropathy with monoclonal gammopathy: glycolipids are frequently antigens for IgM paraproteins. Proc. Natl. Acad. Sci. 82, 6697. 1985.
36. Imawari, M., Akatsuka, N., Ishibashi, M., Beppu, H., Suzuki, H. Syndrome of plasma cell dyscrasia, polyneuropathy and endocrine disturbance. Report of a case. Ann. Intern. Med. 81, 590. 1974.
37. Iwashita, H., Ohnishi, A., Asada, M., Kanazama, Y. Polyneuropathy skin hyperpigmentation, edema and hypertrichosis in localised osteosclerotic myeloma. Neurology. 27, 675. 1977.
38. Johnston, J. B., Weinerman, B., Cooney, T., Bowman, D. M., Pettigrew, N. M., Orr, K. IgD Kappa plasma cell dyscrasia. Extraosseous manifestation including isolated leptomeningitis. Amer. J. Clin. Path. 77, 60. 1982.
39. Jullen, J., Vital, C., Vallat, J. M., Lagueny, A., Deminier, C., Darriet, D. Polyneuropathy in Waldenström's macroglobulinaemia. Arch. Neurol. 35, 423. 1978.
40. Kahn, S. N., Riches, P. G., Kohn, J. Paraproteinaemias in neurological disease: incidence, associations and classification of monoclonal globulins. J. N. N. P. 43, 617. 1980.
41. Kamin, D. F., Helper, R. S. Solitary intracranial plasmocytoma mistaken for retrobulbar neuritis. Amer. J. Ophthal. 73, 584. 1972.
42. Kelly, J. J., Kyle, R. A., Miles, J. M., O'Brien, P. C., Dyck, P. J. The spectrum of peripheral neuropathy in myeloma. Neurology. 31, 24. 1981.
43. Kelly, J. J., Kyle, R. A., Miles, J. M., Dyck, P. J. Osteosclerotic myeloma and peripheral neuropathy. Neurology. 33, 202. 1983.
44. Knisely, M. H. Intravascular erythrocyte aggregation (blood sludge). In: Handbook of Physiology. Ed: by Hamilton, W. F., Dow, F. American Physiological Society. Washington, D. C. 3, 2249. 1965.
45. Lindsley, H., Teller, D., Noonan, B., Peterson, M., Mannik, M. Hyperviscosity syndrome in multiple myeloma: a reversible, concentration-dependent aggregation of the myeloma protein. Amer. J. Med. 54, 682. 1973.
46. Logothetis, J., Silverstein, P., Coe, J. Neurological aspects of Waldenström's macroglobulinaemia. Report of case. Arch. Neurol. 3, 564. 1960.
47. Lundsgaard-Hansen, P. Volume limitations of plasmapheresis. Vox. Sang. 32, 20. 1977.
48. Luxenburg, M. N., Mausolf, F. A. Retinal circulation in hyperviscosity syndrome. Amer. J. Ophthal. 70, 588. 1970.
49. MacKenzie, M. R., Brown, E., Fudenberg, H. H., Goodenday, L. Waldenström's macroglobulinaemia: correlation between expanded plasma volume and increased serum viscosity. Blood. 35, 394. 1970.
50. MacKenzie, M. R., Fudenberg, H. H., O'Reilly, R. A. The hyperviscosity syndrome. I. In IgG myeloma. The role of protein concentration and molecular shape. J. Clin. Invest. 49, 15. 1970.
51. MacKenzie, M., Fudenberg, N. H. Macroglobulinaemia an analysis of 40 patients. Blood. 39, 874. 1972.
52. MacKenzie, M. R., Babcock, J. Studies of the hyperviscosity syndrome. II. Macroglobulinemia. J. Lab. Clin. Med. 85, 227. 1975.
53. MacKenzie, R. M., Lee, Th. K. Blood viscosity in Waldenström makroglobulinemia. Blood. 49, 507. 1977.
54. Mancilla-Jimenez, R., Tavassoli, F. A. Solitary meningeal plasmocytoma. Report of a case with action microscopic and immune histologic observation. Cancer. 38, 798. 1976.
55. Mayo, C. M., Daniels, A., Barron, K. D. Polyneuropathy in osteosclerotic form of multiple myeloma. Trans. Amer. Neurol. Assoc. 93, 240. 1968.
56. McLeod, J. G., Walsh, J. C., Pollard, J. D. Neuropathies associated with paraprotenaemias and dysproteinaemias. In: Peripheral Neuropathy. Ed: by Dyck, D. J., Thomas, P. K., Lambert, E., Bunge, R. W. B. Saunders, New York. 80. 1984.
57. Melmed, C., Frail, D., Duncan, I., Braun, P., Danoff, D., Finlayson, M., Stewart, J. Peripheral neuropathy with IgM kappa monoclonal immunoglobulin directed against myelin associated glucoprotein. Neurology. 33, 1397. 1983.
58. Meltzer, M., Franklin, E. C., Elias, K. Cryoglobulinemia – A clinical and laboratory study. II. Cryoglobulins with rheumatioid factor activity. Amer. J. Med. 40, 837. 1966.
59. Moosey, J., Wilson, C. B. Solitary intracranial plasmacytoma. Arch. Neurol. 16, 212. 1967.
60. Moya-Mir, M. S., Martin-Martin, F., Barbadillo, R., Cuervas, M. V. Plasma cell dyscrasia with polyneuropathy and dermatoendocrine alteration. Postgrad. Med. Jour. June: 427. 1980.
61. Panitch, H. S. Antimyelin antibodies in patients with peripheral neuropathy and benign monoclonal gammopathy. J. Neuropath. Exp. Neurol. 41, 372. 1982.
62. Pope, R. M., Mannik, M., Gilliland, B. C. The hyperviscosity syndrome in rheumatoid arthritis due to intermediate complexes formed by self-association of IgG-rheumatoid factors. Arth. Rheum. 18, 97. 1975.
63. Propp, R. P., Means, E., Deibel, R., Sherer, G. Waldenström's macroglobulinaemia and neuropathy. Deposition of M component on myelin sheats. Neurology. 25, 980. 1975.
64. Pruzansky, W., Watt, J. G. Serum viscosity and hyperviscosity syndrome in IgG multiple myeloma: Report on 10 patients and a review of the literature. Ann. Intern. Med. 77, 853. 1972.
65. Pruzanski, W. Hyperviscosity and immunoglobulin complexes.

Ann. Intern. Med. 80, 107. 1974.
66. Read, D. J., Vanhegan, R. I., Matthews, W. B.
Peripheral neuropathy and benign IgG paraproteinaemias.
J. N. N. P. 41, 215. 1978.
67. Read, D. J., Warlow, C.
Peripheral neuropathy and solitary myeloma.
J. N. N. P. 41, 177. 1978.
68. Reitan, J. B., Pape, E., Fossra, S. D., Julsrod, O. J.
Osteosclerotic myeloma with neuropathy.
Acta Med. Scand. 208, 137. 1980.
69. Rosenblum, W.
Vasoconstriction, blood viscosity and erythrocyte aggregation in macroglobulinaemic and polycythaemic mice.
J. Lab. Clin. Med. 73, 359. 1969.
70. Rousseau, J. J., Franck, J., Grisar, G. T., Reznik, M., Heynen, G.
Osteosclerotic myeloma with polyneuropathy and ectopic secretion of calcitonin.
Europ. J. Cancer. 14, 133. 1978.
71. Ruben, R., Distenfeld, A., Berg, P., Carr, R.
Sudden sequential deafness as the presenting symptom of macroglobulinaemia.
J. Amer. Med. Ass. 209, 1364. 1969.
72. Rubies-Pral, J., Gallart, M. T., Frison, J. C., Caralp, A., Schwartz, S., Bacardi, R.
IgG myeloma Sia test, and serum hyperviscosity.
Acta Haematol. 52, 107. 1974.
73. Russell, J. A., Powles, R. L.
The relationship between serum viscosity, hypervolaemia and clinical manifestations associated with circulating paraprotein.
Brit. J. Haematol. 39, 163. 1978.
74. Schmid-Schönbein, H., Gaehtgens, P., Hirsch, H.
On the shear rate dependence of red cell aggregation in vitro.
J. Clin. Invest. 47, 1447. 1968.
75. Schulman, P., Sun, T., Sharer, L., Hyman, L., Vinciguerra, V., Feinstein, M., Blanck, R., Susin, M., Degan, J.
Meningeal involvement in IgD myeloma with CSF paraprotein analysis.
Cancer. 46, 152. 1980.
76. Schwab, D. J., Okon, E., Fahey, J. L.
Reversal of retinopathy in Waldenström's macroglobulinaemia by plasmapheresis.
Arch. Ophthal. 64, 515. 1960.
77. Sheldon, R., Levin, S. R., Spauldin, A. G., Wirman, J. A.
Multiple myeloma orbital involvement in a youth.
Arch. Ophthal. 95, 642. 1977.
78. Silverstein, A., Doniger, D. E.
Neurological complications of myelomatosis.
Arch. Neurol. 9, 534. 1963.
79. Skoog, W. A., Adams, W., Coburn, J. W.
Metabolic balance study of plasmapheresis in a case of Waldenström's macroglobulinaemia.
Blood. 19, 425. 1962.
80. Smith, I. S., Kahn, S. N., Lacey, B. W., King, R. H. M., Eames, R. A., Whybrew, D. J., Thomas, P. K.
Chronic demyelinating neuropathy associated with benign IgM paraproteinaemia.
Brain. 106, 169. 1983.
81. Sod, L. M., Weiner, L. M.
Intradural extramedullary plasmacytoma.
J. Neurosurg. 16, 107. 1959.
82. Soloman, A., Fahey, J. L.
Plasmapheresis therapy in macroglobulinaemia.
Ann. Intern. Med. 58, 789. 1963.
83. Solomons, R. E. B., Gibbs, D. D.
Plasma cell dyscrasia with polyneuropathy, monoclonal gammopathy and skin changes. POEMS syndrome.
J. Royal Soc. Med. 75, 553. 1982.
84. Somer, T.
Hyperviscosity syndrome in plasma cell dyscrasias. Advances in microcirculation.
Karger, Basel. 6, 1. 1975.
85. Someren, A., Osgood, L. P., Brylski, J.
Solitary posterior fossa plasmocytoma.
J. Neurosurg. 35, 223. 1971.
86. Spaar, F. W.
Handbook of clinical neurology.
Vinken Bryn Elsevier, North Holland. Biomedical Press. 39. 1980.
87. Spiers, A. S. D., Halpern, R., Ross, S. C., Neiman, R. S., Harawi, S., Zipoli, T. E.
Meningeal myelomatosis.
Arch. Intern. Med. 140, 256. 1980.
88. Stark, R. J., Henson, R.
Cerebral compression by myeloma.
J. N. N. P. 44, 833. 1981.
89. Truong, L. D., Kim, H. S., Estrada, R.
Meningeal myeloma.
Amer. J. Clin. Path. 78, 532. 1982.
90. Urbanski, S. J., Bilbao, J. M., Horvath, E., Kovacs, K., So, W., Ward, J. V.
Intrasellar solitary plasmocytoma terminating in multiple myeloma. A report of a case including electromicroscopical study.
Surg. Neurol. 14, 233. 1980.
91. Victor, M., Banker, B. Q., Adams, R. D.
The neuropathy of multiple myeloma.
J. N. N. P. 21, 73. 1958.
92. Waldenström, J. G., Adner, A., Gydell, K., Zettervall, O.
Osteosclerotic „plasmacytoma" polyneuropathy, hypertrichosis and diabetes.
Acta Med. Scand. 203, 297. 1978.
93. Weiner, L. P., Anderson, P. N.
Cerebral plasmacytoma with myeloma protein in CSF.
Neurology. 16, 615. 1966.
94. Whittauer, J. A., Tuddenham, E. G. D., Bradley, J.
Hyperviscosity syndrome in IgA multiple myeloma.
Lancet. ii, 572. 1973.

# XIV. Chromosome studies in monoclonal gammopathies

Györgyi Bujdosó and János Jákó

The chromosome aberrations in the background of the plasma cell tumors have been studied for an almost 30 year period (2, 7, 8, 11, 16, 17, 20, 33, 35, 36, 41, 42, 43, 48, 56).

Though the publications accept the existence of aberrations, no correlation could be observed between pathologic plasma cell proliferation, the character of the monoclonal immunoglobins, or other clinical symptoms. The studies are also hampered by technical difficulties, since a reproducable medulloculture and a proper karyogram could not be obtained for a long time (16, 58). This obstacle has recently been overcome, so cultures and chromosomes can be easily identified by the different methods of strip identification.

The chromosome anomalies are not specific, the identification of the so-called marker chromosomas (large chromosomes, like the chromosome 2 and 3, or acrocentric ones) is not always possible, but if it is, the results are informative.The changes in the chromosomes of groups C and G are frequently found. In monozygote male twins an extra chromosome of the similar size as the Group G one was identified (46). In neoplastic cells the occurrence of chromosomes exceeding karyotype, as well as aberrations is not surprising (1, 10, 34, 37, 53, 54), nevertheless their specificity has to be questioned. Moreover, some pathological changes can not be always detected. Philadelphia chromosome is of course an exception (18, 27, 50, 61).

The $V_L$ and $C_L$ genes responsible for the synthesis of immunoglobulins are localized on the chromosome 22 (23, 61). IgA formation is bound to chromosome 18. (See chapter on the heavy chain disease.) The site of the kappa light chains synthesis is not yet defined. Though in 1983 the examination of a child with lymphoid leukemia revealed a translocation which might be made responsible for the monoclonal formation of the kappa light chain and the attached heavy chain occurring in plasma (52). It'san evidence that abnormalities involve far more chromosomes than would follow from the above description. These observation are authentic, but to explain them a complicated system should be assumed.

MM can be defined as an irregularity of an early haemopoietic primordial cell which can be revealed during the maturation of "B" clone cells (5). Therefore, the mature plasma cells do not continue to proliferate, in other words, the multitude of malignant cells is not due to the spontaneously repeated mitosis of the plasma cells, but are the products of the independent malignant primordial cells (9). It is, therefore, not surprising that certain malignant plasma cell populations can also carry early B, T, or myelomonocytic phenotype markers (22, 26). It should be emphasized that the abnormality of the pluripotent primordial cells is not more than a continuous pathologic response to the otherwise normal antigen stimulus. Both of the pathologic continuity of the antigen presentation or the antigen itself can act as a trigger mechanism. The decision, which alternative holds are true is not easy and can not be reached with the methods of classical haematology. The cytogenetic attempts cannot be successful, as most of the myelomatoid elements are the finally differentiated B cells which are not sensitive to the applied mitogens. Aneuploidity and the chaotic karyotype changes still make difficult myeloma multiplex to define, comparing to other haematologic diseases. Cytogenetic analysis involving quantitative study of DNA showed that in 80% of all the plasmocytoma patients aneuploid, in 70% hyperdiploid, in 20% diploid and in 10% hypodiploid anomaly were present (3, 40). The latter one suggests the lack of one of the sexual chromosomes. Hypodiploid DNA is often accompanied by LCD and resistance to standard therapy (25, 51). A surprising result was obtained in the study of the tumour cells carrying DNA aneoploidity, as despite of the fact that the membrane carries surface IgG-λ isotype, it expresses κ and λ light chains as well (4). The deletion of the long arm of chromosome 6 indicates extensive bone lesions (21). Multidrug resistance (MDR) is found to be due to a break in the long arm of chromosome 7 (14) which is otherwise a site of glycoprotein p (6). Furthermore, broken chromosome 11 q, generally the involvement chromosomes 5, 12, 13, 16 and 17, as well as the formation of 6q 7q marker chromosomes are reported. Irregularity of the 6p short arm are frequently observed. Partial trisomia of the short arm of the chromosome 4 and the anomalies of the chromosomes 11 and 17 are also reported. The study of the carefully drawn maps of the chromosomal breaking points and their possibility has focused the attention on the anomalies of the short arms of the chromosomes 1 and 11 as important factors since the Kristen-ras and Harvey-ras genes are localized there. Furthermore, with the use of pan-ras and H-ras monoclonal antibodies cell-sorter could reveal high titre of p 21 expression in 17 of 23 myeloma patients (59). The inverse correlation between the q 21 levels and the 11th chromosome trisomia suggest that suppressor genes can be localized on the chromosome 11 (29). At the same time the elevated q 21 level might signal the participation of the oncogenic H-ras into the pathogenesis of myeloma (59). Chromosomal translocations are probably due to the errors of the recombinant rearrangement of the immunoglobulin genes causing the activation of the oncogens (38). Consequently, the cause of the self-maintaining inevitably progressive overproduction of the plasma cells is the transformation of the normal plasma cells into abnormal ones, which can be induced by an error of the gene rearrangement (mutation) or by the lack of the effect of the suppressor gene the scavenging function of the (anti-oncogens) (29). We have studied 7 IMG and 17 OMG (untreated) patients. Bone marrow gained by the sternum punction simultaneously performed with taking a venous blood was used for the cultures. The chromosomes were gained with and without the PHA stimulation of the nucleated cells, of the blood as well as from the bone marrow using a direct method (44, 55, 58). The karyograms were composed from preparations of the quality shown in Fig. XIV/1, XIV/2 and XIV/3.

Despite a substantial amount of observations indicating various chromosome aberrations in monoclonal gammopathies, our experience strengthen (Table XIV/I) the general relationship between carcinomas as malignant

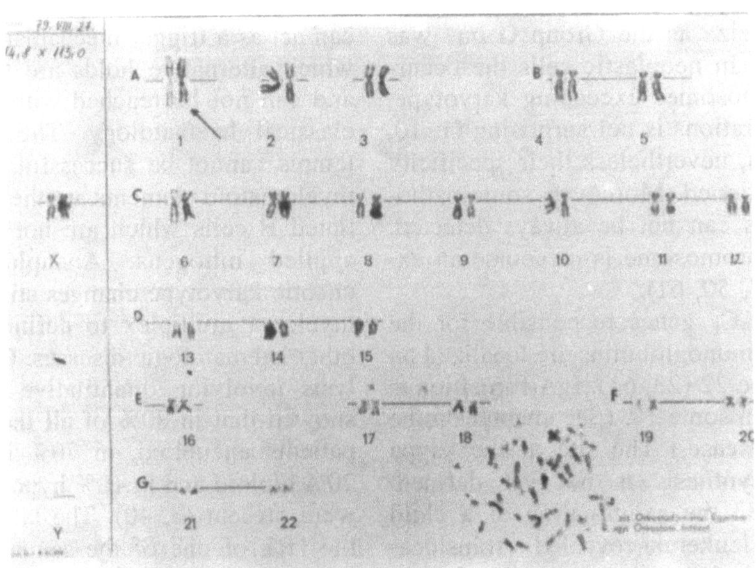


**2n = 46 xx/1q terminal G-stripe, bone marrow**

**Fig. XIV/1**

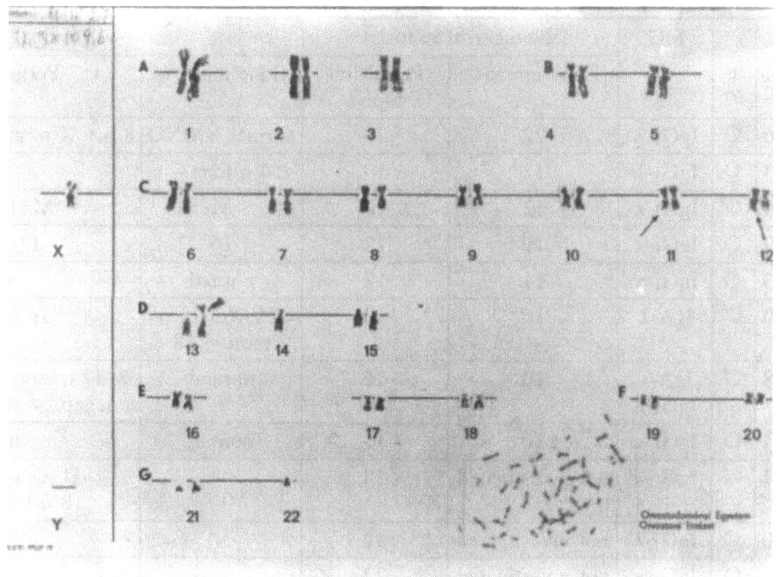

2n = 45 xo/minute G-stripe, blood

Fig. XIV/2

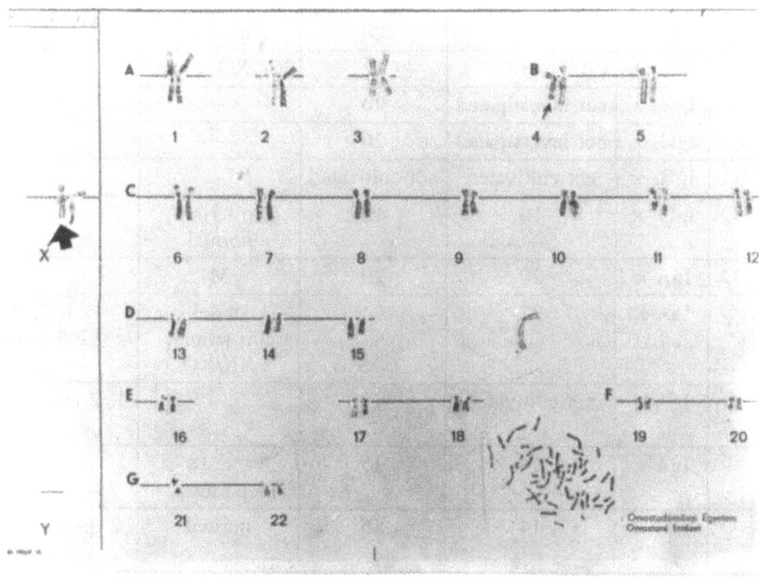

2n = 47 xxi/ + M4q terminal G-stripe, blood

Fig. XIV/3

diseases and the breakpoints of chromosomes is summarized in Table XIV/II (30).

As a final conclusion, two mutual features can be observed in the publications in this field.

First: in the examined preparations aberrations occur only sporadically, the karyogram of cells of cultures prepared from peripheral blood or from the medulla seldom exhibit any differences. The reason of the higher ratio of the normal cells in the studied samples: only a smaller percentage of the myeloma population can be identified (24, 45) Furthermore, it is also true that only a small percentage of the pathologic cells are involved in the synthesis of DNA and conse-

| | Name | Age | Sex | MG | Number of mitosis | | Results | |
|---|---|---|---|---|---|---|---|---|
| | | | | | Bone morrow | Peripheral blood | Bone morrow | Peripheral blood |
| 1. | H. I. | 56 | ♀ | IgG1-λ | 22 | 26 | normal XX/XO | 8 p.+ X term.-, 18, 16 q term.- |
| 2. | B. E. | 83 | ♀ | IgG1-λ | 12 | 19 | 4 q term.- | M |
| 3. | M. F. | 80 | ♂ | IgG1-κ | 12 | 15 | M | M+11 q term.- |
| 4. | L. L. | 89 | ♀ | IgG1-κ | 10 | 12 | 16 | 18-, XX/XO |
| 5. | L. J. | 83 | ♀ | IgG1-κ | 11 | 12 | normal | normal |
| 6. | K. S. | 70 | ♂ | IgA-λ | 10 | 10 | XY/XO 18 p term.-, 18 | 1p term.-, 16- |
| 7 | L. J. | 78 | ♂ | IgA-κ IgG1-λ | 10 | 16 | normal | 2, 12 q term.-, 7 g term. + 18-, 1 gap., 4 chrom. bridge, M |
| 8. | B. P. | 44 | ♂ | IgG1-κ | 10 | 11 | normal | minute's |
| 9. | K. E. | 57 | ♂ | IgG1-κ | not investigated | 18 | - | D, G very large sat., 18 q term.-, minute's |
| 10. | M. J. | 70 | ♀ | IgG1-λ | not investigated | 12 | - | 18 |
| 11. | O. K. | 64 | ♂ | IgG1-κ | not investigated | 23 | - | normal |
| 12. | S. GY. | 53 | ♀ | IgG1-κ | not evaluable | 15 | - | 2, 11, 14 q term.-, 3q term.+, 17 v 18- |
| 13. | M. P. | 59 | ♀ | IgG1-λ | 20 | 15 | normal | 8, 11 q term.-, 1 translocatios cross |
| 14. | S. I. | 63 | ♀ | IgG1-κ | 15 | 20 | 5 q term.- | normal |
| 15. | ZS. M. | 46 | ♂ | IgG2-λ | 13 | 12 | XY/XO 8 gap. | 11 gap. |
| 16. | B. A. | 52 | ♀ | IgG3-λ | not investigated | 20 | - | 8 pq term.-, 16 del., M |
| 17. | F. E. | 48 | ♀ | IgG3-λ | not investigated | 26 | - | normal |
| 18. | T. J. | 65 | ♂ | IgG4-κ | not cultivated | not cultivated | - | - |
| 19. | SZ. T. | 52 | ♀ | IgG4-κ | 10 | 12 | 2 p term.+, normal | normal |
| 20. | W. J. | 64 | ♀ | IgA-κ | 20 | 20 | M | minute's |
| 21. | N. GY. | 72 | ♀ | IgA-λ | 22 | 34 | 18 del, 8 q term.-, XX/XO | 1, 5, 8, 18 q term.-, 2 q term.+6 pq term.-, XX/XO, 13 gap |
| 22. | S. F. | 63 | ♂ | IgA-κ | not cultivated | 18 | - | 1, 2 term.-, 3 q term.+, 1 del., 5 chrom. bridge. |
| 23. | V. L. | 52 | ♂ | IgA-κ | 14 | 15 | 17 v. 18-, 18 pq term.- | 4 q term. |
| 24. | SZ. J. | 56 | ♀ | IgD-λ | 14 | 15 | normal | 2 q term.-, 7 q term.+, minute's |

**Table XIV/I**

quently, a low incidence of mitosis is usually insufficient for the determination of karyotype abnormalities (19, 25).

Second: Multiplex and general pathologic form of manifestation in contrast to the specific Philadelphia chromosome further complicates the situation in myeloid leukaemia.

So far Philadelphia chromosome was found in four cases of multiplex myeloma (60).

A higher incidence of acute leukaemia or solid tumour is observed following Alkeran therapy (39). The damaging effect of the alkylating agent on the chromosomes is assumed in the pathogenesis of this process. Cytogenetic studies have shown that the aneuploid alterations of the chromosomes in leukaemia complicated with multiple myeloma occur more frequently in cases of "de novo" leukaemias (31). On the other hand in several cases when no cytostatic therapy had been

| Chrom.: | Break-points: | Tumor: | Notes: |
|---|---|---|---|
| 1. | -q25 | ANNL-M$_3$<br>Burkitt lymphoma | t (1; 17)<br>duplicatio, triplicatio |
| 2. | -q11 | CML | |
| 5. | q31 | Preleukaemia, other malignities | del (5) (q12; q31) |
| 6. | q26 | Non-Hodgkin lymphoma<br>ALL | del (6) (q21; q term) |
| 7. | q22 | ANLL-M$_1$, M$_2$, M$_{4\text{-}6}$<br>Immunoblastic lymphoma<br>Preleukaemia + other malignity | del (7) (q22; q26)<br>t (5; 7) (q33; q22)<br>del (7) (q22; q term) |
| 9. | p21 | ANLL-M$_5$<br>(Acute monoblastic leukaemia) | t (9; 11) (p21; q23) |
| 11. | p13<br>q13<br><br><br><br><br>q23 | Wilms tumor + aniridia<br>Congenital ALL<br>CLL<br>B-cellular lymphoproliferative disease<br>ANLL-M$_4$-M$_5$<br>ANLL-M$_4$<br>ANLL-M$_5$ (acute monocyte leukaemia)<br>ANLL-M$_5$ | del (11) (p13)<br><br><br>t (11; 14) (q13; q32)<br>del (11) (q13; q25)<br>t (4; 11) (21; q23)<br>t (9; 11) (p22; q23)<br>t (11; 19) (q23; p13) |
| 12. | q13 | Diffuse mixed T-type lymphoma<br>CLL + lymphoma | t (12; 14) (q13; q32)<br>Intrachromosomc rearrangement |
| 14. | | Blood syndrome | t (14, 14) |
| 17. | p12 | CML, ANLL, malignant lymphoma, other malignitis, multiple myeloma | Isodia (17) (p12) |
| 18. | q21 | Follicular histiocyte lymphoma | t (14; 18) (q32.3; q21.3) |
| 20. | p11 | CML | Complex Philadelphia translocatio |
| 22. | q12 | CML<br>ANLL-M$_1$<br>ALL-L$_1$, L$_2$<br>Burkitt lymphoma | t (9; 22) (q34; q12)<br><br><br>t (8; 22) (q24; q11) |
| X. | p22 | CML<br><br>Malignant non-Hodgkin lymphoma<br>diffuse mixed lymphocyte-histiocyte form. | t (X; 22) unusal<br>Philadelphia translocatio<br>t (X; 1q) |

**Table XIV/II**

used, leukaemia nevertheless developed in addition to plasmacytoma (12, 57).

It seems that the treatment alone causes no chromosomal change in the already developed disease.

It must, however, be emphasised that it was not possible to observe any correlation between the quality of the paraprotein, the type of the disease and the abnormality of the chromosome so far (13). However, certain controversy exists in the literature (15, 18, 32, 49). These papers were published between 1967 and 1969 and later these data have not been supported probably because the stripe technique has not yet been introduced, so it was impossible to get adequarely reroducible conclusions.

We can, therefore, state that there is no agreement between the data published by different laboratories and thus no definite conclusions can be made. Though in our material the lack of one sexual chromosome as well as the involvement of 11th and 12th chromosomes were characteristic findings, supporting the data of the literature (37). Still, only the occurrance of pathologic chromosomal changes in certain plasma cell tumours is proved. Recently some important results were obtained in this field too.

| | Abberation | Literature |
|---|---|---|
| Bence Jones in the presence if pritein | Hypodiploidity | 22 |
| Primery drug resistance (cytostaticum) | 7q- | 12, 46 |
| IgA MG | + (8; 14) (q24; q32) | 22 |
| Osteolysis Ap21 (H-ras) level | 6q- | 19 |
| Decrease | Trisomia 11 | 54 |

**Table XIV/III**

It has been found that the repeated examinations of the patients and the parallel detection of their karyotype changes are suitable for the follow-up studies: the secondary aberrations indicate the progression of the disease, in CML there is a danger of transformation, and in MDS leukaemic transformation can happen (47).

It was found that the chromatine particles found in the tumour cells do not form (as we have learned from the earlier literature) a uniform group, and other central cells besides the acentric "m", can contain a centromere. For example the Robertson chromosomes (in mice) can be formed not only by centromere fusion, but also by a real translocation when one of the centromeres disappears.

The new immunocytogenetic methods of chromosome and nucleus analysis will reform the existing views (28), since it can localize certain macromolecules in situ (for example proteins, protein-DNA, and protein-RNA complexes).

In summary: the recent results provide a deeper insight into the mechanisms of the chromosome changes occurring in the plasma cell tumours, as shown in our table (Table XIV/III.) So the primary and the gradually developing karyotype anomalies can be identified and this prognostic information might be quite valuable. We already know that DNA hypodiploid patients are resistant to both melphalan + Prednisolone and VAD. As everywhere else the choice of therapy is the key question.

**References**


1. Baikie, A. G., Court, Brown, W. M., Jacobs, P. A., Milne, J. S.
Chromosome studies in human leukaemia.
Lancet. ii, 425. 1959.
2. Balogh, E., Kaszás, I., Krizsa, F., Fleischmann, T.
Chromosoma vizsgálatok myeloma multiplexben.
Orv. Hetil. 123, 35. 1982.
3. Barlogie, B., Latreille, J., Swartzengruber, D., Smallwood, L., Maddox, A., Raber, M., Drewinko, B., Alexanian, R.
Quantitative cytology in myeloma research.
In: Clinincs in Haematology.
Ed: by Salmon, S. E.
Saunders. London 19. 1982.
4. Barlogie, B., Alexanian, R., Pershouse, M., Smallwood, L., Smith, L.
Cytoplasmic immunoglobulin content in multiple myeloma.
J. Clin. Invest. 76, 765. 1985.
5. Barlogie, B., Epstein, J., Selvanayagam, P., Alexanian, R.
Plasma cell myeloma. New biological insights and advances in therapy.
Blood. 73, 865. 1989.
6. Bell, D. R., Trent, J. M., Willard, H. F., Riordan, J. R., Ling, V.
Chromosomal location of human p-glycoprotein sequences.
Cancer Genet Cytogenet. 25, 141. 1987.
7. Bottura, C., Ferrari, I., Veiga, A. A.
Chromosome abnormalities in Waldenström's macroglobulinaemia.
Lancet. i, 1170. 1961.
8. Bottura, C.
Chromosome abnormalities in multiple myeloma.
Acta Haematol. 30, 274. 1963.
9. Caligaris-Cappio, F., Bergui, L., Tesio, L.
Identification of malignant plasma cell precursors in the bone marrow of multiple myeloma.
J. Clin. Invest. 76, 1243. 1985.
10. Carbonell, F., Benitez, J., Prieto, F., Badia, L., Sanchez-Fayos, J.
Chromosome banding patterns in patients with chronic myelocytic leukemia.
Cancer Genet. Cytogenet. 7, 287. 1982.
11. Castoldi, G. L., Ricci, N., Punturieri, E., Bosi, L.
Chromosomal imbalance in plasmacytoma.
Lancet. i, 829. 1963.
12. Cleary, B., Binder, A., Kales, A. N., Weltry, B. J.
Simultaneous presentation of acute myelomonocytic leukemia and multiple myeloma.
Cancer. 41, 1381. 1978.
13. Coltman, M. C. A.
Multiple myeloma without a paraprotein.
Arch. Intern. Med. 120, 687. 1967.
14. Dalton, W. S., Dure, B. G. M., Alberts, D. S., Gerlach, J. H., Cress, A. E.
Characterization of a new drug-resistant human myeloma cell line that expresses p-glycoprotein.
Cancer Res. 46, 5125. 1986.
15. Dammacco, F., Trizio, D., Bonomo, L.
A case of IgAK myelomatosis with two urinary Bence Jones proteins (BJK and BJL) and multiple chromosomal abnormalities.
Acta Haematol. 41, 309. 1969.
16. Dartnall, J. A., Mundy, G. R., Baikie, A. G.
Cytogenetic studies in myeloma.
Blood. 42, 229. 1973.
17. Das, K. C., Aikat, B. K.
Chromosomal abnormalities in multiple myeloma.
Blood. 30, 738. 1967.
18. De Grouchy, J., de Nava, C., Bilski-Pasquier, G., Zitton, R., Bernadou, A.
Endoréduplication sélective d'un chromosome supernuméraire dans un cas de myélome multiple (Maladie de Kahler).
Ann. Genet. 10, 43. 1967.

19. DeWald, G. W., Kyle, R. A., Hicks, G. A., Greipp, P. R.
The clinical significance of cytogenetic studies in 100 patients with multiple myeloma, plasma cell leukemia or amyloidosis.
Blood. 66, 380. 1985.

20. Dubrova, C. E., Dycin, V. P., Ushkova, E. A.
The cytogenetic and clinico-hematological changes in myeloma disease.
Cytologica. 8, 241. 1966.

21. Durie, B., Baum, V., Vela, E., Mundy, G.
Abnormalities of chromosome 6q and osteoclast activating factor production in multiple myeloma.
Blood. 68, 208. 1986.

22. Epstein, J., Barlogie, B., Katzmann, J., Alexanian, R.
Phenotypic heterogeneity in aneuploid multiple myeloma indicates pre-B cell involvement.
Blood. 71, 861. 1988.

23. Erikson, J., Martinis, J., Crose, C. M.
Assignment of the genes for human immunoglobulin chains to chromosome 22.
Nature. 294, 173. 1981.

24. Ginsberg, D. M.
Circulating plasma cells in multiple myeloma: A method for detection and review of the problem.
Ann. Intern. Med. 57, 843. 1962.

25. Gould, J., Alexanian, R., Goodacre, A., Pathak, S., Hecht, B., Barlogie, B.
Plasma cell karyotype in multiple myeloma.
Blood. 71, 453. 1988.

26. Grogan, T. M., Durie, B. G., Lomen, C.
Delineation of a novel pre-B cell component in plasma cell myeloma: Immunochemical, immunophenotypic, genotypic, cytologic, cell culture and kinetic features.
Blood. 70, 932. 1987.

27. Gropp, A.
Karyogramm und Krebs.
Selecta. 51, 4740. 1982.

28. Haaf, T.
Immunzytogenetik: Neue Wege zur strukturellen und funktionellen Analyse von Chromosom und Zellkern.
Med. Gen. 2/3, 9. 1990.

29. Harris, H.
The Analysis of malignancy by cell fusion: The position in 1988. Perspective in Cancer Research.
Cancer Res. 48, 3302. 1988.

30. Hecht, F., Sutherland, G. R.
Fragile sites and cancer breakpoints.
Cancer Genet. Cytogenet. 12, 179. 1984.

31. Hossfeld, D. K., Holland, J. F., Cooper, R. G., Ellison, R. R.
Chromosomal studies in acute leukemia a developint in Multiple Myeloma.
Cancer Res. 35, 2803. 1971.

32. Houston, E. W., Ritzmann, S. E., Levin, W. C.
Chromosome aberrations common to three types of monoclonal gammopathy.
Blood. 29, 214. 1967.

33. Itani, S., Hoshino, T., Kawasaki, S., Nakayama, S.
Chromosome abnormality and its significance in human multiple myeloma.
Acta Haematol. 33, 54. 1970.

34. Jacobs, P. A., Tough, I. M., Wright, D. H.
Cytogenetic studies in Burkitt's lymphoma.
Lancet. ii, 1144. 1963.

35. Kanzow, U., Lange, B., Niederalt, G., Groop, A.
Chromosomenuntersuchungen bei Paraproteinämien.
Klin. Wschr. 45, 1065. 1967.

36. Karpas, A., Fischer, P., Swirsky, D.
Human plasmacytoma with an unusual karyotype growing in vitro and producing light-chain immunoglobulin.
Lancet. i. 931. 1982.

37. Klein, G.
Lymphoma development in mice and human: diversity of initiation is followed by convergent cytogenetic.
Proc. Natl. Acad. Sci. 76, 2442. 1979.

38. Klein, G., Klein, E.
Myc/Ig juxtaposition by chromosomal translocation: Some new insights, puzzles and paradoxes.
Immunol. Today. 6, 208. 1985.

39. Kyle, R. A., Pierre, R. V., Bayrd, E. D.
Multiple myeloma and acute leukamia associated with alkylated agents.
Arch. Intern. Med. 135, 185. 1975.

40. Latreille, J., Barlogie, B., Gohde, W., Johnston, D., Drewinko, B., Alexanian, R.
Cellular DNA content as a marker of human multiple myeloma.
Blood. 55, 403. 1980.

41. Lewis, F. J. W., MacTaggart, M., Crow, R. S., Wills, M. R.
Chromosomal abnormalities in multiple myeloma.
Lancet. i, 829. 1963.

42. Lewis, F. J. W., Fraser, I. L., MacTaggert, M.
An abnormal chromosomal pattern in myelomatosis.
Lancet. ii, 1013. 1963.

43. Manicienelli, S., Durant, J. R., Hammack, W. J.
Cytogenetic abnormalities in a plasmacytoma.
Blood. 33, 225. 1969.

44. Moorhead, P. S., Nowell, P. C., Mellman, W. J., Battips, D. M., Hungerford, D. A.
Chromosome preparations of leukocytes cultured from human peripheral blood.
Exp. Cell. Res. 20, 613. 1960.

45. Morissette, L., Watkins, C. H.
Multiple myeloma: Diagnostic value of the blood smear.
Proc. Staff. Meet. Mayo Clin. 17, 433. 1942.

46. Ogawa, M., Wurster, H. D., McIntyre, R. O.
Multiple myeloma in one of a pair of monozygotic twins.
Acta Haemat. 44, 295. 1970.

47. Oláh, É.
Primer és szekunder kromoszóma rendellenességek rosszindulatú megbetegedésekben.
Doktori értekezés, Debrecen. 1989.

48. Siebner, H., Spengler, C. A., Bütler, R., Heni, F., Riva, G.
Chromosomenanomalien bei Paraproteinämie.
Schweiz med. Wschr. 95, 1967. 1965.

49. Siebner, H., Aly, F. W., Braun, H. H.
Chromosomenbefund bei D-Plasmozytom.
Klin. Wschr. 47, 884. 1969.

50. Smadja, N., James, J., Krulik, M., Zittoun, R., Debray, J.
Chronic myelogenous leukemia with a Philadelphia chromosome resulting from a complex translocation (2, 9, 22), following an undifferentiated acute leukemia.
Cancer Genet. Cytogenet. 8, 1. 1983.

51. Smith, L., Barlogie, B., Alexanian, R.
Biclonal and hypodiploid multiple myeloma.
Am. J. Med. 80, 841. 1986.

52. Sonnier, J. A., Buchanan, G. R., Howard-Peebles, P. H., Rutledge, J., Graham Smith, R.
Chromosomal translocation involving the immunoglobulin Kappa-chain and heavy-chain loci in a child with chronic lymphocytic leukemia.
N. Engl. J. Med. 309, 590. 1983.

53. Spiers, A. S. D., Baikie, A. G.
A special role of the group 17, 18 chromosomes in reticuloendothelial neoplasia.
Br. J. Cancer. 24, 77. 1970.

54. Spiers, A. S. D., Baikie, A. G.
Anomalies of the small acrocentric chromosomes in human tumour cells.
Aust. NZ. J. Med. 21, 188. 1972.

55. Sumner, A. T., Evans, H. J., Buckland, R.
New technique for distinguishing between human chromosomes.
Nature New Biology. 31, 232. 1971.

56. Tassoni, E. M., Durant, J. R., Becker, S., Kravitz, B.
Cytogenetic studies in multiple myeloma. A study of fourteen cases.
Cancer Res. 27, 806. 1967.

57. Thiagarajan, P.
Simultaneous presentation of multiple myeloma and acute leukemia in the absence of previous chemotherapy–report of a case.
The Mount Sinai Journal of Medicine. 46, 360. 1979.
58. Tjio, J. H., Whang, J.
Chromosome preparations of bone marrow cells without prior in vitro culture or in vivo colchicine administration.
Stain Technol. 37, 17. 1962.
59. Tsuchiya, H., Epstein, J., Selvanayagam, P., Dedman, J., Gallick, G., Alexanian, R., Barlogie, B.
Correlated flow cytometric analysis of H-ras p21 and nuclear DNA in multiple myeloma.
Blood. 72, 796. 1988.
60. Van der Berghe, H., Louwagle, A., Broeckaert-Van Orshoven, A.
Philadelphia chromosome in human multiple myeloma.
J. Natl. Cancer Inst. 63, 11. 1979.
61. Van Loghem, E.
Genetics and diagnostics of immunoglobulin subclasses.
International Congress ISH-IS BT. Budapest, 1982.

# XV. Therapy of the plasma cell tumors

Human mind can not keep up with the dynamic development of medical science. The awareness of insufficient knowledge is directly or indirectly experienced by physicians during their every-day work with the patients.

Physicians encounter only certain aspects of human life. Instead of stimulation, the increasing amount of information in the professional literature rather elicits a feeling of suffocation and useless efforts, hence the impossibile turns into helplessness. Instead of gaining a "horizontal" outlook, specialists who do not devote their entire time to their profession try to read "vertically". Such narrow specialization induces an avalanche which could ruin the deep professional outlook required for prominent specialists.

Moreover, the "vertical" approach is split further into smaller fields; clear and objective understanding even in these fields - for example in the area of therapy - is hampered by divergent opinions of professionals directly influenced by the pharmacological industry. Individual experience is of value only if ample statistical data are obtained. In a small country, howerer, these are futile hopes.

In 1900 the average life span in the USA was 47 years; in 1973 it reached 73 years, and by 2010 this value will be 82.4 years (85, 230). Such an increasing tendency is mainly attributed to social measures. In Hungary, however, the average life span does not show such extremely high values. Still the "ageing" observed in the civilized countries is worth to be dealt with; the ratio of elderly people over 65 years comprises about 10% of the population. The diseases, including the plasma cell malignant tumors, mainly occur in this group and rarely develop in the third-fourth decade. The dosage of drugs is usually expressed per kg-body weight or per unit of body surface, absolutely disregarding the age of the patient. The latter is, however, of importance in both cytotoxic and supporting therapy.

The aged are more susceptible to drug therapy partly because receptors incolved in the action and the side effects of drugs, become more sensitive, leading to an altered reaction of the organism (59, 179). At the same time, in elderly people plasma concentration of drugs is also altered due to changes in their absorption (in oral administration), distribution, metabolism and the excretion. These mechanisms greatly contribute to the individual differences in the expected therapeutical effect (59, 100). The pharmacodynamic (at the receptor site) and pharmacokinetic (due to drug distribution) regulation are responsible for the enhanced reaction (58). The pharmacodynamic changes are the sequelae of a modified receptor function due to the altered function of the organelles at subcellular level. A typical pharmacokinetic sign is the decreased albumin level with subsequent lower carrier protein capacity, due to the decreased renal (between 20 and 90 years 50%); cardiac (after 30 years 1% every year) and liver (after 25 years 1% every year) function (145, 156).

The serum creatinin level is not an accurate index of an impaired renal function as the ageing organism reduces its production.

Drugs are excreted from the organism via three different pathways;

1. by the kidney in an unchanged form,
2. metabolized by the liver, thus the effect of chemotherapeutic agents can be activated or inactivated and the metabolites are excreted with the urine or,
3. are excreted with the bile and feces.

The first phase of the metabolic activity of the liver (oxydation-reduction or hydroxylation and demethylation) alters with age (158), but the second phase (conjugation-acetylation, formation of glucuronides or sulphates) remains unchanged (102, 157).

The above data indicate that due to the decreased metabolic activity of the liver, drug levels in the blood might be higher than in young patients. This is worth pointing out, because the absorption rate shows no changes with age.

Furthermore, it is important to emphasize that elderly patients are usually hypoacidic.

The distribution of drugs depends on the concentration and functional capacity of carrier proteins; the ratio between the extracellular, intravasal and intracellular spaces and that of the aqueous and lipid phases; and the competitive relationship between different drugs if taken simultaneously.

The above mentioned general considerations, which help to understand the present chapter provide certain guidelines for therapy of the elderly. They hold true not only for the cytotoxic but also for the supporting therapeutical modalities (such as glucocorticoids, $H_2$ receptor blocking agents, nonsteroid antiinflammatory drugs and tranquillizers).

Though levels of ACTH and endogenous glycocorticoids show no marked changes with age, their elimination is reduced and the sensitivity of the adrenals to ACTH increases (12, 13). Hence in long-lasting therapy the side-effects might be more pronounced. For example if used in plasmocytoma therapy, these hormones impair glucose tolerance, suppress the immune response and promote osteoporosis (30). In young patients, 30% of cimetidine or ranitidine is metabolized, and 70% excreted in unchanged form. In the elderly patients, only 10–15% of these drugs are metabolized and their clearance is decreased (106). Therefore, doses of these drugs should be carefully selected, all the more, they can bind and inhibit the $H_2$ receptors of T suppressor cells, leading to a relative predominance of helper cells and a consequent enhancement of paraprotein production.

The non-steroid antiinflammatory drugs are excreted in the urine more slowly, therefore, their half-life period is longer and the occurrence of side effects (headaches, hypertension, nausea, diarhea, and intestinal bleeding) higher (389).

During the inevitable administration of morphine derivatives during the terminal stage of the disease, the increased sensitivity of the elderly to opiates should also be considered (40).

Before going into the details of therapy of plasmocytomas, an overview on the cytokines and especially interferon, is presented.

## Interferons and clinical practice

Clinicians have recognized the importance of active molecules which regulate the homeostasis. Alteration of concentrations and ratios of these molecules is used for therapeutic purposes not only in rare diseases, but also in commonly occurring entities. After the discovery of these intrinsic regulators, laboratory experiments were soon replaced by industrial production of these biologically active molecules, some of which are now used in the clinical practice.

These biologically active molecules are referred to as cytokines. Cytokines are protein or polypeptide cellular products which are produced either constitutively or as a result of a certain signal. Subsequently, these highly active molecules bind to high affinity membrane receptors or subcellular stuctures.

Interferons (IFN) can be regarded as one of the subclasses of cytokines. They do not function alone, but require complex cytokine networks. This recognition also means that the cytokine-producing cells neither do function alone and products of one cell type would regulate the activity of the other cell populations. Though all cytokines possess individual biological activity, the same molecule can act as a synergist, an antagonist or a modulator in different cell systems. Tissue culture experiments show that IFNs are capable of all three functions. The in vivo effect of IFNs is a result of an interaction between multiple effector mechanisms. The cytokine network operates as a thoroughly controlled system, which regulates and maintains a certain balance between the inflammatory, immunological, cell proliferative, and differentiation events.

Apart from the antigen presentation and antibody production, macrophages and B lymphocytes contribute to the homeostasis by secreting soluble cytokines (15). These mediators, however, are produced not only by normal cells, but also by malignant B lymphocytes. In contrast to immunoglobulin (monoclonal antibodies, paraprotein) production, which seems to play no role in the proliferation of malignant cells, cytokines can mediate the expression of the malignant phenotype.

Following stimulation by antigen, B lymphocytes undergo activation, proliferation and differentiation until they are transformed into plasma cells (209). The sequence, rate, and final result of this transformation is controlled by the cytokines that are produced mainly by T cells. Some of these mediators can also be secreted by B lymphocytes. Many cytokines are inactive on their own, but elicit their effect indirectly by "forcing" the cell to express the necessary membrane receptors. The effects of cytokines hierarchically bind to each other and regulate the synthesis of the next mediator molecule. This effect is elicited in spatial-temporal networks.

IL-1 exists at least in two molecular forms - alpha and beta - which are the products of two related genes located on the second chromosome (229, 412). IL-1-alpha predominantly binds to the membranes of cells that produce this molecule, while IL-1-beta is secreted into the intercellular space. Altough only 25% of the amino acid sequences of these two molecules is homologous they bind with equally high affinity to the surface receptors of target cells. (46). IL-1 induces the expression of the IL-2 gene in T cells when IL-1 receptors appear on the surface of IL-4 producing T helper cells (237). Apart from triggering B cell proliferation, receptor expression is a result of a cooperative interaction between IL-4, IL-2 and IL-5. IL-1 has other functions, for example, it stimulates fibroblasts and monocytes in a paracrine fashion to secrete IL-6 (258).

IL-2 is the key cytokine in the entire immune system. It was first discovered as a T-cell growth factor, although it also markedly promotes the proliferation and differentiation of B cells (193,290).

The IL-2 receptor, a basic molecule for different important processes, consists of a light and a heavy chain. IL-2 is produced by T helper lymphocytes; mainly the $T_H$-1 subpopulation (296), though such subdivision is debated. Some B cell malignomas were also found to secrete IL-2 under special conditions (29). This cytokine is active during the early maturation phase of B lymphocytes and is stimulated by IFN-gamma. It also stimulates the synthesis of TNF-alpha, which enhances the anti-tumor activity of macrophages (166, 251).

IL-3 is required for the proliferation and final differentiation of plasma cell precursors (189). Its cellular effect is elicited in cooperation with IL-6 (39). IL-3, on its own, promotes the expression of surface immunoglobulins of lymphocytes. IL-3 acts not only on B cells but also plays a part in pro-T cell transformation (369) and affects basophils and mast cells (39, 45a).

IL-4 is a protein with MW of 20 kD and is produced by a $T_H$-2 subpopulation of T helper lymphocytes (296). Its most important function is the stimulation of B lymphocytes to express homologous receptors on their surfaces (317). B cell differention, induction of the S phase and synthesis of other cytokines all depend on the presence of this molecule, therefore, IL-4 can be reagarded as a factor controlling the competence of resting cells. Furthermore, IL-4 induces and enhances expression of the MHC-II molecule which plays a crucial role in T-B cell collaboration and is necessary for the differentiation of IgE-producing B cells. The term "growth factor" indicates the effect of the cytokines on T cells. High doses of IL-4 also stimulate proliferation of B cells (387).

Under normal conditions, similar to IL-4, IL-5 is also secreted by the $T_H$-2 subpopulation. Earlier it has been termed TRF or BCGF-II. IL-5 can send mitogenic signals that initiate proliferation and differentiation of B cells, which develop high-affinity receptors on their surface (293). Moreover, it promotes differentiation of granulocytes and IgA-producing B cells. It is still not known, whether both effects occur simultaneously in the same cell or functionally different cells respond diverqently to stimuli and differentiate into new subpopulations (178).

IL-6 is a 26 kD protein, originally described as a product of B cells, is an autocrine differentiation factor of these cells (300). Data of recent studies indicate that the most important function of IL-6 is the differential splicing of Ig specific RNA that result in the amplification of secretion (357). Moreover, IL-6 can act as a growth factor in certain systems (45, 401), though it is important that this effect never occurs in normal B cell population (300).

Under certain conditions, IL-6 strongly inhibits cell proliferation (213, 259).

Apart from its immunological functions, similar to IL-1, IL-6 plays an important part in the acute phase reaction and is also recognized to be a hepatocyte stimulating factor (HSF), (144). Furthermore, it increases the secretion of ACTH, corticosteroids and the tissue factor produced by endothelial cells.

Finally, TNF should also be mentioned. It exists in two forms; TNF-alpha and TNF-beta, the latter generally known as LT. These two partially homologous cytokines elicit their effect via the same receptor (1). Though this molecule is generally known as a macrophage-derived factor, TNF-alpha can also be produced by activated T-lymphocytes (89). TNF-alpha has a strong cytotoxic effect on tumors carrying TNF receptors.

Recent publications indicate that IL-6 promotes the growth of B cells (205). The above listed data suggest that cytotoxicity and stimulation of cell growth depend not only on receptor binding. Both IL-1 and TNF-alpha are active mediators of inflammatory reactions, including the inhibition of the lipoprotein lipase activity. The latter contributes to cachexia (43) characteristic of neoplastic and inflammatory diseases (302). TNF can also affect the hemostatic system and by involving another mechanism; it activates OAF (96) that increases bone resorption (42). The latter requires synthesis of a substantial amount of IL-1, like for example, in multiple myeloma (374).

In different malignant hematological diseases, due to the expression of TNF gene by tumor cells and assuming the existence of an autocrine mechanism, TNF maintains cell proliferation (77). It is assumed, that similar reactions develop in MM, however, the cytokine - receptor interactions take place on the membranes of the endoplasmatic reticulum and not on the cell membrane !

Endotoxins elevate circulating TNF levels (287), however the measured TNF level shows no correlation with its biological activity (129). The reason for this is that the molecule maintains its antigenic properties for a longer period as compared to its necrotizing function (269). Studies in this field have revealed that TNF is a rather unstable dimeric molecule which is split into monomers at 37 °C (298). The latter property might explain this virtual contradiction (108).

Though there is still no evidence regarding the direct role of cytokines in the pathomechanism of different diseases, their inadequate production can aggravate certain clinical conditious (101).

Proliferating malignant cells produce cytokines. Most frequently, these are growth factors or cofactors utilized by the cells in an autocrine fashion (372). Such processes never occur under normal conditions, as the cells possessing certain types of receptors would never secrete cytokines capable to interact with them, and vice versa. Only two exceptions are known; IL-2 (286) and fibroblast growth factor (FGF) (406) produced by endothelial cells. Moreover, normally the amount of the produced cytokines is thoroughly regulated and controlled these mechanisms turn to be insufficient in the neoplastic cells, so the high secretion rate of the cytokines will maintain malignancy. According to the recent theory concerning the role of the IL-6, this molecule plays a key role in the self-reproducing process in multiple myeloma. Surviving plasma cells in cell culture carry a substantial amount of the IL-6 surface receptors and, at the same time, high titers of the IL-6 secreted by the same plasma cells were detected in the medium.

As a result, uncontrolled cell proliferation occurs (203). A similar autocrine mechanism has been described for IL-5 with a myeloma cell line (211). IL-1 produced by plasma cells (84) is the most potent stimulator of the IL-6 gene transcription (258, 259). These two cytokines should functionally interact to stimulate the myeloma cell proliferation. It is, however, still not known, whether the IL-1 autocrine mechanism primarily induces the expression of the IL-6 cytokine, or its receptor, or both? It is also worth to point out that in normal B lymphocytes IL-6 exhibits no growth factor activity (300).

IFN-alpha decreases the sensitivity of the tumor cells to cytotoxic NK cells (256, 390, 419). At the same time, IL-2 alone, or with IFN as a synergist, is a potent activator of NK cells (397). The activation is a result of a specific binding to the surface receptors, which transmit an activating signal. The receptors involved proved to be identical to the Tac-antigens (low affinity receptor-alpha) (236); pretreatment with anti-Tac MoAb abolishes the activity of IL-2 or IFN-alpha.

Various IFN-alpha effects on the immune system have been described (128, 133, 361), but one of the most dramatic is the normalization of a markedly decreased NK function (127, 233, 342).

Compared to the combined effect of IFN and IL-2, the effect of IFN alone takes several months excluding the possibility of direct NK cell activation by IFN. Additional evidence favoring this view is the sustained NK activation for 2-3 months, following the cessation of IFN treatment (180). It is, however, noteworthy, that IFN-alpha has a direct cytostatic effect on neoplastic cells, though the underlying mechanism is not known.

In summary, observations regarding the participation of cytokines in malignant cell proliferation support the hypothesis that the insufficient surveillance of autocrine growth stimulation is one of the major determinants of neoplastic transformation. The role of the paracrine interaction should not be neglected; after a while the cultured tumor cells do not proliferate any more, probably due to the lack of the intact paracrine network system of the host (212, 322, 423).

The recent finding of a chromosome damage in the 9p 21-22 region (219) suggest the elimination of a suppressor gene (136). As IFN-alpha and beta genes also belong to this region (98), it can be assumed that the antiproliferative effect of IFN is attributed to its suppressor gene function (99).

Since the first publication on the effect of IFN on tumors (278), the phenomenon has been studied in considerable detail (254). Despite of this, only few data are available about the whole process. Changes in the plasma cell count and levels of monoclonal gammaglobulins give no insight into the mechanism as these changes can be multicausal. The exact amount of the administered IFN, its combination with other substances, the microenvironment, and its interactions with other cytokines still remains unknown. This means that the local effective concentration of the biologically active substance could not yet been assessed.

In the case of synergistic cytokines, two or more cytokines, which would be ineffective on their own, exert their effect in combinations. IFN-alpha or beta combined with interferon-gamma (352) exhibit antitumor synergism in vitro and in experimental animals in vivo (95). Though the biochemical background of the synergism is not yet understood, it is assumed that the simultaneous induction of production of these molecules might comprise the background of this observation (32). IFN-gamma contributes to the antiumor effect by potentiating the cytotoxic effect of TNF (20, 134). Due to the increased expression on the target cell level, more TNF receptors are produced (343) and more cytokines secreted (32). Therefore, it is hypothesized that a cascade mechanism is in effect in which one cytokine induces the production of the other and so forth. Hence, one would never know for sure that a given biological effect is directly elicited by the administered cytokine or is rather due to an indirect interaction of active substances expressed by the cytokine. For example, the inter- or intracellular presence of IFN-beta induced by a viral infection triggers the formation of TNF and IL-1, and, finally, the elimination of the infection (400). It is also well known that the increase of the IFN-alpha concentration elicits the production of IFN-gamma (187).

All interferons stimulate the expression of MHC I and MHC II. The strong interaction between IFN-gamma and TNF affects not only the hemopoietic, but also other cells, for example vascular endothelium. (377).

Antagonism has been observed between PDGF and interferons (116). It is assumed that IFN has a negative feedback effect on PDGF (420) or TNF. Still, based on the above data, it can not be explained by what mechanism TNF facilitates the growth of normal fibroblasts and why is it simultaneously cytotoxic to neoplastic cells.

Both the proliferative (growh promoting) and cytostatic effects of TNF are enhanced by an anti-IL-6 MoAb (213); at the same time, TNF stimulates the IL-6 production. This finding reflects not a complex process but rather assumes a simple negative feed back control.

IFN-alpha and beta increase CSF-1 stimulated IL-1 production by macrophages, but inhibit the differentiation and proliferation of macrophage precursors and progenitors (294, 332). Aside from promoting the immune reaction, IFN-gamma elicits a differentiating and antiproliferative effect, for example, by strongly abolishing the formation of myeloid colonies. In an in vivo mouse model IFN induces leukopenia, while simultaneous administration of CSF-1 abolishes this effect (216). Furthermore, antogonism is also observed between IFN-gamma and IL-4. Evidence exists that IL-4 stimulates MHC II expression and increases the number of IgE receptors (CD 23) on Burkitt lymphoma cells, this effect, however, is inhibited by minimal concentrations of IFN-gamma (340).

Mitotic activity of CLL cells is enhanced by the simultaneous administration of IL-4 and phorbol esters and can be abolished by IFN-gamma (148). Similarly, IFN-gamma inhibits the antiproliferative effect of IFN-alpha of human neuroblastoma cells in culture (147). Such antagonism can also be observed in macrophage cultures (422). It is assumed that the antagonism between the IFN-alpha, beta and gamma takes place at the receptor level.

## Interferons

Interferons are dealt with a special emphasis within the group of cytokines. Their full classification is presented below:

Interferon-alpha is a protein induced by viral infection. It has 25 subtypes while those prduced with recombinant techniques contain only one subtype. The effect of natural IFN-alpha is three times higher than that of the recombinant form.

The mechanism of action of IFN-alpha is summarized in Table XV/I. Multiple and divergent effects of IFN-alpha that are partly morphological and partly functional can be allocated into four groups:

I. Alteration of membrane structure,
II. Stimulation or inhibition of induction of protein synthesis,
III. Modification of the cytoskeleton,
IV. Induction of enzyme systems.

It is evident that not all mechanisms are activated simultaneously and the exact nature of the response depends on the type of IFN.

Another, so-called "priming" effect of IFN might be independent of those mentioned above (376). Pretreatment of a cell culture with IFN-alpha at a low final concentration with subsequent addition of an "inducer" (for example a virus) elicits a higher immune-IFN-alpha production as compared to untreated cultures (415).

The problem of low doses is especially relevant in therapy as they would rather stimulate than inhibit cell growth (240). A detailed analysis of an ALL case supports the above view (306).

It is also interesting, that pretreatment with IFN-alpha increases the susceptibility of patients to cytostatic therapy (254).

Though different recombinant interferons (IFN-alpha2, IFN-alpha2-A, IFN-alpha2-B, IFN-alpha2-C, IFN-alpha-C) are already commercially available, their effect is not different from each other (16).

## Interferon beta

IFN-beta is a glycoprotein that has two subtypes. It is produced by fibroblasts, and its production is induced by virus. In contrast to IFN-alpha, following parenteral administration interferon-beta is rapidly eliminated from the circulation (44, 122, 292) and is accumulated in the skin. Therefore, in some viral infections with cutaneous localization, local IFN-beta therapy might be beneficial (151, 353)

## Interferon-gamma

IFN-gamma is one of the cytokines that regulate processes of immune responses resulting from antigen stimuli. This glycoprotein is produced by thymus-dependent T-cells. It shows slight structural resemblance to IFN-alpha and beta. The molecule specifically binds to cell membrane receptors. Other cytokines do not compete for the same receptors with IFN-

| | IFN-α | IFN-β | IFN-γ |
|---|---|---|---|
| Antiviral effect | + + + | + + + | + |
| Mitogenic effect on different cell types | – | – | + |
| Cytostatic effect | depending on type of tumor + ( + + ) | + | ( + )? |
| Increases expression of TNF receptor | + + | + | – |
| Expression of MHC I | + | + | + + + |
| Expression MHC II | + | + | + |
| Increases expression of HLA-DR on keratinocytes | + + | + | – |
| Expression of IgG Fc receptor on myelo-monocytes | + + | + + | – |
| Induces differentiation of myelo-monocytes and B lymphocytes | – | – | + + |
| Decreases the number of transferrin receptors | + + | – | – |
| Increases cytotoxicity of T lymphocytes and NK cells | + | + | - ( + ) |
| Activates macrophages | + | – | + + + |
| Enhances phagocytosis | – | – | + + + |
| Enhances the ADCC reaction | + | + | – |
| Enchances activity of LAK cells | + | + | + |
| Induces expression of ICAM-1 | – | – | + |
| Incudes alterations of cytosceleton | + | – | – |
| Pyrogenic effect | + | + | + |
| Enchances synthesis of acute phase protein | – | – | + |
| Enchances chemiluminescence | + + | + | – |
| Increases CD4/CD8 ratio | + + | + | – |
| Induces new enzyme systems<br>- triglyceride level of cells increases<br>- activity of 2-SA synthase increases<br>- activity of alpha-naphthyl-acetate increases | + + | + | – |

**Table XV/1 (108, 181, 186, 238, 354, 355, 373, 390, 413, 414)**

gamma (391). Production of IFN-gamma is stimulated by IL-1 and IL-2. Its antiviral effect is much less pronounced than that of other interferons. Much more important is the immunomodulatory effect of IFN-gamma. IFN-gamma activates macrophages (391) that lead to their higher antitumor effectiveness. Another immunomodulatory effect is the induction of MHC-7 expression on both normal and malignant cells (104, 163).

Activated macrophages have higher activity also against microorganisms, especially against intracellular species (i.e. toxoplasmosis, leishmaniasis, rickettsiosis). This cytokine is applied in the therapy of chronic granulomatosis because it stimulates the formation of free redicals thus enhancing phagocytosis.

Apart from the already mentioned indirect or immunomodulatory antitumor effect of

IFN-gamma, it also elicits a direct antiproliferative effect, although its clinical relevance needs further investigation (163).

Other fields of its therapeutic use, though very important, have failed to show significant results (330, 341).

Both IFN-alpha and beta are immunogenic, while no report is available on the production of anti-IFN-gamma antibodies (142, 375, 398, 408). It is worth mentioning that the antigenity of ACTH, endorphins and IFN-alpha are similar (48, 49, 50, 51). Anti-IFN antibodies were identified in 1% of treated patients some untreated patients also produced autoantibodies (393). In the clinical practice, antibody production is suspected if the appetitie of the patient normalizes and the fatigue decreases. The ineffectiveness of IFN is not always due to the presence of the antibodies, though the methods of investigation are far from being perfect (194, 202).

Antibody formation against recombinant interferons is more often observed. According to early investigations its suggested incidence was about 10%, more recent data, however, indicate that this value is closer to 30% (372) or 60% (408).

A low dose of rembinant IFN is less immunogenic than a high dose (408). Subcutaneous application is the most immunizing, whereas administration through the intravenuous route is the least effective (215).

The formation of anti-IFN antibodies depends on;
- the nature of the IFN product used (biological, recombinant),
- the dose of IFN,
- the period of IFN treatment,
- the mode of administration,
- the genetic background of the patient.

## Clinical applications of interferon

Data from the literature indicate that, similar to other new methods, therapeutical indications are not yet clearly defined. Interferons have been used many times on an accidental or " it might be beneficial" basis rather than based on the available data derived from in vivo and in vitro results. Our retrospective, though not complete, study summarizes the most important clinical entities in which benefit of IFN treatment can be hypothesized.

***Lymph node diseases (tumors of the immune apparatus)***

Hodgkin's lymphoma
Non-Hodgkin's lymphoma (131, 132) including;
- T-cell lymphomas (66),
- HCL (183, 327),
- MM (80, 326, 330).

***Diseases of hemopoietic organs (171)***

CML (304, 305, 383, 384),
Thrombocytemia (149).

***Solid tumors***

Renal cell carcinoma (220, 301, 324, 328),
Melanoma (86, 87, 208),
Transitional cell carcinoma,
Osteosarcoma (378).

***Cutaneous diseases with skin lesions***

Psoriasis (47),
Henoch-Schönlein syndrome.

***Viral infections***

Herpes zooster,
Herpes simplex,
Papilloma virus,
- verucca vulgaris,
- primary verucca plana,
- juvenile larynx papilloma,

Hepatitis B virus carriers (94, 319),
AIDS,
Kaposi's sarcoma (97, 231).

In the above listed diseases, IFN-alpha is primarily recommended as there are no sufficient data on the effect of IFN-beta and gamma.

In summary, despite the experience and the experimental evidence of several decades, the four most important questions concerning the application of IFN-s still remain unanswered; 1. the proper dose, 2. the indication of therapy, 3. the mode of administration, and 4. the duration of therapy.

(1) In the case of i.v. administered IFN-alpha a peak concentration of 500-600 IU/ml is reached whithin 15-30 min. The half-life period is 2 hours, and its level is normalized in 4 hours.

If the substance is administered subcutaneously or i.m., the maximum concentration is reached in 4 hours; 8–12 hours later the concentration of IFN decreases to 100–200 IU/ml (170, 362).

Administration of IFN-alpha, in a dose of 1-9 x $10^6$ IU/dose per day i.m., results in tolerable side effects.

A dose of 10–18 × $10^6$ is severely toxic and further elevation of IFN concetration to 20–30 × $10^6$ IU/dose has ireversible sequelae. Currently, independent of the indications, a dose of 1 × $10^6$ - 3 × $10^6$ IU/day is applied. Recombinant IFN-alpha is administered at 2–3 fold higher concentrations.

It has to be emphasized, that in malignant disease administration of IFN-alpha plays a dual role;

It activates the effector system and promotes the MHC II and MHC I expression on malignant target cells in order to make them available for the cytotoxic and NK cells (164).

(2) The timing of IFN-alpha treatment in viral and malignant diseases is selected as follows;

Administration in viral infections should be as early as possibile. The reason for this is, on one hand, to minimize the deleterious effects of complications, i.e. by binding to opiate receptors. Another reason is that IFN can prevent pain in herpes zoster (333), and, simultaneously, it diminishes the dissemination of the disease (Fig. XV/1).

According to our experience, in malignant disease (152) IFN can be given as a component of induction therapy (200, 284) alone, or in combination with corticosteroids (60, 63) in order to stabilize the process.

However, its use as a component of a post-relapse combination seems to be not beneficial (68).

(3) With regard to the route of administration (370) it has to be noted that the short half-life period, the possible complications, and the properties of the commercially available forms of IFN preclude the rapid i.v. administration of a single dose. IFN-alpha is rather administered in a slow infusion (337). Subcutaneous infusion (in analogy with the insulin pump) of IFN-alpha can prevent many side effects including those affecting the central nervous system (105, 241). This is the preferred way of administration when prolonged or repeated treatment is needed. When shorter or unrepeated treatment is anticipated, the intramuscular route is preferred.

Intra-lesional applications have also been

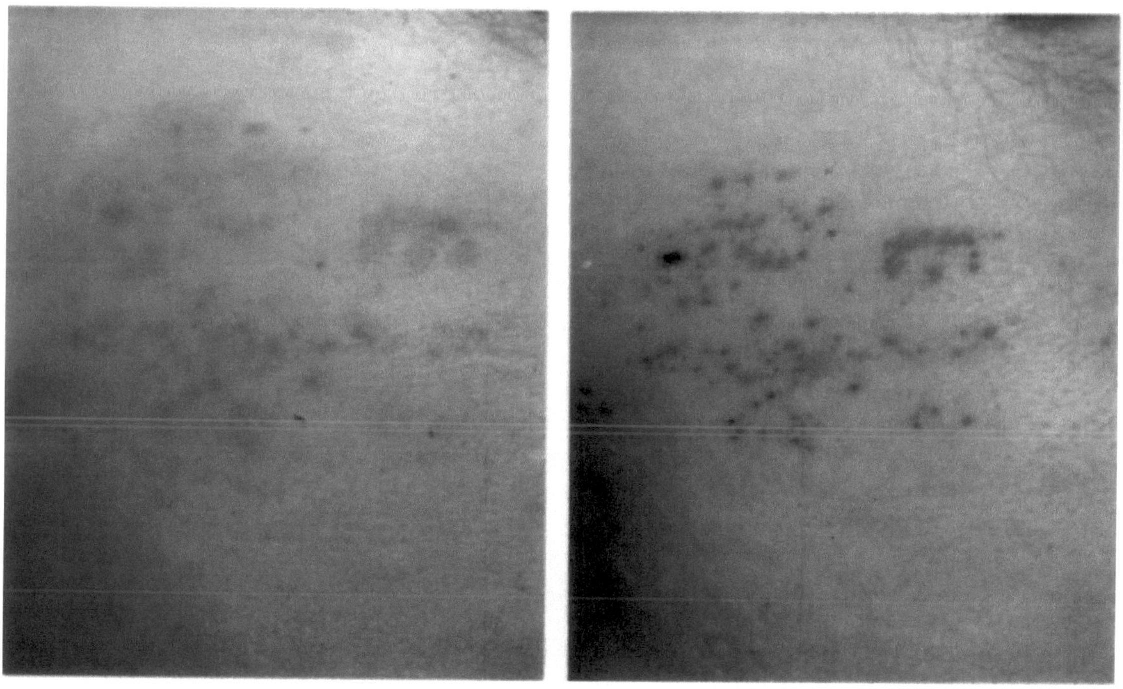

**Fig. XV/1**

described, however these data are insufficient to draw any definite conclusion.

There is debate about the principles of combination with other drugs and the ratio between the different components and IFN, therefore the role of IFN (synergist, adjuvant, potentiating agent etc.) is not determined. It is known that IFN sensitizes cells to cytotoxic drugs, parallel to this, it inhibits hepatic microsomal enzymes in vitro and in amimal experiments. These data focus the attention to possible desirable and undesirable therapeutical interactions. For example, Placicyl (Abbott Laboratories, Chicago) given simultaneously with IFN, induced a reversible comatose state. Interferon also enhances the neurotoxic effect of Vinblastin. Irradiation + IFN cause severe mucovitis (Fig. XV/2).

The above data show that none of the mentioned questions have been answered experimentally or empirically. The following considerations are, however, to be emphasized:

- Unambiguous, "clear" effects of anti-tumor drugs (if IFN is one?) can not be expected at the experimental or early clinical phase, since at the time of diagnosis, malignant tumors have been developing for several years or decades. It is a very bad routine that only immediate therapeutic effects are traditionally accepted and slow or delayed effects draw only minor attention or are insufficiently observed and recorded.
- The species specificity of IFN might also restrict the field of its experimental application. If species specificity exists, testing of human interferons in animals is of questionable value.
- Determination of cytotoxicity of biological substances in the presently existing systems might not always yield reliable results.
- It has been observed that in cases with similar diagnosis, phase, and specifications, IFN therapy of different individuals can be either effective or ineffective. Therefore, definition of sensitivity and insensitivity, as well as studies of intracellular and membrane effects of IFN are necessary. Furthermore, IFN resistance might not necessarily indicate proliferation of the malignant clone treated with this cytokine, but growth of a new one!
- Though we still do not have methods to investigate the in vivo effects of IFN, in the future IFN-alpha will be applied in combination with other cytokines.

(4) How long should IFN therapy last? The answer is rather subjective and empiric and is based on only few objective facts. Usually, patients are treated until remission occurs, as for example, in the case of CAH patients, carrying HBV virus (319). In 10–30% of cases the HBV antigen disappears from the circulation (94). Therapy is maintained for 1-6 months. In a recently published protocol, therapy is initiated with a prednisone boost for two weeks, and is continued with recombinant IFN-alpha ($5 \times 10^6$ IU/day s.c. for 90 days). The efficacy of this protocol has been proved by the disappearance of $HB_c$-Ag, HBV-DNA and DNA polymerase from the circulation in the treated group, as compared to controls. In some patients, $HB_c$-Ag could not be detected in liver cells, indicating that infection had been eliminated.

In malignant disease, even a (A) full remission due to IFN therapy does not lead to recovery, (B) the incidence of relapses and survival rate are not changed; (C) the direct and indirect (long-term) dangers of therapy are not negligible.

Following the administration of IFN, im-

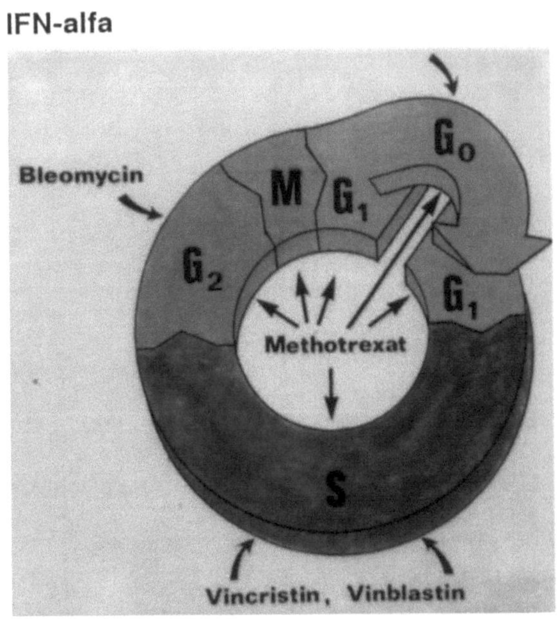


Fig. XV/2

mediate and delayed (or late) symptoms develop. At the present state of knowledge, these changes can not be identfied as true effects, side-effects or toxic effects. They must, however, be accurately recorded and physicians must be aware of their existence.

Immediate symptoms resemble those of influenza, including fever, chills, myalgia, arthralgia, temporary hypotension, headache, dizziness, nausea, and rarely, syncope. This might indicate that the symptoms of influenza infection and those following administration of IFN might be induced by the same mechanism, i.e. the direct effect of IFN. The analysis of the above symptoms suggests that all of them are caused by the fever induced by the IFN-dependent prostaglandin production of the hypothalamus. The central nervous system response, as well as the non-specific IL-1 release by the immune system, are both caused by IFN.

Late symptoms affect different organs and are, therefore, classified according to organ systems.

## Neurotoxic effects

Neurotoxic effects include fatigue, impaired memory, loss of appetite, transient impairment of tasting and smelling, lethargy, confusion, and EEG anomalies (diffuse deceleration of alpha rhythm, appearance of beta and delta waves) similar to that seen in diffuse encephalitis, diffuse mental and motoric deceleration, in extreme cases dysphagia, loss of spontaneous speech and motoric functions without alterations of personality. These changes might probably be due to the impaired function of the frontal lobe caused by IFN itself. There are no available data, whether these changes reflect the dose-dependent physiologic or toxic effect of IFN. According to published data, in two patients who presented with visual and auditory hallucunations coma and cerebral hemorrhage developed (coincidence?). These cases must be considered with special attention. The relationship of neurotransmitters and IFN is also to be addressed. At present, however, no available data exist on the relationship between IFN and neurotransmitters. Nevertheless, it has to be mentioned that depression has never occurred following IFN treatment. In animal experimental models IFN enhaced neuronal activity (70) probably due to its binding to some neurotropic receptors (250). Intracerebral aplication of IFN in mice induced endorphin-like reactions; analgesia, decreased motoric activity, catalepsy. These reactions could be reversed with naloxone (48, 51).

## Cardiovascular effects

Tachycardia, vasoconstriction, cyanosis of the lower extremities and hypotension is probably due to the induced fever and dehydration. Because of such impairment in the cardiac performance, IFN must be administered with special care to cardiac patients. Patients should be thoroughly monitored and treated only symptomatically. Prolonged IFN treatment, on the other hand, can cause congestive cardiomyopathy of unknown origin. Interferon enhances the contraction of fibers and increases the contraction-rate of cultured cardiac muscle cells (333). Occurence of arrhythmias, pericardial effusion, and perimyocarditis have also been reported (247, 310).

## Hemopoietic effect

IFN treatment always results in leukopenia of variable severity - the leukocyte count is sometimes only 40-60% of the original value - which affects both the polymorphonuclear neutrophils and lymphocytes. Changes mainly affect the ratio between the circulating and stored pools, as bone marrow studies did not show any alterations in hemopoiesis or maturation (120). It is, however, known that, under in vitro conditions, IFN inhibits the growth of bone marrow colonies. It is also important that leukopenia is reversible. The effect of IFN also depends on the exact nature of the underlying hemopoietic disease that requires IFN treatment.

Anemia is a regular companion to IFN therapy. In contrast to leukopenia, its normalization takes more time. Anemias are always monochromic and normocytic. In some cases, development of hemolytic anemia has been reported and repeated therapy has led to repeated hemolysis. Anti-RBC antibodies are not involved in the pathomechanism of hemolysis (188).

Development of thrombocytopenia takes more time. Some data indicate that some cases of antibody-induced thrombocytopenias are related to IFN therapy (268).

Though pathological alterations of hemostasis are not characteristic for IFN therapy, they sometimes occur and are related to vitamin K-dependent factors. Prolonged Quick and PTI time are the main features of these canges (291)

*Gastrointestinal symptoms*

Loss of appetite, vomiting, and diarrhea are typical side-effects of IFN therpy. While the former always occurs, nausea and vomiting do not develop regularly, and the appearance of diarrhea is more characteristic for high doses (325, 326). Abdominal colic, bleeding or steatorrhea never occur. It is believed that IFN and choleratoxin bind to the same membrane receptors of the intestinal mucosa. According to the current hypothesis, the cAMP system is involved in the pathomechanism of diarrhea (214).

Gastrointestinal bleeding, caused by IFN-induced mucositis, develops very rarely.

The increased SGOT and SGPT levels, as well as the rarely occuring pathological elavation of ALP and LDH indicate hepatotoxicity (325). Fatty change of hepatocytes is the predominant histological feature (330). High dose IFN treatment induced liver necrosis is experimental mice (162).

*Renal symptoms*

Proteinuria occurs in 15–20% of treated individuals. In contrast to mechanisms of liver injury, renal changes are dose independent. Proteinuria is usually mild, not exceeding 1 g protein/day, and never leads to hypoproteinemia. Two extreme cases of acute renal insufficiency and nephrosis syndrome have been reported (19, 358). Renal biopsies usually show interstitial nephritis and minimal change nephropathy without deposits. It is generally accepted that only a small proportion of IFN metabolites are excreted by the kidney, hence the excretion itself is not responsible for this "toxicity" (56). The pathomechanism of renal demages is still not known and IFN-gamma is the most toxic molecule (330). Despite the apparent controversy, it is accepted that high blood urea nitrogen and creatinin do not contraindicate therapy even in the light-chain excreting myeloma patients (326). The albumin/beta-2 microglobulin ratio in the urine remains unchanged following therapy, suggesting that IFN therapy causes no renal damage (379).

*Skin and appandages*

Regular monitoring of the skin and its appangages is mandatory over the course of IFN treatment. Mild loss of hair or allegy, and a paradoxical growth of eye lashes have been reported (130). Maculo-papular rash may appear and histopathological studies of the affected areas indicate vasculitis. It can not be excluded that urticaria and diffuse erythema develop as a result of vasculitis (356). It is also interesting that following IFN treatment the condition of psoriatic patients deteriorted and IFN could be isolated from epidermal psoriatic blisters (47).

*Bone and muscle*

Response of bone and muscle to IFN should be studied in more detail also on the experimental level. Transient myalgia and arthralgia in patients with CML treated with IFN can be so severe that they require treatment with potent analgetics. However, laboratory analysis show no signs of muscle damage. Another interesting finding is the pain of previously painless ostosteolytic lesions of myeloma patients treated with IFN. These data are in contradiction with reports concerning the effect of IFN on endorphins. Increasing pain of bone metastases of other solid tumors has never been observed (330).

*Endocrine system*

The following laboratory findings support the involvement of the endocrine system:
- elevation of the hydroxycorticosteroid level,

- increased insulin requirement in diabetes,
- decreased progesterone and estron levels,
- development of hyperkalemia and hypercalcemia,
- decrease in HDL concentration,
- increased ratio of triglicerides to cholesterol,
- hypo- or hyperthyroidism.

The latter is probably due to the interaction of IFN with TSH receptors.

In summary, an attempt was made to consider the effect of IFN-s and their relationship to other cytokines in order to emphasize the importance of meticulous, long-term monitoring and follow up, and recording of even minor symptoms of patients treated with IFN.

## Cytostatic therapy

The introduction of melphalan (M) and cyclophosphamide (CTX) 25 years ago for therapy of plasma cell tumors has been a milestone in tumor chemotherapy.

Unfortunately, since that time, another step of similar significance has not been made. Despite the intensive, multidirectional research, our knowledge has not inmpoved and, similar to other tumors, therapy of plasmocytomas still remains an enigma.

All patients with plasmocytomas require treatment but not in all phases of the disease. Conventional staging is not applicable to plasmocytomas. If, for example, the MG level is lower than 3 g/dL, the plasmocyte count in the bone marrow does not exceed 10%, and anemia, renal damages or osteoporosis are not present, patients should not be treated but observed thoroughly. (Certainly, exceptions always occur, i.e. patients having severe pain require treatment.)

Early attempts to combine different chemotherapeutics gave no superior results than the usual melphalan + prednisone (MP) protocol.

The recently applied combinations, however, yielded significantly better results with regard to the improvement of laboratory parameters of myeloma patients, as compared to results with the convertional MP protocol (271, 345). Conversely, in 1979 there was no difference between these two groups with regard to survival (265). Therefore further efforts had to be made with respect to the individualization of myeloma therapy.

The first step of improvement was the employment of cycle-active drugs. It has also been observed that absorbtion of orally administered drugs varies but that of i.v. chemotherapeutics yields standard values (4, 264). Therefore, the majority of components in newer combinations can be administered through the intravenous route. Protocols preferntially contained melphalan (L-phenylalanine mustard), cyclophosphamide, Carmustine, Lomustine, doxorubicine, vincristine, hexamethylmelamine, and cisplatine (cis-diamino-dichloroplatinum) combined with corticosteroids. Since the experience obtained with these combinations has been diverse, a method to determine the sensitivity of a particular tumor to cytostatics was needed. For this purpose, in vitro testing of cytostatics on tumor cell cultures cultivated in soft agar has been introduced (347). Fast uptake of $H^3$-thymidine (labeling index of bone marrow plasma cells (BMPCLI) indicated the intrinsic sensitivity of neoplastic plasma cells to cytostatics (109, 335). These plasmocytomas responded to cytostatic therapy with a rapid decrease in the tumor mass. Further studies showed that these are clones of aneuploid plasmocytes prone to a rapid remission, a short remission period, and an early relapse. Therefore, selection of adequate therapy must be based not only on the determination of LI, but it also requires the determination of the rate of DNA synthesis (107, 150). With the aid of the latter method it was possible to estimate the proliferative characteristics of plasma cell clones and to select a therapeutic regimen permitting alternative cytostatic treatment.

The therapeutic sensitivity of resistance of plasma cells (or the possibility to influence the course of the disease) is controlled by,
- the decrease in the serum and urine concentration of the M component,
- the changes (if any) in the blood urea nitrogen (BUN) and serum creatinine values;
- the serum calcium level,
- the WBC count,
- the decrease in the urine level of $\beta_2$-microglobulin,
- and X-ray studies.

The clinical activity of the disease, however, does not correlate with the changes in the tumor mass, as indicated by the studies listed above. Recent studies indicate that monitoring of the light chain isotypes carried by peripheral lymphocytes is more informative, since inhibition of their expression is characteristic to the plateau phase of the disease. Cessation of this suppression indicates activation of the disease (235, 303, 410, 411) Progression needs to be verfied by the detection of proliferative activity of plasma cells, which is evaluated by bromodeoxyuridine (BrdU) incorporation. The uptake by DNA is visualized by an anti-bromodeoxyuridine MoAb (54, 103, 155). Its value is about 1% in resting tumors and is elevated in relapses (115). Similar to resting tumors low values are measured in MGUS and indolent myelomas (159, 160). Independent of the type of tumor, higher values indicate worse survival rate (109, 114, 232). A new, generally applicable, method to determine tumor growth, based on the use of monoclonal antibodies, is under development. With this, a more accurate assessment of the cell fraction that is capable of proliferation can be made (147, 167).

Further kinetic studies could not have been performed unless human plasma cell culture models had not been available. At present, 17 cultures of unusual phenotype exist (257). Using short-term cell culture, a reliable method was developed to evaluate drug sensitivity. Evidence exists that while $CD_8$+ cells inhibit growth, the presence of $CD_4$+ cells is desirable, because they help maintain survival of the culture (288).

Results of recent research heve substantially contributed to our better understanding of the role of cytokines. In summary, these biologically active growth factors specifically affect certain types of cells (209, 210). Following the processes of information transfer and committent induced by IL-1 and IL-2, B cells differentiating in the presence of the IL-3 (187, 405) are activated by IL-4 (331, 421). Their proliferation is mediated by IL-5 (368) and they acquire their final form (plasma cell) due to the action of IL-6 (182). Recent studies have revealed that BSF-2, interferon β-2, and hybridoma growth factor are identical to IL-6 (144, 399). The observation that myeloma cells have surface receptors for IL-6 raises the possibility that IL-6, as an autocrine growth factor, is responsibile for the irreversibility of the neoplastic growth and the disease itself (203). A paracrine mechanism, is also hypothesized as, according to our present knowledge, IL-6 exerts its effect on the adherent cells in the bone marrow (212). Since it has been found that the action of BCGF (IL-5) also involves an autocrine component, it is assumed that autocrine and paracrine stimulation of cells can be active simultaneously or, according to the particular need, either of them predominates (211). Answers to these questions will be given only following the exploration of the exact molecular mechanism of action interleukins and the activity of receptor genes.

These studies might allow a deeper insight into the mechanism of pathologic proliferation of plasmacytoma.

With regard to prognosis, it is important that neither of the patients with hypodiploid plasmocytes responded to Melphalan-Prednisone or VAD therapy. A higher RNA content, on the other hand, indicated a better responsiveness to therapy (22). The ratio of unclear DNA and cytoplasmic RNA remains constant over the course of the disease, therefore, absolute values of these parameters do not reflect the sensitivity to therapy (27, 69, 221).

- Most of the myeloma cells are aneuploid and exhibit a differentiated B cell phenotype. The presence of diploid cells with an immature phenotype in an aneuploid population suggest involvement of different stem cells (117) (experience gained with missile therapy support this hypothesis).

The already mentioned importance of detection of malignant cell precursors should be specially emphasized (221, 280). Under normal conditions these proliferating cells are rarely found in the circulation. While the number of these precursors increases in relapses only few are detected in remissions (53, 320). Empirical evidence exists that chemosensitivity of these proliferating cells is markedly higher than that of cells at rest (55, 321).

Knowledge and detection of factors characteristic for myeloma, are not only of theoretical importance; they also contribute to the precise diagnosis and selectian of adequate therapy (315). Thus:

- using anti-idiotype antibodies, markers of malignant cells i. e., surface immunoglobulins of transformed B cells that are of identical idiotypes to paraproteins, can be detected (effect of uniforming factor!) (279);
- this test also proves clonal B cell predominance (CBE) (282, 314);
- suppression of light chain isotypes (LCIS) can also be monitored (197). More specifically this means that, for example in a lambda monoclonal gammopathy the increase of the lymphocyte subpopulation carrying a kappa light chain indicates a shift towards equlibirium and, simultaneously, improvement of the disease;
- Karyotype analysis reveals identical anomalies in peripheral and bone marrow lymphocytes (248);
- subsequently, the rearrangement of identical clonal genes occurs in peripheral lymphocytes and plasma cells in the bone marrow (31) according to these it is also assumed that on the basis of cell surface structures some of the pluripotent prognitors belong to neoplastic proliferating clones (119).

In myelomas with bone lesions measurement of the concentration of osteoclast activating factor (OAF) seems reasonable as OAF levels close correlation with the dissemination of bone lessions (III). Changes of OAF as a result of therapy deliver important information. It must be, however, mentioned that OAF isolated from the supernatant of myeloma cell cultures is not homogenous (297), i. e., it consists of a high and a low molecular weight component (299). One molecule has been later identified as TNF-β and another as IL-1-β (84, 143, 204). Mautre B cells contain neither of these components while both are detected in neoplastic plasma cells (84). It is hypothesized that these factors induce expression of IL-5 (211) and IL-6 (203). TNF can act as a tumor growth factor not only in plasmacytoma but also in HCL and (77) CLL. IL-5 (BCGF II) and IL-6 receptors on myeloma cells could serve as targets for missile therapy.

Plasma cells also carry vitamin $D_3$, glycocorticoid, estrogen, and progesterone receptors (92, 93, 255, 339). The role of these receptors in biological functions, malignant transformation, and aggressive behavior is unknown.

It is clear that the expression of glycocorticoid receptors shows correlation with the effect of IFN-α (255, 402). Susceptibility to bacterial infections is due to the diminished function of T and B lymphocytes (191, 201). IgBF can suppress normal B cells and this factor might be identical to the Ig-Fc receptor expressed on the T cells (323).

Resistance of malignant plasma cells to cytotoxic agents (MDR) might turn into a predominant property in tumor cell cultures (90, 239). The MDR phenotype depends on the presence of a p-glycoprotein (p-170), which prevents the effect of certain anti-tumor agents (for example anthracyclins and vinca alcaloids) by stimulating their outflow from cells (190, 417). Experiments with an anti P-170 MoAb (199) indicate that high expression of p-170 has led to VAD resistance (118). It is interesting that such drug resistance could be prevented with Ca-entry blocking agents (395). It is assumed that the Ca-blockers would prevent the increased "pumping out" of cytotoxic drugs from the cell (78). This question, though still unanswered, is by all means of high interest (91).

- No systemic study on the development of the disease have been performed. In the malignization period the LDH level increases. Studies on human plasma cell lines prove that IL-5 is an autocrine growth factor (211). Bone marrow monocytes produce IL-6, which binds to the IL-6 receptors of plasma cells, indicating paracrine regulation (212). A parallel autocrine mechanism also exists as proliferating malignant plasma cells are also capable of producing IL-6 (203).

The increasing activity, proliferation, and differentiation of normal B cells, elicited by IL-4, IL-5, and IL-6, gain an important role in maintaining the growth and differentiation anomalies of myeloma.

Production of IL-1 (84, 204), TNF (77, 143), IL-4, and IL-5 (211) maintains autocrine growth, i.e.a. stimulatory loop, and IL-6 production of bone marrow monocytes contributes to the maturation of plasma cells (212).

## Principles of indications of chemotherapy

The course of plasmocytomas shows a broad variety, from asymptomatic forms to those leading to death in a few weeks. It is known that the incidence of lytic lesions, osteopenia, bone pain, increase of the m-gradient, renal isufficiency, amyloidosis, and weight loss in asymptomatic patients is around 2% annually (12). Profilactic cytostatic therapy has no effect, as statistical analysis indicates no improvement of the survival rate (38). Chemotherapy should be reserved for the most dangerous period of progression (225). If therapy should be immediately started after the diagnosis as reguired by the stage of the disease, but the patient has other complications (infections, inflammation, complications requiring orthopedic or neurosurgical treatment), complications have to be solved prior to the onset of chemotherapy. Chemotherapy must not be, however, delayed if patients have pain, azotemia, hypercalcemia, or hyperviscosity syndrome. The decrease of tumor mass (A), the maintenance of the induced remission (B), and effective resumed treatment if relapses occur (C) are the primary goals of chemotherapy. We must always keep in mind that plasmocytomas, or at least some of them, are the anomalies of the progenitor cells. (25). Therefore, the following questions arise: when should we start therapy (?), which drug combination should be selected (?), and what is the optimal period of treatment (?). All these questions, of course, equally refer to items, A, B, and C.

A) Induction phase is the first step of therapy for untreated patients. Analysis of adata concerning the size and behavior of the tumor ($H^3$ thymidine or bromodeoxyuridine uptake, LI, stage classification, flow cytometry results) (115) indicate that the preclinical period is about 3-33 years. In this period the ratio of proliferating cells is the lowest. Elevation of the otherwise normal LDH values suggests progression, sometimes with lymphoma-like clinical picture (26). High LDH liter indicates that the clonal development resulted in a myeloma of a less differentiated tumor phenotype.

In summary:

1. hypodiploid DNA, low plasma cell RNA index, and high $\beta_2$-microglobulin levels mean absolute drug resistance,
2. high serum LDH and low RNA index indicate short remission,
3. low survival rate is expected at high serum LDH and $\beta_2$-microglobulin levels.

High LDH level (26), low CALLA expression (117), high H-ras activity (394), myelocytic phenotype (168), and high proliferative activity (161), are regarded as important prognostic factors. It is noteworthy that high LDH levels frequently accompany renal insufficiency, suggesting an unfavorable prognosis (14).

The exact nature of the disease can be assessed only after the evaluation of the response to induction therapy.

It has to be emphasied that drug-resistant cells that exist among the proliferating tumor cells are also activated by induction therapy (346). This observation underlines the importance of introduction of cycle-active drugs into the starting chemotherapeutic combination. At present no definite rules can determine the exact period of therapy. Though certain therapeutic protocols have been developed, tolerance of patients to therapy shows marked individual changes. Rigid application of any protocol can be of more harm than benefit.

B) With regard to the kinetics of the proliferating tumor cells, the plateau or remission is a rest period. Even in those cases when MG disappeared, tumor markers could still be detected using special MoAbs (141). Therefore, a so called maintenance therapy has been suggested. Unfortunately, the refractory anemia that almost always develops is followed by monomyelogenous leukemia in 2-6% of cases (153, 338). Therefore, one of the basic requirements regarding maintenance therapy is, parallel to the largest possible tumor mass reduction, that it should minimize the risk of the development of a second tumor. In our opinion, cytostatic therapy is unnecessary (110), and our purpose is rather to stabilize the reduction of the tumor mass. Theoretically, several agents are suitable for this purpose, paractically, however the choice is narrow. At present IFN-α is the only agent of choice, however, the dose and period of therapy needs further studies.

Plasma cells of IFN-α-treated patients are resistant to IFN-α which might be a possible consequence of genetic changes (242, 243, 349).

Usually, IFN-α inhibits the clonogenic growth of plasma cells cultured in soft agar (173). In contrast to these findings, stimulatory effect of IFN-α in a final concentration of 120 IU/ml (242), has also been observed, thus the use of low doses in vivo is risky. Human myeloma cells are heterogeneous with respect to their IFN-α sensitivity, hence, higher doses should be applied.

C) The term relapse denotes an increase of the tumor mass with a simultaneous development of drug resistance. This definition, however, does not always hold true as during the first phase of relapse, or during the first relapse, tumor growth is not solely attributed to the drug-resistant cell population; in these cases therapy might be beneficial. Over the course of repeated relapses the population of the drug-resistant cells increases and finally the whole population might consist of drug-resistant cells. Unfortunately, therapy of such patients will not be successful (184).

Precise diagnosis and indication is the first important requirement before starting therapy. Patients who do not require therapy (IMG) or those, who at the time of diagnosis need not be treated (indolent myeloma) most be excluded. All patients, however, must be thoroughly observed.

The second requirement is the selection of an adequate protocol which depend on the type of the tumor (for example less malignant IgG-κ or more malignant LCD) (113, 267).

The third requirement is the evaluation of the results of induction therapy. As a result of this evaluation, independent from staging, we can categorize each particular case and its prognosis.

The main goal of therapy is to alter the course of the disease as the disease itself is incurable.

Though the prognosis is markedly hindered bay the occurence of asymptomatic (lantanic) myeloma, available data indicate that the mean survival rate post-diagnosis ranges between 7–30 months.

Imrpoving areas of supporting therapy, i. e. management of infections, hypercalcemia, anemia, renal and hemorrhagic complications, all contribute to the better clinical results.

Though the traditional M + P therapy or other accepted combinations have led to improved quality of life, diminished symptoms, and prolonged survival, they basically did not alter the course of the disease.

Results with M + P therapy resemble those obtained with the most agressive agents. Analysis of results and complications of trials with more aggressive combinations, claiming them as more efficient, are not very convincing.

Further development of M + P therapy was based on the observation that both murine and human plasma cell tumors resistant to Melphalan were still sensitive to Cyclophosphamide (34, 312). There is however, evidence that there is no significant difference with respect to remission or prolongation of survival (36).

Despite the fact that treatment with VBCMP protocol resulted in remission in 70% of patients as compared to 50% with M + P, we emphasize again that there is no difference in the mean survival of the two groups.

We also failed to confirm those recently published findings that certain drug combinations (VMCP, VCAP, VBAP), used to treat third stage disease, significantly improve remissions and survival rate (350). Moreover, according to some followup studies of patients treated with M + P, survival was longer by 11 months (11).

These results suggest that the real character of plasma cell tumors can be identified only retrospectively and this observation substantially influences the evaluation of theraputical resutls obtained from randomized trials. Therefore, these trials only partially reflect the efficacy of therapy as these randomized groups immanently carry the possibility of variability of results (Tables XV/II, XV/III, XV/IV, XV/V).

These tables unequivocally prove that compared to the cytostatic era, survival rate und quality of life of patients treated with other drugs were significantly inferior.

Different cytostatic combinations, however, failed to deliver different results, thus rating them as better or worse can not be carried out. Despite of this, a critical analysis of the available therapeutical modalities is presented.

1. Melphalan (1-phenylalanine mustard) for plasmocytoma therapy is commercially available in the name of. Sarcolysin* (alternative names: Alkeran, Peptichemio**). Different protocols suggest a daily dose of 3.5–70 mg (see Table). It is usually combined with prednisone or other steroids, the role of which is still obscure. Steroids undoubtedly improve and prolong remission without affecting survival. They probably enhance catabolism of immunoglobulins (165) and, due to their lymphoclastic effect, inhibit the maturation of precursors. Statistical analysis shows that the survival rate following M + P therapy ranges between 24–30 months (7, 113, 275).
2. In 70% of cases combination therapy elicits beneficial symptomatic effect, furthermore it causes a 70% reduction of tumor mass in 45–70% of the patients. Full remission without any clinical or laboratory signs is rare and is observed in 15% of cases.

Different evaluations of results do not alter the inevitability of the relapses (113). Furthermore, if individual complications and the frequently occurring deterioration in the quality of life are considered, the efficacy of therapy will also be doubted (9, 316, 318, 388). The onset of an optimal therapy depends on the extent of dissemination. Combined cytostatic therapy of patients in stage I-II is less efficient than those is stage III (350). It is generally accepted that duration of chemotherapy-induced remission informs on the tumor mass and is a prerequisite of a long survival.

Further analysis of survival indicates that independent of the applied therapy, similar values are obtained at 48 months. Therefore, this finding can not be explained through therapy only, but suggests the involvement of a variety of other unknown factors.

Of the periodically applied combined therapies the M2 protocol seems to be the most promising (311, 316).

The effect of VCAP protocol applied every three weeks also seems beneficial.

Vincristin-Adriamycin infusions combined with oral high doses of dexamethasone and supplemented with oral CTX (VAD-VCAD) efficiently induces remission, however, it does not improve survival (14).

Comparative studies of ABCM combination and M monotherapy are unequivocally in favor of combination protocol (249).

HDM (high dose Mephlan 140 mg/m$^2$) is a highly toxic but efficient therapeutic modality. Because of its short-term action, relapses occur in 11 months (359). HDM applied as induction therapy induces full remission with normalization of bone marrow plasma cells and disappearence of the M component. It should be, however, emphasized that this form of therapy markedly differs from others as it can be applied only in those patients who are able to tolerate high doses of toxic sobstances. In these individuals long remissions of 37–50 months have been described, though, similar to others, HDM therapy is not curative.

With regard to therapy following aspects should be considered.

1. At $Cr^{51}$ EDTA clearance values of 40 ml/min a full dose can be applied; if the claerance ranges between 31–40 ml/min - half dose (70 mg/m$^2$) is recommended; at lower values a dose of 30 mg/m$^2$ should be used.
2. The $Ca^{++}$ level should be meticulously monitored.
3. Adequate hydration of the patient is important.
4. Though immobility due to pathological fractures is not a contraindication, improvement of this condition, as a result of therapy, can not be expected.
5. IgA MG is more sensitive to HDM therapy than IgG MG.
6. High MG shows an exponentially decreasing tendency.
7. Therapy should be supplemented with preventive treatment against Pseudomonas aeruginosa.
8. HDM therapy is less effective if patients have been previously treated with oral M (263).

* Sarcolysin is a "D" nd Melphalan is an "L" form
** Peptichemio is identical to Sarcolysin peptide

**Table XV/II**

| Treatment of multiple myeloma with melphalan and cyclophosphamide | | | | | | |
|---|---|---|---|---|---|---|
| Source | Time of report | Interval of survey | Dose of used drug and interval of therapy | Number of patients | Remission (%) | Reference |
| 1. Korst, D. R., et al. | 1964 | | cyclophosphamide 2 mg/kg/day | 165 | 48 | 217 |
| 2. Finkel, M. E., et al. | 1966 | | melphalan 0.05 mg/kg/day | 28 | 50 | 125 |
| 3. Hoogstraten, B., et al. | 1967 | 1959-1965 | melphalan 0.15 mg/kg/day through 7 days, in repeated cycles | 64 | 45 | 185 |
| 4. Bergsageé. D. E. | 1967 | | melphalan 1-2 mg/kg through 4 days, repeated for 6-8 weeks | 199 | 34 | 33 |
| 5. Alexanian, R. | 1968 | 1959-1965 | melphalan 1 mg/kg 4 days a week, through 6 weeks | 82 | 49 | 3 |
| 6. Alexanian, R., et al. | 1969 | | melphalan 0.025 mg/kg/day continously | 35 | 19 | 6 |
| | | | melphalan 1.0 mg/kg intermittently | 69 | 35 | |
| | | | melphalan + prednisone | 79 | 70 | |
| 7. Mc Arthur, J. R. | 1970 | 1962-1968 | melphalan 4.0 mg/day | 39 | 41 | 260 |
| 8. MRC I. trial | 1971 | 1964 1968 | cyclophosphamide 150 mg/day | 141 | | 270 |
| | | | melphalan 4 mg/day | 133 | | |
| 9. Velez-Garcia, E. et al. | 1971 | 1964-1968 | melphalan 6-10 mg/day for 7-10 days, then 2 mg/day after a 14 day break | 32 | 46 | 404 |
| 10. George, R. P., et al. | 1972 | | melphalan 0.25 mg/kg/day for 4 days + prednisone 2 mg/kg/day | 31 | 74 | 146 |
| 11. Costa, G. et al. | 1973 | 1965-1966 | melphalan 0.15 mg/kg/day for 2 days, then change | 60 | 25 | 79 |
| | | | melphalan 0.15 mg/kg/day for 2 days + prednisone | 71 | 48 | |
| | | | melphalan + prednisone + testosterone | 58 | 54 | |
| 12. Brook, J., et al. | 1973 | 1962-1968 | melphalan 0.05 mg/kg/day then lower dose continously | 45 | 38 | 61 |
| 13. Farhangi, M., et al. | 1973 | | Continous melphalan treatment | 41 | 80 | 123 |
| 14. Bernard, D. B., et al | 1974 | | melphalan 0.15 mg/kg/day for 7 days + prednisone | 52 | 40 | 41 |
| 15. Cuttner, J. | 1975 | | melphalan 0.05 mg/kg/day | 38 | 16 | 88 |
| | | | melphalan 0.60 mg/kg/day | 46 | 50 | |
| | | | melphalan 0.05 mg/kg/day | 41 | 34 | |
| 16. Lee, B. J. | 1976 | 1963-1975 | melphalan | 60 | 30 | 234 |
| 17. Mellstedt, H., et al. | 1977 | 1970-1974 | melphalan 0.25 mg/kg/day for 4 days, through 6 weeks + prednisone | 32 | 75 | 277 |
| 18. Harley, J. B., et al. | 1979 | | melphalan 0.15 mg/kg/day for 7 days + prednisone | 127 | 56 | 176 |
| 19. Bergsagel, D. E., et al. | 1979 | 1973-1977 | melphalan 9 mg/m$^2$/day for 4 days + prednisone | 100 | 40 | 36 |
| 20. MRC II. vizsgálat | 1980 | 1968-1975 | cyclophosphomide 150 mg/day for 7 days, every 6 weeks | 124 | | 272 |
| | | | melphalan 10 mg/days for 7 days, every 6 weeks | 128 | | |
| | | | melphalan 10 mg/day + prednisone for 7 days, every 6 weeks | 120 | | |
| 21. MRC III. trial | 1980 | 1975-1978 | melphalan 100 mg/day for 7 days + prednisone every 3 weeks | 179 | | 273 |
| | | | cyclophosphamide 600 mg/m$^2$ i·v. every 3 weeks | 174 | | 274 |

| Treatment of multiple myeloma with melphalan and cyclophosphamide | | | | | | |
|---|---|---|---|---|---|---|
| Source | Time of report | Interval of survey | Dose of used drug and interval of therapy | Number of patients | Remission (%) | Reference |
| 22. Ahre, A., et al. | 1983 | 1974–1978 | melphalan 0.07 mg/kg/day for 3 weeks | 99 | 45 | 2 |
| | | | melphalan 0.25 mg/kg for 4 days + prednisone 2 mg/kg/day | 120 | 31 | |
| 23. Jákó, J., et al. | 1990 | 1972–1989 | melphalan 6.0 mg/m² + prednisone 60 mg/m² | 275 | 65 | |

**Table XV/III**

| Combined cytostatic therapy of multiple myeloma | | | | | | | |
|---|---|---|---|---|---|---|---|
| Source | Time of report | Interval of survey | Combination used | Number of patients | Remission (%) | Complications | Reference |
| 1. Bergsagel, D. E., et al. | 1972 | | HD-CTX-800 mg/m² | | | | 34 |
| 2. Shaw, M. T., et al. | 1974 | | COAP CTX<br>Oncovin<br>Cytosine-Arabinoside<br>Prednisone | | | | 363 |
| 3. Alexanian, R., et al. | 1977 | 1965–1974 | VAD V - 0.4 mg i. v. for 4 days<br>A - 9.0 mg/m² i. v., 3 days<br>D - 40.0 mg p. o., then break | 305 | | fever, Gram negative sepsis, Gram positive bacteriemia, herpetiform CMV infection | 8 |
| 4. Bergsagel, D. E., et al. | 1979 | | BCNU-CTX-M-P<br>BCNU - 150.0 mg/m² i. v., 1 day<br>CTX - 325.0 mg/m² i. v., 1 day<br>M - 9–12 mg/m², 4 days<br>P - 100.0 mg p. o., 4 days | | | | 36 |
| 5. Harley, J. B., et al. | 1979 | 1973–1978 | BCMP-MP | 252 | 40 | | 176 |
| 6. Riccardi, A., et al. | 1980 | 1972–1979 | PTC-V-P<br>PTC - 40-60 mg/48h<br>V - 2.0 mg<br>P - 30 mg | | | | 334 |
| 7. Ludwig, H., et al. | 1981 | | Alternating administration of BCNU-CTX-M-P | 65 | 29 | | 243 |
| 8. Cohen, H. J. | 1982 | | Hexamethyl melamine - Prednisone | | | | |
| 9. Bonnet, J., et al. | 1982 | | VBAP | | | | |
| 10. Broun, G. D., et al. | 1982 | | Cysplatin-BCNU-CTX-P<br>Cysplatin - 50.0 mg/m² i. v.<br>BCNU - 50.0 mg/m² i. v.<br>CTX - 300.0 mg/m² i. v.<br>P - 75.0 mg day for 1 week, then a 4 weeks break;<br>3–4 cycles altogether. | | | bronchopneumonia M. tuberculosis meningitis, neutropenia, thrombopenia, acute myocardial infarction | 62 |
| 11. Kyle, R. A., et al. | 1982 | | CAP CTX - 400.0 mg/m² i. v.<br>Adriamycin - 30.0 mg/m² i. b. once a week,<br>P - 0.6 mg/kg devide in three equal portions. | | | | 223 |

| Combined cytostatic therapy of multiple myeloma | | | | | | | |
|---|---|---|---|---|---|---|---|
| Source | Time of report | Interval of survey | Combination used | Number of patients | Remission (%) | Complications | Reference |
| 12. Kyle, R. A., et al. | 1982 | | VBCMP V - 1.2 mg/m² i. v.<br>BCNU - 20.0 mg/m² i. v.<br>M - 8.0 mg/m² p. o.<br>CTX - 400.0 mg/m² i. v.<br>P - 40.0 mg/m² p. o. | | | | 223 |
| 13. Ahre, A., et al. | 1984 | | Alternating administration of VMCP and VBAP | | | | 3 |
| 14. Barlogie, B., et al. | 1984 | 1982-1984 | VAD | 20 | 60 | | 21 |
| 15. Alexanian, R. | 1985 | 1974-1980 | VAD | 256 | 30 | | 13 |
| 16. Cohen, H. J., et al. | 1985 | | BCNU-CTX-P<br>CTX - 400.0 mg/m² i. v.<br>P - 75.0 mg p. o.<br>Administration of i. v. drugs every 7 days, repeated 6 times; continous administration of prednisone | | | | 73 |
| 17. Finnish Leukemia Group | 1985 | | MOCCA<br>V - 0.03 mg/kg (max 2 mg), 1 day<br>CTX - 10.0 mg/kg, 1 day<br>CCNU - 40.0 mg p. o., 1 day<br>M - 0.25 mg/kg, 1-4 days<br>P - 0.8 mg/kg, 1-7 days<br>0.4 mg/kg, 8-14 days<br>Repeated on every 35th day. | | | paralytic ileus, depression, alopecia, peripheral neuropathy, oral herpes and candidiasis | 126 |
| 18. Gore, M. E., et al. | 1989 | | VAMP | 50 | 20 | | 154 |
| 19. Jákó, J. | 1990 | 1981-1989 | VAC MP<br>V - 1.0 mg/mg/m² i. v., 1 day<br>A - 8.0 mg/mg/m² i. v., 1 day<br>CTX - 400.0 mg/m² i. v., 1-5 days<br>P - 0.6 mg/kg p. o every day for 7 days, repeated every 6 weeks. | 17 | 42 | | |

**Table XV/IV**

| Combined cytostatic and IFN treatment of multiple myeloma | | | |
|---|---|---|---|
| Source | Time of report | Combination used | Reference |
| 1. Costanzi, J. K., et al. | 1985 | Chemotherapy + Interferon | 81 |
| 2. Cooper, M. R., et al. | 1986 | M-P + Interferon-alpha 2 | 75 |
| 3. Cooper, M. R., et al. | 1987 | M-P + Interferon-alpha 2b | |
| 4. Ludwig. H., et al. | 1987 | Chemotherapy + Interferon | 246 |
| 5. Ludwig, H., et al. | 1987 | Chemotherapy + Interferon | 245 |
| 6. Kyle, R. A., et al. | 1988 | VBMCP + Interferon-alpha 2 | 226 |
| 7. Mellstedt, H. | 1988 | M-P + Human IFN-alpha | 283 |

**Table XV/V**

| Interferon therapy of multiple myeloma | | | |
|---|---|---|---|
| Source | Time of report | Type of interferon used | Reference |
| 1. Mellstedt, H., et al. | 1979 | Interferon-alpha | 278 |
| 2. Gutterman, J. V., et al. | 1980 | Interferon-alpha | 169 |
| 3. Mellstedt, H., et al. | 1982 | Interferon-alpha | 281 |
| 4. Quesada, J. R., et al. | 1984 | Recombinant interferon | 326 |
| 5. Costanzi, J. J., et al. | 1985 | Interferon-alpha 2 | 80 |
| 6. Oken, M. M., et al. | 1985 | Interferon-alpha 2 | 308 |
| 7. Case, D. C. | 1986 | Interferon-alpha 2 (Intron A) | 71 |
| 8. Ludwig, H., et al. | 1986 | Interferon-alpha 2c | 244 |
| 9. Ohno, R., et al. | 1986 | Interferon-alpha 2a | 307 |
| 10. Quesada, J. R., et al. | 1986 | Interferon-alpha 2b | 329 |
| 11. Taguchi, T., et al. | 1986 | Interferon-alpha 2a (Roferon) | 385 |
| 12. Costanzi, J. J., et al. | 1987 | Interferon-alpha 2b | 82 |
| 13. Mandelli, F., et al. | 1988 | Interferon-alpha 2b | 252 |
| 14. Jákó, J. | 1990 | Human interferon (EGIS) | |
| Initial dose of recombinant interferons: 2-100 x $10^6$ IU/$m^2$, 3 times a week i. v. 2-3 x $10^6$ IU/$m^2$, 3 times a weekk, s. c. | | Maintenance dose of interferons: 10 x $10^6$ IU/$m^2$, weekly, s. c. | |

The plateau phase develops as a result of a balance of the cytoreductive effect of chemotherapy and tumor growth. Fourty four per cent of treated patients rearch the plateau phase which can be characterized as a cytokinetically stable interval (112, 196, 410). It is evident that during this phase only very little benefit can be expeected from therapy, therefore, administration of cytotoxic drugs is not necessary.

Combination therapy has also its risks as the incidence of infections increases and acute leukemia develops significantly more often (67). Still, in the group of so-called "instable" patients with high cell turnover and a dynamic tumor growth, a long continuous therapy seems to be of benefit.

Therapy of relapses: Most patients with a 50-70% of tumor regression stay in remission for 2.5-3 years before relapsing to their disease (113).

During this phase, tumor cells tend to be more aggressive. According to the available data their difference from the original cell population is summarized as follows:

1. A more rapid cell growth,
2. susceptibility to mutations: a new paraprotein appears, and MG turns into a bi- or oligoclonal gammopathy,
3. excretion of light chains in the urine,
4. development of plasmoblastoma.

Generally, patients who show a rapid decrease in their tumor mass relapse more rapidly, and the incidence of CNS complications is higher.

IFN-a treatment of plasma cell tumors has already been mentioned as a part of maintenance therapy.

Though several groups use IFN-α in combinations, we do not favor its use as such. From the pharmacological viewpoint, IFN-α is a cytoreductive agent and in low tumor load disease, as a part of inductive therapy, it can also be of benefit in combinations (253, 329). According to some authors, VBMCP combined with IFN-α induces full remission more often than other regimens (309). Another advantege of IFN-α is its limited efficacy in drug-resistant cases. In contrast to themotherapy, in some cases it could restore polyclonal immunoglobulin synthesis in patients with secondary antibody-deficiency syndrome (329).

Combination of IFN-α with M + P has proved to be very advantageous, as reactivity to M + P only was 47% as compared to 76% with the triple combination in stage I. disease. Survival seemed to be also longer (284). Although no difference could be found in stages II-III. combination protocols of IFN-α with cytostatics are now widely used in the clinical practice, and our experience fully supports this practice. Usually IFN-α is combined with VBMCP (228) or high dose steroids (137, 409). The latter is recommended in unfavorable bone marrow parameters and drug-resistant cases.

Though no available data exist on the therapeutic use of IFN-β and IFN-γ, poly-ICLC (a derivative of poly-ribocytidine and poly-riboinosine) had been administered several years ago to activate endogeneous interferons (10). We now repeat our opinion that IFN-α should be applied for the consolidation of chemotherapy-induced remission. Others also support our view (National Cancer Institute of Canada; Central Pennsylvania Oncology Group, USA; Italian Myeloma Group).

Idiotypes carried by paraproteins are the most characteristic markers of plasmocytomas. Cytostatics, linked to individually synthetized monoclonal antibodies can find and bind to the surface of malignant plasma cells (276). This is termed "missile", or specially selective chemotherapy. Studies however have revealed that cytotoxic molecules bound to antibodies are less toxic than free molecules (289, 295). This effect might be due to the small size of the antibody molecule restricting the binding possibilities. Also binding of a cytotoxic drug alters the tertiary structure, thereby reducing the effectiveness of the complex. It is well known that the cytotoxic effect depends on the concentration of the particular agent within the tumor. In this case, the latter depends on the binding properties of the antibody. Therefore, in monoclonal gammopathies optimal result can be achieved only if there are no circulating paraproteins, which would bind the complex prior to its "arrival" to the site of action. Conseqently, a series of plasma exchanges are required before the therapeutic agent is introduced. Antitumor antibodies not only recognize tumor cells but also elcit antitumor effect by themselves. Their binding either inhibits tumor cell growth or contributes to the destruction of cell membranes (121). Complement activation or ADCC reaction are involved in this mechanism. Independent of the exact mechanism, anti-idiotype antibodies are the basic principle of missile therapy. Production. of individual antibodies is however, extremely time-consuming.

Since the role of the immune growth factors is already known, production and clinical use of antibodies forming complexes with cytokines might turn into reality in the near future (135, 206, 285).

Antibody administration causes only few side effects. Patients may present with dispnea, chest pain during or after treatment, fever, chills nausea, vomiting.

Only few patients have been treated with this method, so no serious conclusions can be drawn at present. Still, analysis of results and failures is important, because these data will serve as the basis for future research (198).

The causes of failures of MoAb treatment can be summarized as follows:

1. Antigens of tumor cells are shed into the extracellular and intercellular space, as well as the circulation, therefore antibodies form complexes at these sites do not reach the target cells,
2. As a result of antigen modulation, target cells avoid interaction with antibodies,
3. If the tumor is not homogeneous (monoclonal) but bi- or oligoclonal destruction of one clone does not lead to the destruction of the whole tumor,
4. The antibody response of the host can block exogenously administered antibodies.

Though this classification is far from being complete, it can still indicate the main trends of MoAb therapy in the future:

1. Application of monoclonal cocktails simultaneously against several markers might enchance the effect.
2. Application of genetically engineered antibodies. Production of high affinity anti-

bodies can be carried out with recombinant DNA technology or with hybridoma mutations and selection. For example, combination of a gene encoding the variable region of a murine monoclonal antibody with the one encoding a human constant fragment and indtrocudtion of this molecule into the myeloma cell line leads to the production of high affinity antibodies recognizing even the smallest antigens (218).

Irradiation can also be applied in order to increase the efficacy of induction therapy, prior to bone marrow transplantation, to help releave pain and to treat local plasmacytomas (172, 224).

Irradiation of the whole marrow mass following 12 weeks of M + P therapy was not enough and has led to complications (226). Even the so-called "hemibody" irradiation proved to be extremely toxic if applied

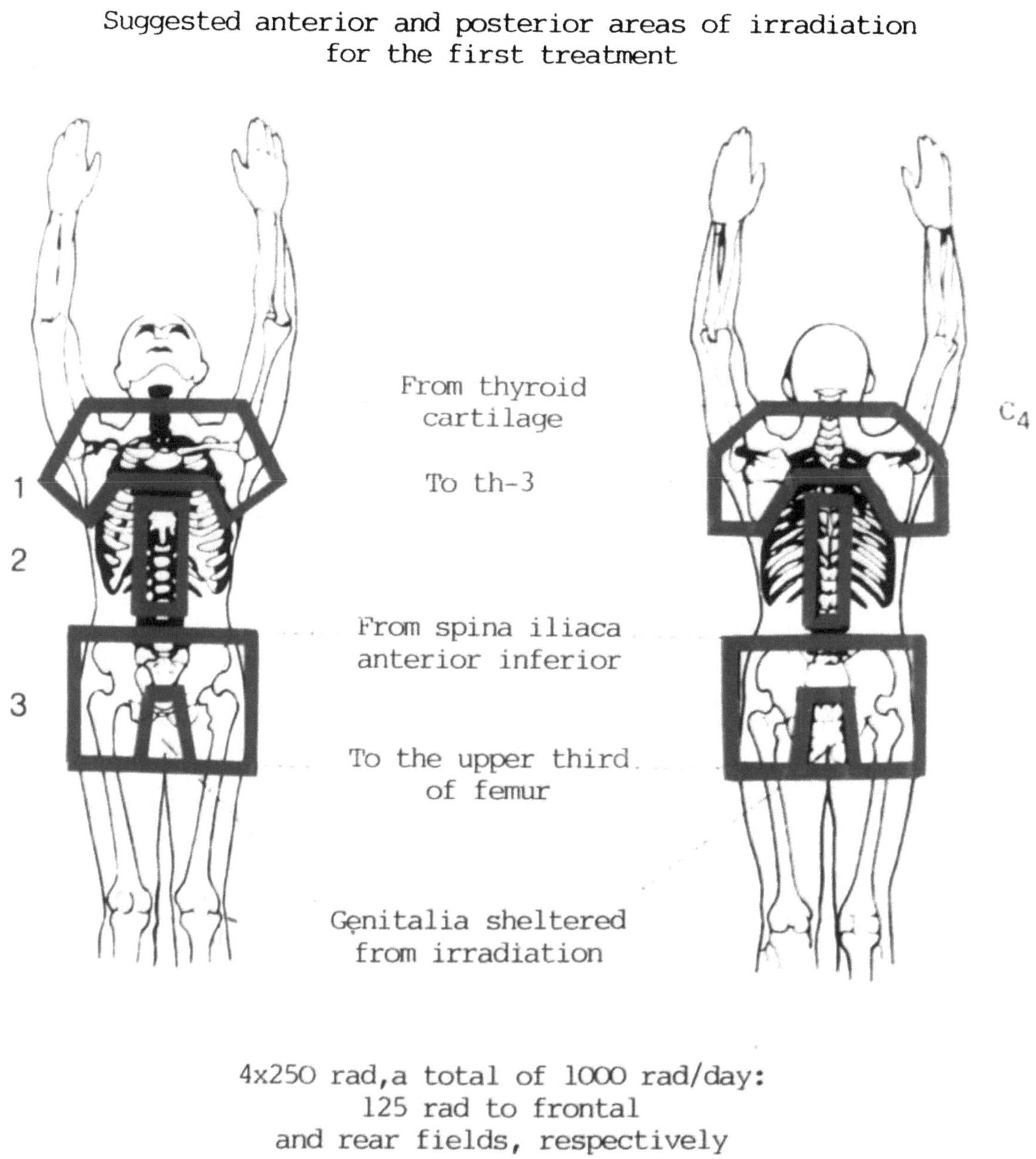


**Fig. XV/3**

together with chemotherapy (351). This infavorable effect might be due to the long term effect of irradiation on the bone marrow (381) preventing regeneration. Doses over 3500 rad usually cause prolonged bone marrow aplsia (382).

Irradation is performed in three steps as shown in Figures XV/3, XV/4, XV/5. The daily dose should be equally distributed. Ventral and dorsal surfaces of the body must be irradiated every day, in the same sequence, in all phases of the disease. As a radiation source cobalt 60 is used, however, a 10 MeV electron beam should be used for irradiation of ribs and skull.

The irradiation in the I phase lasts for 4 days until a final dose of 1000 rad is reached.

Following a 4 weeks rest period, in phase II. a total dose of 900 rad is applied over 6 days.

Suggested anterior and posterior areas of irradiation for the second treatment

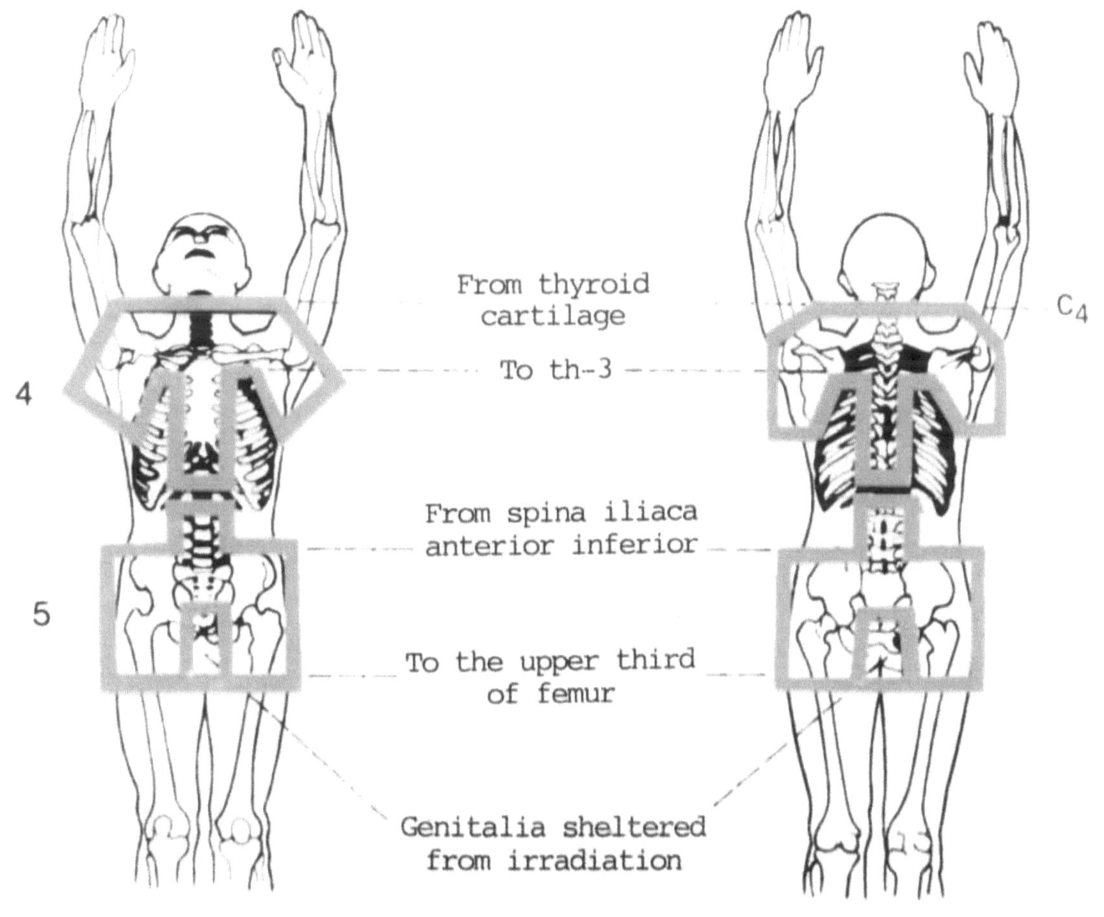


6x150 rad, a total of 900 rad

**Fig. XV/4**

Third irradiation is administered to the right and left side of the skull and the anterior and posterior arches of ribs

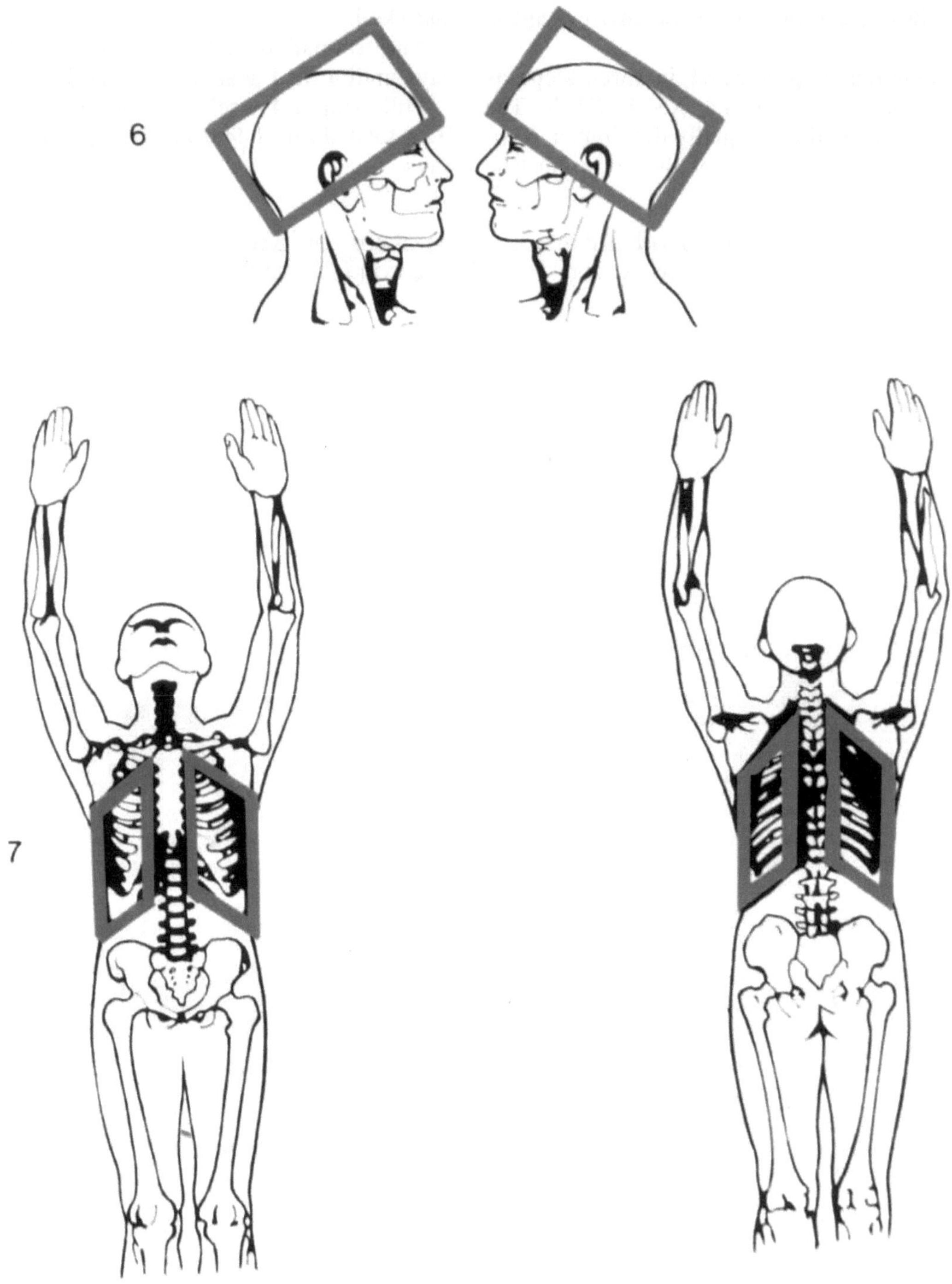


10 MeV electron beam, 8x250 rad, a total of 2000 rad

Fig. XV/5

Irradiation in phase III is also started after a 4 week rest period. The patient receives a total dose of 2000 rad (74) (Fig. XV/3, XV/4, and XV/5).

Using a massive 3000 rad raster irradiation pain may be eliminated and patients can live withont analgesics for several months.

Before bone marrow transplantation, 5 irradiations of 850 rad administered every 12 hours is considered as optimal (Fig. XV/6).

Good results of bone marrow transplantation performed on leukemic and lymphoma patients have persuaded clinicians to test this method also in plasmocytoma. Though indications for this technology are already well defined in plasmocytoma it is less effective than in patients with leukemia, since abdominal destruction of plasma cells still remains an unsolved problem (65).

Survival rates of transplantation have been shorter as compared to similar results of transplantation performed during the remision period of leukemic patients, however, it showed similar results to those obtained ofter transplantation during relapse. Though it is too early to dran any conclutions, it seems that transplantation is worth trying in the II. and III. phases of plasmacytoma. Generally, full remission lasts for 3 years, however, in some patients its duration exceeded 5 years (138). At present the indications for the allogenic bone marrow transplantation are as follows:

- age under 50,
- low $\beta_2$-microglobulin level,
- presence of IgD monoclonal immunoglobulin,
- high LI values (140).

Both allogenic and syngeneic transplantations were performed following CTX treatment and TBI and have led to a marked decrease of the tumor mass. Transplantation itself was performed during partial remission. (124, 138, 313, 396) In many patients monoclonal Ig-s have persisted. Recent data indicate that full remission develops in 50% and GVHD occurs rarely (140). One publication presents a case of a glucocorticoid-resistant GVHD which could be successfully treated with anti-CD MoAb therapy (207).

Of many different forms of therapy applied before bone marrow transplantation, initial treatment with a high dose Melphalan predominates. Following a continuous infusion of VAMP, 200 mg/m$^2$ i. v. Melphalan is administered (262, 359). The latter is followed by BMT. Full remissions were obtained in 30% of the cases; only 10% of fatal outcomes occured shortly after therapy (360).

VAD induction therapy is applied of the tumor mass is medium or large. Patients respondring to induction are then treated with 140 mg/m$^2$ Melphalan and undergo whole body irradiation (850 rad, 5 occasions). Results of this attempt proved to be worse than expected. The higher mean age of the myeloma patients contributes to these unfavorable results. IFN-$\alpha$ and IL-2 therapy would probably improve the responsiveness of such patients (139, 140, 365, 418).

Attempts of autologuous bone marrow transplantation failed to elicit full remission, but have led to symptomatic improvement (23). Preconditions and pretreatment applied in autolognuous transplantation differ from

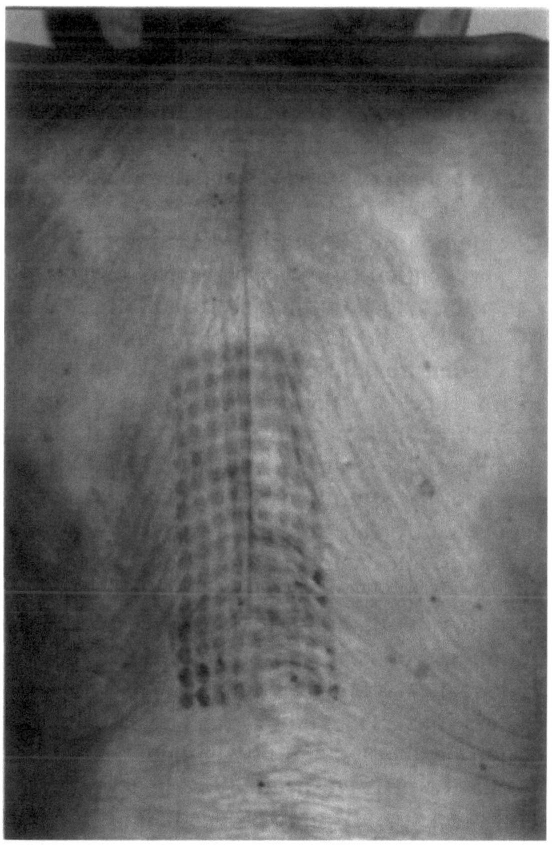

Fig. XV/6

those accepted in allogenic and syngenic transplantation.

The important criteria are as follows:

- plasma cell count about 10%,
- primary irradiation of osteolytic areas followed by
- 70 mg/m$^2$ Melphalan therapy on days 1 and 2 (261) with a subsequent
- fractional, 1200 rad whole body irradiation on day 3 (17),
- finally, the autologuous bone marrow should be purified. Though no successful methods exist, different methods had been tested alone or in combination,
- killer cells activated by lymphokines (LAK),
- B1 and PCA-1 MoAbs,
- a combination of anti-CD9 and anti-CD10 MoAb applied in immunomagnetic purification (these MoAb produced in mice are bound by specific goat anti-murine antibodies which are fixed to the walls of a special dish. In the first step the monoclonal antibodies bind to the membranes of the nondesirable plasma cells which, through these antisera, will be attached to the walls of the dish. This procedure eliminates the malignant plasma cells from the bone marrow) (364).
- Treatment with 60 micromol final concentration of 4-hydroperoxycyclophosphamide (4-HC) is another in vitro method
- Finally, Etoposide (VP-16) can be used for similar purposes (366, 367).

The reason of such thorough preparation for autologuous bone marrow transplantation is the possibility of identification and binding of the small fraction of clonogenous tumor cells as they have a predominant terminal B cell phenotype (348).

The other reason is that 1% of the bone marrow granulocyte-macrophange progenitor colony forming cells (CFU-GM) survives the purification process and this fraction is enough to populate the bone marrow. Available data on autologous bone marrow transplantation indicate that the procedure is favorable however, relapses suggest that both the methods of pretreatment and in vivo purification of bone marrow should be further imroved.

The rare occurence of full remissions following autologuous BMT is probably not due to the reinfusion of tumor cells as a further decrease of MG after transplantation can be observed. The plateau phase of myeloma during remission exhibits a certain resistance even to high doses of Melphalan (110). If the IMG is considered as a „precursor" phase of myeloma (24, 177), the plateau phse analogous to it.

It is also worth pointing out that following allogenic transplantation due to a "graft versus myeloma" reaction the malignant clone is better suppressed as compared to the allogenic clone.

Early complctions that contribute to the failure of transplantations, interstitial preumonitis acute graft versus host disease, bleeding, and ARDS are similar to those seen in autologous bone marrow transplantations performed as a result of other indications.

*Supplementary and supporting therapy*

Apart from myeloma, patients might need treatment for diseases that might have been existing even before the onset of myeloma or which simply accompany old age. Therefore, almost all myeloma patients are treated along three main guidelines:

1. treatment of previous and degenerative diseases,
2. therapy of the plasma cell tumors,
3. prevention of complications developing after cytostatic therapy, irradiation, missile therapy, bone marrow transplantation, and actual treatment of the intercurrent disease.

Pain caused by the osteolytic damages of the vertebrae and extremities is a typical leading symptom. Generaly, it develops without preceding symptoms following a sudden movenent or an unnatural posture. The cause of pain can easily be identified except those cases of rib fraction without dislocation.

Not only the the fractures but also radicular compressions can cause pain. MRI is the most adequate method for diagnosis of soft tissue alterations. Depending on the localization, pain can be successfully eliminated by fractional or massive irradiation. In less successful cases, pain is only reduced and is tolerated by the patient; otherwise it is fully inhibited for months or years.

In the advanced stages of the disease sequential hemibody irradiation can also be used (192).

A localized, focused back pain may not only be due to the already mentioned bone lesions, but also to spinal compression caused by an extradural tumor. In these cases myelography localizes the affected area and laminectomy is performed. It is also worth pointing out that if the histology of frozen intraoperative samples indicates plasmacytoma or lymphoma, radical surgery is not recommended (64), instead, the expansion of the tumor should be determined as it influences both the actual interventions and postoperative radiotherapy (416). The concomitant edema should be treated with an evenly distributed daily dose of 14–18 mg dexamethasone. 25 mg/kg Mithramycin is an efficient analgetic, however, it has a bone marrow destructive side effect. Pyramidone and non-steroid antiinflammatory agents are the drugs for everyday use. During the terminal stage, opiates are indicated; their effect on intestinal passage should be considered, thus proper dosage should be cantiously determined.

The following general principles are recommended for indication of analgetics:

1. Individual titration of analgetics allows the selection of an optimal dose or drug combination that meet the individual needs of the patient and elicit minimum complications. For example, the motoric performance of the patient should not be impaired, otherwise it increases the danger of fractures.
2. Constipation should be prevented or treated with other drugs.
3. If possible, do not let the patient control his or her drug therapy, it should be the task of the environment. Cytostatics also have an analgetic effect.

Remineralization supported by flouride or calcitonin is generally accepted. Fluor can not be administered alone, but should be always combined with other agents. For this purpose we use our own combination Osvital. However, the effect of these substances could be verified by objective methods. Though different methods have been used for assessment of bone corticalization, heither of them could verify any clinical imrpovement. These unsuccesful results urged the introduction of diphosphonates and pyrophosphates which inhibit the liberation of hydroxyvapatite from the bone (344) inhibiting the activity of osteoclasts stimulated by OAF (371).

Unfortunately, the use of these substances induced leukemia in some patients, therefore, an analoguous drug (disodiumethydronate) has recently introduced (195).

The immobilization period should be transient as it promotes bone resorption. Hence, patients must be mobilized. Supporting corsettes are hardly accepted by patients, because they restrict locomotion and cause sleep disturbances. Plaster beds are better tolerated, and patients get used to them.

Hypercalcemia indicates a plasma $Ca^{++}$ level exceeding 2.8 mmol/L. Increased fluid intake (3 liters/day) and restriction of Ca-containing diet is the mainstay of therapy. Letargy and nausea are the leading clinical symptoms. Sodium infusion promotes the renal excretion of calcium. Furosemide, but not thiazides, is the diuretic of choice (222), and steroid therapy is also of benefit. Urgent therapy comprises the following three steps:

1. Mithramycin, 25 microgram/kg i. v. for a short period.
2. Calcitonin, 3-6 Medical Research Council Units/kg (403).
3. Peritoneal or hemodialysis or plasma exchange.

Plasma exchange is the therapy of choice in hyperviscosity syndrome and kryoglobulinemia.

The actual state of the kidney and bone marrow should be always considered when selecting the antibiotic treatment of intercurrent pneumonia.

In the rare cases of hyperurcemia the so-called alcalic diuresis is a selection of choice; 50–100 mEqu of $NaCO_3$ dissolved in 5% dextrose solution is slowly infused (using a pump if possible).

The pH of the urine should be constantly monitored and maintained aroung 7.0. Administration of allopurinol 300–600 mg/day (inhibitor of xanthine oxydase) is also a useful therapeutic agent).

Neuropathy is treated with carbamazepine which is not always efficient and some patients do not respond to it.

Normalization of anemia is another important task (we accept a critical Ht value of 0.30). Washed and donor selected RBC mass is transfused with special care. Viral infection must be tested with all available methods and special premedication (histamine, antipyretics, elevation of the patients head) has to be implemented. Tansfusion is carried out very slowly at low rate per minute.

After this brief overview of supporting therapy, criteria of remission are summarized as follows:

*CR:* complete remission denotes the disapperance of monoclonal gammaglobulins.

A basic diagnostic requirement is that immunelectrophoresis should not detect monoclonal gammaglobulin. A second immunelectrophoresis, repeated at four weeks, has to confirm the negative result of the first test. Our personal opinion about complete remission is controversial since it requires an extremely aggressive, dangerous therapy which does not always justify the achieved results as CR is always followed by a relapse. Furthermore, despite the aggressive therapy paraproteins still persist in the transplanted bone marrow.

*PR:* partial remission denotes a 50% decrease of monoclonal immunglobulins.

*MR:* according to the recommendations of the Myeloma Task Force Study Group, this stage indicates a remission with a 25% decrease of the M component in the urine and blood. There is also an extended version of criteria determined by the Myeloma Task Force Study Group in which, parallel to the disappearance of monoclonal immunoglobulin, CR is characterized by a BJ(LCh) level lower than 0.06 g/24 h. Bone marrow plasma cell count decreases below 10%. The previously palpable plasmoblastoma can not be identified (Fig. XV/7). Hb values and the Ca level are normal, the patient experiences no bone pain and resumes his activity,

*PR:* the level of M component decreases below 50% and protein concentration in the urine, also decreases by 50%. Hb values improve.

*NC* indicates that the condition of the patient did not improve and therapy was ineffective.

*PD* means progression, increase in the M component level and higher concentrations of BJ protein in the urine. The number of osteolytic changes increases.

The analysis of plasmocytoma therapy shows that it is still far from being solved.

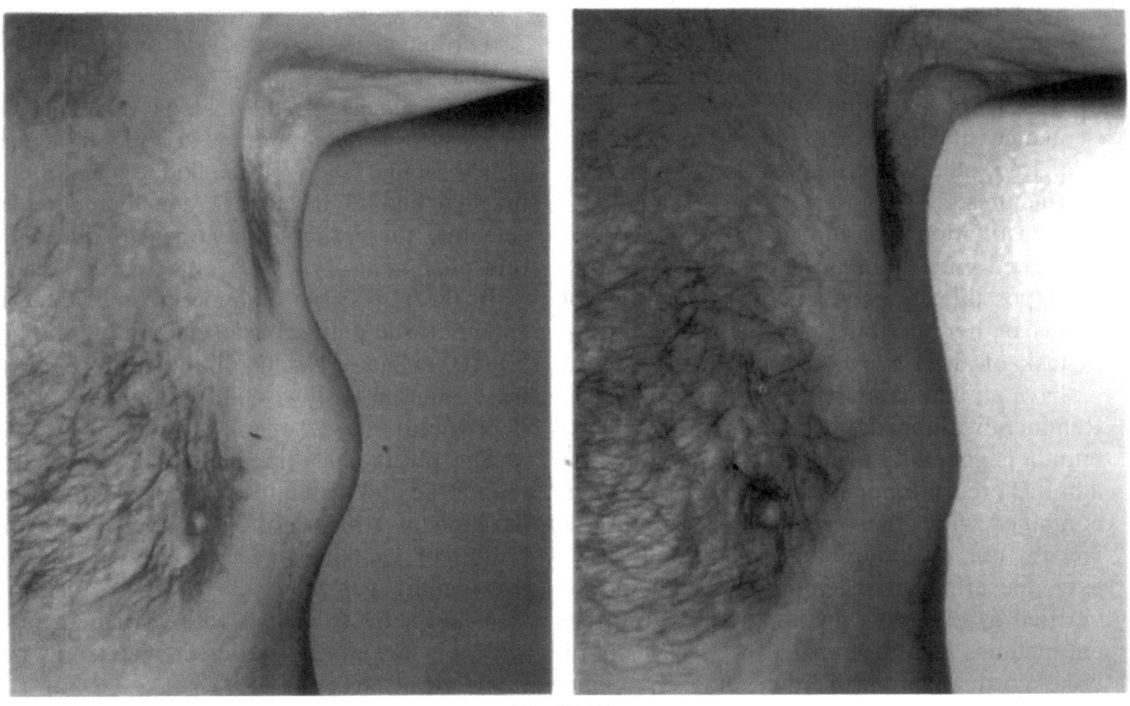

Fig. XV/7

Results achived in the recent decades are insignificant. A dual problem should be solved in the future: a better undestanding the nature of the disease and to create new therapeutic principles which would be capable to prolong the plateau phase. Present data indicate that the duration of this phase ranges between 4 months and 5 years (average 21 months) (28). Repeated remissions, however, get shorter due to the increasing growth rate of myeloma cells during relapse with a concomitant decrease of the doubling time of M protein concentration (35). This high reproduction rate of myeloma cells during relapse makes chemotherapy ineffective, and irreversible bone lesions develop and the disease transfers into the "acute" phase. Pancytopenia and uncontrolled growth of the myeloma cells characterize this phase.

In summary, the survival period depends on the duration of the stable, symptomatic and acute phases of the disease. High growth rate of myeloma cells in the symptomatic phase leads to inevitable death despite modern therapy. Though recovery has never been reported, even traditional therapy has sometimes induced prolinged and stable remission (52, 227, 336, 380, 407).

Therefore, the biological behavior of plasma cell myeloma is much more complicated than expected on the basis of the presence of an uniform malignant cell population. Similar to CML, molecular genetic changes stabilize the disease already in its early phase.

There is a big difference between the phenotype of primordial malignant cells, the target cell population of malignant transformation, and that of the dominant cells characterizing the disease.

Subsequent undrelying processes are obscure, as oligoclonal gammopathy may develop or, conversely, production of MG may diminish. Also exranodal manifestation (plasmoblastoma) with high LDH level and myelodysplastic syndrome (transformation into acute myelogenous leukemia) may result.

Different surface receptors and cytokines produced by myeloma cells are not yet targeted by therapy. The cytokines have a putative autostimulatory effect, thus IL-1, IL-5, IL-6, and OAF are, to some extent, responsible for the manifestations of the disease. The serum $\beta_2$-microglobulin level is an additional information about the tumor mass assessed at stage classification.

A deeper insight into the cellular and molecular biology of, for instance, CALLA and H-ras might reveal some of the special properties of the tumor thus we might understand more about responsen to therapy as well. The fact that the prognosis of myeloma has not changed despite the introduction of different drugs, drug combinations and intensive therapy still justifies the use of the M + P protocol. It is especially indicated in elderly persons, individuals with short expected life span, and those suffering from a variety of other diseases (37). Use of other therapeutic protocols is especially indicated in young patients.

VAD, interferon-α and subsequently HDM therapy should be applied following the diagnosis, the latter parallel to bone marrow transplantation. However, only subsequent prospective studies would be able to evaluate the risks of the early mortality versus the anticipated improvement of survival.

### References


1. Aggarwall, B. B., Eessalu, T. E., Hass, P. E. Characterization of receptors for human tumor necrosis factor and their regulation by γ-interferon. Nature. 318, 665. 1985.
2. Ahre, A., Bjorkholm, M., Mellstedt, H. H., Holm, G., Brenning, G., Engstedt, L., Gahrton, G., Hällen, J., Johansson, B., Johansson, S. G., Karnström, L., Killander, A., Lesner, R., Lockner, D., Lönnquist, B., Sinmonson, B., Stalfelt, A. M., Ternstedt, B., Wadman, B. Intermittent high dose melphalan/prednisone versus continuous low dose melphalan treatment in multiple myeloma. Eur. J. Cancer Clin. Oncol. 19, 499. 1983.
3. Ahre, A., Björkholm, M., Mellstedt, H., Brenning, G., Engstedt, L., Gahston, G., Gallenhammar, H., Holm, G., Johansson, B., Jänmark, M., Karnström, L., Killander, A., Lerner, R., Lockner, D., Lönnquist, B., Nilsson, B., Simonsson, B., Stalfelt, A. M., Strander, H., Svedmyr, E., Wadman, B., Wedelin, C. Human leukocyte interferon and intermittent high-dose Melphalan-Prednisone administration in the treatment of multiple myeloma: A randomized clinical trial from the myeloma group of central Sweden. Cancer. 68, 1331. 1984.
4. Alberts, D. S., Chang, S. Y., Chen, H. S. Variability of melphalan (M) absorption in man. Proc. Am. Assoc. Cancer Res. Am. Soc. Clin. Oncol. 19, 334. 1978.
5. Alexanian, R., Bergsagel, D. E., Migliore, P., Vaughn, W. K., Howe, C. D. Melphalan therapy for plasma cell myeloma. Blood. 31, 1. 1968.
6. Alexanian, R., Haut, A., Khan, A. U., Lane, M., Mc Kelvey, E. U., Migliore, P. J., Stuckey, W. J., Wilson, H. E.

Treatment for multiple myeloma. Combination chemotherapy with different melphalan dose regimens.
JAMA. 208, 1680. 1969.
7. Alexanian, R., Balcerzak, S., Haut, A., Hewlett, J., Gehan, E.
Remission maintainance therapy for multiple myeloma.
A South-west Oncology Group study.
Arch. Intern. Med. 135, 147. 1975.
8. Alexanian, R., Salmon, S., Bonnet, J., Gehan, E., Haut, A., Weick, J.
Combination therapy for multiple myeloma.
Cancer. 40, 2765. 1977.
9. Alexanian, R., Salmon, S., Gutterman, J., Dixon, D., Bonnet, J., Haut, A.
Chemoimmunotherapy for multiple myeloma.
Cancer. 47, 1923. 1981.
10. Alexanian, R., Gutterman, J., Levy, H.
Interferon treatment for multiple myeloma.
Clin. in Haematology. 11, 211. 1982.
11. Alexanian, R., Driecar, R.
Chemotherapy for multiple myeloma.
Cancer. 53, 583. 1984.
12. Alexanian, R.
Diagnosis and management of multiple myeloma.
In: Neoplastic diseases of the blood.
Ed: by Wiermok, P. J., Canellos, G., Kyle, R., Schiffer, C.
Churchill. Livingstone, 529. 1985.
13. Alexanian, R.
Ten-year survival in multiple myeloma.
Arch. Intern. Med. 145, 2073. 1985.
14. Alexanian, R., Barlogie, B.
VCAD-VAD as initial chemotherapy for multiple myeloma.
Blood. 72, (Suppl. 1.). 237. 1988.
15. Ambrus, J. L. Jr., Fauci, A. S.
Current studies examining regulation of the human B cell cycle.
In: Molecular basis of lymphokine action.
Ed: by Webb, D. R., Pierce, C. W., Cohen, S.
Clifton: Humana Press Inc. 137. 1987.
16. Anderka, D., Michalevicz, R., Yair, D., Levo, Y., Douer, D., Ben-Bassat, I., Ramot, B., Shaklai, M., Prokocimer, M., Berrebi, A., Meytes, D., Rosubaum, H., Wallach, D., Revel, M.
Recombinant interferon alpha-C for advanced hairy cell leukemia.
Cancer. 61, 2207. 1988.
17. Anderson, K. C., Barut, B. A., Ritz, J., Freedman, A. S., Nadler, L. M.
Autologous bone marrow transplantation therapy for multiple myeloma.
Eur. J. Haematol. Suppl. 51. 43, 157. 1989.
18. Andres, R., Tobin, J. D.
In: Handbook of the Biology of Aging.
Ed: by Finch, C. E., Hayflick, L.
Van Nostrand Reinhold, N. Y. 357. 1977.
19. Averbuch, S. D., Austin, H. A., Sherwin, S. A., Autonorych, T., Bunn, P. A., Longo, D. L.
Acute interstitial nephritis with the nephrotic syndrome following recombinant leukocyte A interferon therapy for mycosis fungoides.
N. Engl. J. Med. 310, 32. 1984.
20. Balkwill, F. R., Griffin, D., Ward, B. G., Fiers, W.
Antitumour Activity of Interferons alpha and gamma and Tumour Necrosis Factor.
J. Int. Res. 7, 818. 1987.
21. Barlogie, B., Smith, L., Alexanian, R.
Effective treatment of advanced multiple myeloma refractory to alkylating agents.
N. Engl. J. Med. 310, 1353. 1984.
22. Barlogie, B., Alexanian, R., Smith, L., Dixon, D., Smallwood, L., Delasalle, K.
Prognostic implications of tumor cell DNA and RNA content in multiple myeloma.
Blood. 66, 338. 1985.
23. Barlogie, B., Alexanian, R., Dicke, K. A., Zagars, G., Spitzer, G., Jagannath, S., Horwitz, L.
High-dose chemoradiotherapy and autologous bone marrow transplantation for resistant multiple myeloma.
Blood. 70, 869. 1987.
24. Barlogie, B., Alexanian, R., Jagannath, S., Zagars, G., Spitzer, G., Dunphy, F., Yau, J., LeMaistre, F., Dicke, K.
High dose therapy with autologous marrow transplantation for multiple myeloma (MM).
Blood. Abst. Proceeding. 862. 1988.
25. Barlogie, B., Epstein, J., Selvanayagam, P., Alexanian, R.
Plasma cell myeloma – new biological insights and advances in therapy.
Blood. 73, 865. 1989.
26. Barlogie, B., Smallwood, L., Smith, T., Alexanian, R.
High serum levels of lactic dehydrogenase identify a high grade lymphoma-like myeloma.
Ann. Intern. Med. 110, 521. 1989.
27. Bast, B. J. E. G., Boom, S. E., Ballieux, R. E.
Characterization of the colony forming cell in monoclonal gammopathies.
Blood. 60, 608. 1982.
28. Belch, A., Shelley, W., Bergsagel, D. E., Wilson, K., Klimo, P., White, D., Willon, A.
A randomized trial of maintenance versus no maintenance melphalan and prednisone in responding multiple myeloma patients.
Br. J. Cancer. 57, 94. 1988.
29. Benjamin, D., Hartmann, D. P., Bazar, L. S., Jacobson, R. J.
Burkitt's cells can be triggered by teleocidin to secrete interferon-gamma.
Am. J. Hematol. 22, 169. 1986.
30. Bennett, R. W., Gehlhausen, J. A., Poulos, J. T.
Glucocorticoids. A review.
US Pharm. 7, 40. 1982.
31. Berenson, J., Wong, R., Kim, K., Brown, N., Lichtenstein, A.
Evidence for peripheral blood B lymphocyte but not T lymphocyte involvement in multiple myeloma.
Blood. 70, 1550. 1987.
32. Beresini, M. H., Lempert, M. J., Epstein, L. B.
Overlapping polypeptide induction in human fibroblasts in response to treatment with interferon-alpha, interferon-gamma, interleukin 1 alpha, interleukin 1 beta, and tumor necrosis factor.
J. Immunol. 140, 485. 1988.
33. Bergsagel, D. E., Griffith, K. M., Hart, A., Stuckey, W. J.
The treatment of plasma cell myeloma.
Adv. Cancer Res. 10, 311. 1967.
34. Bergsagel, D. E., Cowan, D. H., Hasselback, R.
Plasma cell myeloma: Response of melphalan-resistant patients to high-dose cyclophosphamide.
Can. Med. Assoc. J. 107, 851. 1972.
35. Bergsagel, D. E.
Assessment of the response of mouse and human myeloma to chemotherapy and radiotherapy.
In: Growth kinetics and biochemical regulations of normal and malignant cells.
Ed: by Drewinko, B., Humphreys, R. M.
Williams and Wilkins, Baltimore. 705. 1977.
36. Bergsagel, D. E., Bailey, A. J., Langley, G. R., MacDonald, R. N., White, D. F., Miller, A. B.
The chemotherapy of plasma cell myeloma and the incidence of acute leukemia.
N. Engl. J. Med. 301, 743. 1979.
37. Bergsagel, D. E.
Use a gentle approach for refractory myeloma patients, editorial.
J. Clin. Oncol. 6, 757. 1988.
38. Bergsagel, D. E.
Melphalan/prednisolon versus drug combinations for plasma cell myeloma.
Eur. J. Haematol. Suppl. 51. 43, 117. 1989.
39. Bergui, L., Schena, M., Gaidano, G. L., Riva, M., Caligaris-Cappio, F.

Interleukin-3 and Interleukin-6 synergistically promote the proliferation and differentiation of malignant plasma cell precursors in multiple myeloma.
J. Exp. Med. 170, 613. 1989.
40. Berkowitz, B. A., Ngai, S. H., Yang, J. C.
The disposition of morphine in surgical patients.
Clin. Pharmacol. Ther. 17, 629. 1975.
41. Bernard, D. B., Lynch, S. R., Bothwell, T. H.
Multiple myeloma II. The value of melphalan.
South. African Med. Journ. 48, 1026. 1974.
42. Bertolini, D. R., Nedwin, G. E., Bringman, T. S., Smith, D. D., Mundy, G. R.
Stimulation of bone resorption and inhibition of bone formation in vitro by human tumor necrosis factor.
Nature. 319, 516. 1986.
43. Beutler, B., Cerami, A.
Cachectin and tumour necrosis factor as two sides of the same biological coin.
Nature. 320, 584. 1986.
44. Billiau, A., Bloemmen, J., Bogaerts, M., Claeys, H., Van Damme, J., De Ley, M., de Somer, P., Drochmans, A., Heremans, H., Kriel, A., Schetz, J., Tricot, G., Vermilen, C., Verwilghen, R., Waer, M.
Interferon therapy in multiple myeloma: failure of human fibroblast interferon administration to affect the course of light chain disease.
Eur. J. Cancer. Clin. Oncol. 17, 875. 1981.
45. Billiau, A.
BSF-2 is not just a differentiation factor.
Nature. 324, 415. 1986.
45a. Bischoff, S. C., De Weck, A. L., Dahinden, C. A.
Interleukin-3 and granulocyte/macrophage-colony-stimulating factor render human basophils responsive to low concentrations of complement component C3A.
Proc. Natl. Acad. Sci. 87, 6813. 1990.
46. Bird, T. A., Saklatvala, J.
Identification of a common class of high affinity receptors for both types of porcine interleukin-1 on connective tissue cells.
Nature. 324, 263. 1986.
47. Bjerke, J. R., Livden, J. K., Degre, M., Matre, R.
Interferon in suction blister fluid from psoriatic lesions.
Br. J. Dermatol. 108, 295. 1983.
48. Blalock, J. E., Smith, E. M.
Human leukocyte interferon: structural and biological relatedness to adrenocorticotropine hormone and endorphins.
Proc. Natl. Acad. Sci. USA. 77, 5979. 1980.
49. Blalock, J. E., Stanton, J. D.
Common pathways of interferon hormonal action.
Nature. 283, 406. 1980.
50. Blalock, J. E., Smith, E. M.
Structure and function of interferon (IFN) and neuroendocrine hormone.
In: The Biology of the Interferon System.
Ed: by DeMayer, E., Galasso, G., Schellekans, H.
Elsevier North Holland Biomedical Press, Amsterdam. 93. 1981.
51. Blalock, J. E. Smith, E. M.
Human leukocyte interferon (HuIFN-alfa): potent endorphin-like opioid activity.
Biochem. Biophys. Res. Commun. 101, 472. 1981.
52. Boccadoro, M., Acker, A. V., Carmagnola, A., Conte, F., Pileri, A.
Monitoring of long survival myeloma patient.
Tumori. 67, 571. 1981.
53. Boccadoro, M., Gavarotti, P., Fossati, G., Massaia, M., Pileri, A., Durie, B. G. M.
Kinetics of circulating B-lymphocytes in human myeloma.
Blood. 61, 812. 1983.
54. Boccadoro, M., Redoglia, V., Gavarotti, P., Pileri, A.
Multiple myeloma plasma cell kineticks: rapid and reliable evaluation using 5-bromo-2-deoxyuridine DNA incorporation detected by an anti-BrdUrd monoclonal antibody.
Tumori. 72, 135. 1986.
55. Boccadoro, M., Marmont, F., Tribalto, M., Fossati, G., Redoglia, V., Battaglio, S., Massaia, M., Gallamini, A., Comotti, B., Barbui, T., Campobasso, N., Dammacco, F., Cantonetti, M., Petrucci, M. T., Mandelli, F., Resegotti, L., Pileri, A.
Early responder myeloma: kinetic studies identify a patient subgroup characterized by a very poor prognosis.
J. Clin. Oncol. 7, 119. 1989.
56. Bocci, V., DeFrancesco, P., Pacini, A., Pessina, G. P., Rossi, G. B., Sorrentino, V.
Renal metabolism of homologous serum interferon.
Antiviral Res. 3, 53. 1983.
57. Bonnet, J., Alexanian, R., Salmon, S.
Vincristine, BCNU, doxorubicin, and prednisone (VBAP) combination in the treatment of relapsing or resistant multiple myeloma:
A South-west Oncology Group Study.
Cancer. Treat. Rep. 66, 1267. 1982.
58. Bosanguet, A. G., Gilby, E. D.
Pharmacokinetics of oral and intravenous melphalan during routine treatment of multiple myeloma.
Eur. J. Cancer. Clin. Oncol. 18, 355. 1982.
59. Boss, G. R., Seegmiller, J. E.
Age-related physiologic changes and their clinical significance.
West. J. Med. 135, 434. 1981.
60. Brenning, G., Ahre, A., Nilsson, K.
Correlation between in vitro and in vivo sensitivity to human leukocyte interferon in patients with multiple myeloma.
Scand. J. Haematol. 35, 543. 1985.
61. Brook, J., Bateman, J. R., Gocka, E. F., Nakamura, E., Steinfeld, J. L.
Long-term low dose melphalan treatment of multiple myeloma.
Arch. Intern. Med. 131, 545. 1973.
62. Broun, G. O., Petruska, P. J., Hiramoto, R. N., Cohen, H. J.
Cisplatin, BCNU, Cyclophosphamide, and Prednisone in multiple myeloma.
Cancer. Treat. Rep. 66, 237. 1982.
63. Browning, G.
The in vitro effect of leucocyte alpha-Interferon on human myeloma cells in a semisolid agar culture system.
Scand. J. Haematol. 35, 178. 1985.
64. Bruckman, J. E., Bloomer, W. D.
Management of spinal cord compression.
Seminars in Oncology. 5, 135. 1978.
65. Bruckner, C. D., Fefer, A., Bensinger, W. I., Storb, R., Duire, B. G., Appelbaum, F. R., Petersen, F. B., Weiden, P., Clift, R. A., Sanders, J. E., Sullivan, K. M., Witherss-poon, R. P., Hill, P., Martin, P., Thomas, E. D.
Marrow transplantation for malignant plasma cell disorders: summary of Seattle experience.
Eur. J. Haematol. Suppl. 51. 43, 186. 1989.
66. Brunn, P. A., Foon, K. A., Ihde, D. C., Longo, D. L., Eddy, J., Winkler, C. F., Veach, S. R., Zeffren, J., Sherwin, S., Oddham, R.
Recombinant leucocyte A interferon: an active agent in advanced cutaneous T-cell lymphomas.
Ann. intern. Med. 101, 484. 1984.
67. Buckman, R., Cuzick, J., Galton, D. A. G.
Long term survival in myelomatosis. A report to the MRC Working Party on Leukaemia in Adults.
Brit. J. Haematol. 52, 589. 1982.
68. Buzaid, A. C., Durie, B. G. M.
Management of refractory myeloma: a review.
J. Clin. Oncol. 6, 889. 1988.
69. Caligaris-Cappio, F., Bergui, L., Tesio, L., Pizzolo, G., Malavasi, F., Chilosi, M., Campana, D., van Camp, B., Jánosy G.
Identification of malignant plasma cell precursors in the bone marrow of multiple myeloma.
J. Clin. Invest. 76, 1243. 1985.
70. Calvet, M. C., Gresser, I.
Interferon enhances the excitability of cultured neurons.
Nature. 178, 558. 1979.

71. Case, D. C., Sonneborn, H. L., Paul, St. D.
Phase II study of rDNA alpha-2 interferon (INTRON A) in patients with multiple myeloma utilizing an escalating induction phase.
Cancer Treat. Rep. 70, 1251. 1986.
72. Cohen, H. J., Bartolucci, A. A.
Hexamethylmelamine and prednisone in the treatment of refractory multiple myeloma.
Am. J. Clin. Oncol. 5, 21. 1982.
73. Cohen, H. J., Bartolucci, A.
Age and the treatment of multiple myeloma.
Amer. J. Med. 79, 316. 1985.
74. Coleman, M., Saletan, S., Wolf, D., Nisce, L., Wasser, J., McIntyre, O. R., Tulloh, M.
Whole bone marrow irradiation for the treatment of multiple myeloma.
Cancer. 49, 1328. 1982.
75. Cooper, M. R., Fefer, A., Thompson, J., Case, D. C., Kempf, R., Sacher, R., Neefe, J., Bickers, J., Scarffe, J. H., Spiegel, R., Bonnem, E.
Alpha-2 Interferon/Melphalan/Prednisone in Previously Untreated Patients with Multiple Myeloma: A Phase I–II Trial.
Cancer Treatm. Rep. 70, 473. 1986.
76. Cordingley, F. T., Bianchi, A., Hoffbrand, A. V., Reittie, J. E., Heslop, H. E., Vyakarnam, A., Turner, M., Meager, A., Brenner, M. K.
Tumor necrosis factor as an autocrine tumor growth factor for chronic B-cell malignancies.
Lancet. i, 969. 1988.
77. Cordingley, F. T., Hoffbrand, A. V., Heslop, H. E., Turner, M., Bianchi, A., Reittie, J. E., Vyakarnam, A., Meager, A., Brenner, M. K.
Tumour necrosis factor as an autocrine tumour growth factor for chronic B-cell malignancies.
Lancet. 8592, 969. 1988.
78. Cornwell, M., Pastan, I., Gottesman, M. M.
Certain calcium channel blockers bind specifically to multidrug-resistant human KB carcinoma membrane vesicles and inhibit drug binding to p-glycoprotein.
J. Biol. Chem. 262, 2166. 1987.
79. Costa, G., Engle, R. L., Schilling, A., Carbone, P., Kochwa, S., Nachman, R. L., Glidewell, O., For the acute leukemia gròup „B" and the eastern cooperative oncology group.
Melphalen and Prednisone. An effective combination for the treatment of multiple myeloma.
Amer. J. med. 54, 589. 1973.
80. Costanzi, J. J., Cooper, M. R., Scarffe, J. H., Ozer, H., Gubbs, S. S., Ferraresi, R. W., Pollard, R. B., Spiegel, R. J.
Phase II study of recombinant alpha-2 interferon in resistant multiple myeloma.
J. Clin. Oncol. 3, 654. 1985.
81. Costanzi, J. J., Cao, J., Spiegel, R. J.
The efficacy of chemotherapy following interferon for relapsing multiple myeloma.
In: Proceedings of the serono symposium.
Raven Press. 1985.
82. Costanzi, J. J., Pollard, R. B.
The Use of Interferon in the Treatment of Multiple Myeloma.
Seminars in Oncology. 14. 2. Suppl. 2. 24. 1987.
83. Courtenay-Luck, N. S., Epenetos, A. A., Sivolapenco, G. B., Larche, M., Barkans, J. R., Ritter, M. A.
Development of anti-idiotypic antibodies against tumour antigens and autoantigens in ovarian cancer patients treated intraperitoneally with mouse monoclonal antibodies.
Lancet. ii. 894. 1988.
84. Cozzolino, F., Torcia, M., Aldinucci, D., Rubartelli, A., Miliani, A., Shaw, R., Lansdorp, A. R., Di Guglielmo, R.
Production of interleukin-1 by bone marrow myeloma cells: Its role in the pathogenesis of lytic bone lesions.
Blood. 74, 380. 1989.
85. Crawford, J., Cohen, H. J.
Aging and neoplasia.
Annu. Rev. Gerontol. Geriatr. 4, 3. 1984.
86. Creagan, E. T., Ahmann, D. L., Green, S. J., Long, H. J., Rubin, J., Schutt, A. J., Dziewanowski, Z. E.
Phase II study of recombinant leukocyte A interferon (rIFNαA) in disseminated malignant melanoma.
Cancer. 54, 2844. 1984.
87. Creagan, E. T., Ahmann, D. L., Frytak, St., Long, H. J., Chang, M. N., Itri, L. M.
The consecutive phase II studies of recombinant interferon alfa-2a in advanced malignant melanoma.
Cancer. 59, 638. 1987.
88. Cuttner, J., Wasserman, L. R., Martz, G., Sonntag, R. W., Kyle, R. A., Silver, R. T., Spurr, C., Harley, J. B., Wiernik, P. M., Cornwell, G. G., Falkson, G., Glidewell, O., Holland, J. F.
The use of low dose prednisone and melphalan in the treatment of poor risk patients with multiple myeloma.
Med. Pediat. Oncol. 1, 207. 1975.
89. Cuturi, M. C., Murphy, M., Costa-Giomi, M. P., Weinmann, R., Perussia, B., Trinchieri, G.
Independent regulation of tumor necrosis factor and lymphotoxin production by human peripheral blood lymphocytes.
J. Exp. Med. 165, 1581. 1987.
90. Dalton, W. S., Durie, B. G. M., Alberts, D. S., Gerlach, J. H., Cress, A. E.
Characterization of a new drug-resistant human myeloma cell line that expresses p-glycoprotein.
Cancer. Res. 46, 5125. 1986.
91. Dalton, W. S., Durie, B. G. M., Salmon, S. E., Grogan, T. M., Scheper, R. J., Meltzer, P. S.
Multidrug resistance in multiple myeloma. Detection of p-glycoprotein in clinical specimens and reversal of resistance with verapamil.
Blood. 70, 834A. 1987.
92. Danel, L., Menouni, M., Cohen, J. H., Magaud, J. P., Lenoid, G., Revillard, P., Saez, S.
Distribution of androgen and estrogen receptors among lymphoid and haemopoietic cell lines.
Leuk. Res. 9, 1337. 1985.
93. Danel, L., Vincent, C., Rousset, F., Klein, B., Bataille, R., Flacher, M., Durie, B. G. M., Revillard, P.
Estrogen and progesterone receptors in some human myeloma cell lines and murine hybridomas.
J. Steroid. Biochem. 1989.
94. Davis, G. L., Hoofnagel, J. H.
Interferon in viral hepatitis: role in pathogenesis and treatment.
Hepatology. 6, 1038. 1986.
95. DeClerq, E., Zang, Z. X., Huygen, K.
Synergism in the antitumor effects of type I and type II Interferon in mice inoculated with leukemia L1210 cells.
Cancer Lett. 15, 223. 1982.
96. Dewhirst, F. E., Stasbenko, P. O., Mole, J. E., Tsurumachi, T.
Purification and partial sequence of human osteoclast-activating factor: identity with interleukin 1-beta.
J. Immunol. 135, 2562. 1985.
97. DeWit, R., Schattenkerk, J. K. M. E., Boucher, C. A. B., Bakker, J. M., Veenhof, K. H. N., Danner, S. A.
Clinical and virological effects of high-dose recombinant Interferon-alpha in disseminated AIDS-related Kaposi's sarcoma.
Lancet. i, 1214, 1988.
98. Diaz, M. O., Le Beau, M. M., Pitha, P., Rowley, J. D.
Interferon and c-ets-1 genes in the translocation (9; 11) (p22:q23) in human acute monocytic leukemia.
Science. 231, 265. 1986.
99. Diaz, M. O., Ziemin, S., Le Beau, M. M., Pitha, P., Smith, S. D., Chilcote, R. R., Rowley, J. D.
Homozygous deletion of the alpha and $beta_1$-interferon genes in human leukemia and derived cell lines.
Proc. Natl. Acad. Sci. USA. 85, 5259. 1988.
100. Dillon, N., Chung, S., Kelly, J., O'Melley, K.
Age and beta adrenoceptor-mediated function.

Clin. Pharmac. Ther. 27, 769. 1980.
101. Dinarello, C. A., Mier, J. W.
Current concepts: Lymphokines.
N. Engl. J. Med. 317, 940. 1987.
102. Divoll, M., Abernethy, D. R., Ameer, B., Greenblatt, D. J.
Acetaminophen kinetics in the elderly.
Clin. Pharmac. Ther. 31, 151. 1982.
103. Dolbeare, F., Gratzner, H., Pallavicini, M., Gray, J. W.
Flow cytometric measurement of total DNA content and incorporated bromodeoxyuridine.
Proc. Natl. Acad. Sci. USA. 80, 5573. 1983.
104. Dolei, A., Capobianchi, M. R., Ameglio, F.
Human IFN-gamma enhances the expression of class I and class II major histocompatibility complex products in neoplastic cells more effectively than interferon-alpha and interferon-beta.
Infect. Immun. 40, 172. 1983.
105. Dorr, R. T., Salmon, S. E., Robertone, A., Bonnem, E.
Phase I–II trial of interferon-alfa2b by continuous subcutaneous infusion over 28 days.
J. Interferon Res. 8, 717. 1988.
106. Drayer, D. E., Romankiewicz, J. A., Lorenzo, B., Reidenber, M. M.
Age and renal clearance of cimetidine.
Clin. Pharm. Ther. 31, 45. 1982.
107. Drewinko, B., Alexanian, R., Boyer, H., Barlogie, B., Rubinow, S. I.
The growth fraction of human myeloma cells.
Blood. 57, 333. 1982.
108. Duncombe, A. S., Brenner, M. K.
Is circulating tumor necrosis factor bioactive?
N. Engl. J. Med. 319, 1227. 1988.
109. Durie, B. G. M., Salmon, S. E., Moon, T. E.
Pretreatment tumor mass, cell kinetics and prognosis in multiple myeloma.
Blood. 55, 364. 1980.
110. Durie, B. G. M., Russell, D. H., Salmon, S. E.
Reappraisal of plateau phase in myeloma.
Lancet. ii, 65. 1980.
111. Durie, B. G., Salmon, S. E., Mundy, G. R.
Relation of osteoclast activating factor production to extent of bone disease in multiple myeloma.
Brit. J. Haematol. 47, 21. 1981.
112. Durie, B. G. M.
Staging and kinetics of multiple myeloma.
Clin. in Haematol. 11, 3. 1982.
113. Durie, B. G. M., Salmon, S. E.
The current status and future prospects of treatment for multiple myeloma.
Clin. in Haematol. 11, 181. 1982.
114. Durie, B. G. M., Young, L. A., Salmon, S. E.
Human myeloma in vitro colony growth. Interrelationship between drug sensitivity, cell kinetics, and patients survival duration.
Blood. 61, 929. 1983.
115. Durie, B. G. M., Salmon, S. E.
Staging, kinetics and flow cytometry of multiple myeloma.
In: Neoplastic diseases of the blood.
Ed: by Wiernik, P. J., Canellos, G., Kyle, R., Schiffer, C.
Church. Livingstone. 513. 1985.
116. Einat, M., Resnitzky, D., Kimchi, A.
Inhibitory effects of interferon on the expression of genes regulated by platelet-derived growth factor.
Proc. Natl. Acad. Sci. USA. 82, 7608. 1985.
117. Epstein, J., Barlogie, B., Katzmann, J., Alexanian, R.
Phenotypic heterogeneity in aneuploid multiple myeloma indicates pre-B cell involvement.
Blood. 71, 861. 1988.
118. Epstein, J., Barlogie, B., Ling, V., Alexanian, R.
P-glycoprotein expression in plasma cell myeloma is associated with resistance to VAD.
Blood. 15, 74. 1989.
119. Epstein, J., He X-Y, Xiao, H., Barlogie, B.
Multi-lineage involvement in multiple myeloma (MM).
Proc. ASCO. 8, 256. 1989.
120. Ernstoff, M. S., Kirkwood, J. M.
Changes in the bone marrow of cancer patients treated with recombinant interferon alpha-2.
Am. J. Med. 76, 593. 1984.
121. Etoh, T., Takashi, H., Maie, M., Ohnuma, N., Tanabe, M.
Tumor imaging by antineuroblastoma monoclonal antibody and its application to treatment.
Cancer. 62, 1282. 1988.
122. Ezaki, K., Ogawa, M., Okabe, K., Abe, K., Jnone, K., Horikashi, N., Inagaki, J.
Clinical and immunological studies of human fibroblastoid interferon.
Cancer Chemother. Pharmacol. 8, 145. 1982.
123. Farhangi, M., Osserman, E. F.
The treatment of multiple myeloma.
Seminars in Haematol. 10, 149. 1973.
124. Fefer, A., Cheever, M. A., Greenberg, P. D.
Identical twin (syngeneic) marrow transplantation for hematologic cancers.
JNCI. 76, 1269. 1986.
125. Finkel, H. E., Young, W. J., Salmon, S. E., Schilling, A.
Current concepts in the therapy of multiple myeloma.
Med. Clin. of N. America. 50, 1569. 1966.
126. Finnish Leukaemia Group.
Aggressiv combination chemotherapy in multiple myeloma. A multicentre trial.
Scand. J. Haematol. 35, 205. 1985.
127. Foă, R., Lauria, F., Lusso, P., Raspadori, D., Fierro, M. T., Tazzari, P. L., Caudana, L., Matera, L.
Phenotypic and functional characterization of the circulating NK compartment in hairy-cell leukaemia.
Clin. Exp. Immunol. 64, 392. 1986.
128. Foă, R., Fierro, M. T., Lusso, P., Bonferroni, M., Raspadori, D., Buzzi, M., Zinzani, P. L., Resegotti, L., Lauria, F.
Effect of alpha-interferon on the immune system of patients with hairy cell leukemia.
Leukemia. 1, 377. 1987.
129. Fomsgaard, A., Worsaae, H., Bendtzen, K.
Detection of tumour necrosis factor from lipopolysaccharide-stimulated human mononuclear cells by enzymelinked immunosorbent assay and cytotoxicity bioassay.
Scand. J. Immunol. 27, 143. 1988.
130. Foon, K. A., Dougher, G.
Increased growth of eyelashes in a patient given leukocyte A interferon.
N. Engl. J. Med. 311, 1259. 1984.
131. Foon, K. A., Sherwin, S. A., Abrams, P. G., Longo, D. L., Fer, M. F., Stevenson, H. C., Ochs, J. J., Bottino, G. C., Schoenberger, C. S., Zeffren, J., Jaffe, E. S., Oldham, R. K.
Treatment of advanced 'non-Hodgkin's lymphoma with recombinant leukocyte A interferon.
N. Engl. J. Med. 311, 1148. 1984.
132. Foon, K. A., Bottino, G. C., Abrams, P. G., Fer, M. F., Longo, D. L., Schoenberger, C. S., Oldham, R. K.
Phase II trial of recombinant leukocyte A interferon in patients with advanced chronic lymphocytic leukemia.
Am. J. Med. 78, 216. 1985.
133. Foon, K. A., Maluish, A. E., Abrams, P. G., Wrightington, S., Stevenson, H. C., Atarif, A., Fer, M. F., Overton, W. R., Poole, M., Schnipper, E. F., Jaffe, E. S., Herberman, R. B.
Recombinant leucoyte A interferon therapy for advanced hairy cell leukemia. Therapeutic and immunologic response.
Am. J. Med. 80, 351. 1986.
134. Fransen, L., Van der Heyden, J., Ruysschaert, R., Fiers, W.
Recombinant tumor necrosis factor: its effect and its sinergism with interferon-gamma on a variety of normal and transformed human cell lines.
Eur. J. Can. Clin. Oncol. 22, 419. 1986.
135. Frasca, D., Adorini, L., Landolfo, S., Doria, G.
Enhancement of suppressor T cell activity by injection of anti-IFN-$\gamma$-monoclonal antibody.
J. Immunol. 140, 4103. 1988.

136. Friend, S. H., Dryja, T. P., Weinberg, R. A.
Oncogenes and tumor-suppressing genes.
N. Engl. J. Med. 318, 618. 1988.
137. Fuhuzawa, Y., Kawamura, T., Ohno, R.
Clinical trial of a combination of human lymphoblastoid interferon (HLBI) and methyl-prednisolone (HLBI-mP) in multiple myeloma.
Gan-To-Kagaku-Ryoho. 16, 3581. 1989.
138. Gahrton, G., Tura, S., Flesch, M., Gratwohl, A., Gravett, P., Lucarelli, G., Michallet, M., Reiffers, J., Ringden, O., van Lint, M. T., Vernant, J. P., Zwaan, F. E.
Bone marrow transplantation in multiple myeloma. Report from the European Cooperative Group for Bone Marrow Transplantation.
Blood. 69, 126. 1987.
139. Gahrton, G., Tura, S., Flesch, M., Gratwohl, A., Gravett, P., Lindenberg, A., Lucarelli, G., Michallet, M., Reiffers, J., Ringden, O., Nikoskelainen, J., Van Lint, M. T., Vernant, J. P., Znaan, F. E.
Allogeneic bone marrow transplantation in 24 patients with multiple myeloma reported to the EBMT Registry.
Hematol. Oncol. 6, 181. 1988.
140. Gahrton, G., Tura, S., Belanger, C., Caro, M., Chapvis, B., Ferrant, A., Flesch, M., Gore, M., Gratwohl, A., Gravett, P. J., Harrouseau, J. L., Lindeberg, A., Lingman, P., Löwenberg, B., Lucarelli, G., Michallet, M., Reiffers, J., Ringden, O., Van Lint, M. T., Vernant, J. P., Sellerfors, B., Simonsson, B., Toivanen, A., Troussard, X., Verdonck, L. F., Volin L., Zwaan, F. E.
Allogeneic bone marrow transplantation in patients with multiple myeloma.
Eur. J. Haematol. Suppl. 51. 43, 182. 1989.
141. Gailani, S., Seon, B. K., Nussbaum, A., Henderson, E. S., Pressman, D.
Radioimmunoassay for myeloma idiotype.
J. Natl. Cancer Inst. 58, 1553. 1977.
142. Galton, J. E., Bedford, Ph., Scott, J. E., Brand, C. M., Nethersell, A. B. W.
Antibodies to lymphoblastoid interferon.
Lancet. ii, 572. 1989.
143. Gerrett, L. R., Durie, B. G. M., Nedwin, G. E., Gillespie, A., Bringman, T., Sabatini, M., Bertolini, D. R., Mundy, G. R.
Production of the bone resorbing cytokine lymphotoxin by cultured human myeloma cells.
N. Engl. J. Med. 317. 526. 1987.
144. Gauldie, J., Richards, C., Harnish, D., Landsdorp, P., Baumann, H.
Interferon beta-2/B cell stimulatory factor type 2 shares monocyte-derived hepatocyte stimulating factors and the major acute phase protein response in liver cells.
Proc. Natl. Acad. Sci. USA. 84, 7241. 1987.
145. Geokas, M. C., Haverback, B. J.
The aging gastrointestinal tract.
Am. J. Surg. 117, 881. 1969.
146. George, R. P., Poth, J. L., Gordon, D., Schrier, S. L.
Multiple myeloma – Intermittent combination chemotherapy compared to continuous therapy.
Cancer. 29, 1665. 1972.
147. Gerdes, J., Dallenbach, F., Lennert, K.
Growth fraction in malignant non-Hodgkin's lymphomas (NHL) as determined in situ with the monoclonal antibody Ki-67.
Hematol. Oncol. 2, 365. 1984.
148. Ghaderi, A. A., Richardson, P., Cardona, C., Millsum, M. J., Ling, N., Gillis, S., Ledbetter, J., Gordon, J.
Stimulation of B-chronic lymphocytic leukemia populations by recombinant interleukin-4 and other defined growth-promoting agents.
Leukemia. 2, 165. 1988.
149. Giles, F. J., Singer, C. R. J., Gray, A. G., Yong, K. L., Brozovic, M., Daries, S. C., Grant, I. R., Hoffbrand, A. V., Machin, S. J., Mehta, A. B.,
Alpha-interferon therapy for essential thrombocythaemia.
Lancet. ii, 70. 1988.
150. Giordano, P., Mazzini, G., Riccardi, A.
A simple procedure for DNA cytofluorometry simultaneously with autoradiographic grain count.
In: Proceedings of the Combined International Conference on analytical cytology and with International Symposium on flow cytometry.
Elmenau, West Germany. 75. 1982.
151. Glezerman, M., Lunenfeld, E., Cohen, V., Sarow, I., Movshovitz, M., Doerner, T., Shoham, J., Revel, M.
Placebo-controlled trial of topical interferon in labial and genital herpes.
Lancet. i, 150. 1988.
152. Goldstein, D., László, J.
Interferon therapy in cancer: from imagination to interferon.
Cancer Res. 46, 4315. 1986.
153. Gonzalez, F., Trujillo, J. M., Alexanian, R.
Acute leukemia in multiple myeloma.
Ann. Intern. Med. 86, 440. 1977.
154. Gore, M. E., Viner, C., Meldrum, M., Bell, J., Milan, S., Zuiable, A., Slevin, M., Selby, P. J., Clark, P. I., Millar, B., Maitland, J. A., Judson, I. R., Tillyer, C., Malpas, J. S., McElwain, T. J.
Intensive treatment of multiple myeloma and criteria for complete remission.
Lancet. ii, 879. 1989.
155. Gratzner, H.
Monoclonal antibody to 5-bromo- and 5-iododeoxyuridine: A new reagent for detection of DNA replication.
Science. 218, 474. 1982.
156. Greenblatt, D. J.
Reduced serum albumin concentration in the elderly.
A report from the Boston Collaborative Drug Surveillance Program.
J. Am. Geriatr. Soc. 27, 20. 1979.
157. Greenblatt, D. J., Shader, R. I.
Effects of age and other drugs on benzodiazepine kinetics.
Arzneimittelforsch. 30, 886. 1980.
158. Greenblatt, D. J., Divoll, M., Abernethy, D. R., Harmatz, J. S., Shader, R. J.,
Antipyrine kinetics in the elderly: Prediction of age-related changes in benzodiazepine oxidizing capacity.
J. Pharmacol. Exp. Ther. 220, 120. 1982.
159. Greipp, P. R., Kyle, R. A.
Clinical morphological and cell kinetic differences among multiple myeloma, monoclonal gammopathy of undetermined significance, and smoldering multiple myeloma.
Blood. 62, 166. 1983.
160. Greipp, P. R., Witzing, T. E., Gonchoroff, N. J., Habermann, T. M., Katzmann, J. A., O'Fallon, W. M., Kyle, R. A.
Immunofluorescence labeling indices in myeloma and related monoclonal gammopathies.
Mayo. Clin. Proc. 62, 969. 1987.
161. Greipp, P. R., Katzmann, J. A., O'Fallan, W. M., Kyle, R. A.
Value of beta-2 microglobulin level and plasma cell labeling indices as prognostic factors in patients with newly diagnosed myeloma.
Blood. 72, 219. 1988.
162. Gresser, I., Tovey, M. G., Maury, C., Chouroulinkov, I.
Lethality of interferon preparations for newborn mice.
Nature. 258, 76. 1975.
163. Gresser, I.
Antitumor effects of Interferon.
In: Cancer – A Comprehensive Treatise.
Ed: by Becker, F.
Plenum Press, New York/London. 521. 1977.
164. Gressler, V. H., Weinkauff, R. E., Franklin, W. A., Golomb, H. M.
Modulation of the expression of major histocompatibility antigens on splenic hairy cells – differential effect upon in vitro treatment with alpha-2b-interferon, gamma-interferon, and interleukin-2.
Blood. 72, 1048. 1988.

165. Griggs, R. C., Condemi, J. J., Vaughan, J. H.
Effect of therapeutic dosages of prednisone on human immunoglobulin G metabolism.
J. Allerg. Clin. Immunol. 49, 267. 1972.
166. Grimm, E. A., Ramsey, K. M., Mazumder, A., Wilson, D. J., Djeu, J. Y., Rosenberg, S. A.
Lymphokine-activated killer cell phenomenon. II. Precursor phenotype is serologically distinct from peripheral T lymphocytes, memory cytotoxic thymus-derived lymphocytes, and natural killer cells.
J. Exp. Med. 157, 884. 1983.
167. Grogan, T. M., Lippman, S. M., Spier, C. M., Slyman, D. J., Rybski, J. A., Rangel, C. S., Richter, L. C., Miller, T. P.
Independent prognostic significance of a nuclear proliferative antigen in diffuse large cell lymphomas as determined by the monoclonal antibody Ki-67.
Blood. 71, 1157. 1988.
168 .Grogan, T. M., Durie B. G. M., Spier, C. M., Richter, L., Vela, E.
Myelomonocytic antigen positive multiple myeloma.
Blood 73, 41. 1989.
169. Gutterman, J. U., Blumenschein, G. R., Alexanian, R., Yap, H. W. Y., Buzdar, A. U., Cabaillas, F., Hortobagyi, G. N., Hersh, E. M., Rasmussen, S. L., Harmon, M., Kramer, M., Pestka, S.
Leukocyte interferon-induced tumor regression in human metastatic breast cancer, multiple myeloma, and malignant lymphoma.
Ann. Intern. Med. 93, 399. 1980.
170. Gutterman, J. U., Fein, S., Quesada, J., Horning, S. J., Levine, J. F., Alexanian, R., Bernhardt, L., Kramer, M., Spiegel, H., Colburn, W., Trown, P., Merigan, T., Dziewanowski, Z.
Recombinant leukocyte A interferon: Pharmacokinetics, single-dose tolerance, and biologic effects in cancer patients.
Ann. Intern. Med. 96, 549. 1982.
171. Gutterman, J. U.
The role of interferons in the treatment of hematologic malignancies.
Seminars in Hematol. 25, 3. 1988.
172. Hagelstam, H.
Solitary vertebral plasmacytoma causing paraplegia.
Acta Chir. Scand. 109, 384. 1955.
173. Hamburger, A. W., Salmon, S. E.
Primary bioassay of human myeloma stem cells.
J. Clin. Invest. 60, 846. 1977.
174. Hannigan, G. E., Fish, E. N., Williams, B. R. G.
Modulation of human interferon-alpha receptor expression by human interferon-gamma.
J. Biol. Chem. 259, 8084. 1984.
175. Harley, J. B., Schilling, A., Glidewell, O.
Ineffectiveness of fluoride therapy in multiple myeloma.
N. Engl. J. Med. 286, 1283. 1972.
176. Harley, J. B., Pajak, Th. F., McIntyre, O. R., Kochwa, Sh., Cooper, M. R., Coleman, M., Cuttner, J.
Improved survival of increased-risk myeloma patients on combined triple-alkylating-agent therapy: a study of the CALGB.
Blood. 54, 13. 1979.
177. Harousseau, J. L., Milpied, N., Garand, R., Bourhis, J. H.
High dose melphalan and autologous bone marrow transplantation in high risk myeloma.
Br. J. Haematol. 67, 493. 1987.
178. Harriman, G. R., Kunimoto, D. Y., Elliot, J. F., Paetkau, V., Strober, W.
The role of IL-5 in IgA B cell differentiation.
J. Immunol. 140, 3033. 1988.
179. Health care in the elderly: (WHO).
Report of the technical group on use of medicaments by the elderly.
Drugs. 22, 279. 1981.
180. Herberman, R. B., Djeu, J. Y., Kay, H. D., Ortaldo, J. R., Riccardi, C., Bonnard, G. D., Holden, H. T., Fagnani, R., Santoni, A., Ducetti, P.
Natural killer cells: characteristics and regulation of activity.
Immunol. Rev. 44, 43. 1978.
181. Herberman, R., Ortaldo, J. R., Bonnard, G. D.
Augmentation by interferon of human natural and antibody-dependent cell-mediated cytotoxicity.
Nature. 227, 221. 1979.
182. Hirano, T., Yasakuwa, K., Hanada, H., Tagar, T., Watanabe, Y., Matsuda, T., Kashiwamura, S., Nakajima, K., Koyama, K., Iwanatsu, A., Tsumasawa, S., Sakijama, F., Matsui, H., Takahara, Y., Taniguchia, T., Kishimoto, T.
Complementary DNA for a novel human interleukin (BSF-2) that induces B lymphocytes to produce immunoglobulin.
Nature. 324, 73. 1986.
183. Hoffmann, V., Scheel, A., Schwarzmeier, J., Talstad, I., Varini, M., Kaplan, S., Sauter, Chr., Fehr, J.
Wirksamkeit von Interferon-alpha2 (E. coli) bei der Haarzell-Leukämie.
Schweiz. med. Wschr. 115, 235. 1985.
184. Hokanson, J. A., Brown, B. W., Thompson, J. R., Drewinko, B., Alexanian, R.
Tumour growth patterns in multiple myeloma.
Cancer. 39, 1077. 1977.
185. Hoogstraten, B., Sheehe, P. R., Cuttner, J., Cooper, T., Kyle, R. A., Oberfield, R. A., Townsend, S. R., Harley, J. B., Hayes, D. M., Costa, G., Holland, J. F.
Melphalan in multiple myeloma.
Blood. 30, 74. 1967.
186. Huang, K. Y., Donahoe, R. M., Gordon, F. B., Dressler, H. R.
Enhancement of phagocytosis by interferon-containing preparations.
Infect. Immun. 4, 581. 1971.
187. Hughes, T. K., Baron, S.
Tumour necrosis factor.
J. Biol. Reg. Homeos. Agents. 1, 29. 1987.
188. Hui, S. K., Lichtiger, B., Quesada, J. R.
Cross-reactive red blood cell antigenic substance in human leukocyte alpha interferon.
Arch. Pathol. Lab. Med. 110(2), 128. 1986.
189. Ikebucki, K., Wong, G. C., Clark, S. C., Ihle, J. N., Hirai, Y., Ogawa, M.
Interleukin 6 enhancement of interleukin 3 – dependent proliferation of multipotential hemopoietic precursors.
Proc. Natl. Acad. Sci. 84, 9035. 1987.
190. Inaba, M., Watanabe, T., Sugiyama, Y.
Kinetic analysis of active efflux of vincristine from multidrug-resistant P388 leukemia cells.
Jpn. J. Cancer. Res. 78, 397. 1987.
191. Jacobson, D. R., Zolla-Pazner, S.
Immunosuppression and infection in multiple myeloma.
Semin. Oncol. 8, 282. 1986.
192. Jaffe, J. P., Bosch, A., Raich, P. C.
Sequential hemibody radiotherapy in advanced multiple myeloma.
Cancer. 43, 124. 1979.
193. Jelinek, D., Lipsky, P. E.
Regulation of human B cell proliferation and differentiation by interleukin 2.
In: The molecular basis of B-cell differentiation and function.
Ed.: by Ferrarini, M., Pernis, B.
New York, London: Plenum Press. 63. 1986.
194. Joller-Jemelka, I. H., Fischer, A. E.
Detection of antibodies of interferon alpha in various patient groups without prior administration of interferon: The 1986 ISIR-TNO meeting on the interferon system.
Espoo, Finland, September 7–12. 1986.
195. Jong, A., Ouwenaller, C., van Chantraine, A., Courvoisier, B.
Parenteral diphosphonates for treating malignant hypertension.
Cancer. 48, 1922. 1981.
196. Joshua, D. E., Wearne, A., Kronenberg, H.

Indication for therapy in multiple myeloma: should it be stage or stability.
Lancet. ii, 201. 1985.
197. Joshua, D. E.
Biology of multiple myeloma – Host-tumour interactions and immune regulation of disease activity.
Hematol. Oncol. 6, 83. 1988.
198. Kanellos, J., Pietersz, G. A., Cunningham, Z., McKenzie, I. F. C.
Antitumor activity of aminopterin-monoclonal antibody coniugates; in vitro and in vivo comparison with methotrexate-monoclonal antibody coniugates.
Immunol. Cell. Biol. 65, 483. 1987.
199. Kartner, N., Evernden-Porelle, D., Bradley, G., Ling, V.
Detection of p-glycoprotein in multidrug-resistant cell lines by monoclonal antibodies.
Nature. 316, 820. 1985.
200. Kataoka, T., Fusiko, O. H., Sakura, Y.
Enhancement of antiproliferative activity of vincristine and adrymycin by interferon.
Gann. 75, 548. 1984.
201. Katzmann, J. A.
Myeloma-induced immunosuppression: A multistep mechanism.
J. Immunol. 121, 1405. 1978.
202. Kawade, Y.
Quantitation of neutralisation of interferon by antibody.
Methods Enzymol. 119, 558. 1986.
203. Kawano, N., Hirano, T., Matsuda, T., Taga, T., Horii, Y., Jwato, K., Asaoku, H., Tang, B., Tanabe, O., Tanaka, H., Kuramoto, A., Kishimoto, T.
Autocrine generation and requirement of BSF-2/IL-6 for human multiple myelomas.
Nature. 332, 83. 1988.
204. Kawano, M., Yamamoto, I., Iwato, K., Tanaka, H., Kideki, A., Tanabe, O., Ishikawa, H., Nobuyoshi, M., Ohmoto, Y., Hirai, Y., Kuramoto, A.
Myeloma cells produce IL-1, having bone-resorbing activity.
Blood 72, 145. 1988.
205. Kehrl, J. H., Alvarez-Mon, M., Delsing, G. A., Fauci, A. S.
Lymphotoxin is an important T cell-derived growth factor for human B cells.
Science. 238, 1144. 1987.
206. Kelley, V. E., Goulton, G. N., Hattori, M., Ikegami, H., Eisenboth, G., Strom, T. B.
Anti-Interleukin 2 receptor antibody suppresses murine diabetic insulitis and lupus nephritis.
J. Immunol. 140, 59. 1988.
207. Kernan, N. A., Byers, V., Scannon, P. J., Mischak, R. P., Brochstein, J., Flomenberg, N., Dupont, B., O'Reilly, R. J.
Treatment of steroid-resistant acute graft-vs-host disease by in vivo administration of an anti-T-cell ricin A chain immunotoxin.
JAMA. 259, 3154. 1988.
208. Kirkwood, J. M., Ernstoff, M. S., Davis, C. A., Reiss, M., Ferraresi, R., Rudnick, S. A.
Comparison of intramuscular and intravenous recombinant alpha-2 interferon in melanoma and other cancers.
Ann. Int. Med. 103, 32. 1985.
209. Kishimoto, T.
Factors affecting B cell growth and differentiation.
Ann. Rev. Immunol. 3, 133. 1985.
210. Kishimoto, T.
B-cell stimulating factors (BSFs): Molecular structure, biological function, and regulation of expression.
J. Clin. Immunol. 7, 343. 1987.
211. Klein, B., Jourdan, M., Vasquez, A., Dugas, B., Bataille, R.
Production of growth factors by human myeloma cells.
Cancer Res. 47, 4856. 1987.
212. Klein, B., Zhang, X-G., Jourdan, M., Content, J., Houssiau, F., Regis, L. M., Aarden, L., Piechaczyk, M., Bataille, R.
Paracrine rather than autocrine regulation of myeloma-cell growth and differentiation by interleukin-6.
Blood. 73, 517. 1989.
213. Kohase, M., Henriksen-DeStefano, D., May, L. T., Vilcek, J., Sehgal, P. B.
Induction of beta 2-interferon by tumor necrosis factor: a homeostatic mechanism in the control of cell proliferation.
Cell. 45, 659. 1986.
214. Kohn, L. D., Friedman, R. M., Holmes, J. M., Lee, G.
Use of thyrotropin and cholera toxin to probe the mechanism by which interferon initiates it's antiviral activity.
Proc. Natl. Acad. Sci. USA. 73, 3695. 1976.
215. Konrad, M. W., Childs, A. L., Merigan, T. C., Borden, E. C.
Assessment of the antigenic response in humans to recombinant mutant interferon beta.
J. Clin. Immunol. 7, 365. 1987.
216. Koren, S., Klimpel, G. R., Fleischman, Jr. R.
Treatment of mice with macrophage colony stimulating factor (CSF-1) prevents the in vivo myelosuppression induced by murine alpha, beta and gamma interferons.
J. Biol. Resp. Mod. 5, 481. 1986.
217. Korst, D. R., Mich, A. A., Clifford, G. C., Fowler, W. M., Louis, J., Will, J., Wilson, H. E.
Multiple Myeloma II Analysis of cyclophosphamide therapy in 165 patients.
JAMA 189, 758. 1964.
218. Kozbor, D., Tripputi, P., Roder, J. C., Croce, C. M.
A human hybrid myeloma for production of human monoclonal antibodies.
J. Immunol. 133, 3001. 1984.
219. Kowalczyk, J., Sandberg, A. A.
A possible subgroup of ALL with 9p.
Cancer Genet. Cytogenet. 9, 383. 1983.
220. Krown, S. E.
Interferon treatment of renal cell carcinoma: current status and future prospects.
Cancer. 59, 647. 1987.
221. Kubagawa, H., Vogler, L. B., Capra, J. D., Conrad, M. E., Lawton, A. R., Cooper, M. D.
Studies on clonal origin of multiple myeloma. Use of individual specific (idiotype) antibodies to trace the oncogenic event to it's earliest point of expression in B cell differentiation.
J. Exp. Med. 150, 792. 1979.
222. Kyle, R. A., Bayrd, E. D.
The monoclonal gammophaties multiple myeloma and related disorders.
Thomas. Springfield. 188. 1976.
223. Kyle, R. A., Pajak, Th. F., Henderson, E. S., Nawabi, I. V., Brunner, K., Henry, P. H., McIntyre, O. R., Holland, J. F.
Multiple myeloma resistant to Melphalan: treatment with Doxorubicin, Cyclopharphamid Carmustine (BCNU), and Prednisone.
Cancer. Treat. Rep. 66, 451. 1982.
224. Kyle, R. A.
Treatment of multiple myeloma.
N. Engl. J. Med. 310, 1382. 1984.
225. Kyle, R. A., Greipp, P. R.
Monoclonal gammopathies of undetermined significance.
In: Neoplastic diseases of the blood.
Ed.: by Wiernik, P. H., Canellos, C. P., Kyle, R. A., Schitter, C. A.
Church. Livingstone N. Y. 653. 1985.
226. Kyle, R. A., Oken, M. M., Greipp, P. R.
Alternating Cycles of VBMCP with Recombinant Alpha-Interferon (Intron A) in the Treatment of Multiple Myeloma.
XXII. Congress of the International Society of Hematology, University of Milan, Italy. 1988.
227. Kyle, R. A.
IgD multiple myeloma: A cure of 21 years.
Am. J. Hematol. 29, 41. 1988.
228. Kyle, R. A.
Multiple myleoma. An update an diagnosis and management.
Acta. Oncol. 1, 29. 1990.
229. Lafage, M., Maroc, N., Dubreuil, P., de Waal Malciijt, R.,

Pébusque, M. J., Carcasonne, T., Mannoni, P.
The human interleukin-1 a gene is located on the long arm of chromosome 2 at band q13.
Blood. 73, 104. 1989.

230. Lamy, P. P.
Prescribing for the Elderly.
Littleton. MA. PSG. Pub. 1980.

231. Lane, H. C., Kovacs, J. A., Feinberg, J., Herpin, B., Davay, V., Wallar, R., Deyton, L., Metcalf, J. A., Baseier, M., Salzman, N.
Anti-retroviral effects of interferon-alpha in AIDS-associated Kaposi's sarcoma.
Lancet. i, 1218. 1988.

232. Latreille, J., Barlogie, B., Dosik, G., Johnston, D. A., Drewinko, B., Alexanian, R.
Ploidy and proliferative characteristics in monoclonal gammopathies.
Blood. 59, 43. 1982.

233. Lauria, F., Foǎ, R., Matera, L., Raspadori, D., Tura, S.
Membrane phenotype and functional behaviour of T-lymphocytes in hairy cell leukemia (HCL).
Sem. Oncol. 11, 409. 1984.

234. Lee, B. J.
Melphalan in the treatment of multiple myeloma.
Clinical Bulletin. 6, 142. 1976.

235. Leonard, R. C. F., Maclennan, I. C. M., Smart, Y., Van Hegan, R. I., Cuzick, P.
In conjunction with the MRC Working Party for Leukaemia in Adults and the Oxford Lymphoma Group. Light chain isotypeassociated suppression of normal plasma cell numbers in patients with multiple myeloma.
Int. J. Cancer. 24, 385. 1979.

236. Le thi Bich-Thuy, Dukovich, M. Teffer, N. J., Fauci, A. S., Kehrl, J. H., Greene, W. C.
Direct activation of human resting T cells by IL-2: The role of an IL-2 receptor distict from the Tac protein.
J. Immunol. 139, 1550. 1987.

237. Lichtman, A. H., Chin, J., Schmidt, J. A., Abbas, A. K.
Role of interleukin 1 in the activation of T lymphocytes.
Proc. Natl. Acad. Sci. USA. 85, 9699. 1988.

238. Lindahl, P., Leary, P., Gresser, I.
Enhancement by interferon of the specific cytotoxicity of sensitized lymphocytes.
Proc. Natl. Acad. Sci. USA. 69, 721. 1972.

239. Ling, V.
Multi-drug-resistant mutants.
In: Molecular cell genetics.
Ed.: by Gottesman, M. M.
Willey, N. Y. 773. 1985.

240. Ludwig, Ch. U., Durie, B. G. M., Salmon, S. E., Moon, Th. E.
Tumor growth stimulation in vitro by interferons.
Eur. J. Cancer. Clin. Oncol. 19, 1625. 1983.

241. Ludwig, Ch. U., Stoll, H. R., Obrist, R., Sutter, C., Obrecht, J. P.
Improved tolerance of interferon by continuous subcutaneous infusion.
Proc. Amer. Soc. clin. Oncol. 22, 915. 1986.

242. Ludwig, H., Swetly, P.
In vitro inhibitory effect of interferon on colony formation of myeloma stem cells.
Cancer Immunol. Immunother. 93, 139. 1980.

243. Ludwig, H., Fritz, E.
Individualized chemotherapy in multiple myeloma by cytostatic drug sensitivity testing of colony-forming stem cells.
Anticancer. Res. 1, 329. 1981.

244. Ludwig, H., Cortelezzi, A., Scheithauer, W., van Camp, B. G. K., Kuzmits, R., Fillet, G., Peetermans, M., Polli, E., Flener, R.
Recombinant interferon alfa-2c versus polychemotherapy (VMCP) for treatment of multiple myeloma: A prospective randomized trial.
Eur. J. Cancer. Clin. Oncol. 22, 1111. 1986.

245. Ludwig, H., Cortelezzi, A., Kührer, I.
Combined interferon-polychemotherapy versus polychemotherapy in multiple myeloma: A phase III study.
In: The biology of the interferon system.
Ed.: by Cantell, K., Schellekens, H.
M. Nijhoff Publishers, Dordrecht, The Netherlands. 363. 1987.

246. Ludwig, H., Cortelezzi, A., Fritz, E.
A prospective and randomized clinical trial of combined interferon polychemotherapy versus polychemotherapy in multiple myeloma (MM).
J. Clin. Oncol. 6, 149. 1987.

247. Lutton, C. W., Gauntt, C. J.
Ameliorating effect of IFN-beta and anti-IFN-beta on coxsackievirus B3-induced myocarditis in mice.
J. Interferon Res. 5, 137. 1985.

248. MacKenzie, M. R., Lewis, J. P.
Cytogenetic evidence that the malignant event in multiple myeloma occurs in a precursor lymphocyte.
Cancer. Genet. Cytogenet. 17, 13. 1985.

249. MacLennan, I. C. M., Kelly, K., Crockson, R. A., Cooper, E. H., Cuzick, J., Chapman, C.
Results of the MRC myelomatosis trials for patients entered since 1980.
Haematol. Oncol. 6, 145. 1988.

250. Maheshware, R. K., Lazo, P. S., Friedman, R. M.
Enhancement of interferon activity by a membrane glycoprotein related to the thyrotropin receptor.
In: Interferon: properties and clinical uses.
Ed.: by Khan, S., Hill, N. O., Dorn, G. L.
Leland Fikes Foundation. 387. 1979.

251. Malkovsky, M., Loveland, B., North, M., Asherson, G. L., Gao, L., Ward, P., Fiers W.
Recombinant interleukin-2 directly augments the cytotoxicity of human monocytes.
Nature. 325, 262. 1987.

252. Mandelli, F., Tribalto, M., Avvisati, G.
Recombinant alpha-2b interferon as maintenance therapy in responding multiple myeloma patients: Report of the Italian multiple myeloma study group (MMSG). (Preliminary results). XXII Congress of the International Society of Hematology.
University of Milan, Italy. 1988.

253. Mandelli, F., Tribalto, M., Avvisati, G., Cantonetti, M., Petrucci, M. T., Boccadoro, M., Pileri, A., Marmont, F., Resengotti, L., Lauta, V., Dammacco, F.
Recombinant interferon alfa-2b (Intron A) as postinduction therapy for responding multiple myleoma patients. M84 protocol.
Cancer. Treat. Rev. 15, (Suppl. A.) 43. 1988.

254. Mandelli, F., Avvisati, G., Petrucci, M. T.
Interferon: a new strategy in the treatment of multiple myeloma.
Eur. J. Haematol. Supp. I. 51. 43, 129. 1989.

255. Marchetti, D., Van, N. T., Gametchu, B., Thompson, E. B., Kobayashi, Y., Watanabe, F., Barlogie, B.
Flow cytometric analysis of glucocorticoid receptor using monoclonal antibody and fluoresceinated ligand probes.
Cancer Res. 49, 863. 1989.

256. Marini, S., Guadagni, F., Bonmassar, E., Potenza, P., Giuliani, A.
Influence of interferon on the functional expression of NK target structures of murine lymphoma cells.
Cell Immunol. 102, 113. 1986.

257. Matuzaki, H., Hata, H., Takeya, M., Takatsuki, K.
Establishment and characterization of an a amylase producing human myeloma cell line.
Blood 72, 211. 1988.

258. May, L. T., Ghrayeb, J., Santhanam, U., Tatter, S. B., Sthoeger, Z., Helfgott, D. C., Chiorazzi, N., Grieninger, G., Sehgal, P. B.
Synthesis and secretion of multiple forms of $\beta_2$-interferon/B-cell differentiation factor 2/hepatocyte-stimulating factor by human fibroblast and monocytes.
J. Biol. Chem. 263, 7760. 1988.

259. May, L. T., Torcia, M. G., Cozzolino, F., Ray, A., Tatter,

S. B., Santhanam, U., Schgal, P. B., Stern, D.
Interleukin-6 gene expression in human endothelial cells: RNA start sites, multiple. IL-6 proteins and inhibition of proliferation.
Biochem. Biophys. Res. Commun. 159, 991. 1989.

260. McArthur, J. R., Athens, J. W., Wintrobe, M. M., Cartwright, G. E.
Melphalan and myeloma: Experience with a low-dose continouos regime.
Ann. Intern. Med. 72, 665. 1970.

261. McElwain, T. J., Hedley, D. W., Burton, G., Clink, H. M., Gordon, M. Y., Jazman, M., Tuttner, C. A., Millar, J. L., Milsted, R. A. V., Prentice, G., Smith, I. E., Spence, D., Woods, M.
Morrow autotransplantation accelerates haematologicall recovery in patients treated with high dose melphalan.
Br. J. Cancer. 40, 72. 1979.

262. McElwain, T. J., Powles, R. L.
High-dose intravenous melphalan for plasma-cell leukaemia and myeloma.
Lancet. i, 822. 1983.

263. McElwain, T. J., Selby, P. J., Core, M. E., Viner, C., Meldrum, M., Millar, B. C., Malpas, J. S.
High-dose chemotherapy and autologous bone marrow transplantation for myeloma.
Eur. J. Haematol. Suppl. 51. 43, 152. 1989.

264. McIntyre, O. R., Leone, L., Pajak, T. F.
The use of intravenous melphalan (L-PAM) in the treatment of multiple myeloma.
Blood. Suppl. 1. 52, 274. 1978.

265. McIntyre, O. R.
Current concepts in cancer: Multiple myeloma.
N. Engl. J. Med. 301, 193. 1979.

266. McIntyre, O. R., Tefft, M., Propert, K., Wolf, D. J., Coleman, M., Leone, L., Cooper, M. R., Eaton, W., Zimmer, B.
Melphalan and prednisone plus total bone marrow irradiation as initial treatment for multiple myeloma.
J. Rad. Oncol. Biol. Phys. 15, 1007. 1988.

267. McLaughlin, P., Alexanian, R.
Myeloma protein kinetics following chemotherapy.
Blood. 60, 851. 1982.

268. McLaughlin, P., Talpaz, M., Quesada, J. R., Saleem, A., Barlogie, B., Gutterman, J. U.
Immune thrombocytopenia is induced by alpha interferon in cancer patients.
JAMA. 254, 1353. 1985.

269. Meager, A., Parti, S., Leung, H., Wolley, J., Peil, E., Sighu, S., Roberts, T.
A two-site sandwich immunoradiometric assay of human lymphotoxin with monoclonal antibodies and its applications.
J. Immunol. Methods. 104, 31. 1987.

270. Medical research council's working party on leukaemia in adults. Myelomatosis. Comparison of melphalan and cyclophosphamide therapy.
Brit. Med. J. 1, 640. 1971.

271. Medical Research Council's Working Party for Therapeutic Trials in Leukaemia: Report on the first myelomatosis trial. Analysis of presenting features of prognostic importance.
Brit. J. Haematol. 24, 123. 1973.

272. Medical Research Council Working Party on Leukaemia in Adults.
Report on second myelomatosis trial after five years follow up.
Brit. J. Cancer. 42, 813. 1980.

273. Medical Research Council Working Party on Leukaemia in Adults.
Treatment comparisons in the 3rd MRC myelomatosis trial.
Brit. J. Cancer. 42, 823. 1980.

274. Medical Research Council Working Party on Leukaemia in Adults.
Prognostic features in the 3rd MRC Myelomatosis trial.
Brit. J. Cancer. 42, 831. 1980.

275. Medical Research Council Working Party on Leukaemia in Adults. Objective evaluation of the role of vincristine in induction and maintainance therapy for myelomatosis.
Brit. J. Cancer. 52, 153. 1985.

276. Meeker, T., Lowder, J., Cleary, M. L., Stewart, S., Warnke, R., Sklar, J., Levy, R.
Emergence of idiotype variants during treatment of B-cell lymphoma with anti-idiotype antibodies.
N. Engl. J. Med. 312, 1658. 1985.

277. Mellstedt, H., Bjorkholm, M., Holm, G.
Intermittent melphalan and prednisolone therapy in plasma cell myeloma.
Acta Med. Scand. 202, 5. 1977.

278. Mellstedt, H., Ahre, E., Bjorkholm, M., Holm, G., Johanson, B., Strander, H.
Interferon therapy in myelomatosis.
Lancet. i, 247. 1979.

279. Mellstedt, H., Pettersson, D., Holm, G.
Monoclonal B lymphocytes in myelomatosis. Relation to activity of the disease.
In: Prevention and Detection of Cancer. Part II. Detection. Vol. 2. Cancer Detection in Specific Sites.
Macel and Dekker, Inc. N. Y. 2439. 1980.

280. Mellstedt, H., Holm, G., Petterson, D., Peest, D.
Idiotype bearing lymphoid cells in plasma cell neoplasia.
In: Clinical Haematology.
Ed.: by Salmon, S. E.
W. B. Saunders Ltd. London. 65. 1982.

281. Mellstedt, H., Aahre, A., Bjorkholm, M.
Interferon therapy of patients with myeloma.
In: Immunotherapy of human cancer.
Ed.: by Terry, H., Rosenberg, G.
Elsevier North Holland, Inc. 387. 1982.

282. Mellstedt, H., Holm, G., Björkholm, M.
Multiple myeloma, Waldenström's macroglobulinemia and benign monoclonal gammopathy. Characteristics of the B cell clone, immunoregulatory cell populations and clinical implications.
Adv. Cancer Res. 41, 257. 1984.

283. Mellstedt, H.
Melphalan/prednisone versus melphalan/prednisone plus human alpha-interferon therapy in patients with multiple myeloma, stage II and III.
XXII Congress of the International Society of Hematology, University of Milan, Italy. 1988.

284. Mellstedt, H., Österborg, A., Björkholm, M., Björeman, M., Brenning, G., Gahrton, G., Grimfors, G., Gyllenhammar, H., Hast, R., Johansson, B., Juliusson, G., Järnmark, M., Killander, A., Kimby, E., Lerner, R., Merk, K., Paul, C., Simonsson, B., Smedmyr, B., Stalfelt, A. M., Strander, H., Svedmyr, E., Udén, A. M., Wadman, B., Ösby, E.
Induction treatment with alpha-interferon in multiple myeloma: An interim report from MGCS.
Eur. J. Haematol. Suppl. 51. 43. 124. 1989.

285. Merz, B.
Antibodies to tumor growth factors herald new era in chemotherapy.
JAMA. 259, 7. 1988.

286. Meuer, S. C., Hassey, R. E., Cantrell, D. A., Hodgdon, J. C., Schlossman, S. F., Smith, K. A., Reinherz, E. L.
Triggering of the T3-Ti antigen-receptor complex results in clonal T cell proliferation through an interleukin-2-dependent autocrine pathway.
Proc. Natl. Acad. Sci. USA. 81, 1509. 1984.

287. Michie, H. R., Manogue, K. R., Spriggs, D. R., Revhang, A., O'Dwyer, S., Dinarella, C. A., Gerami, A., Wolff, S. M., Wilmore, D. W.
Detection of circulating tumor necrosis factor after endotoxin administration.
N. Engl. J. Med. 318, 1481. 1988.

288. Millar, B. C., Bell, J. B. G., Lakhani, A., Ayliffe, M. J., Selby, P. J., McElwain, T. J.
A simple method for culturing myeloma cells from human bone marrow aspirates and pheripheral blood in vitro.

Br. J. Haematol. 69, 197. 1988.
289. Miller, R. A., Maloney, D. G., Warnke, R., Levy, R. Treatment of B-cell lymphoma with monoclonal anti-idiotype antibody. N. Engl. J. Med. 306, 517. 1982.
290. Mingari, M. C., Gerosa, F., Carra, G., Accolla, R. S., Moretta, A., Zubler, R. H., Waldmann, T. A., Moretta, L. Human interleukin-2 promotes proliferation of activated B cells via surface receptors similar to those of activated T cells. Nature. 312, 641. 1984.
291. Mirro, J., Kalwinsky, J., Whisnant, P. Coagulopathy induced by continuous infusion of high doses of human lymphoblastoid interferon. Cancer. Treat. Rep. 69, 315. 1985.
292. Misset, J. L., Mathé, J., Gastiaburu, A., Goutner, A., Dorval, T., Gouveia, J., Hayat, M., Jasmin, C., Schwarzenberg, L., Machover, D., Ribaud, P., De Vassal, F., Horoszevicz, J. S. Treatment of lymphoid neoplasias with interferon I. Human fibroblastic beta-interferon in malignant gammapathies. Phase II trial. Anticancer Res. 2, 63. 1982.
293. Mita, S., Harada, N., Naomi, S., Hitashi, Y., Sakamoto, K., Akagi, M., Tominaga, A., Takatsu, K. Receptors for T cell-replacing factor/interleukin 5. Specificity, quantitation and its implication. J. Exp. Med. 168, 863. 1988.
294. Moore, R. N., Pitnizzello, F. J., Robinson, R. M., Rouse, B. T. Interferon produced endogenously in respanse to CSF-1 augments the functional differentiation of progeny marrophages. J. Leuk. Biol. 37, 659. 1985.
295. Morrow, K. J. Monoclonal antibodies in the treatment of B-cell malignancies. Ann. Intern. Med. 104, 263. 1988.
296. Mosmann, T. R., Cherwinski, H., Bond, M. W., Giedlin, M. A., Coffman, R. L. Two types of murine helper T cell clone. I. Definition according to profiles of lymphokine activities and secreted proteins. J. Immunol. 136, 2348. 1986.
297. Mundy, G. R., Cooper, L. G., Schecter, R. A., Salmon, S. E. Evidence for the secretion of an osteoclast stimulating factor in myeloma. N. Engl. J. med. 291, 1041. 1974.
298. Munck Petersen, C., Moller, B. K. Immunological reactivity and bioactivity of tumour necrosis factor. Lancet. i, 934. 1988.
299. Mundy, G. R., Raisz, L. G. Big and little forms of osteoclast activating factor. J. Clin. Invest. 60, 122. 1977.
300. Muraguchi, A., Hirano, T., Tang, B., Matsuda, T., Horii, Y., Nakajima, K., Kishimoto, T. The essential role of B cell stimulatory factor 2 (BSF-2/IL-6) for the terminal differentiation of B cells. J. Exp. Med. 167, 332. 1988.
301. Muss, H. B., Constanzi, J. J., Leavitt, R., Williams, R. D., Kempf, R. A., Pollard, R., Ozer, H., Zekan, P. J., Grunberg, S. M., Mitchell, M. S. Recombinant alfa interferon in renal cell carcinoma: a randomized trial of two routes of administration. J. Clin. Oncol. 5, 286. 1987.
302. Nawroth, P. P., Bank, I., Handley, D., Cassimeris, J., Chess, L., Stern, D. Tumor necrosis factor/cachectin interacts with endothelial cell receptors to induce release of interleukin I. J. Exp. Med. 163, 1363. 1986.
303. Nicholls, M., Vincent, P. C., Repka, E., Saunders, J., Gunz, F. W. Isotypic-discordance of paraproteins and lymphocyte surface immunoglobulins in myeloma. Blood. 57, 192. 1981.
304. Niederle, N., Kloke, O., Doberauer, C., Becher, R., Beelen, D. W., Schmidt, C. G. Alpha Interferon: erste Behandlungsergebnisse bei der chronischen myeloischen Leukämie. Dtsch. med. Wschr. 111, 767. 1986.
305. Niederle, N., Kloke, O., May, D., Becher, R., Osieka, R., Schmidt, C. G. Treatment of chronic myelogenous leukemia with recombinant interferon alfa-2b. Invest. new Drugs. Suppl. 5, 19. 1987.
306. Ochs, J., Abromowith, M., Rudnick, S., Murphy, S. Phase I–II study of recombinant alpha-2 interferon against advanced leukemia and lymphoma in children. J. Clin. Oncol. 4, 883. 1986.
307. Ohno, R., Kimura, K. Treatment of multiple myeloma with recombinant interferon alpha-2a. Cancer. Suppl. 8. 57, 1685. 1986.
308. Oken, M. M., Kyle, R. A., Kay, N. E. A Phase II trial of interferon alpha$_2$ (rIFN$\alpha_2$) in the treatment of resistant multiple myeloma (MM). Proc. Am. Soc. Clin. Oncol. 4, 215. 1985.
309. Oken, M. M., Kyle, R. A., Greipp, P. R., Kay, N. E., Tsiatis, A., O'Connell, M. J. Alternating cycles of VBMCP with interferon (rIFN-alpha 2) in the treatment of multiple myeloma. Proc. Am. Soc. Clin. Oncol. 7, 868. 1988.
310. Oldham, R. Toxic effects of interferon. Science. 219, 902. 1982.
311. Osby, E., Reizenstein, P. Differential treatment of myeloma. A non-randomized phase II study. Anticancer Res. 4, 163. 1984.
312. Osgood, E. E. The survival time of patients with plasmacytic myeloma. Cancer Chemother. Rep. 9, 1. 1960.
313. Osserman, E. F., DiRe, L. B., DiRe, J., Sherman, W. H., Hersman, J. A., Storb, R. Identical twin marrow transplantation in multiple myeloma. Acta Haematol. 68, 215. 1982.
314. Österborg, A., Nilsson, B., Björkholm, M., Holm, G., Johansson, B., Lindelman, C., Pettersos, D., Ahre, A., Mellstedt, H. Blood clonal B cell excess at diagnosis in multiple myeloma. Relation to prognosis. Eur. J. Haematol. 38, 173. 1987.
315. Österborg, A., Mellstedt, H. Monoclonal and biclonal immunoglobulinproducing disorders. Eur. J. Haematol. Suppl. 51. 43, 11. 1989.
316. Paccagnella, A., Carter, G., Fosser, V., Salvagno, L., Bolzonella, S., Sileni, V. C., Fiorentino, M. V. Treatment of multiple myeloma with M-2 protocol and without maintainance therapy. Eur. J. Cancer. Clin. Oncol. 19, 1345. 1982.
317. Paul, W. E., Ohara, J. B-cell stimulatory factor-1/Interleukin. Annu. Rev. Immunol. 5, 429. 1987.
318. Pavlovsky, S., Saslavsky, J., Tezanos-Pinto, M. A randomized trial of melphalan and prednisone versus melphalan, prednisone, cyclophosphamide. MeCCNU and vincristine in multiple myeloma. J. Clin. Oncol. 2, 836. 1984.
319. Perrillo, R. P. Interferon therapy for chronic type B hepatitis: the promise comes of age. Gastroenterol. 96, 532. 1989.
320. Pileri, A., Carmagnola, A., Boccadoro, M. Human myeloma: kinetics of the remission phase. Acta Haemat. 68, 237. 1982.
321. Pileri, A., Ferrero, D., Massaisa, M., Dianzi, U., Boccadoro, M.

Advances in biology of multiple myeloma: cell kinetics, molecular biology and immunology.
Eur. J. Haematol. Suppl. 51. 43, 30. 1989.
322. Pistoia, V., Cozzolino, F., Rubartelli, A., Torcia, M., Roncella, S., Ferrarini, M.
In vitro production of interleukin 1 by normal and malignant B lymphocytes.
J. Immunol. 136, 1688. 1986.
323. Preud'homme, J. L., Brouet, J. C., Seligmann, M.
Membrane-bound IgD on human lymphoid cells, with special reference to immunodeficiency and immunoproliferative diseases.
Immunol. Rev. 37, 127. 1977.
324. Quesada, J. R., Gutterman, J. U., Swanson, D., Tundade, A.
Renal cell carcinoma: Antitumor effects of leukocyte interferon.
Cancer Res. 43, 940. 1983.
325. Quesada, J. R., Gutterman, J. U.
Clinical study of recombinant DNA produced leukocyte interferon in an intermittent schedule in cancer patients.
JNCI. 70, 1041. 1983.
326. Quesada, J. R., Hawkins, M. Merigan, T. C., Adams, F., Gutterman, J. U.
A collaborative phase I–II study of recombinant DNA-produced leukocyte interferon (clone A) in metastatic breast cancer, malignant lymphoma and multiple myeloma.
Am. J. Med. 77, 427. 1984.
327. Quesada, J. R., Reuben, J., Manning, J. T., Hersch, E. M., Guttermann, J. U.
Alpha interferon for induction of remission in hairy cell leukemia.
New. Engl. J. Med. 310, 15. 1984.
328. Quesada, J. R., Rios, A., Swanson, D., Trown, P., Gutterman, J. U.
Antitumor activity of recombinant-derived interferon alpha (rIFNαA) in metastatic renal cell carcinoma.
J. Clin. Oncol. 3, 1522. 1985.
329. Quesada, J. R., Alexanian, R., Hawkins, M., Barlogie, B., Borden, E., Itri, L., Gutterman, J. U.
Treatment of multiple myeloma with recombinant alpha-interferon.
Blood. 67, 275. 1986.
330. Quesada, J. R., Talpaz, M., Rios, A., Kurzrock R., Guttermann, J. U.
Clinical toxicity of interferons in cancer patients: a review.
J. Clin. Oncol. 4, 234. 1986.
331. Rabin, E. M., Mond, J. J., Ohara, J., Paul, W. E.
B cell stimulatory factor (BSF)-1 prepares resting B cells to enter S phase in response to anti-IgM and to lipopolysaccharide.
J. Exp. Med. 164, 517. 1986.
332. Raefsky, E. L., Platanias, L. C., Zoumbos, N. C., Young, N. S.
Studies of interferon as a regulator of hematopoietic cell proliferation.
J. Immunol. 135, 2507. 1985.
333. Reyes-Vasquez, C., Weisbrodt, N., Dafny, N.
Does interferon exert it's actions through opiate receptors.
Life Sciences. 35, 1015. 1984.
334. Riccardi, A., Merlini, G., Montecucco, C., Perugini, S.,
Vincristine in the treatment of multiple myeloma.
Haematol. 65, 595. 1980.
335. Riccardi, A., Montecucco, C., Danova, M., Ucci, G., Merlini, G., Ascari, E.
Rate of M-component changes and plasma cell labeling index in 25 patients with multiple myeloma treated with Peptichemio.
Cancer Treatment Reports. 69, 971. 1985.
336. Robert, J. L., Tubiana, N., Lejeune, C., Carcassone, Y.
Multiple myeloma: A case of a cure?
Nov. Rev. Fr. Hematol. 31, 35. 1989.
337. Rohatiner, A. S. Z., Balkwill, F. R., Griffin, D. B., Malpas, J. S., Lister, T. A.
A phase I study of human lymphoblastoid interferon administered by continuous intravenous infusion.
Cancer Chemother. Pharmacol. 9, 97. 1982.
338. Rosner, F., Grünwald, H.
Multiple myeloma terminating in acute leukemia: report of 12 cases and review of the literature.
Am. J. Med. 57, 927. 1974.
339. Rossi, J-F., Durie, B. G. M., Duperray, C., Braich, T., Marion, S. L., Pike, J. W., Haussler, M. R., Janbon, C., Bataille, R.
Phenotypic and functional analysis of 1,25-dihydroxyvitamin D3 receptor mediated modulation of the human myeloma cell line RPMI 8266.
Cancer Res. 48, 1213. 1988.
340. Rousset, F., Malefijt, R. W., Slierendregt, B., Aubry, J. P., Bonnefoy, J. Y., Defrance, T., Banchereau, J., DeVries, J. E.
Regulation of Fc-receptor for IgE (CD 23) and class II MHC antigen expression Burkitt's lymphoma cell lines by human IL-4 and IFN-gamma.
J. Immunol. 140, 2625. 1988.
341. Rubin, B. Y., Gupta, S. L.
Different efficacies of human type I and type II interferons as antiviral and antiproliferative agents.
Proc. Natl. Acad. Sci. USA. 77, 5928. 1980.
342. Ruco, L. P., Procopio, A., Maccalini, V., Calogero, A., Uccini, S., Annino, L., Mandelli, F., Baroni, C. D.
Severe deficiency of natural killer activity in the peripheral blood of patients with hairy cell leukemia.
Blood. 61, 1132. 1983.
343. Ruggiero, V., Tavernier, J., Fiers, W., Baglioni, C.
Induction of the synthesis of tumor necrosis factor receptors by interferon.
J. Immunol. 136, 2445. 1986.
344. Russel, R. G. G., Fleisch, H.
Pyrophosphates and disphosphonates in skeletal metabolism: physiological, clinical and therapeutic aspects.
Clin. Orthop. 108, 241. 1975.
345. Salmon, S. E., Smith, B. A.
Immunoglobulin synthesis and total body tumor cell number in IgG multiple myeloma.
J. Clin. Invest. 49, 1114. 1970.
346. Salmon, S. E.
Expansion of the growth fraction in multiple myeloma with alkylating agents.
Blood. 45, 119. 1975.
347. Salmon, S. E., Hamburger, A. W., Soehnlen, B., Durie, B. G. M., Alberts, D. S., Moon, T. E.
Quantitation of differential sensitivity of human-tumor stem cells to anticancer drugs.
N. Engl. J. Med. 298, 1321. 1978.
348. Salmon, S. E.
In vitro cloning and chemosensitivity of marrow myeloma stem cells.
In: Clinics in Hematology.
Ed: by Salmon, S. E.
Saunders. 11. 1982.
349. Salmon, S. E., Durie, B. G. M., Young, L., Liu, R. M., Trown, P. W., Stebbing, N.
Effects of cloned human leukocyte interferons in the human tumor stem cell assay.
J. Clin. Oncol. 1, 217. 1983.
350. Salmon, S. E., Haut, A., Bonnet, J. D.
Alternating combination chemotherapy and levamisole improves survival in multiple myeloma: A South-west Oncology Group Study.
J. Clin. Oncol. 1, 453. 1983.
351. Salmon, S., Tesh, D., Crowley, J., Saeed, S., Fonley, P., Stock-Novak, D.
Combination chemotherapy is superior to hemibody irradiation (HXRT) for remission consolidation in multiple myeloma (MM): A South-west Oncology Group (SWOG) study.

Proc. Am. Soc. Clin. Oncol. 8, 250. 1989.
352. Schiller, J. H., Groveman, D. S., Schmid, S. M., Willson, J. K. V., Cummings, K. B., Borden, E. C.
Synergistic antiproliferative effects of human recombinant alpha54 – or $\beta_{ser}$ – interferon with $\gamma$-interferon on human cell lines of various histogenesis.
Cancer Res. 46, 483. 1986.
353. Schonfeld, A., Nitke, S., Schattner, A., Wallach, D., Crespi, M., Hahn, T., Levari, H., Yarden, O., Shoham, J., Doerner, T.
Intramuscular human interferon-beta injections in treatment of condylomata acuminata.
Lancet. i, 1038. 1984.
354. Schulze, H. J., Mahrle, G.
Effect of interferons (rIFN-alpha 2, rIFN-gamma) on DNA'synthesis and HLA-DR expression in psoriasis.
Arch. Dermatol. Res. 278, 416. 1986.
355. Schulze, H. J., Mahrle, G.
Gamma interferon and psoriasis.
Lancet. ii, 926. 1986.
356. Scott, G. M.
The toxic effects of interferon in man.
In: Interferon.
Ed: by Gresser, I.
Orlando Fla, Academic. 5, 86. 1983.
357. Sehgal, P., May, L. T., Tamm, I., Vilcek, J.
Human $\beta_2$ interferon and B-cell differentiation factor BSF-2 are identical.
Science. 235, 731. 1987.
358. Selby, P. F., Kohn, J., Raymond, J., Judsou, I., McElwain, T.
Nephrotic syndrome during treatment with interferon.
Br. Med. J. 290, 1180. 1985.
359. Selby, P. F., McElwain, T. J., Nandi, A. C., Perren, T. J., Powles, R. L., Tillyer, C. R., Osborne, R. J., Slevin, M. L.,. Malpas, J. S.
Multiple myeloma treated with high dose intravenous melphalan.
Br. J. Haematol. 66, 55. 1987.
360. Selby, P. F., Zulian, G., Forgeson, G., Nandi, A., Milan, S., Meldrum, M., Osborne, R., Malpas, J. S., McElwain, T. J.
The development of high dose melphalan and of autologous bone marrow transplantation in the treatment of multiple myeloma. Royal Marsden and St. Bartholomew's Hospital studies.
Hematol. Oncol. 6, 173. 1988.
361. Semenzato, G., Pizzollo, G., Agostini, C., Ambrosetti, A., Zambello, R., Trentin, R., Luca, M., Masciarelli, M., Chilosi, M., Vinante, F., Penona, G., Cetto, G.
Alpha-interferon activates the natural killer system in patients with hairy-cell leukemia.
Blood. 68, 293. 1986.
362. Shah, I., Band, J., Samson, M., Young, J., Robinson, R., Bailey, B., Lerner, A. M., Prasad, A. S.
Pharmacokinetics and tolerance of intravenous and intramuscular recombinant alpha-interferon in patients with malignancies.
Am. J. Hematol. 17, 363. 1984.
363. Shaw, M. T., Twele, T. W., Nordquist, R. E.
Plasma cell leukemia: detailed studies and response to chemotherapy.
Cancer. 33, 619. 1974.
364. Shimazaki, C., Wisniewski, D., Scheinberg, D. A., Atzpodien, J., Strife, A., Gulati, S., Fried, J., Wisniewolski, R., Wang, C. Y., Clarkson, B. D.
Elimination of myeloma cells from bone marrow by using monoclonal antibodies and magnetic immunobeads.
Blood. 72, 1248. 1988.
365. Shimazaki, C., Atzpodien, J., Wisniewski, D., Gulati, S. C., Kalitz, J. E., Fried, J., Clarkson, B.
Cellmediated toxicity of interleukin-2-activated lymphocytes against autologous and allogeneic human myeloma cells.
Acta Haemat. 80, 203. 1988.
366. Shimazaki, C., Gulati, S. C., Atzpodien, J., Fried, J., Colvin, O. M., Clarkson, B. D.
Ex vivo treatment of myeloma cells by 4-hydroperoxycyclophosphamide and VP-16-213.
Acta Haematol. 80, 17. 1988.
367. Shimazaki, C., Inaba, T., Murakami, S.
Purging of myeloma cells from bone marrow using monoclonal antibodies and magnetic immunobeads in combination with 4-hydroperoxycyclophosphamide.
Leukemia. 1989.
368. Shimizu, K., Hirano, T., Tshibashi, K., Nakano, N., Taga, T., Sugamura, K., Yamamura, Y., Kishimoto, T.
Immortalization of BGDF (BCGF II)-and BGDF-producing T cells by human T cell leukemia virus (HTLV) and characterization of human BGDF (BCGF II).
J. Immunol. 134, 1728. 1985.
369. Sideras, P., Palacios, R.
Bone marrow pro-T and pro-B lymphocyte clones express functional receptors for interleukin (IL) 3 and IL-4/BSF-1 and non-functional receptors for IL-2.
Eur. J. Immunol. 17, 217. 1987.
370. Spiegel, R. J.
Clinical overview of alpha interferon.
Cancer. 59, 626. 1987.
371. Spiro, F. P., Mundy, G. R.
Effect of the diphosphonates APD Aminohydroxypropane disphosphonate) and C12 MDP Dichloromethylene diphosphanate) on bone resorption.
American Society for Bone and Mineral Research. 18. 1980.
372. Sporn, M. B., Roberts, A. B.
Autocrine growth factors and cancer.
Nature. 313, 745. 1985.
373. Stanton, G. J., Weigent, D. A., Langford, M. P., Blalock, J. E.
Human leukocyte transfer of viral resistance to heterologous cells.
In: Interferons properties and clinical uses.
Ed: by Kahn, A., Hill, N. O., Dorn, G.
Wadley Institute of Molecular Medicine. i, 355. 1980.
374. Stashenko, P., Dewhirst, F. E., Peros, W. J., Kent, R. L., Ago, J. M.
Synergistic interaction between interleukin 1, tumor necrosis factor, and lymphotoxin in bone resorption.
J. Immunol. 138, 1464. 1987.
375. Steis, R. G., Smith II, J. W., Urba, W. J., Clark, J. W., Itri, L. M., Evans, L. M., Schoenberger, C., Longo, D. L.
Resistance to recombinant interferon alfa-2a in hairy-cell leukemia associated with neutralizing anti-interferon antibodies.
N. Engl. J. Med. 318, 1988.
376. Stewart, W. E. X.
II Non-antiviral actions of interferons.
In: The interferon system.
Ed: by Stewart, W. E.
Springer-Verlag. N. Y. 134. 1979.
377. Stolpen, A. H., Guinan, E. C., Fiers, W., Pober, J. S.
Recombinant tumor necrosis factor and immune interferon act singly and in combination to reorganize human vascular endothelial cell monolayers.
Amer. J. Path. 123, 16. 1986.
378. Strander, H., Cantell, K., Inginarison, S.
Interferon treatment of osteogenic sarcoma: a clinical trial. In Conference on modulation of host immune resistance in the prevention and treatment of induced neoplasms.
United States Government printing office, Washington. 28, 377. 1980.
379. Sumpio, B. E., Ernstoff, M. S., Kirkwood, J. M.
Urinary excretion of interferon, albumin, and beta-2 microglobulin during interferon treatment.
Cancer Res. 44, 3599. 1984.
380. Suyehira, L. A., Lawrence, H. J.
Seventeen year survival in multiple myeloma.
Am. J. Haematol. 30, 192. 1989.

381. Sykes, M. P., Savel, H., Chu, F. C. H., Bonadonna, G., Farrow, J., Mathis, H.
Long-term effects of therapeutic irradiation upon bone marrow.
Cancer. 17, 1144. 1964.
382. Sykes, M. P., Chu, F. C. H., Savel, H., Bonadonna, G., Mathis, H.
The effects of varying dosages of irradiation upon sternal-marrow regeneration.
Radiology. 83, 1084. 1964.
383. Talpaz, M., McCredie, K. D., Mayligit, G. M., Guttermann, J. U.
Leukocyte interferon-induced myeloid cytoreduction in chronic myelogenous leukemia.
Blood. 62, 689. 1983.
384. Talpaz, M., Kantarijan, H. M., McCredie, K., Trujillo, J. M., Keating, M. J., Guttermann, J. U.
Hematologic remission and cytogenetic improvement induced by recombinant human interferon alpha A in chronic myelogenous leukemia.
New Engl. J. Med. 314, 1065. 1986.
385. Taguchi, T.
Clinical studies of recombinant interferon alfa-2a (Roferon[R]-A) in cancer patients.
Cancer. Suppl. 57, 1705. 1986.
386. Thomas, E. D., Buckner, C. D., Clift, R. A., Fefer, A., Johnson, F. L., Neiman, P. E., Sale, G. E., Sanders, J. E., Singer, J. W., Shulman, H., Storb, R., Weiden, P. L.
Marrow transplantation for acute nonlymphoblastic leukemia in first remission.
N. Engl. J. Med. 301, 597. 1979.
387. Thompson, C. B., Schaefer, M. E., Finkelman, F. D., Scher, I., Farrar, J., Mond, J. J.
T cell-derived B Cell growth factor(s) can induce stimulation of both resting and activated B cells.
J. Immunol. 134, 369. 1985.
388. Tierelli, V., Crivellari, D., Carbone, A., Verosesi, A., Galligioni, E., Trovo, M. G., Tumolo, S., Gigoletto, E.
Combination chemotherapy for multiple myeloma with melphalan, prednisone, cyclophosphamide, vincristine and carmustine (M-2 protocol).
.Cancer Treat. Rep. 66, 1971. 1982.
389. Traeger, A., Kunze, M., Stein, G., Ackermann, H.
Zur Pharmacokinetik von Indomethacin bei alten Menschen.
Z. Alterusforsch. 27, 151. 1973.
390. Trinchieri, G., Santoli, D.
Antiviral activity by culturing lymphocytes with tumor-derived or virus-transformed cells: Enhancement of human natural killer cell activity by interferon and antagonistic inhibition of susceptibility of target cells by lysis.
J. Exp. Med. 147, 1314. 1978.
391. Trinchieri, G., Perussia, B.
Immune interferon: a pleiotropic lymphokine with multiple effects.
Immunology today. 6, 131. 1985.
392. Trinchieri, G., Kobayashi, M., Rosen, M., Loudon, R., Murphy, M., Perussia, B.
Tumor necrosis factor and lymphotoxin induce differentiation of human myeloid cell lines in synergy with immune interferon.
J. Exp. med. 164, 1206. 1986.
393. Torwn, P., Kramer, M., Dennin, R.
Antibodies to human leukocyte interferons in cancer patients.
Lancet. i, 81. 1983.
394. Tsuchiya, H., Epstein, J., Selvanayagam, P., Dedman, J., Gallick, G., Alexanian, R., Barlogie, B.
Correlated flow cytometric analisis of H-ras p21 and nuclear DNA in multiple myeloma.
Blood. 72, 796. 1988.
395. Tsuro, T., Iida, H., Tsukagoshi, S., Sakurai, Y.
Overcoming vincristine resistance in P388 leukemia in vivo and in vitro through enhanced cytotoxicity of vincristine and vinblastine by verapamil.
Cancer Res. 41, 1969. 1981.
396. Tura, S., Cavo, M., Baccarani, M., Ricci, P., Gobbi, M.
Bone marrow transplantation in multiple myeloma.
Scand. J. Haematol. 36, 176. 1986.
397. Ueno, Y., Myawaki, T., Seki, H., Matsuda, A., Taga, K., Sato, H., Taniguchi, N.
Differential effects of recombinant human interferon-gamma and IL-2 on NK cell activity of peripheral blood in early human development.
J. Immunol. 135, 180. 1985.
398. Vallbracht, A., Treuner, J., Flehmig, B.
Interferon-neutralizing antibodies in a patient treated with human fibroblast interferon.
Nature. 289, 496. 1981.
399. Van Damme, J., Opdenakker, G., Simpson, R. J., Rubira, M. R., Cayphas, S., Vink, A., Billiau, A., van Snick, J.
Identification of the human 26 kd protein, interferon beta-2 (IFN-beta-2), as a B-cell hybridoma/plasmacytoma growth factor induced by interleukin and tumor necrosis factor.
J. Exp. Med. 165, 914. 1987.
400. Van Damme, J., De Ley, M., Van Snick, J., Dinarello, C. A., Billiau, A.
The role of interferon-$beta_1$ and the 26-kDa protein (interferon-$beta_2$) as mediators of the antiviral effect of interleukin-1 and tumor necrosis factor.
J. Immunol. 139, 1867. 1987.
401. Van Damme, J., Van Beeumen, J., Decock K., van Snick, J., De Ley, M., Billiau, A.
Separation and comparison of two monokines with lymphocyte-activating factor activity: IL-1$\beta$ and hybridoma growth factor (HGF). Identification of Leukocyte-derived HGF as IL-6.
J. Immunol. 140, 1534. 1988.
402. Van, N. T., Raber, M. N., Barrows, G., Barlogie, B.
Estrogen receptor analysis by flow cytometry.
Science. 224, 876. 1984.
403. Vaughn, C. B., Vaitkevicius, K.
The effects of calcitonin in hypercalcaemia in patients with malignancy.
Cancer. 34, 1268. 1974.
404. Velez-Garcia, E., Maldonado, N.
Long-term follow up and therapy for multiple myeloma.
Cancer. 27, 44. 1971.
405. Vink, A., Coulie, P. G., Wauters, P., Nordan, R. P., van Snick, J.
B-cell growth and differentiation activity of interleukin-1. HP 1 and related plasmacytoma growth factors. Synergy with Interleukin-1.
Eur. J. Immunol. 18, 607. 1988.
406. Vlodavsky, I., Folkman, J., Sullivan, R., Fridman, R., Ishai-Michaeli, R., Sasse, J., Klagsbrun, M.
Endothelial cell-derived basic fibroblast growth factor: synthesis and deposition into subendothelial extracellular matrix.
Proc. Natl. Acad. Sci. USA. 84, 2292. 1987.
407. von Scheele, C.
Light chain myeloma with features of the adult Fanconi syndrome: Six years remission following one course of melphalan.
Acta Med. Scand. 199, 533. 1976.
408. von Wussow, P., Frunc, M., Black, B., Deidrich, H., Poliwoda, H, Deicher, H.
Clinical significance of anti-IFN-alfa antibody titres during interferon therapy.
Lancet. ii, 635. 1987.
409. Wahlin, A., Holm, J.
Rapid response to recombinant interferon alpha 2B and high-dose prednisone in multiple drug-resistant multiple myeloma.
Eur. J. Haematol. 352, 43. 1989.
410. Warne, A., Joshua, D. E., Kronenberg, H.
Light chain isotype associated suppression of surface im-

munoglobulin expression on peripheral blood lymphocytes in myeloma during plateau phase.
Br. J. Haematol. 58, 483. 1984.
411. Wearne, A., Joshua, D. E., Kronenberg, H.
Monitoring myeloma: light chain isotype suppression. A new parameter.
Aust. N. Z. J. Med. 15, 629. 1985.
412. Webb, A. C., Collins, K. L., Auron, P. E., Eddy, R. L., Nakai, H., Byers, M. G., Haley, L. L., Henry, L. L., Henry, W. N., Shows, T. B.
Interleukin-1 gene (IL-1) assigned to long arm of human chromosome 2.
Lymphokine Res. 5, 77. 1986.
413. Weigent, D. A., Langford, M. P., Stanton, G. J., Blalock, J. E.
Interferon-induced transfer of viral resistance by human B and T lymphocytes.
Cell Immunol. 87, 678. 1984.
414. Weigent, D. A., Lloyd, R. E., Blalock, J. E., Stanton, G. J.
Synergism of antiviral activity in cell cultures treated with low concentrations of interferon-treated lymphocytes.
Proc. Soc. Exp. Biol. Med. 177, 257. 1984.
415. Weigent, D. A., Stanton, G. J.
Pretreatment of human lymphocytes with interferon enhances the synthesis of interferon in cocultures with allogeneic cells.
J. Leukoc. Biol. 40, 459. 1986.
416. Wild, W. O., Porter, R. W.
Metastatic epidural tumour of spine. A study of 45 cases.
Arch. Surgery. 87, 137. 1963.
417. Willingham, M. C., Cornwell, C., Cardarelli, C., Gottesman, M. M., Pastan, I.
Single cell analysis of daunomycin uptake and efflux in multidrug-resistant and sensitive KB cells: effects of verpamil and other drugs.
Cancer Res. 46, 5941. 1986.
418. Wolff, S. N., McCurley, T. L., Giannone, L.
High-dose chemoradiotherapy with syngeneic bone marrow transplantation for multiple myeloma: A case report and literature review.
Am. J. Hematol. 26, 191. 1987.
419. Wright, S. C., Bonavida, B.
Studies on the mechanism of natural killer mediated cytotoxicity. IV. Interferon-induced inhibition of NK target cell susceptibility to lysis is due to a defect in their ability to stimulate release of natural killer cytotoxic factors (NKCF).
J. Immunol. 130, 2965. 1983.
420. Zullo, J. N., Cochran, B. H., Huang, A. S., Stiles, C. D.
Platelet-derived growth factor and double-stranded ribonuclein acids stimulate expression of the same genes in 3T3 cells.
Cell. 43, 793. 1985.
421. Yokota, T., Otsuka, T., Mosmann, T., Banchereau, J., DeFrance, T., Blanchard, D., Vries, J. E. D., Lee, F., Arai, K.
Isolation and characterization of a human interleukin cDNA clone, homologous to mouse B-cell stimulating activites.
Proc. Natl. Acad. Sci. USA. 83, 5894. 1986.
422. Yoshida, R., Murray, H. W., Nathan, C. F.
Agonist and antagonist effect of interferon alpha and beta on activation of human macrophages. Two classes of interferon gamma receptors and blockade of the high-affinity sites by interferon alpha or beta.
J. Exp. Med. 167, 1171. 1988.
423. Young, D. S., Griffin, J. D.
Autocrine secretion of GM-CSF in acute myeloblastic leukemia.
Blood. 68, 1171. 1986.

# XVI. The possibilities and limits of the plasma exchange in the treatment of monoclonal gammopathies

*Life is not a quantitative but a qualitative notion*
*(Géza Hetényi)*

About a quartermillion of plasmaphoreses are performed yearly in the world (19). After the initial euphoria when the method had been used without criticism for 17 years, definite indications for plasma exchange has been introduced. The history of plasma exchange goes back to Hippocrates since bleeding can be considered as a one-side version of it. Venesection was widely applied in the second half of the 18th century, when it had more victimes than successful results. In his work Anatomy of melancholy Brunton was the first to call the attention to the dangers of venesection though emphasising its importance after having answered the questions to whom; how much; and when. There was still debate about the problem of taking blood, but Brunton had the most sober attitude stressing that a controlled venesection applied when necessary will not only relieve the symptoms but can even save life (7). The first true plasma pheresis was mentioned by Abel (1) who had described the method and proved that animals can survive the removal of the total estimated plasma volume with adequate fluid and cellular replacement.

Prior to the use of separators plasma exchange was a time-consuming, dangerous intervention often resulting in infections and other complications. The continuous or intermittent separation of blood components has become by now an accepted method used not only for theraapeutic needs.

Though Brunton had put his questions (to whom, how much and when) 400 years ago, they still need a correct answer at present.

The analysis of the question where in the literaturecomprises about 40 diseases and 20 additional indications when the intervention is necessary. Nevertheless, besides the diagnosis and the symptoms requiring plasmapheresis, individual tolerance of the patients also affects the conditions and parameters of the intervention. The frequency, intensity and duration of intensive plasmapheresis are very important. For example, partial removal of the monoclonally produced paraproteins by plasma exchange, though decreases their concentration, enhances their production. However, when repeated it will delay the reproduction process and make the pathologic plasma cells producing these proteins more sensitive to cytostatics. It should be emphasized that in addition to the removal of the undisirable plasma components the decrease of other biologically active agents has a great stimulating effect, so the organism trying to replace the loss, will produce increased amounts of these substances.

The quality of the fluid chosen for the maintenance of normovolaemia should be carefully considered.

Saline solution (0.9 %)
Sugar solution (5 %)
Albumin preparation
Gelifundol
Gelfusin
Haemacell
Rheomacrodex
HES, and
FFP are used for plasma replacement.

he latter is rarely used in case of haemorrhagic complications. Since the MG patients almost always have hyperproteinaemia, saline solution or Gelifundol is administered. The question what and when are very important, as the osmoreceptors of the portal veins of the liver (15, 16) are very sensitive. Hypotonic solutions raise the diuresis and raise haematocrit, while hypertonic solutions elicit hypertension during the plasma exchange, by raising the arginin-vasopressin level. The above possibilties can endanger, for example, the treatment of HVS (11). Moreover, albumin solution and FFP can be sources of infection (HBV, HBC, HIV, etc.)

The problem of fluid volume is very important, as there should be no changes in the portal venous pressure since its decrease stimulates diuresis, while higher pressure definitely reduces it (31). After plasma exchange the removed protein is substituted in 24 to 36 hours, at the same time the half-life of HES, Gelifundol, Gelfusin , and Haemaecel are 12, 3-3, and 5 hours ( 2 hours) respectively.

Theoretically during the pheresis the albumin (32) and IgG content of the extravascular and intercellular space in the skin and muscles remains unchainged, although their concentration decreases because it is diluted by the low protein liquid that flows from the intravasal space. The protein content of the lymph will also decrease as a consequence of the above described events, and also due to the protein retaining effect of the intercellular collagen matrix. Parallel to the tissue processes, substitution of proteins and fluids in the intravasal space begins (12, 28). Refilling is realized from the two sources: increased protein synthesis and proteins comming from the interstitium via the lymphatics. However, the protein concentration in the lymph continuously decreases indicating its retaining (2, 24). Experimental evidence indicates that the homogenous distribution of albumin (4, 33) occurs in three days.

Almost half of the plasma volume of the organism is localized in the skin and extravascular space of the skeletal musculs (29, 30). This quantity and localization are extremely stable. Neither the decrease of the total protein concentration, nor the low plasma volume can cause a significant transfer of the extravasal plasma proteins from the interstitium into the plasma (2, 24, 26, 28, 46) across the vein walls. This transfer can only occur via the lymphatics (48).

According to our measurements (described later) this can be observed already during the plasma exchange. Considering the above mentioned events the process is as follows: because of the decreased plasma protein concentration fluid flows into the extravascular space, diluting the extravascular fluid; the increasing pressure enhances the lymph flow (27). The lymph thus formed contains plasma proteins at lower concentration. At the same time fluid also flows from the interstitium into the capillaries (14). The two directional flow continues until an equilibrium between the intravascular and extravascular fluid is reached (47). Permeability is never increased, the fluid transport is due to the pressure gradients. As it has already been mentioned, collagen and glycosamines as the most important substances of the interstitium matrix regulate the presence of other macromolecules in this space, since two macromolecules cannot occupy the same place at the same time. Hence the concentration of plasma proteins in the interstitium is higher than that theoretically calculated from the extravascular protein mass. Both albumin and IgG do not occur in at least 50% of the interstitial space (2, 3, 29). The interstitium consists of two compartments, one containing collagen and glycosaminoglycan as interstitial matrix, the other - plasma proteins. Under normal conditions their concentration correlates with the protein level of the lymph. After plasma exchange the concentration of albumin and IgG in the lymph decreases by 60%. However, calculations indicate that the true interstitial protein concentration (the ratio between extravascular protein amount quantity and the intracellular volume) would only decrease by 40%. Surprisingly enough experimental evidence suggests that the increase of the interstitial volume is responsible for this difference. According to histologic studies intercellular compartment increases from 0.156 ml/g to 0.503 ml/g (dry tissue weight) (30). In a special model it was found that the cells and proteins that are still present in the vascular bed after removal of plasma without fluid replacement elicit a strong sucking effect gaining fluid fromm the extravasal space. Consequently fluid from the intracellular space flows into the more concentrated interstitial space. The sucking effect of the interstitial collagen contributes to this process. Therefore, the cell volume decreases and the extracellular space expands due to the dilution from the intracellular space. Thus the protein concentration of the lymph will be 60% lower. In the case of a forced liquid infusion the reversed process takes place (29). Finally, appropriate fluid replacement performed in normovolaemia with a solution maintaining the oncotic pressure might overcome the above mentioned pulsation of the extracellular space. Adequate selection of the periods between the cycles might promote the protein production.

Removal of the plasma is accompanied by the AT III and fibrinogen loss, while the II, V, X factors, the Quick and the reptilase time hardly change. Appropriate technique can also maintain the platelet count (21, 49).

*Where should plasmapheresis be used?*

Nowdays plasma exchange is no longer used as a general, widely applied symptomatic therapy, its indication became more rational, but still no final criteria for its use exist. Plasma exchange can be effectively used in certain diseases (a) and less effectively applied in others (b). In the case of monoclinal gammopathies HVS and drug resistant plasmocytoma belong to group A and cryoglobulinaemia, pyroglobulinaemia and renal lesion - to group B (30, 37).

*When should this procedure be used?*

In hyperproteinaemia, in cases when the viscosity exceeds 4cP, in cryoglobulinaemia causing the Raynaud syndrome and in all cases, in which monoclonally produced immunoglobulins aggravate the symptoms. Furthermore, it is also used for the removal of the A and B type antigens from the blood when the recipients are prepared for the transplantation of the ABO incompatible bone marrow. This intervention decreases the risk of the haemolytic transfusion reaction due to the lysis of the donor's erythrocytes and also prevents the destruction of stem cells.

*How much of the circulating plasma can be changed and how often?*

The data of the literature are very divergent, though it is generally accepted that it can be performed up to the point where beneficial result is achieved. Usually it means 40 ml/kg/session. It is however, important to emphasise that even in case of a planned intervention the individual tolerance of the patients should be considered. Neither the weight nor the surface of the body can give a direct or an indirect answer to the question how much. Factors affecting the actual succes of the plasma exchange depend on the patientès tolerance. We perform a 35-40 percent plasma exchange, for according to our calculations this is a rational intervention. The answer to the question how often is very simple: certainly more than once.

The answer to the question how obviously implies the minimum risk and maximum efficacy. We have, therefore, introduced a computer program based on series of calculations. Before the onset of the intervention the software uses the input parameters accepted in the literature and by us to select the size of the separating centrifuge rotor, the amount of replacement fluid, and the number of cycles required (it displays the amount of the exchanged plasma per cycle in millilitres and % of the calculated initial plasma volume) (Fig. XVI/1).

For example, if the patients total circulating blood volume is 5000 ml, haematocrit value is 40% and the protein content of the plasma is 110 g/l, and we want to reduce the latter to 70 g, then (See Table) three cycles are needed with a 375 ml centrifuge rotor each cycle replacing 750 ml of liquid until 36.4% of the plasma is removed (Fig. XVI/2).

If the procedure is accurately realized plasmapheresis practically does not lead to leukocyte or platelet loss. At the same time it provides an adequate control of the replaced fluid avoiding unnecessary saving (at the expense of the patient), or administration of the excess fluid (at the expense of the cost-efficiency).

The calculations make up a separate book, for they contain all the relevant data necessary for a start, thus all the patients can be treated individually.

The answer to the question with what is that obviously with disposable instruments made from insoluble material. There is evidence proving the pathogenetic role of the plastic particles and other chemical agents used in the production of the plastics in the development of the late complications.

If possible, only otherwise physiologically uccurring (?), but pathologically accumulated plasma components should be removed. Avoid overshooting the goal, removing other

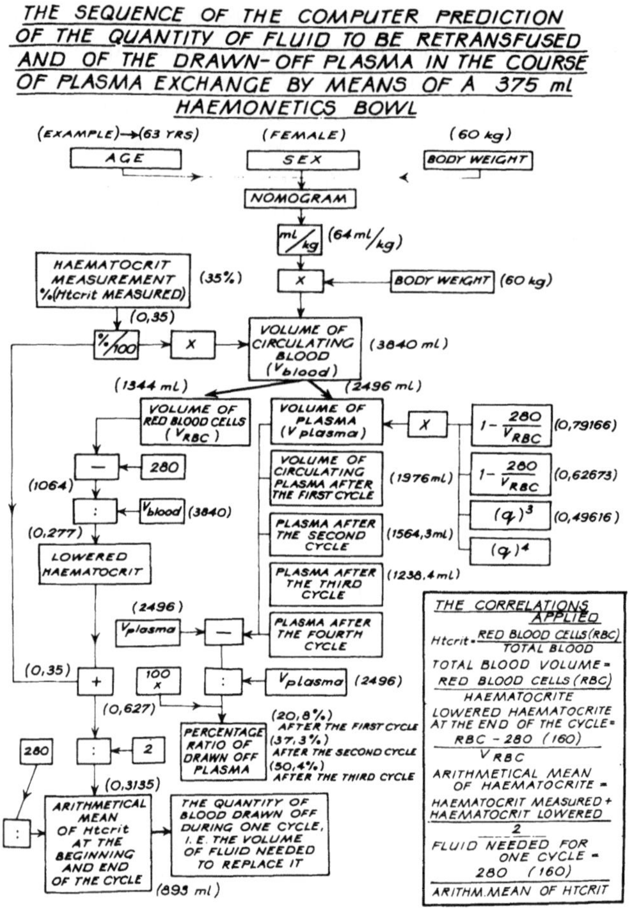


**Computerized evaluation of fluid replacement volume and the removed plasma volume in a plasma exchange performed with a 378 ml haemonetics equipment.**

**Fig. XVI/1**

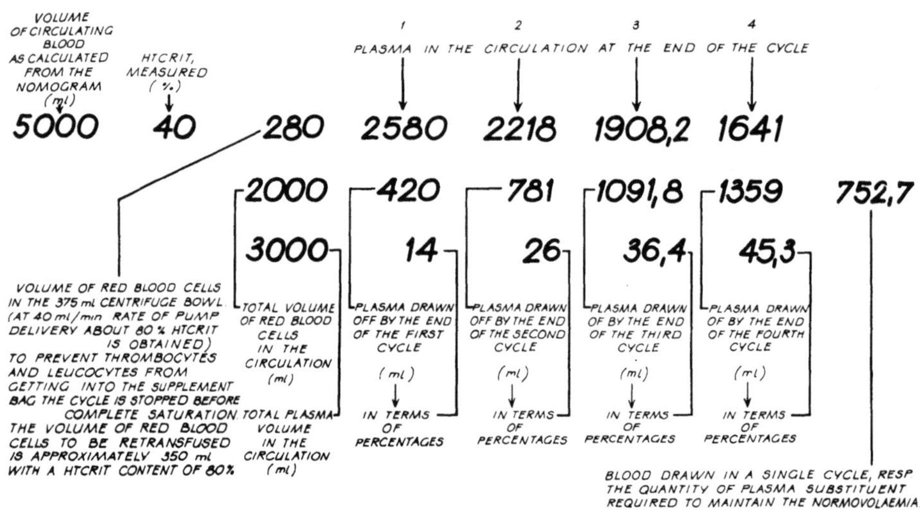


**Fig. XVI/2**

valuable components (cascade filtration, plasma absorption (9, 25, 44).

Centrifuges (41) or filters (39) are used for plasma exchange. Two types of centrifuge procedures exist: a continuous and a fractional system. We are using the latter and our experiences are mainly gained with this system, though we have acquired some generally valid statements (Fig. XVI/3a, b).

In the course of our studies we found that by the end of the plasma exchange procedure

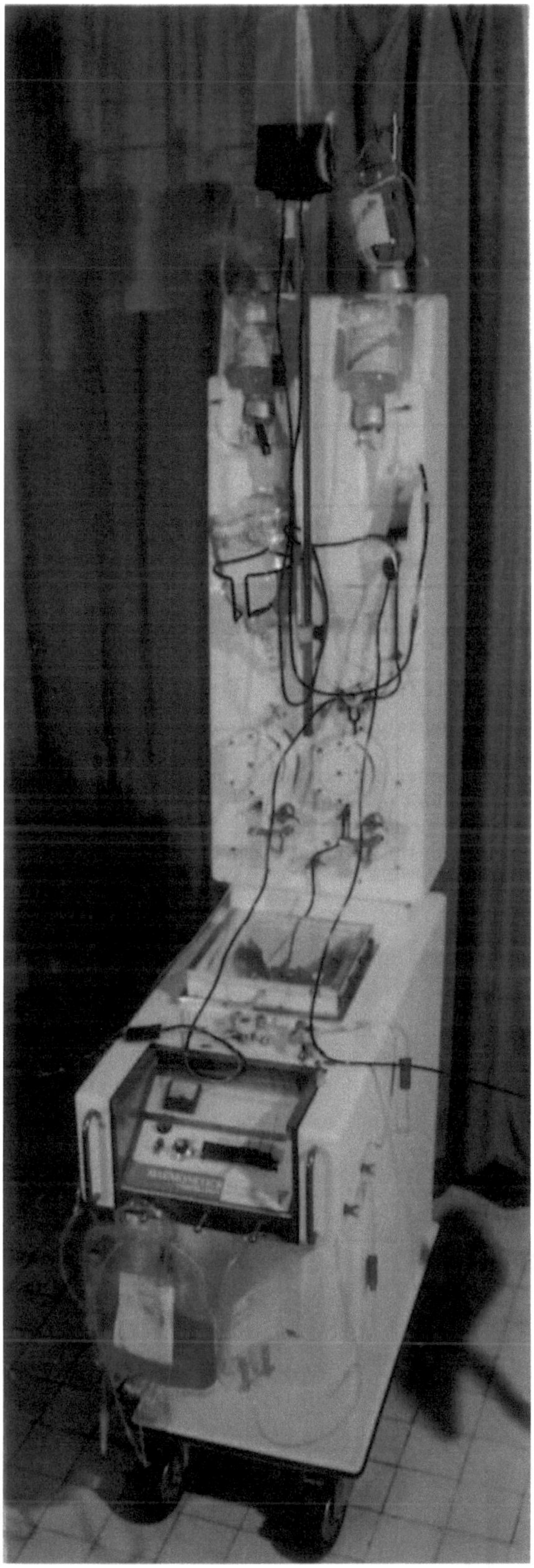

**Haemonetics M 30**

**Fig. XVI/3/a**

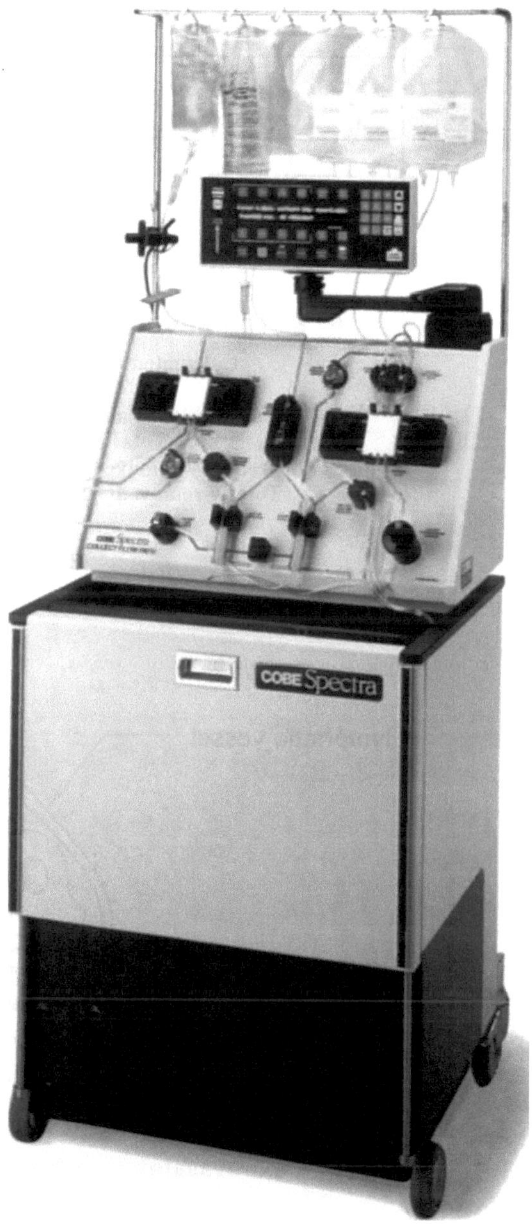


**Cobe Spectra**

**Fig. XVI/3/b**

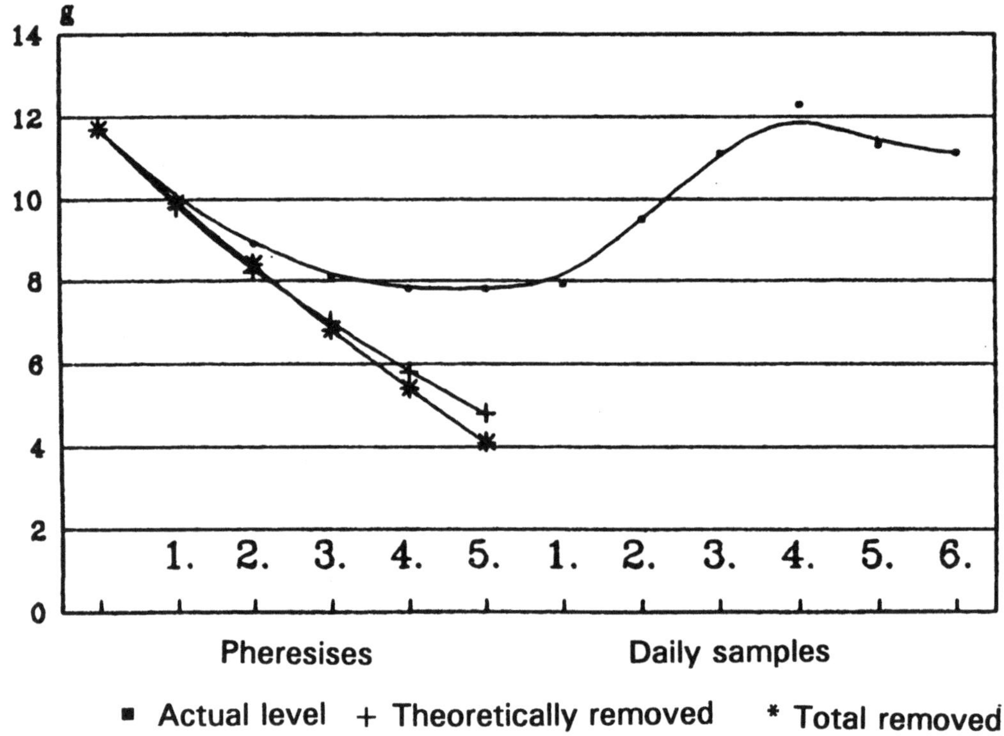


**Calculated blood volume of an albino rabbit is 250 ml**
**Number of cycles: 5**
**Analysis of blood samples taken for 6 days after pheresis Actual level:**
**Total protein value of the rabbit plasma**
**Theoretically removed: calculated protein concentration of the plasma**
**Totally removed: protein concentration of the removed plasma**

**Fig. XVI/4**

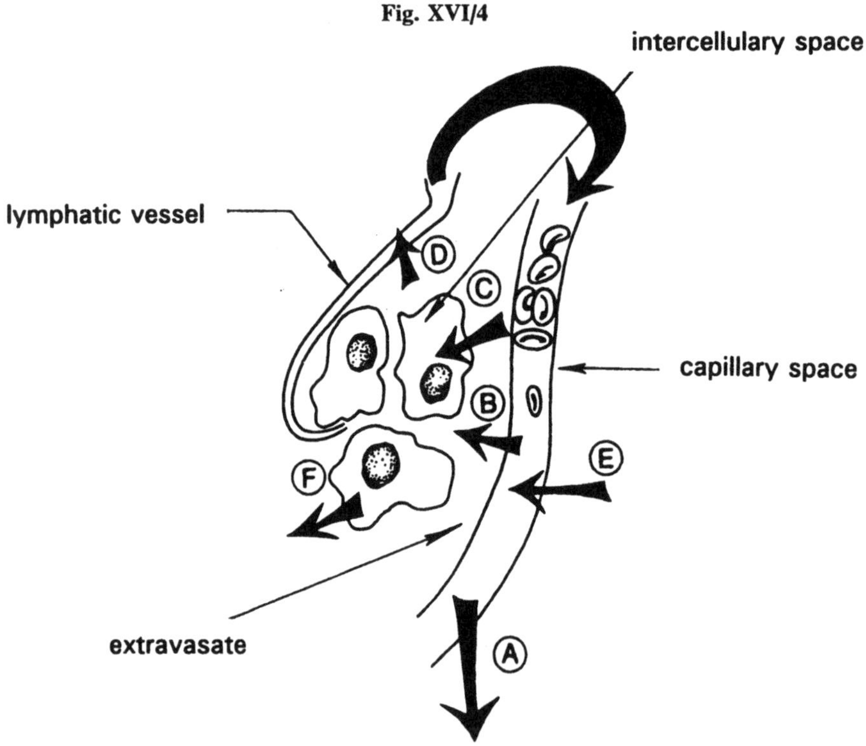


**Fig. XVI/5**

protein concentration was always higher than that expected from our calculations. Moreover, we found that the amount of protein in the collecting sack was alsohigher than theoretically expected, thereby the blood protein level was lower than the calculated one (Fig. XVI/4).

Similar studies were performed in animal experiments perfectly confirming the above results. Furthermore, our measurements of albumin and IgG changes supported the above mentioned difference between the measured and calculated values of total protein. Our findings are presented in Fig. XVI/5.

Removal of protein reduces the intravasal protein concentration (A) and decreases the oncotic pressure. Fluids used for the maintenance of normovolaemia affect the ion composition of blood, due to the diluting effect of the administered solution.

The latter causes fluid transsudation from the capillaries into the extravasal space (B), and as a consequence, the intracellular space is diluted. The concentrations of the ions and their ratios would also change. The fluid uptake by the cells (C) though not significant (the swelling of the cells is minute), still would decrease the extracellular space increasing the pressure and enhancing the lymph flow. The ion exchange between the extracellular and intracellular spaces occurs as dictated by the new conditions. The resulting increased lymph flow enhances the fluid return into the vascular bed (D); its protein content is, however, lower than that before the intervention, since it has been diluted. The flow rate increases, though the ion content isaltered. Plasma flows directly from the interstitium into the capillaries when blood with high haematocrit returns to the vascular bed by the end of the cycle (E). Consequently the flow is accelerated, and the circulating blood volume exceeds the transfused one. In the last step - emphasising that these events actually occur simultaneously and not one by one - the outflow from the cells (F) restores the equilibrium. The protein concentration of the fluid returning into the blood stream shows a reciprocal relationship with the amount of the protein loss. This fluid originates from the interstitium and returns to the capillaries either directly or via the lymphatics. The results of our studies prove that a substancial amount of proteins return into the circulation via the lymphatics. Our findings indicate that the protein concentration measured at the end of the

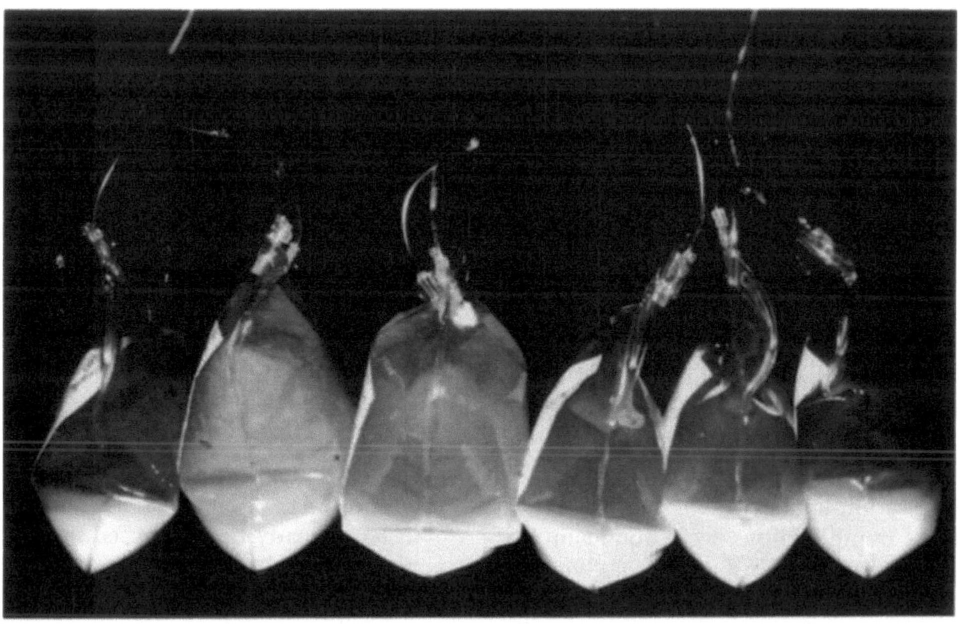

**Plasma removed from a cryoglobulinaemic patient (on 6 occasions). Precipate can be clearly seen in the cooled samples.**

**Fig. XVI/6**

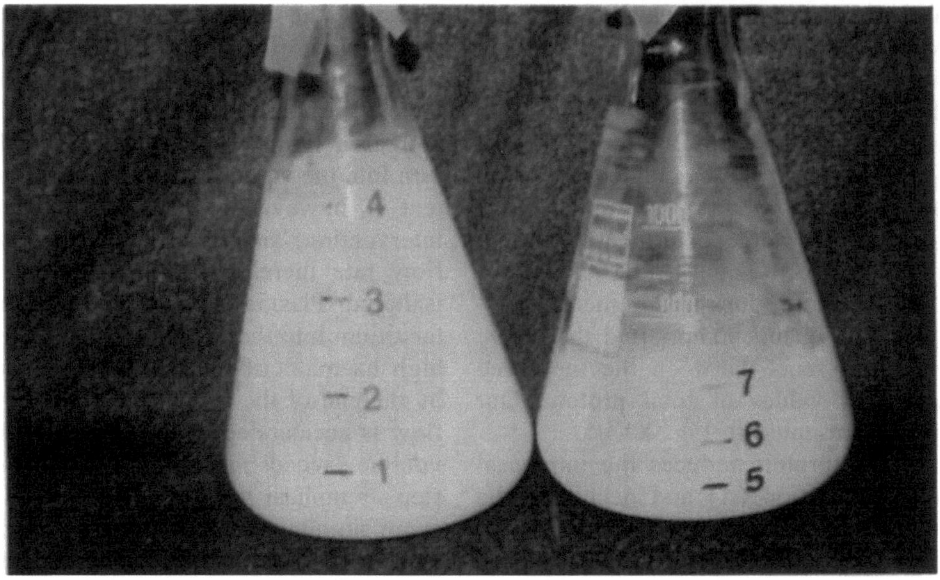


**The quantity of the collected washed precipitate remains almost unchanged from one occasion to the other. The sample forms a thick mass.**

**Fig. XVI/6**

plasma exchange exceeded the calculated value. Though it might seem that the effecacy of plasma exchange is worse than excepted, since more protein remains in the blood stream, the total amount of the removed protein is, however, higher than the calculated one.

The above mentioned characteristics of the flow can explain this apparent contradiction, furthermore, a case of cryoglobulinaemia has excellently proved the correctness of our explanation: after a series of plasma exchanges the amount of cryoprecipitate remained unchanged (Fig. XVI/6, XVI/7), indicating continuous refilling from the tissues.

Plasma exchange is rather dangerous (36). According to international statistical data the incidence of lethal complications is 0.03%. The analysis of these results has revealed that 83% of the deaths occurred after the administration of FFP as fluid replacement (18).

Minor complaints which can not be regarded as complications were found in 1/5 - 1/3 of the interventions (22, 35). Symptoms such as shivering fever, allergic reactions, nausea, transient hypotension (5) might all be due to certain technical problems, as for example, a too rapid infusion rate of a not properly warmed albumin solution. These minor complaints should be differentiated from the group of early complications related to the administration of inadequate volume of fluid or incorrect doses of anticoagulants.

Generally haemodynamic complications occur. Hypotension develops in 0.6% (35) but acute cardiac insufficiency, ARDS, myocardial ischemia and arrhytmia have also been mentioned (6, 22, 43). Non-cardiogenic pulmonary edema developed after the administration of FFP, though its cause still cannot be explained (8, 38). Hyperexcitability, perioral paresthesia and tetania can occur as the side-effects of ACD (17). It was found that the PTH level rises rapidly during plasma exchange probably for prompt compensation of the serum $Ca^{++}$ level (40).

Oliguric patients should be observed with special care, for ACD causes alkalosis by metabolizing citrate into bicarbonate that cannot be excreted because of the impaired renal function (34).

It has already been stressed that plasma exchange deprives the organism of some valuable proteins, including coagulation factors, nevertheless, bleeding complications after this intervention are rare probably because the coagulants and the anti-coagulants are proportionally removed (23). It seems paradoxical that in contrast to rarely occurring bleedings, pulmonary embolism or thrombosis are not rare complications (42). It is assumed that in such cases the lack of AT III is predominant thus leading to clot formation. This view still holds true despite the fact that the AT III is substituted within 24 hours and in 13% of the plasma exchanges its biologic activity remains unchanged (13). In cases of pulmonary micro-embolisms emboli might develop from the leukocytes or platelets (45). Late complications indicate infections.

*Summary*

Scientific outlook can scarcely keep up with the rapidly developing medicine. It is partly due to a certain inertia of ideas, and partly to a growing body of publications spreading half-finished information instead of true statements. The logical question arises: which parameters prove truth? Perhaps careful observations and explanations of the phenomena can be considered truthful? The latter is, however, is a rather dangerous attempt, since on most occasions it serves only apologias instead of true explanation. The controversy between potentia pura the possibility of Aristoteles, versus the actus purus (action) further complicates the situation characteristic of our medium-developed health care. Unfortunately the gap between possibility and reality has not yet decreased but is still growing deeper.. I think that revealing the truth is our duty.

### References

1. Abel, J. J., Rowntree, L. G., Turner, B. B.
Plasma removal with return of corpuscles (Plasmapheresis).
J. Phar. Exp. Ther. 5, 625. 1914.
2. Aukland, K., Nicolaysen, G.
Interstital fluid volume: local regulatory mechanisms.
Physiol. Rev. 61, 556. 1981.
3. Bert, J. L., Pearce, R. H.
The interstitium and microvascular exchange.
In: Handbook of Physiology: The Cardiovascular System. Microcirculation.
Ed: by Bethesda, MD.
Am. Physiol. Soc. sect 2, vol. IV. chapt. 12, 521. 1984.
4. Bill, A.
Plasma protein dynamics: albumin and IgG capillary permeability, extravascular movement and regional blood flow in unanesthetized rabbits.
Acta Physiol. Scand. 101, 28. 1977.
5. Bland, J. H. L., Laver, M. B., Lowenstein, E.
Vasodilator effect of commercial 5% plasma protein fraction solutions.
JAMA. 224, 1721. 1973.
6. Boogaerts, M. A., Roelant, C., Goosens, W., Verwilghen, R. L.
Complement activation and adult respiratory distress syndrome during intermittent flow apheresis procedure.
Transfusion. 26, 82. 1986.
7. Brunton, L.
Therapeutics of the circulation.
J. Murray, London, p. 172. 1908.
8. Buckner, C. D., Clift, R. A., Volwiler, W., Donolme, D., Burnell, J. M., Saunders, F. C., Thomas, E. D.
Plasma exchange in patients with fulminant hepatic failure.
Arch. Intern. Med. 132, 487. 1973.
9. Burgstaler, E. A., Pineda, A. A., Ellefson, R. D.
Removal of plasma lipoproteins from circulating blood with a heparin-agarose column.
Mayo. Clin. Proc. 55, 180. 1980.
10. Burton, R.
Anatomy of melancholy.
PT. 2, Sect. 4. Memb. 3. 1601.
11. Chwalbinska-Moneta, J.
Role of hepatic portal osmoreception in the control of ADH release.
Am. J. Physiol. 236, 603. 1979.
12. Ebert, R. V., Stead, E. A., Warren, V., Watts, W. E.
Plasma protein replacement after hemorrhage in dogs with and without shock.
Am. J. Physiol. 136, 299. 1942.
13. Flaum, M. A., Cuneo, R. A., Appelbaum, F. R., Deisseroth, A. B., Engel, W. K., Gralnik, H. R.
The hemostatic imbalance of plasma-exchange transfusion.
Blood. 54, 694. 1979.
14. Granger, H. J.
Role of the interstitial matrix and lymphatic pump in regulation of transcapillary fluid balance.
Microvas. Res. 18, 209. 1979.
15. Haberich, F. J.
Osmoreception in the portal circulation.
Fed. Proc. 27, 1137. 1968.
16. Haberich, F. J., Aziz, O., Nowacki, P. E., Ohm, W.
Zur Spezifität der Osmoreceptoren in der Leber.
Pflügers Arch. 313, 289. 1969.
17. Hamblin, T. J.
Plasmapheresis and plasma exchange.
Eden Press. 1, 13. 1979.
18. Huestis, D. W.
Mortality in therapeutic haemapheresis.
Lancet. i, 1043. 1983.
19. Ivanovich, P., Schmitt, E., Klinkmann, H.
Plasmapheresis an international determination of status.
Z. Gesamte Inn. Med. 43, 693. 1988.
20. Johnson, W. J., Kyle, R. A., Pineda, A. A., O'Brien, P. C., Holley, K. E.
Treatment of renal failure associated with multiple myeloma. Plasmapheresis, haemodialysis and chemotherapy.
Arch. Intern. Med. 150, 863. 1990.
21. Keller, A. J., Chirnside, A., Urbaniak, S. J.
Coagulation abnormalities produced by plasma exchange on the cell separator with special reference to fibrinogen and platelets.
Br. J. Haematol. 42, 593. 1979.
22. Keller, F., Schultze, G., Offermann, G., Molzahn, M.
Complications during therapeutic plasma exchange.
Acute Care. 11, 100. 1985.
23. Kiprov, D. D.
An overview of therapeutic apheresis.
Dial. Trans. 14, 195. 1985.
24. Kramer, G. C., Harms, B. A., Bodai, B. I., Demling, R. H., Renkin, E. M.
Mechanisms for redistribution of plasma protein following acute protein depletion.
Am. J. Physiol. 243, 803. 1982.

25. Lauterburg, B. H., Dixon, E. R., Pineda, A. A., Carlson, G. L., Taswell, H. F.
Removal of bile acids and bilirubin by plasmaperfusion of U. S. P. charcoal-coated glass beads.
J. Lab. Clin. Med. 94, 585. 1979.
26. Liljedahl, S. O., Rieger, A.
Plasma volume and plasma protein. IV. Importance of thoracic duct lymph in restitution of plasma volume and plasma proteins after bleeding and immediate substitution in the splenectomized dog.
Acta Clin. Scand. Suppl. 379, 39. 1967.
27. Manning, R. D. Jr., Guyton, A. C.
Dynamics of fluid distribution between the blood and interstitium during overhydration.
Am. J. Physiol. 238, 645. 1980.
28. Matthews, C. M. E.
Effects of plasmapheresis on albumin pools in rabbits.
J. Clin. Invest. 40, 603. 1961.
29. Mullins, R. J., Bell, D. R.
Changes in interstitial volume and masses of albumin and IgG in rabbit skin and skeletal muscle after saline volume loading.
Circ. Res. 51, 305. 1982.
30. Mullins, R. J., Powers, M. R., Bell, D. R.
Albumin and IgG in skin and skeletal muscle after plasmapheresis with saline loading.
Amer. J. Physiol. 252, 71. 1987.
31. Ohm, W., Haberich, F. J.
Über den Einfluss des Druckes im Portalkreislauf auf die Diurese der wachen Ratte.
Pflügers Arch. 306, 227. 1969.
32. Oratz, M., Rothschild, M. A., Schreiber, S. S.
Albumin-osmotic function.
In: Albumin Structure, Function and Uses.
Ed. by Rosenoer, V. W., Oratz, M., Rothschild, M. A.
New York, Pergamon, p. 275. 1977.
33. Owen, M., Triffitt, J. T.
Extravascular albumin in bone tissue.
J. Physiol. Lond. 257, 293. 1976.
34. Pearl, R. G., Rosenthal, M. H.
Metabolic alkaloids due to plasmapheresis.
Am. J. Med. 79, 391. 1985.
35. Puig, L. I., Mazzara, R., Torras, A., Castillo, R.
Adverse effects secondary to the treatment with plasma exchange.
Int. J. Artif. Organs. 8, 155. 1985.
36. Reimann, P. M., Mason, P. D.
Plasmapheresis: technique and complications.
Intensiv. Care Med. 16, 3. 1990.
37. Rollino, C., Coppo, R., Mazzucco, G., Roccatello, D., Amore, A., Basolo, B., Martina, G., Gianoglio, B., Piccoli, G.
Monoclonal gammopathy and glomerulonephritis with organized micro tubular deposits.
Am. J. Kidney Dis. 15, 276. 1990.
38. Rubenstein, M. D., Wall, R. T., Wood, G. S., Edwards, M. A.
Complications of therapeutic apheresis including a fatal case with pulmonary vascular occlusion.
Am. J. Med. 75, 171. 1983.
39. Samtleben, W., Randerson, D. H., Blumenstein, M., Habersetzer, R., Schmidt, B., Gurland, H. J.
Membrane plasma exchange: principles and application techniques.
J. Clin. Apheresis. 2, 163. 1984.
40. Silberstein, L. E., Naryshkin, S., Haddad, J. J., Strauss III, J. F.
Calcium homeostasis during therapeutic plasma exchange.
Transfusion. 26, 151. 1986.
41. Sibinga, C. Th. S.
Apheresis machines and techniques.
In: Therapeutic hemapheresis.
Ed. by Valbonesi, M., Pineda, A. A., Briggs, J. C.
Wichtig, Milano, p 1. 1986.
42. Sultan, Y., Bussel, A., Maisonneuve, P., Poupeney, M., Sitty, X., Gajdos, P.
Potential danger of thrombosis after plasma exchange in the treatment of patients with immune disease.
Transfusion. 19, 588. 1979.
43. Sutton, D. M. C., Nairn, R. C., Rock, G. and the Canadian Apheresis Study Group.
Complications of plasma exchange.
Transfusion. 29, 124. 1989.
44. Terman, D. S., Buffaloe, G., Mattioli, C., Cook, G., Tillquist, R., Sullivan, H., Ayus, H. C.
Extracorporeal immunoabsorption: initial experience in human systemic lupus erythematosus.
Lancet. ii, 824. 1979.
45. Toren, M., Goffinet, J. A., Kaplow, L. S.
Pulmonary bed sequestration of neutrophils during hemodialysis.
Blood. 36, 337. 1970.
46. Wasserman, K., Mayerson, H. S.
Mechanism of plasma protein changes following saline infusions.
Am. J. Physiol. 170, 1. 1952.
47. Wasserman, K., Joseph, J. D., Mayerson, H. S.
Kinetics of vascular and extravascular protein exchange in unbled and bled dogs.
Am. j. Physiol. 184, 175. 1956.
48. Yuile, C. L., Lucas, F. V., Neubecker, R. D., Cochran, C. G., Whippel, G. H.
Depletion of reserve protein from extravascular extracellular fluid.
J. Exp. Med. 109, 165. 1959.
49. Ziemer, S., Roeseler, E., Monike, A., Gromnica-Ihle, E.
Therapeutic plasmapheresis and haemostatic system.
Folia Haematol. 115, 563. 1988.

MIX
Papier aus verantwortungsvollen Quellen
Paper from responsible sources
FSC® C105338

If you have any concerns about our products,
you can contact us on
**ProductSafety@springernature.com**

In case Publisher is established outside the EU,
the EU authorized representative is:
**Springer Nature Customer Service Center GmbH**
**Europaplatz 3, 69115 Heidelberg, Germany**

Printed by Libri Plureos GmbH
in Hamburg, Germany